CIRCUIT ANALYSIS FOR ENGINEERS

CIRCUIT ANALYSIS FOR ENGINEERS

Continuous and Discrete Time Systems

Dwight F. Mix
Neil M. Schmitt
University of Arkansas

JOHN WILEY & SONS
New York *Chichester* *Brisbane* *Toronto* *Singapore*

Library of Congress Cataloging in Publication Data:

Mix, Dwight F., 1932–
Circuit analysis for engineers.

Includes indexes.
1. Electric circuit analysis. 2. Discrete-time systems.
I. Schmitt, Neil M. II. Title.
TK454.M49 1985 621.31′92 84-11916
ISBN 0-471-08432-8

Printed in the United States of America

10 9 8 7 6 5 4 3 2 1

PREFACE

Circuit analysis forms the foundation of nearly all curricula in Electrical Engineering education. This text is intended to assist the professor in providing the student with both a broad and a deep understanding of this topic. Students enter engineering programs with a wide range of mathematical capabilities. This text is written so that students can take the first course in circuit analysis while they are taking their first course in calculus. In the early portions of the text any necessary calculus or differential equation concepts needed are introduced as part of the text material. In the latter portions of the text we have assumed that the student has completed a course in calculus and the mathematical concepts presented are considerably more sophisticated.

There is a letter on record from one fifteenth-century Italian nobleman to another advising the recipient to send his son to a certain German university "because they teach long division there." Considerable progress in our educational system has been made since then, and this progress will continue in the future. Discrete-time signal processing was first introduced into the graduate curriculum of many universities some 10 to 20 years ago. Since that time this topic has gradually filtered into the undergraduate curriculum. The advent of very sophisticated but inexpensive digital hardware has now dictated that this topic be a required component of the undergraduate curriculum. In this text we introduce this topic in the beginning circuits sequence and make it a natural part of beginning circuit analysis.

It is our contention that unfamiliar topics are neither inherently difficult nor inherently easy to understand. The manner in which they are presented, coupled with the student's background, primarily determines the level of difficulty experienced in comprehending new ideas. With this in mind, we have employed the following tactics to augment the learning process.

1. **Tell the student what he is expected to learn.** Each section of the text begins with a set of objectives that establishes the learning goals for the student for that section. We assume that these will be augmented or modified as desired by the professor.
2. **Help the student determine whether she has grasped the concepts.** Each section concludes with a set of drill problems that measures how well the student has met the learning objectives. The problems at the end of the chapter

are keyed to the sections in the chapter to further enhance the evaluation process. The end-of-chapter problems are more difficult than the drill problems and some are open-ended or extend the material presented in the chapter.

3. **Present the most basic concepts first and then build on these.** Essentially all concepts and terminology used in the text are introduced in the text as they are encountered rather than relying on previous experiences of the student in physics or similar courses. The meaning of circuit analysis is explained and illustrated as much as possible through the use of the resistive circuit. The addition of other circuit elements then leads to expansion of these concepts to include analysis for any deterministic input signal and analysis in the frequency domain.
4. **Use examples frequently.** For any student encountering a subject for the first time, illustration of principles and concepts by means of carefully chosen examples is important. You will find that examples are used generously as part of the text and they are easily identified within the text material.

We have included two other topics often omitted in a beginning circuits text: linearity and the convolution operation. Linearity is assumed in all beginning texts, but the exploration of this and the time invariance concept allows its use in deriving the transfer function and convolution operation. Under certain conditions (controllability and observability), there are three equivalent methods of finding the response of linear, time-invariant circuits: namely, convolution, transforms, and differential or difference equations. We give complete treatment to all three methods. Convolution has attained additional importance because of its ease of use in digital computer analysis of circuits. Also, more material on transforms than is found in the average two-semester circuits text is included.

The concept of state and the state-variable approach are not introduced. There are several reasons for this.

It is confusing to the beginner to learn two closely related procedures at the same time. This is well documented in a famous article by D. R. Entwisle and W. H. Huggins, "Interference in the Learning of Circuit Theory," (*Proc. IEEE*, July 1963, 986–990). We felt it would be especially confusing to introduce the closely related but distinct concepts of zero-state, zero-input response along with the traditional homogeneous (transient) and particular (steady-state) responses. The state-variable concepts are usually introduced in higher-level courses.

Approximately three-fourths of the text is devoted to traditional topics (continuous-time circuits) and one-fourth is devoted to discrete-time analysis. The new material is integrated with traditional topics as much as possible. For example, discrete convolution is presented in one section, followed by continuous-time convolution in the next section. Whenever a principle applies to both continuous and discrete systems, examples of both are given. Since discrete-time systems are so prevalent in industry, it is our belief that discrete-time topics will eventually occupy a major portion of the typical electrical engineering undergraduate curriculum. Since this text is among the first to introduce this topic in the beginning course, however, we have chosen to be cautious rather than bold. Also, there is some evidence that continuous-time signal processing will make a comeback because of the time limitations of discrete signal processing. It may take seconds to find the two-dimensional Fourier transform of a large array. This same processing can be done optically in microseconds.

It has been our intent to write this book for the students. We hope that they find the style less formal than that of many texts, the historical insights enlightening and amusing, and the order and method of presentation helpful in assimilating the myriad concepts and principles contained within these pages. Likewise, we hope this text is of value in assisting the professor in the educational process.

Each author acknowledges that all errors or omissions are the fault of the other author. Seriously, we would like to acknowledge the helpful review of an earlier version of this text by Professor Edwin C. Jones at Iowa State University. None of the faults in this text should be attributed to him.

Dwight F. Mix
Neil M. Schmitt

CONTENTS

CIRCUIT ANALYSIS FOR ENGINEERS

CHAPTER 1 FUNDAMENTAL DEFINITIONS AND LAWS

1.1 Circuit Analysis: What Is It?

You are about to embark on a journey through the world of circuit analysis. Naturally you must be curious about this topic and want to know why it is so fundamental to the engineering profession. A primary function of an engineer is to apply fundamental laws of science to problems facing the human race in order to effect a solution or at least to minimize the effect of the problem. A significant portion of these laws embrace the area of electricity. We all know that the atom is the basic building block of things that exist in our universe. Atoms are made up in part of charged particles (electrons and protons). The movement of these charged particles is what we commonly call electricity.

An electrical circuit is a grouping of certain devices in order to control the flow of electrons in such a manner as to achieve a desired outcome. The display of a picture on a television screen is the result of controlled movement of electrons by electrical circuits. The uncontrolled flow of electrons, such as in a bolt of lightning, can have devastating effects.

One function of the engineer is to develop devices to be used in electrical circuits, but that is a subject for other texts. A second important function is circuit design—the interconnection of devices with the intent of achieving some goal. Another function is to analyze a group of devices that have been interconnected in some way to determine their effect on the flow of electrons. This function is called circuit analysis, and is the topic of this book. Circuit analysis and circuit design are obviously interrelated; indeed, analysis exists primarily to support the design task.

There are many ways to analyze electric circuits, and the choice of the best method of analysis is made by the engineer based on knowledge and experience. It is the objective of this text to provide you with a "bag of circuit analysis tools" from which you may choose the most appropriate tool for the task you face. It is also the purpose of this text to provide you with experience in using these tools to solve circuit analysis problems.

We will begin by explaining the system of units to be used in the book and by defining basic elements of electricity. Two of the most fundamental laws of circuit analysis will also be introduced. Finally, conventions to be followed in the text will be agreed upon. It is our sincere desire that the subject matter introduced through this text will launch you into a rewarding and satisfying lifelong career as an engineer.

1.2 System for Describing Circuit Values

LEARNING OBJECTIVES

After completing this section you should be able to do the following:

1. Express measures of units in scientific notation.
2. Convert values to the international system of units.

In this book we are interested in conversing with (and in training you to converse with) technically oriented individuals. Consequently we use a system of units that has been adopted by all the professional engineering societies in the United States and by most professional technical societies in the world. It is called the International System of Units and is abbreviated "SI."

There are six basic elements involved in this system of units. The five that are applicable to circuit analysis provide measures of time, weight, distance, temperature, and current. They are respectively: second (s), kilogram (kg), meter (m), degree Kelvin (°K), and ampere (A). The sixth is luminous intensity measured in candela (cd). The accepted abbreviations are shown in parentheses.

Each unit has been accurately defined so that it is both permanent and reproducible. For instance the meter is defined as:

$1.65076373 \times 10^6 \times$ the wavelength of radiation of the orange line of krypton 86

We encounter both very large and very small numbers with respect to the standard units of measure in our study of electric circuits. In order to easily accommodate these numbers we will use a combination of scientific notation (with which we assume the reader is familiar) and standard prefixes. Table 1.1 gives prefixes commonly used, along with their symbol and meaning. Hence, the following statements are equivalent:

1,280,000 seconds

1.28×10^6 seconds

1.28 megaseconds

1.28 Ms

TABLE 1.1
COMMONLY USED PREFIXES

Prefix	Symbol	Meaning
pico	p	10^{-12}
nano	n	10^{-9}
micro	μ	10^{-6}
milli	m	10^{-3}
centi	c	10^{-2}
deci	d	10^{-1}
kilo	k	10^{3}
mega	M	10^{6}
giga	G	10^{9}
tera	T	10^{12}

Calculations involving large or small numbers are very handily performed using scientific notation. In fact, most electronic calculators have the capability to use this method to accept input and to display results.

Additional units of measure related to the SI standard will be introduced as they are encountered in the text.

LEARNING EVALUATIONS

1. Express each of the following terms in three additional ways.
   - **a.** 0.00168 meters
   - **b.** 3.577×10^{-2} s
   - **c.** 14 G °K
   - **d.** 0.877 A
2. Convert the following values to SI values.
   - **a.** 281 ft-lb
   - **b.** 40 °F
   - **c.** 6 in.
   - **d.** 422 s

1.3 Charge and Current

LEARNING OBJECTIVES

After completing this section you should be able to do the following:

1. Find the charge in coulombs of elementary particles.
2. Derive current $i(t)$ from charge $q(t)$.

Early investigators in electricity had few devices to work with, either to produce electric phenomena or to measure electric quantities. They rubbed amber with cat's fur or glass with silk in order to produce electricity. A number of other substances were found to be capable of being "electrified," including gems and rock crystal. These substances were called "electrics."

There were a number of substances, including metal, that could not be electrified, and hence were called "nonelectrics." Today we call these substances conductors, and the electrics are called insulators.

Electricity was considered somehow to be a fluid, for it could be demonstrated that if a glass rod was electrified, and if another electric such as cork was attached to the rod, then it too would become electrified. This was demonstrated in 1729 by an English electrician, Stephen Grey (1696–1736). Thus the fluid spread throughout the configuration. If a nonelectric (conductor) was connected between the electric and ground, the electric fluid would flow rapidly to ground.

It was recognized that there were two types of electricity. When glass was rubbed and the charge transferred to two cork balls, they repelled each other. The same thing happened when the charge from resin was used. But if one cork ball was charged with glass while the other was charged with resin they attracted each other. The glass induced charge was called "vitreous electricity" (from a Latin word for glass), while the resin induced charge was called "resinous electricity."

Benjamin Franklin (1706–1790) was one of the pioneers in this young science. In the 1740s he conducted experiments that led him to speculate that every substance contained electric fluid, but that a deficiency or excess of this fluid could be induced by rubbing electrics. It was known that when both glass and resin were rubbed they attracted each other, hence, one must contain an excess of fluid while the other had a deficiency. Franklin called vitreous electricity positive, and resinous electricity negative. This is the source of our present convention of calling electrons "negative" and protons "positive," for it is now known that vitreous electricity has a deficiency of electrons, while resinous electricity has a surplus of electrons.

Franklin intended for current to be the flow of excess fluid from positive regions to negative regions, but he had no way of knowing which was which. He was forced to guess, and he guessed wrong, for electric current is primarily the flow of electrons in conductors. But he could hardly have known, for it was not for another century and a half that electricity came to be associated with subatomic particles. We still use his convention today for several reasons. First, no harm is done (despite the horror of laymen who first discover this error). Second, current flow in many devices (especially semiconductors) is considered to be the flow of positive charges. Third, and most important, is long standing convention. The scientific community has persisted in using Franklin's convention, and they are not about to change now.

Current is the flow of charge. Measurements show that negatively charged electrons drift through a metal conductor under the influence of an electric field, but the effect of this charge movement travels at the speed of light down a conductor. Thus we get the picture of charge moving in unison down the length of a conductor, much like soldiers marching in single file. The last one moves almost simultaneously with the first one, the slight delay being due to the finite velocity of light.

The fundamental unit of charge is called the coulomb (C) and 1 C of negative charge has been internationally defined as the net charge of 6.24×10^{18} electrons. The symbol for charge is q and the charge of one electron then is

$$q = \frac{-1}{6.24 \times 10^{18}} \text{ C} = -1.602 \times 10^{-19} \text{ C} \tag{1.3.1}$$

Likewise, the charge of one proton is $+1.602 \times 10^{-19}$ C. The charge on an object can be a function of time and would be shown as $q(t)$.

Charge, however, is not of primary interest in circuit analysis. Rather, it is the movement of charge that demands our attention. Through control of the movement of charge we are able to transmit (move from one point to another) energy and information. The movement of charge is called current and is measured in terms of the ampere, after Andre Ampere, the French mathematician who published a treatise explaining the concept of current in 1820. The symbol for current is i and the abbreviation for the ampere (commonly called amp) is A. We normally are interested only in those situations where the electrons' movement is confined, such as in a copper wire. The ampere is defined as follows:

Definition 1.3.1. One ampere of current at a given point is the movement of one coulomb of charge through that point in exactly one second.

Let us return to our copper wire and choose a point on the wire. If an impartial official counts the number of electrons moving past that point in the same direction for a 1 s interval and gets an answer of 6.24×10^{18}, then 1 A of current has passed through that point.

Several questions should arise in your mind at this point. What if the charge were positive instead of negative? What if the charge were going in the other direction? An understanding of these concepts form the basis of circuit theory and is worthy of your efforts to conquer the subject matter. The answers to the two questions are not universally accepted, but the majority of engineering professionals assume the following convention:

> Positive current results from the movement of positive charge. The direction of the positive current flow is the direction of the movement of the positive charge.

This seems simple enough but is deceptively difficult to apply because the movement of charge is usually caused by the movement of electrons (negative charges). Consider the wire in Fig. 1.1. If we observe that there is a current of 1 A flowing from point A to point B, this can be caused by any of three situations:

1. A positive charge moving from A to B at the rate of 1 C/s.
2. A negative charge moving from B to A at the rate of 1 C/s.
3. A movement of positive charge from A to B and negative charge from B to A in such a manner that the *net* rate of charge moving from A to B is 1 C/s.

It is also equivalent to state that a current of $-i$ is flowing from B to A. Mathematically we see that current is the time rate of change of charge

$$i = \frac{\Delta q}{\Delta t} \tag{1.3.2}$$

Consider the plot of charge as a function of time given in Fig. 1.2. To calculate the current for the interval $0 \le t \le 2$ we simply calculate $\Delta q/\Delta t$ and note that the slope is constant over the time interval. Hence,

$$i(t) = \frac{(-1)-(2)}{2} = -1.5 \text{ A} \qquad 0 < t < 2$$

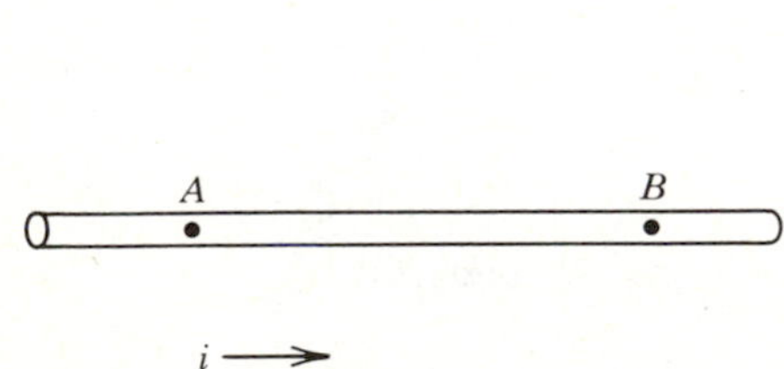

Figure 1.1. Direction of current.

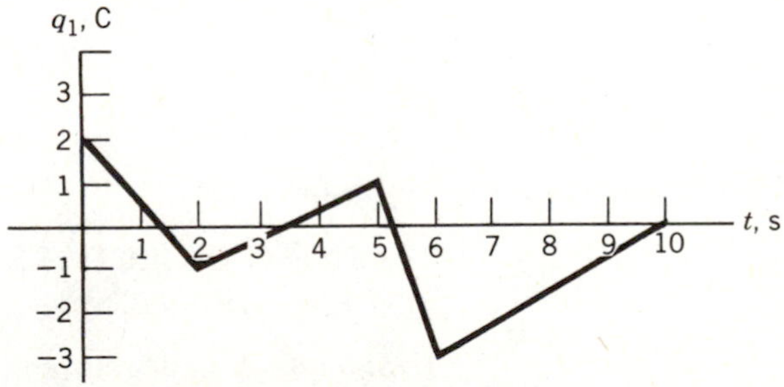

Figure 1.2. Charge as a function of time.

In the same manner,

$$i(t) = \frac{1-(-1)}{3} = 0.67 \text{ A} \qquad 2 \le t \le 5$$

$$i(t) = \frac{(-3)-(1)}{1} = -4 \text{ A} \qquad 5 \le t \le 6$$

$$i(t) = \frac{0-(-3)}{4} = 0.75 \text{ A} \qquad 6 \le t \le 10$$

We can plot $i(t)$ as shown in Fig. 1.3.

It is emphasized that to specify a current correctly we must specify a magnitude *and* a direction. Usually in circuit analysis we will not know beforehand the magnitude or direction of currents in our circuit. We will *assume* a direction of positive current flow and indicate this direction in our circuit by an arrow. If later calculations show the value of current to be negative we know that we assumed the direction of positive current flow in the direction opposite to the actual positive current flow. Return to Fig. 1.1. If we find that $i = -5$ A, then we say -5 A flows from A to B *or* $+5$ A flows from B to A.

LEARNING EVALUATIONS

1. Find the charge associated with

a. 10^{12} protons.
b. 10^{14} electrons and 10^{12} protons.

2. Find and plot $i(t)$ if $q(t)$ is given by the graph in Fig. 1.4.

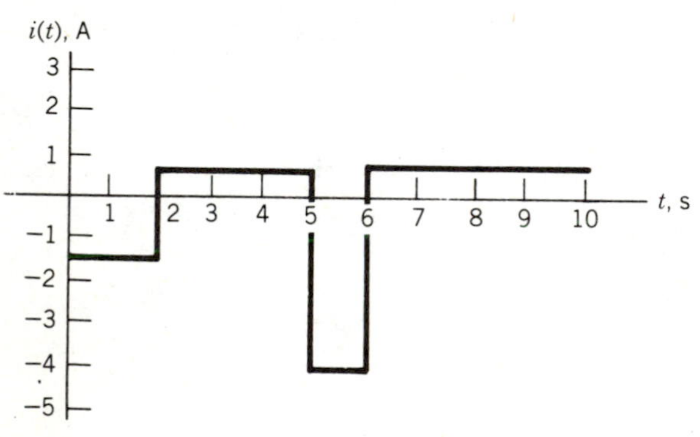

Figure 1.3. Current versus time.

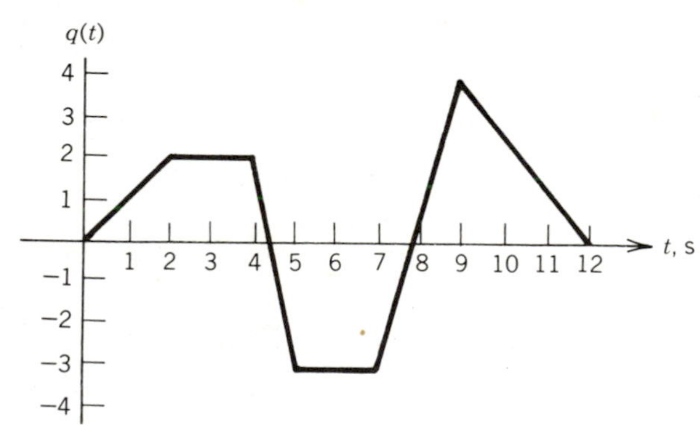

Figure 1.4

1.4 Kirchhoff's Current Law

LEARNING OBJECTIVES

After completing this section you should be able to do the following:

1. State Kirchhoff's current law.

2. Use Kirchhoff's current law to find the current entering or leaving a node.

Current flow in conductors is analogous to water flow in plumbing. Any water that enters a junction must leave. Thus the net quantity of water entering a junction must be zero.

Around 1870 Gustav Kirchhoff discovered that this same principle applied to current in conductors (wires) that are connected to the same point. This point is called a *node*. This principle is called *Kirchhoff's current law* (*KCL*) and can be stated as:

> The sum of currents entering a node equals the sum of currents leaving a node.

This is illustrated in Fig. 1.5 where by KCL at node A,

$$i_1 + i_3 = i_2 + i_4 + i_5 \tag{1.4.1}$$

If any four currents are known, the fifth can easily be calculated. However, if only three values are known, no conjecture can be made concerning the value of the two unknown currents.

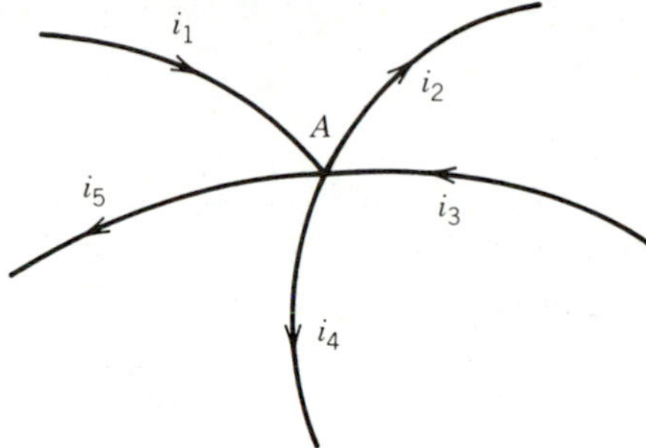

Figure 1.5. Node with five currents.

EXAMPLE 1.4.1 For the node shown in Fig. 1.6, find the value of current i.

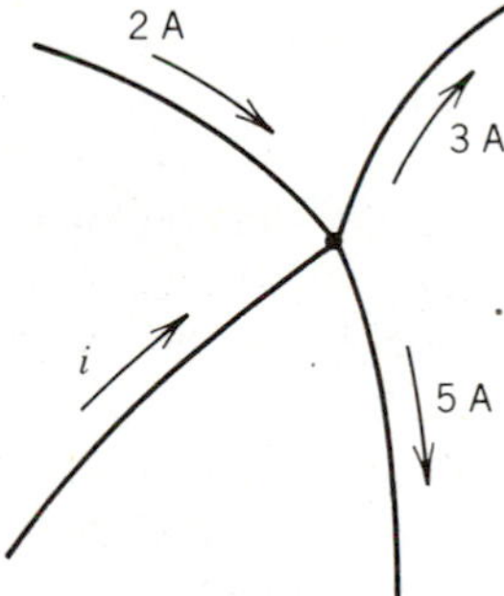

Figure 1.6

Solution

By KCL,

$$i + 2\text{ A} = 3\text{ A} + 5\text{ A}$$
$$i = 8\text{ A} - 2\text{ A}$$
$$i = 6\text{ A}$$

EXAMPLE 1.4.2 For the node shown in Fig. 1.7, find i.

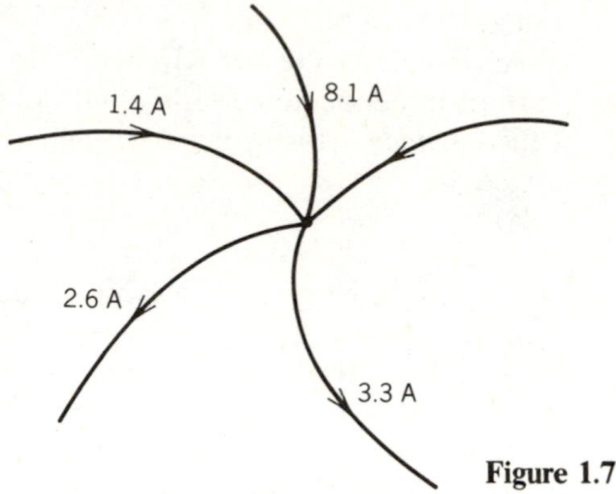

Figure 1.7

Solution

By KCL,

$$i + 8.1\text{ A} + 1.4\text{ A} = 2.6\text{ A} + 3.3\text{ A}$$

$$i = 5.9\text{ A} - 8.1\text{ A} - 1.4\text{ A}$$

$$i = -3.6\text{ A}$$

In Example 1.4.2, the determination of i as a negative number means that the positive direction of current flow of i is away from the node instead of toward the node. That is, we say that a current of -3.6 A is flowing toward the node or that $+3.6$ A is flowing away from the node.

Equation 1.4.1 can be rewritten as

$$i_1 + i_3 - i_2 - i_4 - i_5 = 0 \tag{1.4.2}$$

Equations 1.4.1 and 1.4.2 are obviously identical. Equation 1.4.2 is a mathematical representation of another, more formal, statement of Kirchhoff's current law:

> The algebraic sum of all currents entering a node is zero.

In using this latter version of the law you must assume that the currents entering a node are positive and those leaving are therefore negative or vice versa. You should convince yourself that these two statements of Kirchhoff's current law are identical.

EXAMPLE 1.4.3 Consider again the node shown in Fig. 1.6. If we assume that currents entering a node are positive and use the latter version of KCL, we have

$$i + 2\text{ A} - 3\text{ A} - 5\text{ A} = 0$$

or

$$i = 6\text{ A}$$

If we assume that the currents leaving a node are positive we have

$$-i - 2\text{ A} + 3\text{ A} + 5\text{ A} = 0$$

or

$$i = 6\text{ A}$$

In all cases we get the same answer which indicates that our initial assumptions serve only as a point of reference in writing the KCL equation. Since the answer is positive, we have correctly chosen the direction of the current i. The key to success is to be consistent. Once you make some assumptions on the direction of current flow you cannot change them. If your answer is negative it simply means that positive current flows in the opposite direction.

The version of Kirchhoff's current law that you use should be determined by which version you feel most comfortable with.

LEARNING EVALUATION

For each of the nodes in Fig. 1.8, find the value of i.

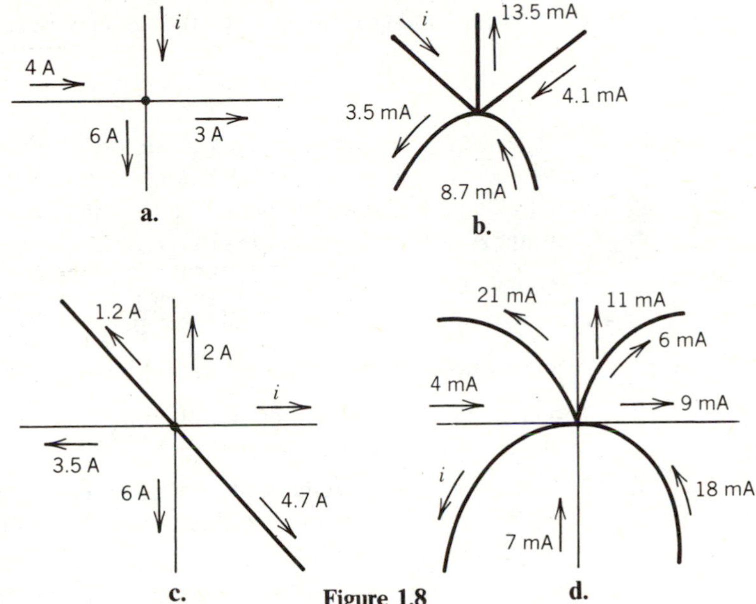

Figure 1.8

1.5 Voltage and Power

LEARNING OBJECTIVES

After completing this section you should be able to do the following:

1. Define voltage and give its unit of measure and abbreviation.
2. Define a circuit element.
3. Define power and give its unit of measurement and abbreviation.
4. Describe the difference between power being absorbed by a circuit element and power being delivered by a circuit element.

Voltage is simply electric potential energy. To grasp this idea, consider for a moment gravitational potential energy. When a mass is lifted against gravity its potential energy is increased by the increase in height. This gain in potential energy is supplied by the lifting device (our muscles, perhaps). The quantity of energy that must be added depends on the difference in height through which the mass is lifted, not on its absolute height. This increase in energy is called gravitational potential difference. The joule (J) is the unit of energy and the kilogram (kg) is the unit of mass. Therefore, the unit of gravitational potential difference is a joule per kilogram (J/kg).

In an electrical system, energy must be supplied to separate two unlike charges (or to move two like charges closer together). The energy supplied is converted into electric potential energy, and the difference between the two energy states before and after moving the charge is called electric potential difference. The unit of charge is the coulomb (C); therefore, the unit of electric potential difference (voltage) is a joule per coulomb (J/C).

If w is the work (or energy), q is the charge, and v is the voltage, then the relationship among these three quantities may be expressed as

$$v = \frac{\Delta w}{\Delta q}$$

Thus, the voltage between points A and B is the amount of energy required to move 1 C of positive charge from point A to point B. The unit of voltage is the volt (V) and is named in honor of the Italian physicist Alessandro Giuseppe Antonio Anastasio Volta (1745–1827). Let us formalize this by a definition.

Definition 1.5.1. The voltage between points A and B is the work expended in moving a unit charge from point A to point B.

Some interesting observations should be made at this point.

1. The voltage between point A and point B is independent of the path taken by the charge as it moves from A to B.
2. Voltage is a measurement of how much energy *would* be required to move a charge from A to B and can therefore exist between two points even if there is no charge movement.
3. The movement of charge (current) can only take place if there is a closed path along which the movement can occur.
4. Voltage must have reference directions assigned just as we did with current. This is done by marking one point with a "+" sign and the other with a "−" sign.

Before we discuss voltage and current and their relation any further, it is to our advantage to define a *circuit element*. We said at the beginning of this chapter that an electrical circuit is a grouping of certain devices for a particular purpose. Any one device is called a circuit element and can be represented as shown in Fig. 1.9.

Points A and B are called the terminals of the circuit element. Every circuit

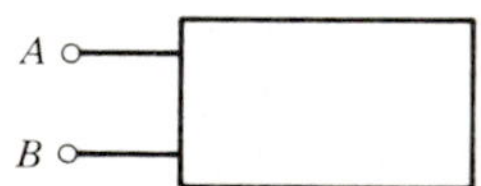

Figure 1.9. A circuit element.

element must have at least two terminals. An electrical circuit with three elements is shown in Fig. 1.10.

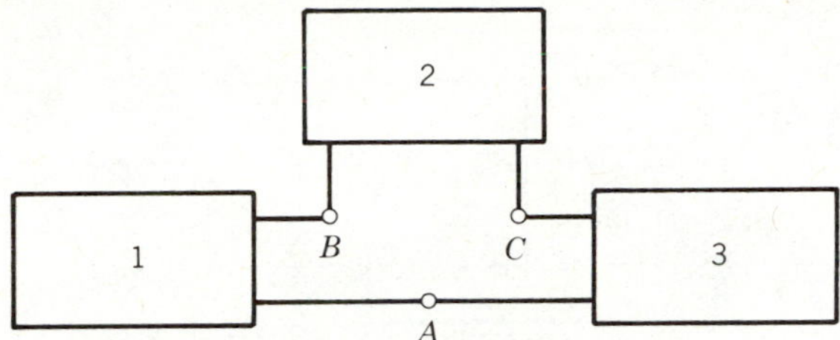

Figure 1.10. A three-element circuit.

In Fig. 1.10 the terminals associated with element 1 are A and B; with element 2 are B and C; and with element 3 are A and C. You will note that points A, B, and C are nodes, according to our previous definition.

Now we can establish the relationship between current, voltage, and a circuit element. A voltage between terminals A and B in Fig. 1.9 will cause a current to flow through the circuit element. By our definition of voltage, if a positive sign is placed at A and a negative sign at B, then current will flow into the element at point A. See Fig. 1.11.

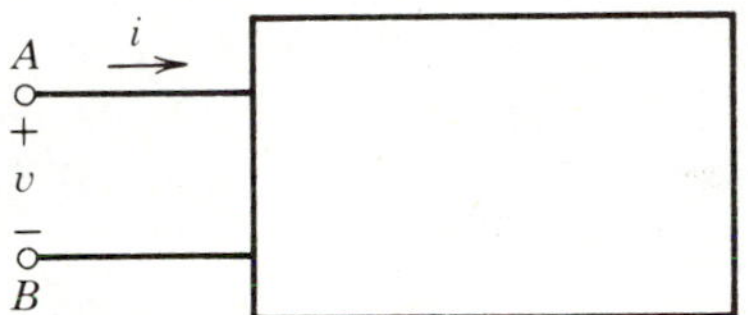

Figure 1.11. A circuit element showing voltage–current relationship.

In order for current to flow through an ordinary circuit element, a voltage *must* be present at its terminals. However, a voltage can exist without any actual current flow. Voltage is analogous to water pressure and current to water flow. The pressure exists at the nozzle of a hose regardless of whether any water flows through the nozzle (that is, we can have the nozzle shut off). Conversely, if there is no water pressure, water will not flow from the nozzle even if it is open.

The amount of current flowing through a circuit element for a given voltage is determined by the characteristics of the element and is the subject of later chapters.

The *power*, p, absorbed by the element is defined as the current flowing through the element from the positive terminal times the voltage across the terminals of the element:

$$p = vi \tag{1.5.1}$$

The unit of power is the watt (W). If the product of Eq. 1.5.1 is negative, then the element is delivering power instead of absorbing it.

EXAMPLE 1.5.1 Find the power absorbed by the element in each of the cases in Fig. 1.12.

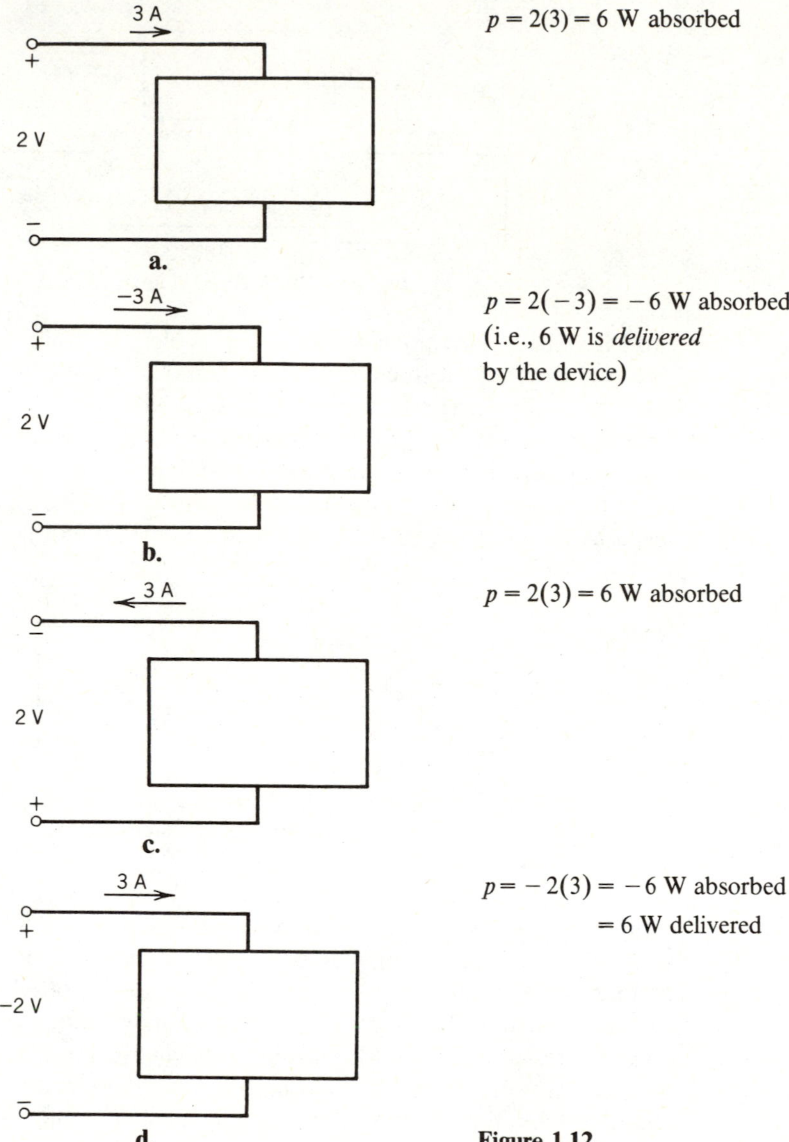

Figure 1.12.

Here is a short review. We have thus far introduced five quantities, their symbols, units, and abbreviations as listed in Table 1.2. We have established three relationships among the five quantities:

$$i = \frac{\Delta q}{\Delta t}, \qquad v = \frac{\Delta w}{\Delta q}$$

and

$$p = vi$$

TABLE 1.2
BASIC ELECTRICAL TERMS

Quantity	Symbol	Unit	Abbreviation
Charge	q	Coulomb	C
Current	i	Ampere	A
Voltage	v	Volt	V
Energy	w	Joule	J
Power	p	Watt	W

We have defined two terms:

Node. A connection of two or more conductors.

Circuit element. A device with at least two terminals.

And we have stated one law:

Kirchhoff's current law. The algebraic sum of the currents entering a node is zero.

Before proceeding to the last sections of this chapter, you should be sure you understand the concepts introduced thus far.

LEARNING EVALUATION
Find the power *absorbed* by each of the circuit elements in Fig. 1.13.

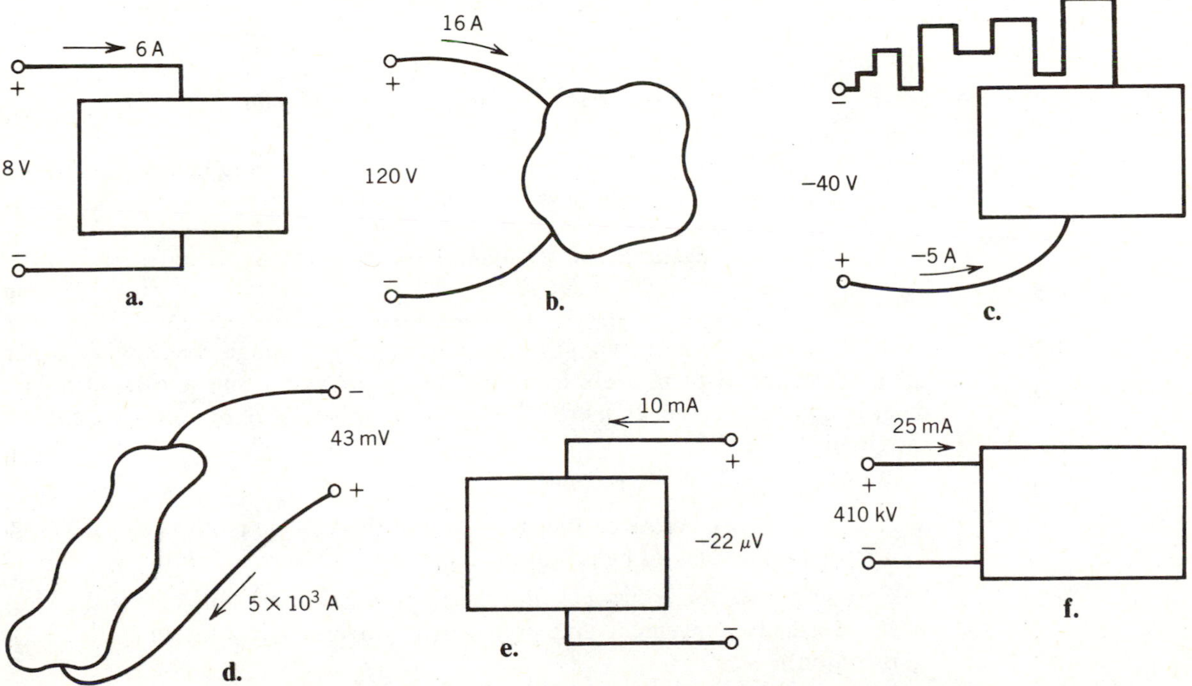

Figure 1.13

1.6 Kirchhoff's Voltage Law

LEARNING OBJECTIVES

After completing this section you should be able to do the following:

1. State Kirchhoff's voltage law.
2. Use Kirchhoff's voltage law to determine an unknown voltage around a closed path.

Refer again to our basic circuit element in Fig. 1.11. The voltage v is referred to as a voltage drop across the circuit element; that is, a voltage drop from terminal A to terminal B. Energy is expended in forcing a charge to travel from A to B so the energy remaining at B is less than the energy at A if measured with respect to some neutral point of reference.

Often two or more elements are connected together to form a closed path (loop). Three such circuits are shown in Fig. 1.14.

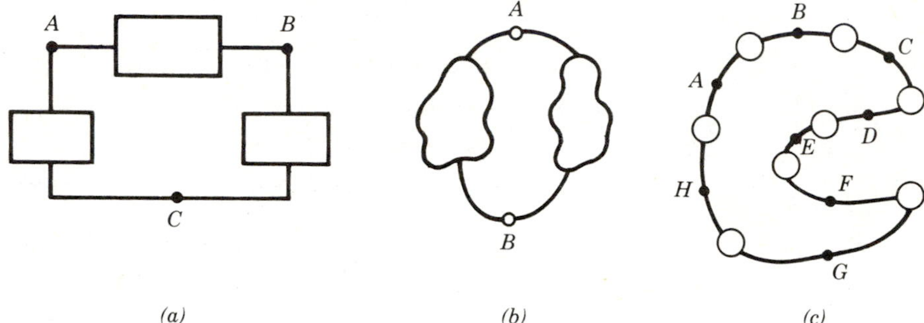

Figure 1.14. Circuits with one closed path.

Gustav Kirchhoff has a second circuit law to his credit called *Kirchhoff's voltage law* (*KVL*).[1] This law says:

The algebraic sum of all voltages around any closed path is zero.

Unless voltages are known, arbitrary polarities for voltage are assigned along the circuit path. If there are n elements then n voltages occur and $n-1$ values must be known in order to use KVL to find the other value of voltage. Consider the circuit in Fig. 1.15.

In order to apply KVL we must:

a. Assign a voltage across each element where the voltage is unspecified. Which terminal we assume to be positive is arbitrary.
b. Choose a direction to traverse the closed path and mark this on the drawing. (It is usually convenient and practical to traverse all paths in a clockwise direction.)

[1]Kirchhoff (1824–1887) is best known for his work with light. He invented the spectroscope and pioneered its use in astronomy.

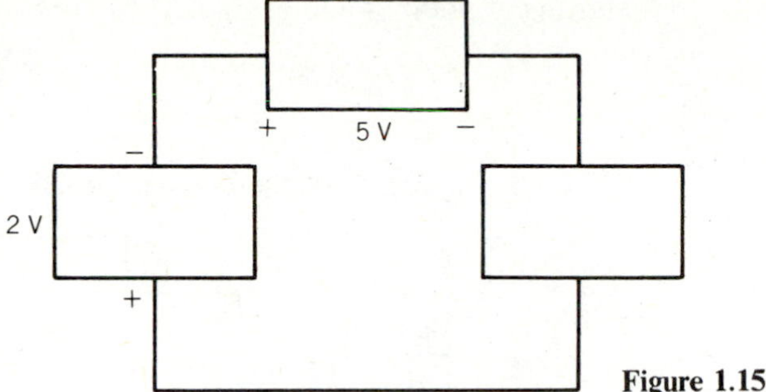

Figure 1.15

c. As we start at any point and traverse the path in the direction chosen, enter all voltages for which we encounter the positive terminal first as positive voltages in our equation; enter all values for which we encounter the negative terminal first as negative.

In Fig. 1.16 we have added a voltage reference and direction arbitrarily to the circuit of Fig. 1.15. This satisfies step (a) in the preceding list.

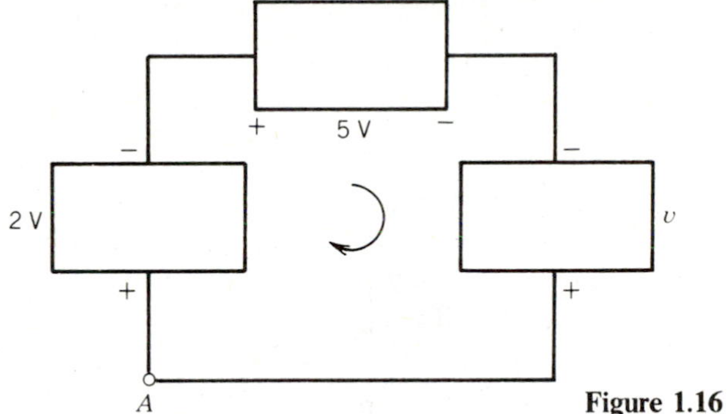

Figure 1.16

Starting at point A and going clockwise (cw) the KVL equation is

$$+2\text{ V} + 5\text{ V} - v = 0$$

or

$$v = 7\text{ V}$$

In Fig. 1.17 the direction has been changed to counterclockwise (ccw). Again,

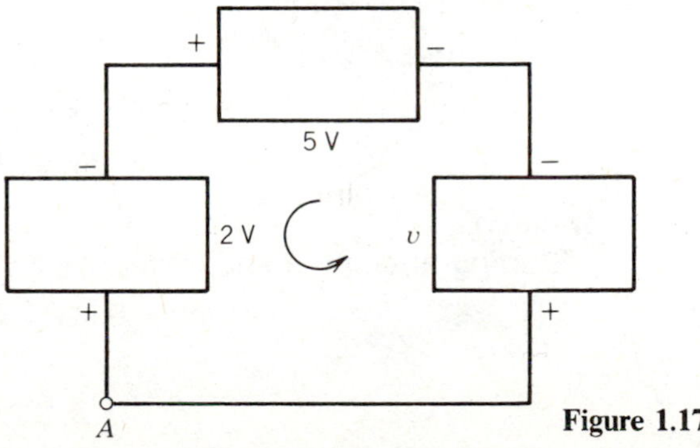

Figure 1.17

starting at point A and using KVL we obtain

$$+v - 5\text{ V} - 2\text{ V} = 0$$

or

$$v = 7\text{ V}$$

Finally, in Fig. 1.18 the assumed polarity of v is reversed. Starting at A and

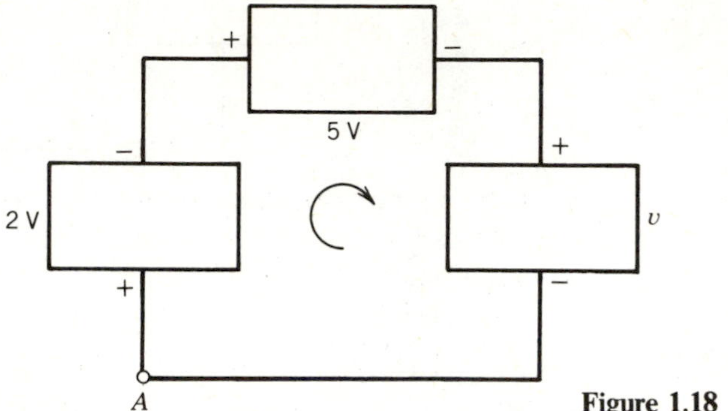

Figure 1.18

going cw, KVL gives us

$$+2\text{ V} + 5\text{ V} + v = 0$$

or

$$v = -7\text{ V}$$

The negative sign in the answer indicates we have chosen the opposite polarity and therefore all answers agree. We should now all agree that the direction of travel around the closed path and the arbitrary assignment of voltage polarity for unknown voltages will not affect the outcome of applying KVL.

It seems that the most common practice is to always traverse the path in the cw direction, but this is a matter you should settle for yourself. It is our experience that students who choose a particular direction and stick with it on all problems make far fewer mistakes than those who vacillate between cw and ccw.

EXAMPLE 1.6.1 For the circuit shown in Fig. 1.19, find v.

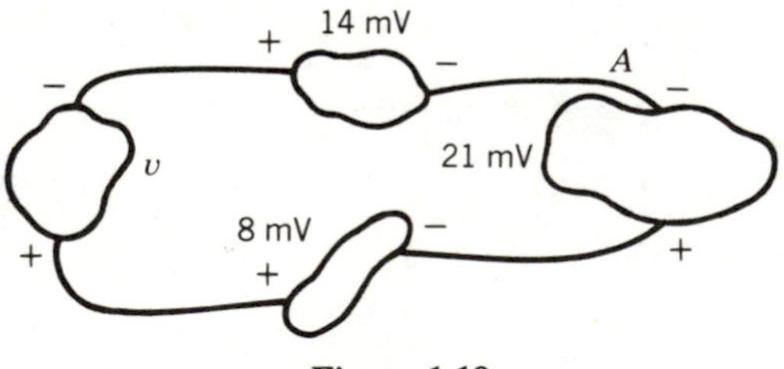

Figure 1.19

Solution

Starting at point A and going cw, KVL yields

$$-21\text{ mV} - 8\text{ mV} + v + 14\text{ mV} = 0$$

or

$$v = 15\text{ mV}$$

EXAMPLE 1.6.2 Find the value of v in the circuit shown in Fig. 1.20 by using KVL.

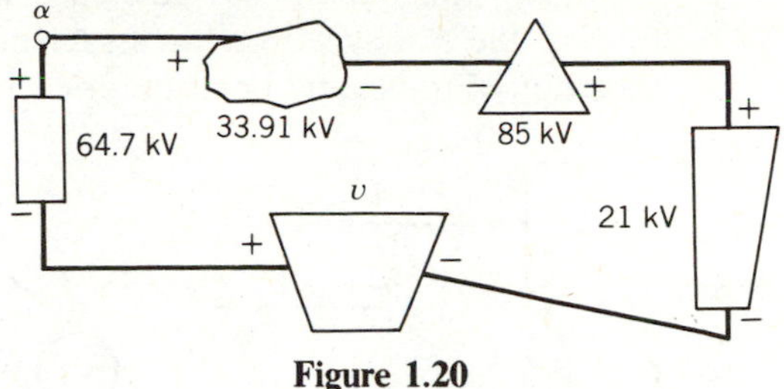

Figure 1.20

Solution

Applying KVL in a cw direction from point α,

$$33.91 \text{ kV} - 85 \text{ kV} + 21 \text{ kV} - v - 64.7 \text{ kV} = 0$$

or

$$v = -94.79 \text{ kV}$$

Here the negative sign on the answer means that the actual polarity of v is opposite of that shown in the circuit.

We will have hundreds of occasions to use Kirchhoff's two circuit laws as we introduce new concepts and methods throughout the book.

LEARNING EVALUATIONS

1. Use KVL to find v for each of the circuits in Fig. 1.21:

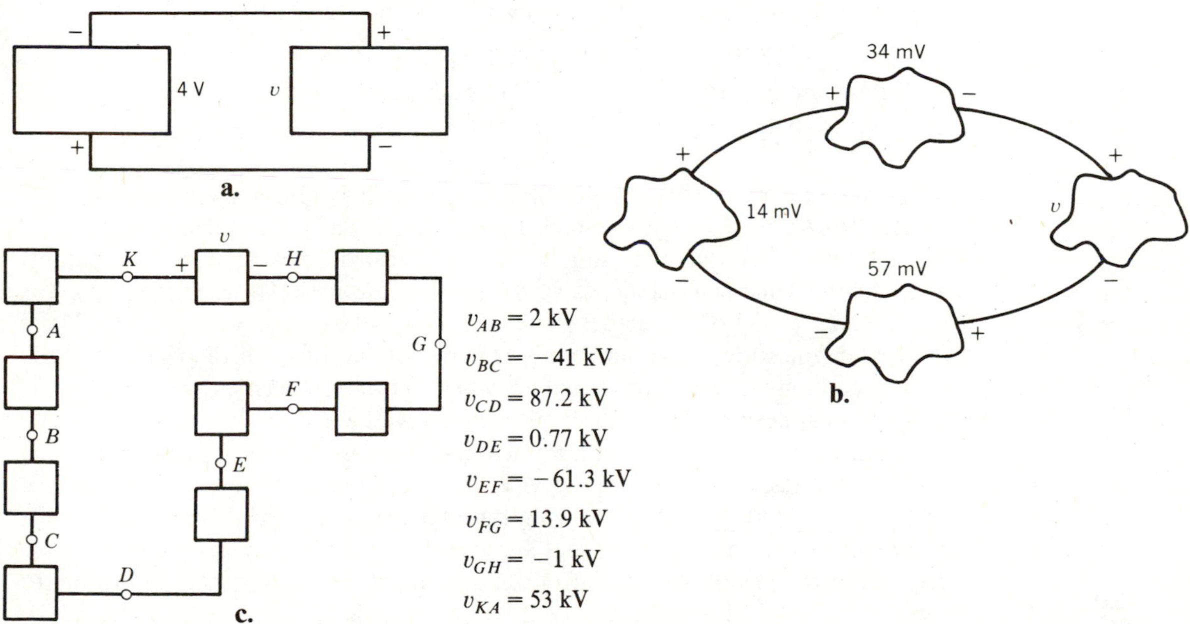

Figure 1.21

2. Use KVL in each circuit in Fig. 1.22 to show that any chosen direction of traversing the circuit and any chosen polarity give the same answer when finding the voltage drop across element β (i.e., use four combinations of direction of travel and polarity for each circuit).

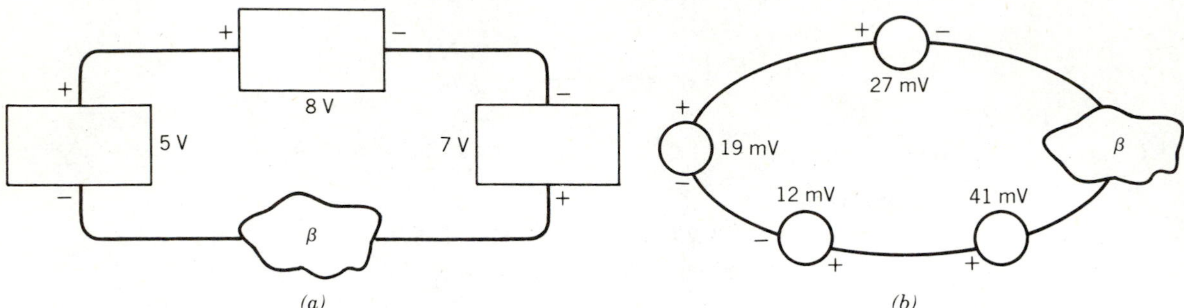

Figure 1.22

1.7 Circuit Element Concepts

LEARNING OBJECTIVES

After completing this section you should be able to do the following:

1. Define and give the circuit model symbol for an independent voltage source.
2. Describe what is meant by an "ideal" voltage source.
3. Define and give the circuit model symbol for an independent current source.
4. Determine whether a source is delivering or absorbing power.
5. Relate energy to power when the power is not a function of time.
6. Define and give the circuit model for a dependent voltage source.
7. Define and give the circuit model for a dependent current source.

We are all familiar with the electronic games such as Ping-Pong played on a TV screen. We recognize that the ball, paddles, and walls are not actual physical devices but rather that they simulate the physical devices; that is, they exhibit some of the same characteristics as the physical devices. The electronic game is said to be a model of the actual game.

Most engineering design and analysis is performed through the use of models. This is certainly true in circuit analysis where our circuit element will be a *model* of a physical device. The basic circuit element we have described so far has been an arbitrarily shaped device with no defining mathematical relationships between the variables associated with the device; that is, between voltage at the terminals of the device and the current flowing through the device. In this section we will introduce four circuit elements that are models of physical energy sources. In the remainder of the text we will introduce additional circuit elements from time to time.

The law of conservation of energy says that energy can neither be created nor destroyed. If we are going to expend energy in a circuit we must then have an

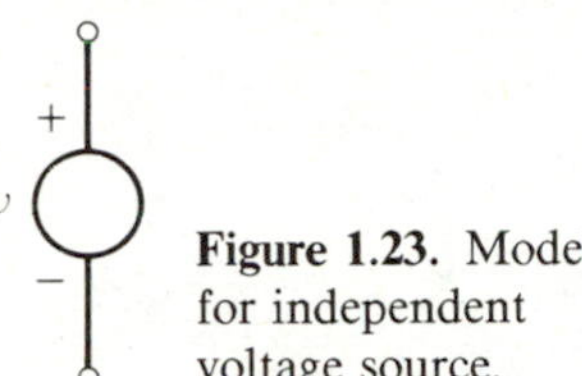

Figure 1.23. Model for independent voltage source.

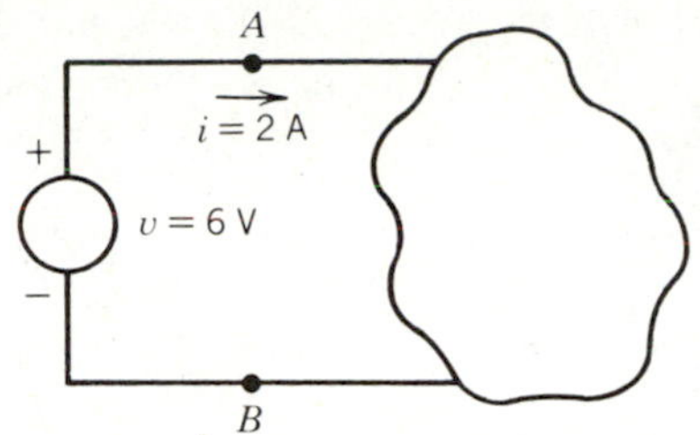

Figure 1.24

equivalent energy source. The first energy source we will examine is called the *independent voltage source* and is represented in the circuit model by the symbol shown in Fig. 1.23.

This source is assumed to supply a voltage v (which is often called the terminal voltage) regardless of the amount of current flowing out of (or into) the voltage source. The amount of power supplied by the source is given by Eq. 1.5.1 as the product of the voltage and the current.

The independent voltage source in Fig. 1.24 is supplying 12 W of power, assuming some other element or group of elements is connected to terminals A and B. This source is called an "ideal" source and does not model any physical source exactly for all values of v and i. However, the ideal source approximates the actual source closely enough to be very useful in circuit analysis. In general, v is represented by a time varying function. Initially we will make use of a special case where v is a constant value for all time. This is often called a direct current, or "dc," voltage source.

The second energy source we describe is called the "independent current source" and is represented in the circuit model by the symbol shown in Fig. 1.25.

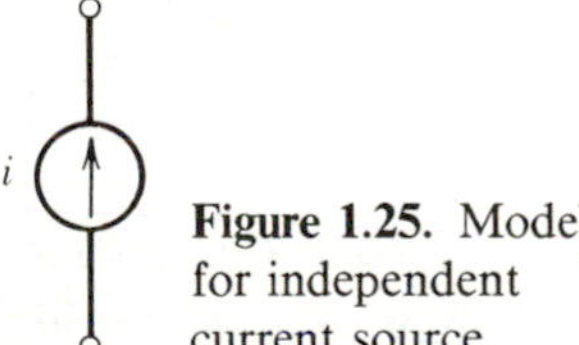

Figure 1.25. Model for independent current source.

This source is defined as being able to supply a current i independent of whatever other circuit elements may be attached to its terminals. In other words, the magnitude of the current supplied by the source is not affected by the voltage that may appear across its terminals. The direction of current flow is indicated by the arrow inside the circle. As before, an important special case exists where i is not a function of time. (i.e., the current has a constant value for all time of interest). This is usually called a dc current source.

As in the case of the independent voltage source, the power supplied is given by Eq. 1.5.1. The point should be made here that power can either be "supplied" or "absorbed" by a source. Basically if the current flows out of the positive voltage terminal of a source, power is supplied by the source; otherwise, it is said to be absorbed by the source. This subject will be revisited and the definition formalized later in the text.

EXAMPLE 1.7.1 For each of the sources in Fig. 1.26, give the type of source and specify whether power is absorbed or supplied by the source.

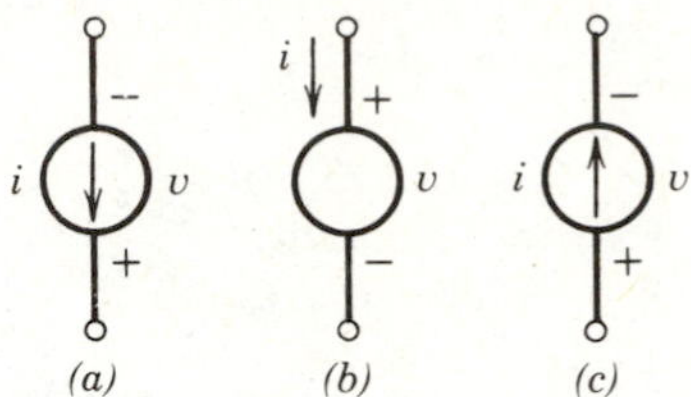

Figure 1.26

Solution

Parts (*a*) and (*c*) are current sources, since an arrow appears inside the circle. Part (*b*) is a voltage source since no arrow appears inside the circle. Part (*a*) supplies power since the current is shown as flowing out of the positive voltage terminal; (*b*) and (*c*) absorb power because the current is shown as flowing into the positive voltage terminal.

By now you may be asking why we started describing energy sources but have only talked about voltage and current sources. The answer lies in the fact that both of these *are* energy sources. If the voltage and current are both independent of time (dc voltage and dc current), then the energy supplied (absorbed) by the source is equal to the power supplied (absorbed) by the source multiplied by the length of time, t, the power is supplied (absorbed).

$$w = pt \tag{1.7.1}$$

Here w is used as the symbol for energy and this equation is only true if p is not a function of time. As we did elsewhere in this chapter, we promise that later developments in the text will demonstrate a much more general form of Eq. 1.7.1.

EXAMPLE 1.7.2 An ordinary car battery can be modeled as a dc independent voltage source with a satisfactory degree of accuracy. We assume the electrical system of the car is attached to its terminals. A model of this physical system is shown in Fig. 1.27.

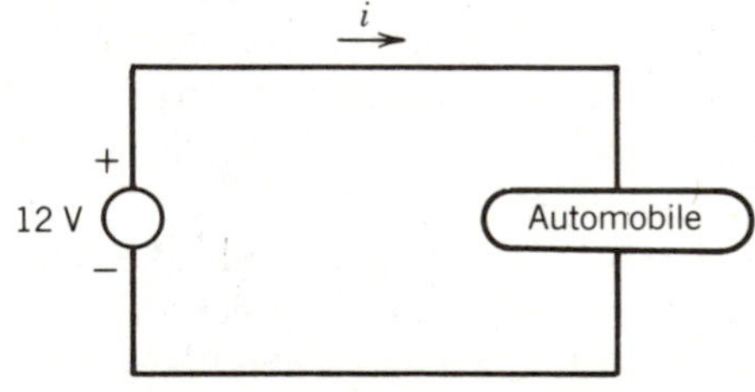

Figure 1.27

Assume that we read on the battery that if it is fully charged it can supply 90 A of current for 1 hour (h). How much energy can the battery supply?

Solution

From Eq. 1.7.1

$$w = pt$$

but using Eq. 1.5.1

$$w = vit$$

for our system

$$\begin{aligned} w &= (12)(90)(1) \\ &= 1080 \text{ W-h} \\ &= 1.08 \text{ (kilowatt-hours) kWh} \end{aligned}$$

Some additional observations about this example are in order. After the automobile is started, the electrical system reverses the indicated direction of current so that current actually flows into the positive terminal of the battery. Under these conditions, the battery is said to be "charging," or absorbing energy. Charge is actually stored inside the battery, and when fully charged, enough charge exists in the battery to supply a current of 90 A for 1 hour. Remember charge is measured in coulombs and current is measured in the number of coulombs flowing past the measurement point in 1 s. A "brand x" battery may be more cheaply built and may not be able to store as much charge, and could therefore only supply 90 A for a time less than 1 hour. It is also possible for the electrical system of the auto to be damaged or altered in a way such that the battery cannot maintain a voltage of 12 V. Therefore, our model of the system is only true under certain conditions. Fortunately these conditions hold for sufficient numbers of cases to make the model useful.

The last two circuit elements we will introduce here are the dependent voltage and dependent current sources. The circuit models for these two sources are shown in Fig. 1.28.

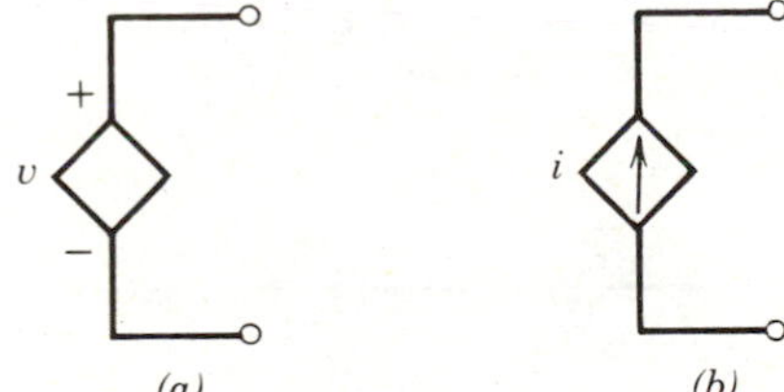

Figure 1.28. Models for dependent voltage and current sources. (*a*) Dependent voltage source. (*b*) Dependent current source.

The "circle" symbol has been changed to the "diamond" symbol to indicate that the source is dependent instead of independent. By "dependent" we mean that the voltage or current supplied is not predetermined but rather is a function of some other circuit parameter. We have not yet introduced enough circuit concepts to explain completely this situation but perhaps it can be illustrated by using an analogy of a burner of a gas stove. The burner can be considered a source of heat energy but the amount of energy supplied is a function of the setting of the control knob on the stove. For a given knob setting the energy output of the burner is constant regardless of the size of pan placed on the burner. In the same way, a dependent energy source in an electrical circuit is a function of some other circuit value but acts as an independent source to the remaining circuit elements. Dependent sources play a very important role in modeling some physical electronic devices such as transistors.

With the understanding of charge, current, voltage, power, and energy, the enlightenment given by Kirchhoff's voltage and current laws, and the descriptions of the energy sources as circuit elements, we are now ready to consider the application of these concepts in electric circuits involving the resistor. This is the subject of Chapter 2.

LEARNING EVALUATIONS

1. For each of the sources shown in Fig. 1.29, specify the type of source and determine whether power is absorbed or delivered.

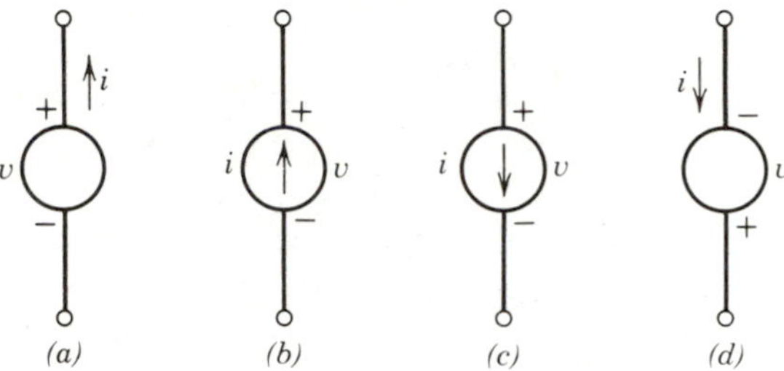

Figure 1.29

2. An air conditioner in a home can be considered a circuit element. Assume that 220 V is connected across the terminals of a special air conditioner and that 15 A flows through the air conditioner from the positive terminal.
   a. Calculate the energy used in one month if the air conditioner runs 9.6 h/day
   b. If electrical energy costs 9 cents per kilowatt-hour, determine the monthly charge for using the air conditioner.

PROBLEMS

SECTION 1.2

1. The speed of light is close to a foot per nanosecond. Assuming the speed of light is 3×10^8 m/s, determine the correct value in terms of feet per nanosecond.
2. In computer technology one limit on the speed (clock rate) is the response time of the gates, or elementary switches. In the early computers that employed vacuum tubes the cycle times were about 10^{-5} s. With the advent of transistors the cycle times were reduced to about 10^{-7} s. Integrated circuit technologies have further cut the cycle times to less than 10^{-9} s. Express each of these times in nanoseconds.

3. Express each of the given values in terms of scientific notation and basic SI units.

   a. 0.0001 grams
   b. 432×10^6 centiseconds
   c. 1432 °K
   d. 1760 yards

SECTION 1.3

4. For the copper wires in Fig. 1.30 specify the magnitude and direction of positive current flow. [In parts (c), (d), and (e), one charge = 0.5 C.]

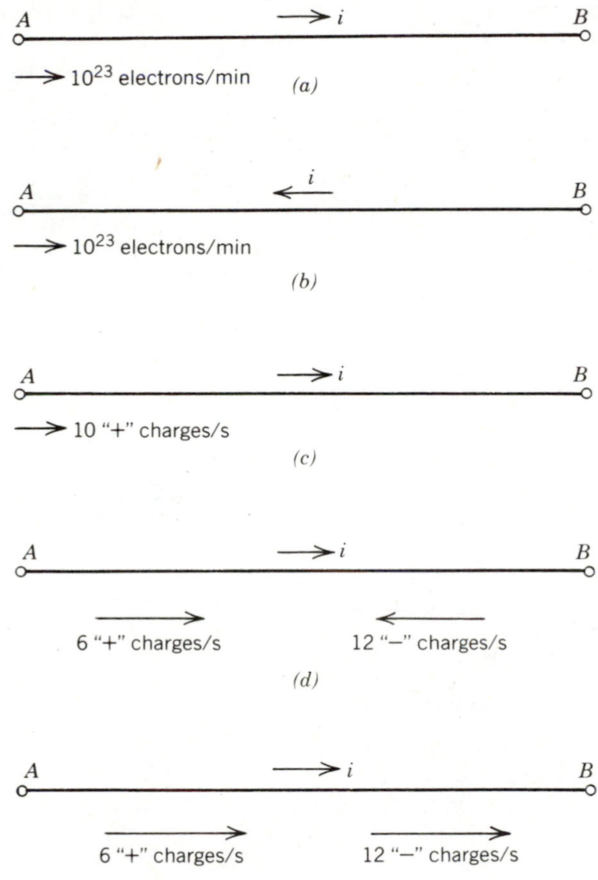

Figure 1.30

5. The current through a circuit element is monitored for 10 ms, and the results are given as

$$
\begin{array}{ll}
25 \text{ mA} & 0 \le t \le 2 \text{ ms} \\
0 \text{ mA} & 2 \le t \le 2.5 \text{ ms} \\
200 \text{ mA} & 2.5 \le t \le 8 \text{ ms} \\
-50 \text{ mA} & 8 \le t \le 10 \text{ ms}
\end{array}
$$

Also, $i = 0$ for $t < 0$ and $t > 10$ ms.

a. Plot the current from $t = 0$ ms to $t = 10$ ms.

b. Calculate and plot the charge (in coulombs) transferred during each time period.

6. A battery delivers a constant current of 6 A for 30 min. How many coulombs of charge pass through the positive terminal of the battery?

SECTION 1.4

7. For the following situations, use the first statement of Kirchhoff's current law to find the unknown current(s) in Fig. 1.31.

a. $a = +2$ A	**b.** ar $= +2$ A	**c.** $a = 160\ \mu$A
$b = -2$ A	$c = -5$ A	$b = +3.2$ mA
$d = +7$ A	$d = 5$ A	$c = 3$ mA

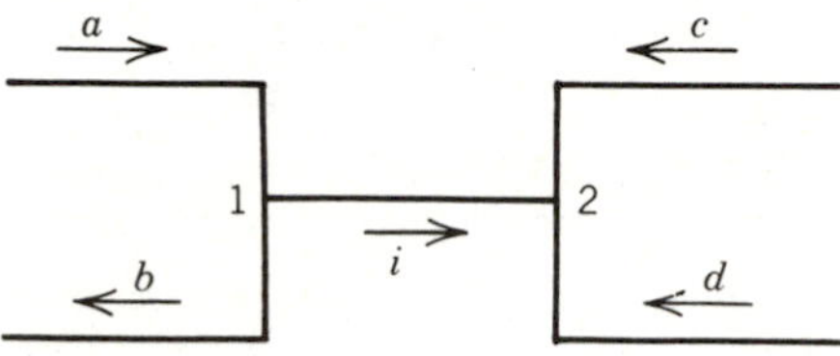

Figure 1.31. Problem 7(a) to (c).

d. See Fig. 1.32.

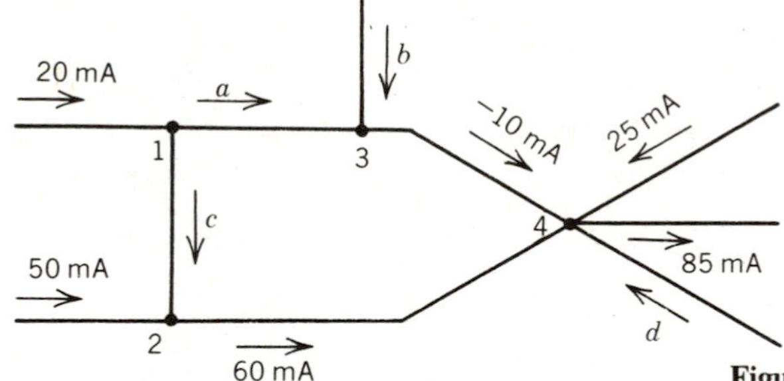

Figure 1.32. Problem 7(d).

e. See Fig. 1.33.

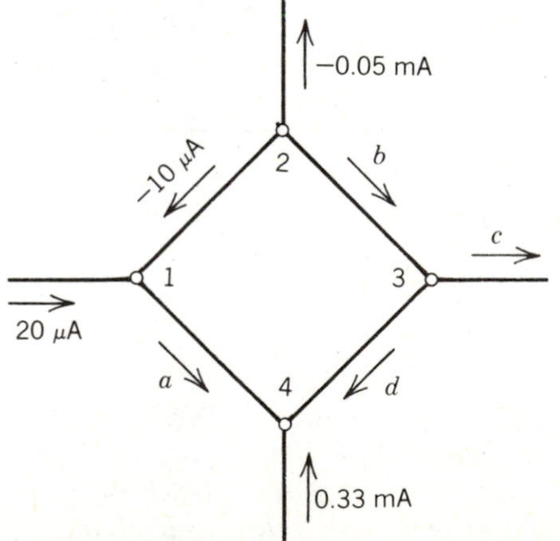

Figure 1.33. Problem 7(e).

8. Reverse the direction of the arrows of each current of Problem 7. Find the unknown currents by using KCL. Compare your results to those of Problem 7.

9. Use the second version of KCL to find the unknown current(s) in the diagrams of Fig. 1.34. (Assume current flow into a node is "+" current.)

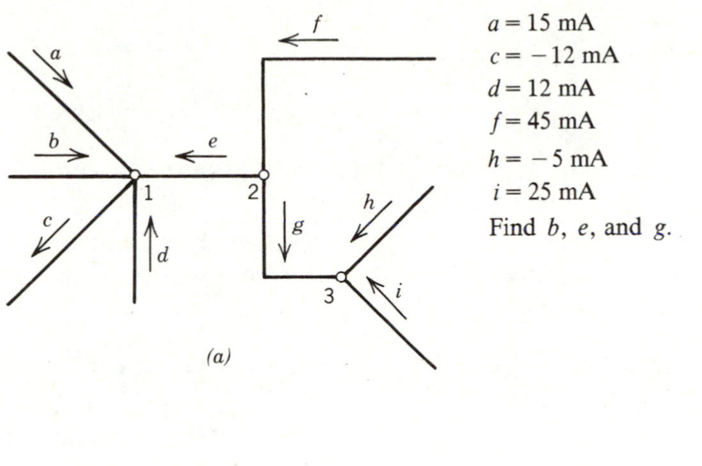

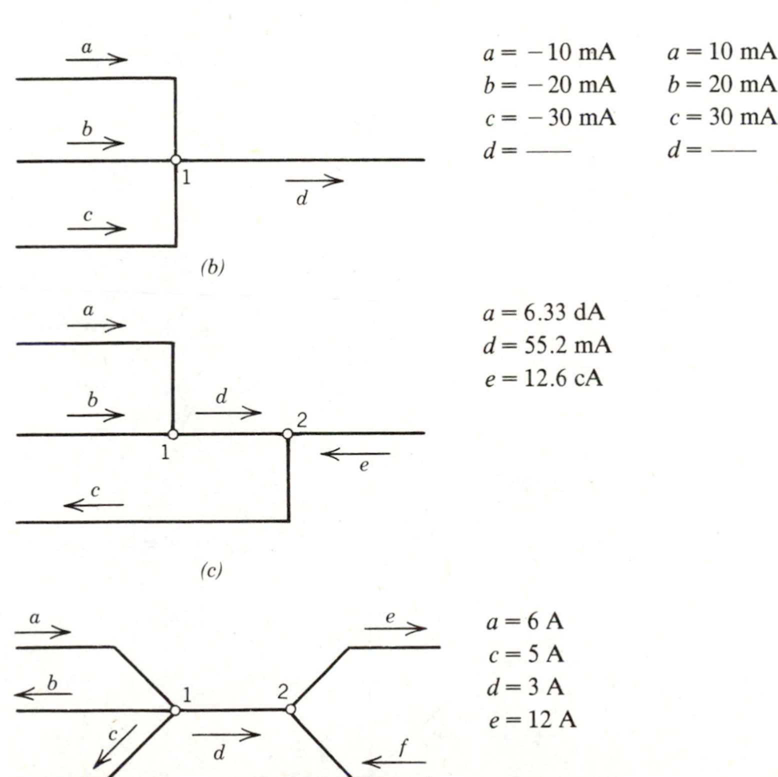

Figure 1.34.

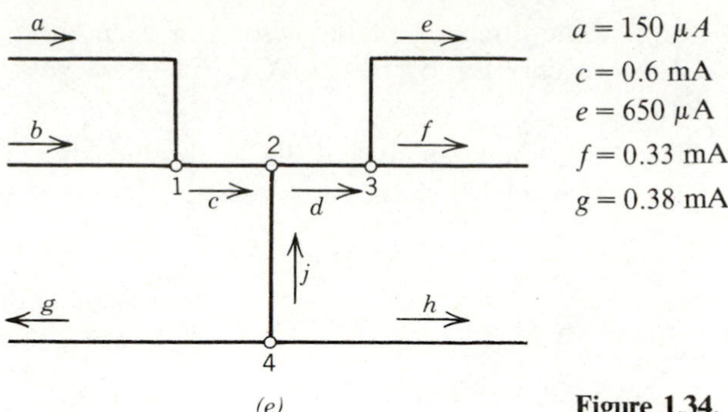

Figure 1.34. (*Continued*).

10. For each part of Problem 9, assume that current flow into a node is "−" current

11. Using KCL, determine whether the current i is correct in Fig. 1.35. If i is incorrect, calculate a value consistent with the other given currents in the problem.

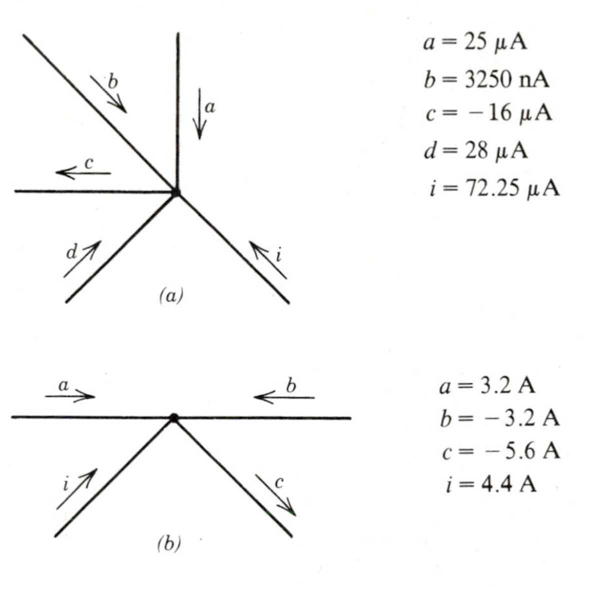

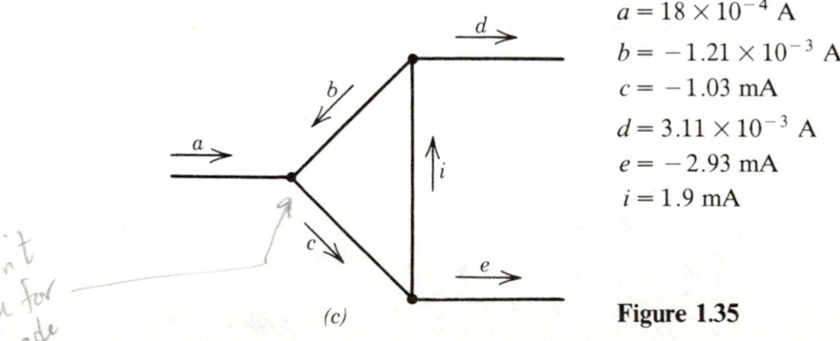

Figure 1.35

SECTION 1.5

12. A certain battery has energy that will transfer 74.98×10^{18} elementary positive charges between two points. Determine the voltage across the "+" and "−" terminals of the battery.

13. For the circuit element in Fig. 1.36 choose a voltage polarity convention. Show the current flowing into the "+" terminal of the circuit element and calculate the power absorbed by the element.

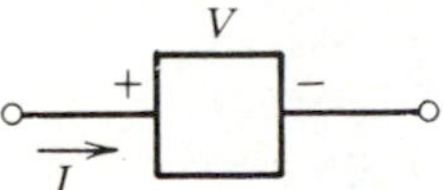

Figure 1.36

a. $V = 6.5$ mV
$i = 4$ mA

b. $V = 66.5$ kV
$i = -21.4$ mA

c. $V = -18$ V
$i = -12$ A

d. $V = -42.1$ mV
$i = 16.4\ \mu$A

14. Use the relation $p = iv$, and KCL to find the power associated with the circuit element in Fig. 1.37. Is the power absorbed or delivered?

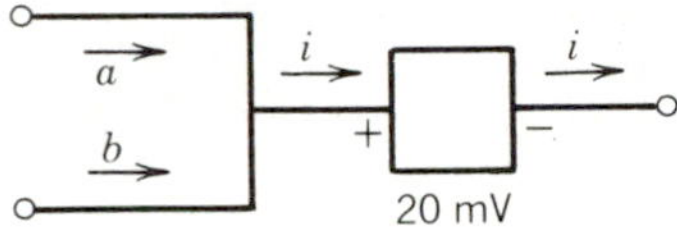

Figure 1.37

a. $a = 95$ mA; $b = -43$ mA
b. $a = 40\ \mu$A; $b = -40\ \mu$A
c. $a = 50$ mA; $b = -85$ mA

15. If the power delivered in Fig. 1.37 is 350 mW, calculate the current required for the following voltage drops across the circuit element.

a. 45 mV
b. 2 V
c. − 16 mV

16. An electrical circuit is represented by the circuit elements in Fig. 1.38.

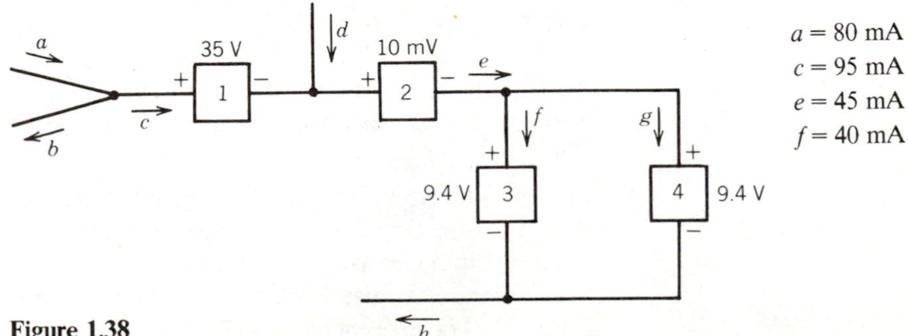

Figure 1.38

Assume the current going into a circuit element equals the current coming out of the element.

a. Find the unknown currents.
b. Calculate the power absorbed by each element.
c. As an electrical engineer, your employer authorizes you to replace one circuit element with a more efficient circuit element that absorbs 20% less power. Which element would you replace and why?

SECTION 1.6

17. The device in Fig. 1.39 consists of five circuit elements. Use KVL to find the unknown voltage drop for each unspecified element.

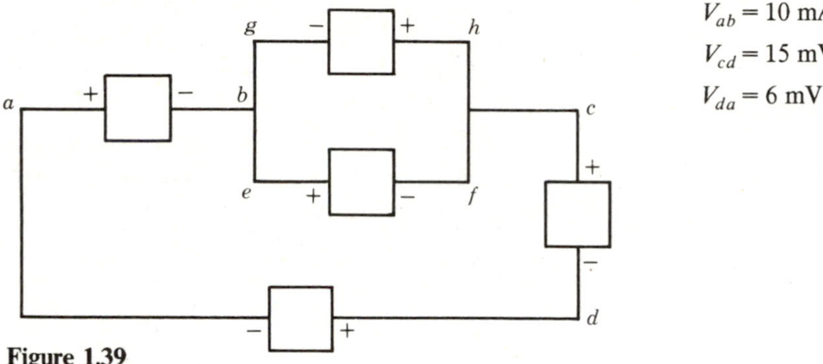

Figure 1.39

18. Four circuit elements are connected as shown in Fig. 1.40.

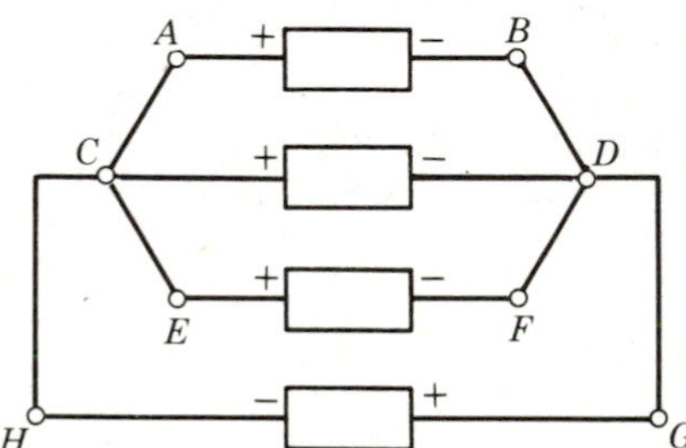

Figure 1.40

If $v_{AB} = 5$ V, what are v_{CD}, v_{EF}, and v_{GH}? If N circuit elements were connected in similar manner, what would be the voltage drop across each element?

19. In practical situations, the voltage drop across one element can often be thought of in terms of the voltage drop across another circuit element. This relationship depends upon the physical characteristics of the circuit elements. Considering this, use KVL to find the voltage drop across each element in Fig. 1.41. Traverse each circuit clockwise.

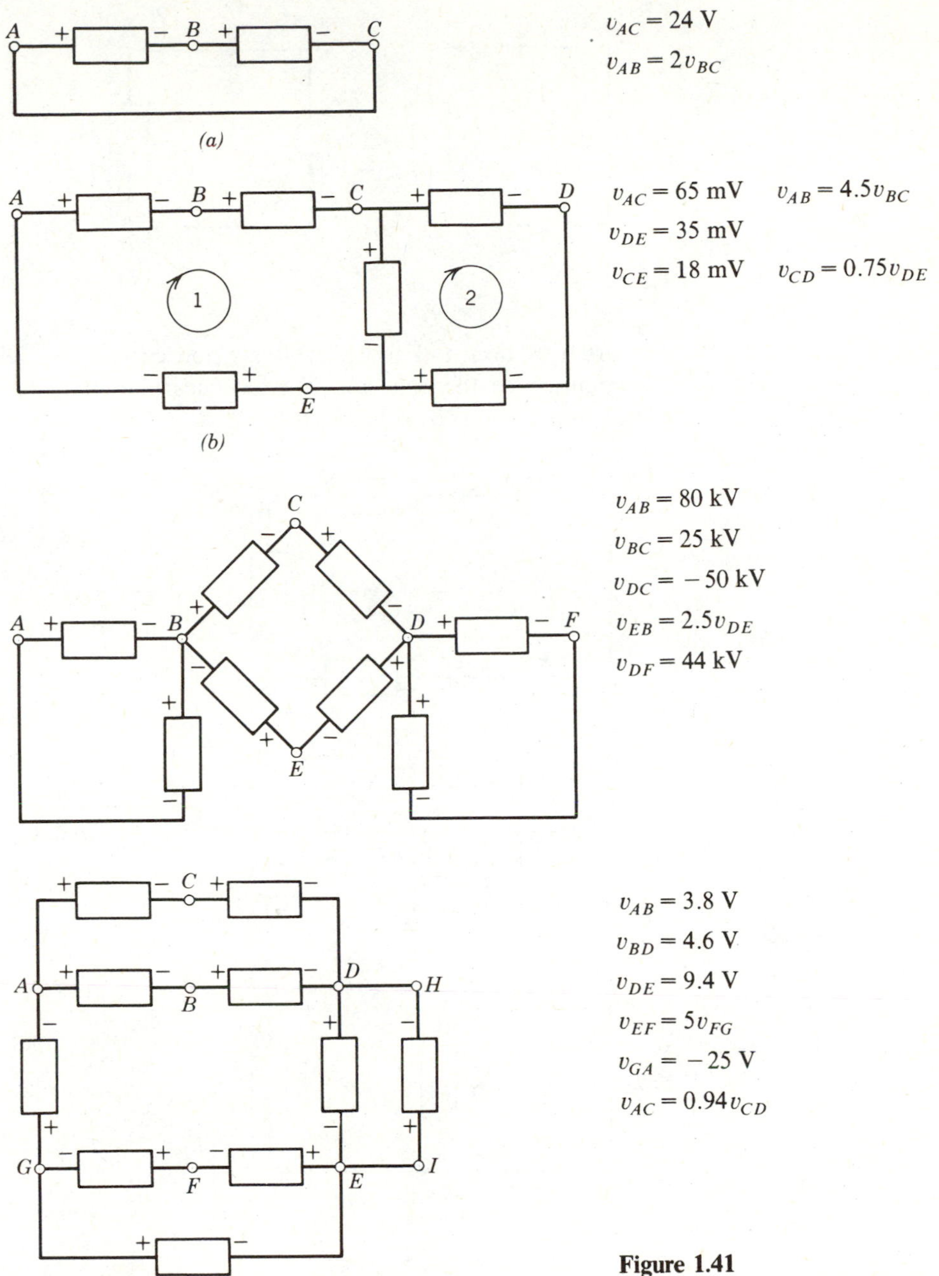

Figure 1.41

SECTION 1.7

20. Use KCL or KVL to find the unknown current/voltage for each of the situations shown in Fig. 1.42. State whether power is absorbed or delivered by the source and each element. Compare the total power absorbed and delivered in the circuit.

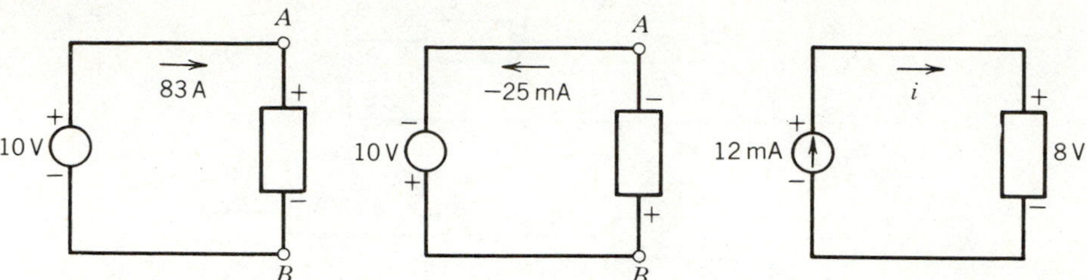

Figure 1.42

21. Often we need to know all voltages and currents associated with an electrical circuit. Use the concepts you have learned thus far to determine any unknown currents or voltages in Fig. 1.43.

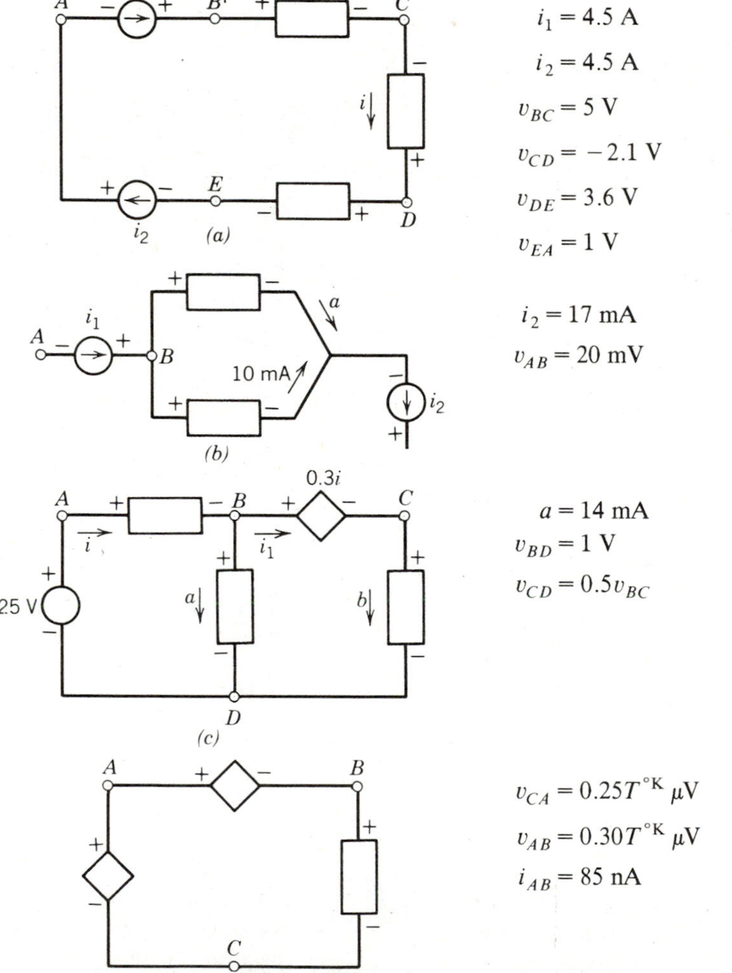

$i_1 = 4.5$ A
$i_2 = 4.5$ A
$v_{BC} = 5$ V
$v_{CD} = -2.1$ V
$v_{DE} = 3.6$ V
$v_{EA} = 1$ V

$i_2 = 17$ mA
$v_{AB} = 20$ mV

$a = 14$ mA
$v_{BD} = 1$ V
$v_{CD} = 0.5 v_{BC}$

$v_{CA} = 0.25 T^{\circ K}$ μV
$v_{AB} = 0.30 T^{\circ K}$ μV
$i_{AB} = 85$ nA

Figure 1.43. Find v_{BC} for $T = 300$ °K.

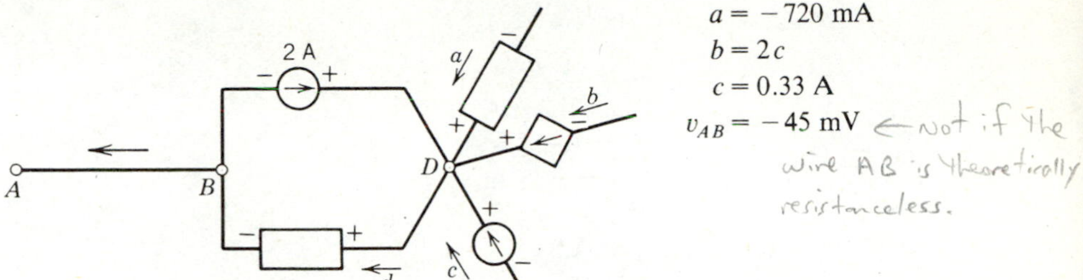

Figure 1.43. (*Continued*)

22. Refer to the circuit elements in Problem 21 that are between points A and B. Determine the amount of power absorbed/delivered and the total energy released for a time period of 15 s.

LEARNING EVALUATION ANSWERS

Section 1.2

1. a. 0.00168 m = 0.168 cm, 1.68 mm, and 1680 μm
b. 3.577×10^{-2} s = 3.577 cs, 35.77 ms, and 35770 μs
c. 14 G °K = 14×10^9 °K, 0.014 T °K, and 14000 M °K
d. 0.877 A = 8.77 dA, 87.7 cA, and 877 mA

2. a. 281 ft − lb = 38.85 m-kg
b. 40 °F = 277.68 °K
c. 6 in. = 0.1524 m
d. 422 s = 422 s (already in SI units)

Section 1.3

1. a. $q = 1.602 \times 10^{-7}$ C
b. $q = -1.586 \times 10^{-5}$ C

2. See Fig. 1.44.

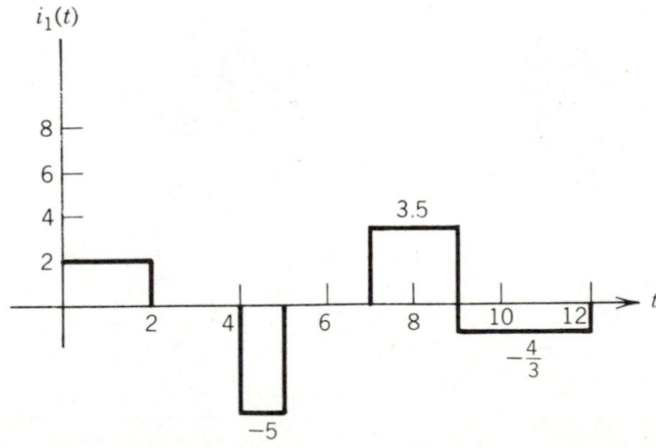

Figure 1.44

Section 1.4

a. $i = 5$ A
b. $i = 4.2$ mA
c. $i = -17.4$ A
d. $i = -18$ mA

Section 1.5

a. $p = 48$ W
b. $p = 1.92$ kW
c. $p = 200$ W
d. $p = 215$ W
e. $p = -2.20(10)^{-7}$ W
f. $p = 10.25$ kW

Section 1.6

1. **a.** $v = -4$ V
b. $v = -77$ mV
c. $v = 53.57$ kV

2. **a.** $\beta = 4$ V if + terminal is on right-hand side of β.
b. $\beta = 21$ mV if + terminal is on top of β.

Section 1.7

1. **a.** Voltage, delivered.
b. Current, delivered.
c. Current, absorbed.
d. Voltage, delivered.

2. **a.** 950.4 kWh
b. $85.54

CHAPTER 2 RESISTIVE CIRCUIT ANALYSIS

This chapter introduces a new circuit element called the resistor. Circuits made up of one or more resistors are called "resistive circuits." These circuits usually also contain one or more energy sources since circuits containing resistors only are of little practical value. We will limit the voltage and current sources used in the chapter to dc sources. Therefore, we will be investigating the subject called dc resistive circuit analysis. General dc voltage and current sources are usually designated by V and I, respectively. That is, capital letters indicate dc sources. This is sometimes confusing because V is also the symbol for the unit of volts.

In its broadest form, "dc resistive circuit analysis" can be defined as finding the voltage across and current through *every* element of a circuit for *all time*. Through the use of Eq. 1.5.1, we also know the power absorbed by every element. Often we are interested in a subset of this general solution, that is, the voltage across and/or current through a specific circuit element. Techniques for finding the solution to this subset will also be addressed in this chapter.

2.1 The Resistor

LEARNING OBJECTIVES

After completing this section you should be able to do the following:

1. Define and give the symbol for a resistor.
2. Define resistivity.
3. Find the resistance of a device of uniform cross-sectional area, given its resistivity.
4. State the difference between insulators, semiconductors, and conductors.
5. Define conductance, give its relationship to resistance, amd give its unit and symbol.

Resistance is defined as the opposition to current flow. A physical device that possesses this resistance property is called a *resistor*. All physical devices have some value of resistance, but the term "resistor" is usually reserved for a device that is specifically designed and constructed to provide a given amount of resistance to current flow.

The most common commercial type of resistor is the carbon resistor. This device is usually cylindrical and contains a mixture of carbon and another material, called a "binder." The binder binds the carbon to a given form. A

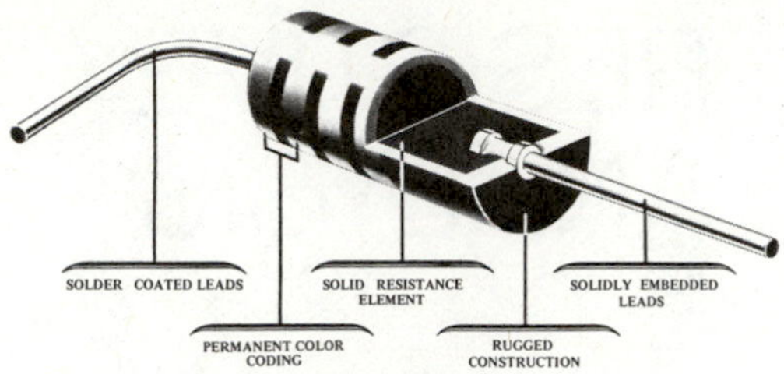

Figure 2.1

protective outer coating of a plastic-type material is used to ensure the integrity of the carbon cylinder. Finally, a wire is imbedded in each end of the cylinder (Fig. 2.1).

The physical dimensions of the cylinder and the amount of carbon used determines the resistance of the device. We see from Fig. 2.1 that the resistor fits the description of a two-terminal circuit element.

The symbol used to model a physical resistor in an electric circuit is shown in Fig. 2.2.

R

Figure 2.2. Circuit model of a resistor.

R indicates the value of resistance associated with the circuit model. Figure 2.3 shows a resistor and an independent voltage source combined to form an electrical circuit.

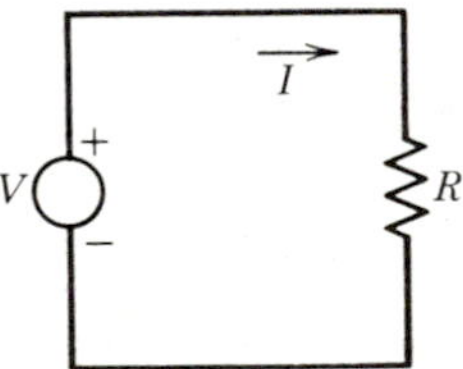

Figure 2.3. Electrical circuit containing resistor and voltage source.

What if we had used a material other than carbon in our resistor? The resistance would have been altered based on the material used. A material is classified as an insulator, semiconductor, or conductor based on the number of valence electrons it contains in its atomic structure. Many handbooks give the value of resistivity, ρ, associated with different material. Table 2.1 lists a few values of interest. The values shown are a function of temperature and purity.

The resistance, R, of a device made up of an element is given by

$$R = \frac{\rho l}{A} \tag{2.1.1}$$

where

ρ = the resistivity of the element
l = the length of the device
A = the cross-sectional area of the device

TABLE 2.1
RESISTIVITY VALUES FOR SELECTED MATERIALS

	Materials	Resistivity (Ω-m)
Insulators	Glass	10^{14}
	Hard rubber	10^{12}
	Ceramic	10^{14}
Semiconductors	Silicon	100
	Germanium	0.46
Conductors	Aluminum	2.83×10^{-8}
	Copper	1.77×10^{-8}
	Gold	2.97×10^{-8}
	Silver	1.63×10^{-8}

Equation 2.1.1 only applies to structures of uniform cross-sectional area. The units associated with ρ must be consistent with those used for l and A. The unit of resistance is the ohm, named after Georg Simon Ohm, the nineteenth-century German physicist who developed Ohm's law, which we will encounter shortly. The symbol used for ohm is the Greek letter omega, Ω. If resistivity is measured in ohm-meters, then l must be expressed in meters and A in square meters.

EXAMPLE 2.1.1 A gold wire ($\rho = 2.97 \times 10^{-8}$ Ω-m) 0.75 m long has a cross-sectional area of 1.6 cm^2. Find the resistance of the wire.

Solution

From Eq. 2.1.1,

$$R = \frac{\rho l}{A}$$

$$= \frac{(2.97 \times 10^{-8})(0.75)}{1.6 \times 10^{-4}}$$

$$= 1.358(10)^{-4}\ \Omega$$

EXAMPLE 2.1.2 A copper stereo speaker connector ($\rho = 1.77 \times 10^{-8}$) has a resistance of 5.5×10^{-4} Ω. The cross-sectional area is 9 mm^2. What is the length of the connector?

Solution

From Eq. 2.1.1,

$$l = \frac{RA}{\rho}$$

$$= (5.5 \times 10^{-4}\ \Omega)\,(9 \times 10^{-6}\ \text{m}^2)/(1.77 \times 10^{-8}\ \Omega\text{-m})$$

$$= 0.280\ \text{m}$$

Instead of discussing the resistance of a device to current flow, we could discuss the ability of the device to allow current flow. This ability is expressed as the *conductance* of a device. Conductance is denoted by the letter G and the conductance is defined as the reciprocal of the resistance.

$$G = \frac{1}{R} \tag{2.1.2}$$

$$R = \frac{1}{G} \tag{2.1.3}$$

The unit for conductance was "mho" (ohm spelled backward) and the symbol was ℧ (upside-down omega) until about 1970. The name siemen and the symbol S was then adopted. You will likely see it both ways for some time to come, although we will use the modern version.

EXAMPLE 2.1.3 A certain circuit element has a resistance of 400 Ω. What is its conductance?

Solution

$$R = 400\ \Omega$$

and

$$G = \frac{1}{R} = \frac{1}{400} = 0.0025\ \text{S}$$

or

$$G = 2.5\ \text{mS}$$

We express this as a conductance of 2.5 millisiemens.

LEARNING EVALUATIONS

1. A rectangular bus bar (a current-carrying bar) has dimensions of 3×6 cm. Its length is 4 m. Determine the resistance if the bar is made:

   a. Of silver. **b.** Of silicon.

2. What is the conductance of each bar?
3. A copper wire has circular cross-sectional area of 1 mm^2. If the conductance of the wire is 2 S, find the length of the wire.

2.2 Ohm's Law

LEARNING OBJECTIVES

After completing this section you should be able to do the following:

1. State Ohm's law.
2. Express Ohm's law with six different equations.
3. Use Ohm's law to find the voltage across, current through, or resistance of an element in a circuit.
4. Calculate the power absorbed by a resistor in a given circuit.

We stated that we are interested in finding the voltage across and current through a circuit element. This is called the "volt–ampere" characteristic of a two-terminal circuit element. Assume we are to conduct an experiment using the circuit model in Fig. 2.3 in the following way:

1. Choose a resistor of value R.
2. Choose a voltage source of value V_1.
3. Connect these as shown in Fig. 2.3.
4. Measure the value of the current and call it I_1.
5. Repeat steps 2 to 4 for nine other values of voltage.
6. Graph the points (V_i, I_i) for $i = 1, 2, \ldots, 10$.

The result of our experiment would be a graph such as the one shown in Fig. 2.4. Several conclusions are noted from Fig. 2.4:

1. If the voltage is zero, the current is also zero.
2. If the voltage is positive, the current is positive.
3. If the voltage is negative, the current is negative.
4. The slope of the line connecting the points (V_i, I_i) is constant (i.e., a straight line).

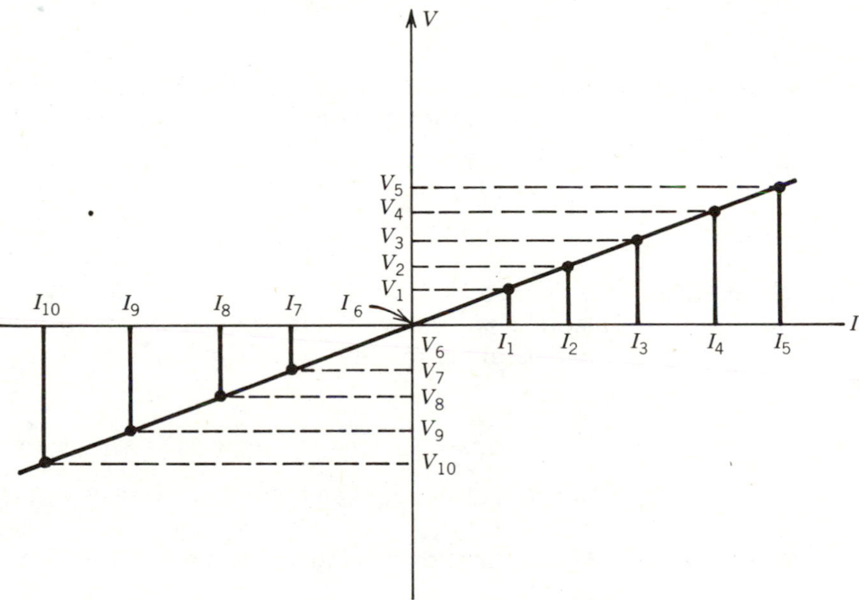

Figure 2.4. Volt–ampere characteristic for resistor of value R.

In general, the equation for a straight line is $y = ax + b$, where a is the slope of the line and b is the y intercept. From our conclusion in Fig. 2.4, we can write $V = kI$ since the V intercept is zero (conclusions 1 and 4). Also from conclusions 2 and 3 it is obvious that k is a positive constant and can be expressed as

$$k = \frac{V}{I}$$

Evaluation of k for any set of points V_i, I_i shows k to be the value of resistance, R. That is,

$$R = \frac{V}{I}$$

or

$$V = IR \tag{2.2.1}$$

Georg Simon Ohm (1787–1854) discovered this relationship in the early nineteenth century and it is known as *Ohm's law*. This law (Eq. 2.2.1) is a fundamental fact used in circuit analysis and takes its place alongside Kirchhoff's voltage and current laws to provide a solid foundation on which we will build and expand our circuit analysis capabilities. Ohm's law can be expressed in any of the following forms:

$$V = IR \qquad I = \frac{V}{R} \qquad R = \frac{V}{I}$$

$$V = \frac{I}{G} \qquad I = GV \qquad G = \frac{I}{V} \tag{2.2.2}$$

In these equations, the units on V, I, G, and R are volts, amperes, siemens, and ohms, respectively.

EXAMPLE 2.2.1 A resistive circuit element of value 32 Ω has a voltage of value 83 V connected across its terminals. Calculate the current in the resistor.

Solution

$$I = \frac{V}{R} = \frac{83}{32} = 2.59 \text{ A}$$

EXAMPLE 2.2.2 It is desired to have a current of 5 mA flowing through an 8-kΩ resistor. Calculate the value of the voltage that must be connected across the terminals of the resistor.

Solution

$$V = IR = (5 \times 10^{-3})(8 \times 10^{3}) = 40 \text{ V}$$

Notice that we must be careful to express V, I, and R in the proper units.

EXAMPLE 2.2.3 In the circuit shown in Fig. 2.5 find the values R and G.

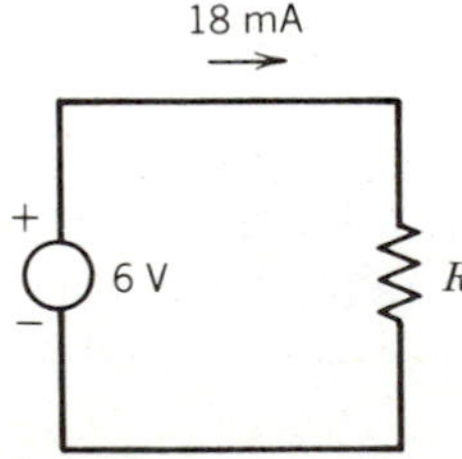

Figure 2.5

Solution

$$R = \frac{V}{I} = \frac{6}{18 \times 10^{-3}} = 0.333 \times 10^3 = 333\ \Omega$$

$$G = \frac{I}{R} = \frac{1}{333} = 3 \times 10^{-3}\ \text{S}$$

Recall Eq. 1.5.1, which states that power is expressed as voltage times current. Using Eq. 2.2.1, we can find the expression for power absorbed in a resistor:

$$P = VI = (IR)I = I^2R \tag{2.2.3}$$

or

$$P = VI = V\left(\frac{V}{R}\right) = \frac{V^2}{R} \tag{2.2.4}$$

Here the capital P is used to indicate we are concerned with power that is not a function of time, since V and I are not functions of time.

The power dissipated (absorbed) in a resistor is an important quantity, since resistor values can be permanently changed or the resistor destroyed by absorbing too much power. Since energy is power multiplied by time, and energy is neither created nor destroyed, it follows that all power absorbed by a resistor must be dissipated or stored. A resistor does not have the ability to store energy so all power must be dissipated in the form of heat. Resistors are rated according to the amount of power they can absorb without damage or significant resistance change. A 2-W resistor, for instance, can absorb 2 W of power indefinitely without damage to the resistor under normal environmental conditions.

A light bulb is a good example of these concepts. The filament in a light bulb is simply a piece of wire (usually made of tungsten) with a certain resistance. The current flowing through the light bulb is inversely proportional to the resistance of the wire (Ohm's law.) The power absorbed by the filament causes it to heat up and glow, thereby emitting a quantity of light. Anyone who has touched a bulb while the lamp is turned on knows that it also dissipates a considerable amount of heat. The light emitted is a function of the power absorbed.

EXAMPLE 2.2.4 Does a 100-W light bulb have a higher or lower resistance than a 40-W bulb?

Solution

By Eq. 2.2.4,

$$P = \frac{V^2}{R}$$

If V is constant and P increases then R must decrease.

$$P_{100} = 2.5P_{40}$$

$$\frac{V^2}{R_{100}} = 2.5\frac{V^2}{R_{40}}$$

$$R_{40} = 2.5R_{100}$$

Therefore, the resistance of a 40-W bulb must be 2.5 times the resistance of 100-W bulb.

EXAMPLE 2.2.5 If 50 V is applied across the terminals of a 10-kΩ resistor, find the current I, the power absorbed by the resistor, and the power delivered by the voltage source.

Solution

The circuit model is given in Fig. 2.6.

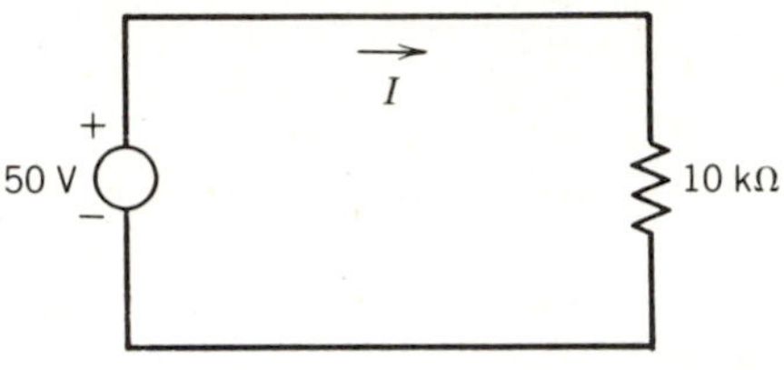

Figure 2.6

By Ohm's law,

$$I = \frac{V}{R} = \frac{50}{10 \times 10^3} = 5 \times 10^{-3}\ \text{A} = 5\ \text{mA}$$

By Eq. 2.2.4,

$$P_R = \frac{(50)^2}{10 \times 10^3} = \frac{2.5 \times 10^3}{10 \times 10^3} = 0.25\ \text{W} = 250\ \text{mW}$$

By Eq. 1.5.1, we find the power delivered by the source is

$$P_S = VI = (50)(5 \times 10^{-3}) = 250 \times 10^{-3}\ \text{W} = 250\ \text{mW}$$

Example 2.2.5 illustrates another conclusion that can be drawn from the law of conservation of energy: The power absorbed by elements of a circuit must equal the power delivered by elements of the circuit.

We have now defined five terms by the equations

$$P = VI$$
$$V = IR$$
$$G = \frac{1}{R} \tag{2.2.5}$$

so that given the values of any two terms (as long as resistance and conductance are not both included), the values of the other three terms can be calculated. At least 24 different combinations of Eqs. 2.2.5 are possible to generate in order to solve for a given term. These are too many for you to memorize. You should commit the Eqs. 2.2.5 to memory and then be able to derive any other form of these equations you might need by algebraic manipulation.

LEARNING EVALUATIONS

1. Complete the values in the following table:

	Voltage	Current	Power	Resistance	Conductance
a.		2.0 A	16 W		
b.	87 V			8 kΩ	
c.			931 mW		283 mS
d.	57 mV	16 mA			
e.			43 kW	6 Ω	
f.		241 A		650 mS	
g.	1.73 V		2.88 mW		
h.		89 μA		47 MΩ	
i.	538 V				14 μS

2. Plot the volt-ampere characteristic for a 75-Ω resistor
3. Find the relationship between the resistance of a 75-W light bulb and the resistance of a 150-W light bulb.
4. The current through a 25-MΩ resistor is 86 μA. Find the voltage that must be across the terminals of the resistor.

2.3 A First Look at Network Analysis

LEARNING OBJECTIVES

After completing this section you should be able to do the following:

1. Be able to determine whether two elements are in series.
2. Write Kirchhoff's voltage law around a series circuit.
3. Find the voltage drop across and power absorbed or delivered by each element of a series circuit.
4. Show that the power delivered in a circuit is equal to the power absorbed in a circuit.
5. Determine if two elements are connected in parallel. Find the voltage across, current through, and power absorbed or delivered by each element in a parallel circuit.
6. Find the voltage across, current through, and power absorbed or delivered by each element in a series-parallel circuit.

Now that we have some laws to depend on and a few circuit elements to use, let us investigate some simple circuits and try to analyze (find the voltage across and current through one or more elements) them using this newly acquired knowledge. We will first try the series circuit, then the parallel circuit, and finally attack the series-parallel circuit.

The Series Circuit

Two circuit elements are said to be in "series" if the same current *always* flows through both elements regardless of the construction of the rest of the circuit. Figure 2.7 shows a circuit in which all elements are in series. The voltage source is shown as a dc source since a capital V is used with the source.

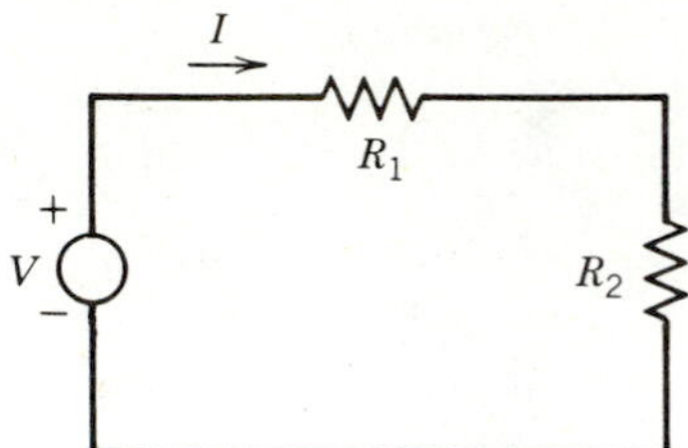

Figure 2.7. A series circuit.

Consequently the current is also dc and is shown with a capital I. The current is confined to the path so all current that passes through R_1 must also pass through R_2. Therefore R_1 and R_2 are connected in series. By the same argument the voltage source is in series with both R_1 and R_2. Now we will write Kirchhoff's voltage law around this circuit. The current direction has already been given so we only need to assign voltage drops across each resistor (after one gains experience with this procedure, this step can be eliminated). These assignments are shown in Fig. 2.8.

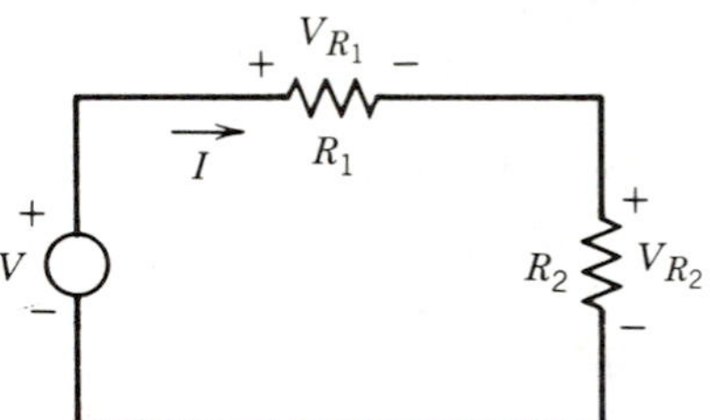

Figure 2.8. Voltage assignments.

Recall that the positive side of the voltage is assumed to be at the terminal where the current enters the circuit element. Application of KVL gives

$$-V + V_{R_1} + V_{R_2} = 0$$

or

$$V = V_{R_1} + V_{R_2} \tag{2.3.1}$$

But by Ohm's law we know

$$V_{R_1} = IR_1 \tag{2.3.2}$$

and

$$V_{R_2} = IR_2 \tag{2.3.3}$$

Substituting Eqs. 2.3.2 and 2.3.3 into 2.3.1 gives

$$V = IR_1 + IR_2 \tag{2.3.4}$$

If we know the voltage of the independent source and the values of R_1 and R_2 we can use Eq. 2.3.4 to find the current I. We can then use Eqs. 2.3.2 and 2.3.3 to find the voltage drop across each resistor. Finally we can use Eq. 2.2.4 to find the power absorbed by each resistor and our circuit analysis task is completed.

EXAMPLE 2.3.1 Find I_1, V_{R_1}, V_{R_2}, P_{R_1}, P_{R_2}, P_S in the circuit in Fig. 2.9.

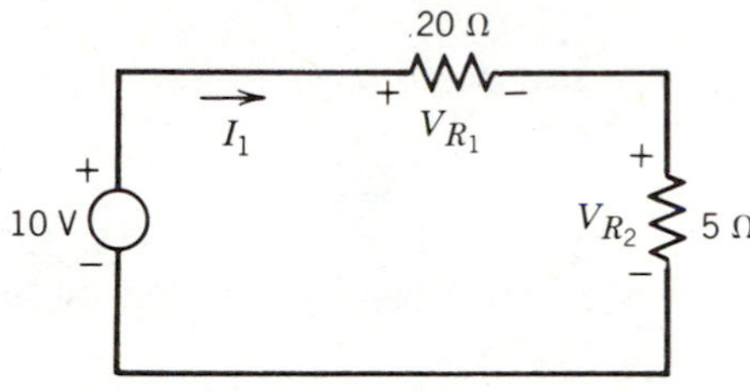

Figure 2.9

Solution

By KVL,

$$-10 + V_{R_1} + V_{R_2} = 0 \tag{2.3.5}$$

By Ohm's law,

$$V_{R_1} = 20I_1$$
$$V_{R_2} = 5I_1$$

so

$$-10 + 20I_1 + 5I_1 = 0$$
$$25I_1 = 10$$
$$I_1 = \tfrac{10}{25} = 0.40 \text{ A}$$

and

$$V_{R_1} = 20(0.4) = 8 \text{ V}$$
$$V_{R_2} = 5(0.4) = 2 \text{ V}$$

Using Eq. 2.2.4,

$$P_{R_1} = \frac{V_{R_1}^2}{R_1} = \frac{(8)^2}{20} = \frac{64}{20} = 3.2 \text{ W}$$

$$P_{R_2} = \frac{V_{R_2}^2}{R_1} = \frac{(2)^2}{5} = \frac{4}{5} = 0.8 \text{ W}$$

$$P_S = VI_1 = (10)(0.4) = 4.0 \text{ W}$$

As a check we place our calculated values back in Eq. 2.3.5:

$$-10 + 8 + 2 = 0$$
$$0 = 0 \text{ (checks)}$$

Also remember the power delivered in a circuit must equal the power absorbed so

$$P_S = P_{R_1} + P_{R_2} \tag{2.3.6}$$

Substituting in our calculated values

$$4.0 = 3.2 + 0.8$$
$$= 4.0$$

We see that this is indeed the case, so our calculated values should be correct.

EXAMPLE 2.3.2 For the circuit shown in Fig. 2.10 find the voltage drops across each resistor, the power absorbed by each resistor, and the power delivered by each source.

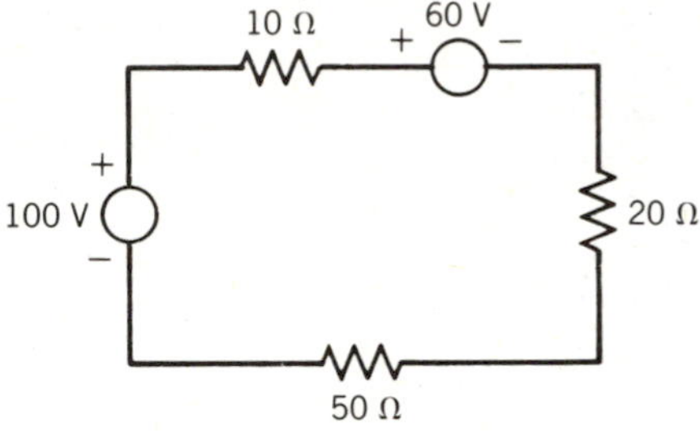

Figure 2.10

Solution

First we must assume a direction of current (note that all elements are in series). Since it makes no difference which direction we choose, let us pick ccw. Next we assign voltage drops across each resistor. The application of these actions results in the circuit of Fig. 2.11.

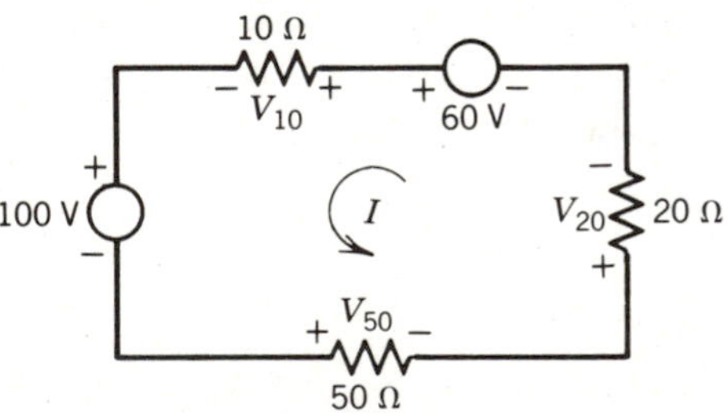

Figure 2.11

Note: Picking the ccw direction violated our stated consistency condition of Chapter 1. We are violating it here to illustrate that the same answer is obtained no matter which way we assume the current to flow. Now back to the solution.

By KVL,

$$V_{20} - 60 + V_{10} + 100 + V_{50} = 0$$

But by Ohm's law,

$$V_{10} = 10I$$
$$V_{20} = 20I$$
$$V_{50} = 50I$$

Substituting these values into Eq. 2.3.6 and simplifying gives

$$-80I = 40$$

or

$$I = -0.5 \text{ A}$$

The negative value of I simply means that I actually flows in the cw direction with a magnitude of 0.5 A. (You should verify this by assuming a cw current in Fig. 2.10 and resolving for I.) The voltages across and power through each resistor can now be calculated as

$$V_{10} = 10I = -5 \text{ V} \qquad P_{10} = V_{10}I = (-5)(-0.5) = 2.5 \text{ W}$$
$$V_{20} = 20I = -10 \text{ V} \qquad P_{20} = V_{20}I = (-10)(-0.5) = 5 \text{ W}$$
$$V_{50} = 50I = -25 \text{ V} \qquad P_{50} = V_{50}I = (-25)(-0.5) = 12.5 \text{ W}$$

We must be careful when we calculate the power delivered by each voltage source. Remember, to *deliver* power, a *positive* current must flow *out* of the *positive* side of the source. In examining Fig. 2.11 and keeping in mind that the *positive* current was found to be flowing cw, we find that the 100-V source is delivering power while the 60-V source is absorbing power. The values of each are

$$P_{100V} = 100(0.5) = 50 \text{ W delivered}$$
$$P_{60V} = 60(0.5) = 30 \text{ W absorbed}$$

As a check we see if the total power absorbed in the circuit equals the total power delivered.

$$P_{\text{total abs}} = 2.5 + 5 + 12.5 + 30 = 50 \text{ W}$$
$$P_{\text{total del}} = 50 \text{ W}$$

In working problems of this nature, the fact these two agree does not guarantee that your calculations are correct, but does allow you to go on to the next problem with increased confidence.

EXAMPLE 2.3.3 The circuit in Fig. 2.12 contains a *dependent* voltage source. The value of this source is 3.2 times the voltage across R_2. Find the current I.

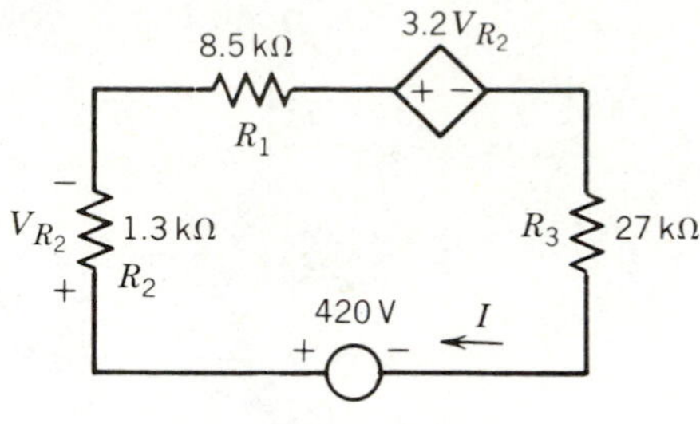

Figure 2.12

Solution

First we assume a direction of current (let us pick cw) then assign the appropriate voltage drops across the resistors. These results are shown in Fig. 2.13.

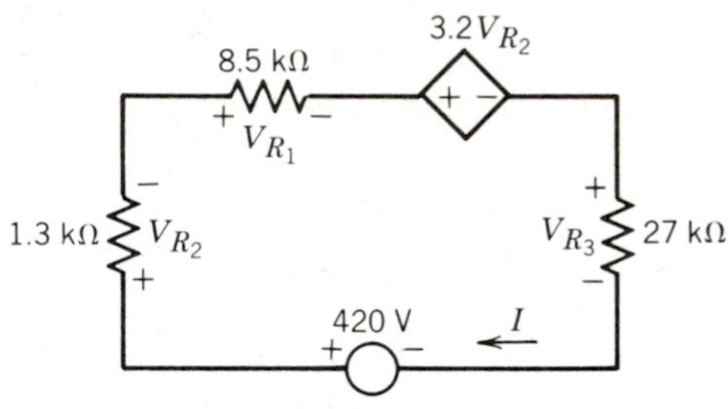

Figure 2.13

Now apply KVL starting at the lower right-hand corner and proceeding cw:

$$-420 + V_{R_2} + V_{R_1} + 3.2V_{R_2} + V_{R_3} = 0$$

Using Ohm's law to find the resistor voltages gives us

$$-420 + (1.3 \times 10^3 I) + (8.5 \times 10^3 I) + 3.2(1.3 \times 10^3 I) + (27 \times 10^3 I) = 0$$

or

$$(40.96 \times 10^3)I = 420$$

$$I = 10.25 \times 10^{-3}\text{ A}$$

$$I = 10.25\text{ mA}$$

So we see that solving for values in a circuit containing a dependent source presents no special problems.

The Parallel Circuit

Two circuit elements are said to be connected in "parallel" if the *same voltage* appears across both elements regardless of the values of the elements. Consider the circuit in Fig. 2.14. The two resistors R_1 and R_2 are connected in parallel. To

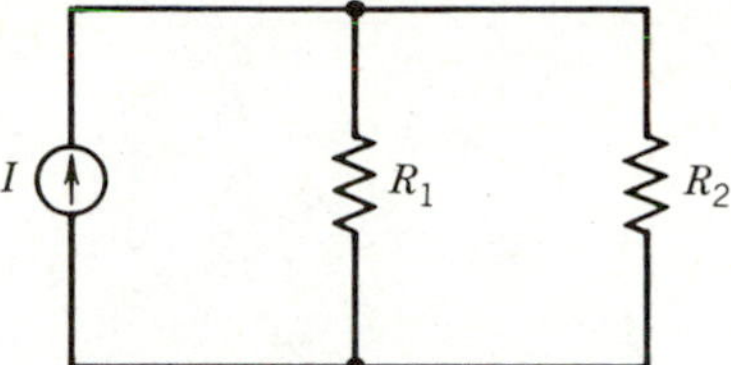

Figure 2.14. A parallel circuit.

prove this we must show that the voltage drops across the two resistors are the same. Assume two arbitrary voltage drops V_{R_1} and V_{R_2}, as shown in Fig. 2.15. Now write KVL around the closed path $ABCDA$ starting at A.

$$-V_{R_1} + V_{R_2} = 0$$

or

$$V_{R_1} = V_{R_2}$$

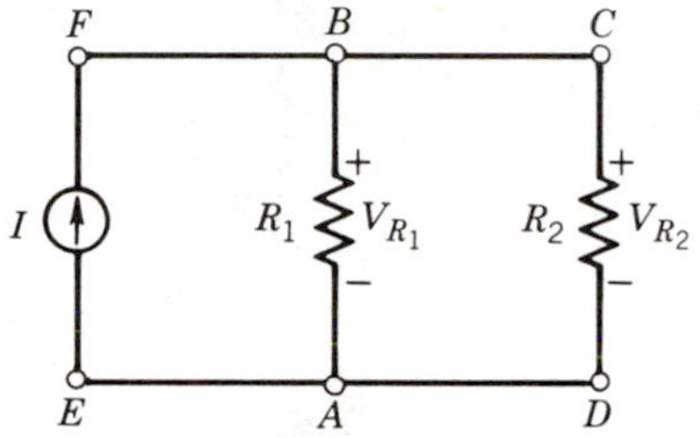

Figure 2.15. Voltage references for a parallel circuit.

So R_1 and R_2 are indeed connected in parallel. If the current source has a value of 20 A and $R_1 = 60\ \Omega$ while $R_2 = 80\ \Omega$, what are the values of V_{R_1} and V_{R_2}? Since R_1 and R_2 are in parallel we know

$$V_{R_1} = V_{R_2} = V_R$$

The 20-A current enters node B. Assume current I_1 flows through R_1 and I_2 flows through R_2 as shown in Fig. 2.16. By KCL at node B,

$$20 = I_1 + I_2$$

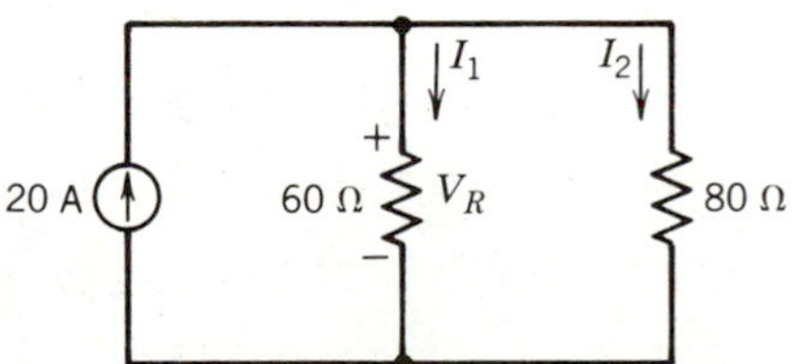

Figure 2.16. Current reference directions.

By Ohm's law,

$$I_1 = \frac{V_R}{60}$$

$$I_2 = \frac{V_R}{80}$$

so

$$20 = \frac{V_R}{60} + \frac{V_R}{80}$$

or

$$V_R = 685.71V = V_{R_1} = V_{R_2}$$

In general, KCL is used to solve for unknown variables in parallel circuits while KVL is used in series circuit. Ohm's law is used in both types of circuits to supply the additional equations necessary to solve for the variables in question.

EXAMPLE 2.3.4 Find the value of V_R in the circuit shown in Fig. 2.17.

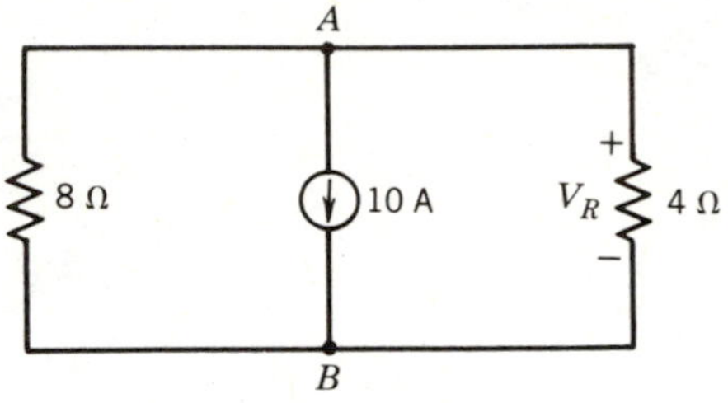

Figure 2.17

Solution

First we assume current directions through the resistors. One choice is indicated in Fig. 2.18.

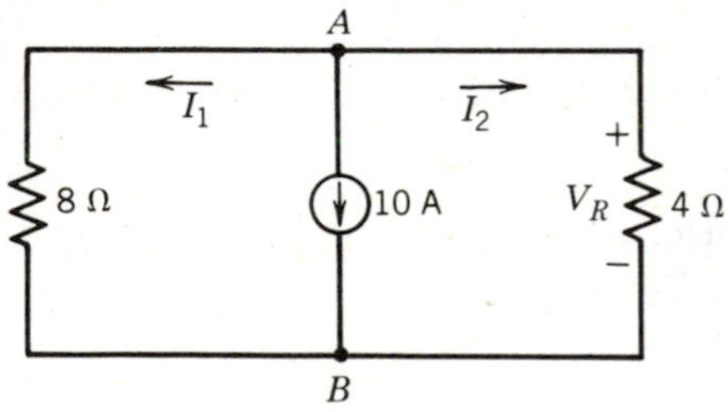

Figure 2.18

Now apply KCL at point B:

$$10 + I_1 + I_2 = 0$$

Also since the 8- and 4-Ω resistors are in parallel,

$$V_8 = V_4 = V_R$$

By Ohm's law,

$$I_1 = \frac{V_R}{8}$$

$$I_2 = \frac{V_R}{4}$$

Substituting,

$$10 + \frac{V_R}{8} + \frac{V_R}{4} = 0$$

or

$$V_R = -26.67 \text{ V}$$

EXAMPLE 2.3.5 Find the value of I for the circuit in Fig. 2.19.

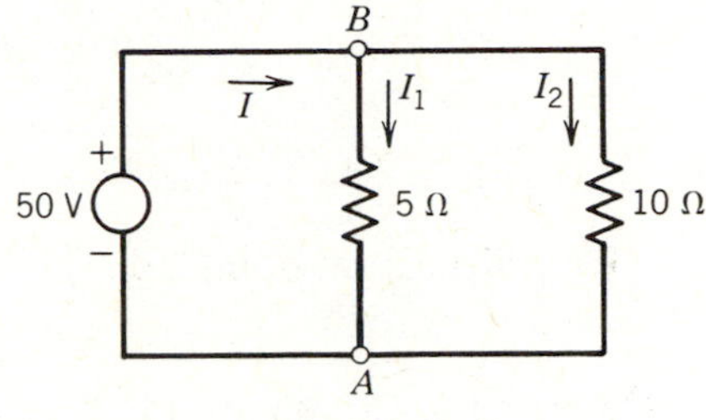

Figure 2.19

Solution

First assign the currents through each resistor arbitrarily. These are shown as I_1 and I_2 in Fig. 2.19. Next realize that the two resistors are in parallel and the voltage across each resistor is 50 V. Then by Ohm's law,

$$I_1 = \tfrac{50}{5} = 10 \text{ A}$$

$$I_2 = \tfrac{50}{10} = 5 \text{ A}$$

Then using KCL at node B,

$$\begin{aligned} I &= I_1 + I_2 \\ &= 10 + 5 \\ I &= 15 \text{ A} \end{aligned}$$

EXAMPLE 2.3.6 In the circuit of Fig. 2.20, which contains a dependent current source, find I.

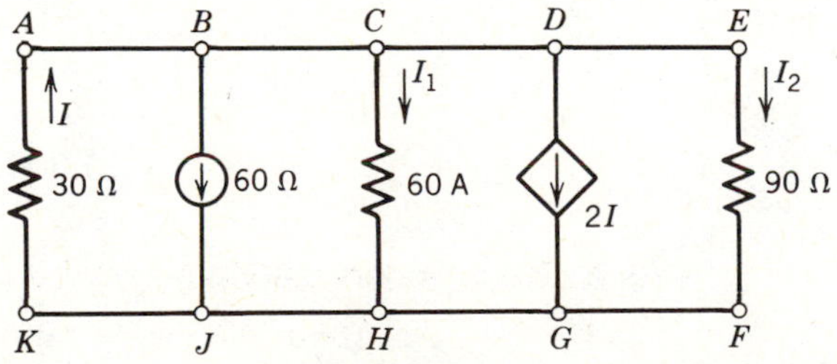

Figure 2.20

Solution

The first step here is to realize that points A, B, C, D, E are all the same point electrically. Likewise, points F, G, H, J, K are all the same point electrically. (That is, there is no voltage drop between points A and B, B and C, etc., since there is no resistance between these points.) So the circuit can be redrawn as shown in Fig. 2.21.

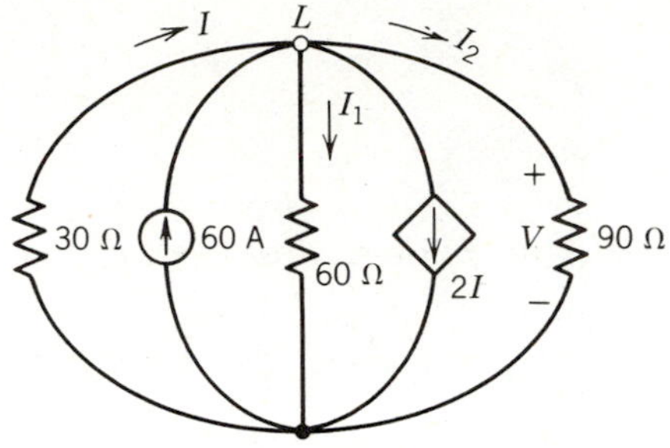

Figure 2.21

Applying KCL at point L gives

$$I + 60 = I_1 + 2I + I_2 \tag{2.3.7}$$

Since all elements are in parallel, the same voltage V appears across the elements. Again, using Ohm's law,

$$I = \frac{-V}{30}$$

$$I_1 = \frac{V}{60}$$

$$I_2 = \frac{V}{90}$$

Substituting

$$\frac{-V}{30} + 60 = \frac{V}{60} + 2\left(-\frac{V}{30}\right) + \frac{V}{90}$$

Solving for V,

$$V = -10{,}800 \text{ V}$$

Then

$$I = \frac{-(-10{,}800)}{30} = 360 \text{ A}$$

$$I_1 = \frac{-10{,}800}{60} = -180 \text{ A}$$

$$I_2 = \frac{-10{,}800}{90} = -120 \text{ A}$$

As a check, insert the values for I, I_1, I_2 into Eq. 2.3.7:

$$(360) + 60 = (-180) + 2(360) + (-120)$$

$$420 = 420$$

So our results are correct.

As in the case of the series circuit, the power absorbed in the parallel circuit must equal the power delivered. In the above example, the two circuit elements capable of delivering power are the independent and dependent current sources. (We must be careful here! The current must flow *out* of the source at the *positive* voltage terminal in order for the source to *deliver* power.) This example is really tricky. The voltage V is the voltage across each of the sources since all elements are in parallel. But V was found to be a negative value for the reference voltage chosen. Consequently the actual reference direction of V is opposite that shown. It follows that the dependent source delivers energy since $2I$ is positive while the independent source absorbs energy.

$$P_{\text{del}_{2I}} = VI = (10{,}800)(2)(360) = 7.776 \text{ MW}$$

$$P_{\text{del}_{60}} = VI = -(10{,}800)(60) = -648 \text{ kW}$$

The power absorbed by the resistors is

$$P_{30} = I^2R = (360)^2(30) = 3.88 \text{ MW}$$

$$P_{60} = I_1^2R_2 = (-180)^2(60) = 1.944 \text{ MW}$$

$$P_{90} = I_2^2R = (-120)^2(90) = 1.296 \text{ MW}$$

Does

$$P_{\text{del}_{60}} + P_{\text{del}_{2I}} = P_{30} + P_{60} + P_{90}$$

$$7776 \text{ kW} + (-648 \text{ kW}) = 3888 \text{ kW} + 1944 \text{ kW} + 1296 \text{ kW}$$

$$7128 \text{ kW} = 7128 \text{ kW}$$

Yes! The power delivered equals the power absorbed.

As you gain experience in circuit analysis, you will quickly learn to determine if two circuit elements are in parallel by inspection.

Series-Parallel Circuits

As the name implies, a series-parallel circuit contains elements in series as well as elements in parallel. Consider Fig. 2.22. The same current passes through R_1 and R_2 so they are in series. Likewise R_4 and the voltage source are in series. R_3, R_5 and R_6 are in parallel because the same voltage appears across each element.

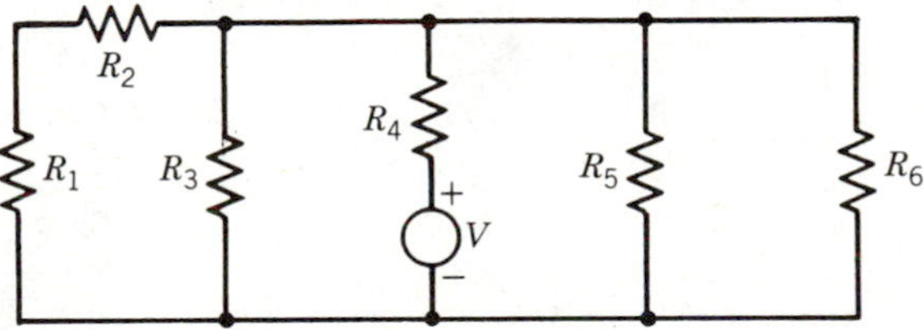

Figure 2.22. A series-parallel circuit.

How do we proceed to find voltages and currents in series-parallel circuits? We use the same tools as we did before: Kirchhoff's current and voltage laws, Ohm's law, and the power equation.

EXAMPLE 2.3.7 Find I in the circuit shown in Fig. 2.23.

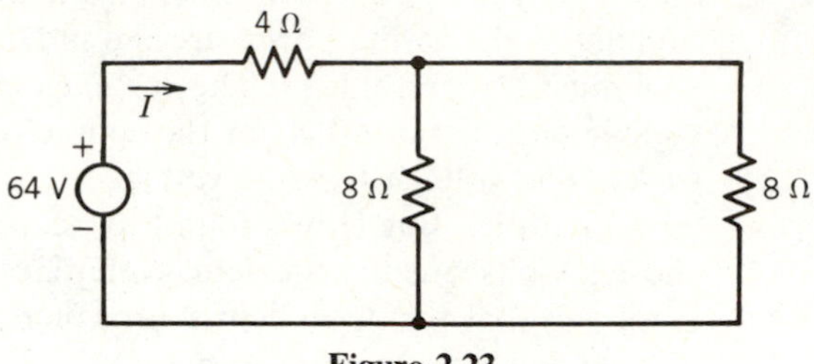

Figure 2.23

Solution

First we *assume* currents through the elements not already specified as shown in Fig. 2.24.

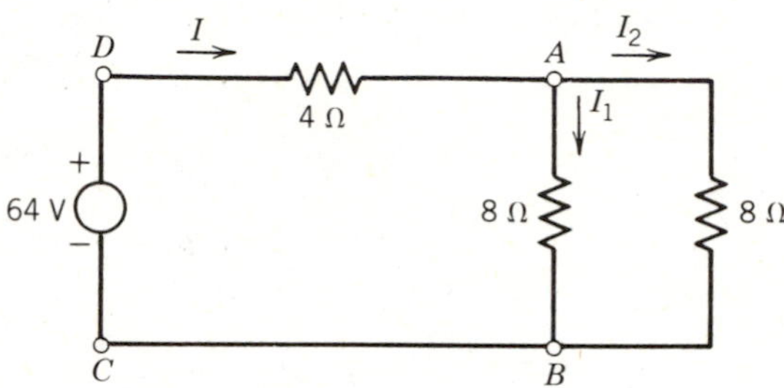

Figure 2.24

We have three unknowns (I, I_1, I_2) so we need three independent equations in order to solve for I.

By KCL at node A,

$$I = I_1 + I_2 \tag{2.3.8}$$

By KVL around path $ABCDA$,

$$4I + 8I_1 - 64 = 0 \tag{2.3.9}$$

By inspection the 8-Ω resistors are in parallel so the voltage across each is the same and is given by Ohm's law as

$$8I_1 = 8I_2$$

or

$$I_1 = I_2 \tag{2.3.10}$$

Equations 2.3.8 through 2.3.10 may be solved for I. Substituting Eq. 2.3.10 into Eq. 2.3.8 gives

$$I = 2I_1 = 2I_2$$

or

$$I_1 = I_2 = 0.5I \tag{2.3.11}$$

Now putting Eq. 2.3.11 into Eq. 2.3.9 yields

$$4I + 8(0.5I) - 64 = 0$$

or

$$I = 8 \text{ A}$$

EXAMPLE 2.3.8 Find the voltage across, current through, and power absorbed by each resistor in the circuit of Fig. 2.25.

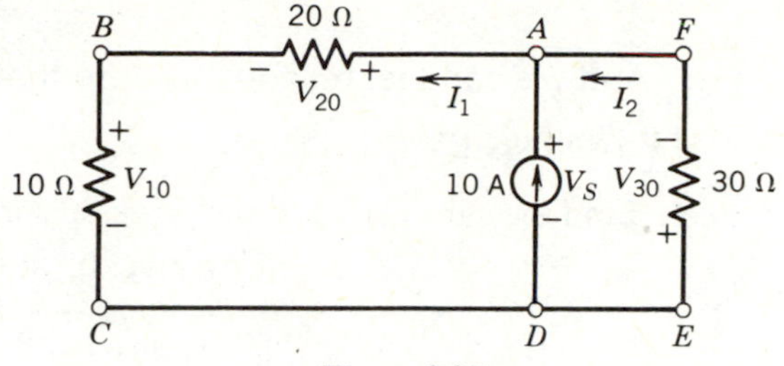

Figure 2.25

Solution

Writing KCL at node A,

$$10 + I_2 = I_1 \tag{2.3.12}$$

Writing KVL around the loop $ABCD$,

$$-V_S + V_{20} + V_{10} = 0$$

or using Ohm's law,

$$-V_S + 20I_1 + 10I_1 = 0$$

$$V_S = 30I_1 \tag{2.3.13}$$

Writing KVL around loop $AFED$,

$$-V_{30} - V_S = 0$$

or

$$V_S = -V_{30} = 30I_2 \tag{2.3.14}$$

Combining Eqs. 2.3.12 through 2.3.14 yields

$$I_1 = 5\ \text{A}$$

$$I_2 = -5\ \text{A}$$

Then

$$V_{10} = 10I_1 = 50\ \text{V}$$

$$V_{20} = 20I_1 = 100\ \text{V}$$

$$V_{30} = -(30)(-5) = -150\ \text{V}$$

and

$$P_{10} = (5)^2(10) = 250\ \text{W}$$

$$P_{20} = (5)^2(20) = 500\ \text{W}$$

$$P_{30} = (-5)^2(30) = 750\ \text{W}$$

As a check the total power absorbed is

$$P_{abs} = P_{10} + P_{20} + P_{30} = 1500\ \text{W}$$

while the power delivered is

$$P_{del} = 10V_S = (10)(-30)(-5) = 1500 \text{ W}$$

so

$$P_{abs} = P_{del}$$

And we can feel reasonably sure that our answers are correct.

LEARNING EVALUATIONS

1. Find the current I, and the power absorbed by each element in Fig. 2.26.

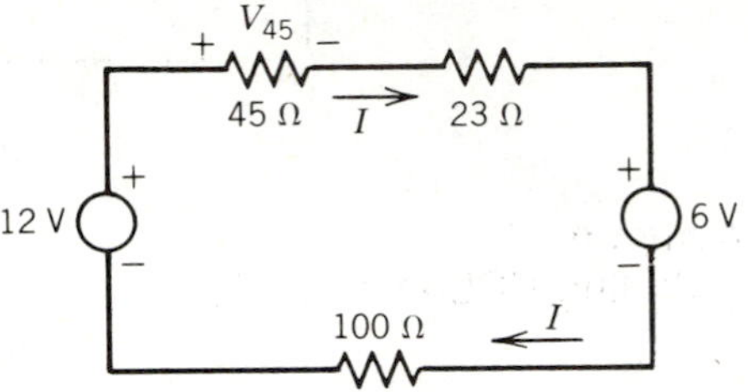

Figure 2.26

2. Repeat Problem 1 with the 6-V independent source replaced by a dependent source of value $4.5V_{45}$.
3. Determine the voltage drop across the source and current through each resistor in Fig. 2.27.

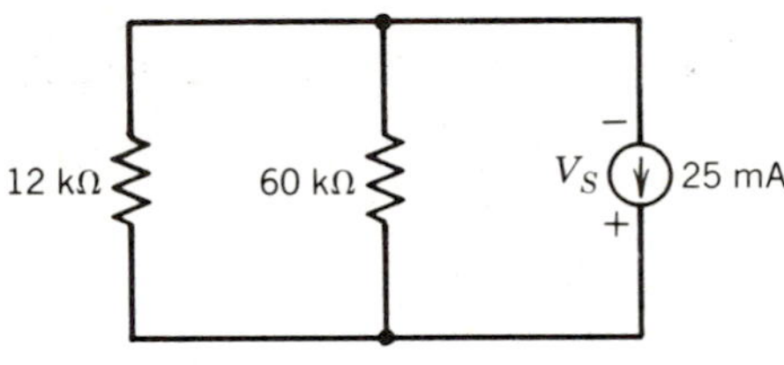

Figure 2.27

4. Using series-parallel analysis methods, find the voltage across and current through each element in Fig. 2.28.

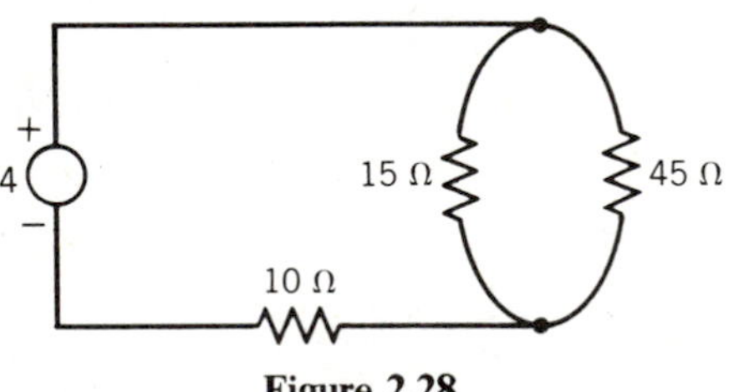

Figure 2.28

2.4 Circuit Simplification Through Element Combination

LEARNING OBJECTIVES

After completing this section you should be able to do the following:

1. Prove that N resistors in series can be replaced by one equivalent resistor whose value is equal to the sum of the values of the individual resistors.

2. Prove that N voltage sources in series may be replaced by a single source whose value equals the algebraic sum of the individual sources.
3. Find the equivalent resistance of N resistors in parallel.
4. Find the equivalent current source for N current sources in parallel.
5. Use circuit reduction techniques to simplify problems.

Often we can save ourselves considerable time and effort by replacing many circuit elements with one or a few elements that have the same effect on the remainder of the circuit. Consider the two circuits shown in Fig. 2.29.

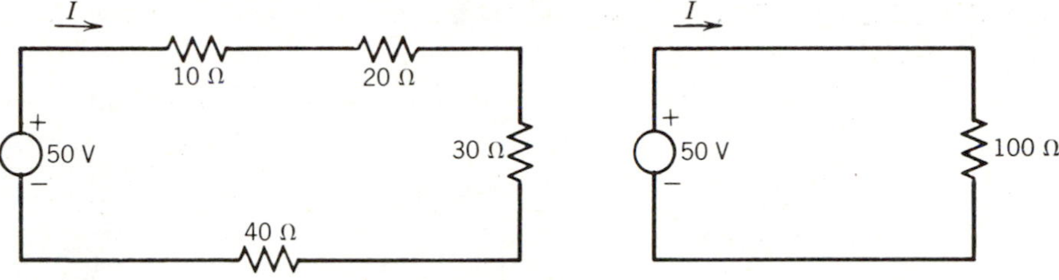

Figure 2.29

If we calculate the current flowing from the 50-V source, we find it is 0.5 A in both circuits. That is, the four resistors in the first circuit can be replaced by the single 100-Ω resistor in the second circuit and the voltage source does not know the difference! A current of 0.5 A will be supplied by the source in either case. The secret is in knowing the value of the single resistor that will replace the four resistors. The problem can be stated in a general form as follows: If a network contains N series resistors as shown in Fig. 2.30*a*, find the value of single resistor, R_{eq}, as seen in Fig. 2.30*b*, which can replace the N resistors such that the current I is unaffected. To solve this problem, write KVL around each circuit.

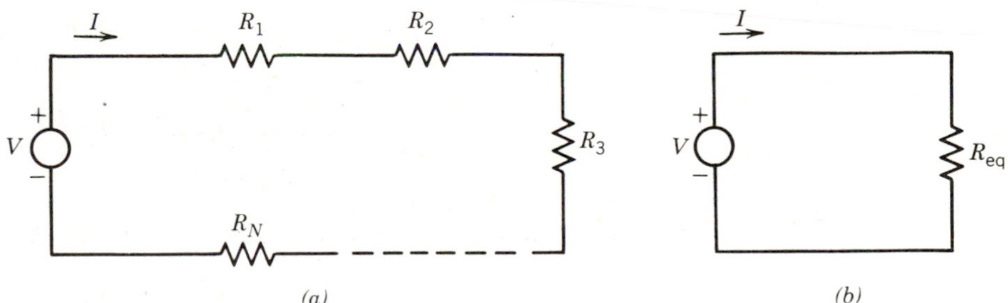

Figure 2.30

$$V = R_1 I + R_2 I + R_3 I + \cdots + R_N I \tag{2.4.1}$$

and

$$V = R_{eq} I \tag{2.4.2}$$

If Eq. 2.4.1 is simplified as

$$V = (R_1 + R_2 + R_3 + \cdots + R_N) I$$

and compared to Eq. 2.4.2, we find

$$R_{eq} = R_1 + R_2 + R_3 + \cdots + R_N \tag{2.4.3}$$

This result says that:

> N resistors in series can be replaced by a single resistor whose value is the sum of the resistances of the N resistors without affecting the remainder of the circuit.

EXAMPLE 2.4.1 Find the value of the current I in the circuit of Fig. 2.31.

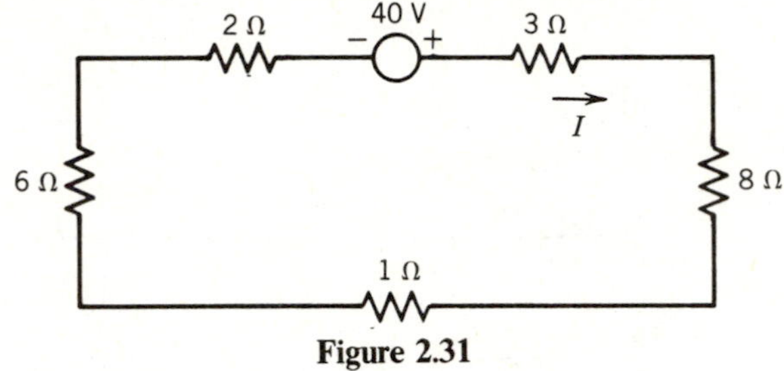

Figure 2.31

Solution

Using Eq. 2.4.3,

$$R_{eq} = 3 + 8 + 1 + 6 + 2 = 20\ \Omega$$

so an equivalent circuit is shown in Fig. 2.32.

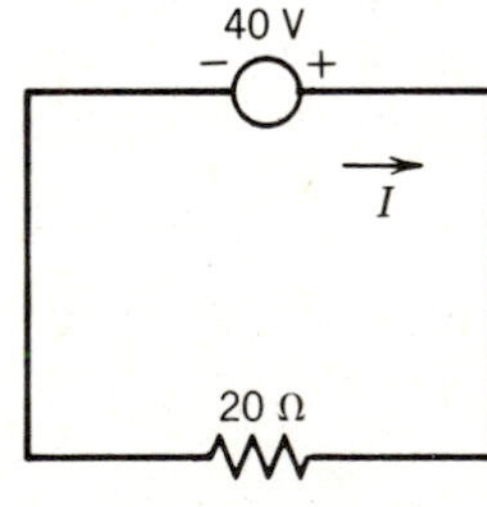

Figure 2.32

By Ohm's law the current is immediately found to be

$$I = \frac{V}{R_{eq}} = \frac{40}{20} = 2\ \text{A}$$

We can use exactly the same arguments to substantiate the following law:

> N voltage sources in series may be replaced by a single source whose value equals the algebraic sum of the N sources.

Consider the circuit in Fig. 2.33.

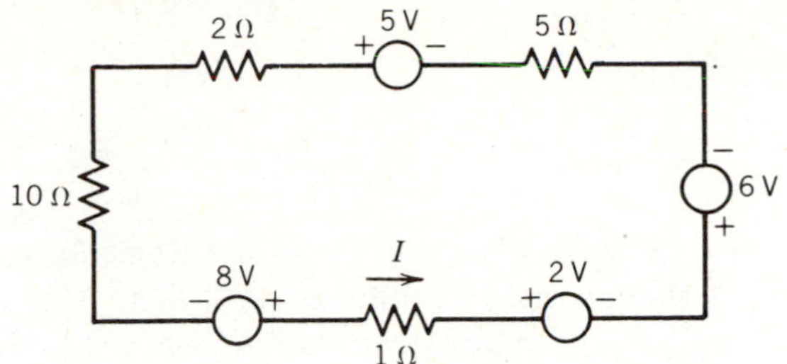

Figure 2.33. Circuit with four voltage sources.

We calculate

$$R_{eq} = 2 + 5 + 1 + 10 = 18\ \Omega$$

In order to calculate V_{eq} we must choose a reference direction. In order to be consistent with the reference direction established for Ohm's law we will assume that the current I flows out of the positive value for V_{eq}. We then calculate V_{eq} as

$$V_{eq} = 8 - 2 - 6 + 5 = 5\ V$$

Our new circuit is shown in Fig. 2.34, and the current is

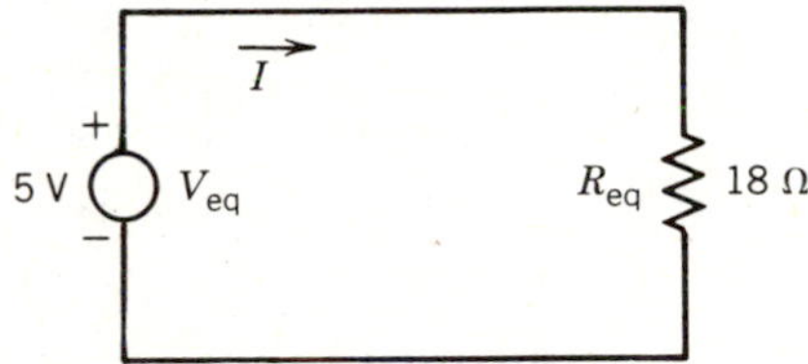

Figure 2.34. Equivalent circuit to Fig. 2.33.

$$I = \tfrac{5}{18} = 0.28\ A$$

We note here that the positive reference direction of V_{eq} must be chosen and then the *algebraic sum of the voltages determined.*

EXAMPLE 2.4.2 The circuit in Fig. 2.35 contains two voltage sources of equal value but opposite polarity. What will be the new current flow?

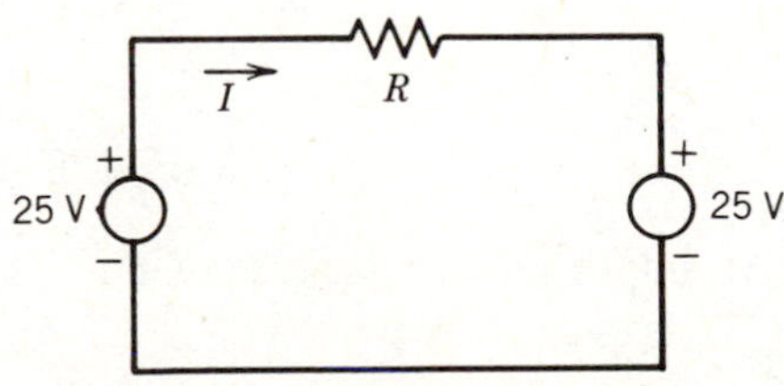

Figure 2.35

Solution

$$V_{eq} = 25 - 25 = 0 \text{ V}$$

so

$$I = \frac{V_{eq}}{R} = 0$$

Can this be true? Certainly! Remember voltage is a pressure and current flows along a pressure differential. If there is no pressure difference, there is no current.

We have now seen that voltage sources and resistors in series can be replaced by equivalent circuit elements. The next logical question would concern elements in a parallel circuit. N resistors in parallel with a current source, such as in Fig. 2.36*a* can be examined with KCL to determine the value of R_{eq} of Fig. 2.36*b*.

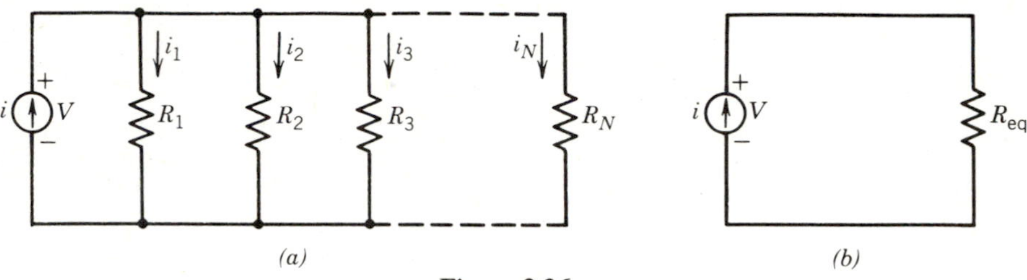

Figure 2.36

KCL yields

$$\begin{aligned} i &= i_1 + i_2 + i_3 + \cdots + i_N \\ &= \frac{V}{R_1} + \frac{V}{R_2} + \frac{V}{R_3} + \ldots + \frac{V}{R_N} \end{aligned}$$

or

$$i = \left(\frac{1}{R_1} + \frac{1}{R_2} + \frac{1}{R_3} + \cdots + \frac{1}{R_N} \right) V \tag{2.4.4}$$

By Ohm's law,

$$i = \frac{V}{R_{eq}} \tag{2.4.5}$$

Comparing Eqs. 2.4.4 and 2.4.5 reveals

$$\frac{1}{R_{eq}} = \frac{1}{R_1} + \frac{1}{R_2} + \frac{1}{R_3} + \cdots + \frac{1}{R_N}$$

or

$$R_{eq} = \frac{1}{1/R_1 + 1/R_2 + 1/R_3 + \cdots + 1/R_N} \tag{2.4.6}$$

An important special case of Eq. 2.4.6 occurs when $N = 2$. In this situation

$$R_{eq} = \frac{R_1 R_2}{R_1 + R_2} \tag{2.4.7}$$

Equations 2.4.3, 2.4.6, and 2.4.7 are fundamental to circuit analysis and should be memorized.

EXAMPLE 2.4.3 Find the equivalent resistance of the parallel combination shown in Fig. 2.37*a*.

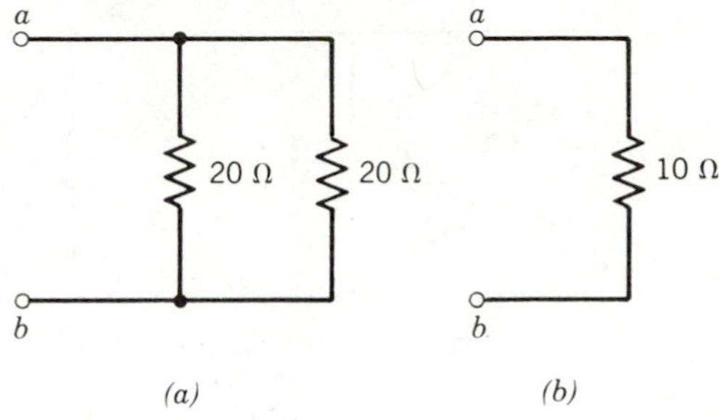

Figure 2.37

Solution

By Eq. 2.4.7,

$$R_{eq} = \frac{(20)(20)}{20 + 20} = 10\ \Omega$$

That is, the two 20-Ω resistors can be replaced by the equivalent circuit shown in Fig. 2.37*b*.

EXAMPLE 2.4.4 Find the equivalent resistance of the network in Fig. 2.38.

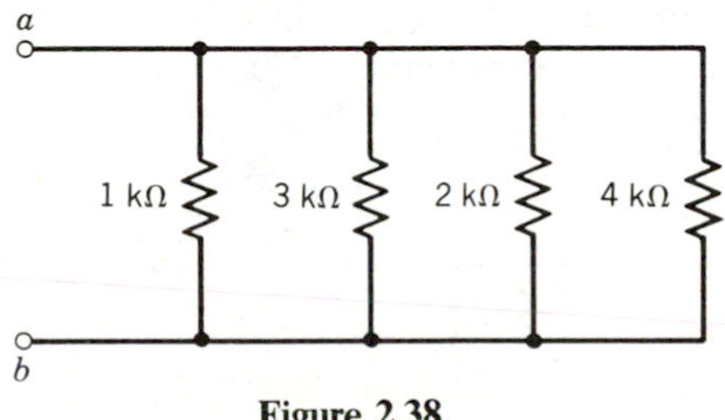

Figure 2.38

Solution

We have at least two choices of approach. First we could apply Eq. 2.4.6 directly:

$$\begin{aligned} R_{eq} &= \frac{1}{\frac{1}{1000} + \frac{1}{3000} + \frac{1}{2000} + \frac{1}{4000}} \\ &= \frac{12{,}000}{12 + 4 + 6 + 3} \\ &= \frac{12{,}000}{25} \\ &= 480\ \Omega \end{aligned}$$

Second, we could combine the 1- and 3-kΩ resistors using Eq. 2.4.7 and do likewise with the 2- and 4-kΩ resistors to obtain the intermediate result in Fig. 2.39, then combine R_1 and R_2 again using Eq. 2.4.7 to find R_{eq}.

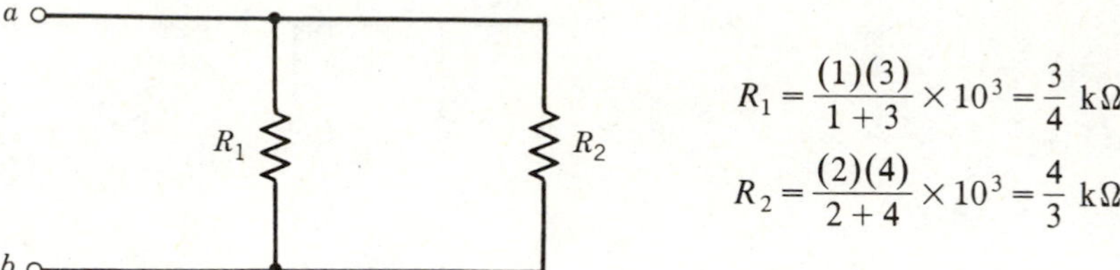

Figure 2.39. Two resistors in parallel.

$$R_{eq} = \frac{(\frac{3}{4})(\frac{4}{3})}{\frac{3}{4} + \frac{4}{3}} \times 10^3 = \frac{1}{\frac{25}{12}} \times 10^3 = \frac{12}{25} \times 10^3 = 480\ \Omega$$

Finally, current sources in parallel add algebraically. If two parallel sources are supplying current in the same direction their equivalent source is the sum of the two individual sources. If the two sources are opposite in direction of current the equivalent source supplies a new current equal to the difference of the two sources.

In Fig. 2.40, circuits (a) and (b) are equivalent and circuits (c), (d), and (e) are all equivalent. You can prove this to yourself by connecting a 1-Ω resistor across the terminals X and Y and calculating the voltage across the resistor in each case using KCL and Ohm's law.

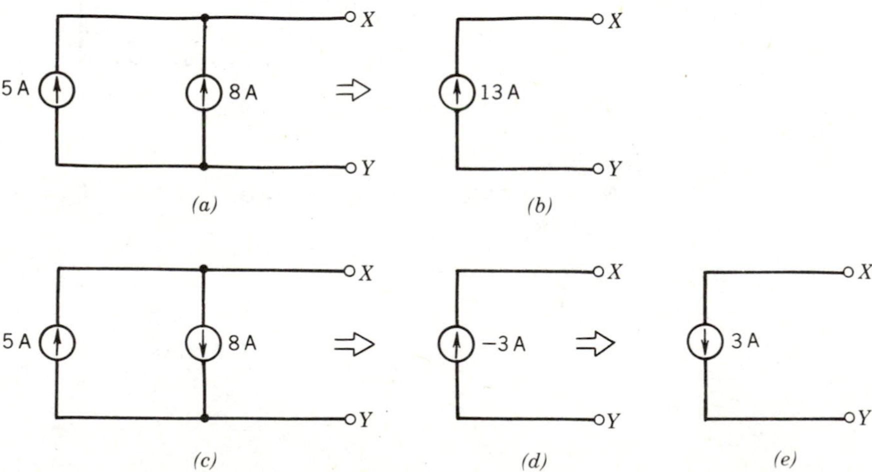

Figure 2.40. Circuit reduction of two parallel current sources.

EXAMPLE 2.4.5 For the circuit in Fig. 2.41 find I_{eq} and R_{eq}.

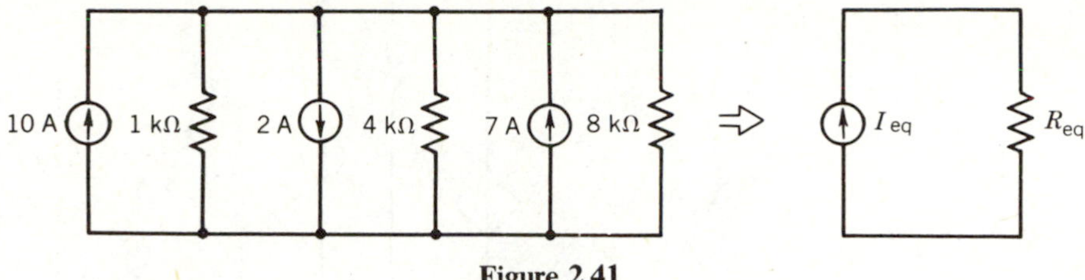

Figure 2.41

Solution

For the reference direction shown

$$I_{eq} = 10 - 2 + 7 = 15 \text{ A}$$

By Eq. 2.4.4,

$$R_{eq} = \frac{1}{\frac{1}{1000} + \frac{1}{4000} + \frac{1}{8000}} = 727.27\ \Omega$$

Several observations and comments can be made about combining circuit elements as a means of analyzing circuit behavior.

1. The combination of resistors in parallel will always result in an equivalent resistance smaller than the smallest value of resistance of the original resistors.
2. The combination of resistors in series will always result in an equivalent resistance larger than the largest of the original resistors.
3. If three or more resistors are in parallel, the equivalent resistance can be found by the repeated application of Eq. 2.4.5.
4. Dependent sources are usually not combined with any other sources in order to achieve network reduction. Dependent sources should remain identifiable. Also the circuit parameter that determines the value of a dependent source is usually not combined with other circuit elements.
5. Once two circuit elements are combined, we lose our ability to find the voltage across, current through, and power delivered or absorbed by that element. This can be accomplished, however, by reconstructing a portion of (or all of) the original circuit after the equivalent circuit is used to calculate its values of interest.
6. From our definition of ideal voltage and current sources, two voltage sources of different values cannot be connected in parallel and two current sources of different values cannot be connected in series. We will relax this rule later in the text under the discussion of practical voltage and current sources.

In our final examples of the use of equivalent circuits we will consider circuits with both parallel and series components as well as a dependent source.

EXAMPLE 2.4.6 Find the value of the current in Fig. 2.42.

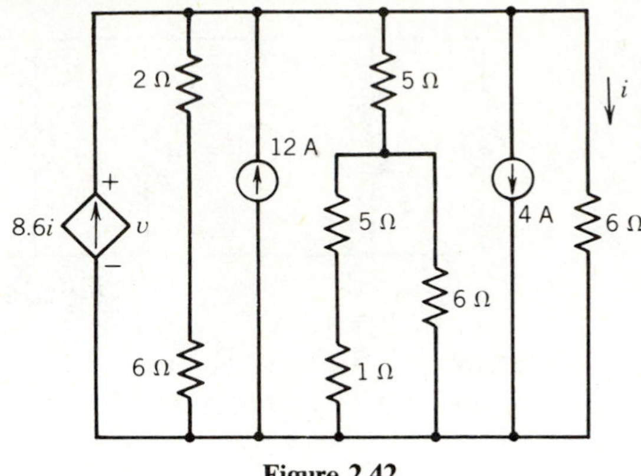

Figure 2.42

Solution

The elements of the circuit are combined in the series of steps shown in Fig. 2.43.

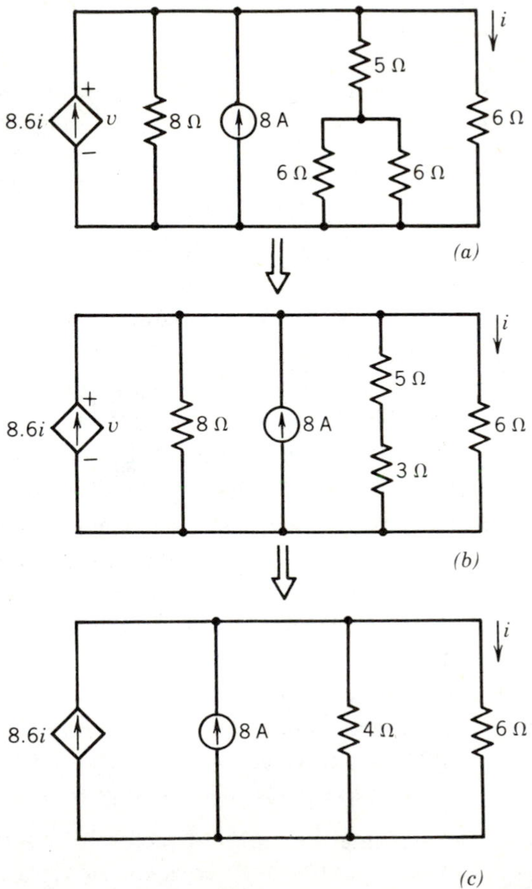

Figure 2.43

Now by KCL,

$$8.6i + 8 = \frac{V}{4} + \frac{V}{6}$$

and by Ohm's law,

$$V = 6i$$

so

$$8.6i + 8 = \tfrac{6}{4}i + i$$

or

$$i = -1.31 \text{ A}$$

EXAMPLE 2.4.7 Find I in the circuit of Fig. 2.44 by using network reduction.

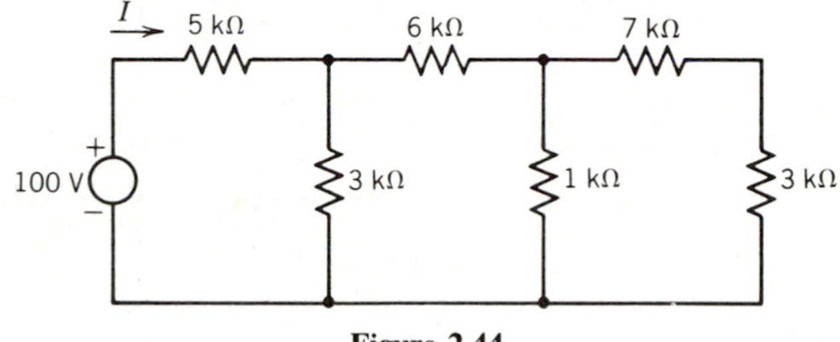

Figure 2.44

Solution

See Fig. 2.45. By Ohm's law from Fig. 2.45*e* we have

$$I = \frac{100}{7.092 \times 10^3} = 14.1 \text{ mA}$$

EXAMPLE 2.4.8 A 5-V battery is connected in series with two resistors of values 2 and 5 kΩ. Determine the voltage drop across each resistor.

Solution

$$R_{eq} = 2k + 5k = 7k$$

$$I = \frac{V}{R_{eq}} = \frac{5}{7(10)^3} = \frac{5}{7}(10)^{-3}\text{A}$$

$$V_{2k} = 2(10)^3 \times \tfrac{5}{7}(10)^{-3} = \tfrac{10}{7}V$$

$$V_{5k} = 5(10)^3 \times \tfrac{5}{7}(10)^{-3} = \tfrac{25}{7}V$$

As a check,

$$V = V_{2k} + V_{5k}$$

$$5 = \tfrac{10}{7} + \tfrac{25}{7} = \tfrac{35}{7} = 5$$

LEARNING EVALUATION

Transform each of the circuits in Fig. 2.46 to a circuit with a single equivalent independent source and resistance.

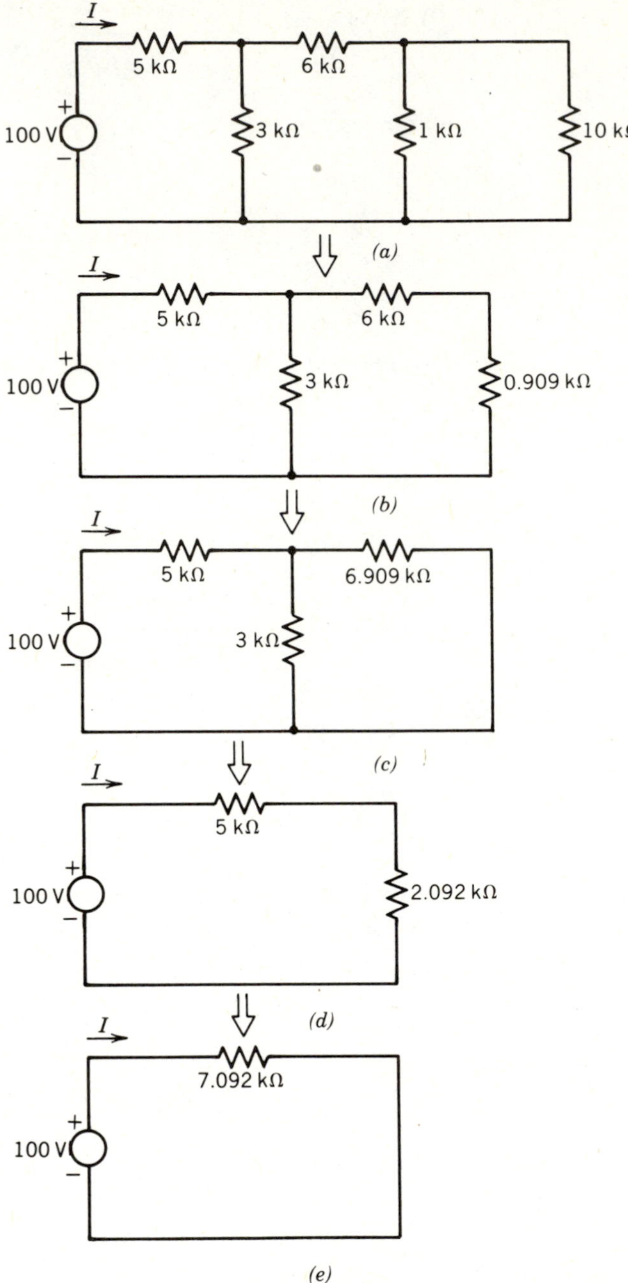

Figure 2.45

2.5 Voltage and Current Division

LEARNING OBJECTIVES

After completing this section you should be able to do the following:

1. Derive the equation for the division of voltage between two resistors in series.

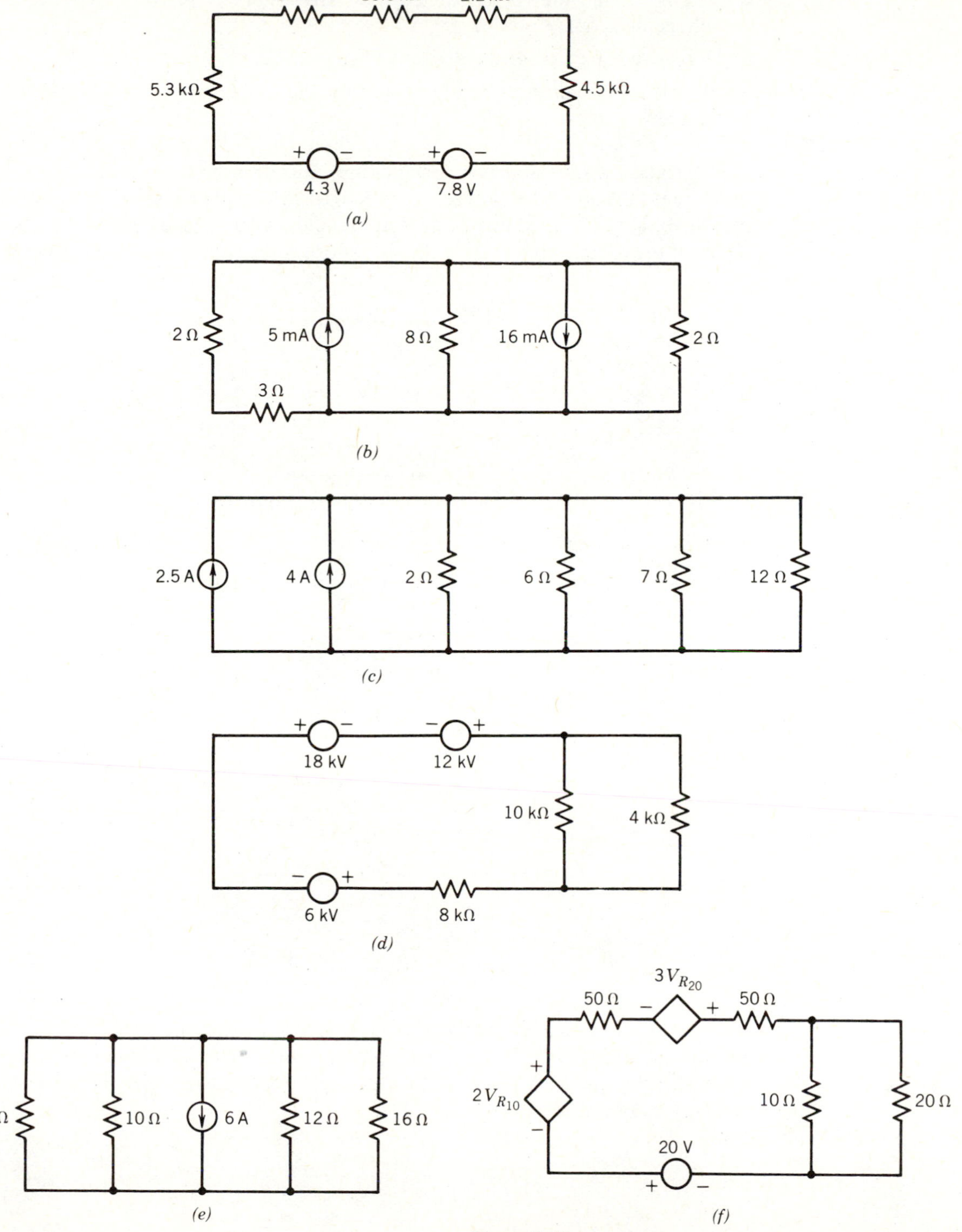
8.4 kΩ
13.5 kΩ
2.2 kΩ
5.3 kΩ
4.5 kΩ
4.3 V
7.8 V
(a)
2 Ω
5 mA
8 Ω
16 mA
2 Ω
3 Ω
(b)
2.5 A
4 A
2 Ω
6 Ω
7 Ω
12 Ω
(c)
18 kV
12 kV
10 kΩ
4 kΩ
6 kV
8 kΩ
(d)
15 Ω
10 Ω
6 A
12 Ω
16 Ω
(e)
3 $V_{R_{20}}$
50 Ω
50 Ω
2 $V_{R_{10}}$
10 Ω
20 Ω
20 V
(f)

Figure 2.46

2. Be able to find the voltage drop across any one of N resistors in series using voltage division.
3. Derive an equation for the division of current between two resistors in parallel.
4. Be able to use current division to find the current in any one of N resistors in parallel.

If a voltage source and two resistors are all connected in series (Fig. 2.47) we know that the sum of the voltage drop across each of the resistors must equal the source voltage (KVL). How can we determine the value of the voltage drop across either of the resistors if we know the values of source voltage and the resistors?

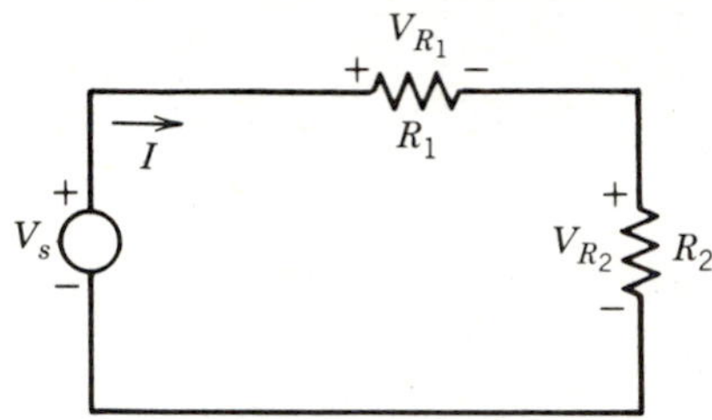

Figure 2.47. A simple voltage divider circuit.

Begin by writing KVL around the circuit,

$$V_s = V_{R_1} + V_{R_2} = IR_1 + IR_2 = I(R_1 + R_2)$$

$$I = \frac{V_s}{R_1 + R_2}$$

but

$$V_{R_1} = IR_1$$

so

$$\boxed{V_{R_1} = \frac{R_1}{R_1 + R_2} V_s} \tag{2.5.1}$$

Likewise,

$$\boxed{V_{R_2} = \frac{R_2}{R_1 + R_2} V_s} \tag{2.5.2}$$

This result says that the voltage drop across a resistor in series with another resistor is simply the ratio of that resistor value to the sum of the resistances multiplied by the source voltage. This can be extended to the general formula for N resistors in series:

$$\boxed{V_{R_i} = \frac{R_i}{R_i + R_2 + \cdots + R_N} V_s} \tag{2.5.3}$$

where R_i is one of the resistors in series.

Figure 2.47 is often called a voltage divider network (circuit) because the source voltage does divide across the two resistors as a function of the resistor values. The design of voltage divider networks is a very important function of engineering.

EXAMPLE 2.5.1 A calculator can be modeled as a resistor with a resistance of 100 kΩ. In order to operate properly the calculator needs 9 V supplied to it (across the 100-kΩ resistor.) If we have only a 12-V battery, what value of resistance must we put in series with the calculator to ensure a 9-V supply to the calculator?

Solution

We can model our system with an electric circuit as shown in Fig. 2.48.

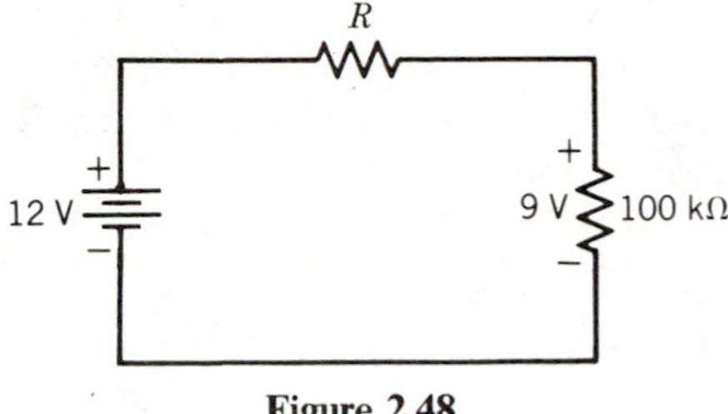

Figure 2.48

By Eq. 2.5.2,

$$9 = \frac{100\ \text{k}\Omega}{100\ \text{k}\Omega + R}(12)$$

Solving for R, we find

$$R = 33.33\ \text{k}\Omega$$

Note that we used a new symbol for the 12-V battery in the model. This symbol is quite common and is most often used to model batteries but may be used to model general dc voltage sources. The "long" lines indicate the positive side of the battery and the "short" lines the negative side.

EXAMPLE 2.5.2 Use voltage division to find the voltage across the 5-Ω resistor in Fig. 2.49.

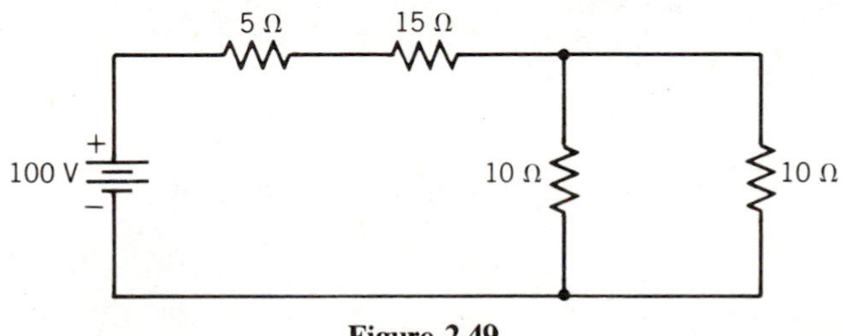

Figure 2.49

Solution

We can use the general formula for N resistors in series, Eq. 2.5.3, if we first replace the two 10-Ω resistors by their equivalent parallel resistance.

This is shown in Fig. 2.50. Applying Eq. 2.5.3 to Fig. 2.50 gives us

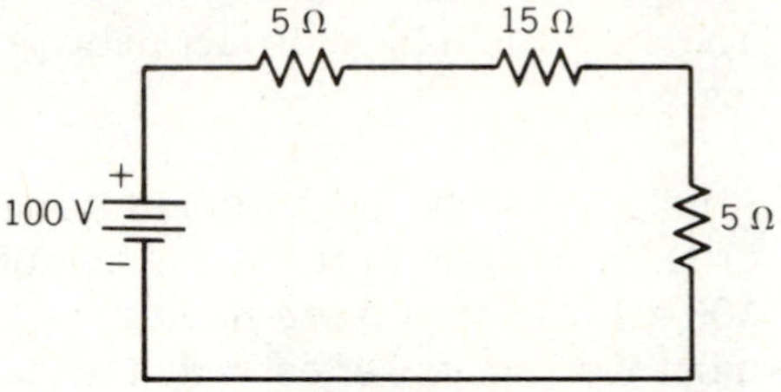

Figure 2.50

$$V_5 = \frac{5}{5 + 15 + 5}(100) = 20 \text{ V}$$

Now let us consider current division. A current source in parallel with two conductances G_1 and G_2 is shown in Fig. 2.51. Since all three elements are in parallel the voltage V appears across all the elements. Using KCL and Ohm's law we may say

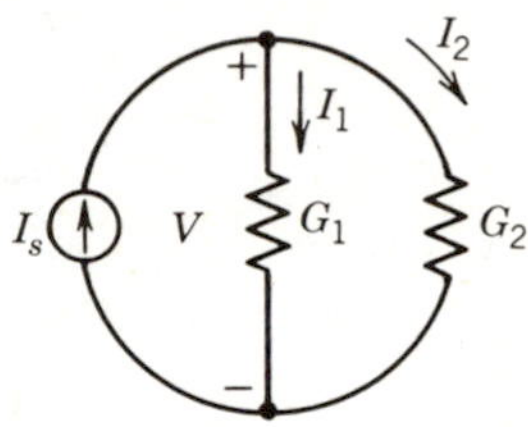

Figure 2.51

$$\begin{aligned} I_s &= I_1 + I_2 \\ &= VG_1 + VG_2 = (G_1 + G_2)V \end{aligned}$$

or

$$V = \frac{I_s}{G_1 + G_2} \tag{2.5.4}$$

but

$$I_1 = VG_1$$

so if we multiply both sides of Eq. 2.5.4 by G_1, we obtain

$$\boxed{I_1 = \frac{G_1}{G_1 + G_2} I_s} \tag{2.5.5}$$

By a similar argument we can show that I_2 is related to I_s by

$$\boxed{I_2 = \frac{G_2}{G_1 + G_2} I_s} \tag{2.5.6}$$

Equations 2.5.5 and 2.5.6 are dangerously similar to Eqs. 2.5.1 and 2.5.2. This has caused much gnashing of teeth by students who have failed to recall the difference during exams. If you can just remember that conductance goes with current division, and that resistance goes with voltage division, you should have no trouble.

These current division formulas can be extended to the general formula for N conductances in parallel:

$$I_{G_i} = \frac{G_i}{G_1 + G_2 + \cdots + G_N} I_s \qquad (2.5.7)$$

EXAMPLE 2.5.3 For the circuit in Fig. 2.52, find I_1 and I_2.

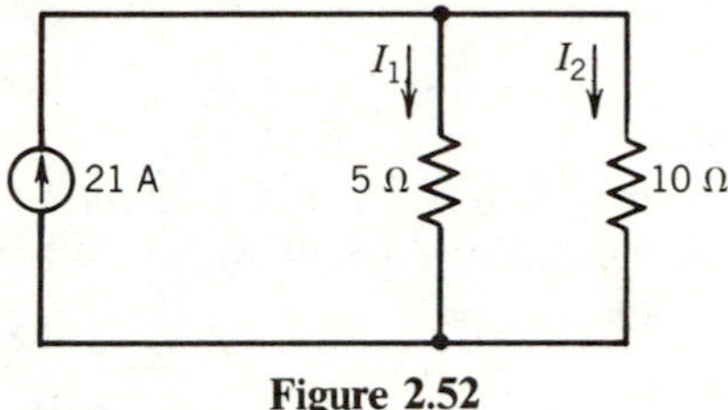

Figure 2.52

Solution

From Eq. 2.5.5 with $G_1 = \frac{1}{5}$ S,

$$I_1 = \frac{\frac{1}{5}}{\frac{1}{5} + \frac{1}{10}} 21 = 14 \text{ A}$$

From Eq. 2.5.6,

$$I_2 = \frac{\frac{1}{10}}{\frac{1}{5} + \frac{1}{10}} = 7 \text{ A}$$

EXAMPLE 2.5.4 Find I_1 in the circuit of Fig. 2.53.

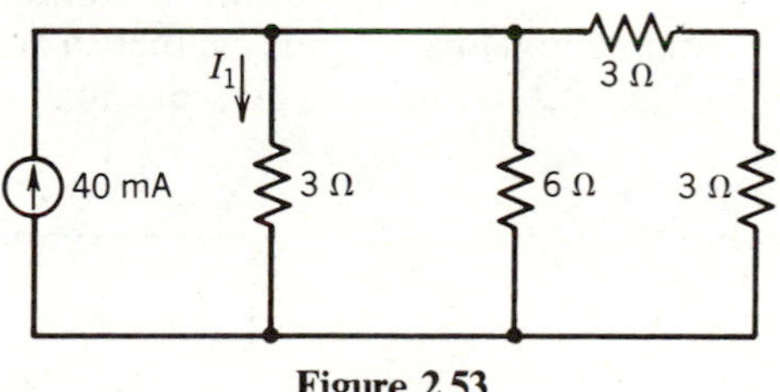

Figure 2.53

Solution

First combine the two series 3-Ω resistors on the right to obtain one 6-Ω resistor. Then apply Eq. 2.5.7.

$$I_1 = \frac{\frac{1}{3}(40)}{\frac{1}{3} + \frac{1}{6} + \frac{1}{6}} = 20 \text{ mA}$$

LEARNING EVALUATIONS

1. Using the voltage divider concept, determine the voltage drop across the 200-Ω resistor in Fig. 2.54.

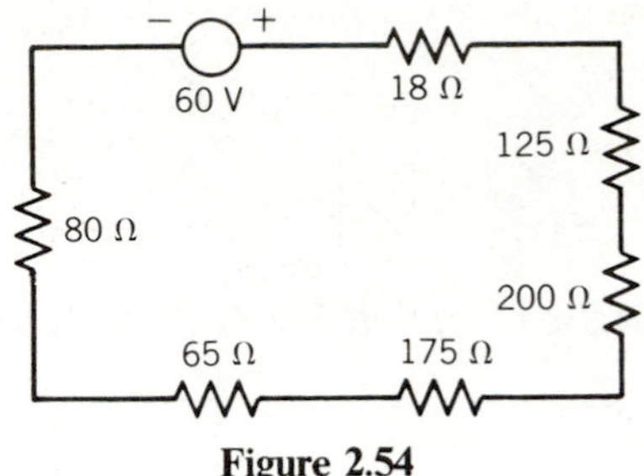

Figure 2.54

2. A 50-kΩ resistor and a 150-kΩ resistor are placed in parallel across a 4-mA independent current source. What are the voltage across and current through each resistor?

2.6 Delta-Wye Transformations

LEARNING OBJECTIVES

After completing this section you should be able to do the following:

1. State the need for delta-wye transformation.
2. Be able to use the delta-wye transformations to reduce a circuit to a series-parallel circuit.

We have now learned to combine some circuit elements that are in series or parallel in order to simplify network analysis. But what if these elements are neither in series nor in parallel? General techniques for this problem will be developed in Chapter 3 but if the elements are resistors a special technique called the delta-wye or π-t transformation may be useful. The circuit in Fig. 2.55 contains resistors that are neither in series nor parallel and the calculation of the current I is beyond our present capabilities.

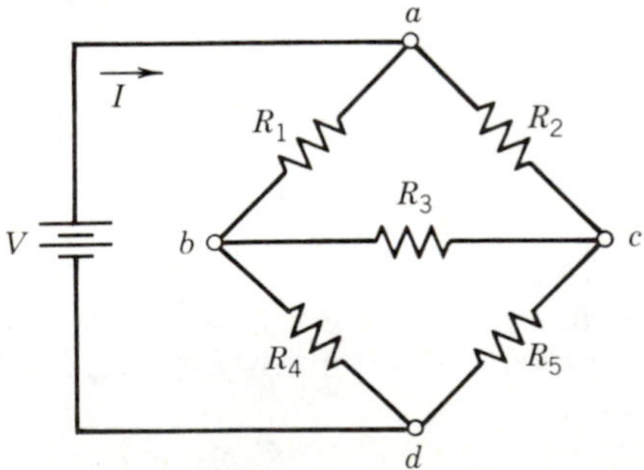

Figure 2.55. A resistive circuit.

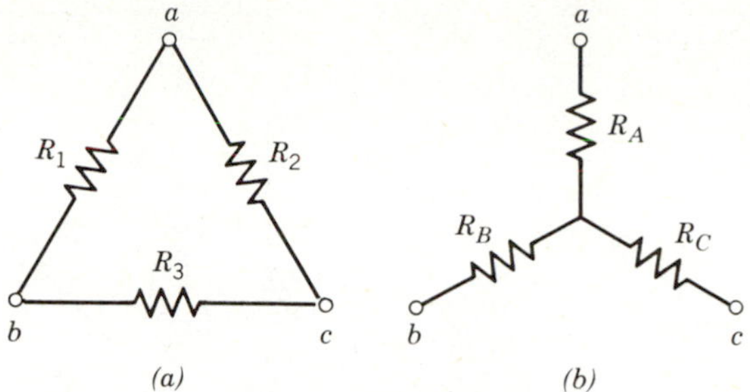

Figure 2.56. Resistors connected in delta (a) and wye (b).

The delta-wye transformation states that with respect to terminals a, b, and c the two circuits in Fig. 2.56 are equivalent if the relationships given in Eqs. 2.6.1 and 2.6.2 are true. When we say the two circuits are equivalent with respect to terminals a, b, and c, we mean that if one of these circuits is inside a box with only three terminals exposed, there is no way short of opening the box to tell whether the circuit inside is delta or wye.

$$R_A = \frac{R_1 R_2}{R_1 + R_2 + R_3} \tag{2.6.1a}$$

$$R_B = \frac{R_1 R_3}{R_1 + R_2 + R_3} \tag{2.6.1b}$$

$$R_C = \frac{R_2 R_3}{R_1 + R_2 + R_3} \tag{2.6.1c}$$

or

$$R_1 = \frac{R_A R_B + R_B R_C + R_C R_A}{R_C} \tag{2.6.2a}$$

$$R_2 = \frac{R_A R_B + R_B R_C + R_C R_A}{R_B} \tag{2.6.2b}$$

$$R_3 = \frac{R_A R_B + R_B R_C + R_C R_A}{R_A} \tag{2.6.2c}$$

These equations seem formidable at first glance, but a brief study reveals a logical pattern that aids in remembering them. For example, the denominators of Eqs. 2.6.1 are the same and are simply the sum of the three delta resistors. The numerator of the wye resistor value being calculated is the product of the two delta resistors connected to the node also common to the wye resistor. Similar observations can be made for Eqs. 2.6.2.

Now we will return to Fig. 2.55 and replace the delta resistor arrangement connected to nodes a, b, c by its wye equivalent to obtain the circuit of Fig. 2.57.

The values of R_A, R_B, and R_C are given by Eqs. 2.6.1. An equivalent resistance between terminals a and d may now be calculated as

$$R_{\text{eq}} = R_A + \frac{(R_B + R_4)(R_C + R_5)}{R_B + R_4 + R_C + R_5}$$

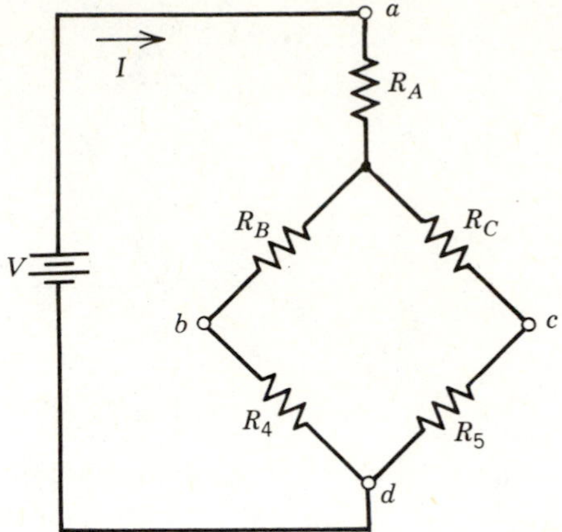

Figure 2.57. Equivalent circuit to Fig. 2.55.

and finally the current I would then be

$$I = \frac{V}{R_{eq}}$$

EXAMPLE 2.6.1 Find the equivalent resistance between terminals X and Y in Fig. 2.58.

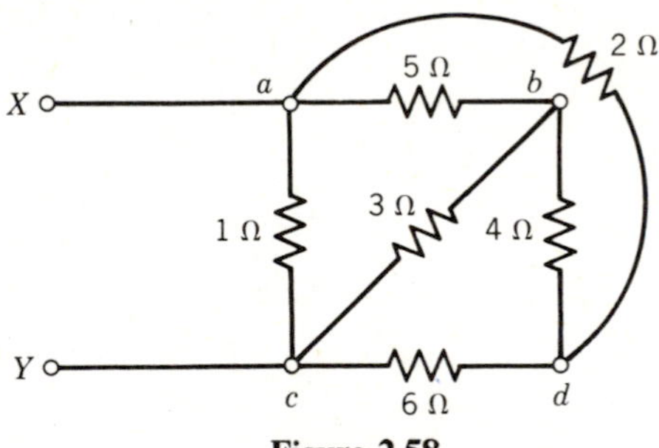

Figure 2.58

Solution

Use delta-to-wye transformation on the 4-, 5-, and 2-Ω resistors to obtain the circuit in Fig. 2.59*a*.

$$R_A = \frac{(2)(5)}{2+4+5} = \frac{10}{11}\ \Omega$$

$$R_B = \frac{(4)(5)}{11} = \frac{20}{11}\ \Omega$$

$$R_C = \frac{(2)(4)}{11} = \frac{8}{11}\ \Omega$$

Using previous techniques of combining series and parallel resistors the circuit can be further reduced in the series of steps indicated in Figs. 2.59*b* through *e*.

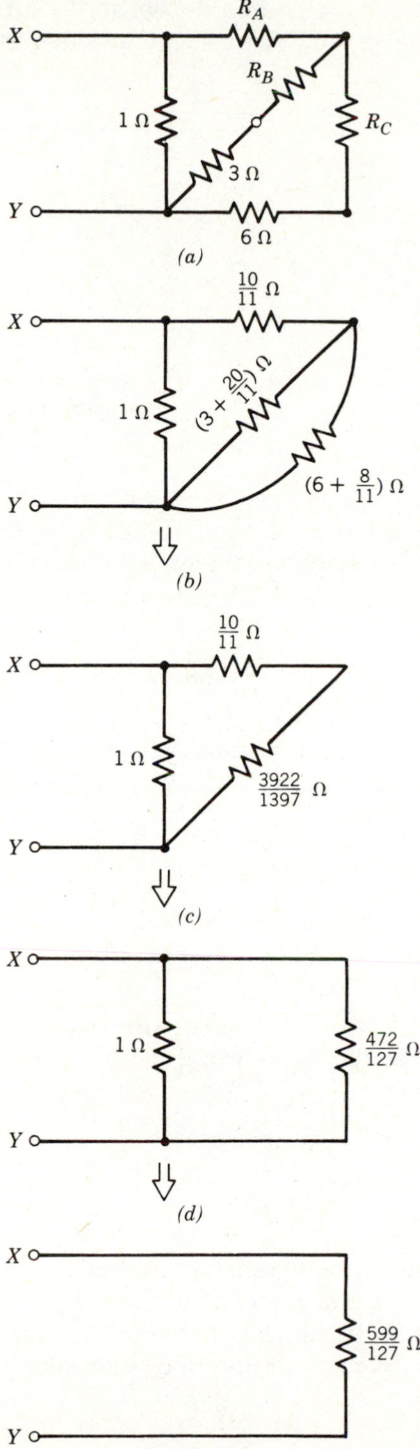

X
Y
R_A
R_B
R_C
1 Ω
3 Ω
6 Ω
(a)
$\frac{10}{11}$ Ω
$(3 + \frac{20}{11})$ Ω
$(6 + \frac{8}{11})$ Ω
(b)
$\frac{3922}{1397}$ Ω
(c)
$\frac{472}{127}$ Ω
(d)
$\frac{599}{127}$ Ω
(e)

Figure 2.59

EXAMPLE 2.6.2 Find the voltage source necessary to supply a current, I, of 500 μA to the circuit in Fig. 2.60. All resistance values are in megohms.

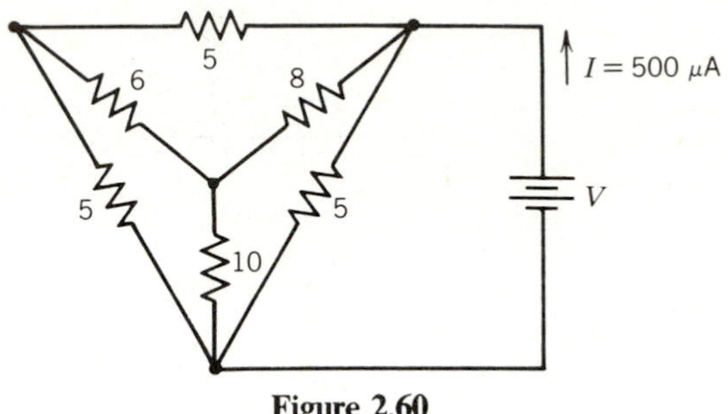

Figure 2.60

Solution

Replace the 6-, 8-, and 10-MΩ wye-connected resistors with their delta equivalents, then reduce the circuit. The steps are shown in Fig. 2.61.

$$V = IR = (500 \times 10^{-6})(2.803 \times 10^{6}) = 1401.5 \text{ V}$$

Delta-wye transformations will prove to be particularly useful in later chapters where electrical power systems and substation loads are discussed.

LEARNING EVALUATIONS

1. Draw a delta connection with $R_1 = 2.5$ kΩ, $R_2 = 3.0$ kΩ, and $R_3 = 4.5$ kΩ. Calculate the equivalent wye resistors, R_A, R_B, and R_C and draw equivalent wye diagram.
2. Draw a wye connection with $R_A = 2.5$ kΩ, $R_B = 3.0$ kΩ, $R_C = 4.5$ kΩ. Calculate the equivalent delta resistors, R_1, R_2, and R_3 and draw the delta diagram.

2.7 The Forest and the Trees

One of the most difficult and challenging tasks facing a student of any complex subject is keeping the details of all newly discovered knowledge in proper perspective. In this section we will summarize our efforts of these first two chapters in order to help you develop your perspective.

$$R_1 = \frac{(6)(8)+(6)(10)+(8)(10)}{10} = \frac{188}{10}\ \mathrm{M\Omega}$$

$$R_2 = \frac{188}{8}\ \mathrm{M\Omega}$$

$$R_3 = \frac{188}{6}\ \mathrm{M\Omega}$$

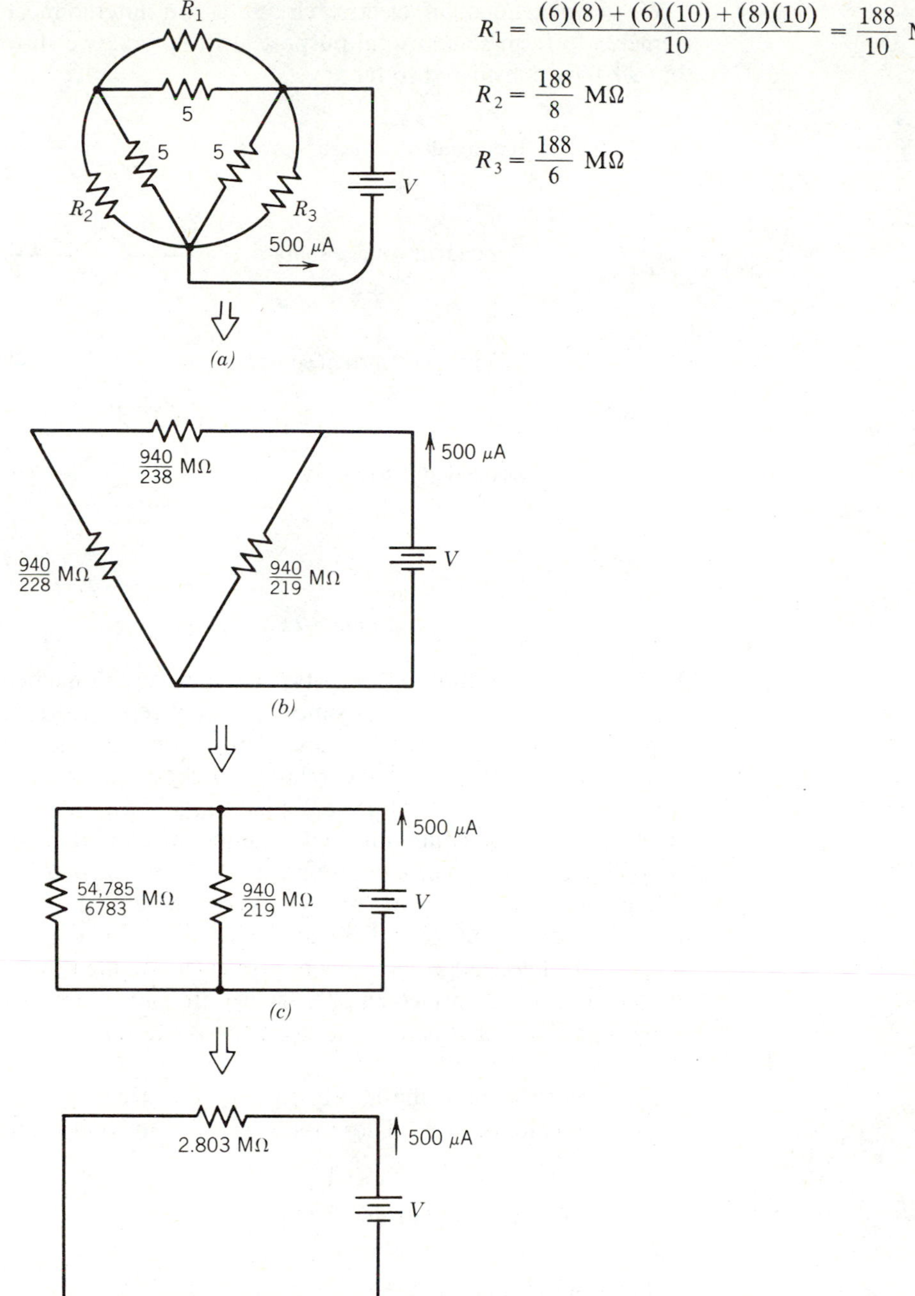

Figure 2.61

We have defined an electric circuit as an interconnection of two or more elements to form some useful purpose. In Fig. 2.62 we show the circuit elements that we have considered so far.

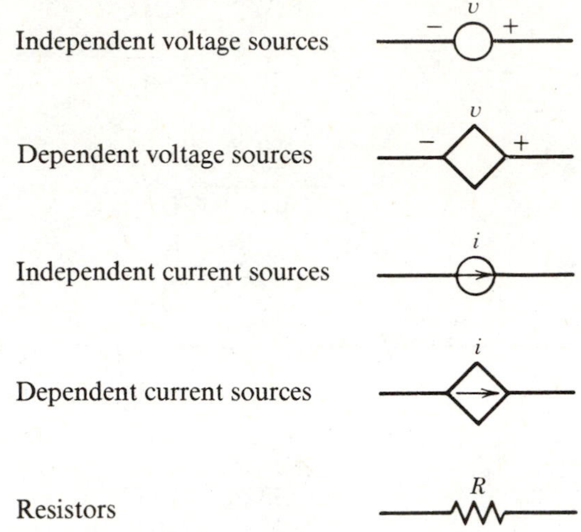

Figure 2.62. Circuit symbols.

A *node* is a connection between two or more circuit elements. Remember that a "circuit" drawn in this text is simply a *model* representing a physical system.

We have defined terms for describing network behavior as current, voltage, power, and energy and have related these terms to the atomic structure of physical devices. We have also assigned units and symbols to these terms so that their values can be communicated to others in an understandable fashion.

We have limited our definitions of "circuit analysis" so far to finding the voltage across and current through every element of an electric circuit and we have limited our voltage and current sources to being constant (time-independent) values. This knowledge allows us to also calculate the power and energy absorbed or delivered by each element. Often we are only interested in a subset of this general problem, that is, we determine these relations for only one or a few of the elements of the circuit.

In order to acquire the ability to perform circuit analysis we are developing various tools to assist us. Those tools we have introduced thus far include:

Kirchhoff's current law (KCL)
Kirchhoff's voltage law (KVL)
Ohm's law
The power equation
The energy equation
Network reduction
  Combining resistors in series

Combining resistors in parallel
Combining voltage sources in series
Combining current sources in parallel
Voltage division
Current division
Delta-wye or π-t transformation

When confronted with a circuit analysis problem you should be able to look through your "bag of tools" and select the tool or tools that will allow you to solve your problem in the most expeditious manner. In problems in the text (and on exams) the tool you are to use may be specified simply to give you experience with, or test your knowledge of, that tool. As a practicing professional engineer you will not be under this restraint and the "tool selection" will be entirely yours.

PROBLEMS

SECTION 2.1

1. Aluminum wire is often used to carry current in power distribution systems because it is cheaper and lighter than copper. The disadvantage of aluminum is that it has higher resistivity than copper. Design an aluminum conductor to replace a copper conductor whose area is 10 cm^2. If the resistance must be the same, which is cheaper and which is lighter?
2. A piece of material measures 10 mm $\times$ 5 mm $\times$ 10 m. The resistance measured across the 10-m length is 250 Ω. Is this a conductor, semiconductor, or insulator?
3. Which is longer, a wire whose conductance is 2 S or a wire whose conductance is 1 S? Both wires are made of the same material and have the same cross-sectional area.

SECTION 2.2

4. Which light bulb contains the largest resistance, a 60-W bulb or a 100-W bulb?
5. The power absorbed by a 330-kΩ resistor is 0.2 W. Determine the voltage across and the current through the resistor.
6. An ammeter (current measuring device) has a voltage drop of 45 mV for a full-scale deflection of 15 A. What is the conductance of the meter, and how much power does it absorb?
7. The voltage across a 150-W light bulb is 110 V. Find the conductance and resistance of the filament and the current through the filament. If electric energy costs 9.5 cents/kWh, determine the cost for 10 days of continuous operation.

8. Find the maximum allowable voltage or current for the resistors with the following power ratings.

a. $\frac{1}{2}$ W, $I = 10\ \mu A$
b. 10 W, $V = 50$ mV
c. $R_1 = \frac{1}{4}$ W and $R_2 = \frac{1}{2}$ W in series: $V_{R_1} = 10$ V, $V_{R_2} = 5$ V
d. $R_1 = \frac{1}{4}$ W and $R_2 = \frac{1}{4}$ W in parallel: $I_{R_1} = 18$ mA, $I_{R_2} = 30$ mA

SECTION 2.3

9. Find the current through, voltage across, and power absorbed by each resistor in Fig. 2.63.

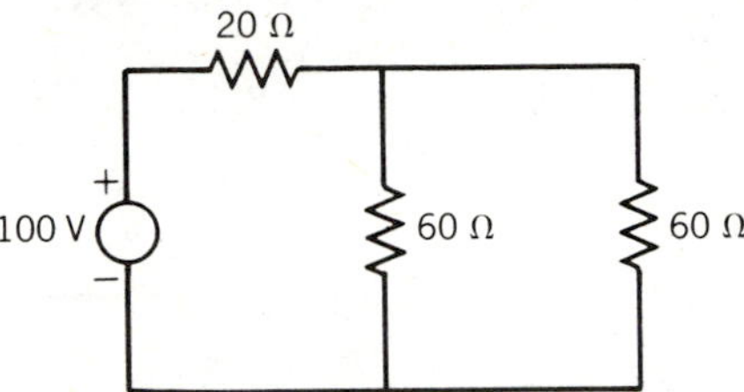

Figure 2.63

10. Find the current in the network of Fig. 2.64.

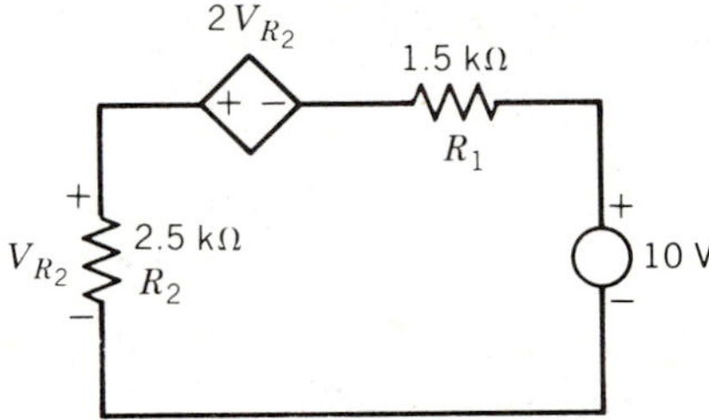

Figure 2.64

11. Find the value of R in Fig. 2.65.

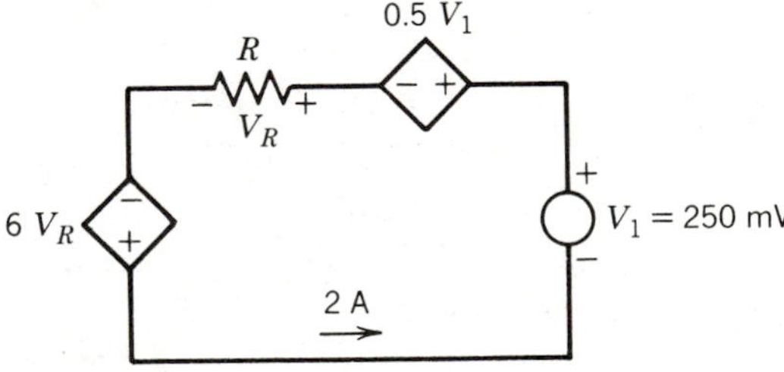

Figure 2.65

12. Two resistors are placed in parallel with a 24-V source. Current of 6.5 A flows through R_1, and R_2 dissipates 750 mW. Find the resistance of both resistors, and the current through R_2.

13. An electric motor requires an input of 50 mA and a terminal voltage of 30 V. This is illustrated in Fig. 2.66, where the available source has a value of 145 V. Find the value of R in the circuit.

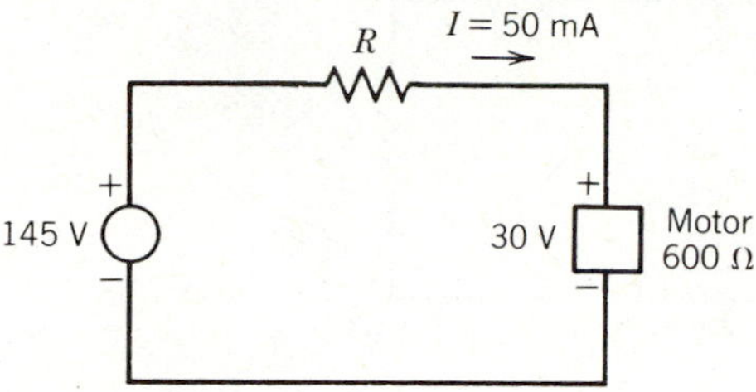

Figure 2.66

14. Find the current, voltage, and power associated with the 50-Ω resistor in Fig. 2.67.

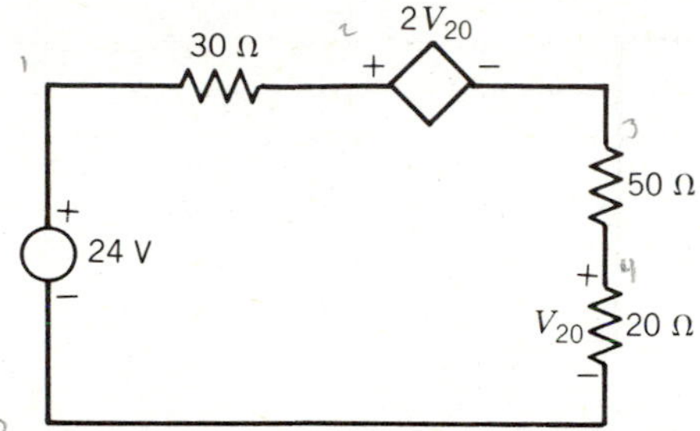

Figure 2.67

15. Find the current, voltage, and power associated with each 40-Ω resistor in Fig. 2.68. Also find the voltage across each dependent source.

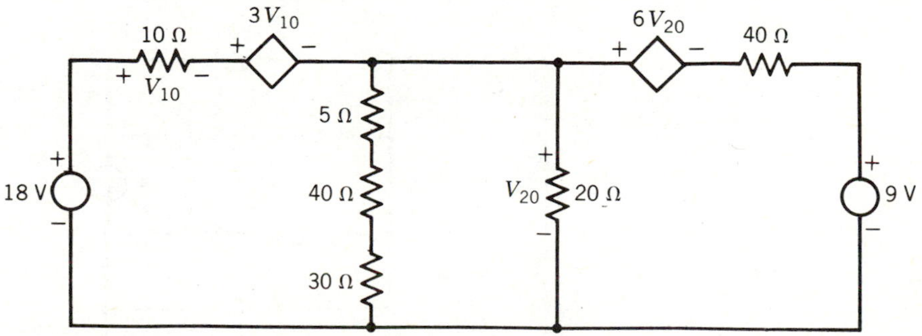

Figure 2.68

SECTION 2.4

16. Consider the circuit in Fig. 2.69. Find the current I by:

a. KVL
b. Element combination

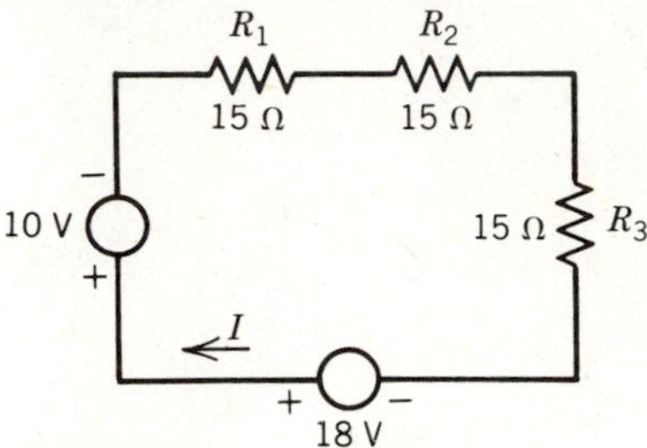

Figure 2.69

17. Consider the circuit in Fig. 2.70 with $G_1 = 0.25$ S, $G_2 = 0.5$ S, $G_3 = 1$ S, and $G_4 = 0.01$ S. Find voltage V by:

a. KCL
b. Element combination

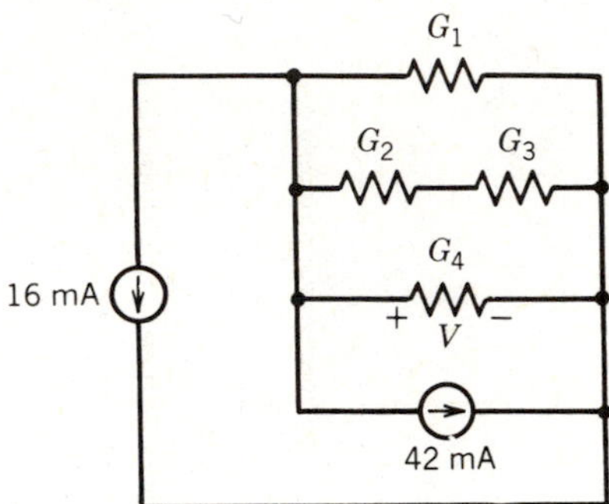

Figure 2.70

18. In Fig. 2.71 what is the current through and voltage across the 12- and 21-Ω resistors? What is the voltage from point C to D?

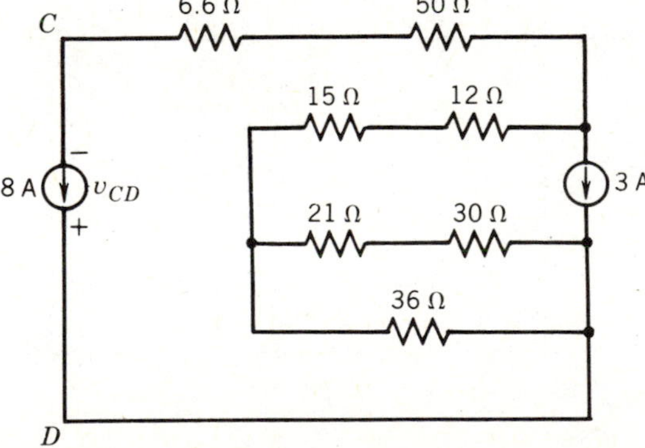

Figure 2.71

19. Determine the current in the loop, the voltage of each source, and show that the power absorbed equals the power delivered in Fig. 2.72.

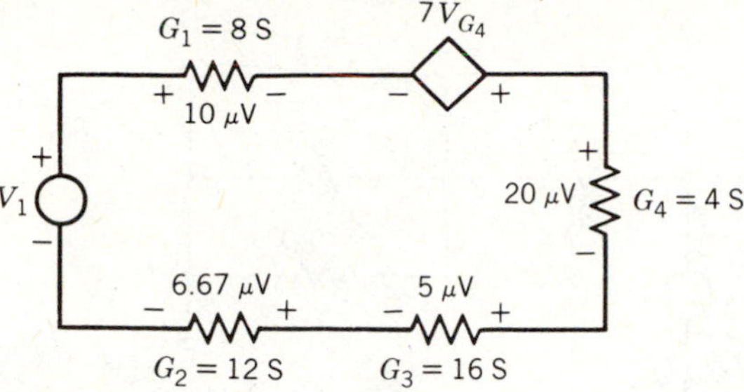

Figure 2.72

20. Find the current through each element in Fig. 2.73.

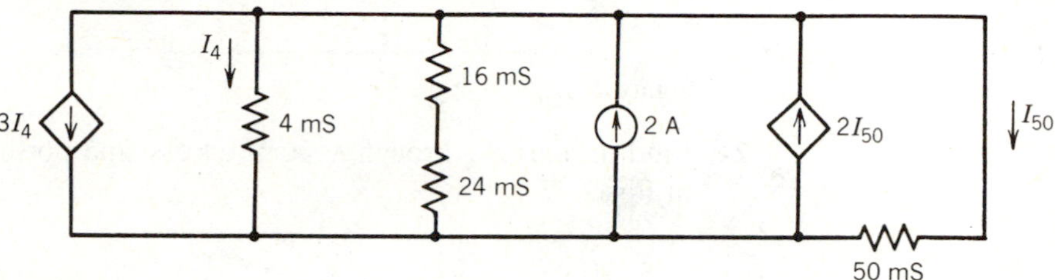

Figure 2.73

SECTION 2.5

21. Derive a formula for voltage division of two series elements in terms of conductance G, that is, in Fig. 2.74 express the voltage V_2 in terms of G_1, G_2, and V.

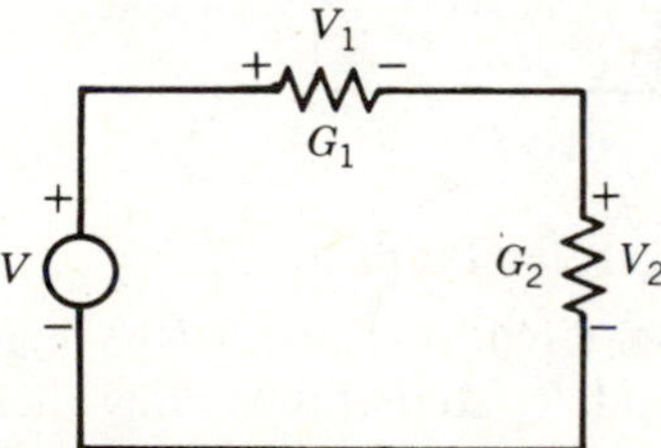

Figure 2.74

22. Derive a formula for current division between two parallel elements in terms of resistance R, that is, in Fig. 2.75 express the current I_2 in terms of R_1, R_2, and I.

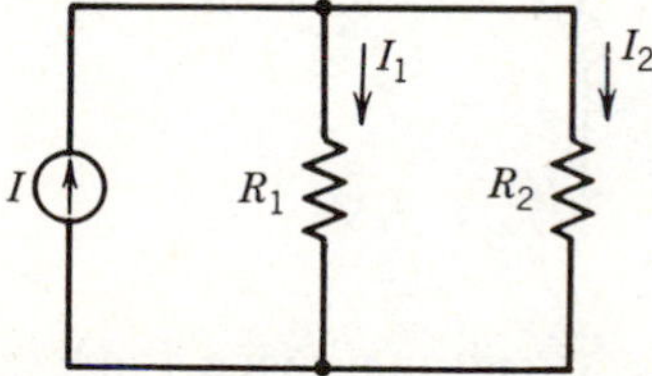

Figure 2.75

23. How much power is absorbed by the 35-Ω resistor in Fig. 2.76? What is the current through the 112-Ω resistor?

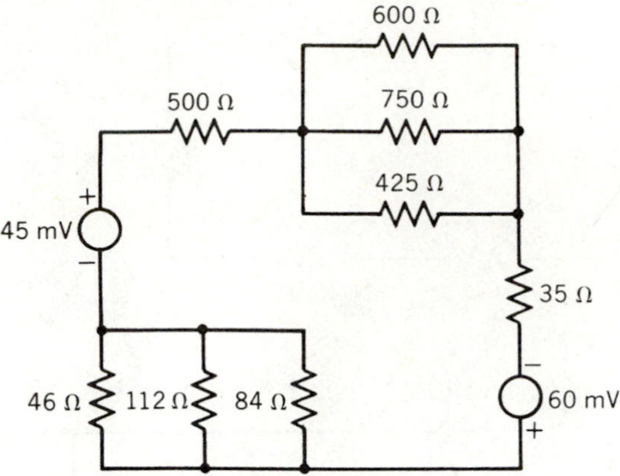

Figure 2.76

24. Find the current through, voltage across, and power absorbed by each resistor in Fig. 2.77.

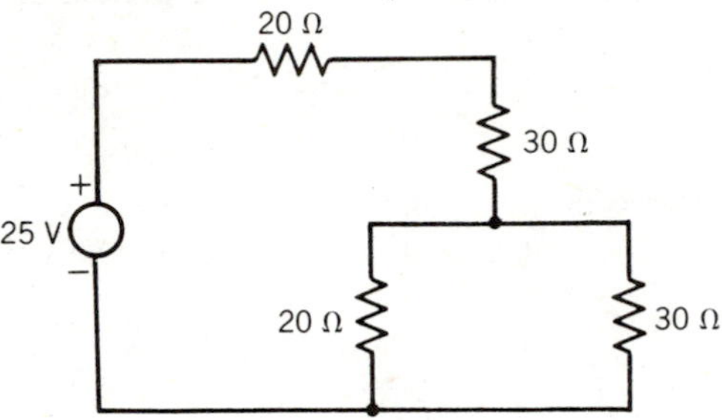

Figure 2.77

25. Use current division to solve Problem 17.

26. Design a voltage divider circuit to provide 40 V across R_3 in Fig. 2.78, that is, choose values for R_1 and R_2 so that the voltage across R_3 is 40 V.

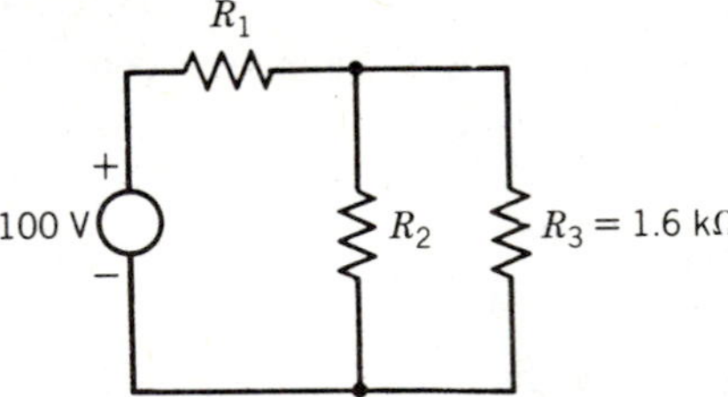

Figure 2.78

27. Design a current divider circuit to provide 4 mA through G_3 in Fig. 2.79, that is, choose G_1 and G_2 so that the current through G_3 is 4 mA.

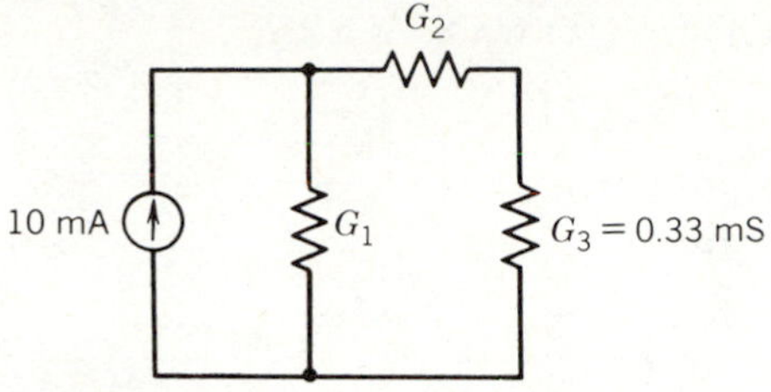

Figure 2.79

SECTION 2.6

28. Find the current I in Fig. 2.80.

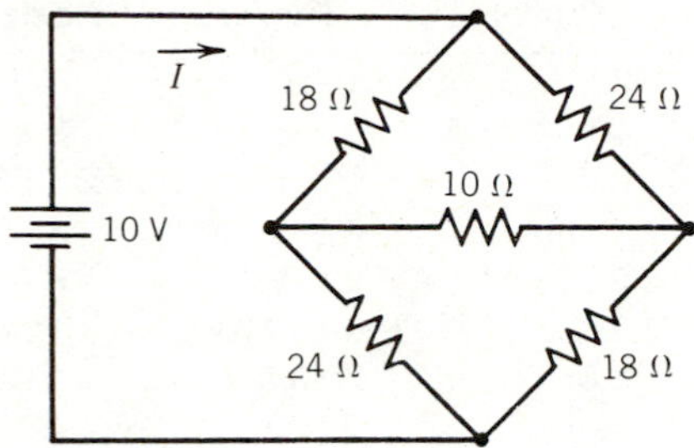

Figure 2.80

29. Find the voltage source necessary to supply a current I of 20 mA in Fig. 2.81.

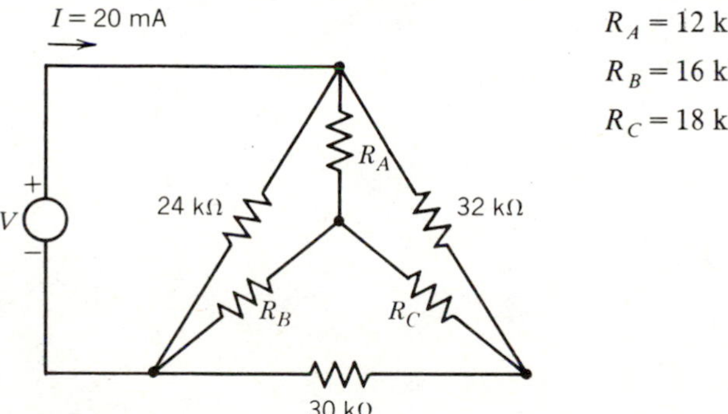

Figure 2.81

30. Find the power delivered by the source in Fig. 2.82.

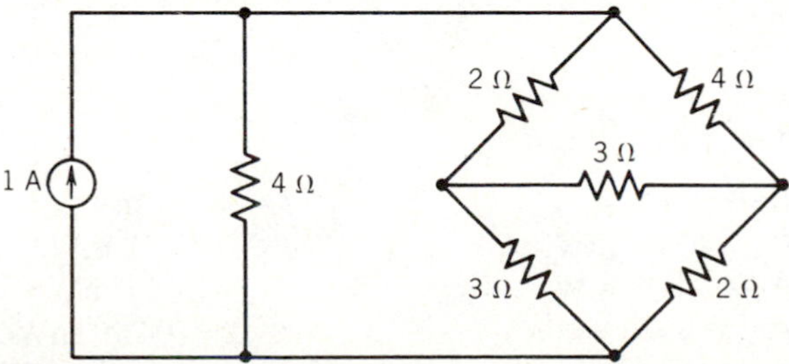

Figure 2.82

LEARNING EVALUATION ANSWERS

Section 2.1

1. a. $R = 3.62 \times 10^{-5}\ \Omega$
b. $R = 2.22 \times 10^{5}\ \Omega$

2. a. $G = 2.76 \times 10^{4}$ S
b. $G = 4.50 \times 10^{-6}$ S

3. $l = 28.25$ m

Section 2.2

	Voltage	Current	Power	Resistance	Conductance
1. a.	8.0 V	2.0 A	16 W	4.0 Ω	0.25 S
b.	87 V	10.875 mA	0.946 W	8 kΩ	0.125 S
c.	1.813 V	0.513 A	931 mW	3.534 Ω	283 mS
d.	57 mV	16 mA	0.912 mW	3.56 Ω	0.281 S
e.	507.9 V	84.66 A	43 kW	6 Ω	0.166 S
f.	371 kV	241 A	89.4 MW	1.54 kΩ	650 μS
g.	1.73 V	1.66 mA	2.88 mW	1.04 kΩ	0.962 mS
h.	4.18 kV	89 μA	0.372 W	47 MΩ	0.021 μS
i.	538 V	7.53 mA	4.05 W	71.4 kΩ	14 μS

2. See Fig. 2.83.

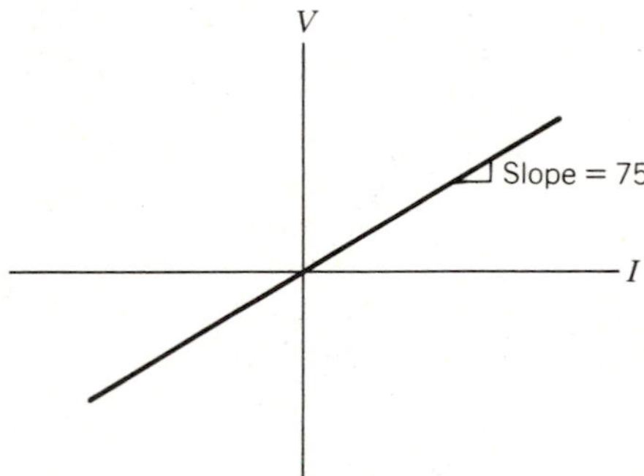

Figure 2.83

3. $\dfrac{R_{75}}{R_{150}} = \dfrac{150}{75}$

4. $V = 2.15$ kV

Section 2.3

1. $I = 35.71$ mA
$P_{45} = 57.40$ mW
$P_{23} = 29.34$ mW
$P_{100} = 127.55$ mW
$P_{12V} = 428.57$ mW
$P_{6V} = -214.29$ mW

2. $I = 32.39$ mA
$P_{45} = 47.21$ mW
$P_{23} = 24.13$ mW
$P_{100} = 104.90$ mW
$P_{12V} = 388.66$ mW
$P_{dependent} = -212.43$ mW

3. $V_S = 250$ mV
$I_{12} = 20.83$ mA
$I_{60} = 4.17$ mA

4. $V_{15} = V_{45} = 2.117$ V
$V_{10} = 1.882$ V
$I_{10} = 188.24$ mA
$I_{15} = 141.18$ mA
$I_{45} = 47.06$ mA

Section 2.4

See Fig. 2.84.

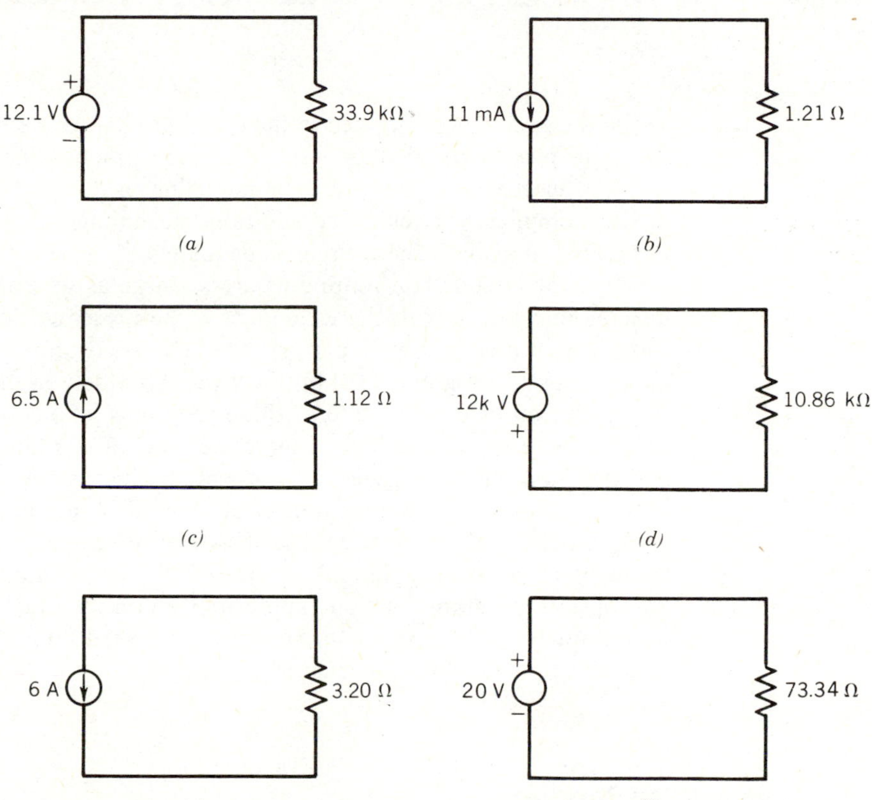

Figure 2.84

Section 2.5

1. $V_{200} = 18.10$ V

2. $V = 150$ V, $I_{50} = 3$ mA, $I_{150} = 1$ mA

Section 2.6

1. $R_A = 0.75$ kΩ
$R_B = 1.125$ kΩ
$R_C = 1.35$ kΩ

2. $R_1 = 7.167$ kΩ
$R_2 = 10.75$ kΩ
$R_3 = 12.90$ kΩ

CHAPTER

3 ANALYSIS OF THE MORE COMPLEX RESISTIVE CIRCUIT

While the circuits we analyzed in the previous chapter were useful to demonstrate the principles involved, they were of limited practical application in the broad scope of electrical engineering. The modeling of physical systems usually results in very complicated circuits and additional techniques are required in order for us to be able to analyze these circuits adequately.

Often the circuits confronting us are so large as to make manual calculations uneconomical even with the techniques to be presented in this chapter. In these cases, use of a computer to perform the analysis is appropriate. Many computer circuit analysis programs exist but they vary so widely in their use and application that inclusion of one or more of these programs in this text would be useful to only a few of the readers and is therefore precluded. Many hand-held calculators and desk-top microcomputers also provide limited circuit analysis capabilities.

It is important to realize, however, that all computer programs for circuit analysis are based on techniques such as those discussed in this text and you must be aware of the limitations and advantages of each method in order to choose the appropriate program for your application. You should be convinced that it is important for you to know the methods presented here regardless of the circuit analysis approach you use as an engineer.

3.1 The Superposition Principle

LEARNING OBJECTIVES

After completing this section you should be able to do the following:

1. State the principle of superposition.
2. Solve for all circuit parameters in each element of a circuit using the principle of superposition.

A circuit such as that in Fig. 3-1 presents us with an analysis problem in which none of the tools we have learned to use so far are applicable.

The voltage sources cannot be combined because they are not in series. Likewise the resistance cannot be combined and a delta-to-wye transformation

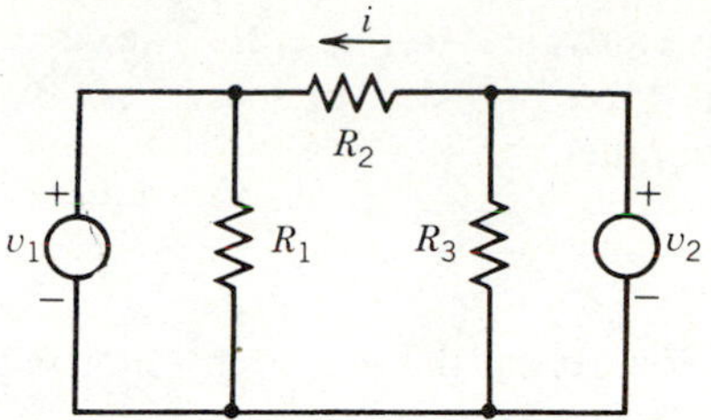

Figure 3.1. A circuit with two voltage sources.

does not improve the situation. For linear circuits there exists a theorem called the *superposition principle*, which states:

> For linear circuits containing more than one energy source, the voltage across or current through any circuit element is the algebraic sum of the voltage or current from each of the independent sources applied separately with all other independent voltage sources replaced by short circuits and all other independent current sources replaced by open circuits.

Linearity will be defined in Chapter 8, but all circuits discussed in this text are linear.

Let us apply this superposition theorem to the circuit in Fig. 3.1 to see if it helps us determine the current, i. First we will short circuit voltage source v_2. The resulting circuit is shown in Fig. 3.2a. Notice that R_3 is in parallel with the short circuit (zero resistance). This combination results in a zero resistance so Fig. 3.2a

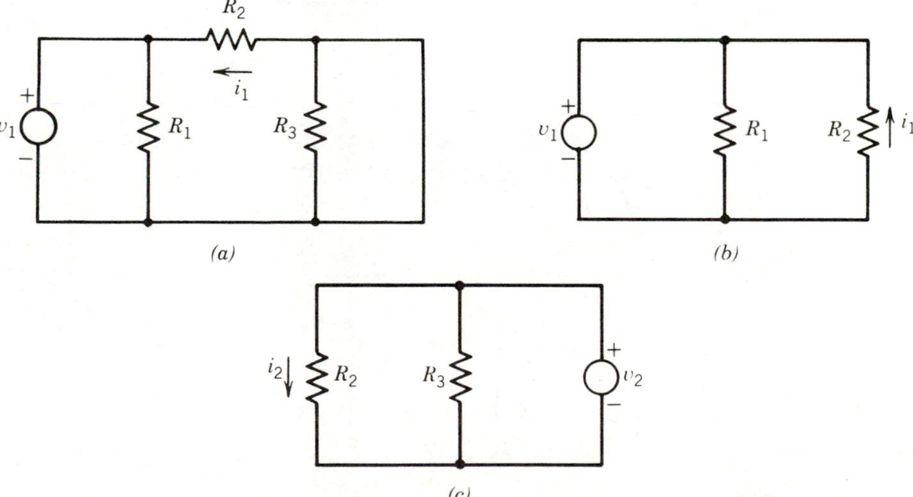

Figure 3.2. (a) Circuit of Fig. 3.1 with v_2 short circuited. (b) Equivalent circuit for (a). (c) Circuit of Fig. 3.1 with v_1 short circuited.

is equivalent to Fig. 3.2*b*. The current i has been designated i_1 to indicate it is the portion of i due to source v_1. By Ohm's law, since v_1, R_1, and R_2 are all in parallel,

$$i_1 = -\frac{v_1}{R_2} \text{ A}$$

Now repeat the process by shorting v_1 to obtain the circuit of Fig. 3.2*c*. Again by Ohm's law,

$$i_2 = \frac{v_2}{R_2} \text{ A}$$

And by the principle of superposition,

$$i = i_1 + i_2$$

$$i = -\frac{v_1}{R_2} + \frac{v_2}{R_2}$$

$$i = \frac{v_2 - v_1}{R_2} \text{ A}$$

EXAMPLE 3.1.1 Use superposition to find the voltage v_R in the circuit of Fig. 3.3.

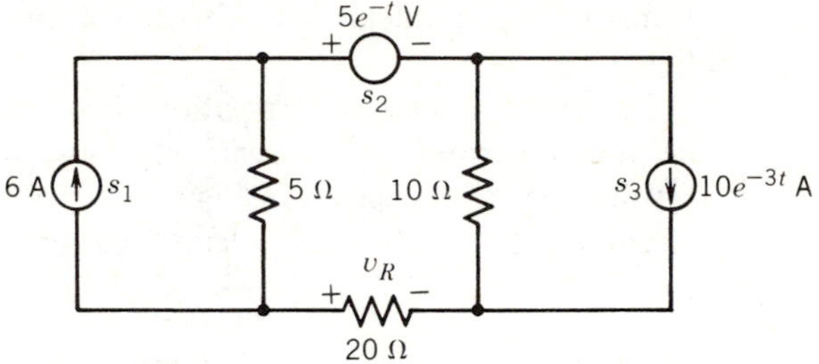

Figure 3.3

Solution

1. Find the v_{s_1} due to the 6-A current source. Shorting the voltage source and opening the other current source leaves the circuit of Fig. 3.4*a*.

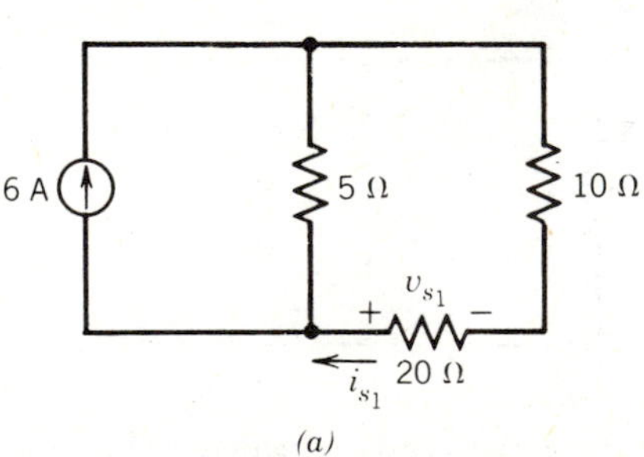

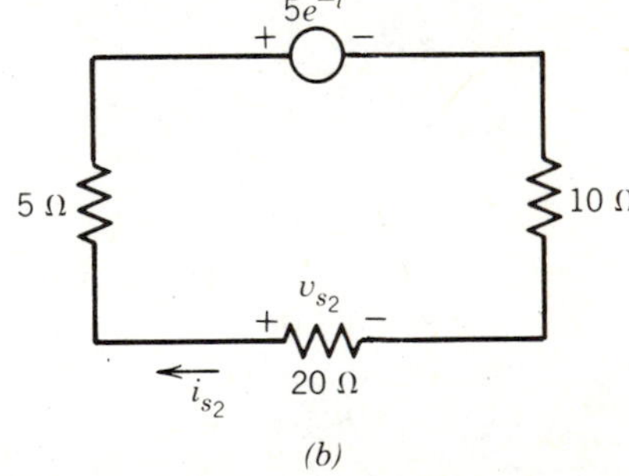

Figure 3.4

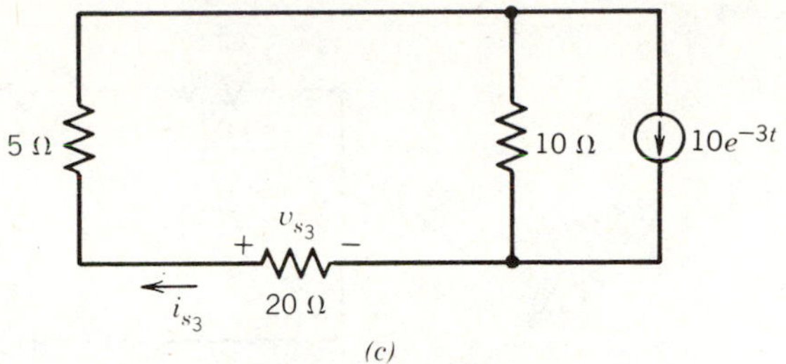

Figure 3.4. (*Continued*).

By current division,

$$i_{s_1} = \frac{5(6)}{5 + (20 + 10)} = \frac{6}{7}\ \text{A}$$

so

$$v_{s_1} = -20 i_{s_1} = -20(\tfrac{6}{7})$$

$$v_{s_1} = -\tfrac{120}{7}\ \text{V}$$

2. Find v_{s_2}. Opening s_1 and s_3 gives the circuit of Fig. 3.4*b*. By voltage division,

$$v_{s_2} = \frac{20}{20 + 5 + 10} 5e^{-t}$$

$$v_{s_2} = \tfrac{20}{7} e^{-t}$$

3. Find v_{s_3} by opening s_1 and shorting s_2. This gives Fig. 3.4*c*. By current division,

$$i_{s_3} = \frac{\frac{1}{25}}{\frac{1}{25} + \frac{1}{10}} 10e^{-3t}$$

$$= \tfrac{20}{7} e^{-3t}$$

so

$$v_{s_3} = -20 i_{s_3}$$

$$= -20(\tfrac{20}{7} e^{-3t})$$

$$= -\tfrac{400}{7} e^{-3t}$$

4. Finally by superposition,

$$v_R = v_{s_1} + v_{s_2} + v_{s_3}$$

$$= -\tfrac{120}{7} + \tfrac{20}{7} e^{-t} - \tfrac{400}{7} e^{-3t}$$

$$= \tfrac{1}{7}(20e^{-t} - 400e^{-3t} - 120)\ \text{V}$$

What do we do if our circuit contains dependent sources? The superposition principle applies only to *independent* sources. We may treat the dependent source as just another circuit element and include its effect in all calculations. Consider

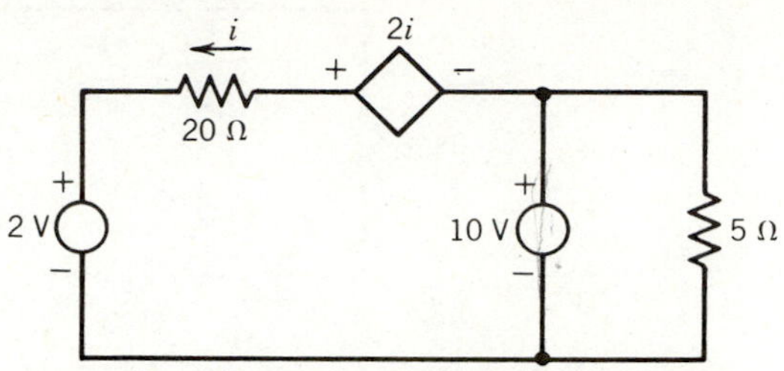

Figure 3.5. Circuit with dependent source.

the circuit in Fig. 3.5. This circuit contains two independent sources and one dependent source. We are to determine i. By shorting the 10-V source we obtain a single closed path where KVL yields

$$-2 - 20i_2 + 2i_2 = 0$$

or

$$i_2 = -\tfrac{1}{9}\text{ A}$$

By shorting the 2-V source and writing KVL around the path containing the dependent source we obtain

$$-10 - 2i_{10} + 20i_{10} = 0$$

or

$$i_{10} = \tfrac{5}{9}\text{ A}$$

so by superposition

$$i = i_2 + i_{10} = -\tfrac{1}{9} + \tfrac{5}{9}$$

$$i = \tfrac{4}{9}\text{ A}$$

LEARNING EVALUATION

Use superposition to find the current i for the circuits in Fig. 3.6.

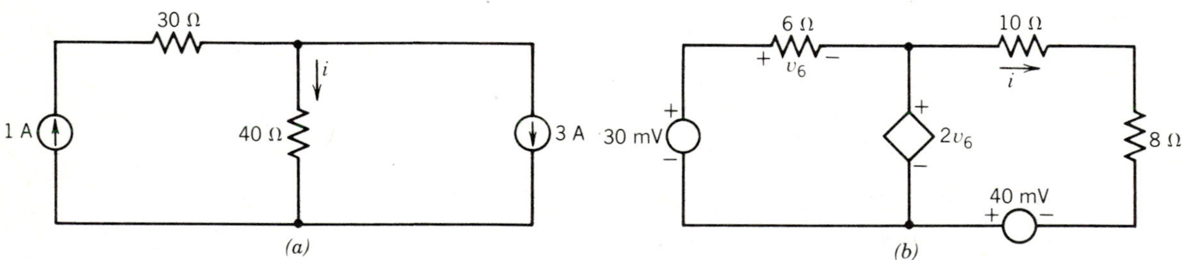

Figure 3.6

3.2 Source Conversions

LEARNING OBJECTIVES

After completing this section you should be able to do the following:

1. Define a practical source.
2. Demonstrate why the internal resistance of a current source should be large and that of a voltage source should be small.

3. Change practical voltage sources to practical current sources and vice versa.

A practical energy source, such as a battery, may be modeled as either a voltage source or as a current source. Practical models consist of an ideal source coupled with a source resistance or conductance. Before studying nodal analysis and mesh analysis in the next two sections, we need to be able to convert voltage sources to current sources and vice versa. In this section we will develop a method for making such conversions.

First recall that an ideal voltage source produces a voltage that is not a function of the current flowing through it. Likewise an ideal current source produces a current that is independent of the voltage across the terminals of the current source. Unfortunately, such devices cannot be built because they would be required to produce infinite energy. Practically all voltage sources have a series resistance called the "source resistance." A model for a practical voltage source connected in series with another resistor that is often called the "load resistance" is shown in Fig. 3.7. By voltage division,

$$v_L = \frac{R_L}{R_L + R_v} v_s \tag{3.2.1}$$

If

$$R_L \gg R_v$$

then

$$v_L \approx v_s$$

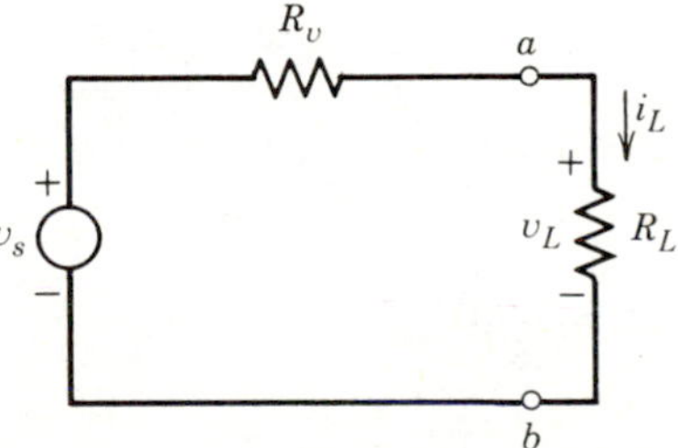

Figure 3.7. A practical voltage source connected to a load resistance.

In other words, if R_v is very small compared to R_L then the practical source can be *approximated* by an ideal source of value v_s. This approximation is only good over a certain range of values for R_L. Generally, if

$$R_L \geq 100 R_v$$

then

$$V_L \geq 0.99 v_s$$

and the approximation is valid within 1%. The voltage across the load resistance, v_L, is called the "load voltage."

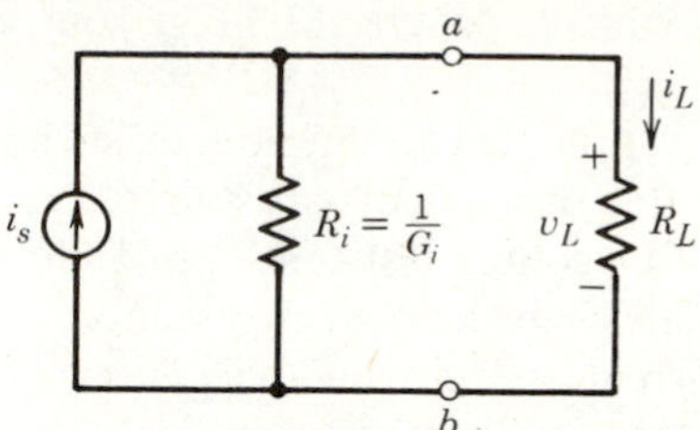

Figure 3.8. Practical current source connected to a load resistance.

A practical current source always contains a parallel resistance, as shown to the left of the terminals a–b in Fig. 3.8. With $R_i = 1/G_i$ and $R_L = 1/G_L$, current division gives

$$i_L = \frac{G_L}{G_L + G_i} i_s = \frac{R_i}{R_L + R_i} i_s \tag{3.2.2}$$

If

$$R_i \gg R_L \quad \text{or} \quad G_L \gg G_i$$

then

$$i_L \approx i_s$$

As before, if $R_i \geq 100R_L$ then the practical source can be approximated by an ideal current source of value i_s.

Now we can restate our original question as follows: Can we replace the practical voltage source in Fig. 3.7 with the practical current source in Fig. 3.8 and if so, what values of i_s and R_i do we use? The replacement can be legitimately made only if the two circuits appear the same to R_L; that is, if v_L and i_L are unchanged by the replacement for any given value of R_L. So we can make the comparison, rename the voltage and current across R_L due to the voltage source, i_1 and v_1, and due to the current source, i_2 and v_2. Then for the replacement to be valid,

$$i_1 = i_2 \tag{3.2.3}$$

and

$$v_1 = v_2 \tag{3.2.4}$$

By Ohm's law,

$$i_1 = \frac{v_1}{R_L} = \left(\frac{1}{R_L}\right)\left(\frac{R_L}{R_L + R_v}\right) v_s$$

$$i_1 = \frac{v_s}{R_L + R_v} \tag{3.2.5}$$

By Eq. 3.2.3, Eqs. 3.2.2 and 3.2.5 must be equal so

$$\frac{v_s}{R_L + R_v} = \frac{R_i}{R_L + R_i} i_s \tag{3.2.6}$$

Also by Ohm's law,

$$v_2 = R_L i_2 = (R_L)\frac{R_{is}}{R_L + R_{is}} i_s \tag{3.2.7}$$

By Eq. 3.2.4, Eqs. 3.2.1 and 3.2.7 must be equal so

$$(R_L)\left(\frac{R_i}{R_L + R_i} i_s\right) = \frac{R_L}{R_L + R_v} v_s$$

or

$$\frac{R_i}{R_L + R_i} i_s = \frac{v_s}{R_L + R_v} \tag{3.2.8}$$

By inspection, the only way both Eqs. 3.2.6 and 3.2.8 can be true is for

$$R_i = R_v \tag{3.2.9}$$

then

$$v_s = R_i i_s \tag{3.2.10}$$

The theorem for source conversion can therefore be stated as:

> With respect to the rest of the circuit, a voltage source v_s and series resistance R_v can be interchanged with a current source i_s and parallel resistor R_i as long as the following relationships hold:
>
> $$R_i = R_v = R_s$$
> $$v_s = R_s i_s$$

By examining Eqs. 3.2.3 through 3.2.10, we see that this theorem is valid for any values of R_i or R_v.

EXAMPLE 3.2.1 Change the voltage source in Fig. 3.9 to a current source.

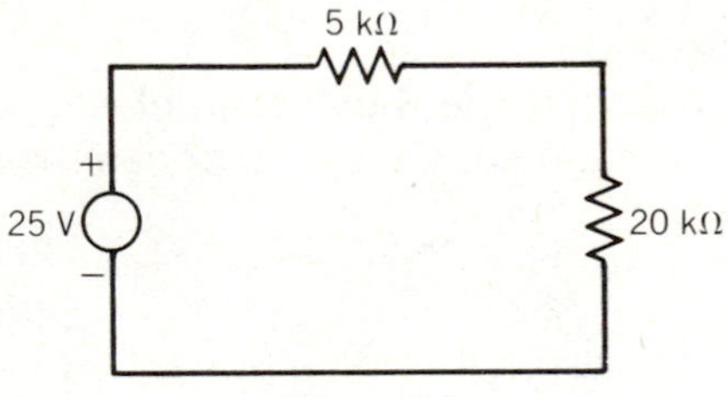

Figure 3.9

Solution

Let $R_s = 5\ \text{k}\Omega$ and $V_s = 25$ V; then

$$I_s = \frac{V_s}{R_s} = \frac{25}{5 \times 10^3} = 5 \times 10^{-3} = 5\ \text{mA}$$

so the circuit of Fig. 3.9 is replaced by the one shown in Fig. 3.10.

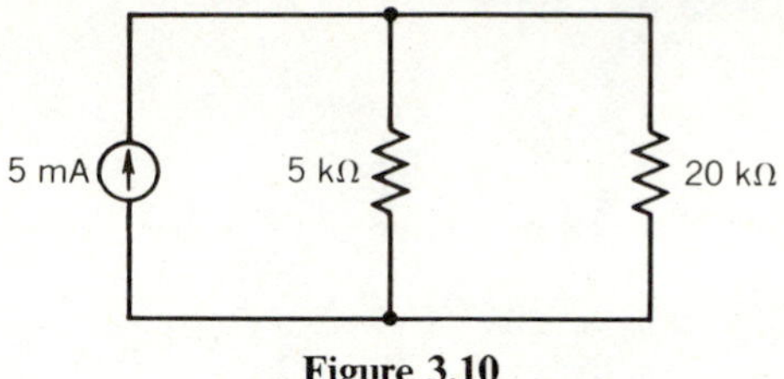

Figure 3.10

Some observations can be made from this example:

1. The two circuits are equivalent *only* with respect to the 20-kΩ resistor; that is, the voltage across and current through the 20-kΩ resistor is the same in either circuit.
2. Current flows from the current source to the terminal originally connected to the positive side of the voltage source.
3. The value of the resistor in series with the voltage source can be any value and the source conversion will still be valid.

EXAMPLE 3.2.2 For the circuit in Fig. 3.11, use source conversion to find I.

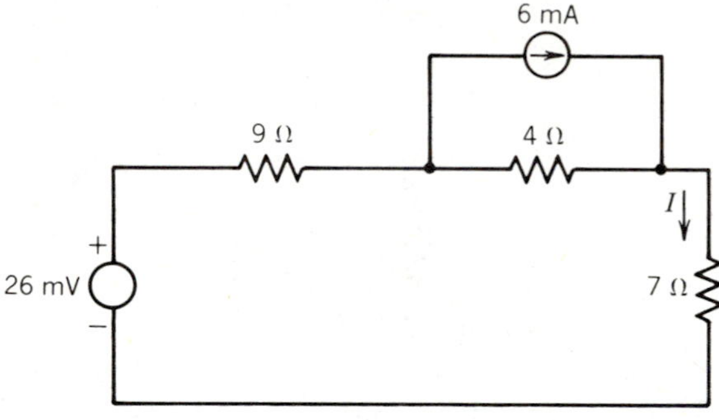

Figure 3.11

Solution

Convert the 6-mA current source in parallel with the 4-Ω resistor to a voltage source V_s and series resistance R_s where

$$V_s = R_i i_s = (4)(6 \times 10^{-3})$$
$$= 24 \text{ mV}$$
$$R_s = R_i = 4\ \Omega$$

So our new circuit is given in Fig. 3.12.

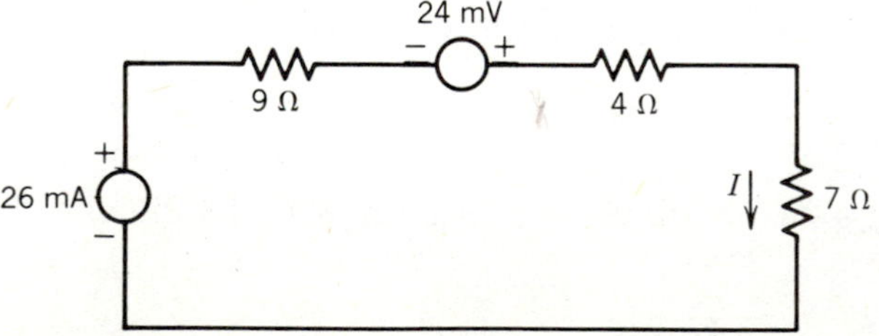

Figure 3.12

By KVL,

$$-26 \times 10^{-3} + 9I - 24 \times 10^{-3} + 4I + 7I = 0$$
$$20I = 50 \times 10^{-3}$$
$$I = 2.5 \text{ mA}$$

Source conversion now becomes another tool in our bag of circuit analysis techniques. This tool is normally used to simplify or alter circuits so that other circuit analysis methods can be more easily applied.

LEARNING EVALUATIONS

1. In the circuits shown in Fig. 3.13, transform each voltage source to an equivalent current source. In each case, the source resistance is the resistance closest to the positive side of the voltage source.

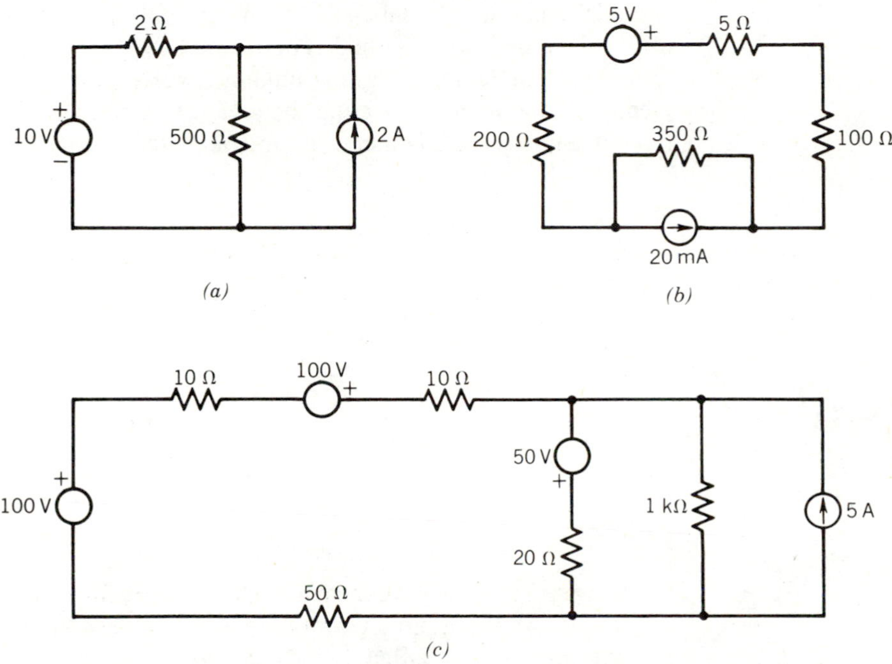

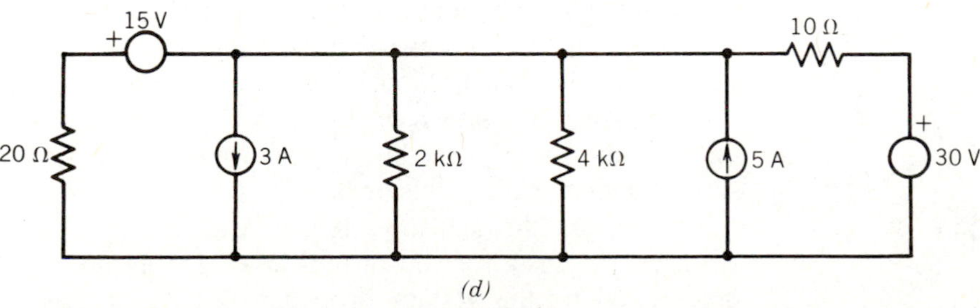

Figure 3.13

2. For each circuit in Fig. 3.13, transform each current source to its equivalent voltage source.

3.3 Nodal Analysis

LEARNING OBJECTIVES

After completing this section you should be able to do the following:

1. Solve for all circuit parameters in each element of a circuit using the method of nodal analysis.
2. Find the conductance matrix of a circuit.
3. Apply the method of nodal analysis to circuits containing voltage sources by creating "super nodes."

Nodal analysis is widely used in circuit analysis. In fact, it probably has more utility than any other method. The method consists of choosing a "common" or "reference" node and assigning unknown voltages with respect to this reference. Generally, the node connecting the greatest number of elements is chosen as the reference node. KCL is then used to complete the analysis.

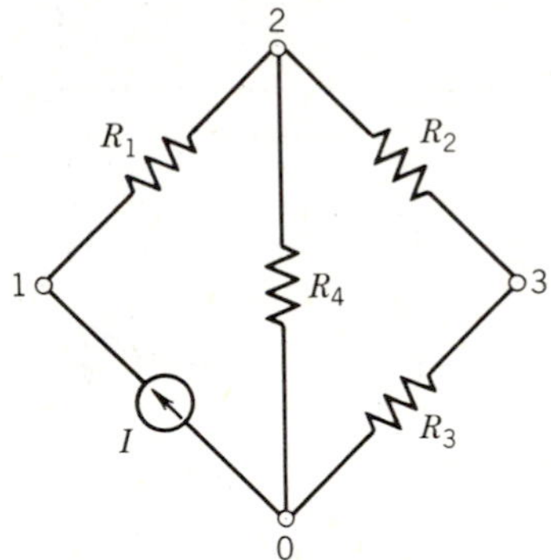

Figure 3.14. A four-node circuit.

We have previously defined a node as a junction of two or more elements. Figure 3.14 shows a circuit containing four nodes. Nodes 0 and 2 each connect three elements and either could be chosen as the reference node. We will arbitrarily choose node 0. (Note: *Any* node can be chosen as the reference node, but our computational efforts are normally minimized by choosing a node connecting a large number of elements.) We will now define node voltage v_1 to be the voltage that exists between node 1 and the reference node with node 1 positive with respect to node 0. Similar definitions are made for node voltages v_2 and v_3. These voltages are shown in Fig. 3.15. The node voltage reference directions are understood and are generally not shown on the circuit diagram and will not be shown from now on in this book.

The current through any element can now be stated in terms of the node voltages. For instance, the current flowing from node 1 to node 2 through R_1 is determined by the voltage drop across R_1 which is $(v_1 - v_2)$ for the current

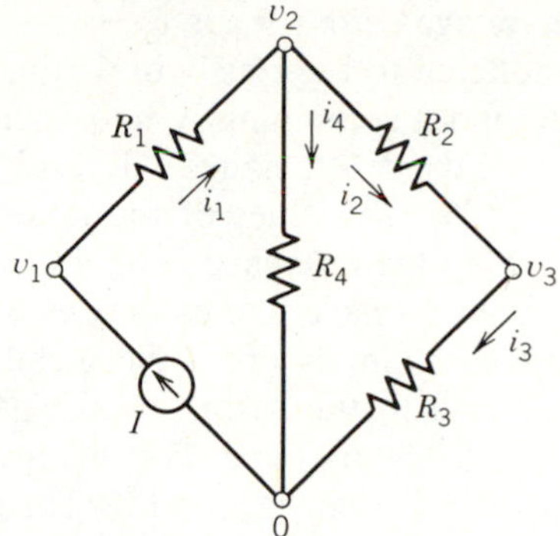

Figure 3.15. Assignment of node voltages.

direction assumed. Therefore, we can write KCL equations at each node to which a node voltage has been assigned.

Node 1

$$I = i_1 \tag{3.3.1}$$

Node 2

$$i_1 = i_2 + i_4 \tag{3.3.2}$$

Node 3

$$i_2 = i_3 \tag{3.3.3}$$

Substitute Ohm's law into each node equation to obtain

Node 1

$$I = \frac{v_1 - v_2}{R_1} \tag{3.3.4}$$

Node 2

$$\frac{v_1 - v_2}{R_1} = \frac{v_2 - v_3}{R_2} + \frac{v_2}{R_4} \tag{3.3.5}$$

Node 3

$$\frac{v_2 - v_3}{R_2} = \frac{v_3}{R_3} \tag{3.3.6}$$

If all resistor values and sources are known, then we have three equations with three unknowns, v_1, v_2, and v_3. Once these voltages are determined, all other circuit parameters (current, power, etc.) can be calculated.

It is important that you understand how to assign polarities on the voltage terms in Eqs. 3.3.4 through 3.3.5. First, the directions we chose for i_1, i_2, i_3, and i_4 in Fig. 3.15 were arbitrary. Once these are chosen, however, we are forced to choose voltage polarities that are consistent with our assumption. For example, if i_1 flows from node 1 to node 2, then the voltage across R_1 must be $v_1 - v_2$ and not $v_2 - v_1$ in Fig. 3.15. Since i_4 is assumed to flow from node 2 to node 0, then the voltage across R_4 is $v_2 - 0 = v_2$ and not $0 - v_2 = -v_2$. This convention must be adhered to rigorously in writing Eqs. 3.3.4 through 3.3.6.

A circuit containing N nodes will have $N-1$ node voltages. Applying KCL at each of the $N-1$ nodes will result in $N-1$ independent equations which can be solved for the values of the node voltages. An additional comment is in order regarding the reference node. In construction of an actual electrical circuit, many components have one of their two terminals connected to a common point. This common point is often the metal chassis of the circuit which may in turn be connected to the earth through a water pipe, large conductor, or the electrical system of the building. For this reason this common point is referred to as "earth ground" or simply "ground." The ground node of such circuits is a logical choice for the reference node when using nodal analysis. This point is often marked on circuit schematic diagrams with the symbol ⏚ . For instance, in Fig. 3.16, one terminal of the current source, R_2, and R_4 are all connected to the same point and node voltages v_1, v_2, and v_3 can be assigned to the three remaining nodes.

EXAMPLE 3.3.1 For the circuit in Fig. 3.16, find the current through R_4 if $R_1 = 2\ \Omega$, $R_2 = 4\ \Omega$, $R_3 = 1\ \Omega$, $R_4 = 3\ \Omega$.

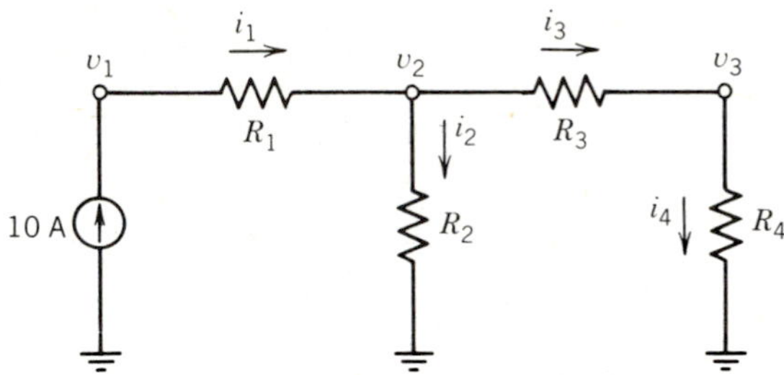

Figure 3.16. A circuit using the symbol for ground.

Solution

We first assume currents i_1 through i_4 as shown in Fig. 3.16. Writing KCL at each node gives

Node 1

$$10 = i_1$$

or

$$10 = \frac{V_1 - V_2}{R_1}$$

Node 2

$$i_1 = i_2 + i_3$$

or

$$\frac{V_1 - V_2}{R_1} = \frac{V_2}{R_2} + \frac{V_2 - V_3}{R_3}$$

Node 3

$$i_3 = i_4$$

or

$$\frac{v_2 - v_3}{R_3} = \frac{v_3}{R_4}$$

Substituting the given values of resistance and simplifying yields

$$\tfrac{1}{2}v_1 - \tfrac{1}{2}v_2 = 10 \qquad (3.3.7)$$

$$\tfrac{1}{2}v_1 + \tfrac{7}{4}v_2 - v_3 = 0 \qquad (3.3.8)$$

$$-v_2 + \tfrac{4}{3}v_3 = 0 \qquad (3.3.9)$$

Notice that we need to find only v_3 in order to calculate the current i_4. Solving Eqs. 3.3.9 and 3.3.7 for v_1 and v_2 in terms of v_3 and inserting these values in Eq. 3.3.8 results in

$$v_3 = 15 \text{ V}$$

Consequently,

$$i_4 = \frac{v_3}{3} = 5 \text{ A}$$

EXAMPLE 3.3.2 Circuits with dependent sources can be handled using these same procedures. Suppose that in Fig. 3.16 the resistor R_2 is replaced by a dependent current source as shown in Fig. 3.17. Find the current through R_4.

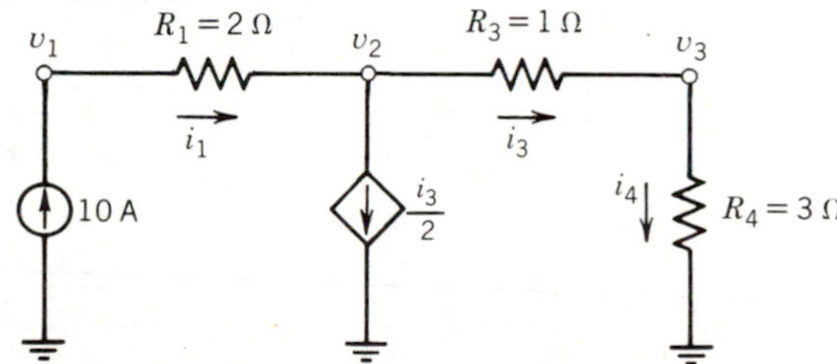

Figure 3.17. A circuit with a dependent current source.

Solution

Writing KCL at each node gives

Node 1

$$10 = i_1$$

or

$$10 = \frac{v_1 - v_2}{R_1}$$

Node 2

$$i_1 = i_3 + \frac{i_3}{2} = \frac{3}{2}i_3$$

or

$$\frac{v_1 - v_2}{R_1} = \frac{3}{2}\left(\frac{v_2 - v_3}{R_3}\right)$$

Node 3

$$i_3 = i_4$$

or

$$\frac{v_2 - v_3}{R_3} = \frac{v_3}{R_4}$$

Substituting the given values of resistance and simplifying gives

$$v_1 - v_2 = 20$$

$$\frac{v_1}{2} - 2v_2 + \frac{3}{2}v_3 = 0$$

$$v_2 - \tfrac{4}{3}v_3 = 0$$

Solving for v_3 gives $v_3 = 20$ V. Therefore,

$$i_4 = \frac{v_3}{3} = \frac{20}{3} \text{ A}$$

EXAMPLE 3.3.3 Suppose a dependent voltage source is present in the center branch of the circuit, as shown in Fig. 3.18. Find the current v_4.

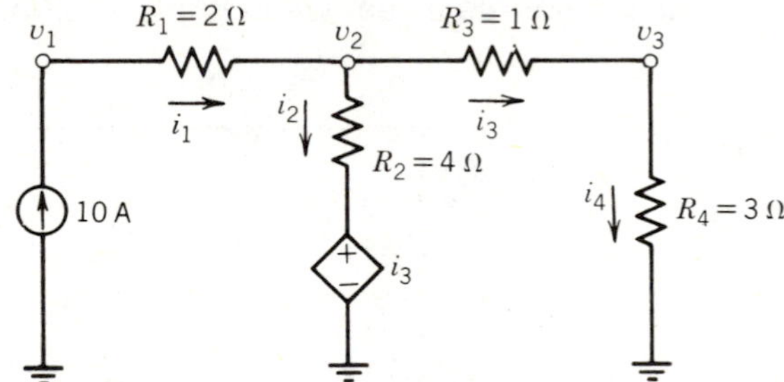

Figure 3.18. A circuit with a dependent voltage source.

Solution

Writing KCL at each node gives

Node 1

$$10 = i_1$$

or

$$10 = \frac{v_1 - v_2}{R}$$

Node 2

$$i_1 = i_2 + i_3$$

or

$$\frac{v_1 - v_2}{R_1} = \frac{v_2 - i_3}{R_2} + \frac{v_2 - v_3}{R_3}$$

and since

$$i_3 = \frac{v_2 - v_3}{R_3}$$

we substitute to get

$$\frac{v_1 - v_2}{R_1} = \frac{v_2 - [(v_2 - v_3)/R_3]}{R_2} + \frac{v_2 - v_3}{R_3}$$

Node 3

$$i_3 = i_4$$

or

$$\frac{v_2 - v_3}{R_3} = \frac{v_3}{R_4}$$

Substituting values and solving for v_3 gives

$$v_3 = \tfrac{120}{7}$$

so

$$i_3 = \frac{v_3}{3} = \frac{40}{7} \text{ A}$$

Matrix Equations

The simultaneous equations that are derived by nodal analysis may be displayed in matrix form. For example, Eqs. 3.3.7 through 3.3.9 are given in matrix form by

$$\begin{bmatrix} \frac{1}{2} & -\frac{1}{2} & 0 \\ \frac{1}{2} & \frac{7}{4} & -1 \\ 0 & -1 & \frac{4}{3} \end{bmatrix} \begin{bmatrix} v_1 \\ v_2 \\ v_3 \end{bmatrix} = \begin{bmatrix} 10 \\ 0 \\ 0 \end{bmatrix} \tag{3.3.10}$$

There is a systematic procedure for arriving at the matrix form of such equations when there are no dependent sources. This procedure bypasses the tedious and error-prone steps of assuming currents, writing KCL equations, and then substituting Ohm's law. The matrix equation can be written directly by inspection of the circuit diagram.

In order to present the procedure, let us begin with the three-node network in Fig. 3.19. The two independent nodes are labeled v_1 and v_2, and conductance is used instead of resistance for the values of the circuit elements. Upon assuming branch currents, writing KCL, and substituting Ohm's law, the following matrix equation results:

$$\begin{bmatrix} G_1 + G_2 & -G_2 \\ -G_2 & G_2 + G_3 \end{bmatrix} \begin{bmatrix} v_1 \\ v_2 \end{bmatrix} = \begin{bmatrix} i_1 \\ 0 \end{bmatrix}$$

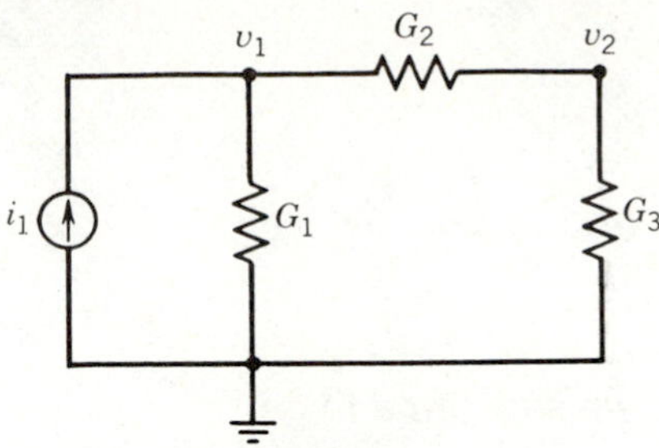

Figure 3.19

This equation has the general form given by

$$\begin{bmatrix} g_{11} & g_{12} \\ g_{21} & g_{22} \end{bmatrix} \begin{bmatrix} v_1 \\ v_2 \end{bmatrix} = \begin{bmatrix} i_1 \\ i_2 \end{bmatrix} \tag{3.3.11}$$

where g_{11} is the sum of all conductances attached to node 1, $g_{12} = g_{21}$ is the negative of the conductance between node 1 and 2, and g_{22} is the conductance attached to node 2. The variables v_1 and v_2 are the unknown node voltages, and i_1 and i_2 represent the sum of all current sources that force current into node 1 and node 2, respectively.

EXAMPLE 3.3.4 Write the node equations in matrix form for the circuit in Fig. 3.20.

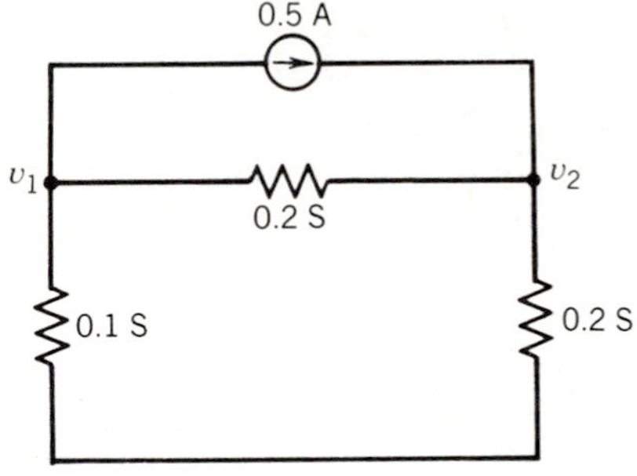

Figure 3.20

Solution

$$\begin{bmatrix} 0.3 & -0.2 \\ -0.2 & 0.4 \end{bmatrix} \begin{bmatrix} V_1 \\ V_2 \end{bmatrix} = \begin{bmatrix} -0.5 \\ 0.5 \end{bmatrix}$$

Notice that the forcing function at node 1 is -0.5 A because the current source forces current away from node 1. Similarly, the forcing function at node 2 is positive because 0.5 A is forced into node 2.

For n independent nodes, the general form of the node equations is given by

$$\begin{bmatrix} g_{11} & g_{12} & \cdots & g_{1n} \\ g_{21} & g_{22} & \cdots & g_{2n} \\ \vdots & \vdots & \ddots & \vdots \\ g_{n1} & g_{n2} & \cdots & g_{nn} \end{bmatrix} \begin{bmatrix} v_1 \\ v_2 \\ \vdots \\ v_n \end{bmatrix} = \begin{bmatrix} i_1 \\ i_2 \\ \vdots \\ i_n \end{bmatrix} \tag{3.3.12}$$

where g_{ii} is the conductance common to node i, and g_{ij} is the negative of the conductance between nodes i and j. The voltage terms are the unknown variables, and the current terms represent sources that force current into the respective nodes. Notice that all off-diagonal terms in the conductance matrix are negative, whereas all diagonal terms are positive. This is a general feature of conductance matrices and is a result of our assumption that all node voltages are positive with respect to ground. The conductance matrix is symmetrical about the diagonal if there are no dependent sources in the circuit.

EXAMPLE 3.3.5 A four-node circuit is shown in Fig. 3.21*a*. The circuit is redrawn with conductance values and node voltages assigned in Fig. 3.21*b*. Write the node equations in matrix form.

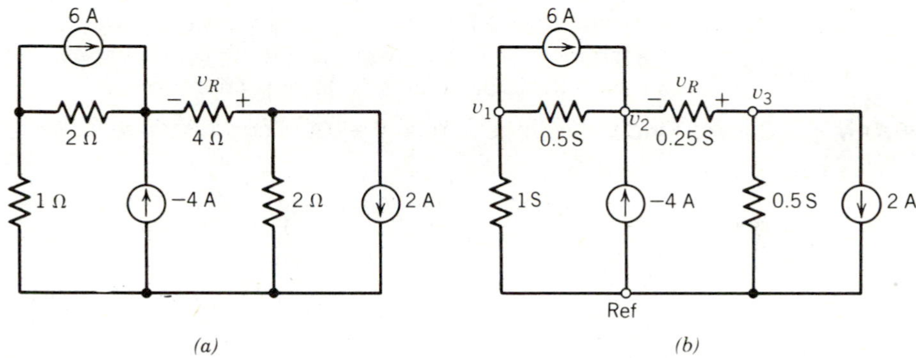

Figure 3.21. (*a*) A four-node circuit. (*b*) Node voltages assigned and resistances changed to conductances.

Solution

The matrix equation is given by

$$\begin{bmatrix} 1.5 & -0.5 & 0 \\ -0.5 & 0.75 & -0.25 \\ 0 & -0.25 & 0.75 \end{bmatrix} \begin{bmatrix} V_1 \\ V_2 \\ V_3 \end{bmatrix} = \begin{bmatrix} -6 \\ 2 \\ -2 \end{bmatrix} \tag{3.3.13}$$

The forcing vector terms are found as follows: There is one current source connected to node 1 and that source has 6 A flowing *away* from node 1. Thus the first term is -6. Node 2 has $+6$ A and -4 A forced into it, so the second term is $6-4=2$. The third node has 2 A forced away from it, giving the -2 term.

The Solution of Matrix Equations

For matrix equations of the form of Eq. 3.3.12 the solution is given

$$v = G^{-1}i$$

where G is the conductance matrix, v is the vector of unknown voltages, and i is the forcing function vector.

EXAMPLE 3.3.6 Find the node voltages v_1, v_2, and v_3 in Fig. 3.21.

Solution

The inverse of the conductance matrix in Eq. 3.3.13 is

$$G^{-1} = \begin{bmatrix} \frac{8}{9} & \frac{2}{3} & \frac{2}{9} \\ \frac{2}{3} & 2 & \frac{2}{3} \\ \frac{2}{9} & \frac{2}{3} & \frac{14}{9} \end{bmatrix}$$

so

$$\begin{bmatrix} v_1 \\ v_2 \\ v_3 \end{bmatrix} = \begin{bmatrix} \frac{8}{9} & \frac{2}{3} & \frac{2}{9} \\ \frac{2}{3} & 2 & \frac{2}{3} \\ \frac{2}{9} & \frac{2}{3} & \frac{14}{9} \end{bmatrix} \begin{bmatrix} -6 \\ 2 \\ -2 \end{bmatrix} = \begin{bmatrix} -4.444 \\ -1.333 \\ -3.111 \end{bmatrix}$$

If our circuit model contains practical voltage sources (those with series resistors) we need to convert them to an equivalent current source first. Then we can write the node equations. Here is an example.

EXAMPLE 3.3.7 Find the voltage across the 0.1-S conductance in Fig. 3.22*a*.

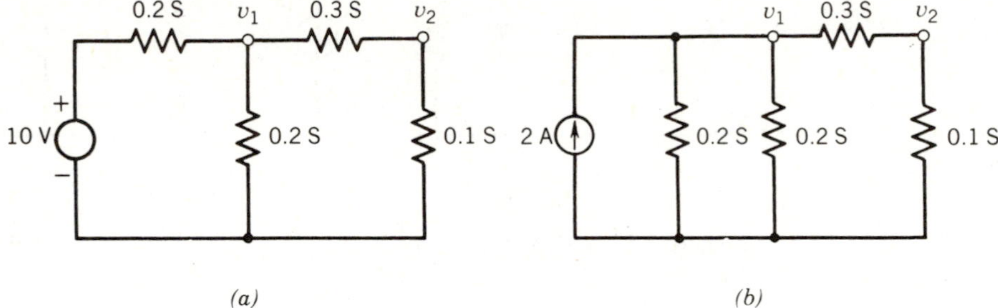

Figure 3.22

Solution

Using procedures introduced in Section 3.2 we first convert the voltage source in series with the 0.2-S conductance to the equivalent current source in parallel with the 0.2-S conductance as shown in Fig. 3.22*b*. The matrix equations for v_1 and v_2 can now be written as

$$\begin{bmatrix} 0.7 & -0.3 \\ -0.3 & 0.4 \end{bmatrix} \begin{bmatrix} v_1 \\ v_2 \end{bmatrix} = \begin{bmatrix} 2 \\ 0 \end{bmatrix}$$

Using Cramer's rule, (Appendix I)

$$v_2 = \frac{\begin{vmatrix} 0.7 & 2 \\ -0.3 & 0 \end{vmatrix}}{\begin{vmatrix} 0.7 & -0.3 \\ -0.3 & 0.4 \end{vmatrix}} = 3.16 \text{ V}$$

If our circuit contains one or more ideal voltage sources we must alter the above procedure in order to accommodate this condition. In examining Fig. 3.23

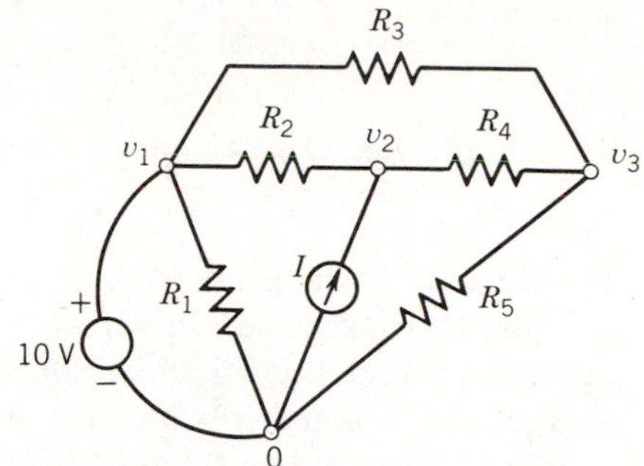

Figure 3.23. Nodal analysis with voltage source between one node and the reference node.

we observe that a 10-V source is connected between node 1 and the reference node. We cannot apply KCL to node 1 because we do not know what current is flowing through the 10-V source and we cannot express it in terms of v_1. However, remember that our goal is *not* to simply write KCL equations but to obtain three equations that are independent and are expressed in terms of the node voltages. By observation we can state that

$$v_1 = 10 \text{ V}$$

We can apply KCL at nodes 2 and 3 to obtain our other two independent equations. In this case at node 2,

$$\frac{v_2 - v_1}{R_2} + \frac{v_2 - v_3}{R_4} = I$$

At node 3,

$$\frac{v_3 - v_1}{R_3} + \frac{v_3 - v_2}{R_4} + \frac{v_3}{R_5} = 0$$

If we know the values of the resistors and the current source we can solve for the node voltages v_2 and v_3.

When the voltage source is connected between two nodes where neither is the reference node, we need to resort to a little subterfuge. Consider the circuit shown in Fig. 3.24. If we apply KCL at each node we obtain

Node 1

$$i_1 + i_2 - i_3 - i_7 = 0 \tag{3.3.14}$$

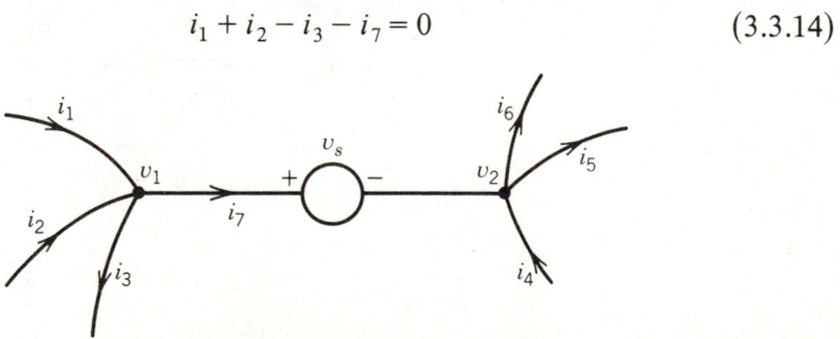

Figure 3.24. Two nodes connected by a voltage source.

Node 2

$$i_4 + i_7 - i_6 - i_5 = 0 \tag{3.3.15}$$

We can add Eqs. 3.3.14 and 3.3.15 to find

$$i_1 + i_2 - i_3 + i_4 - i_5 - i_6 = 0 \tag{3.3.16}$$

The current i_7 disappears from the equation, which is exactly what we want, since i_7 cannot be expressed in terms of v_1 and v_2. Is there some way we could legitimately find Eq. 3.3.16 without having to write Eqs. 3.3.14 and 3.3.15 and then add them together? Of course there is or we would not pose the question. If we consider nodes 1 and 2 as one entity with regard to KCL (this is shown within the dashed lines in Fig. 3.25) and write the KCL equation, we get

$$i_1 + i_2 - i_3 + i_4 - i_5 - i_6 = 0 \tag{3.3.17}$$

Equations 3.3.16 and 3.3.17 are identical, showing that our method is appropriate. This entity shown in Fig. 3.23 is often called a "super node." This method reduces the number of nodal equations that can be written by one so an additional equation will be needed. This is generated by relating the voltage source to the node voltage it connects. In our example here,

$$v_1 - v_2 = v_s$$

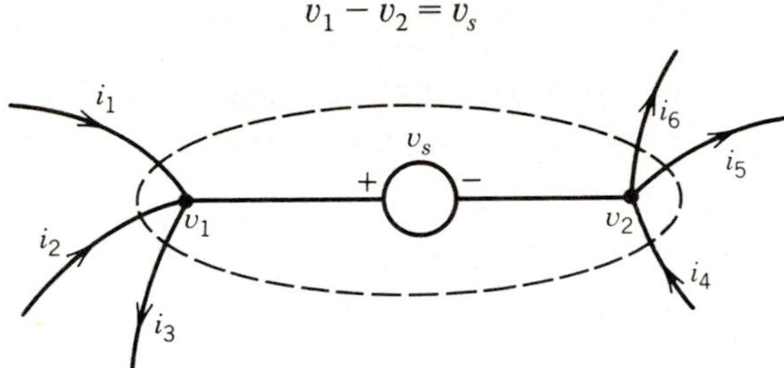

Figure 3.25. Combining nodes in nodal analysis.

EXAMPLE 3.3.8 Write the node voltage equations for the circuit in Fig. 3.26.

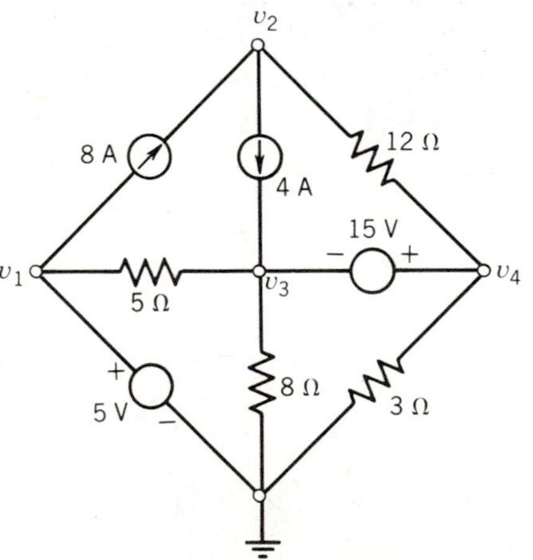

Figure 3.26. Nodal analysis with independent voltage sources.

Solution

Since a voltage source is connected between node 1 and the reference node, we can immediately state

$$v_1 = 5 \text{ V} \tag{3.3.18}$$

Another voltage source is connected between nodes 3 and 4 so we consider these two nodes as one entity and apply KCL to get

$$\frac{v_3 - v_1}{5} + \frac{v_3}{8} - 4 + \frac{v_4 - v_2}{12} + \frac{v_4}{3} = 0 \tag{3.3.19}$$

Also we observe that

$$v_4 - v_3 = 15 \tag{3.3.20}$$

Finally at node 2,

$$-8 + 4 + \frac{v_2 - v_4}{12} = 0 \tag{3.3.21}$$

Equations 3.3.18 through 3.3.21 can be solved for the four unknown node voltages.

If our circuit contains dependent sources, we proceed as before and simply relate the dependent variables to the node voltages. If the source is a dependent voltage source, we still create a super node in order to write KCL equations. If the source is a dependent current source we can apply KCL directly.

EXAMPLE 3.3.9 Modify the circuit in Fig. 3.26 by replacing the 5-V voltage source with a dependent voltage source and the 15-V voltage source with a dependent source as shown in Fig. 3.27. Find the value of v_3.

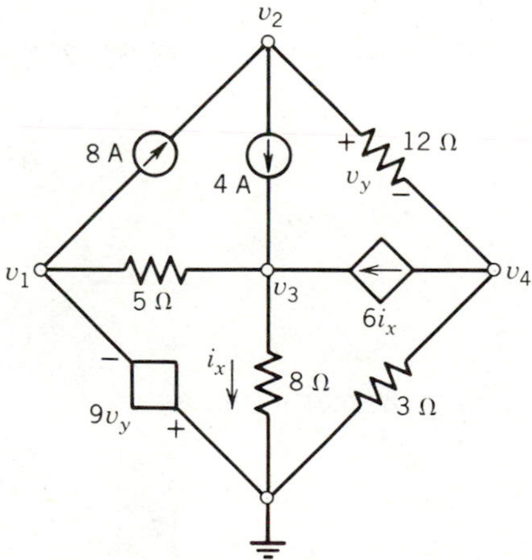

Figure 3.27. Nodal analysis with dependent sources.

Solution

At Node 1

$$v_1 = -9v_y$$

But

$$v_y = v_2 - v_4$$

So

$$v_1 = -9(v_2 - v_4) \tag{3.3.22}$$

At Node 2

$$-8 + 4 + \frac{(v_2 - v_4)}{12} = 0 \tag{3.3.23}$$

At Node 3

$$-4 - 6i_x + \frac{(v_3 - v_1)}{5} + \frac{v_3}{8} = 0 \tag{3.3.24}$$

But

$$i_x = \frac{v_3}{8} \tag{3.3.25}$$

So substituting Eq. 3.3.25 into Eq. 3.3.24,

$$-4 - \frac{6}{8}v_3 + \frac{(v_3 - v_1)}{5} = 0 \tag{3.3.26}$$

At Node 4

$$\frac{(v_4 - v_2)}{12} + \frac{v_4}{3} + 6i_x = 0 \tag{3.3.27}$$

Again using Eq. 3.3.25 in Eq. 3.3.27 we find

$$\frac{v_4 - v_2}{12} + \frac{v_4}{3} + \frac{3}{4}v_3 = 0 \tag{3.3.28}$$

Equations 3.3.22, 3.3.23, 3.3.26, and 3.3.28 can be simplified and arranged as

$$\begin{aligned} v_1 + 9v_2 \qquad\qquad - 9v_4 &= 0 \\ v_2 \qquad\qquad - v_4 &= 48 \\ -4v_1 \qquad - 11v_3 \qquad &= 80 \\ -v_2 + 9v_3 + 5v_4 &= 0 \end{aligned}$$

Using Cramer's rule again,

$$v_3 = \frac{\begin{vmatrix} 1 & 9 & 0 & -9 \\ 0 & 1 & 48 & -1 \\ -4 & 0 & 80 & 0 \\ 0 & -1 & 0 & 5 \end{vmatrix}}{\begin{vmatrix} 1 & 9 & 0 & -9 \\ 0 & 1 & 0 & -1 \\ -4 & 0 & -11 & 0 \\ 0 & -1 & 9 & 5 \end{vmatrix}} = 152.27 \text{ V} \tag{3.3.29}$$

We note that the denominator of Eq. 3.3.29 is no longer symmetric about the major diagonal values. This is due to the presence of the dependent sources.

Circuits containing more than five nodes generate equations that are too cumbersome to solve by hand unless the resulting determinants contain mostly zeros or unless your instructor insists you do it one time so you will appreciate the difficulty. Such circuits are easily solved on large computers where 1024×1024 size determinants are readily evaluated. Analysis of large power systems is often performed using nodal analysis and results in hundreds of independent equations that are solved using computer methods.

LEARNING EVALUATIONS

1. Write the nodal equations in matrix form for the circuit in Fig. 3.28.

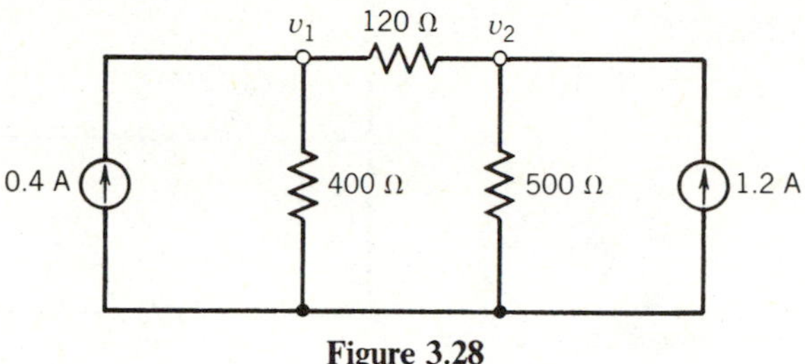

Figure 3.28

2. Use nodal analysis to find v_{300} in Fig. 3.29.

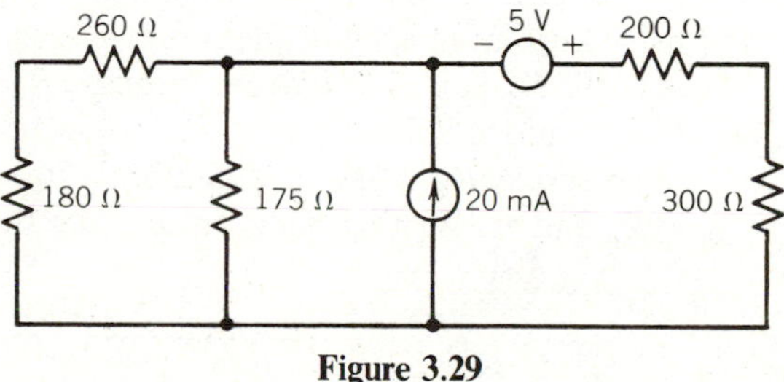

Figure 3.29

3.4 Mesh Analysis

LEARNING OBJECTIVES

After completing this section you should be able to do the following:

1. Define a mesh.
2. Solve for all circuit parameters in each element of a circuit by using the method of mesh analysis.
3. Be able to use the "super mesh" concept in applying mesh analysis to circuits containing current sources.
4. Define a planar circuit and determine if a given circuit is planar.

The more complex circuit has a larger number of independent variables (voltages and currents). If we can write as many independent equations as there are independent variables, then we can solve for any or all of these variables. Mesh analysis is a method for generating the independent equations necessary to solve for the circuit variables, as was nodal analysis.

A *mesh* is defined as a closed path around a circuit that does not contain any other closed paths. The circuit in Fig. 3.30 contains three meshes (a–b–g–f–a), (b–c–d–g–b), (f–g–d–e–f), and other closed paths that are not meshes [e.g., (a–b–c–d–g–f–a), (a–b–g–d–e–f–a)]. In mesh analysis we assign a mesh current to each mesh and assume that the current flows around the mesh in a cw direction.

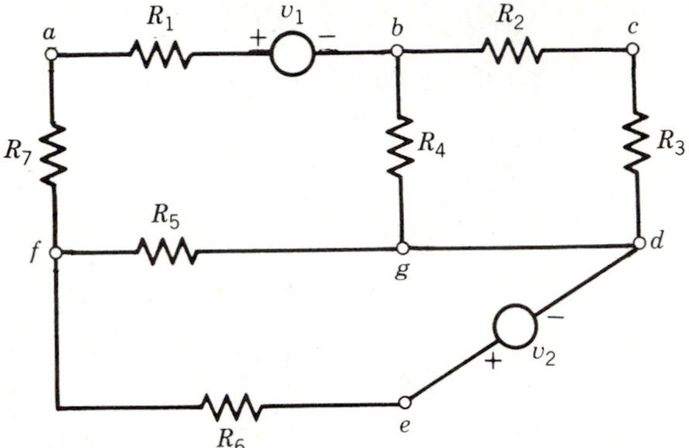

Figure 3.30. Circuit containing three meshes.

The circuit of Fig. 3.30 is redrawn in Fig. 3.31 showing the assignment of a mesh current to each of the three meshes. Notice that the current flowing through R_1 and R_7 is i_1. The current through R_4 is the algebraic sum of mesh currents i_1 and i_2 and is either $(i_1 - i_2)$ or $(i_2 - i_1)$ depending on which direction the voltage drop across R_4 is assumed. Similarly, the current through R_5 is the algebraic sum of i_1 and i_3. Writing KVL around the three meshes gives

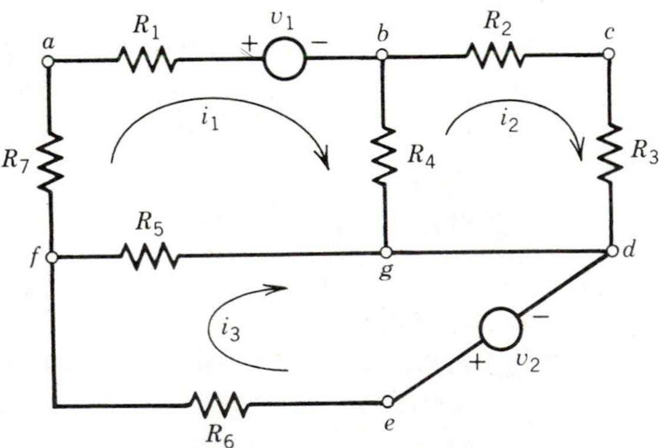

Figure 3.31. Assignment of mesh currents.

Mesh 1

$$R_7 i_1 + R_1 i_1 + v_1 + R_4(i_1 - i_2) + R_5(i_1 - i_3) = 0$$

or

$$(R_7 + R_1 + R_4 + R_5)i_1 - R_4 i_2 - R_5 i_3 = -v_1 \tag{3.4.1}$$

Mesh 2

$$R_2 i_2 + R_3 i_2 + R_4(i_2 - i_1) = 0$$

or

$$-R_4 i_1 + (R_2 + R_3 + R_4)i_2 = 0 \tag{3.4.2}$$

Mesh 3

$$-v_2 + R_6 i_3 + R_5(i_3 - i_1) = 0$$

or

$$-R_5 i_1 + (R_5 + R_6)i_3 = v_2 \tag{3.4.3}$$

Equations 3.4.1 through 3.4.3 form a set of independent equations that can be solved for i_1, i_2, and i_3 if the values of the resistors and voltage sources are known. Once the currents are determined, then all values of voltages, currents, and power in the circuit can be calculated from our elementary laws.

EXAMPLE 3.4.1 Use mesh analysis to find the voltage, V_3, and the power absorbed in the 5-Ω resistor in the circuit of Fig. 3.32.

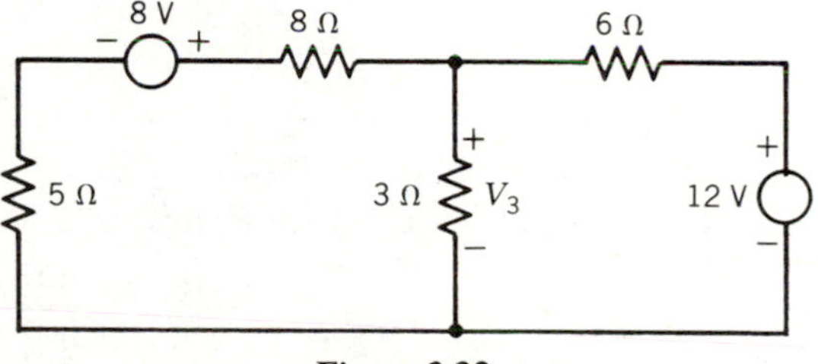

Figure 3.32

Solution

First we observe there are two meshes, so we assign the two mesh currents I_1 and I_2 as shown in Fig. 3.33.

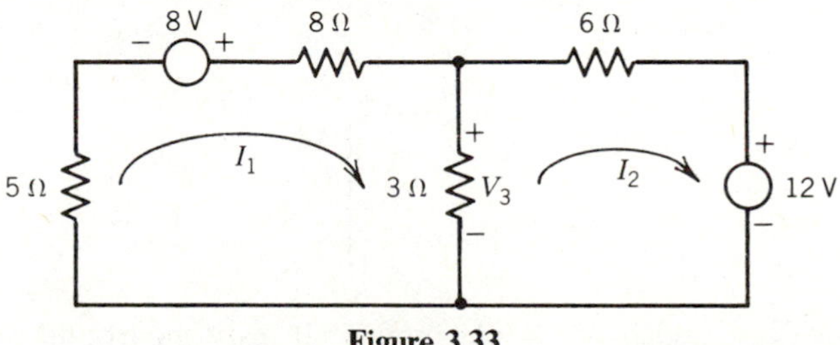

Figure 3.33

Now write KVL around each mesh.

Mesh 1

$$5I_1 - 8 + 8I_1 + 3(I_1 - I_2) = 0$$

or

$$16I_1 - 3I_2 = 8 \tag{3.4.4}$$

Mesh 2

$$6I_2 + 12 + 3(I_2 - I_1) = 0$$

or

$$-3I_1 + 9I_2 = -12 \tag{3.4.5}$$

Equations 3.4.4 and 3.4.5 represent two equations with two unknowns, which may be solved to yield

$$I_1 = \tfrac{4}{15}\text{ A}$$

$$I_2 = -\tfrac{56}{45}\text{ A}$$

The current flowing through the 3-Ω resistor is $(I_1 - I_2)$ because of the voltage direction shown for V_3. So by Ohm's law,

$$V_3 = 3[\tfrac{4}{15} - (-\tfrac{56}{45})] = \tfrac{68}{15}\text{ V}$$

The power absorbed by the 5-Ω resistor is

$$\begin{aligned} P_5 &= I_1^2(5) \\ &= (\tfrac{4}{15})^2(5) \\ &= \tfrac{16}{45}\text{ W} \end{aligned}$$

Matrix Equations

When displayed in matrix form Eqs. 3.4.4 and 3.4.5 become

$$\begin{bmatrix} 16 & -3 \\ -3 & 9 \end{bmatrix} \begin{bmatrix} i_1 \\ i_2 \end{bmatrix} = \begin{bmatrix} 8 \\ -12 \end{bmatrix}$$

There is a systematic way to arrive at this form directly from the circuit diagram, just as there was for nodal equations. The method is applicable to planar networks containing only independent sources. A planar network is one that can be drawn or constructed on a plane surface such that no element crosses over or under any other element. The general form of the matrix equation for an n-mesh network is given by

$$\begin{bmatrix} r_{11} & r_{12} & \cdots & r_{1n} \\ r_{21} & r_{22} & \cdots & r_{2n} \\ \vdots & \vdots & \ddots & \vdots \\ r_{n1} & r_{n2} & \cdots & r_{mn} \end{bmatrix} \begin{bmatrix} i_1 \\ i_2 \\ \vdots \\ i_n \end{bmatrix} = \begin{bmatrix} v_1 \\ v_2 \\ \vdots \\ v_n \end{bmatrix} \tag{3.4.6}$$

where r_{ii} is the sum of all resistors around mesh i, and r_{ij} is the negative of the resistance common to meshes i and j. The current vector is composed of the unknown mesh currents, and the terms in the voltage vector represent the forcing functions in each mesh. Here is an example.

EXAMPLE 3.4.2 Write the mesh equations for Fig. 3.34 in matrix form.

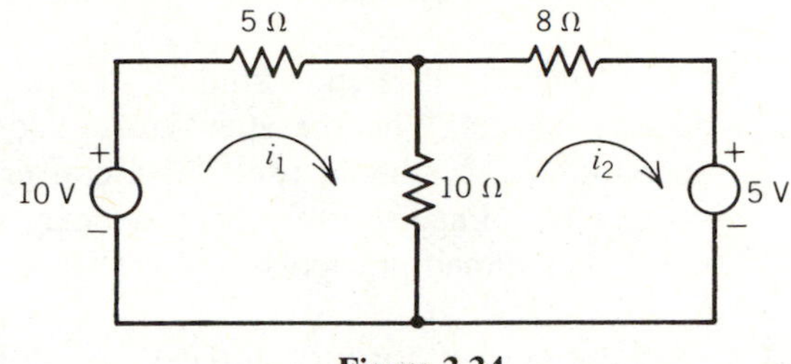

Figure 3.34

Solution

$$\begin{bmatrix} 15 & -10 \\ -10 & 18 \end{bmatrix} \begin{bmatrix} i_1 \\ i_2 \end{bmatrix} = \begin{bmatrix} 10 \\ -5 \end{bmatrix} \tag{3.4.7}$$

The values in the resistance matrix are derived as follows: There is a total of 15 Ω around mesh 1 and 18 Ω around mesh 2. Hence these values appear on the main diagonal. There is 10 Ω common to both mesh 1 and mesh 2, and it has a negative value in the resistance matrix because we assumed both mesh currents cw.

The 10-V forcing function is positive because it is forcing current to move in the direction of i_1. The 5-V source is negative in Eq. 3.4.7 because it is forcing current in the direction opposite to that which we assumed for i_2.

EXAMPLE 3.4.3 Write the mesh equations for Fig. 3.35 in matrix form.

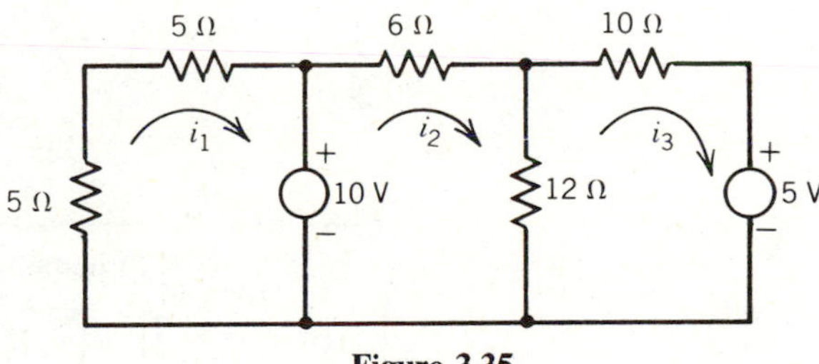

Figure 3.35

Solution

$$\begin{bmatrix} 10 & 0 & 0 \\ 0 & 18 & -12 \\ 0 & -12 & 22 \end{bmatrix} \begin{bmatrix} i_1 \\ i_2 \\ i_3 \end{bmatrix} = \begin{bmatrix} -10 \\ 10 \\ -5 \end{bmatrix}$$

The r_{12} and r_{13} terms are each zero, but for different reasons. The resistance of the source common to mesh 1 and mesh 2 is zero, hence

$r_{12} = 0$. There is no resistance common to mesh 1 and 3, so $r_{13} = 0$.

The sum of all voltage sources around mesh 1 is negative because the 10-V source is opposing i_1. The source in mesh 2 is aiding i_2, so it is positive in the matrix equation. Finally, the 5-V source in mesh 3 opposes i_3 so it is negative.

Notice that all circuits studied in this section until now have contained no current sources. When practical current sources (those with parallel resistors) are encountered, we should use source conversion (from Section 3.2) to convert all sources to voltage sources. Then we may write the KVL equations around each mesh, as outlined previously.

EXAMPLE 3.4.4 Find the current through the 4-Ω resistor in Fig. 3.36.

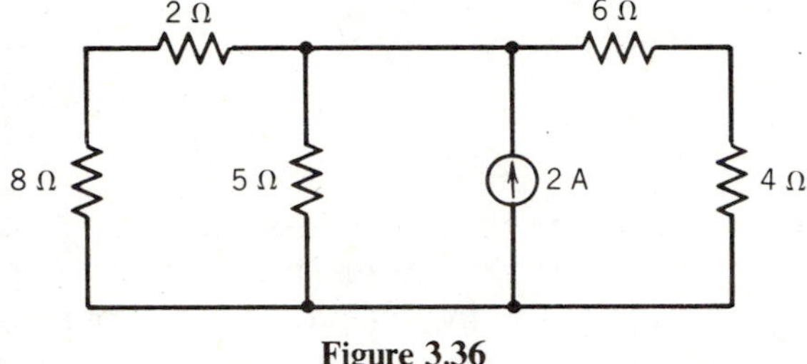

Figure 3.36

Solution

The 2-A source in parallel with the 5-Ω resistor is first converted to the equivalent voltage source shown in Fig. 3.37. Then mesh current i_2 is found as follows.

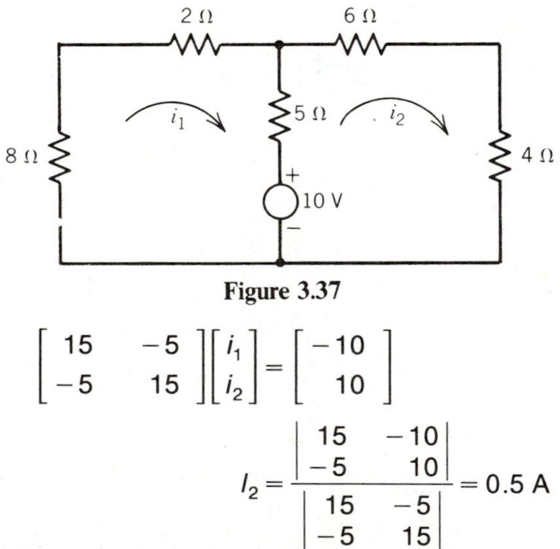

Figure 3.37

$$\begin{bmatrix} 15 & -5 \\ -5 & 15 \end{bmatrix} \begin{bmatrix} i_1 \\ i_2 \end{bmatrix} = \begin{bmatrix} -10 \\ 10 \end{bmatrix}$$

$$I_2 = \frac{\begin{vmatrix} 15 & -10 \\ -5 & 10 \end{vmatrix}}{\begin{vmatrix} 15 & -5 \\ -5 & 15 \end{vmatrix}} = 0.5 \text{ A}$$

We next address the method for handling ideal current sources—those with no resistor in parallel. Remember we are concerned with generating the same number of independent equations as we have variables in our circuit. Up to this point we needed N mesh equations if we had N variables. Consider the portion of a circuit with N meshes shown in Fig. 3.38.

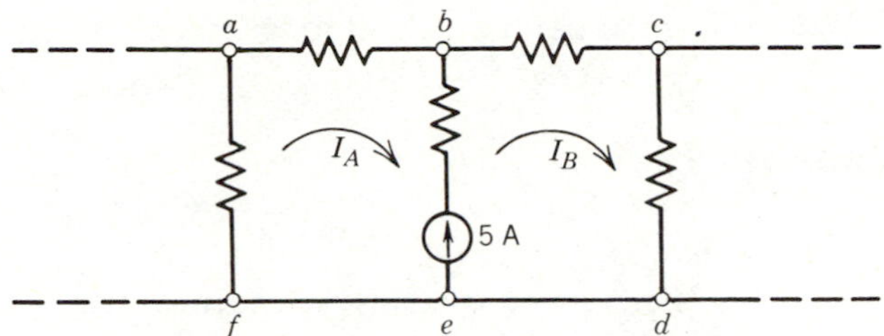

Figure 3.38. Portion of a circuit that contains an ideal current source.

The mesh currents I_A and I_B can be related by inspection to the current source of 5 A as

$$I_B - I_A = 5 \tag{3.4.8}$$

Consequently we now need only $N - 1$ additional equations to solve for N circuit variables. Therefore, our approach will be to write one KVL equation around the loop (a, b, c, d, e, f, a) and use this in conjunction with Eq. 3.4.8 to replace the two mesh equations from mesh A and mesh B. This outer loop is sometimes called a "super mesh."

EXAMPLE 3.4.5 The circuit in Fig. 3.39 contains an ideal current source and already has the mesh currents assigned. Find the power absorbed by the 5-Ω resistor.

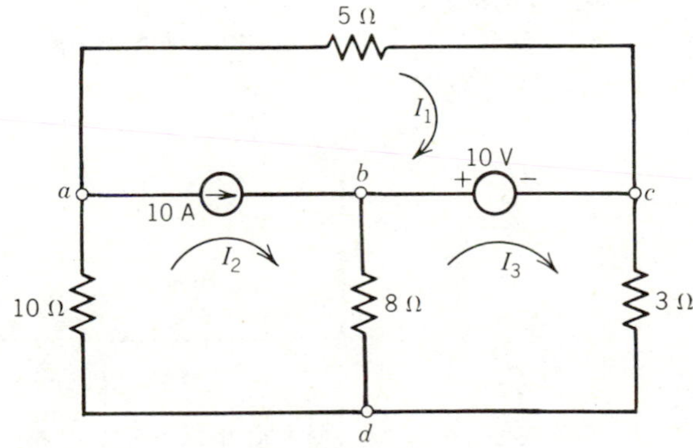

Figure 3.39

Solution

In order to answer the question, we need to find the value of I_1. By inspection of the mesh currents and the current source we note that

$$I_2 - I_1 = 10 \tag{3.4.9}$$

Writing KVL around the mesh $(a\text{–}c\text{–}b\text{–}d\text{–}a)$ gives

$$5I_1 - 10 + 8(I_2 - I_3) + 10I_2 = 0 \tag{3.4.10}$$

and writing KVL around mesh 3 gives

$$8(I_3 - I_2) + 10 + 3I_3 = 0 \tag{3.4.11}$$

Equations 3.4.9 through 3.4.11 are three equations with three unknowns which can be solved for I_1 to obtain

$$I_1 = 6.97 \text{ A}$$

so

$$P_5 = I_1^2(5) = (6.97)^2(5) = 243 \text{ W}$$

If our circuit contains dependent voltage sources we can proceed in exactly the same manner, but we must be careful to relate the dependent source value to one or more of the mesh currents. This technique is best illustrated by means of the following example.

EXAMPLE 3.4.6 Find the value of I in Fig. 3.40 using mesh analysis.

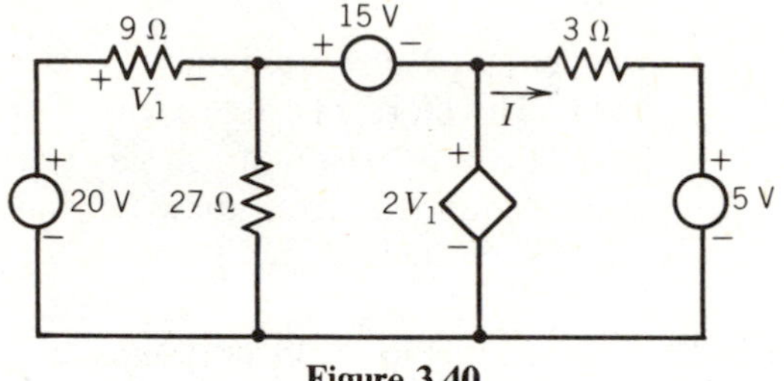

Figure 3.40

Solution

First assign the mesh currents to the three meshes as shown in Fig. 3.41.

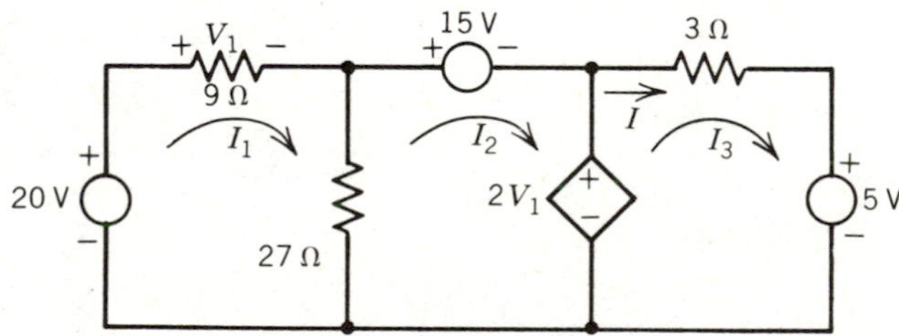

Figure 3.41

By observation we see that the current I is equal to mesh current I_3 and that V_1 is related to mesh current I_1 by $V_1 = 9I_1$, so the dependent voltage source is related to the mesh currents by

$$2V_1 = 18I_1 \tag{3.4.12}$$

Now write the three mesh equations:

Mesh 1

$$-20 + 9I_1 + 27(I_1 - I_2) = 0$$

or

$$36I_1 - 27I_2 = 20 \qquad (3.4.13)$$

Mesh 2

$$15 + 2V_1 + 27(I_2 - I_1) = 0$$

or using Eq. 3.4.12 and combining terms,

$$-9I_1 + 27I_2 = -15 \qquad (3.4.14)$$

Mesh 3

$$3I_3 + 5 - 2V_1 = 0$$

or

$$-18I_1 + 3I_3 = -5 \qquad (3.4.15)$$

Equations 3.4.13 through 3.4.15 can now be solved using Cramer's rule. Solving for I_3 we get

$$I_3 = \frac{\begin{vmatrix} 36 & -27 & 20 \\ -9 & 27 & -15 \\ -18 & 0 & -5 \end{vmatrix}}{\begin{vmatrix} 36 & -27 & 0 \\ -9 & 27 & 0 \\ -18 & 0 & 3 \end{vmatrix}} = -0.56 \text{ A}$$

since $I = I_3$ we have $I = -0.56$ A.

By reading the problem statement carefully and relating the required variable to the mesh currents, we realized that I_3 was the only mesh current that we needed to find. Values of I_1 and I_2 were of no interest to us in this problem. Your should *always* very carefully determine what you are required to find by examining the problem statement and then seek the least difficult method of solution. Your instructor would not be impressed if this example problem were on a test and you found I_1 and I_2 also. Your employer would be even less impressed under similar circumstances.

LEARNING EVALUATIONS

1. Write the mesh equations in matrix form for the circuit in Fig. 3.42.

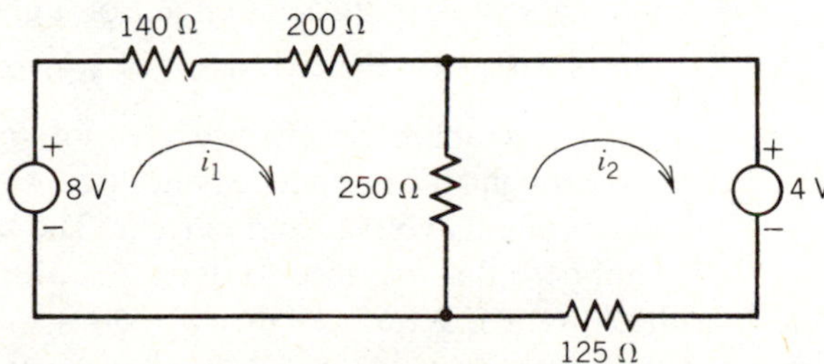

Figure 3.42

2. Find the current through the 20-Ω resistor in Fig. 3.43 by mesh analysis.

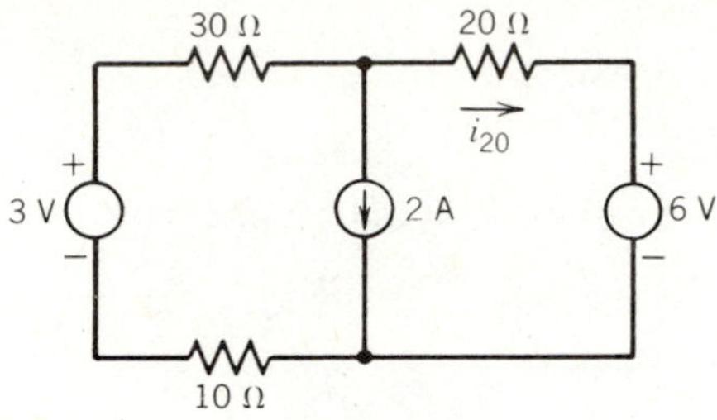

Figure 3.43

3. Find mesh current i_2 in Fig. 3.44.

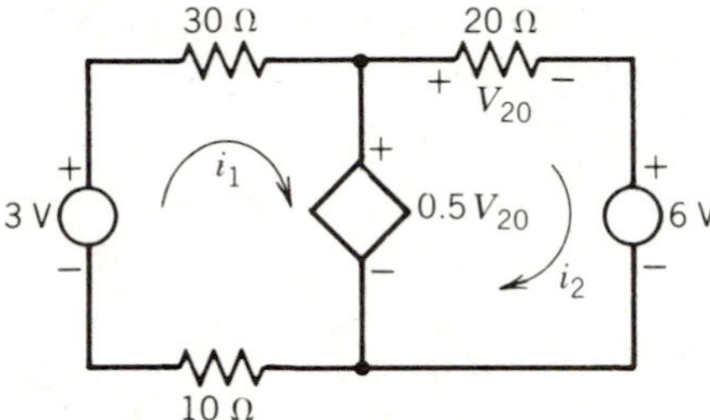

Figure 3.44

3.5 Thévenin and Norton Theorems

LEARNING OBJECTIVES

After completing this section you should be able to do the following:

1. Define the following terms:
   Open circuit voltage
   Short circuit current
   Input resistance
   Thévenin's theorem
   Norton's theorem
2. Solve for the open circuit voltage, short circuit current, and input resistance of a given network.
3. Find Thévenin's equivalent circuit of a given network.
4. Find Norton's equivalent circuit of a given network.

In many complex circuits we are not interested in the voltage across and current through all components, but often want to single out only one component and determine its voltage and current. The circuit in Fig. 3.45 will illustrate this idea. Suppose that we need to determine the effect of changes in the value of R_5 on the current i. With our current tools it would be a long and laborious process to calculate a new value of i for each new value of R_5.

About one hundred years ago (1883), M. Thévenin, an electrical engineer, provided us with a method for simplifying the above task. He proved that all the

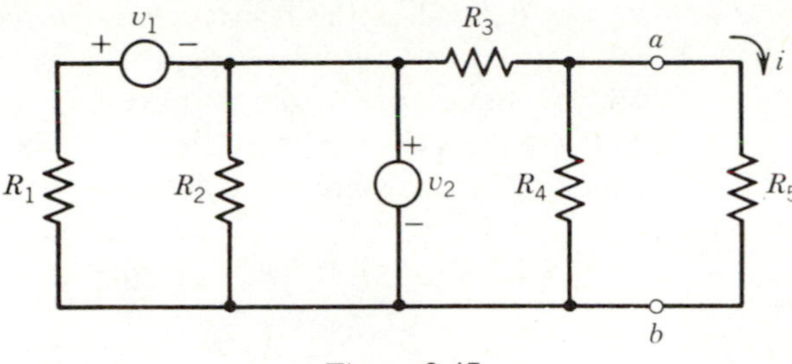

Figure 3.45.

circuit components to the left of terminals a–b could be replaced by a single voltage source (called v_{TH}) in series with a single resistance (called R_{TH}) without altering the current i. This replacement is shown in Fig. 3.46 where v_{TH} and R_{TH} comprise what is called the Thévenin equivalent circuit. Now we find that

$$i = \frac{v_{TH}}{R_{TH} + R_5}$$

and the variation of i as a function of the value of R_5 can easily be calculated. But how can we find the values of v_{TH} and R_{TH}?

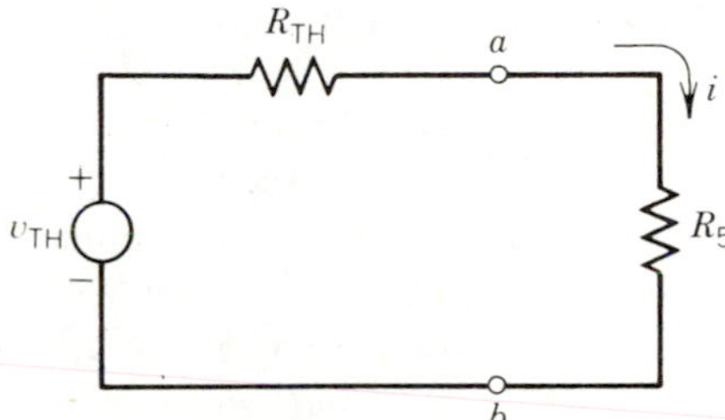

Figure 3.46. Thévenin equivalent circuit of Fig. 3.45.

Thévenin showed that v_{TH} is the open circuit voltage appearing at the terminals a–b after the resistor R_5 had been removed. See Fig. 3.47. By using superposition and voltage division or other analysis techniques in your bag you can show that

$$v_{TH} = \frac{R_4}{R_3 + R_4} V_2$$

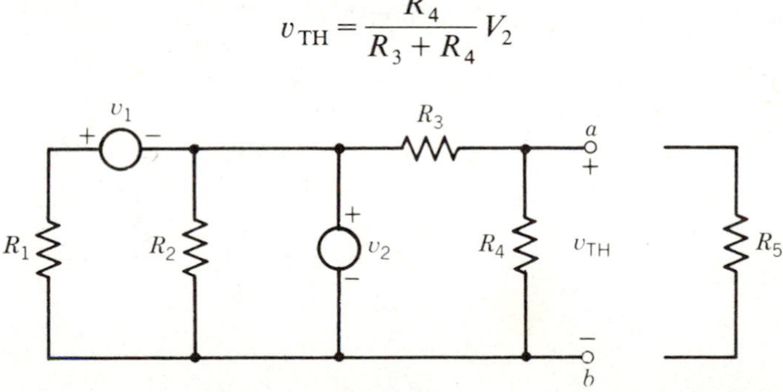

Figure 3.47. Definition of v_{TH} for the circuit of Fig. 3.45.

R_{TH} is defined as the resistance measured at the terminals a–b after R_5 has been removed and all independent voltage sources have been short circuited and all independent current sources have been open circuited. Figure 3.48 shows the circuit of Fig. 3.45 after these steps have been taken. Standing at terminals a–b, and looking into the circuit we observe that

$$R_{TH} = \frac{R_3 R_4}{R_3 + R_4}$$

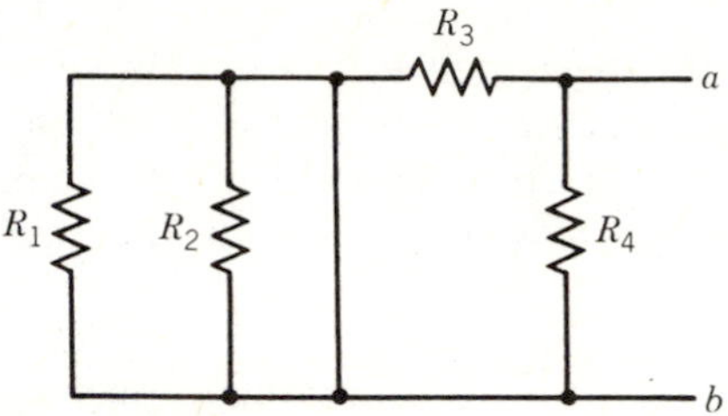

Figure 3.48. Circuit for determining R_{TH}.

We will now formally state Thévenin's theorem:

> Any linear two-terminal circuit can be replaced by a voltage source, v_{TH}, in series with a resistance, R_{TH}, such that a second circuit connected to the two terminals is not affected by the substitution. v_{TH} is the open-circuit voltage appearing across the two terminals when the second circuit has been removed. R_{TH} is the resistance at the two terminals when the second circuit has been removed and all independent voltage sources have been replaced by short circuits and all independent current sources have been replaced by open circuits.

EXAMPLE 3.5.1 Determine Thévenin's equivalent circuit for the circuit shown in Fig. 3.49.

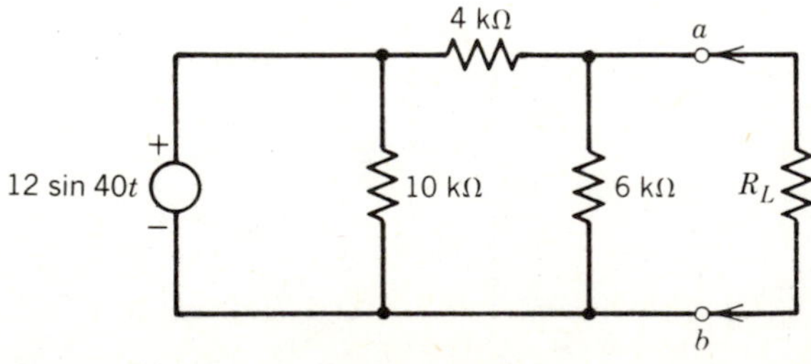

Figure 3.49

Solution

Assume that R_L has been removed. The v_{TH} is the voltage across the 6-kΩ resistor. By voltage division,

$$v_{TH} = \frac{6\text{ k}\Omega}{6\text{ k}\Omega + 4\text{ k}\Omega}(12\sin 40t)$$
$$= 7.2\sin 40t\text{ V}$$

R_{TH} is found by shorting the voltage source, which leaves the 6- and 4-kΩ resistors in parallel so

$$R_{TH} = \frac{(6\text{ k}\Omega)(4\text{ k}\Omega)}{6\text{ k}\Omega + 4\text{ k}\Omega}$$
$$= 2.4\text{ k}\Omega$$

The Thévenin equivalent circuit is shown in Fig. 3.50.

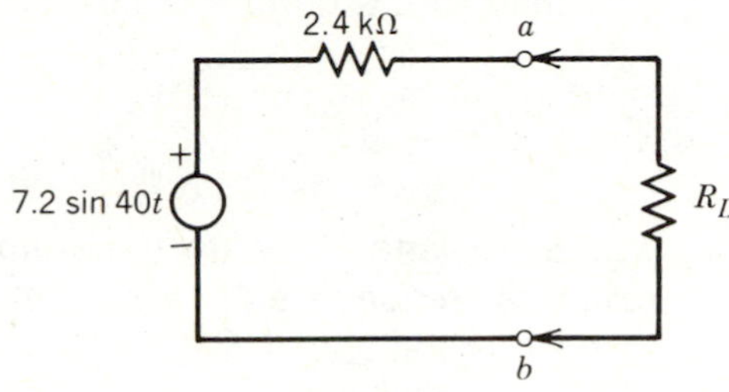

Figure 3.50

EXAMPLE 3.5.2 The bridge circuit in Fig. 3.51 provides a good illustration of the use of Thévenin's theorem.

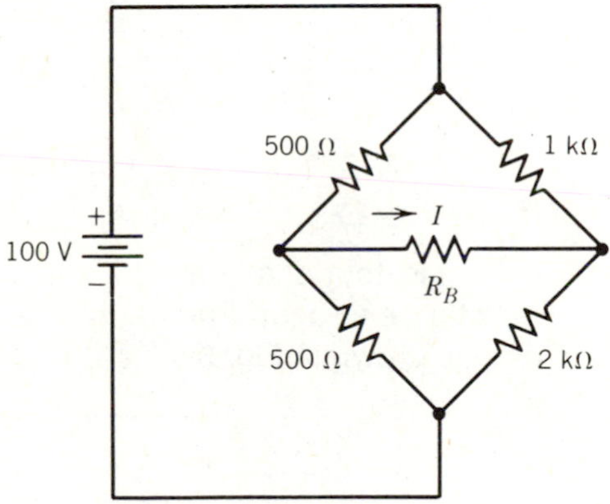

Figure 3.51

Find the current through the bridge resistor R_B for the following values of R_B: 50 Ω, 200 Ω, 1 kΩ, 10 kΩ.

Solution

Since we are interested only in the current through R_B, we can replace the rest of the circuit with its Thévenin's equivalent. Remove R_B and determine V_{TH} and R_{TH}.

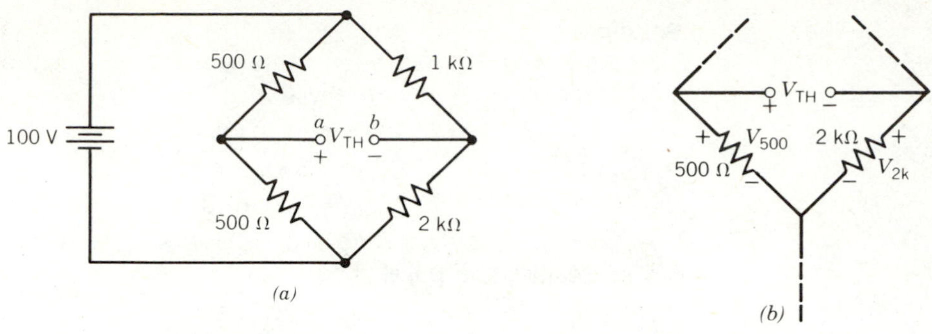

Figure 3.52(*a*) **Figure 3.52**(*b*)

To find V_{TH} we again make use of KVL in Fig. 3.52*a*. Consider the mesh in Fig. 3.52*b*, which is one part of the circuit shown in Fig. 3.51.

Writing KVL around the mesh we have

$$-V_{500} + V_{TH} + V_{2\,k\Omega} = 0$$

or

$$V_{TH} = V_{500} - V_{2\,k\Omega}$$

where V_{500} and $V_{2\,k\Omega}$ are the voltage drops across the 500-Ω and 2-kΩ resistors, respectively. Using voltage division in Fig. 3.52*a* gives the following relations:

$$V_{500} = \frac{500}{500 + 500}100$$

$$= 50 \text{ V}$$

$$V_{2k} = \frac{2000}{2000 + 1000}100$$

$$= 66.67 \text{ V}$$

so

$$V_{TH} = 50 - 66.67$$

$$V_{TH} = -16.67 \text{ V}$$

R_{TH} is found to be shorting the 100-V battery and determining the resistance as seen from terminals a–b. Shorting the battery results in the circuit shown in Fig. 3.53.

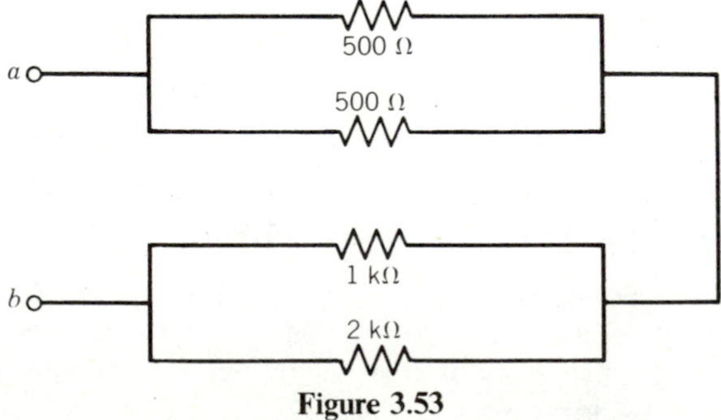

Figure 3.53

so

$$R_{TH} = \frac{(500)(500)}{500 + 500} + \frac{(1000)(2000)}{1000 + 2000}$$

$$R_{TH} = 916.67\ \Omega$$

Our Thévenin equivalent circuit is shown in Fig. 3.54.

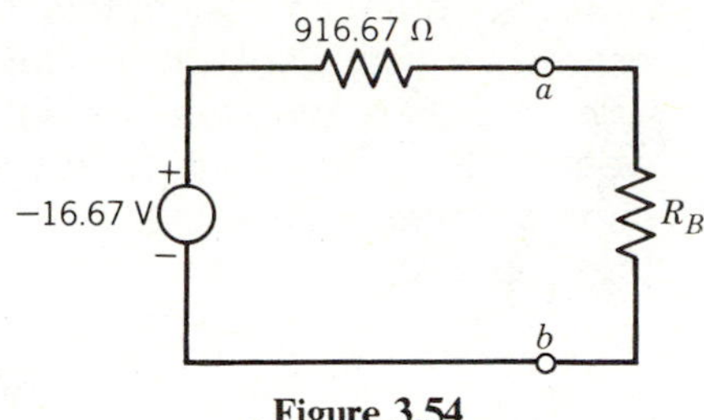

Figure 3.54

and by inspection

$$I = \frac{-16.67}{916.67 + R_B}\ \text{A}$$

The following table can then be completed for the values of R_B of interest.

R_B	50 Ω	200 Ω	1 kΩ	10 kΩ
I	−17.24 mA	−14.92 mA	−8.70 mA	−1.53 mA

This approach results in much less work than writing three node equations or three mesh equations and solving the resulting three algebraic equations every time for the four values of R_B.

Thévenin's theorem may be used for circuits containing dependent sources. Here is an example.

EXAMPLE 3.5.3 Use Thévenin's theorem to find the voltage across the 3-Ω resistor in Fig. 3.55.

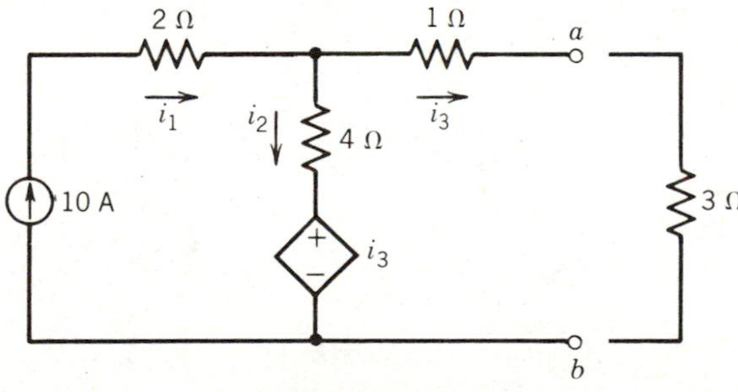

Figure 3.55

Solution

After removing the 3-Ω resistor, the voltage from *a* to *b* may be found by noting that $i_3 = 0$. Therefore, the dependent source has zero voltage across it. The voltage from *a* to *b* is just the voltage across the 4-Ω resistor, giving

$$V_{ab} = V_{TH} = 40 \text{ V}$$

It is not quite so easy to find Thévenin's resistance, however. The best procedure with dependent sources is to apply a voltage source to the terminals *a*–*b* and measure the current through this source. Of course we must first remove all independent sources (open current sources and short voltage sources). This is shown in Fig. 3.56. Writing KVL around the loop gives

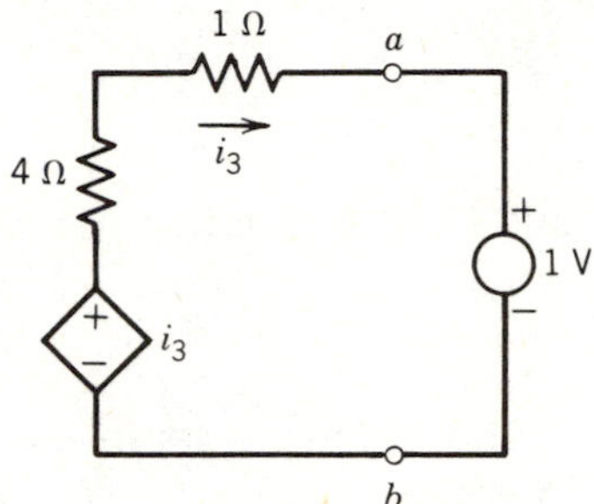

Figure 3.56

$$0 = -i_3 + 4i_3 + i_3 + 1$$

or

$$i_3 = -\tfrac{1}{4}$$

Therefore the 1-V source forces $\frac{1}{4}$ A into node *a*, giving an equivalent resistance of

$$R_{TH} = \frac{1 \text{ V}}{\frac{1}{4} \text{ A}} = 4 \ \Omega$$

Combining these results and attaching the 3-Ω resistor gives (see Fig. 3.57).

$$V_0 = \frac{3}{3+4}(40) = \frac{120}{7} \text{ V}$$

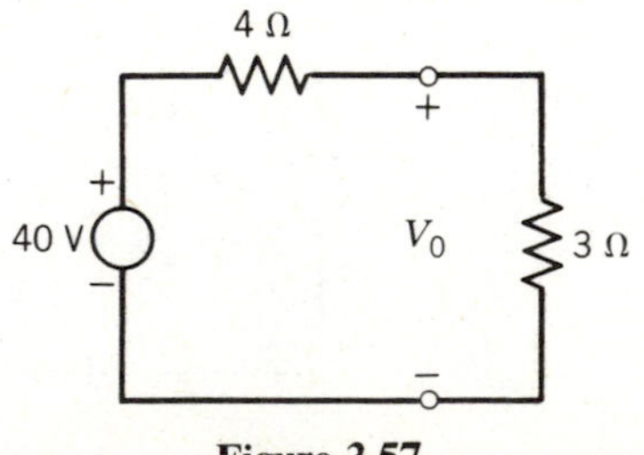

Figure 3.57

Norton's Equivalent Circuit

From our previous discussion of source conversions we know that the circuit of Fig. 3.46 can be converted to that of Fig. 3.58, and R_5 will not know the difference if

$$i_N = \frac{v_{TH}}{R_{TH}}$$

and

$$R_N = R_{TH}$$

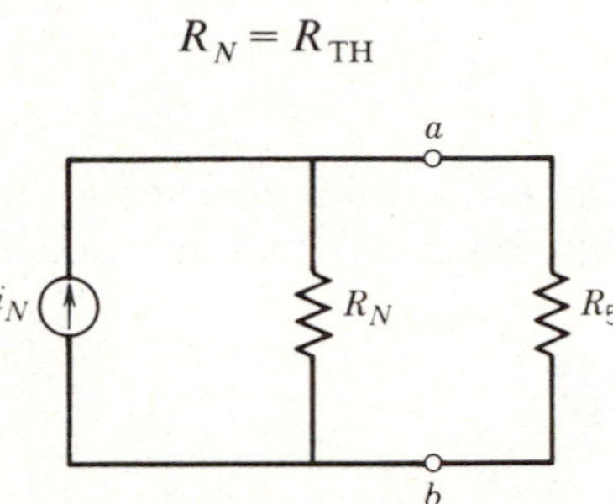

Figure 3.58. Current source equivalent of Fig. 3.46.

Thanks to E. C. Norton, we can find i_N and R_N directly without first finding the Thévenin equivalent circuit. Norton's theorem states:

> Any linear two-terminal circuit can be replaced by a current source, i_N, in parallel with a resistance, R_N, such that a second circuit connected to the two terminals is not affected by the substitution. The current i_N is the short circuit current flowing through the two terminals when the second circuit is replaced by a short circuit. The resistance R_N is the resistance at the two terminals when the second circuit has been removed, all independent voltage sources have been replaced by short circuits, and all independent current sources have been replaced by open circuits.

Equation 3.5.2 shows that R_{TH} and R_N are equivalent and are therefore calculated in exactly the same manner. In addition to the resistance, if either v_{TH} or i_N is known, the other can be found using Eq. 3.5.1. Whether one finds v_{TH} or i_N is strictly a function of the circuit that is to be analyzed. An intuitive judgment of which would be easiest to find must be made.

EXAMPLE 3.5.4 For the circuit used in Example 3.5.1, find the Norton equivalent circuit.

Solution

R_N is the same as R_{TH} found earlier so

$$R_N = 2.4\ \text{k}\Omega$$

To find i_N, short terminals a–b together as illustrated in Fig. 3.59.

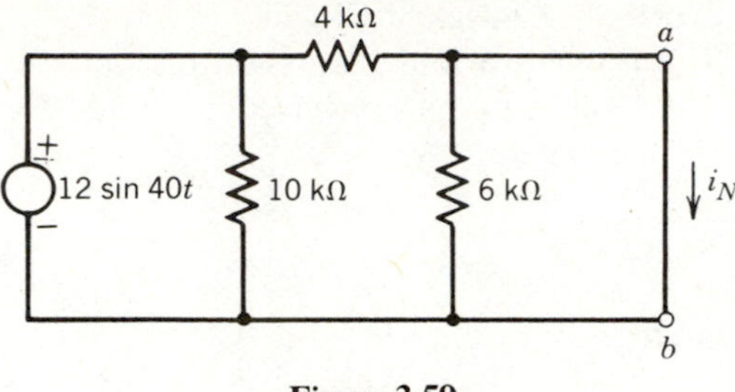

Figure 3.59

Since the 6-kΩ resistor is in parallel with the short circuit it can be ignored and i_N can be calculated from Ohm's law as

$$i_N = \frac{12 \sin 40t}{4 \times 10^3}$$
$$= 3 \sin 40t \text{ mA}$$

As a check we can calculate v_{TH} using Eq. 3.5.1.

$$v_{TH} = R_N i_N$$
$$= (2.4 \times 10^3)(3 \times 10^{-3} \sin 40t)$$
$$= 7.2 \sin 40t \text{ V}$$

which agrees with the value found in Example 3.5.1.

If a circuit contains both dependent and independent sources we cannot find R_{TH} or R_N directly because the dependent sources are not shorted or opened. A careful reading of Thévenin's and Norton's theorems indicates that only *independent* sources are shorted or opened when finding the Thévenin or Norton resistance. So for circuits with both dependent and independent sources one procedure will be to determine v_{TH} and i_N, then use Eq. 3.5.1 to find R_{TH} or R_N.

EXAMPLE 3.5.5 The circuit in Fig. 3.60 contains both independent and dependent sources. Find R_{TH} for the portion of the circuit to the left of terminals a–b.

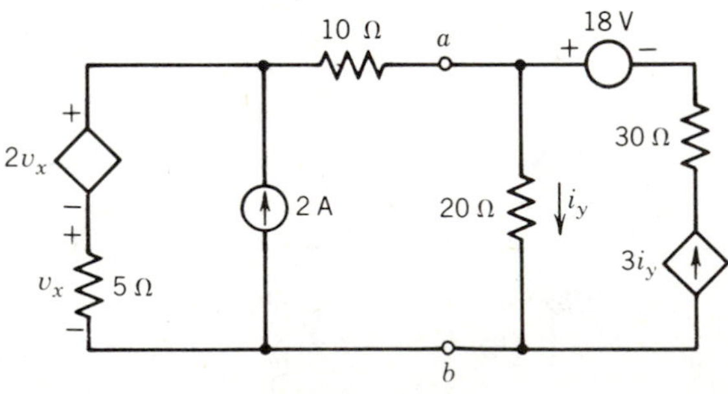

Figure 3.60

Solution

First find v_{TH} by open circuiting terminals a–b (Fig. 3.61).

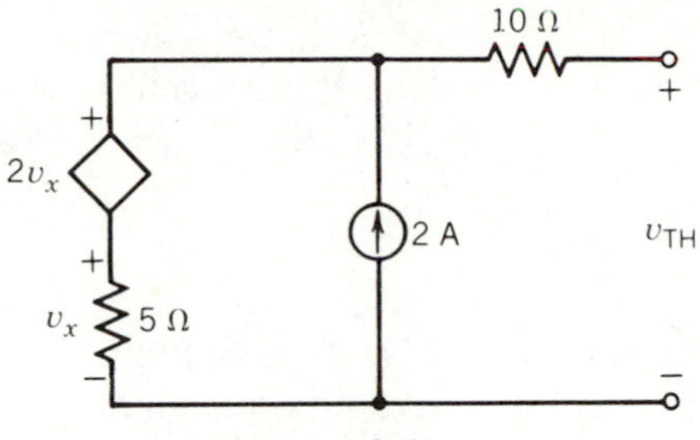

Figure 3.61

We observe that all 2 A of the current source will flow through the 5-Ω resistor, since it represents the only closed path. Writing KVL around the outer loop of the circuit gives

$$-v_x - 2v_x + v_{TH} = 0$$

but

$$v_x = 5(2) = 10$$

so

$$v_{TH} = 30 \text{ V}$$

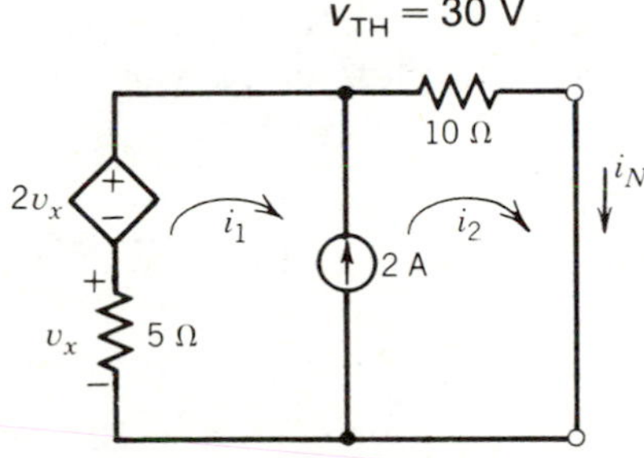

Figure 3.62

To find i_N we short terminals a–b, which yields the circuit in Fig. 3.62. Using mesh analysis and creating a super mesh, we find

a. $i_2 - i_1 = 2$

b. $-v_x - 2v_x + 10i_2 = 0$

c. $v_x = -5i_1$

d. $i_N = i_2$

Solving these yields

$$i_N = 1.2 \text{ A}$$

therefore

$$R_{TH} = \frac{v_{TH}}{i_N} = \frac{30}{1.2} = 25 \ \Omega$$

Finally, if our circuit contains only dependent sources, a slight subterfuge must be used. With only dependent sources, the open circuit or Thévenin voltage will always be zero because there is no independent energy source to activate the circuits. To find R_{TH} we will apply a test current source of 1 A value (a test voltage source of 1 V value could also be used) across terminals a–b and calculate the resulting voltage that appears across a–b, v_{ab}. Then

$$R_{TH} = \frac{v_{ab}}{1}$$

EXAMPLE 3.5.6 The circuit in Fig. 3.63 contains only a dependent current source. We want to find the Thévenin equivalent circuit with respect to terminals x–y.

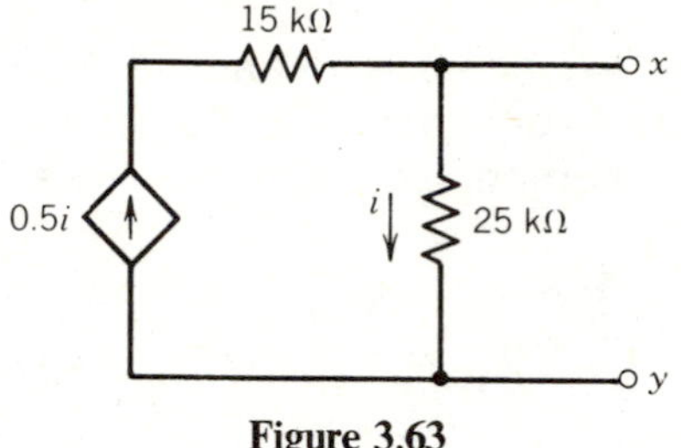

Figure 3.63

Solution

Applying a test current of 1 A to terminals x–y gives the circuit in Fig. 3.64.

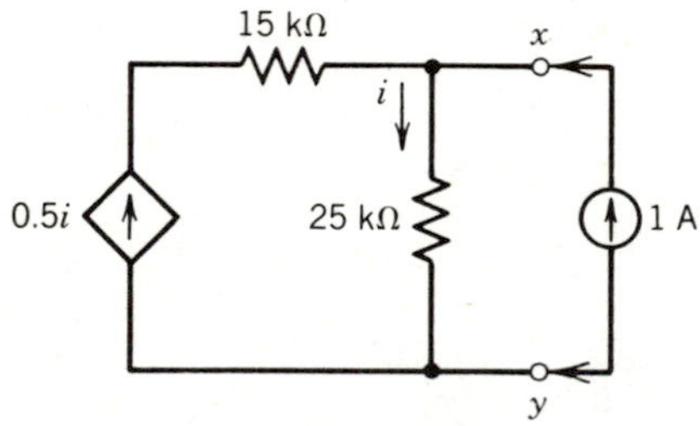

Figure 3.64

Applying KCL,

$$0.5i + 1 = i$$

or

$$i = 0.5 \text{ A}$$

so

$$V_{xy} = (25 \text{ k}\Omega)(0.5)$$
$$= 12.5 \text{ kV}$$

and

$$R_{\text{TH}} = \frac{12.5 \text{ kV}}{1 \text{ A}}$$
$$= 12.5 \text{ k}\Omega$$

The Thévenin equivalent circuit is given by Fig. 3.65.

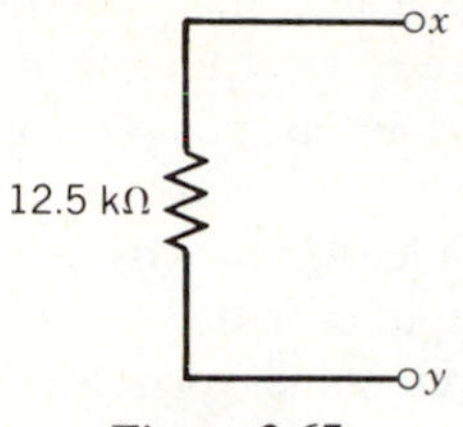

Figure 3.65

Thévenin's and Norton's theorems add new circuit analysis tools to your collection that will prove to be increasingly useful as more complex circuits are encountered in your academic and professional career.

LEARNING EVALUATIONS

1. Use Thévenin's theorem to find the current through the 6-kΩ resistor in Fig. 3.66.

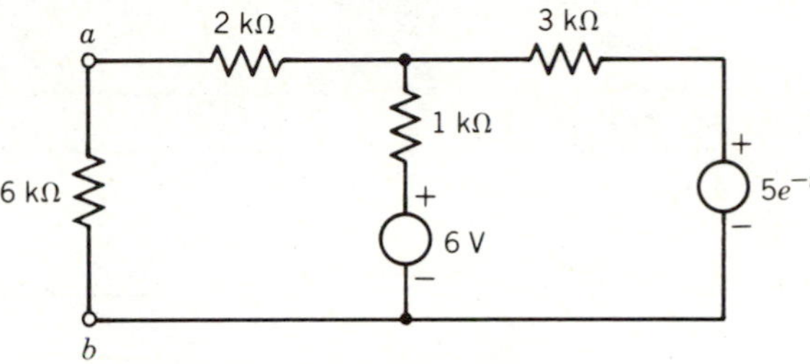

Figure 3.66

2. Use Norton's theorem to find the current through the 2-kΩ resistor in Fig. 3.66.
3. Use either Thévenin's or Norton's theorem to find the current in the 5-Ω resistor in Fig. 3.67.

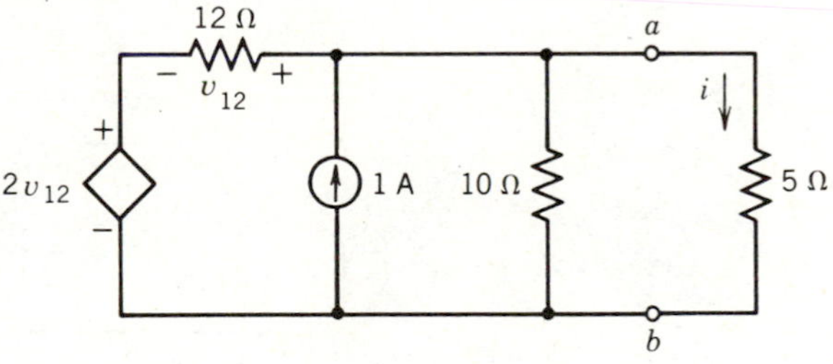

Figure 3.67

3.6 Operational Amplifiers

LEARNING OBJECTIVES

After completing this section you should be able to do the following:

1. Analyze circuits containing op amps.
2. Use op amps to design dependent sources.

To expand your appreciation for the usefulness of the concepts developed in this chapter and to whet your intellectual appetite with a foretaste of things to come in electrical engineering, we are introducing you to an active electronic circuit element in this section. We will see very shortly that circuits containing this element quickly reduce to types of circuits we can investigate with our current (no pun intended) knowledge.

An amplifier is defined as a device whose output is linearly proportional to its input over some predetermined range of operation. Usually the constant of proportionality is greater than 1. The ideal voltage amplifier shown in Fig. 3.68 is called an operational amplifier, or op amp. There are two input ports and one output port. (A port is a pair of terminals.) The input terminal marked "−" is the inverting terminal (meaning that the output is negatively proportional to the input) and the "+" terminal is the noninverting terminal. The input resistance seen between any two input terminals is very high, ideally infinite. This means that if the resistance is measured between terminals 1 and 2 or between 1 and ground, or between 2 and ground, the value will be high. In a practical op amp these resistances may be on the order of tens of megohms.

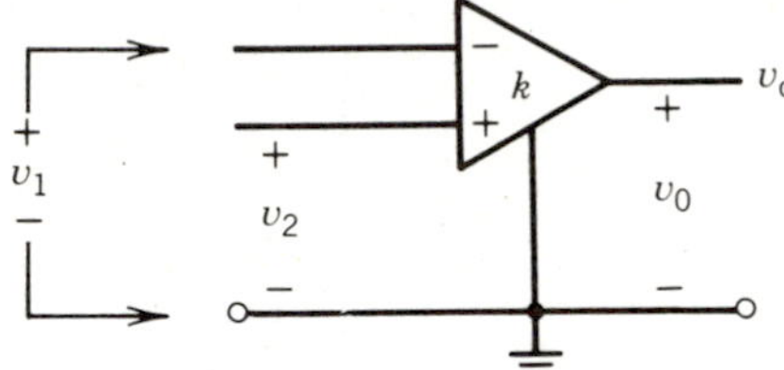

Figure 3.68. An operational amplifier.

The output voltage is proportional to the difference between v_2 and v_1. The proportionality constant k (the gain) is large, on the order of 10^5 or 10^8 in practical op amps. In our analysis we will assume an ideal amplifier with $k = \infty$.

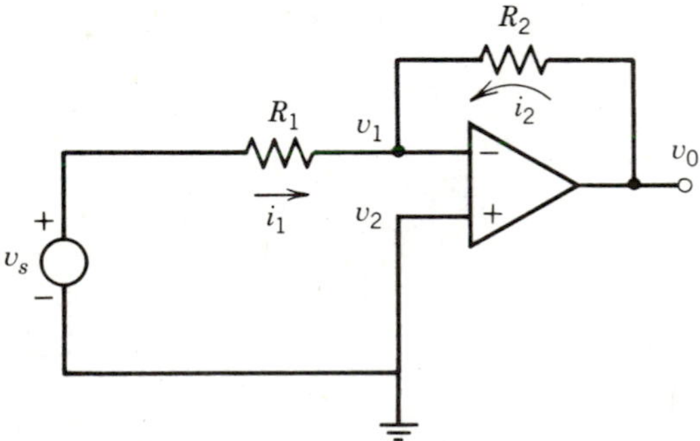

Figure 3.69. An inverting amplifier with gain $-R_2/R_1$.

A circuit using an op amp as shown in Fig. 3.69 is called a "feedback" circuit because the output is connected back to the input through the resistor R_2. The current entering the negative terminal of the amplifier is zero because of the infinite input resistance. Therefore,

$$i_1 + i_2 = 0$$

or

$$\frac{v_s - v_1}{R_1} + \frac{v_0 - v_1}{R_2} = 0 \tag{3.6.1}$$

Since

$$v_0 = k(v_2 - v_1) \tag{3.6.2}$$

and $v_2 = 0$, we have $v_1 = -v_0/k$. Substituting this relationship into Eq. 3.6.1 gives

$$\frac{v_s - (v_0/k)}{R_1} + \frac{v_0 - (v_0/k)}{R_2} = 0 \tag{3.6.3}$$

Typically, v_0 and v_s are on the order of 1 to 10 V, and k is 10^5 or more. Therefore, the v_0/k terms in Eq. 3.6.3 are negligible compared to v_s and v_0. This gives

$$v_0 \simeq -\frac{R_2}{R_1} v_s \tag{3.6.4}$$

This is an inverting amplifier, used for isolation, signal inversion, and amplification. The gain is $-R_2/R_1$, and therefore is adjustable.

Operational amplifiers as common circuit "building blocks" were made possible when solid-state technology allowed the manufacture of integrated circuits containing hundreds of transistors, resistors, and capacitors interconnected on one physical device commonly called a "chip." Normally the input differential $(v_2 - v_1)$ to an op amp must be small, on the order of 0.1 to 1.0 mV. These devices are among those most commonly used by electrical engineers. A discussion of a few applications follows.

The voltages v_1 and v_2 and their difference at the input terminals of the amplifier are approximately zero. Equation 3.6.2 illustrates this point, for v_0 is typically less than 10 V in magnitude, making v_1 less than 10^{-4} to 10^{-7}, depending on k. It will simplify our future analysis to assume $v_1 - v_2 = 0$.

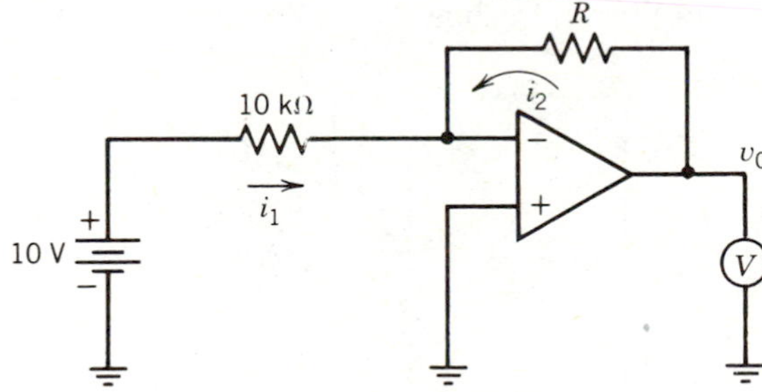

Figure 3.70. An ohmmeter.

The circuit shown in Fig. 3.70 may be used as an ohmmeter. The voltmeter reading will be proportional to the resistance value of R. Summing currents i_1 and i_2 gives

$$\frac{10}{10^4} + \frac{v_0}{R} = 0$$

or

$$v_0 = -10^{-3}R$$

To produce an upscale reading on the voltmeter the positive terminal should be connected to ground. When the voltmeter reads 15 V, the resistor $R = 15$ kΩ. If the voltmeter reads 8 V, $R = 8$ kΩ.

A summing circuit is shown in Fig. 3.71. Figure 3.71*b* shows a practical circuit for realizing the block diagram shown in Fig. 3.71*a*. To analyze this circuit we note that

$$i_1 + i_2 + i_f = 0$$

$$\frac{v_1}{R_1} + \frac{v_2}{R_1} + \frac{v_0}{R_2} = 0$$

or

$$v_0 = -(v_1 + v_2)\frac{R_2}{R_1} \tag{3.6.5}$$

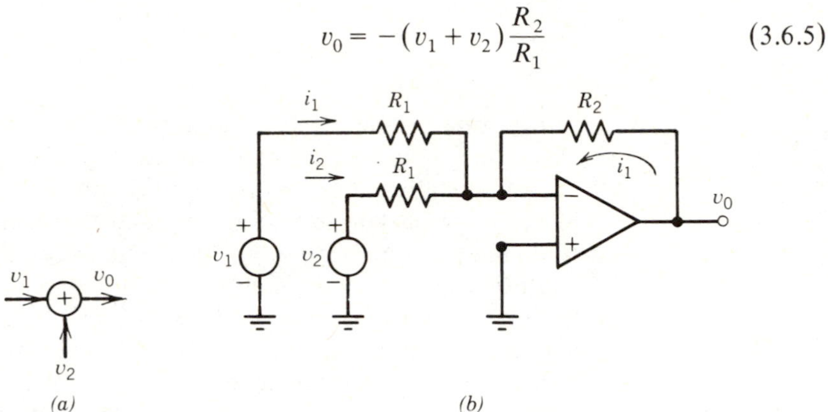

Figure 3.71. A voltage summing circuit.

The output voltage is the negative of the sum multiplied by R_2/R_1. In order to construct the circuit shown in Fig. 3.71*a* we need to have $R_1 = R_2$, and the circuit in Fig. 3.71*b* should be followed by an inverting amplifier (Fig. 3.69).

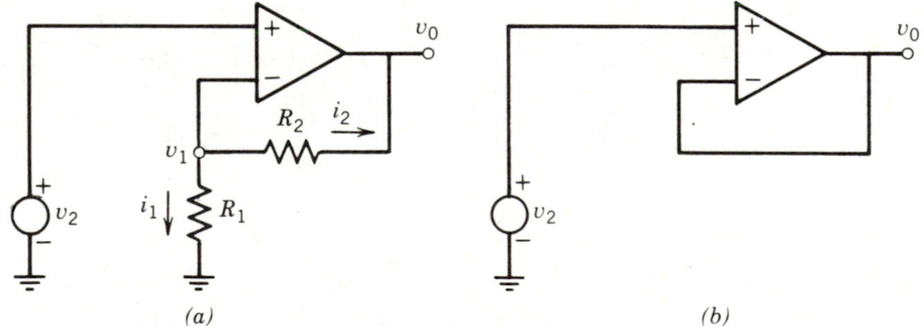

Figure 3.72. A voltage follower.

A voltage follower circuit is shown in Fig. 3.72*b*. To understand this circuit, first consider the circuit shown in Fig. 3.72*a*. With the current into the op amp equal to zero we have

$$i_1 + i_2 = 0$$

$$\frac{v_1}{R_1} + \frac{v_1 - v_0}{R_2} = 0$$

Since

$$v_2 - v_1 = 0$$

or

$$v_2 = v_1$$

$$\frac{v_2}{R_1} + \frac{v_2 - v_0}{R_2} = 0$$

or

$$v_0 = \left(1 + \frac{R_2}{R_1}\right) v_2 \tag{3.6.6}$$

If $R_2 = 0$ then R_1 becomes superfluous and the voltage follower circuit of Fig. 3.72*b* results. In this case

$$v_0 = v_2 \tag{3.6.7}$$

This configuration is used for isolation. To explain what this means, consider the simple circuit shown in Fig. 3.73*a*. If R_L is changed, then the voltage across and current through each resistor in the circuit is changed. On the other hand, changing R_L in Fig. 3.73*b* changes only the voltage across and current through R_L. Thus the circuit portion to the left of the op amp is isolated from the load resistor R_L.

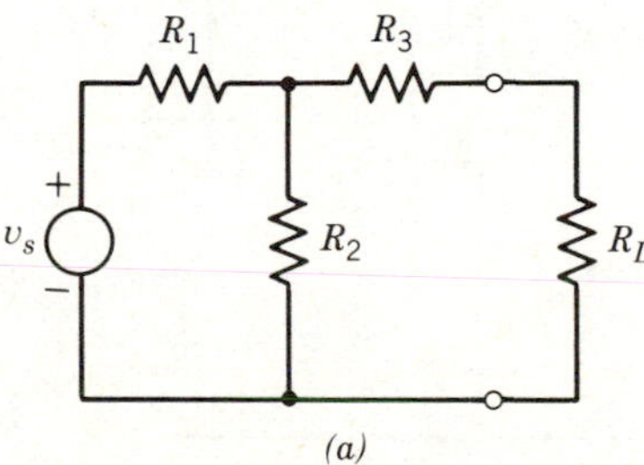

(*a*)

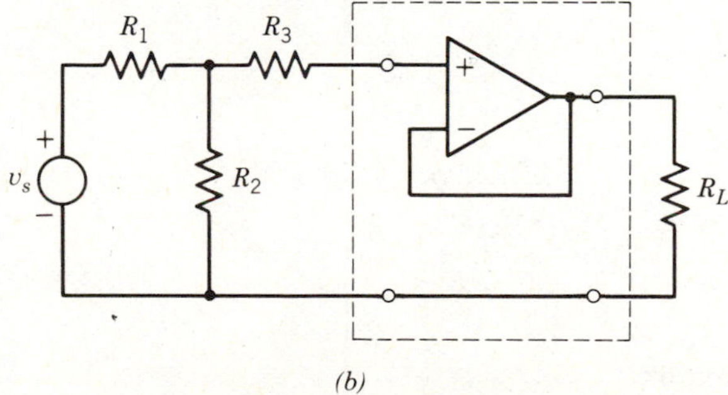

(*b*)

Figure 3.73. The op amp used for isolation.

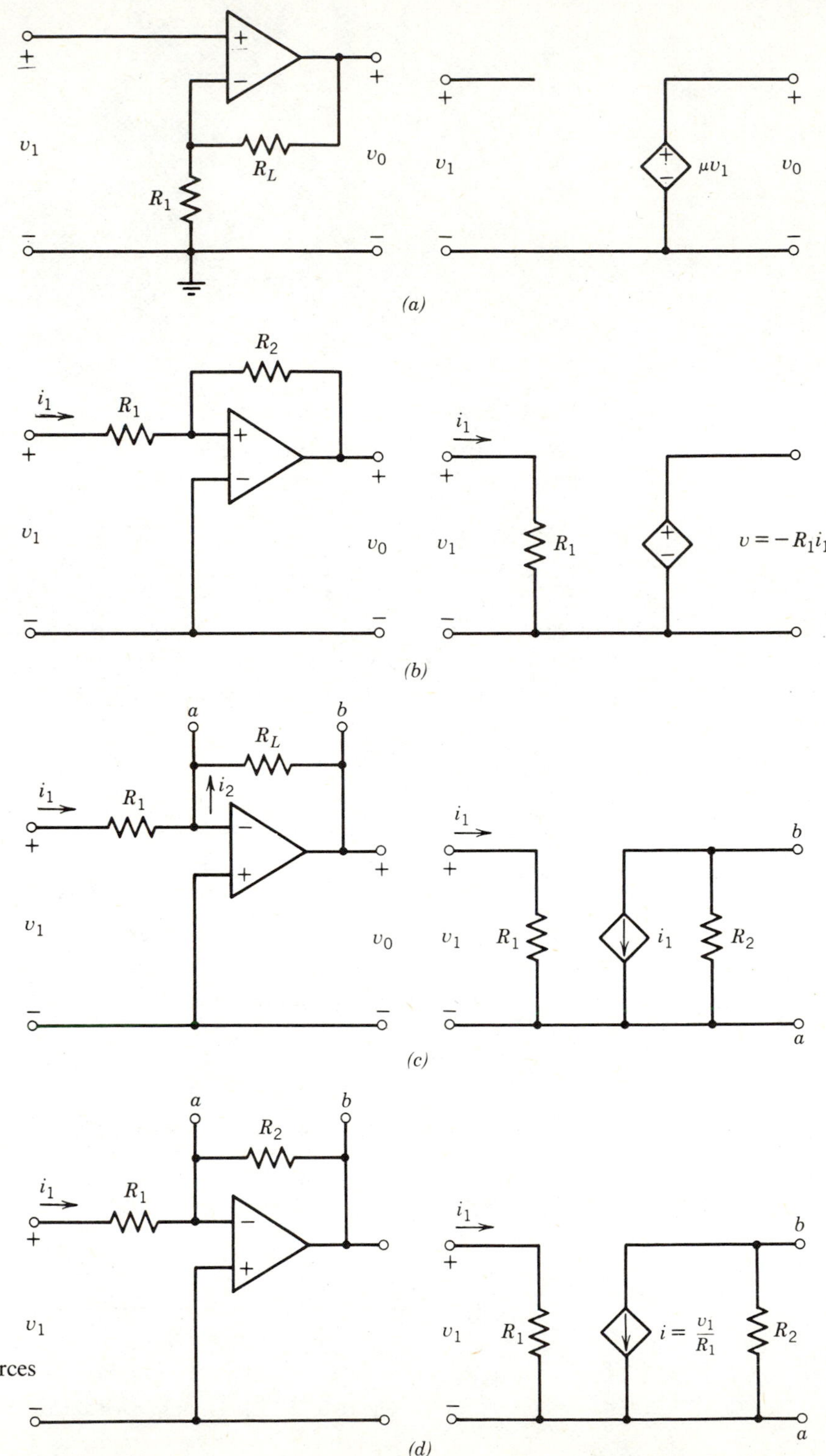

Figure 3.74. Controlled sources using op amps.
(*a*) VCVS. (*b*) ICVS.
(*c*) ICIS. (*d*) VCIS.

Note: All of the preceding analysis is independent of the orientation of the op amp. It does not matter whether the "+" or "−" terminals are connected as shown or if they are reversed. In a practical op amp circuit, however, the inverting terminal must always be connected to the output through the feedback circuit. The amplifier will be unstable otherwise, meaning that the output voltage will increase until it is limited by the available supply voltage. (This condition is called "saturation.")

There are four different varieties of dependent (controlled) sources, a voltage controlled voltage source (VCVS), a current controlled voltage source (ICVS), a voltage controlled current source (VCIS), and a current controlled current source (ICIS). Any of the four can be synthesized by using op amps, as shown in Fig. 3.74.

The VCVS in Fig. 3.74*a* has $\mu = 1 + R_2/R_1$, as our analysis of Fig. 3.74*a* indicates. In Fig. 3.74*b* the inverting amplifier is used to provide a ICVS. From the analysis of Fig. 3.69,

$$v_0 = -\frac{R_2}{R_1}v_1$$

but

$$v_1 = i_1 R_1$$

so

$$v_0 = -R_2 i_1$$

thus providing a voltage source that is controlled by i_1.

Dependent current sources are obtained from the circuit in Figs. 3.74*c* and 3.74*d*. Since the current into the op amp is zero, we have

$$i_1 = i_2 = \frac{v_1}{R_1}$$

This gives the current controlled current source of Fig. 3.74*c* with a gain of 1. This current is also equal to v_1/R_1, giving the controlled source of Fig. 3.74*d*.

LEARNING EVALUATION

The circuit shown in Fig. 3.75 is called a difference or differential amplifier. Show that the output voltage is proportional to $v_1 - v_2$. Find the proportionality constant.

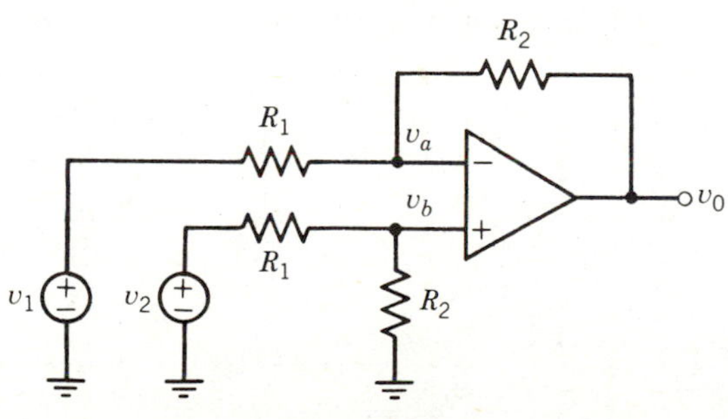

Figure 3.75

3.7 A Beginner's Tool Bag

Including all the circuit analysis techniques listed at the end of Chapter 2 and the methods of mesh analysis, nodal analysis, superposition, and source transformation, as well as Thévenin's and Norton's theorems discussed in this chapter, you have all the tools necessary to perform a complete analysis on any linear circuit containing dependent and independent current and voltage sources and resistors. As we introduce additional circuit elements throughout the remainder of the text these tools will still be applicable and will be used.

PROBLEMS

SECTION 3.1

1. Find the current through the 8-Ω resistor in Fig. 3.76 by superposition.

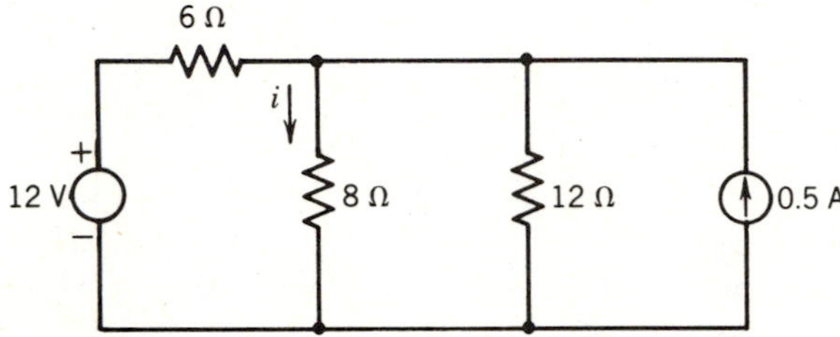

Figure 3.76

2. Find the voltage across the 100-Ω resistor in Fig. 3.77 by superposition.

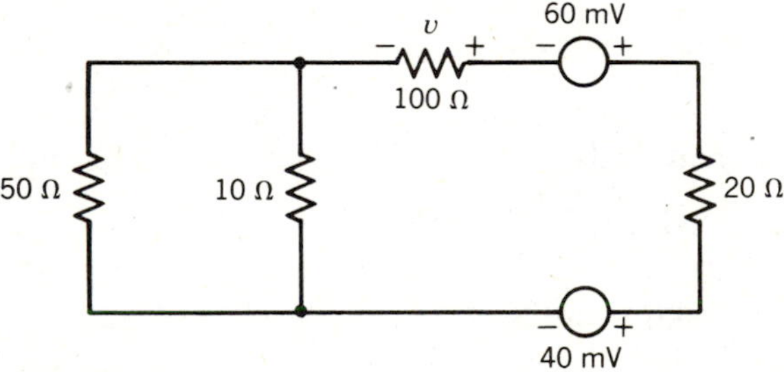

Figure 3.77

3. Use superposition to find i in Fig. 3.78.

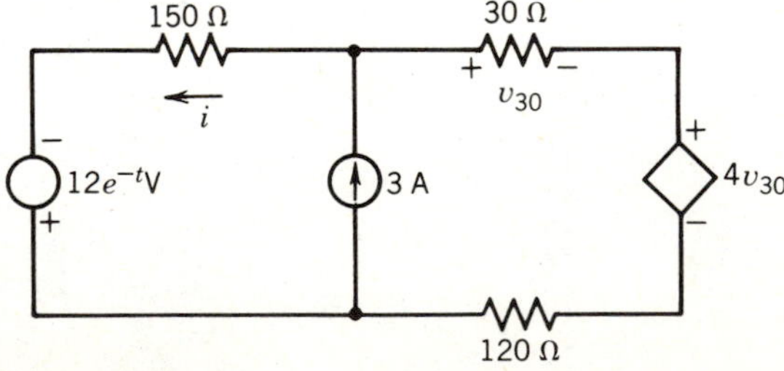

Figure 3.78

4. Use superposition to find i in Fig. 3.79.

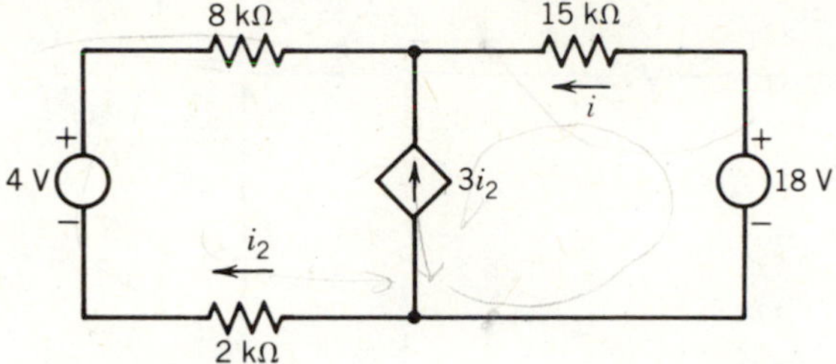

Figure 3.79

SECTION 3.2

5. Use source conversion to find the current through the 5-Ω resistor in Fig. 3.80.

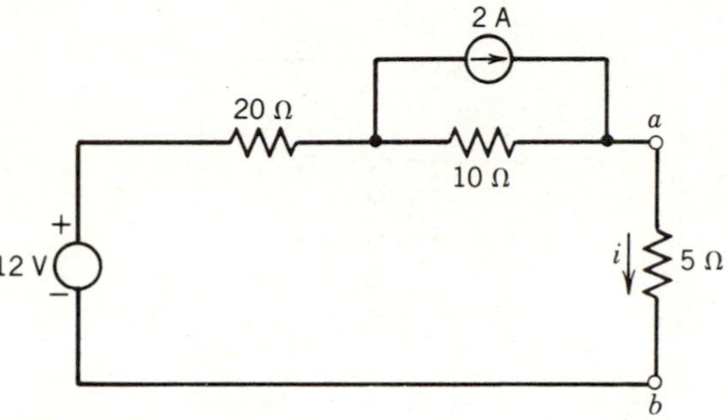

Figure 3.80

6. Use source conversion to find the voltage across the 12-Ω resistor in Fig. 3.81.

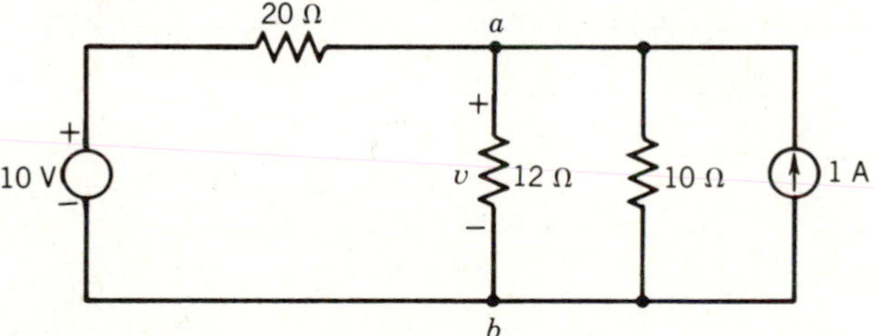

Figure 3.81

SECTION 3.3

7. Find node voltages v_1 and v_2 in Fig. 3.82.

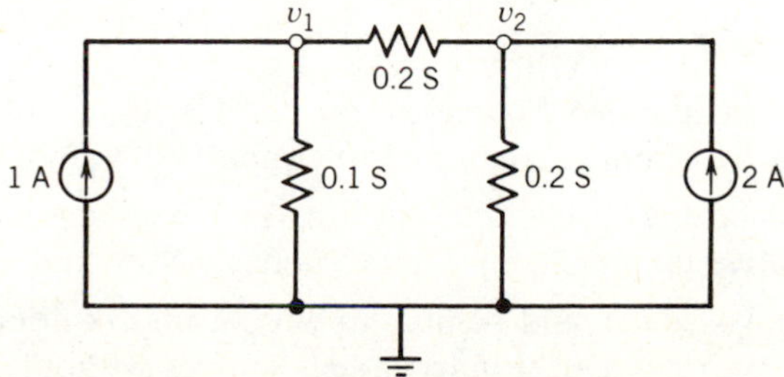

Figure 3.82

8. Make any necessary source transformations and find v_1 and v_2 in Fig. 3.83.

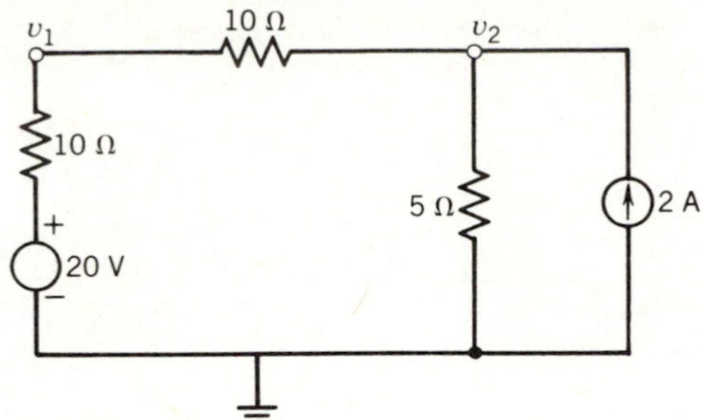

Figure 3.83

9. Set up the matrix nodal equations and solve for v_3 in Fig. 3.84.

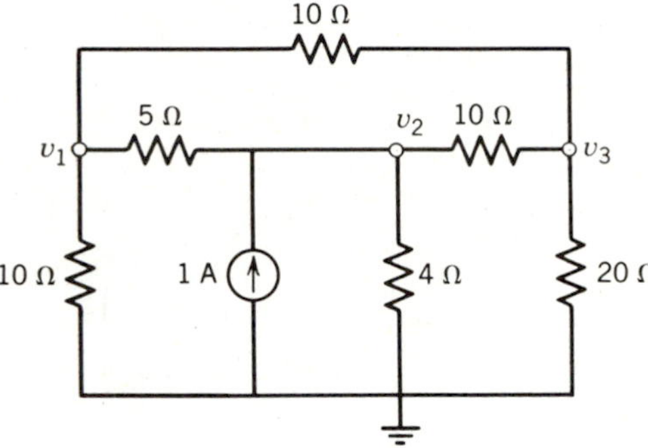

Figure 3.84

10. Set up the matrix nodal equations and solve for v_3 in Fig. 3.85.

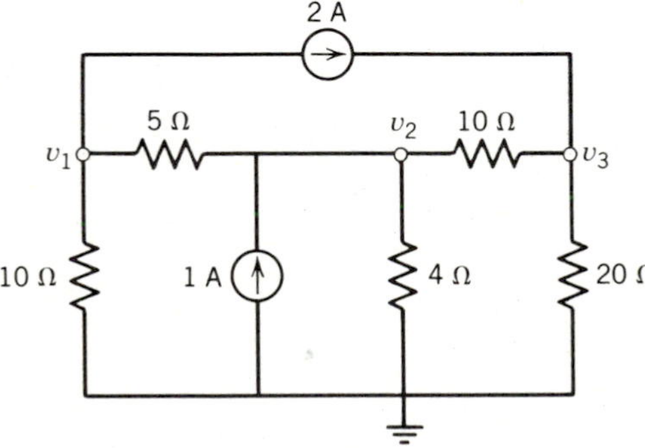

Figure 3.85

11. Replace the 1-A source in Fig. 3.84 with an ideal 10-V source with the negative terminal connected to ground. Now find v_3.

12. Replace the 2-A source in Fig. 3.85 with an ideal 10-V source with the negative terminal connected to node 1. Now find v_3.

13. By now you should be able to write a matrix nodal equation for any circuit containing practical independent sources without assuming currents, writing KCL equations, and then substituting Ohm's law. That is, you should be able to write the matrix equations directly from the circuit. Demonstrate your mastery of this technique by writing directly the single nodal equation to solve for v in Fig. 3.86.

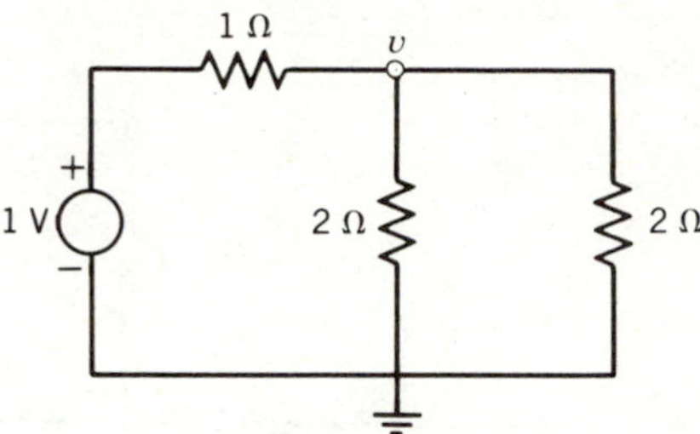

Figure 3.86

14. Use nodal analysis to find v_3 in Fig. 3.87.

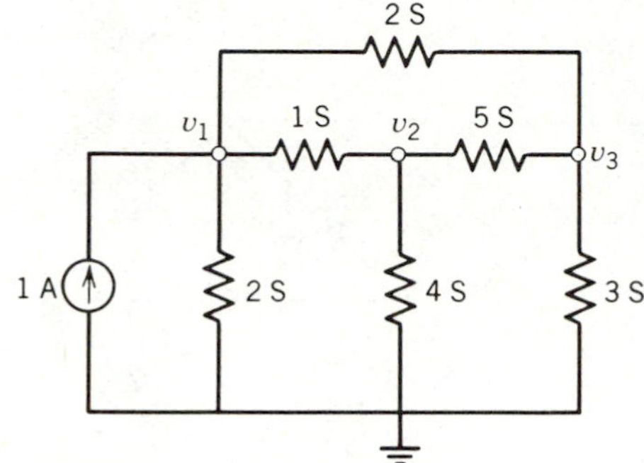

Figure 3.87

15. Use nodal analysis to find i in Fig. 3.88.

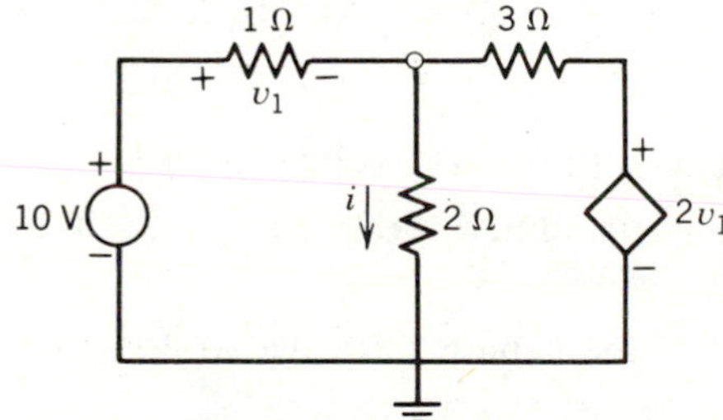

Figure 3.88

16. Use nodal analysis to find i_1 in Fig. 3.89.

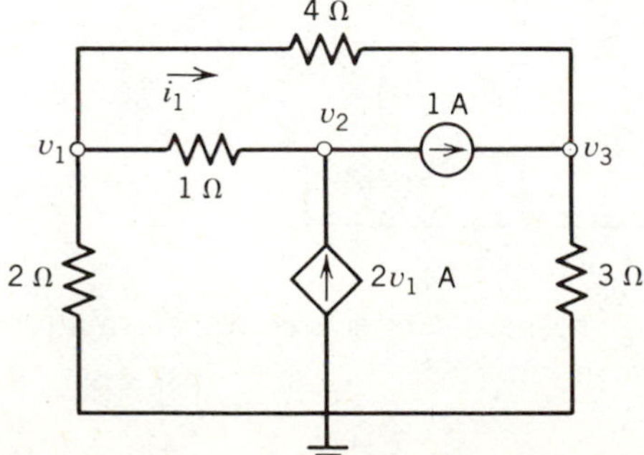

Figure 3.89

17. A typical h-parameter model for a common emitter transistor is shown inside the dashed lines in Fig. 3.90. If $v_1 = 1$ V and $R_L = 10$ kΩ, find v_2 using nodal analysis.

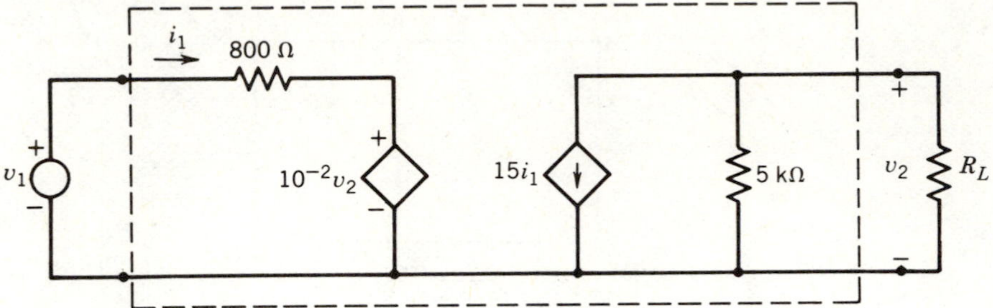

Figure 3.90

18. Use nodal analysis to find the voltage from a to d in Fig. 3.91.

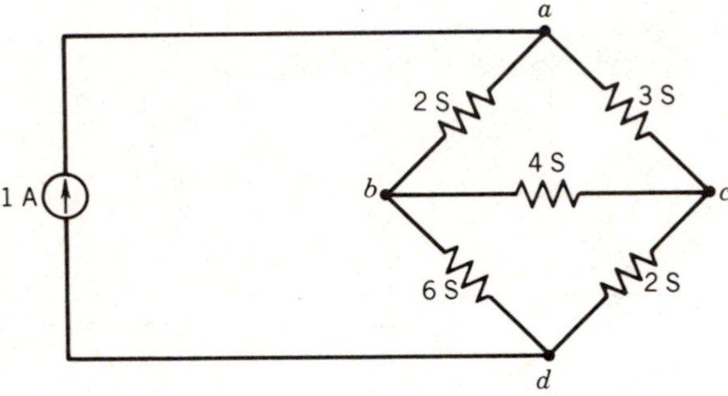

Figure 3.91

SECTION 3.4

19. Use mesh analysis to find the voltage across the 12-Ω resistor in Fig. 3.81.

20. Use source conversion and mesh analysis to find the voltage across the 20-Ω resistor in Fig. 3.84.

21. Write and solve the equations in matrix form for the mesh currents in Fig. 3.92.

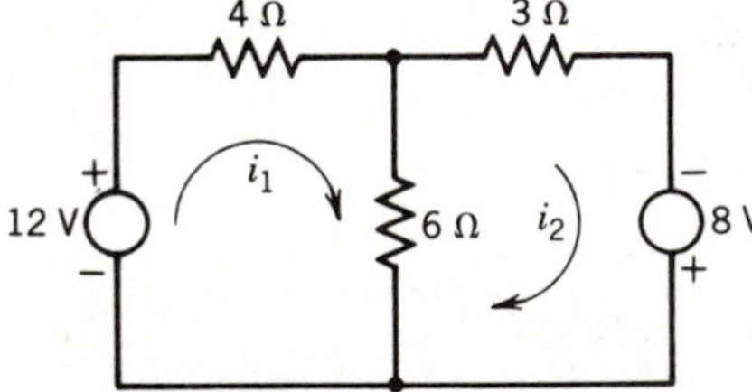

Figure 3.92

22. After making the necessary source transformations, write and solve for the mesh currents in Fig. 3.93.

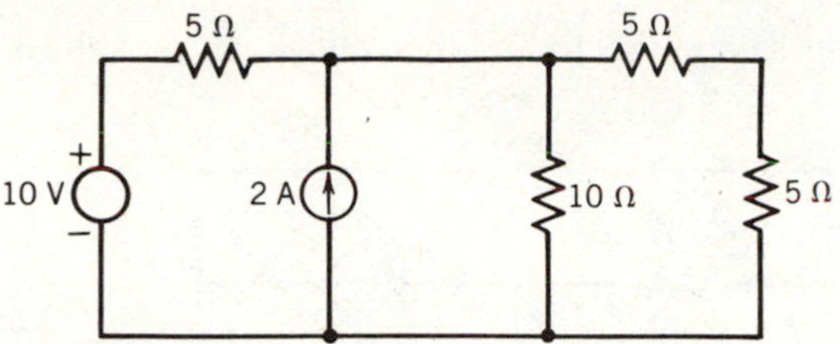

Figure 3.93

23. Set up the mesh equations in matrix form and solve for the voltage across the 12-Ω resistor in Fig. 3.94.

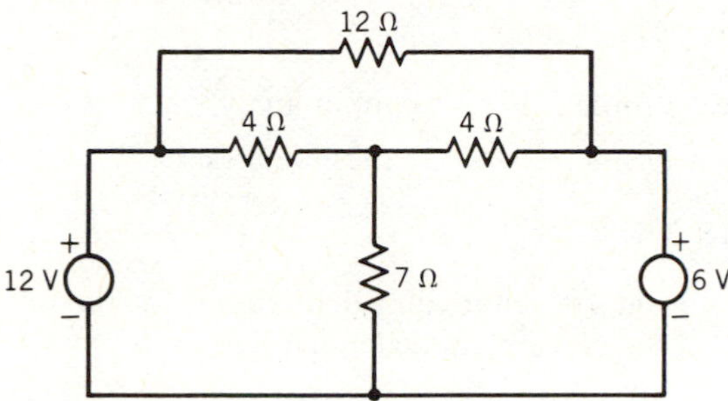

Figure 3.94

24. Write mesh equations and solve for the voltage across the 1-Ω resistor in Fig. 3.95.

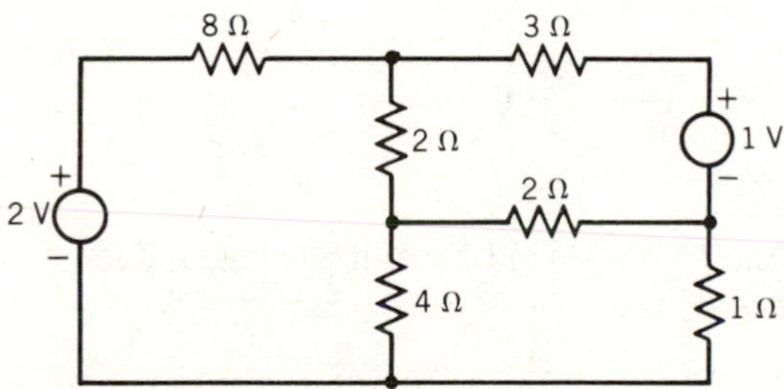

Figure 3.95

25. Use mesh analysis to find v_3 in Fig. 3.96.

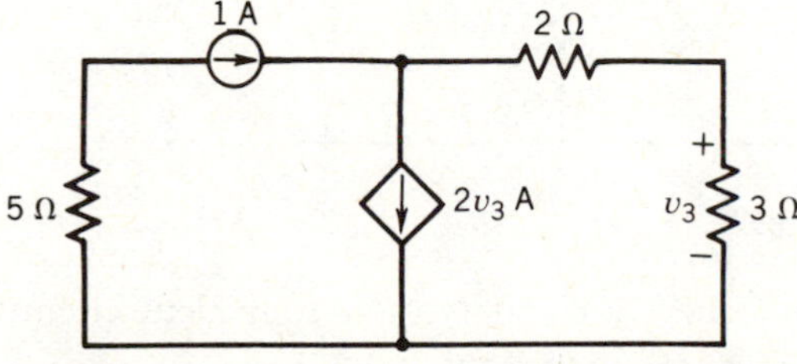

Figure 3.96

26. Use mesh analysis to find the voltage v across the 3-A source in Fig. 3.97.

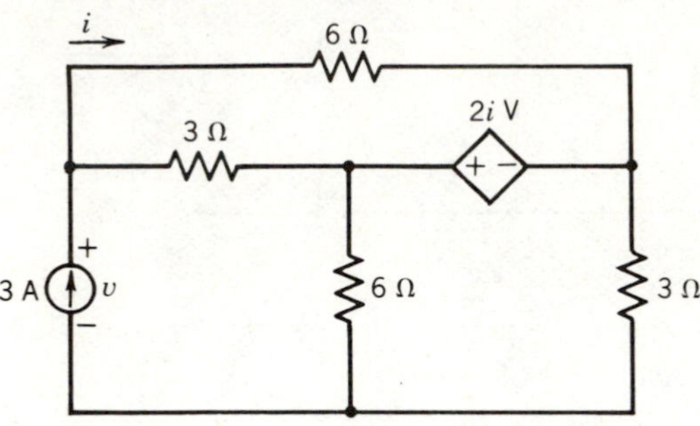

Figure 3.97

27. Repeat Problem 17 using mesh analysis.

28. Repeat Problem 18 using mesh analysis.

SECTION 3.5

29. **a.** Find the Thévenin equivalent circuit for the network in Fig. 3.98.
b. Find the Norton equivalent circuit.

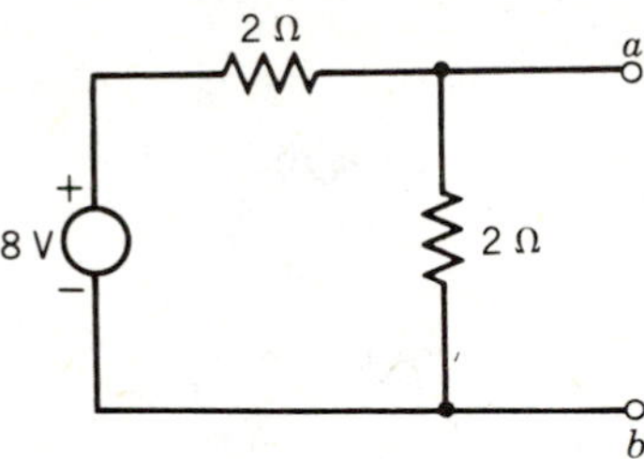

Figure 3.98

30. Find the Thévenin and Norton equivalent circuits for the network in Fig. 3.99 at the terminals a–b.

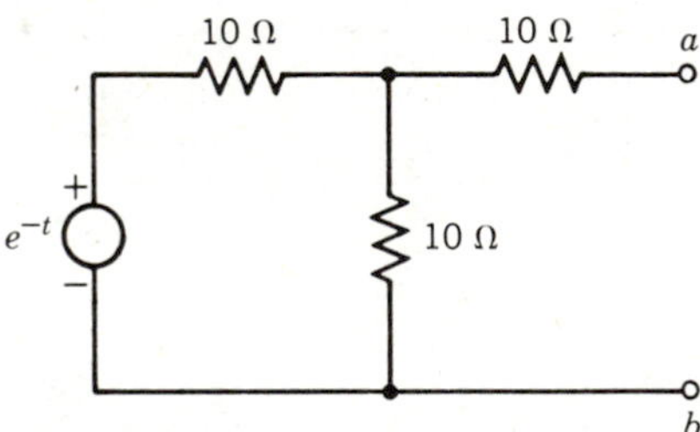

Figure 3.99

31. Find the Thévenin and Norton equivalent circuits for the network in Fig. 3.100 at the terminals a–b.

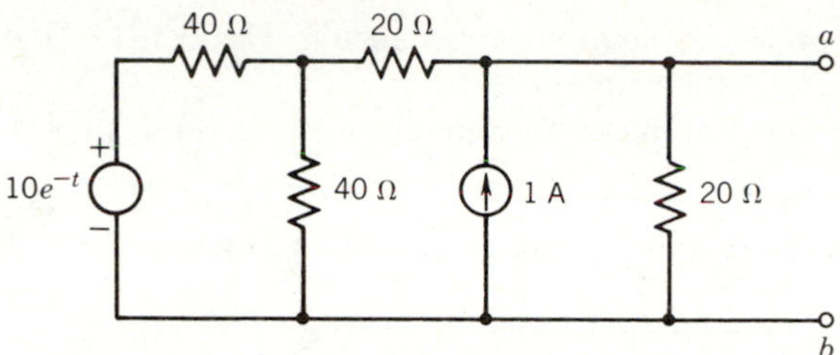

Figure 3.100

32. For the circuit in Fig. 3.80 find the current i by using Thévenin's theorem. To do this, remove the 5-Ω resistor and find Thévenin's equivalent circuit between terminals a–b. Then replace the 5-Ω resistor and solve for current i.

33. Use Norton's theorem to find the voltage across the 12-Ω resistor in Fig. 3.81. To do this, remove the 12-Ω resistor and find Norton's equivalent circuit between terminals a–b. Then replace the 12-Ω resistor and solve for the voltage v.

34. Use Thévenin's theorem to find the voltage across the 20-Ω resistor in Fig. 3.84.

35. Use Norton's theorem to find the voltage across the 20-Ω resistor in Fig. 3.85.

36. Use Thévenin's theorem to find the current through the 2-Ω resistor in Fig. 3.88.

37. Use Norton's theorem to find i_1 in Fig. 3.89.

38. Use Thévenin's theorem to solve Problem 17.

SECTION 3.6

39. Find Thévenin's equivalent circuit for the inverting amplifier of Fig. 3.69.

40. Find Thévenin's equivalent circuit for the voltage follower circuit in Fig. 3.72a.

41. Find the voltage gain v_0/v_1 for the circuit in Fig. 3.101.

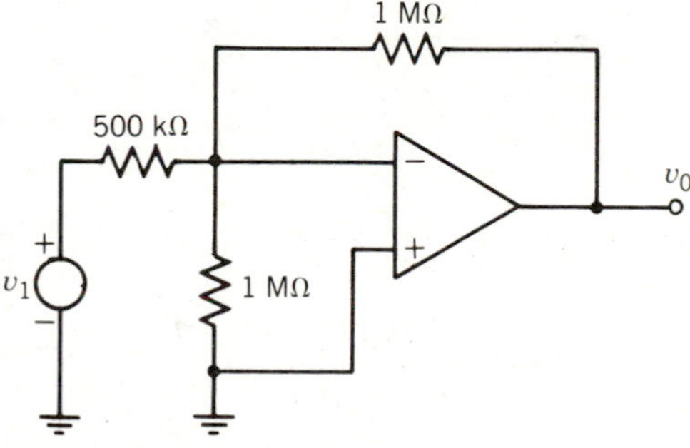

Figure 3.101

42. **a.** What is the input impedance in Fig. 3.101? That is, what is the resistance seen by the source v_1?
b. What is the output impedance (Thévenin's resistance)?

43. Find the ratio v_0/i_s in Fig. 3.102.

44. Find the source resistance and output resistance in Fig. 3.102.

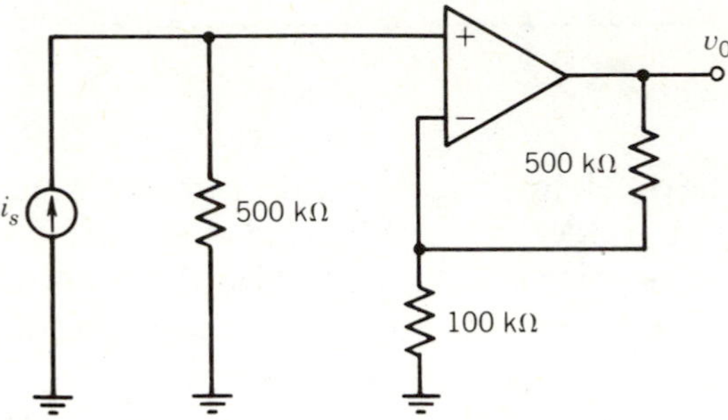

Figure 3.102

LEARNING EVALUATION ANSWERS

Section 3.1

a. $i = -2$ A
b. $i = 3.33$ mA

Section 3.2

1. See Fig. 3.103.

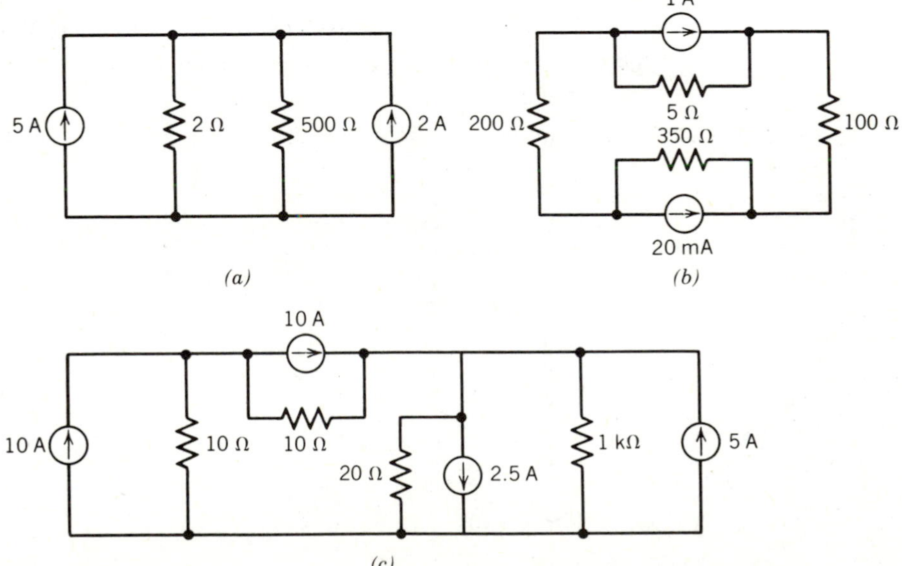

Figure 3.103

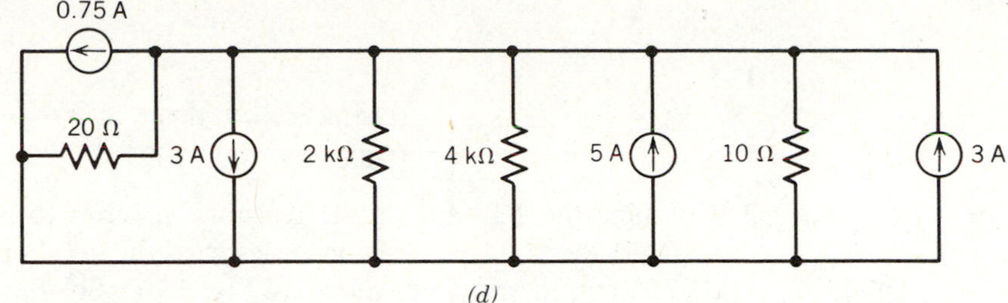

(d)

Figure 3.103. (*Continued*)

2. See Fig. 3.104.

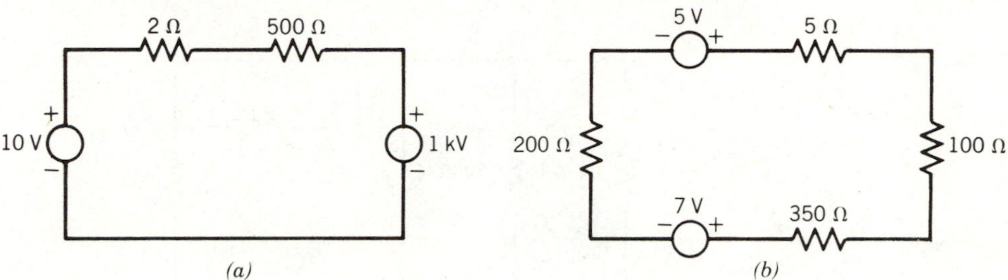

(a) *(b)*

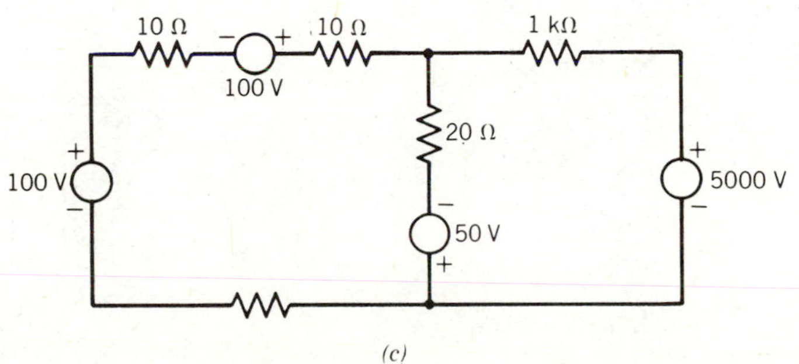

(c)

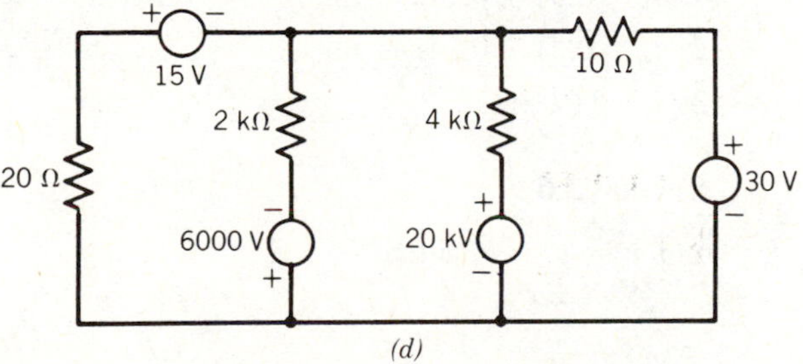

(d)

Figure 3.104

Section 3.3

1.

$$\begin{bmatrix} 0.01083 & -0.0083 \\ -0.0083 & 0.01033 \end{bmatrix} \begin{bmatrix} v_1 \\ v_2 \end{bmatrix} = \begin{bmatrix} 0.4 \\ 1.2 \end{bmatrix}$$

2. Combine the 260- and 180-Ω resistors in series to obtain one 440-Ω resistor. Convert the voltage source in series with the 200-Ω resistor into an equivalent 25-mA current source as shown in Fig. 3.105. Solving for v_2 gives

$$v_2 = -1.198 \text{ V}$$

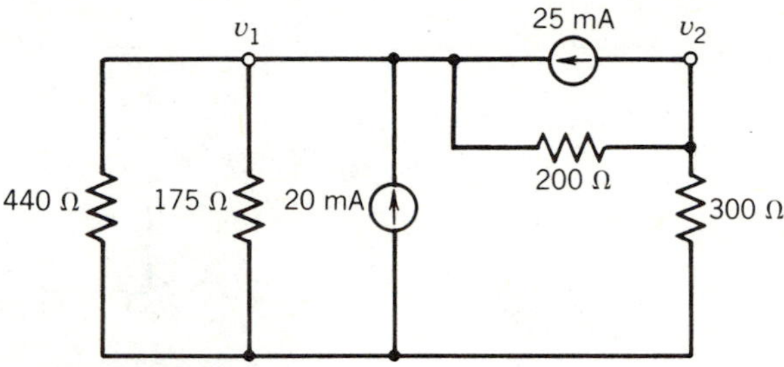

Figure 3.105

Section 3.4

1.

$$\begin{bmatrix} 590 & -250 \\ -250 & 375 \end{bmatrix} \begin{bmatrix} i_1 \\ i_2 \end{bmatrix} = \begin{bmatrix} 8 \\ -4 \end{bmatrix}$$

2. $i_{20} = -1.383$ A

3. $i_2 = -0.6$ A

Section 3.5

1. $i_6 = 0.51 + 0.14e^{-t}$ mA

2. $i_6 = 0.51 + 0.14e^{-t}$ mA

3. $i = 0.61$ A

Section 3.6

Let $v_a = v_b = v$. Then

$$\frac{v_1 - v}{R_1} + \frac{v_0 - v}{R_2} = 0$$

or

$$(v_1 - v)R_2 + (v_0 - v)R_1 = 0 \tag{3.6.8}$$

Also,

$$\frac{v_2 - v}{R_1} - \frac{v}{R_2} = 0$$

or

$$(v_2 - v)R_2 - vR_1 = 0 \tag{3.6.9}$$

Subtracting Eq. 3.6.9 from 3.6.8 gives

$$(v_1 - v_2)R_2 + v_0 R_1 = 0$$

or

$$v_0 = \frac{R_2}{R_1}(v_1 - v_2)$$

CHAPTER

4 ENERGY STORAGE ELEMENTS

We have limited the circuits in our studies so far by including only one circuit element besides energy sources—the resistor. As we have seen, the resistor dissipates energy but does not have the capacity to store energy. The resistor has the desirable feature that the current through it is algebraically related to the voltage across it. In this chapter two additional circuit elements are introduced—the capacitor and the inductor. Each of these has the capacity to store energy but the voltage–current relationship is more complicated and requires the use of differential and integral calculus for the first time in this text.

Also in this chapter some additional waveforms for energy sources are introduced and explained. This allows us to broadly expand the range and complexity of electrical circuits available for analysis.

4.1 The Capacitor

LEARNING OBJECTIVES

After completing this section you should be able to do the following:

1. Derive the voltage–current relationship for a capacitor.
2. Find the current $i(t)$ for given $v(t)$ across a capacitor.
3. Find the voltage $v(t)$ for given $i(t)$ through a capacitor.
4. Calculate the power and energy in a capacitor for given current or voltage waveforms.
5. Find the equivalent capacitance for a given network of capacitors.

The classical picture of a capacitor is a pair of conducting plates separated by an insulating material called a dielectric. Here we will only include a general discussion of the operation of a capacitor and will not derive the electromagnetic equations governing this operation. Figure 4.1 shows two parallel conducting plates of surface area A separated from each other by a distance d and connected to a battery of voltage V.

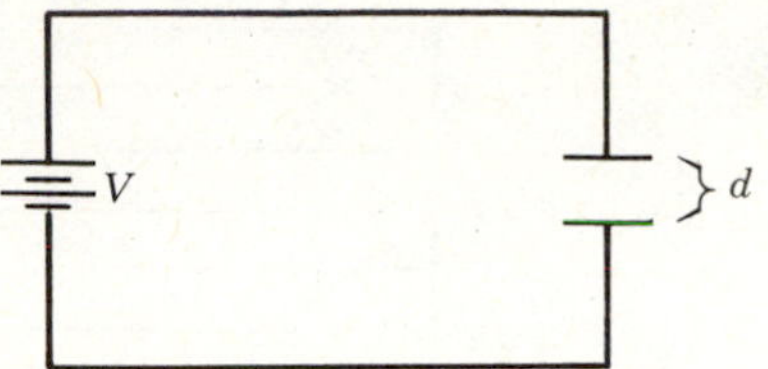

Figure 4.1. Model for a simple parallel plate capacitor.

There can be no current flow, but a positive charge ($+Q$) will exist on the upper plate and a negative charge ($-Q$) on the lower plate due to the voltage from the battery. The electrical characteristics of this capacitor are determined by its "capacitance" which is defined as:

> The *capacitance* of a capacitor is the ratio of the magnitude of the charge on either plate to the voltage difference between the plates, assuming that the conductors have charges equal in magnitude but opposite in polarity.

Therefore capacitance, represented by the symbol C, can be expressed as

$$C = \frac{Q}{V}$$

Q and V are shown here as constants. In general, they will be functions of time. The unit of capacitance is the farad, abbreviated F, and is named after Michael Faraday (1791–1867), an English physicist and chemist. Most capacitors have a capacitance of only a small fraction of a farad and are usually expressed in terms of microfarads (10^{-6} F). Capacitance is a function of the surface area, A, separation, d, and the permittivity, ε, of the dielectric. For the parallel plate capacitor in Fig. 4.1,

$$C = \frac{\varepsilon A}{d}$$

For a dielectric of air,

$$\varepsilon \approx \frac{1}{36\pi} nF/m$$

Capacitors may come in shapes other than that shown in Fig. 4.1 and the relation $Q = CV$ still holds but the expression for C in terms of the physical dimensions of the device will change.

A simple but very useful example of a capacitor is that of a radio tuner. It consists of a set of plates as shown in Fig. 4.2. As you move the tuning knob on your radio you are moving one set of plates into or out of the other set, thereby altering the surface area A, and hence the capacitance, C.

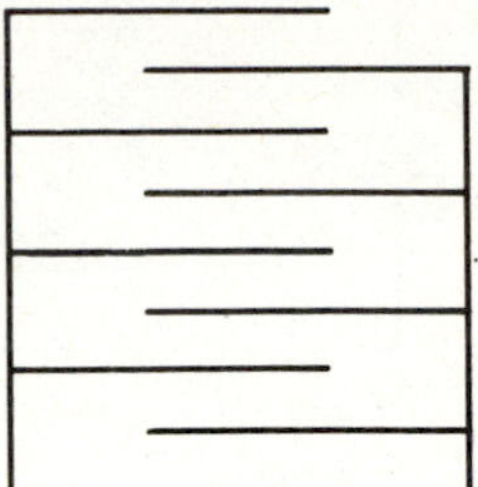

Figure 4.2. Multiplate capacitor used as a tuner.

We are interested in the voltage–current relationship across a capacitor. The circuit model for a capacitor is pictured in Fig. 4.3. We showed in Eq. 1.3.5 that current is the time rate of change of charge or

$$i = \frac{dq}{dt}$$

But since

$$q = Cv$$

we have

$$\boxed{i(t) = C\frac{dv}{dt}} \tag{4.1.1}$$

or

$$v(t) = \frac{1}{C}\int_{-\infty}^{t} i(\lambda)\, d\lambda \tag{4.1.2a}$$

A useful alternate form of Eq. 4.1.2a is

$$v(t) = \frac{1}{C}\int_{0}^{t} i(\lambda)\, d\lambda + v(0) \tag{4.1.2b}$$

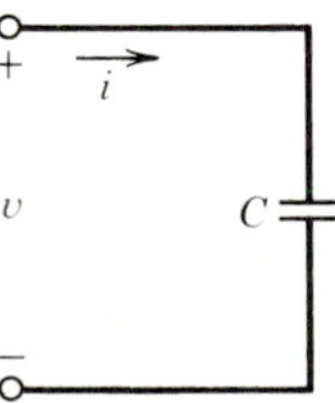

Figure 4.3. The capacitor.

Equations 4.1.1 and 4.1.2 deserve our attention, for they are our first encounter with *operators*. Mathematically, an operator is a function whose domain and range are functions in their own right. This concept is pictured in Fig. 4.4 as a box with input $x(t)$ and output $y(t)$. The derivative operator of Eq. 4.1.1 has input $v(t)$ and output $i(t)$. The integral operator of Eq. 4.1.2 has $i(t)$ as the box input, and $v(t)$ as the output. This box is often called a "black box," indicating that we do not know what is actually contained in the box and do not really care as long as we can completely determine its effect on the input, $x(t)$. Here are some examples that illustrate these concepts.

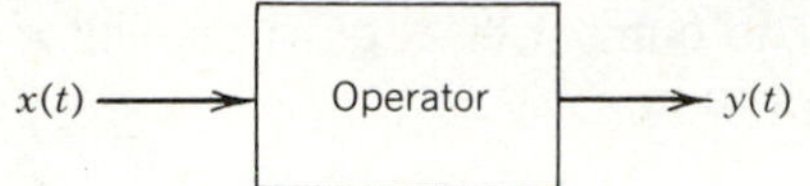

Figure 4.4. An operator is a black box whose input and output are both functions.

EXAMPLE 4.1.1 The voltage across a 100 μF capacitor is shown in Fig. 4.5 as a truncated ramp. Find the current $i(t)$.

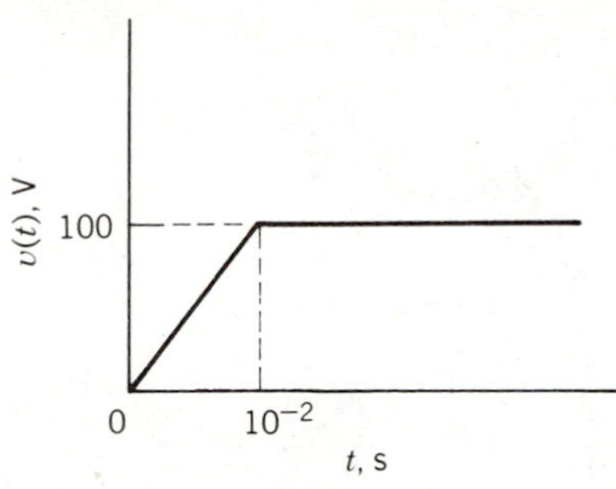

Figure 4.5

Solution

$$i(t) = C\frac{dv}{dt} \qquad \text{from Eq. 4.1.1}$$

We can express $v(t)$ analytically as

$$v(t) = \begin{cases} 10{,}000t & 0 \le t \le 10^{-2} \\ 100 & t \ge 10^{-2} \end{cases}$$

Therefore,

$$\frac{dv(t)}{dt} = \begin{cases} 10{,}000 & 0 \le t \le 10^{-2} \\ 0 & t \ge 10^{-2} \end{cases}$$

Since $C = 10^{-4}$ F, we have

$$i(t) = \begin{cases} 1 & 0 \le t \le 10^{-2} \\ 0 & t \ge 10^{-2} \end{cases}$$

$i(t)$ is shown in Fig. 4.6.

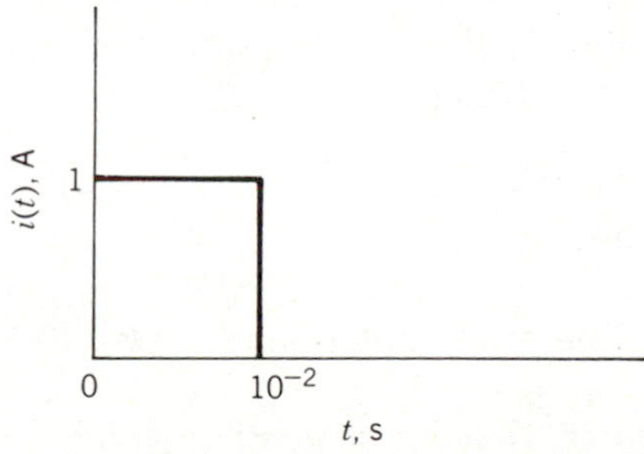

Figure 4.6

EXAMPLE 4.1.2 The current through a capacitor of 100 μF is shown in Fig. 4.7*a*. Find the voltage $v(t)$.

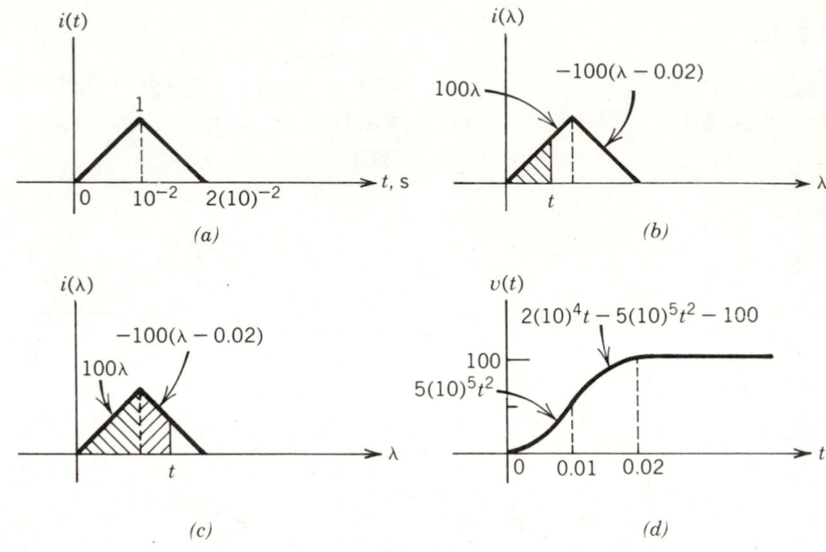

Figure 4.7

Solution

The first step in evaluating Eq. 4.1.2 is to change $i(t)$ to $i(\lambda)$, as shown in Fig. 4.7*b*. Next, a value of t is selected for the upper limit on the integral, and $v(t)$ is calculated. For $t < 0$ the area under $i(\lambda)$ is zero. For $0 < t < 0.01$, as shown in Fig. 4.7*b*, the value of the integral is the area depicted by the shaded region. That is,

$$v(t) = \frac{1}{C}\left(\int_{-\infty}^{0} 0\, d\lambda + \int_{0}^{t} 100\lambda\, d\lambda\right) = 5(10)^5 t^2, \qquad 0 < t < 0.01$$

For $0.01 < t < 0.02$ the area is the sum of the triangle area from zero to 0.01, plus the area under the curve from 0.01 to t.

$$v(t) = \frac{1}{C}\left(\int_{-\infty}^{0} 0\, d\lambda + \int_{0}^{0.01} 100\lambda\, d\lambda + \int_{0.01}^{t} -100(\lambda - 0.02)\, d\lambda\right)$$
$$= 2(10)^4 t - 5(10)^5 t^2 - 100, \qquad 0.01 < t < 0.02$$

Finally, for $t > 0.02$ the capacitor voltage is a constant 100 V. This is found by evaluating Eq. 4.1.2 as follows.

$$v(t) = \frac{1}{C}\left(\int_{-\infty}^{0} 0\, d\lambda + \int_{0}^{0.01} 100\lambda\, d\lambda + \int_{0.01}^{0.02} -100(\lambda - 0.02)\, d\lambda + \int_{0.02}^{t} 0\, d\lambda\right)$$
$$= 100 \text{ V}, \qquad t > 0.02$$

The entire solution is shown in Fig. 4.7*d*.

Let us make some observations about the voltage–current relationship in a capacitor.

1. The relationship between voltage and current in a capacitor is unique. For an initially uncharged capacitor, a given voltage waveform will produce a unique current waveform. This same current waveform will then produce the given voltage. Mathematically this condition is met if the current satisfies the condition $i(t)=0$ for $t<t_0$, where t_0 is *any* finite value of time. In other words, $i(-\infty)=0$.
2. We will use derivatives and integrals similar to those in Eqs. 4.1.1 and 4.1.2 throughout the remainder of this text. We will always assume that the function under the integral has zero value at $t=-\infty$. This will assure us that the relationship between the two functions is unique.
3. The voltage across a capacitor at time t_1 depends only on the historical value of the current through the capacitor until time t_1 and not on the present or future value of current beyond t_1.
4. The current in a capacitor is not a function of the magnitude of the voltage across the capacitor but depends only on the rate of change of that voltage.

Modern solid-state technology makes use of capacitance in the development of large scale and very large-scale integrated circuits. The semiconductor junction found in these circuits has a voltage–charge relationship expressed in terms of a "barrier capacitance," where

$$C_{\text{barrier}}=\frac{\frac{1}{2}(q_0/V_0)}{[1-(v/V_0)]^{1/2}} \qquad v\geq 0$$

This is seen to be a nonlinear relationship. This type of capacitor clearly does not have any physical relationship to that shown in Fig. 4.1. Therefore, we will define a capacitor as any electrical device in which the charge on the device is a function of the voltage across it. That is,

$$q=f(v)$$

The capacitance of the device is the derivative of q with respect to v. The capacitor shown in Fig. 4.1 is a special case of this where we have said that

$$q=Cv$$

This is particularly useful since the capacitance is linear.

Power and Energy

Instantaneous power and instantaneous energy are related by an integral–derivative relationship similar to that for voltage and current in a capacitor. With $p(t)$ denoting power, and $w(t)$ denoting energy, this relationship is expressed by the following formulas.

$$p(t)=\frac{d}{dt}w(t) \tag{4.1.3}$$

$$w(t)=\int_{-\infty}^{t}p(\lambda)\,d\lambda \tag{4.1.4}$$

This instantaneous power supplied to a capacitor is given by

$$p(t) = v(t)i(t) = Cv(t)\frac{dv(t)}{dt} \text{ W} \tag{4.1.5}$$

where the last relationship is determined by using Eqs. 1.5.1 and 4.1.1. The product $v(t)(dv/dt)$ may be positive or negative, meaning that power may be received by the capacitor (positive power), or supplied by the capacitor (negative power).

Substituting Eq. 4.1.5 into 4.1.4 gives

$$w(t) = \int_{-\infty}^{t} Cv(\lambda)\frac{dv}{d\lambda}\,d\lambda = \int_{v(-\infty)}^{v(t)} Cv\,dv$$

$$= \frac{1}{2}Cv^2(t) - \frac{1}{2}Cv^2(-\infty)$$

By our previous agreement, $v(-\infty) = 0$, so

$$\boxed{w(t) = \tfrac{1}{2}Cv^2(t) \text{ J}} \tag{4.1.6}$$

EXAMPLE 4.1.3 The voltage supplied to a 10-μF capacitor is shown in Fig. 4.8. Find the current, power, and energy as a function of time.

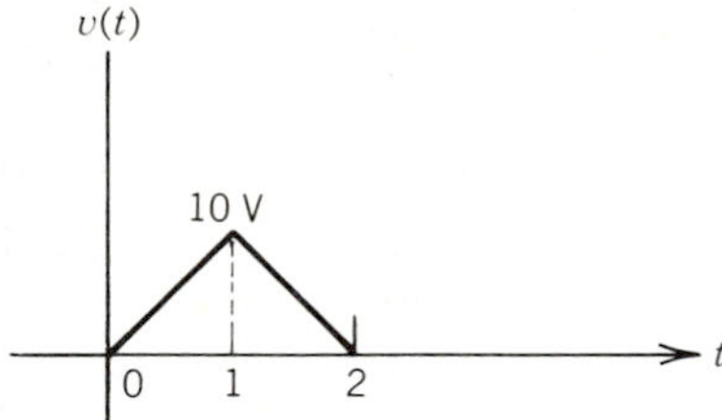

Figure 4.8

Solution

The current $i(t)$ is graphed in Fig. 4.9*a*, and is found by differentiating the given voltage according to Eq. 4.1.1. The power shown in Fig. 4.9*b* is found by multiplying $v(t)i(t)$. One purpose of this example is to illustrate that the energy $w(t)$ may be arrived at by either Eq. 4.1.4 or 4.1.6. First let us integrate according to Eq. 4.1.4.

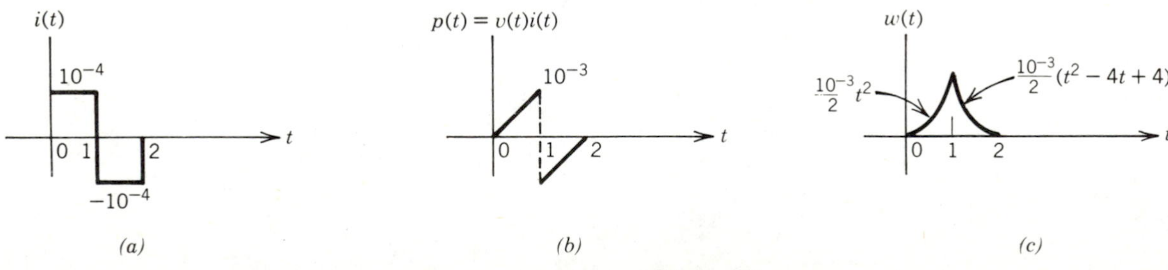

Figure 4.9

$$w(t) = \int_{-\infty}^{t} p(\lambda)\, d\lambda = 0, \qquad t < 0$$

$$= \int_0^t 10^{-3}\lambda\, d\lambda = \frac{10^{-3}}{2} t^2, \qquad 0 < t < 1$$

$$= \frac{10^{-3}}{2} + \int_1^t 10^{-3}(\lambda - 2)\, d\lambda = \frac{10^{-3}}{2}(t^2 - 4t + 4), \qquad 1 < t < 2$$

$$= 0, \qquad t > 2$$

Next, notice that $v^2(t)$ is given by

$$v^2(t) = 0, \qquad t < 0$$

$$= 100t^2, \qquad 0 < t < 1$$

$$= [-10(t-2)]^2, \qquad 1 < t < 2$$

$$= 0, \qquad t > 2$$

Multiplying by $C/2$ yields the same function for $w(t)$ as above. This function is graphed in Fig. 4.9*c*.

Equivalent Capacitance

Two capacitors in parallel, as shown in Fig. 4.10, may be replaced by a single capacitor whose value is given by

$$C_{eq} = C_1 + C_2 \tag{4.1.7}$$

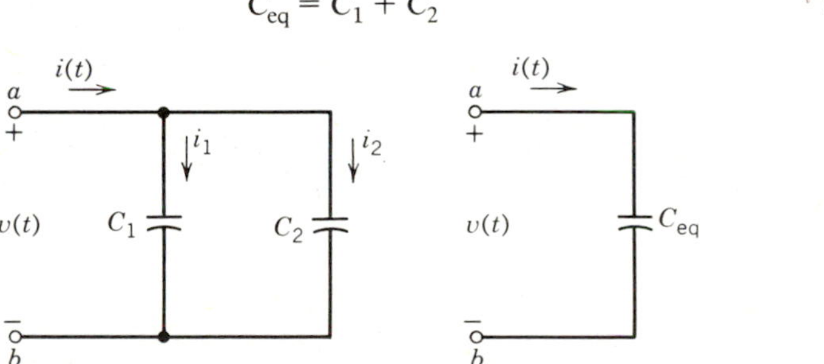

Figure 4.10. Combining capacitors in parallel.

This relationship may be derived by using the v–i characteristic for a capacitor.

$$i_1(t) = C_1 \frac{dv}{dt}$$

$$i_2(t) = C_2 \frac{dv}{dt}$$

By KCL,

$$i(t) = i_1(t) + i_2(t) = (C_1 + C_2)\frac{dv}{dt}$$

This relationship may be extended to any number of capacitors in parallel

$$\boxed{C_{eq} = C_1 + C_2 + \cdots + C_N} \tag{4.1.8}$$

The equivalent value for two series capacitors is given by

$$\frac{1}{C_{eq}} = \frac{1}{C_1} + \frac{1}{C_2} \tag{4.1.9a}$$

or

$$\boxed{C_{eq} = \frac{C_1 C_2}{C_1 + C_2}} \tag{4.1.9b}$$

To derive Eq. 4.1.9, note that the same current $i(t)$ flows through both C_1 and C_2 in Fig. 4.11, and $v(t) = v_1(t) + v_2(t)$. Hence $v(t)$ is given by

$$\begin{aligned} v(t) &= \frac{1}{C_1}\int_{-\infty}^{t} i(\lambda)\, d\lambda + \frac{1}{C_2}\int_{-\infty}^{t} i(\lambda)\, d\lambda \\ &= \left(\frac{1}{C_1} + \frac{1}{C_2}\right)\int_{-\infty}^{t} i(\lambda)\, d\lambda \\ &= \frac{1}{C_{eq}}\int_{-\infty}^{t} i(\lambda)\, d\lambda \end{aligned}$$

Equation 4.1.9a may be extended to n capacitors in series.

$$\boxed{\frac{1}{C} = \frac{1}{C_1} + \frac{1}{C_2} + \cdots + \frac{1}{C_n}} \tag{4.1.10}$$

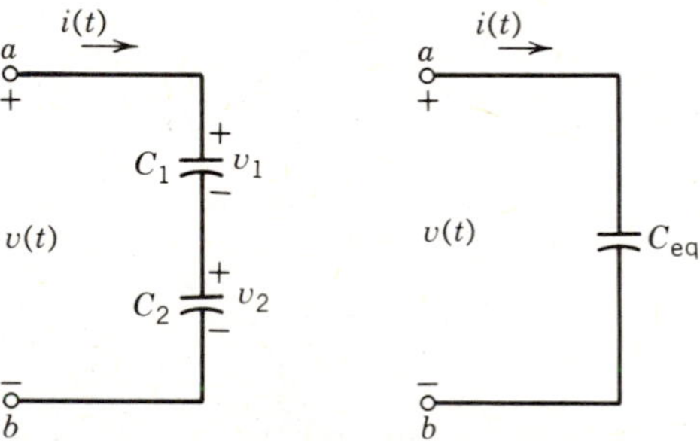

Figure 4.11. Combining capacitors in series.

EXAMPLE 4.1.4 Find the equivalent value of capacitance for the network shown in Fig. 4.12.

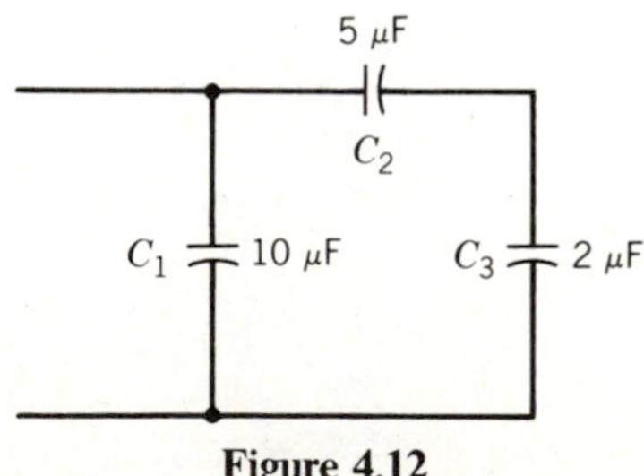

Figure 4.12

Solution

First, replace the C_2, C_3 series combination by

$$C_4 = \frac{C_2C_3}{C_2 + C_3} = \frac{10}{7}\ \mu\text{F}$$

Then

$$C_{eq} = C_1 + C_4 = \frac{80}{7}\ \mu\text{F}$$

LEARNING EVALUATIONS

1. The voltage $v(t)$ across a capacitance of value $C = 80\ \mu\text{F}$ is given by

$$v(t) = \begin{cases} 0 & t < 0 \\ 10 - 10e^{-t/2} & t > 0 \end{cases}$$

Find and plot the current $i(t)$

2. The current $i(t)$ through a capacitance of value $C = 80\ \mu\text{F}$ is plotted in Fig. 4.13. Find and plot the voltage across the capacitor.

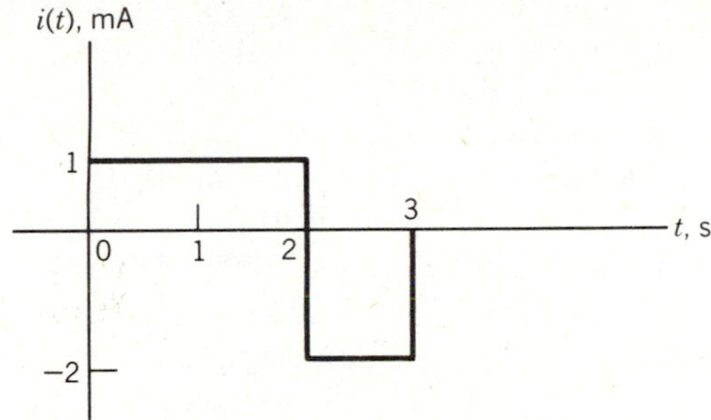

Figure 4.13. Current waveform.

3. Find and plot the power $p(t)$ delivered to the capacitor in Problem 1 above.
4. Find and plot the energy stored in the capacitor in Problem 1 above.
5. Find the equivalent capacitor C for the network in Fig. 4.14.

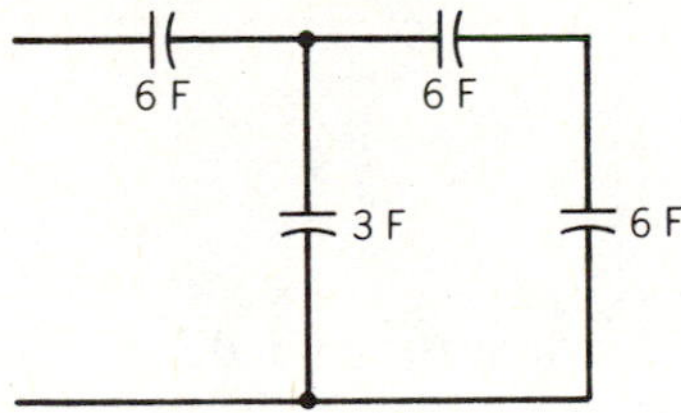

Figure 4.14.

4.2 Waveforms

LEARNING OBJECTIVES

After completing this section you should be able to do the following:

1. Write a mathematical expression for and sketch each of the following waveforms.
   a. A constant.

b. A step function.
c. A ramp function.
d. An impulse function.

2. Describe the relationship between the impulse, step, and ramp signals in terms of the derivative and integral operators.

3. Write equations for signals in terms of the impulse, step, and ramp signals.

In circuit analysis all voltages and currents can be considered functions of time. Up to this point most of the forcing functions used have been constant for all time just to simplify the mathematics involved. We were interested in your learning the circuit analysis concept and not in becoming engrossed (or lost) in the mathematical manipulations. A variety of voltage waveforms are shown in Fig. 4.15. In this figure (a) represents our familiar constant voltage source, such as that available from a battery; (b) represents the waveform of voltage found in the electrical outlet of a house; (c) is a typical waveform received by an AM radio; (d) represents a digital signal (only two values possible), such as might be found in a computer; (e) represents an input to a heart monitor instrument; and (f) represents a signal typical of the output from a sample and hold amplifier.

Waveforms fundamental to the study of electrical circuits include the constant, step, ramp, impulse, exponential, and sinusoidal functions. Nearly all waveforms encountered in circuit analysis can be expressed as combinations of these fundamental waveforms. Exponential waveforms will be investigated in the next chapter and sinusoidal waveforms will be discussed in Chapter 9. The remainder of these waveforms will be introduced here.

Let us begin by defining the unit step.

$$u(t) = \begin{cases} 1 & t \geq 0 \\ 0 & t < 0 \end{cases} \tag{4.2.1}$$

This function is pictured in Fig. 4.16a. If we substitute $t - t_0$ for t in Eq. 4.2.1, we get

$$u(t - t_0) = \begin{cases} 1 & t - t_0 \geq 0 \quad \text{or} \quad t \geq t_0 \\ 0 & t - t_0 < 0 \quad \text{or} \quad t < t_0 \end{cases} \tag{4.2.2}$$

This translated step function is pictured in Fig. 4.16b. There are two important features of the notation involving the argument $(t - t_0)$. First, the function changes value when the argument is zero. That is, $u(t - t_0)$ changes value when $t - t_0 = 0$, or $t = t_0$. Second, the function is zero when the argument is negative, and it is nonzero when the argument is positive. To illustrate this second feature the function $u(-t)$ is shown in Fig. 4.16c.

The main purpose of this section is to introduce you to methods of mathematically expressing waveshapes. Here are some examples using the unit step function.

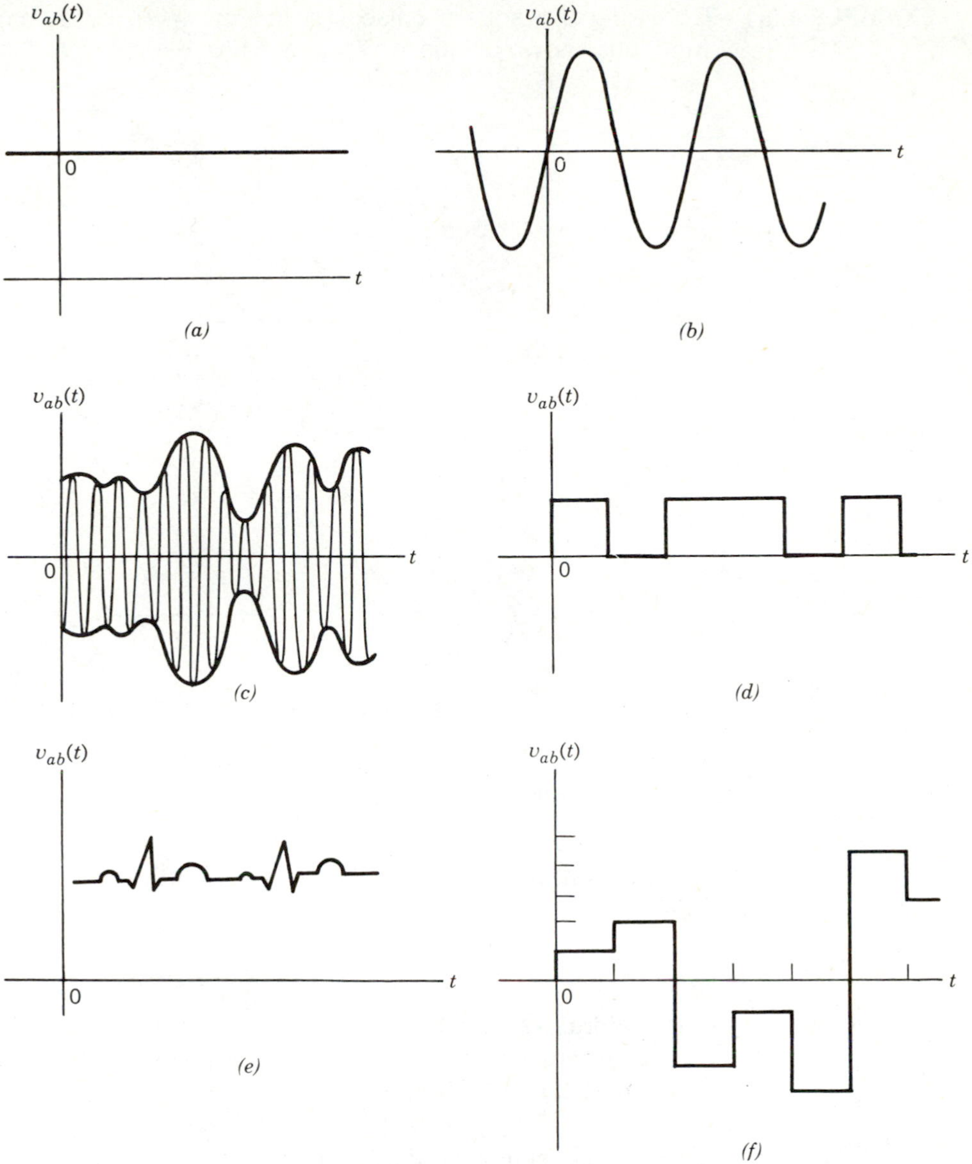

(a) (b) (c) (d) (e) (f)

Figure 4.15. Typical circuit waveforms.

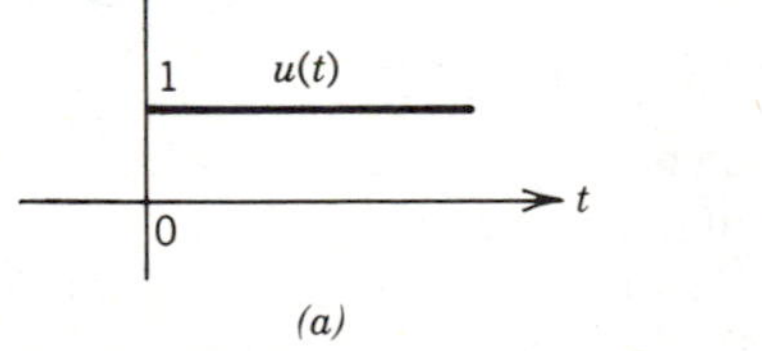

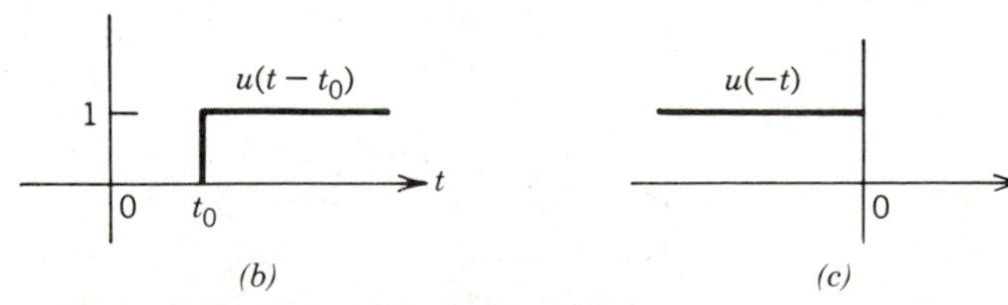

(a) (b) (c)

Figure 4.16. The unit step function.

EXAMPLE 4.2.1 Construct the square pulse $v(t)$ in Fig. 4.17*a* by summing the two step functions shown in Figs. 4.17*b* and 4.17*c*.

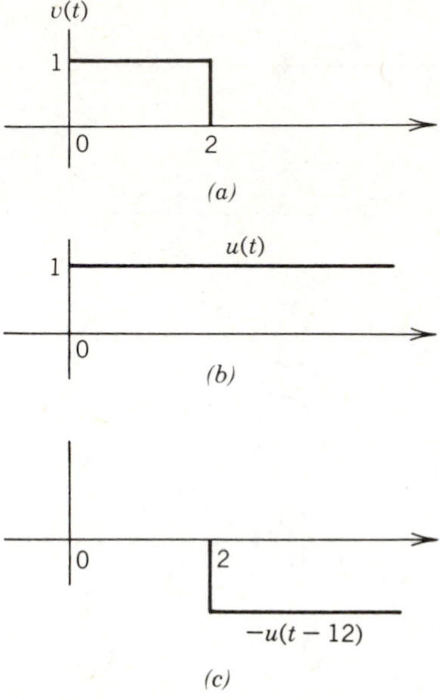

Figure 4.17. Synthesis of pulse $v(t)$.

Solution

We can write

$$v(t) = u(t) - u(t-2)$$

Notice that we can also express $v(t)$ directly

$$v(t) = \begin{cases} 0 & t < 0 \\ 1 & 0 < t < 2 \\ 0 & t > 2 \end{cases}$$

Therefore, we now have two ways to express signals, either in terms of other (more elementary) signals, or by writing the formula directly. We will need both methods.

We should note that there are many ways to express functions. Another representation of the square pulse in Fig. 4.17*a* is as the product of two step functions, where one step function begins at $t = 0$ and the other one ends at $t = 2$.

$$v(t) = u(t) \cdot u(-t+2)$$

We can express waveshapes similar to those encountered in Figs. 4.15*d* and 4.15*f* in terms of the unit step. The typical digital computer signal in Fig. 4.15*d* is repeated with labels along the time axis in Fig. 4.18*a*. When expressed in terms of the unit step we have

$$v_{ab}(t) = Au(t) - Au(t-1) + Au(t-2) - Au(t-4) \\ + Au(t-5) - Au(t-6)$$

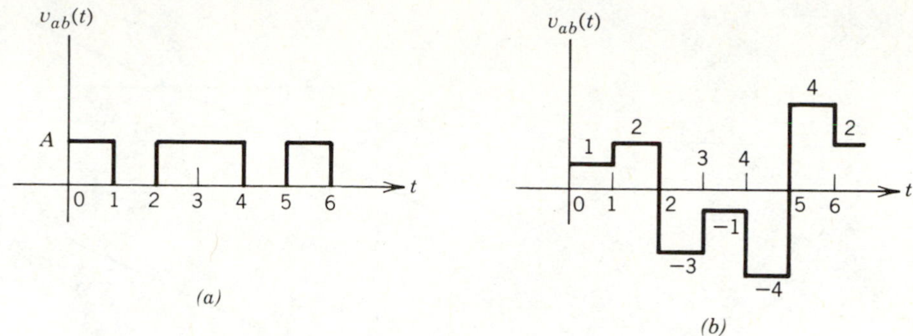

Figure 4.18. The waveforms of Figs. 4.15*d* and 4.15*f*.

The sample and hold signal in Fig. 4.18*b* is given by

$$v_{ab}(t) = u(t) + u(t-1) - 5u(t-2) + 2u(t-3) - 3u(t-4) + 8u(t-5) - 2u(t-6) + \cdots$$

Next let us consider the unit ramp function, which is defined by

$$r(t) = \begin{cases} t & t \geq 0 \\ 0 & t < 0 \end{cases} \tag{4.2.3}$$

This function is pictured in Fig. 4.19*a*. If this ramp is displaced along the time axis as in Fig. 4.19*b*, we have

$$r(t - t_0) = \begin{cases} t - t_0 & t \geq t_0 \\ 0 & t < t_0 \end{cases} \tag{4.2.4}$$

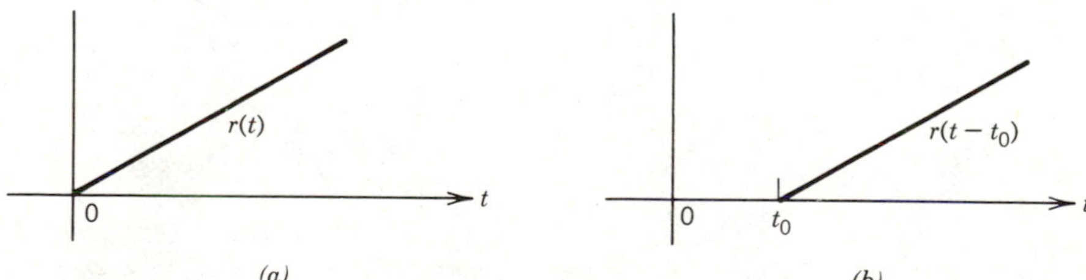

Figure 4.19. The unit ramp.

Notice that the two properties about the argument of the step are also possessed by the ramp. The function changes at $t = t_0$, and the function is zero for negative values of the argument (i.e., for $t < t_0$).

In Fig. 4.20 we illustrate another representation of the ramp function. The product of the line $t - t_0$ and the step $u(t - t_0)$ produces the ramp.

$$r(t - t_0) = (t - t_0)u(t - t_0) \tag{4.2.5}$$

Thus the step function is used to "cut off" the straight line $(t - t_0)$ and produce the ramp $r(t - t_0)$. Here is an example that illustrates the use of the unit ramp to describe a signal.

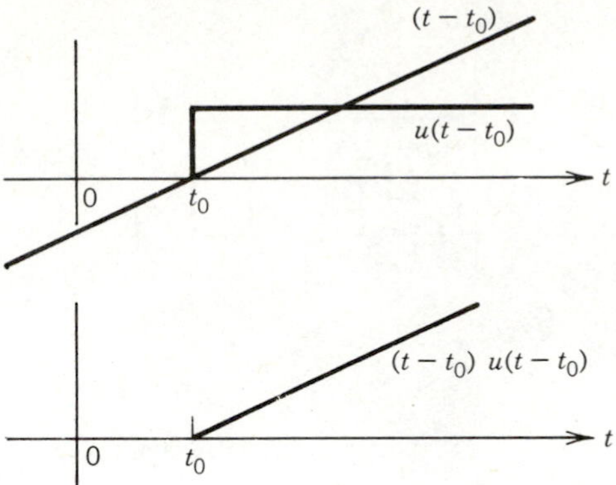

Figure 4.20. The unit ramp expressed as the product $(t - t_0) \cdot u(t - t_0)$.

EXAMPLE 4.2.2 Decompose the triangular pulse $v(t)$ shown in Fig. 4.21*a* into the sum of ramp functions.

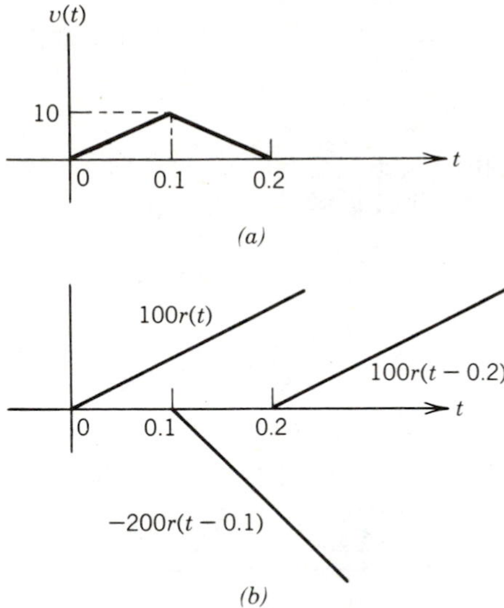

Figure 4.21. A triangular pulse expressed as the sum of three ramp functions.

Solution

Express $v(t)$ as shown in Fig. 4.21*b* as

$$
\begin{aligned}
v(t) &= 100tu(t) - 200(t - 0.1)u(t - 0.1) + 100(t - 0.2)u(t - 0.2) \\
&= 100r(t) - 200r(t - 0.1) + 100r(t - 0.2)
\end{aligned}
$$

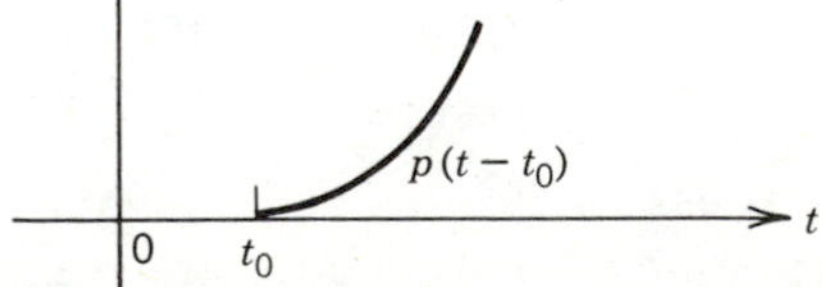

Figure 4.22. The unit parabola.

Notice that this is actually two representations, one in terms of step functions and the other in terms of the unit ramp $r(t)$. A third representation is

$$v(t) = \begin{cases} 0 & t < 0 \\ 100t & 0 < t < 0.1 \\ -100(t-0.2) & 0.1 < t < 0.2 \\ 0 & t > 0.2 \end{cases}$$

Again we emphasize that there are two basic methods of expressing signals, the direct method, exemplified by the last equation, and the decomposition method, in which the signal is expressed in terms of elementary signals such as ramps and steps.

The step and ramp functions are related to each other by a derivative–integral relationship. The step is the derivative of the ramp.

$$u(t) = \frac{d}{dt} r(t) \tag{4.2.6}$$

Conversely, we have

$$r(t) = \int_{-\infty}^{t} u(\lambda)\, d\lambda \tag{4.2.7}$$

It seems natural to ask if this chain can be extended in either direction. That is, consider the relationship

$$? \xrightarrow{\text{Integrate}} \text{Step function} \xrightarrow{\text{Integrate}} \text{Ramp function} \xrightarrow{\text{Integrate}} ?$$

or the converse relationship

$$? \xleftarrow{\text{Differentiate}} \text{Step function} \xleftarrow{\text{Differentiate}} \text{Ramp function} \xleftarrow{\text{Differentiate}} ?$$

What sorts of functions are represented by the question marks?

Let us first integrate the ramp function.

$$\int_{-\infty}^{t} r(\lambda)\, d\lambda = \int_{-\infty}^{0} 0\, d\lambda + \int_{0}^{t} \lambda\, d\lambda = \frac{t^2}{2} u(t)$$

This is called the unit parabola, $p(t)$.

$$p(t) = \begin{cases} t^2/2 & t \geq 0 \\ 0 & t < 0 \end{cases}$$

The parabola $p(t - t_0)$ is shown in Fig. 4.22.

Now we have a chain. Begin with $p(t)$ and successively apply the derivative operator to obtain first $r(t)$ and then $u(t)$. Or we can go the other way. Begin with $u(t)$ and successively apply the integral operator to obtain $r(t)$ and then $p(t)$. This chain could be extended even further by integrating $p(t)$, but we will not do so. Instead, we will discuss the delta (δ) function, which is the derivative of the step function.

The δ function may be approached in several ways. We will take two approaches, first defining it by listing its properties, and then giving a plausibility argument to make it more familiar. By definition, a δ function possesses the following two properties.

$$\int_{-\infty}^{t} \delta(\lambda)\, d\lambda = u(t) \tag{4.2.8}$$

$$\int_{-\infty}^{\infty} f(t)\delta(t-t_0)\, dt = f(t_0) \tag{4.2.9}$$

These two properties serve to define the δ function. The following plausibility argument illustrates how one particular form of the δ function satisfies the above two properties.

Consider the truncated ramp $q_1(t)$ shown in Fig. 4.23*a*. The derivative of this function is the square pulse $h_1(t)$. Notice that the area under $h_1(t)$ is 1. Next consider $q_2(t)$, where the value of a in Fig. 4.23*b* has been reduced. The derivative $h_2(t)$ is a taller, shorter pulse, again with unit area. Finally, a glance at Fig. 4.23*c* illustrates that as $a \to 0$ and $q_3(t)$ approaches a unit step, the derivative $h_3(t)$ approaches a tall thin pulse of unit area, that is, a δ function. This helps explain Eq. 4.2.8.

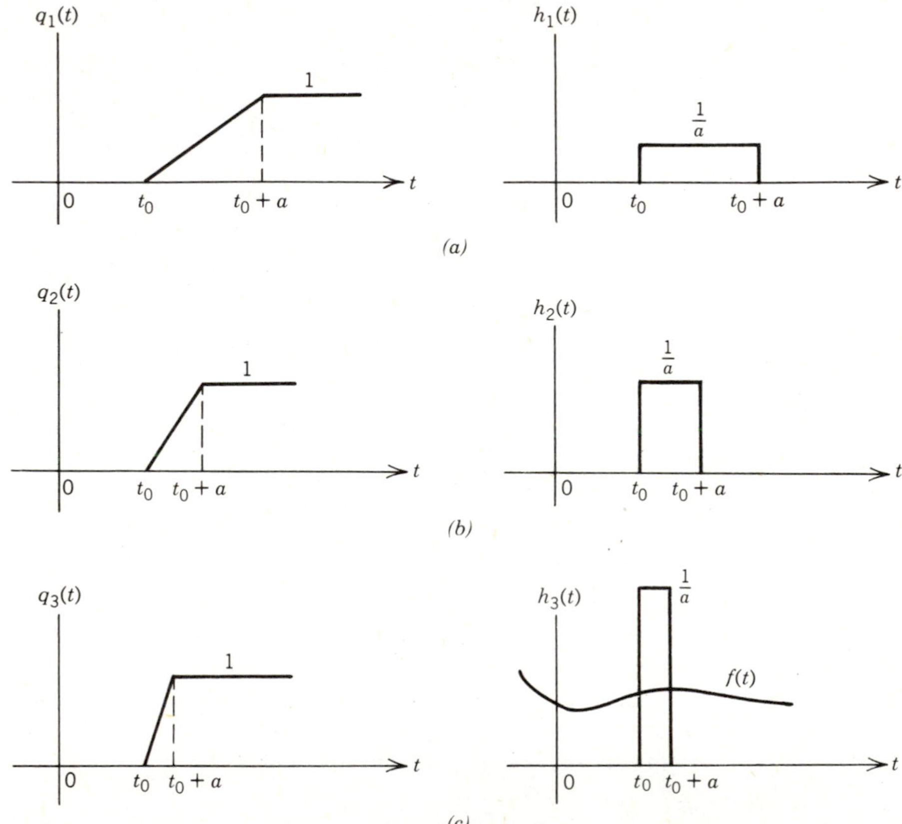

Figure 4.23. Description of the δ function.

Also in Fig. 4.23c, an arbitrary function $f(t)$ is shown superimposed on $h_3(t)$. If $h_3(t)$ is very tall and thin, then $f(t)$ is approximately constant in the interval $t_0 \le t < t_0 + a$. The integral of the product is given by

$$\int_{-\infty}^{\infty} f(t) h_3(t)\, dt \approx \int_{-\infty}^{\infty} f(t_0) h_3(t)\, dt$$

Since $f(t_0)$ is constant, it may be removed from under the integral sign to give

$$f(t_0)\int_{-\infty}^{\infty} h_3(t)\, dt = f(t_0)$$

since the area under $h_3(t)$ is 1. This helps explain Eq. 4.2.9 if the δ function is substituted for $h_3(t)$.

Study the following examples carefully to make certain that you agree with the answers. These examples illustrate the use of Eq. 4.2.8 and 4.2.9 in evaluating integrals that contain δ functions.

a. $\int_{-\infty}^{t} \delta(\lambda - t_0)\, d\lambda = u(t - t_0)$

b. $\int_{-\infty}^{\infty} t\delta(t-2)\, dt = 2$

c. $\int_{-\infty}^{0} t\delta(t-2)\, dt = 0$

d. $\int_{-\infty}^{\infty} 2\delta(t-5)e^{-j\omega t}\, dt = 2e^{-j\omega 5}$

e. $\int_{0}^{4} t^2\delta(t-2)\, dt = 4$

f. $\int_{-4}^{0} t^2\delta(t-2)\, dt = 0$

EXAMPLE 4.2.3 Find a derivative of each function in Fig. 4.24.

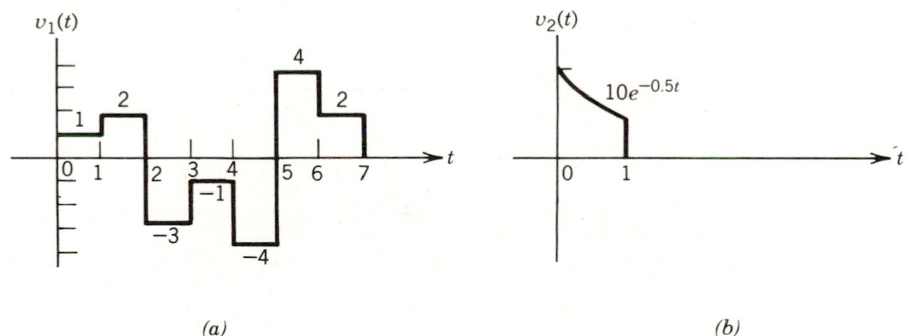

Figure 4.24

Solution

See Fig. 4.25. The derivative of $v_1(t)$ is a sequence of δ functions. Note that the area under each δ function is equal to the height of each step in $v_1(t)$. It is customary to draw all δ functions at uniform height, for supposedly their height is infinite. The differences in the δ functions are indicated by showing the area as a number in parentheses.

The sign on each δ function is determined by the direction of each step in $v_1(t)$. As you trace the curve from left to right, a step up indicates a positive δ function for the derivative. A step down indicates a negative δ function. This is illustrated in the derivative of $v_2(t)$, where two steps are present. The first step at the origin is up 10 units, giving rise to the δ function at the origin in Fig. 4.25*b*. The second step is down at $t=1$ and has height $6.07 = 10e^{-0.5(1)}$, giving rise to the δ function at $t=1$ in the derivative. Analytically, the derivative of $v_2(t)$ is given by

$$\frac{d}{dt}v_2(t) = 10\delta(t) - 6.07\delta(t-1) - 5e^{-0.5t}u(t) \cdot u(-t+1)$$

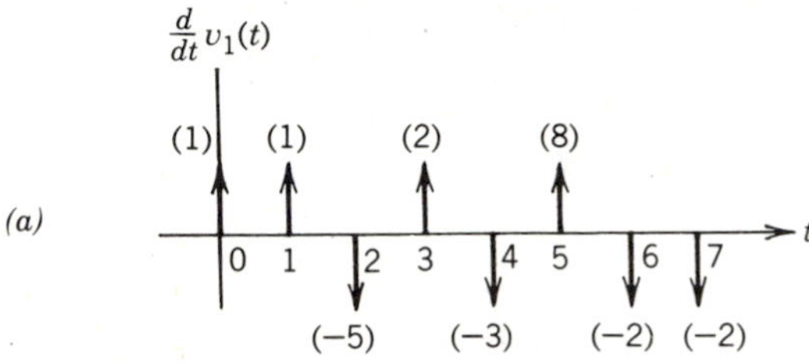

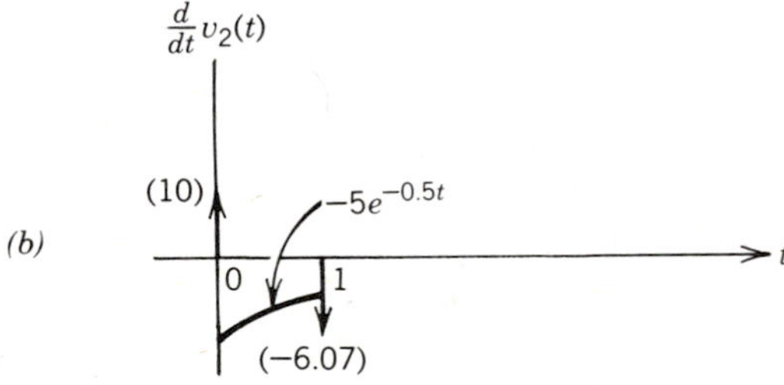

Figure 4.25. Derivatives of the functions in Fig. 4.24.

LEARNING EVALUATIONS

1. Sketch each of the following waveforms.
   a. $x(t) = 2u(t) - 2u(t-3)$
   b. $x(t) = 5u(t+1) - 3u(t)$
   c. $z(t) = \frac{d}{dt}x(t)$ where $x(t)$ is defined in (*b*)
   d. $x(t) = r(t) - r(t-1) - r(t-2) + r(t-3)$
   e. $x(t) = e^{-t}u(t)$
2. Give a mathematical representation for each waveform in Fig. 4.26. Express your representation in terms of steps, ramps, and so on.

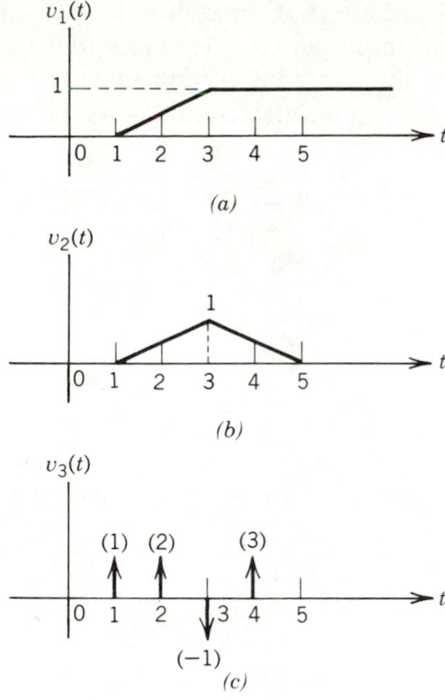

Figure 4.26

3. Find the derivative of $v(t)$ shown in Fig. 4.27.

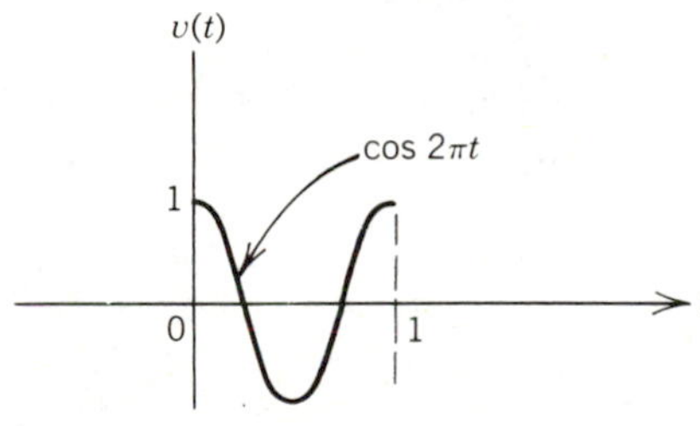

Figure 4.27

4.3 Capacitor Waveforms

LEARNING OBJECTIVES

After completing this section you should be able to do the following:

1. State the continuity condition for capacitors.
2. Find the initial voltage and current in a capacitor.
3. Find the final voltage and current in a capacitor.

In this section we will develop some properties of the capacitor by applying various waveforms. Our purpose is to develop insight into the behavior of the capacitor in the presence of changing voltage and current waveforms. Of particular importance is the ability to determine initial conditions in a capacitor for the solution of differential equations. These initial conditions can be determined through an understanding of the continuity condition, which will be introduced shortly.

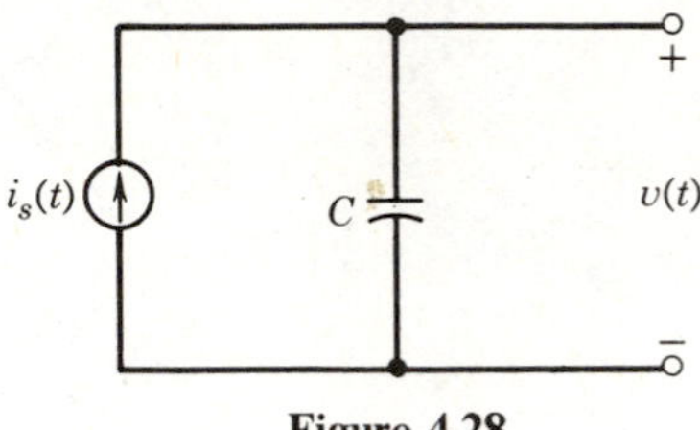

Figure 4.28

We begin by considering the circuit in Fig. 4.28. The forcing function $i_s(t)$ produces a voltage across the capacitor given by

$$\begin{aligned} v(t) &= \frac{1}{C}\int_{-\infty}^{t} i_s(\lambda)\, d\lambda \\ &= \frac{1}{C}\int_{0}^{t} i_s(\lambda)\, d\lambda + v(0) \end{aligned} \tag{4.3.1}$$

The term $v(0)$ summarizes the past history (before $t = 0$) of the capacitor in a single number. It is called the initial state of the capacitor. It is this initial state or initial condition that we must specify in solving for the response of circuits containing capacitors.

The input–output pair $i_s(t) - v(t)$ is shown in Fig. 4.29 for three input signals. An initial voltage of zero is assumed in each case. For a δ-function input in Fig. 4.29*a*, the voltage response is a step function since

$$v(t) = \frac{1}{C}\int_{0}^{t} A\delta(\lambda)\, d\lambda = \frac{A}{C}u(t)$$

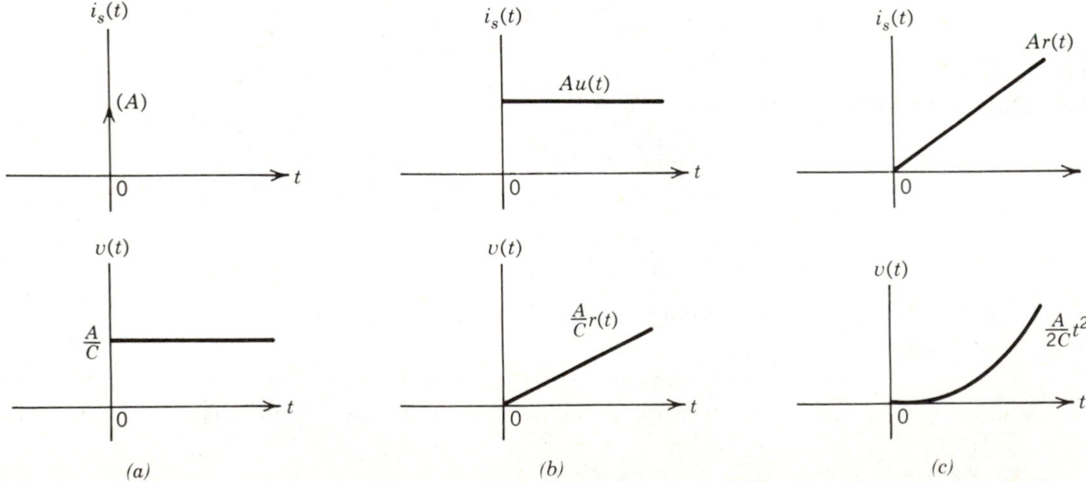

Figure 4.29. Input–output pairs for the capacitor.

For a step input of height A as shown in Fig. 4.29b, the response is a ramp function.

$$v(t) = \frac{1}{C}\int_0^t Au(\lambda)\,d\lambda = \frac{A}{C}r(t)$$

Finally for a ramp input of slope A the response is a parabola given by

$$v(t) = \frac{1}{C}\int_0^t Ar(\lambda)\,d\lambda = \frac{A}{2C}t^2u(t)$$

as shown in Fig. 4.29c.

The presence of an initial voltage on the capacitor does not alter these basic relationships. A δ function produces a step, a step produces a ramp, and a ramp produces a parabola. Consider the input–output relationships established in Fig. 4.30. Each response curve $v(t)$ has an initial voltage $v(0)$. For the δ-function input of strength A the response is given by

$$v(t) = \frac{1}{C}\int_0^t A\delta(\lambda)\,d\lambda + v(0) = \frac{A}{C}u(t) + v(0)$$

For the step input we have

$$v(t) = \frac{1}{C}\int_0^t Au(\lambda)\,d\lambda + v(0) = \frac{A}{C}r(t) + v(0)$$

and finally for the ramp input of slope A we have

$$v(t) = \frac{1}{C}\int_0^t Ar(\lambda)\,d\lambda + v(0) = \frac{A}{2C}t^2u(t) + v(0)$$

These waveforms are all illustrated in Fig. 4.30.

If we are given the voltage $v(t)$ we may determine the current from

$$i(t) = C\frac{dv}{dt} \tag{4.3.2}$$

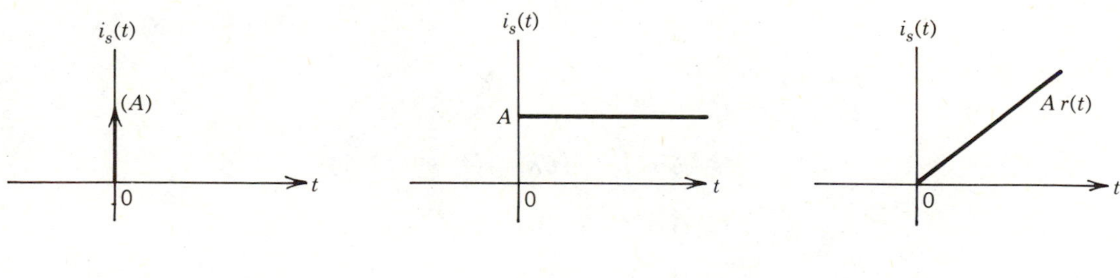

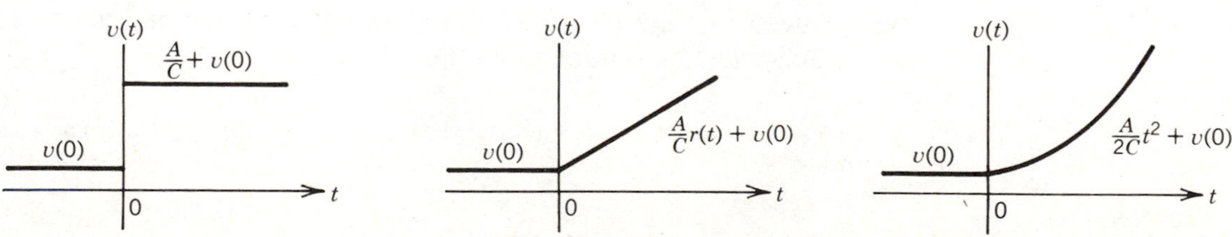

Figure 4.30. Capacitor waveforms with initial conditions.

Application of this relationship to each voltage waveform in Fig. 4.30 results in the corresponding current waveform.

An important observation can be made from Figs. 4.29 and 4.30. Except for the case where $i(t)$ is a δ function, the voltage is continuous for all time. When the step current is applied, the voltage is continuous at $t = 0$; likewise for the ramp forcing function. This is called the continuity condition. Since the voltage waveform is proportional to the integral of the current, this continuity condition will always hold if the current is finite.

> ***Continuity Condition for Capacitors.*** The voltage across a capacitor is a continuous function of time if the current is bounded.[1]

This means that δ functions of current are ruled out as forcing functions. Since it is physically impossible to generate a δ function in the laboratory, this imposes no hardship on us.

EXAMPLE 4.3.1 The current waveform $i(t)$ in the circuit of Fig. 4.31 is a square pulse. Find $v_c(t)$, $v_R(t)$, and $v(t)$ if $v_c(0) = 0$.

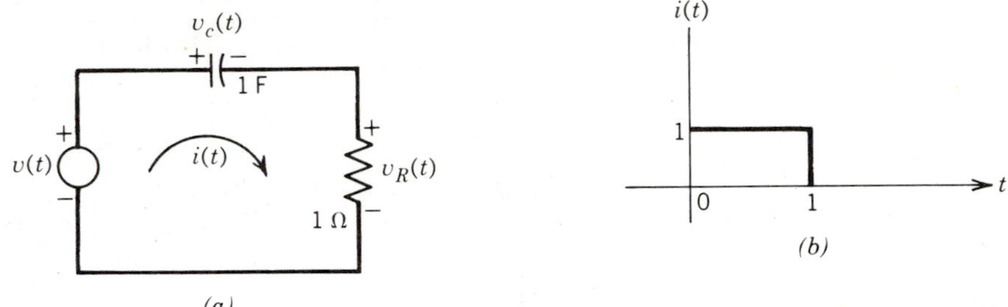

Figure 4.31

Solution

From Eq. 4.3.1,

$$v_c(t) = \int_0^t i(\lambda)\, d\lambda = r(t) - r(t-1)$$

$$v_R(t) = i(t)R = u(t) - u(t-1)$$

$$v(t) = v_c(t) + v_R(t) = u(t) + r(t) - [u(t-1) + r(t-1)]$$

These waveforms are plotted in Fig. 4.32. Notice that the voltage across the capacitor is a continuous function of time.

[1]This condition applies to linear time-invariant capacitors. We will define linearity and time-invariance in Chapter 7.

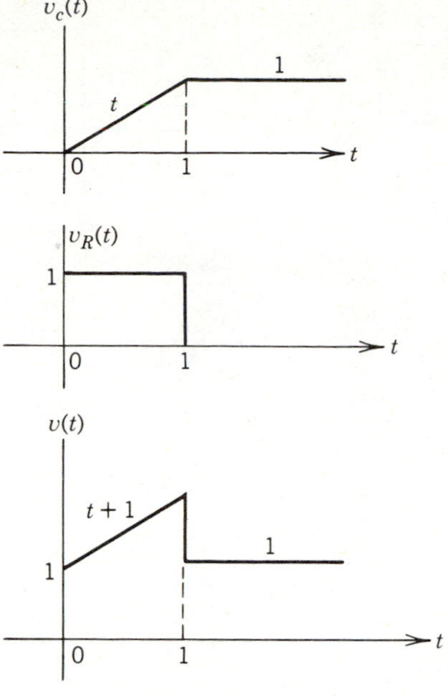

Figure 4.32

EXAMPLE 4.3.2 Find $v_c(t)$ in Fig. 4.31 if the initial voltage $v_c(0)$ is 0.5 V.

Solution

From Eq. 4.3.1 we have

$$\begin{aligned} v(t) &= \int_0^t i(\lambda)\, d\lambda + 0.5 \\ &= 0.5, \qquad t < 0 \\ &= \int_0^t 1\, d\lambda + 0.5 = t + 0.5, \qquad 0 < t < 1 \\ &= 1.5, \qquad 1 < t \end{aligned}$$

This voltage is plotted in Fig. 4.33. Notice once again that the capacitor voltage is a continuous function of time.

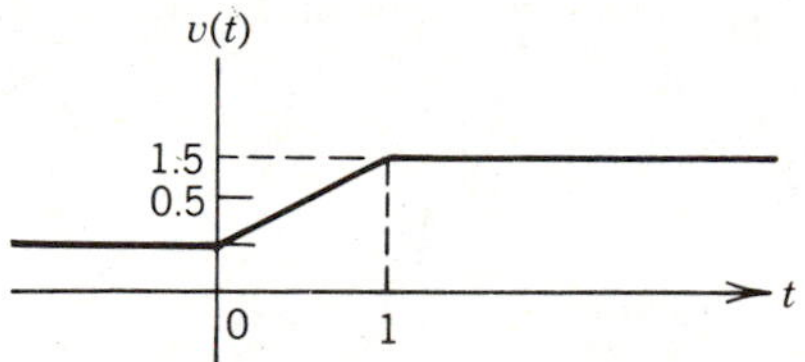

Figure 4.33

The continuity condition for capacitors applies to the voltage across the capacitor. It imposes no corresponding condition on the current through a

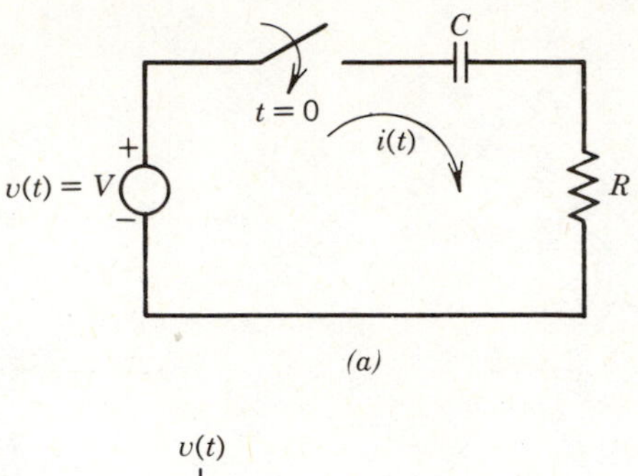

(a)

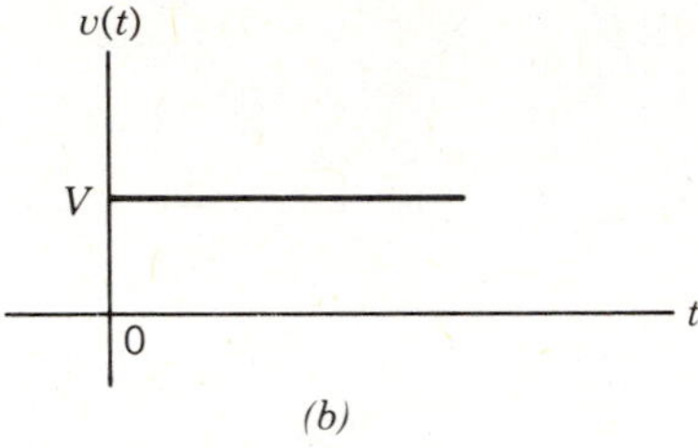

(b)

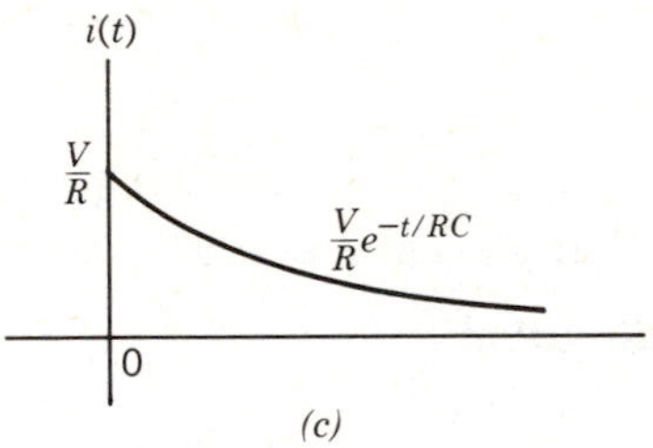

(c)

Figure 4.34

capacitor. In fact, there is no corresponding condition on the current, and it is allowed to be discontinuous. Consider the simple RC circuit in Fig. 4.34a. The constant voltage source in series with the switch that closes at $t=0$ may be replaced by a step voltage $v(t)=Vu(t)$ as shown in Fig. 4.34b. Keep in mind that this is the voltage applied to the circuit, and not the capacitor voltage. If the capacitor voltage is zero before the switch is closed, then the capacitor voltage must be zero immediately after the switch is closed. This is written as

$$\text{if} \quad v_c(0^-)=0$$

$$\text{then} \quad v_c(0^+)=0$$

where the notation means

$$v_c(t_0^-)=\lim_{\substack{t\to t_0\\ t<t_0}} v(t)$$

and

$$v_c(t_0^+)=\lim_{\substack{t\to t_0\\ t>t_0}} v(t)$$

That is, $v_c(0^-)$ is the capacitor voltage just before $t=0$, and $v_c(0^+)$ is the capacitor voltage immediately after $t=0$.

Kirchhoff's voltage law must always be satisfied. The sum of all voltages around any closed path must be zero at any time instant. At $t=0^-$, the applied voltage is zero because the switch is open, the voltage across the capacitor is zero, and so the voltage across the resistor is also zero. At $t=0^+$, however, the situation changes. The applied voltage is V, the capacitor voltage remains zero because of the continuity condition, and so the voltage across the resistor must also equal V. This can occur only if $i(0^+)=V/R$. Thus, for the circuit in Fig. 4.34 we have

$$i(0^-)=0$$

$$i(0^+)=\frac{V}{R}$$

For your information, the current waveform is shown in Fig. 4.34*c*. We will determine how to find this complete waveform in the next chapter. For now, notice that $i(0^-)=0$ and $i(0^+)=V/R$. Also notice that as t becomes large, $i(t)$ approaches zero.

Physically the following events occur. After the switch is closed, charge has mass, and therefore inertia, and so it cannot move instantaneously. Thus the voltage, which is determined by the quantity of charge on the plates, cannot change instantaneously.

On the other hand, current is the rate of change of charge. This rate is limited only by the resistor, so the initial current is V/R.

As time increases the capacitor becomes charged to the maximum voltage V. The current decays to zero because the rate of change of the charge becomes zero. Thus as $t \to \infty$ we have

$$\lim_{t\to\infty} v_c(t)=V$$

$$\lim_{t\to\infty} i(t)=0$$

EXAMPLE 4.3.3 Find the initial current, final current, and final voltage across the capacitor in Fig. 4.35 if $v_c(0^-)=0$. The switch is closed at $t=0$.

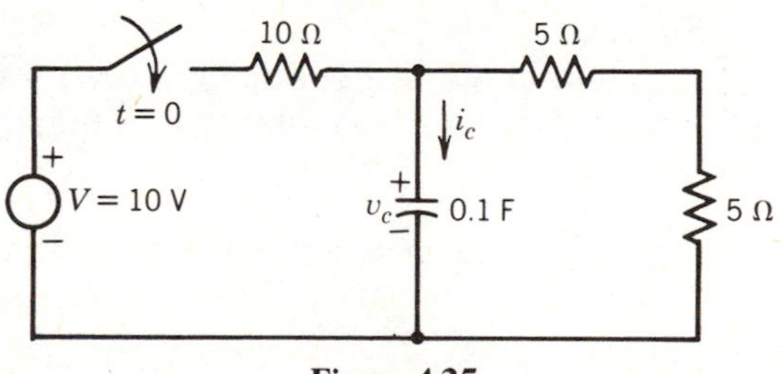

Figure 4.35

Solution

By the continuity condition $v_c(0^-)=v_c(0^+)=0$. Writing Kirchhoff's voltage law around the loop containing the source and capacitor at $t=0^+$ gives

$$V = i(0^+)R + v_c(0^+)$$

Since $v_c(0^+) = 0$ we have

$$i(0^+) = \frac{V}{R} = \frac{10}{10} = 1 \text{ A}$$

At $t = \infty$ the capacitor is charged and the current flow in the capacitor is zero. Thus any current through the 10-Ω resistor also flows through each 5-Ω resistor. Writing Kirchhoff's voltage law around the outside loop at $t = \infty$ gives

$$V = 10i + 5i + 5i$$

or

$$i(\infty) = \frac{V}{20} = \frac{1}{2} \text{ A}$$

The capacitor voltage must equal the voltage across two 5-Ω resistors, giving

$$v_c(\infty) = i(\infty)10 = 5 \text{ V}$$

EXAMPLE 4.3.4 Consider the op amp circuit in Fig. 4.36. If the input voltage $v_s(t)$ is a unit step, find the initial voltage and current through the capacitor. Assume $v_c(0^-) = 0$.

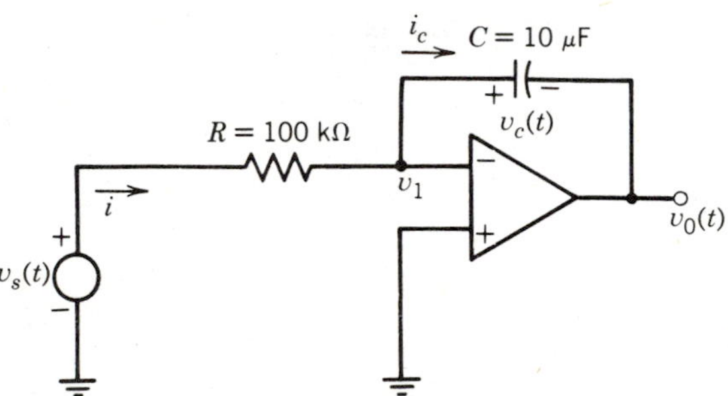

Figure 4.36. Op amp circuit for Example 4.3.4.

Solution

By the continuity condition $v_c(0^+) = v_c(0^-) = 0$. To find the initial capacitor current we begin by noting that

$$i(t) = i_c(t)$$

because there is no current into the op amp. Since $v_1 = 0$ we have by KVL

$$v_s(t) = Ri(t)$$

or

$$i(t) = i_c(t) = \frac{v_s(t)}{R}$$

At $t = 0^+$ this gives

$$i_c(0^+) = \frac{1}{R} = 10^{-5}\text{ A}$$

LEARNING EVALUATION

The switch in Fig. 4.37 has been open for a long time. The switch is closed at $t = 0$. Find the voltage across and current through the capacitor at $t = 0^-$ and at $t = 0^+$.

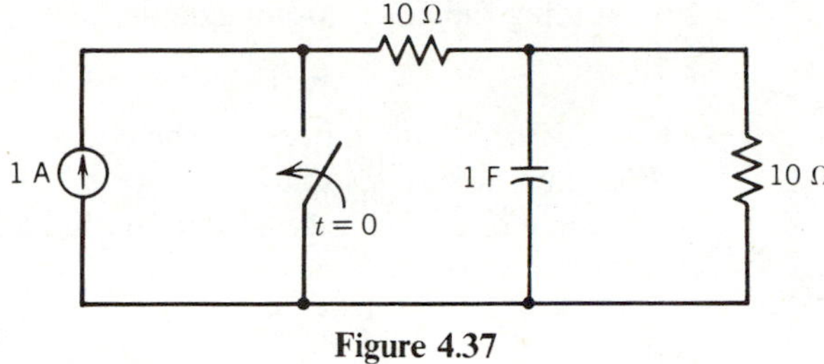

Figure 4.37

4.4 The Inductor

LEARNING OBJECTIVES

After completing this section you should be able to do the following:

1. Find the voltage $v(t)$ for a given current $i(t)$ through an inductor.
2. Find the current $i(t)$ for a given voltage $v(t)$ across an inductor.
3. Calculate the power and energy in an inductor for a given current or voltage waveform.
4. Find the equivalent inductance for a given network of inductors.

Joseph Henry (1797–1878), an American physicist, invented the first telegraph and the first motor based on electromagnetic principles. He and Faraday both performed their investigations in the mid-nineteenth century and independently discovered many of the same principles. Henry's work led to the third circuit element we will consider: the *inductor*. The circuit symbol for the inductor, L, and standard voltage–current reference directions are shown in Fig. 4.38. The *inductance* of the inductor will be defined shortly; the unit of inductance is the henry.

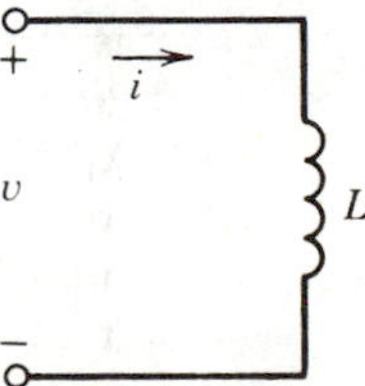

Figure 4.38. The circuit symbol for the inductor.

In order to determine the voltage–current relation in an inductor, we must state a few electromagnetic principles. If we pass a current through a wire, a magnetic field that circles the wire is established. This magnetic field, described by lines of flux, ϕ, is linearly related to the amount of current in the wire. By winding this wire in a cylindrical coil, the flux passes through or links the turns of the coil. The flux linkage, λ, is defined as

$$\lambda = N\phi \tag{4.4.1}$$

where N is the number of turns of the coil linked by the flux, ϕ. If the medium surrounding the coil is nonmagnetic then

$$\lambda = Li \tag{4.4.2}$$

where L is the inductance and the coil is called an inductor.

Faraday's law states that a voltage is "induced" across an inductor that is equal to the time rate of change of the flux linkage. Hence,

$$v(t) = \frac{d\lambda(t)}{dt} \tag{4.4.3}$$

Putting Eq. 4.4.2 into Eq. 4.4.3 results in

$$\boxed{v(t) = L\frac{di(t)}{dt}} \tag{4.4.4}$$

This is the voltage–current relationship across the inductor which we sought. If we express the current as a function of voltage we have

$$i(t) = \frac{1}{L}\int_{-\infty}^{t} v(\alpha)\, d\alpha \tag{4.4.5}$$

or

$$i(t) = \frac{1}{L}\int_{0}^{t} v(\alpha)\, d\alpha + i(0) \tag{4.4.6}$$

Remember that Eqs. 4.4.4 and 4.4.5 hold only if the medium surrounding the coil is nonmagnetic. Figure 4.39*a* shows the λ–i relation (Eq. 4.4.2) for such a case. If we place a magnetic medium such as an iron core in our inductor then the λ–i relationship becomes nonlinear and will have the general shape shown in Fig. 4.39*b*.

For a nonmagnetic medium we see that a zero current always results in zero flux linkage and a given value of i, say I_a, always results in a given flux linkage, λ_A. For a magnetic medium the situation is far different. If the inductor has never had a current pass through it, a current increasing from zero to I_m will generate a curve $0A$. When the current is reduced from I_m to $-I_m$, curve ABC is generated and if the current is then increased back to I_m, curve CDA is generated. This curve is called a *hysteresis curve* and a value of current I_1 will produce a flux linkage between $-\lambda_1$ and λ_1, depending on the portion of the curve currently being traversed. Arrows show the direction of travel on the curve. The hysteresis effect plays a very important role in electrical machine design.

For the remainder of the text we will assume that we are dealing with linear inductors whose v–i relation is expressed by Eqs. 4.4.4 through 4.4.6.

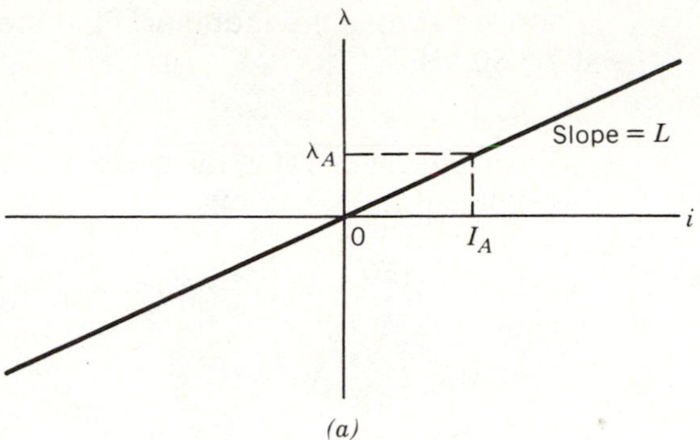

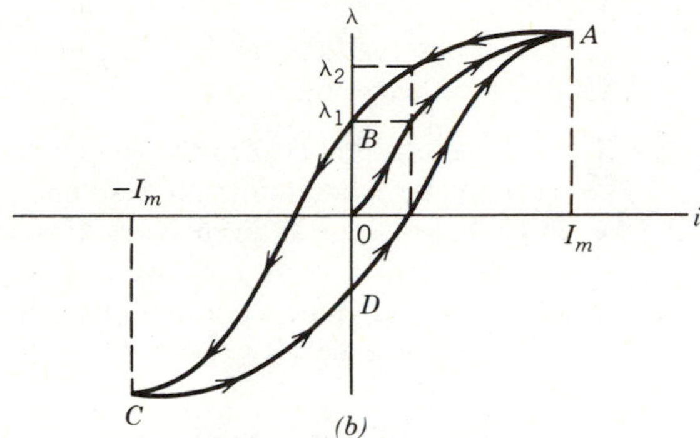

Figure 4.39. Relation for (*a*) nonmagnetic medium and (*b*) magnetic medium.

EXAMPLE 4.4.1 The current through a 50-mH inductor is

$$i(t) = 400 \sin 800t \text{ mA}$$

Find the voltage across the inductor.

Solution

From Eq. 4.4.4,

$$\begin{aligned} v(t) &= L\frac{di(t)}{dt} \\ &= (50 \times 10^{-3})\frac{d}{dt}(400 \times 10^{-3} \sin 800t) \\ &= (50 \times 10^{-3})(400 \times 10^{-3})(800) \cos 800t \\ &= 16 \cos 800t \text{ V} \end{aligned}$$

EXAMPLE 4.4.2 At time $t = 0$, a 200-mH inductor has a current of 5-A flowing through it. At $t = 0$ a voltage

$$v(t) = 60e^{-40t} \text{ V}$$

is applied across the inductor. Find the value of the current in the inductor at $t = 50$ ms.

Solution

Since an initial value of current was given, we will use Eq. 4.4.6 and evaluate it for $t = 50$ ms.

$$
\begin{aligned}
i(50\times 10^{-3}) &= 5 + \frac{1}{200\times 10^{-3}}\int_0^{50\times 10^{-3}} 060e^{-40\alpha}\,d\alpha \\
&= 5 + \frac{60}{200\times 10^{-3}}\left(\frac{1}{-40}\right)(e^{-40\alpha})\Big|_0^{50\times 10^{-3}} \\
&= 5 + (-7.5)(e^{-2} - 1) \\
&= 11.48 \text{ A}
\end{aligned}
$$

Several observations can be made about the voltage–current relationship in an inductor at this point.

1. The relationship between the voltage and current in an inductor is unique. For an inductor containing no residual magnetism (the magnetic field in the inductor is zero) a given current waveform will produce a unique voltage waveform. This voltage waveform would also produce the given current. Mathematically the condition is met if the voltage satisfies the condition $v(t) = 0$ for some $t < t_0$ where t_0 is *any* finite value of time. In other words, $V(-\infty) = 0$.
2. The current in an inductor at time t_1 depends on the historical value of the voltage across the inductor up until time t_1 and not on the present or future value of the voltage beyond t_1.
3. The voltage across an inductor does not depend on the magnitude of the current flowing in the inductor but only on the rate of change of this current.

Power and Energy

As in the case of the capacitor, the power and energy in an inductor are related by an integral–differential relationship. In fact, Eqs. 4.1.3 and 4.1.4 hold for the inductor also. The instantaneous power in an inductor can be found using Eqs. 1.5.1 and 4.4.4 as

$$p(t) = v(t)i(t) = Li(t)\frac{di(t)}{dt} \text{ W} \tag{4.4.7}$$

Since the product $i(t)[di(t)/dt]$ can be positive or negative, the inductor can receive power (positive power) or supply power (negative power).

Substituting Eq. 4.4.7 into Eq. 4.1.4 gives

$$
\begin{aligned}
w(t) &= \int_{-\infty}^{t} Li(\alpha)\frac{di(\alpha)}{d\alpha}\,d\alpha = \int_{i(-\infty)}^{i(t)} Li\,di \\
&= \tfrac{1}{2}Li^2(t) - \tfrac{1}{2}Li^2(-\infty)
\end{aligned}
$$

By our earlier discussion $i(-\infty)$ is assumed to be zero, so

$$\boxed{w(t) = \tfrac{1}{2}Li^2(t)\text{ J}} \tag{4.4.8}$$

EXAMPLE 4.4.3 Find the power delivered and energy stored in a 250-mH inductor carrying a current of 25 A.

Solution

From Eq. 4.4.7,

$$p(t) = (250 \times 10^{-3})(25)\frac{d}{dt}(25) = 0$$

and from Eq. 4.4.8,

$$w(t) = \tfrac{1}{2}(250 \times 10^{-3})(25)^2 = 78.125 \text{ J}$$

This example shows that even though the power delivered or received by an inductor is zero, it can still store energy. This is due to the fact that the energy stored is only a function of current and not of voltage.

EXAMPLE 4.4.4 The current through a 10-mH inductor is shown in Fig. 4.40. Find the voltage, power, and energy as a function of time.

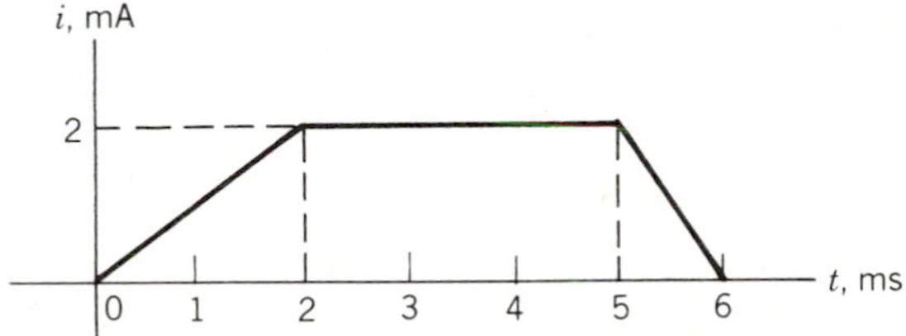

Figure 4.40. Current waveform for Example 4.4.4.

Solution

First notice that $i(t)$ consists of five distinct parts and can be written as

$$i(t) = \begin{cases} 0 & t \le 0 \\ t & 0 \le t \le 2\text{ ms} \\ 2 & 2\text{ ms} \le t \le 5\text{ ms} \\ 2 - 2(t-5) & 5\text{ ms} \le t \le 6\text{ ms} \\ 0 & t \ge 6\text{ ms} \end{cases}$$

Using Eq. 4.4.4, we find

$$v(t) = \begin{cases} 0 & t \le 0 \\ 10^{-2} & 0 \le t \le 2\text{ ms} \\ 0 & 2\text{ ms} \le t \le 5\text{ ms} \\ -2 \times 10^{-2} & 5\text{ ms} \le t \le 6\text{ ms} \\ 0 & t \ge 6\text{ ms} \end{cases}$$

Using Eq. 4.4.7 and the foregoing results, $p(t)$ can be written directly as

$$p(t) = \begin{cases} 0 & t \leq 0 \\ 10^{-2}t & 0 \leq t \leq 2 \text{ ms} \\ 0 & 2 \text{ ms} \leq t \leq 5 \text{ ms} \\ [-4 + 4(t-5)] \times 10^{-2} & 5 \text{ ms} \leq t \leq 6 \text{ ms} \\ 0 & t \geq 6 \text{ ms} \end{cases}$$

Finally Eq. 4.4.8 gives us $w(t)$:

$$w(t) = \begin{cases} 0 & t \leq 0 \\ 5 \times 10^{-3}t^2 & 0 \leq t \leq 2 \text{ ms} \\ 20 \times 10^{-3} & 2 \text{ ms} \leq t \leq 5 \text{ ms} \\ [2 - 4(t-5) + 2(t-5)^2] \times 10^{-2} & 5 \text{ ms} \leq t \leq 6 \text{ ms} \\ 0 & t \geq 6 \text{ ms} \end{cases}$$

Equivalent Inductance

You will often be faced with the task of combining inductors when trying to simplify a circuit. Let us first consider the case of two inductors in series, as shown in Fig. 4.41*a*. We want to determine the value of L in Fig. 4.41*b* such that, with respect to terminals *a–b*, the two circuits are identical. Using KVL around the circuit in Fig. 4.41*a* we obtain

$$-v(t) + L_1\frac{di}{dt} + L_2\frac{di}{dt} = 0$$

or

$$v(t) = (L_1 + L_2)\frac{di}{dt} \tag{4.4.9}$$

Doing the same around Fig. 4.41*b*, we get

$$v(t) = L_{eq}\frac{di}{dt} \tag{4.4.10}$$

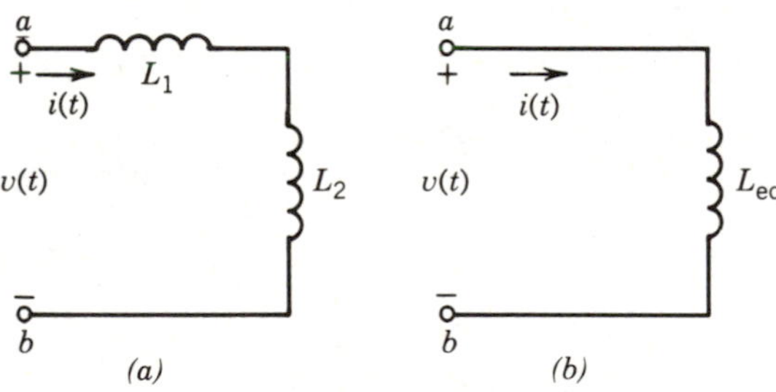

Figure 4.41. Equivalent series inductance.

If the two circuits are to be equivalent at terminals *a–b*, then Eq. 4.4.9 must equal Eq. (4.4.10) so

$$\boxed{L_{eq} = L_1 + L_2} \tag{4.4.11}$$

This reasoning can be extended to say that N inductors in series can be replaced by one equivalent inductance, L_{eq}, of value

$$\boxed{L_{eq} = L_1 + L_2 + \cdots + L_N} \tag{4.4.12}$$

Now investigate two inductors in parallel as shown in Fig. 4.42a and determine the value of L_{eq} so the two circuits are equivalent with respect to terminals a–b. Use KCL on the circuit in Fig. 4.42a to get

$$\begin{aligned} i(t) &= i_1(t) + i_2(t) \\ &= \frac{1}{L_1}\int_{-\infty}^{t} v(\alpha)\,d\alpha + \frac{1}{L_2}\int_{-\infty}^{t} v(\alpha)\,d\alpha \end{aligned} \tag{4.4.13}$$

Now from Fig. 4.42b,

$$i(t) = \frac{1}{L_{eq}}\int_{-\infty}^{t} v(\alpha)\,d\alpha \tag{4.4.14}$$

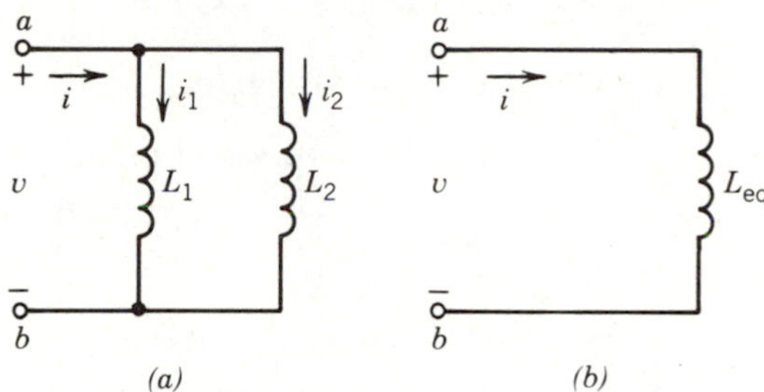

Figure 4.42. Equivalent parallel inductance.

Comparing Eqs. 4.4.13 and 4.4.14 reveals that

$$\boxed{\frac{1}{L_{eq}} = \frac{1}{L_1} + \frac{1}{L_2}} \tag{4.4.15}$$

Extending this argument for N inductors in parallel yields

$$\boxed{\frac{1}{L_{eq}} = \frac{1}{L_1} + \frac{1}{L_2} + \cdots + \frac{1}{L_N}} \tag{4.4.16}$$

Comparing Eq. 4.4.12 and 4.4.16 with Eq. 2.4.3 and 2.4.6 shows that inductors in a circuit combine in exactly the same way as resistors in a circuit.

EXAMPLE 4.4.5 Find the equivalent inductance or the network shown in Fig. 4.43.

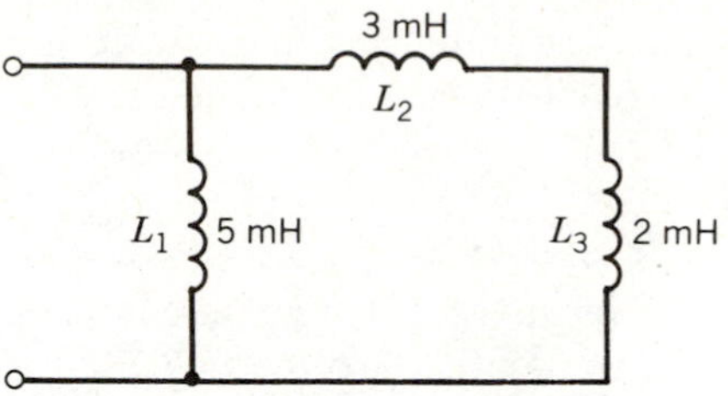

Figure 4.43. Circuit for example 4.5.4.

Solution

First combine L_2 and L_3 to give

$$L_4 = 3 \text{ mH} + 2 \text{ mH} = 5 \text{ mH}$$

Then L_4 and L_1 are in parallel and can be combined to find L_{eq}.

$$L_{eq} = \frac{L_1 L_4}{L_1 + L_4} = \frac{(5 \times 10^{-3})(5 \times 10^{-3})}{(5 \times 10^{-3}) + (5 \times 10^{-3})}$$

$$= 2.5 \text{ mH}$$

LEARNING EVALUATIONS

1. The voltage across a 4-mH inductor is given by the relation

$$v_s(t) = 0, \qquad t \leq 0$$
$$= 4e^{-3t} \text{ mV}, \qquad t > 0$$

Find and plot $i_L(t)$. $i_L(0) = 0$ A.

2. The current through a 15-mH inductor is shown in Fig. 4.44.

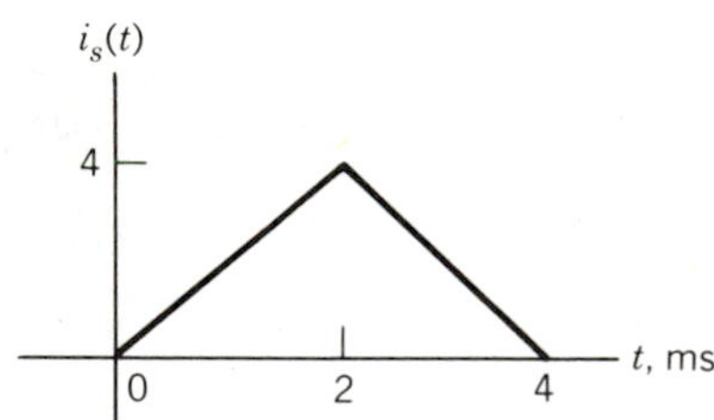

Figure 4.44

Find and plot the resulting voltage.

3. Find and plot the energy stored and power delivered for the inductor circuit in Problem 1.
4. Find the equivalent inductance for the following circuit in Fig. 4.45.

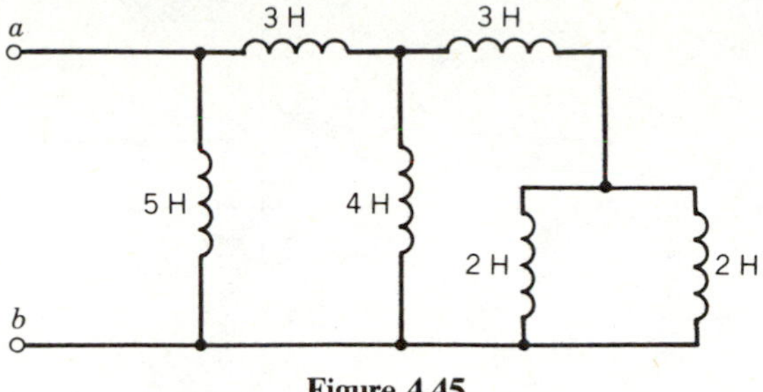

Figure 4.45

4.5 Inductor Waveforms

LEARNING OBJECTIVES

After completing this section you should be able to do the following:

1. State the continuity condition for inductors.
2. Find the initial voltage and current in an inductor.
3. Find the final voltage and current in an inductor.

The inductor has the ability to store energy in a manner that is analogous to the way that energy may be stored in a capacitor. Because of this there is a continuity condition for the inductor. An understanding of this continuity condition will allow us to derive both initial values and final values of current and voltage in an inductor. These initial and final values will then be used in the next chapter as part of the solution for variables in circuits containing resistors, capacitors, and inductors.

Consider the circuit in Fig. 4.46, which contains a voltage source and an inductor. The current in the circuit is given by

$$\begin{aligned} i(t) &= \frac{1}{L}\int_{-\infty}^{t} v(\lambda)\, d\lambda \\ &= \frac{1}{L}\int_{0}^{t} v(\lambda)\, d\lambda + i(0) \end{aligned} \tag{4.5.1}$$

The initial state of the inductor is $i(0)$. This term summarizes the past history (before $t = 0$) of the inductor in a single number. It is this initial state that we must specify when solving for the response of circuits containing inductors.

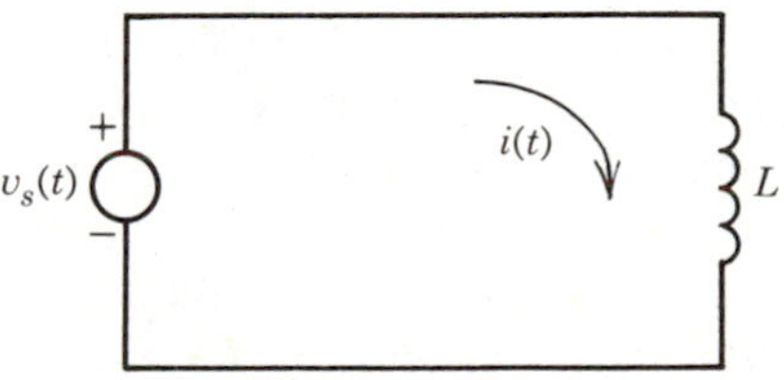

Figure 4.46

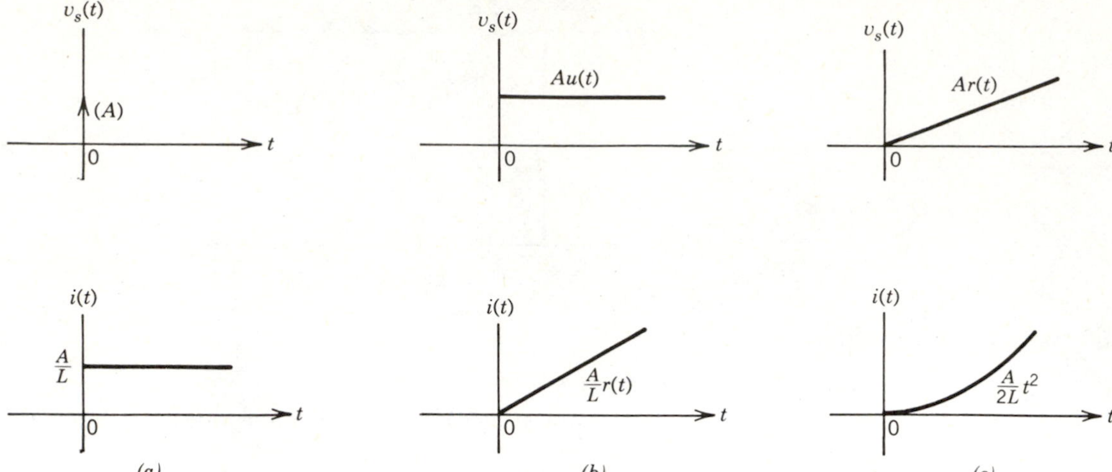

Figure 4.47. Input–output pairs for the inductor.

Several input–output pairs are shown in Fig. 4.47. The input is the source voltage $v_s(t)$ and the output is the current through the inductor $i(t)$. We mean by this that we will control the voltage (input) and then measure the current (output). In each case we have assumed an initial current of zero. That is, $i(0^-) = 0$. Using Eq. 4.5.1 the δ-function input in Fig. 4.47*a* produces a current given by

$$i(t) = \frac{1}{L}\int_0^t A\delta(\lambda)\,d\lambda = \frac{A}{L}u(t)$$

The step input in Fig. 4.47*b* produces the current

$$i(t) = \frac{1}{L}\int_0^t Au(\lambda)\,d\lambda = \frac{A}{L}r(t)$$

Finally, the ramp input produces the parabola response given by

$$i(t) = \frac{1}{L}\int_0^t Ar(\lambda)\,d\lambda = \frac{A}{2L}t^2u(t)$$

The inductor stores energy in the form of flux. The presence of current generates a flux field about the inductor, and this magnetic flux represents stored energy. If the initial current in Fig. 4.47 is not zero, then the current for $t > 0$ is changed by this initial current. This is illustrated in Fig. 4.48, where an impulse, step, and ramp voltage are applied to an inductor with nonzero initial current.

The continuity condition for inductors is similar to that for capacitors (see Section 4.3). Because integration is used to find current for a given voltage in an inductor, the continuity condition states that there are no discontinuities in the current waveform for any voltage that we can generate in the laboratory. This is illustrated in Figs. 4.47 and 4.48. Unless a δ-function voltage waveform is applied to the inductor (an impossibility), the resulting current is continuous. We can state this formally as follows.

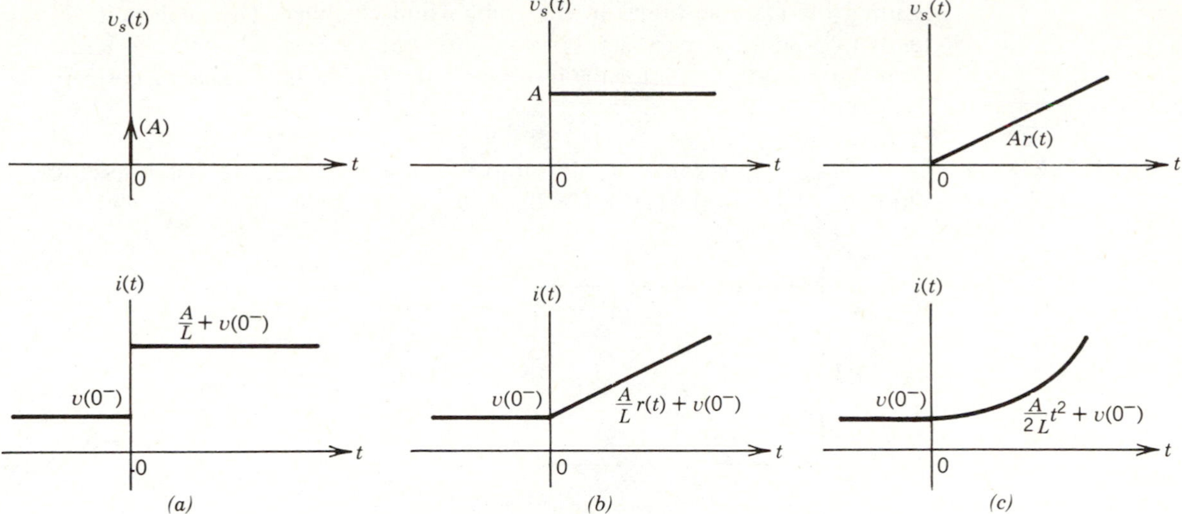

Figure 4.48. Inductor waveforms with initial conditions.

> ***Continuity Condition for Inductors.*** The current through an inductor is a continuous function of time if the voltage is bounded.[2]

An interesting application of the principles that led us to the continuity condition can be found in the method used to generate high voltage in your TV set. If we reverse the input and output so that current $i(t)$ is the inductor input and voltage across the inductor is the output we have

$$v(t) = L\frac{di}{dt}$$

Now consider Fig. 4.48*a*. Here it is possible to approximate the step current through the inductor, so that the resulting voltage will approximate a δ function. A current is established in the inductor and then switched off by an electronic device, such as a transistor. The resulting negative step of current then produces an extremely high voltage across the inductor for a short duration. This voltage waveform is then averaged by a filter circuit to produce the 20,000 or so volts used to accelerate electrons toward the picture tube face. It is the action of the electrons striking the phosphor on the face of the tube that produces the picture.

The concept of an inductor (also often called a "coil" because of the nature of its construction) storing energy is used in many other applications. An automobile ignition system uses the energy from a coil to fire a spark plug in exactly the same manner as the high voltage is generated in a TV system. The "points" in the ignition system serve as the switch to break the current in the coil. The resulting high voltage produces a spark across the narrow gap in the spark plug, thus

[2]As was the case for the capacitor, this applies to a linear time-invariant inductor.

igniting the gasoline fumes in the combustion chamber. This high voltage also tends to produce a spark across the points, but a capacitor is placed across the points to provide a path for the resulting high current, thus reducing corrosion of the points.

EXAMPLE 4.5.1 The voltage across the RL circuit in Fig. 4.49 is a square Pulse as shown. Find $i_L(t)$, $i_R(t)$, and $i(t)$ if $i(0) = 0$.

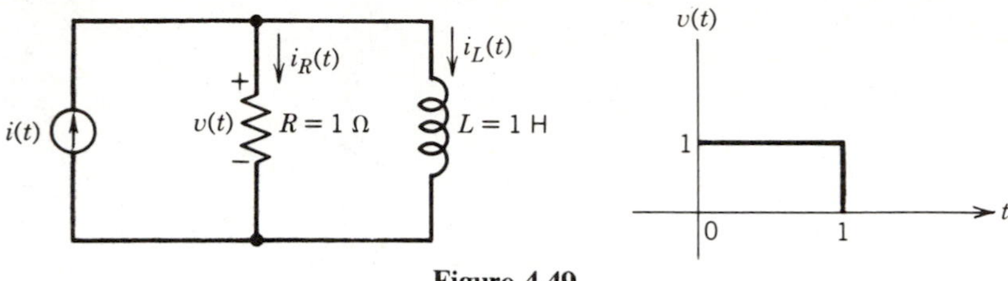

Figure 4.49

Solution

From Eq. 4.5.1,

$$i_L(t) = \int_0^t v(\lambda)\, d\lambda = r(t) - r(t-1)$$

$$i_R(t) = \frac{v(t)}{R} = u(t) - u(t-1)$$

$$i(t) = i_L(t) + i_R(t) = u(t) + r(t) - [u(t-1) + r(t-1)]$$

These waveforms are plotted in Fig. 4.32 if we substitute $i_L(t)$ for $v_c(t)$, $i_R(t)$ for $v_R(t)$, and $i(t)$ for $v(t)$. Notice that the current through the inductor [which has the shape of $v_c(t)$ in Fig. 4.32] is a continuous function of time.

Since any voltage that we can generate must be bounded (i.e., is less than infinity), the continuity condition for inductors means that we must have

$$i_L(0^-) = i_L(0^+)$$

This applies only to the current through an inductor. The continuity condition imposes no restriction on the voltage across an inductor. Consider the parallel RL circuit in Fig. 4.50a. The constant current source in parallel with a switch that opens at $t = 0$ supplies a step current to the circuit, as shown in Fig. 4.50b. Before the switch is open the current I flows through the switch and the branch currents i_R and i_L are both zero. Thus we have

$$i_L(0^-) = 0$$

$$v(0^-) = 0$$

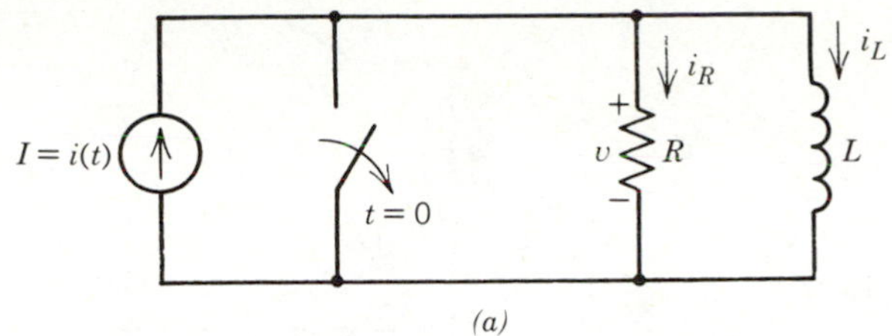

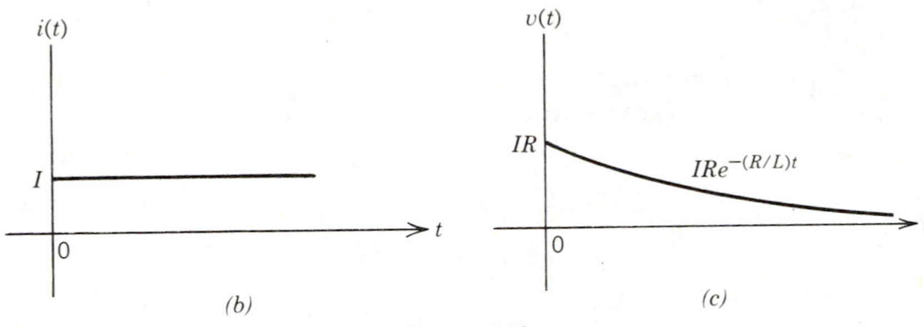

Figure 4.50

At $t = 0^+$ the source current flows into the top node.

$$i_L(0^+) = 0$$

By the continuity condition then, $i_R(0^+) = I$ and the voltage v becomes

$$v(0^+) = IR$$

Thus the voltage changes abruptly at $t = 0$. The complete voltage waveform is shown in Fig. 4.50c. Notice that as t becomes large, $v(t)$ approaches zero.

Physically the following events occur in Fig. 4.50. After the switch is opened the flux field begins to build up about the inductor. This transfer of energy from the source to the field cannot occur instantaneously, so the inductor current must be continuous. Since Kirchhoff's current law must always be satisfied, the source current must initially flow through the resistor, imposing a voltage across the circuit $v(0^+) = IR$. As time increases the inductor current increases until it is limited only by the source. Since all the source current is through the inductor, the resistor current is zero. With $v = Ri_R$ we have $v(\infty) = 0$. That is,

$$\lim_{t \to \infty} i_R(t) = 0$$

$$\lim_{t \to \infty} v(t) = 0$$

EXAMPLE 4.5.2 Find the current through the inductor at $t = 0^+$ and at $t = \infty$ in Fig. 4.51. The applied voltage is a unit step, and $i_L(0^-) = 0$.

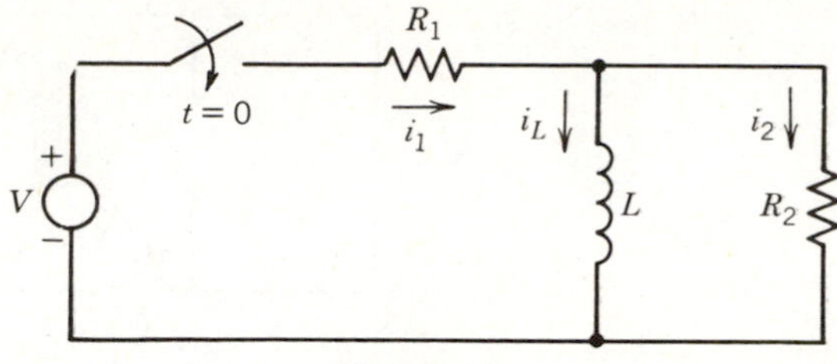

Figure 4.51

Solution

By the continuity condition

$$i_L(0^+) = i_L(0^-) = 0$$

Therefore, we have $i_1(0^+) = i_2(0^+) = V/(R_1 + R_2)$. At $t = \infty$ the inductor is fully energized, and $i_1 = i_L$, or $i_2 = 0$. Therefore,

$$i_L(\infty) = \frac{V}{R_1}$$

Table 4.1 provides a summary of the behavior of all three elements, resistors, capacitors, and inductors, at $t = 0$ and $t = \infty$. The resistor is not affected by initial or final conditions since it does not store energy. As we have seen, however, the capacitor and inductor are a different matter.

An initially uncharged capacitor acts like a short circuit at $t = 0$ because the voltage across it does not change, and the current is not limited by the capacitor. If the capacitor is charged to a voltage V as shown by the third entry in Table 4.1, then the continuity condition assures us that $v(0^+) = V$. Hence, the equivalent circuit for the charged capacitor at $t = 0^+$ is a voltage source of value V. The final condition is derived from the relationship

$$i(t) = C\frac{dv}{dt}$$

If the final voltage is a constant (a fully charged capacitor), then $i(\infty) = 0$. Hence, the equivalent circuit is an open circuit in series with a voltage source whose value is this final voltage.

An inductor with zero initial current acts like an open circuit at $t = 0$ because the current through it does not change, regardless of the nature of the applied voltage, just as long as the voltage is finite. If the initial current is not zero, then the inductor acts like a current source at $t = 0$. The final condition is derived from the relationship

$$v(t) = L\frac{di}{dt}$$

If the final current is constant, then $v(\infty) = 0$. Hence, the equivalent circuit is a short circuit in parallel with a current source (which may be zero) as shown in Table 4.1.

TABLE 4.1
EQUIVALENT CIRCUIT ELEMENTS AT $t = 0^+$ AND AT $t = \infty$

Element	Equivalent Circuit	
	At $t = 0^+$	**At $t = \infty$**
C, $v = 0$	SC	OC
C, $-\ +$, $v = V$	$-\ +$, V	$-\ +$, V, OC
L, $i = 0$	OC	SC
L, $i = I$	I	SC, I

LEARNING EVALUATION
The switch in Fig. 4.52 has been in position a for a long time. The switch is changed to position b at $t = 0$. Find the voltage across and current through the inductor at $t = 0^-$ and at $t = 0^+$.

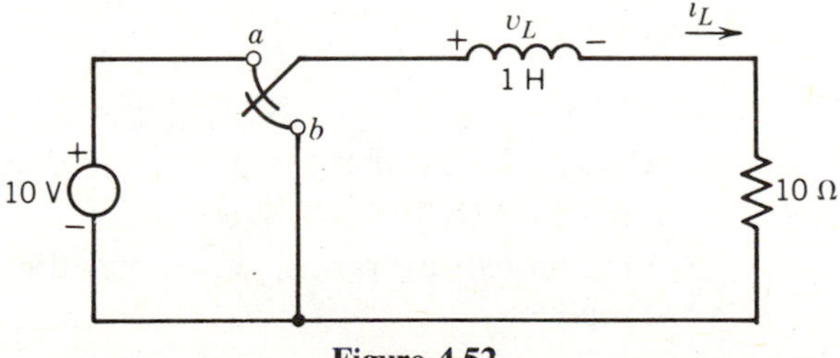

Figure 4.52

PROBLEMS

SECTION 4.1

1. Find the current through a 30-μF capacitor if each waveform in Fig. 4.53 is the capacitor voltage.

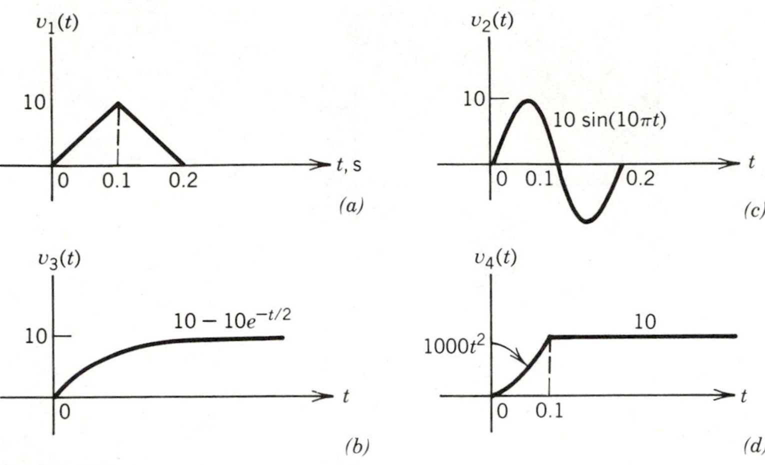

Figure 4.53

2. Find the voltage across an initially uncharged 30-μF capacitor if each waveform in Fig. 4.54 is the capacitor current.

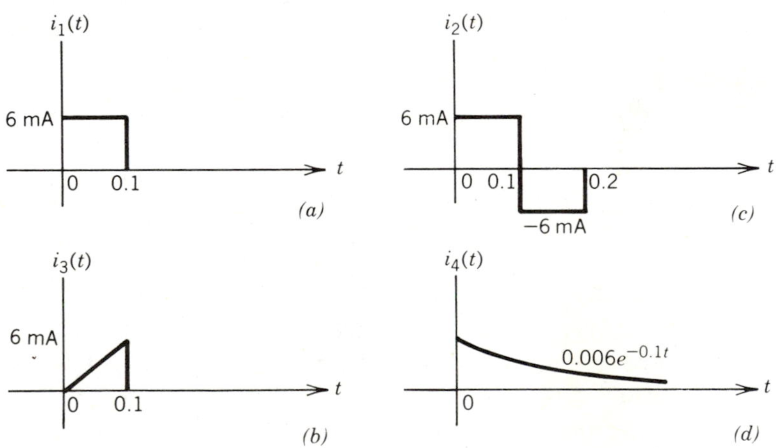

Figure 4.54

3. Repeat Problem 2 if the capacitor has an initial voltage $v(0) = 10$ V.
4. Find and plot the power $p(t)$ supplied to the capacitor for each voltage waveform in Problem 1.
5. Find an plot the power $p(t)$ supplied to the capacitor for each current $i(t)$ in Problem 2.

6. Find and plot the energy $w(t)$ stored in the capacitor for each voltage waveform in Problem 1.

7. Find and plot the energy $w(t)$ stored in the capacitor for each current waveform in Problem 2.

8. Find the equivalent capacitance between terminals a–b in Fig. 4.55.

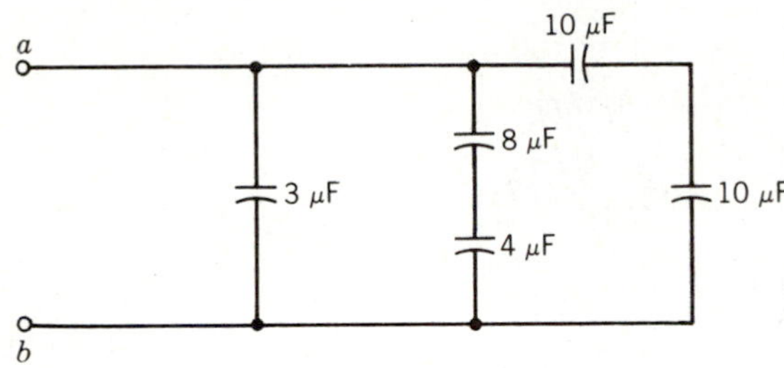

Figure 4.55

SECTION 4.2

9. Sketch the waveforms defined by the following analytical representations.

a. $y(t)=20-10e^{-t},\ t\geq 0$

b. $y(t)=-20-10e^{-t},\ t\geq 0$

c. $x(t)=15\cos 500t,\ t\geq 0$

d. $x(t)=-3+\sin 500\pi t,\ t\geq 0$

e. $y(t)=e^{-t}\sin \pi t,\ t\geq 0$

f.

$$z(t)=\begin{cases} 0 & t<1 \\ 12(t-1) & 1\leq t\leq 2 \\ 3(t-4)^2 & 2\leq t\leq 4 \\ 0 & t>4 \end{cases}$$

g. $x(t)=4\delta(t)-4\delta(t-1)-4\delta(t-3)+2\delta(t-5)-2\delta(t-6)$

$y(t)=\int_0^t x(\lambda)\,d\lambda,\ 0\leq t$

h. $x(t)=u(t)-u(t-3)+4u(t-4)+u(t-6)$

$y(t)=\dfrac{d}{dt}[x(t)]$

10. Give an analytical representation for each waveform in Fig. 4.53 in terms of step, ramp, and parabolic functions.

11. Repeat Problem 10 for Fig. 4.54.

12. Find and plot the derivative of each function shown in Fig. 4.53.

13. Repeat problem 12 for Fig. 4.54.

14. Find and plot the integral of each function shown in Fig. 4.53.

15. Repeat Problem 14 for Fig. 4.54.

16. Evaluate each of the following integrals.

a. $\int_{-\infty}^{t} 2\delta(\lambda - 2)\, d\lambda$

b. $\int_{0}^{t} 2\delta(\lambda - 2)\, d\lambda$

c. $\int_{-\infty}^{\infty} 2\delta(\lambda - 2)\, d\lambda$

17. Evaluate each of the following integrals

a. $\int_{-\infty}^{\infty} \cos(2\pi t)\delta(t-1)\, dt$

b. $\int_{-\infty}^{0} \cos(2\pi t)\delta(t-1)\, dt$

c. $\int_{0}^{2} \cos(2\pi t)\delta(t-1)\, dt$

SECTION 4.3

18. The capacitor in Fig. 4.56 is initially uncharged. The switch is closed at $t = 0$. What are the capacitor current and voltage at $t = 0^+$?

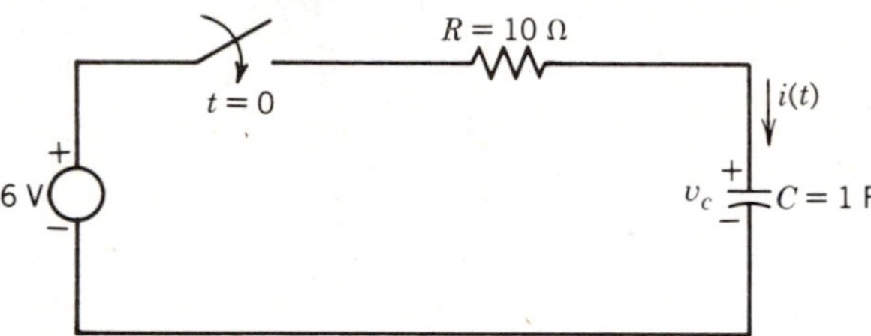

Figure 4.56

19. In Problem 18, what are the final ($t = \infty$) capacitor current and voltage?

20. The switch in Fig. 4.57 has been closed for a long time. It is opened at $t = 0$. Find the capacitor current and voltage at $t = 0^+$.

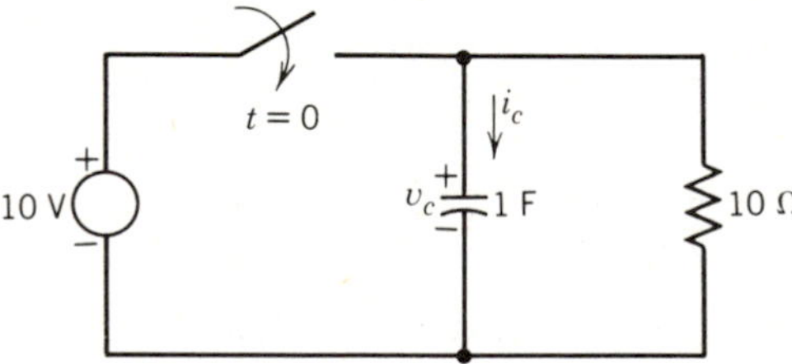

Figure 4.57

21. Find the final ($t = \infty$) capacitor voltage and current in Problem 20.

22. The switch is closed at $t = 0$ in Fig. 4.58. Find the initial current $i(0^+)$ and $i_c(0^+)$ if the capacitor is initially uncharged.

23. Find the final current $i(\infty)$ and the final capacitor voltage $v_c(\infty)$ in Problem 22.

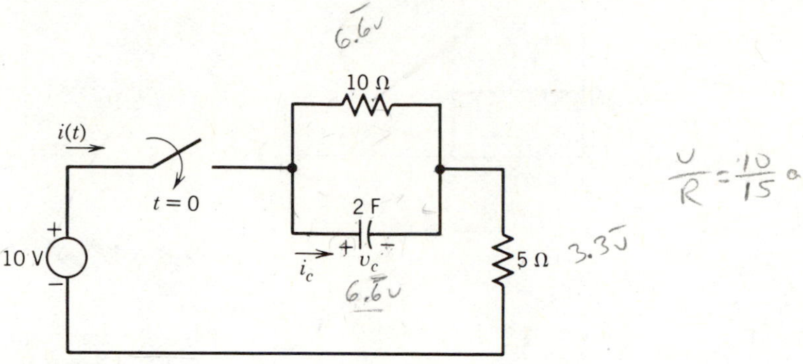

Figure 4.58

SECTION 4.4

24. Find the voltage across a 10-mH inductor if each waveform in Fig. 4.59 is the inductor current.

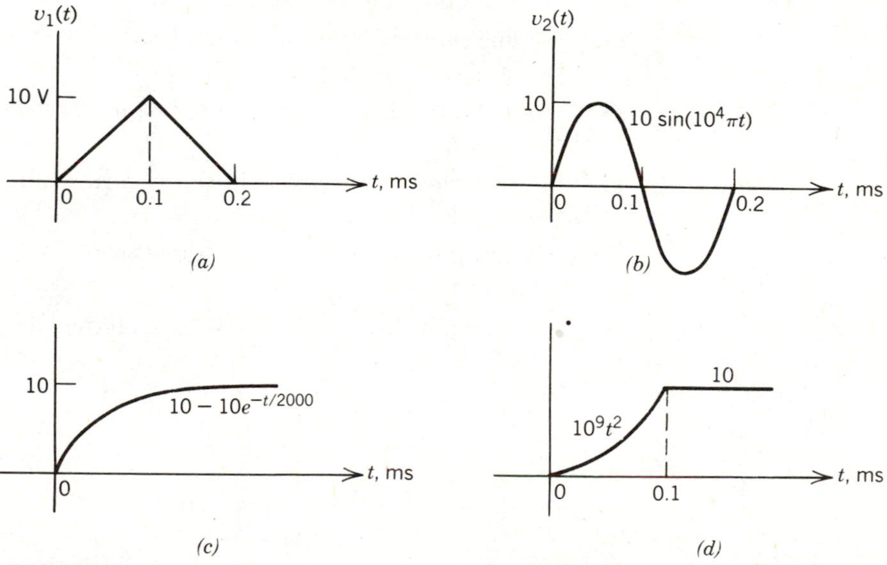

Figure 4.59

25. Find the current through a 10-mH inductor if each waveform in Fig. 4.60 is the inductor voltage. The initial inductor current is zero.

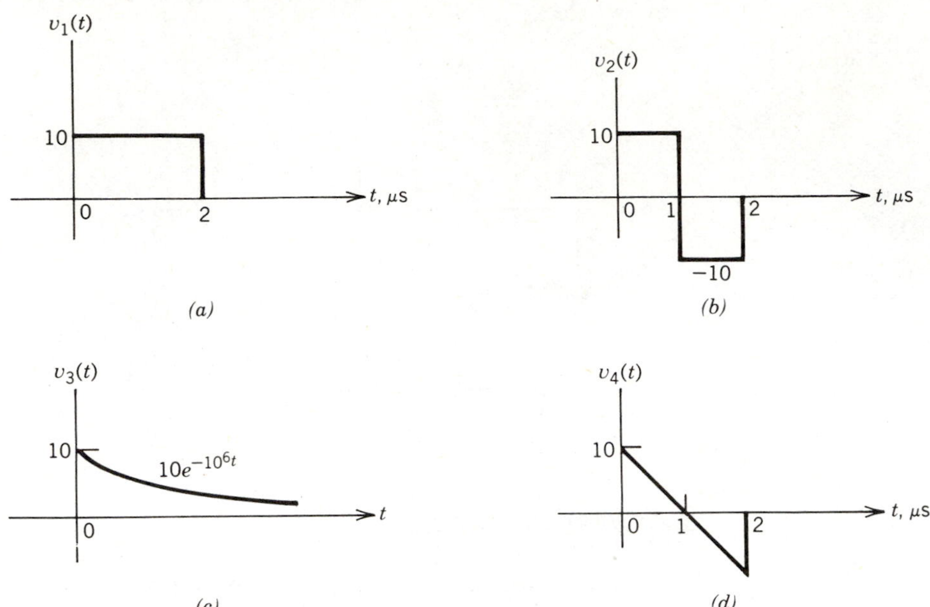

Figure 4.60

26. Repeat Problem 25 if the initial inductor current is 1 ma.

27. Find and plot the power $p(t)$ supplied to the inductor for each waveform in Problem 24.

28. Find and plot the power $p(t)$ supplied to the inductor for each waveform in Problem 25.

29. Find and plot the energy $w(t)$ stored in the inductor for each waveform in Problem 24.

30. Find and plot the energy $w(t)$ stored in the inductor for each waveform in Problem 25.

31. Find the equivalent inductance between terminals a–b in Fig. 4.61.

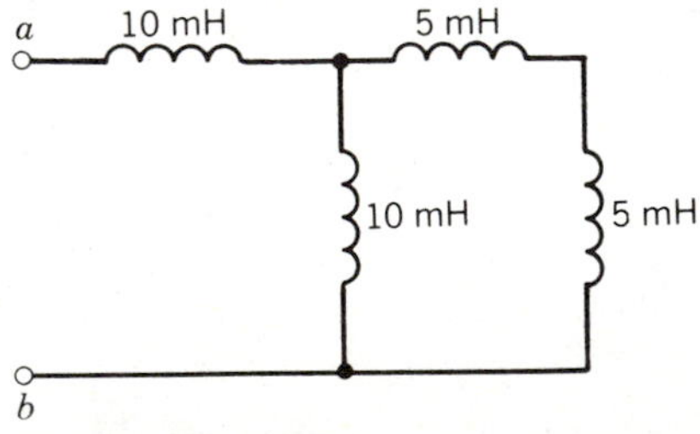

Figure 4.61

SECTION 4.5

32. The switch in Fig. 4.62 is closed at $t = 0$. If the initial inductor current is zero, find $i(0^+)$ and $v_L(0^+)$.

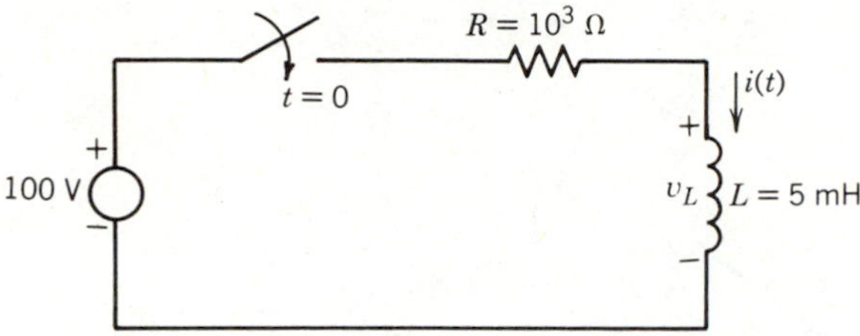

Figure 4.62

33. In Problem 32 what is the final ($t = \infty$) inductor current and voltage?

34. The switch in Fig. 4.63 is opened at $t = 0$. Find $v_L(0^+)$ and $i_L(0^+)$. Also find the final ($t = \infty$) inductor current and voltage.

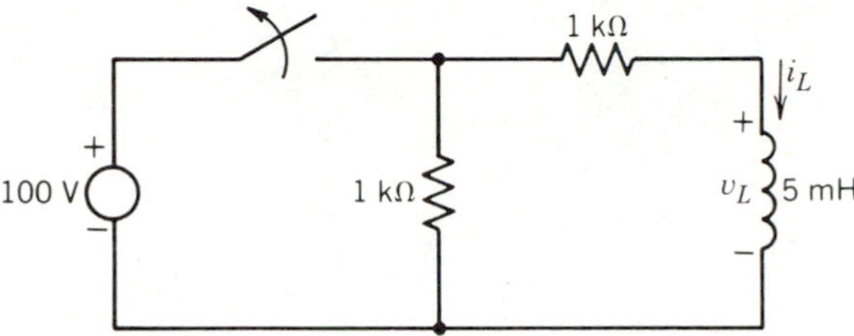

Figure 4.63

35. The switch in Fig. 4.64 is closed at $t = 0$. If the initial capacitor voltage and inductor current are both zero, find $v_R(0^+)$ and $i_R(0^+)$.

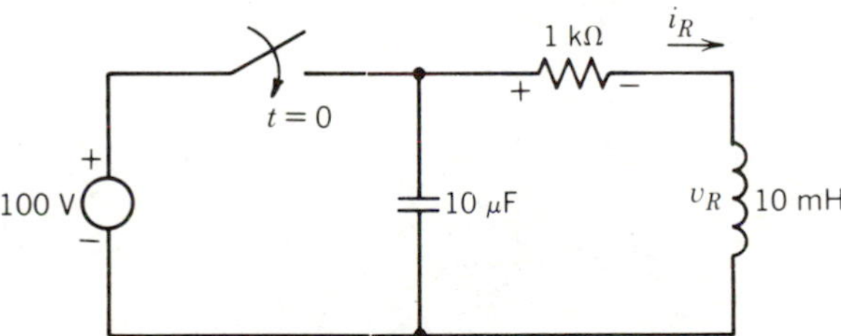

Figure 4.64

36. Suppose the switch in Fig. 4.64 has been closed for a long time when it is opened at $t = 0$. Find $v_R(0^+)$ and $i_R(0^+)$.

LEARNING EVALUATION ANSWERS

Section 4.1

1.

$$i(t) = \begin{cases} 0 & t < 0 \\ 0.4e^{-t/2}\ \text{mA} & t > 0 \end{cases}$$

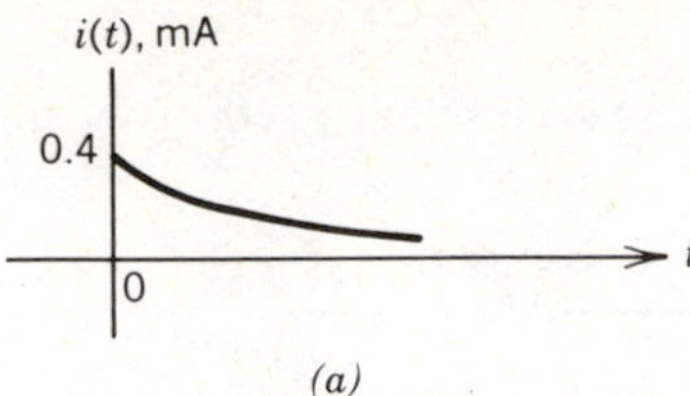

(a)

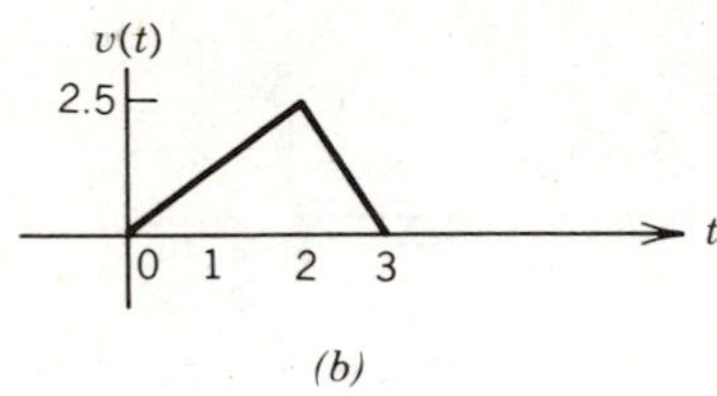

(b)

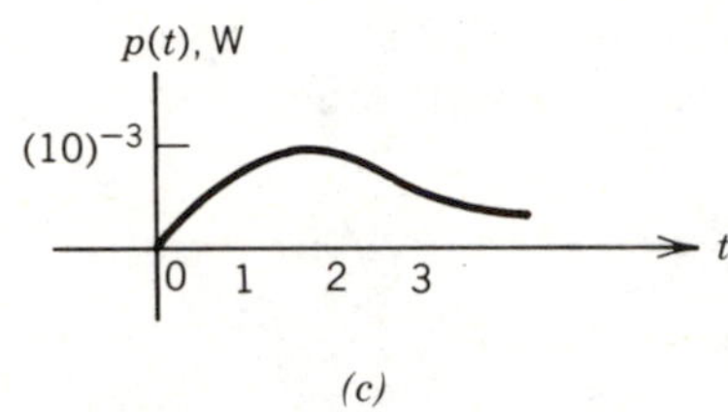

(c)

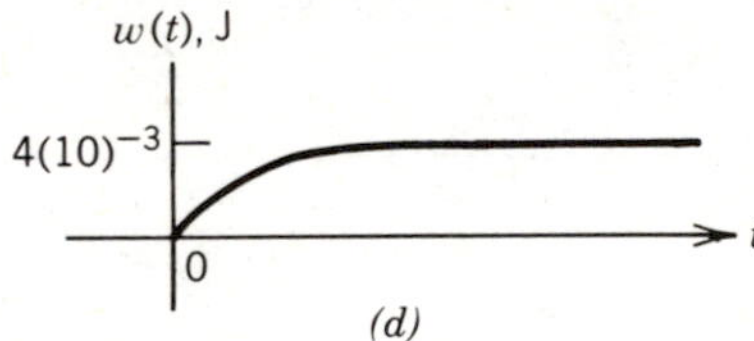

(d)

Figure 4.65

2.

$$v(t) = \begin{cases} 0 & t > 0 \\ \dfrac{25}{2}t & 0 > t > 2 \\ -25(t-3) & 2 < t < 3 \\ 0 & t > 3 \end{cases}$$

(Fig. 4.65*b*)

3.

$$p(t) = \begin{cases} 0 & t < 0 \\ 4(10)^{-3}(e^{-t/2} - e^{-t}) & t > 0 \end{cases}$$

(Fig. 4.65*c*)

4.

$$w(t)=\begin{cases} 0 & t<0 \\ 4(10)^{-3}(1-2e^{-t/2}+e^{-t}) & t>0 \end{cases}$$

(Fig. 4.65d)

5. $C = 3F$.

Section 4.2

1. See Fig. 4.66.

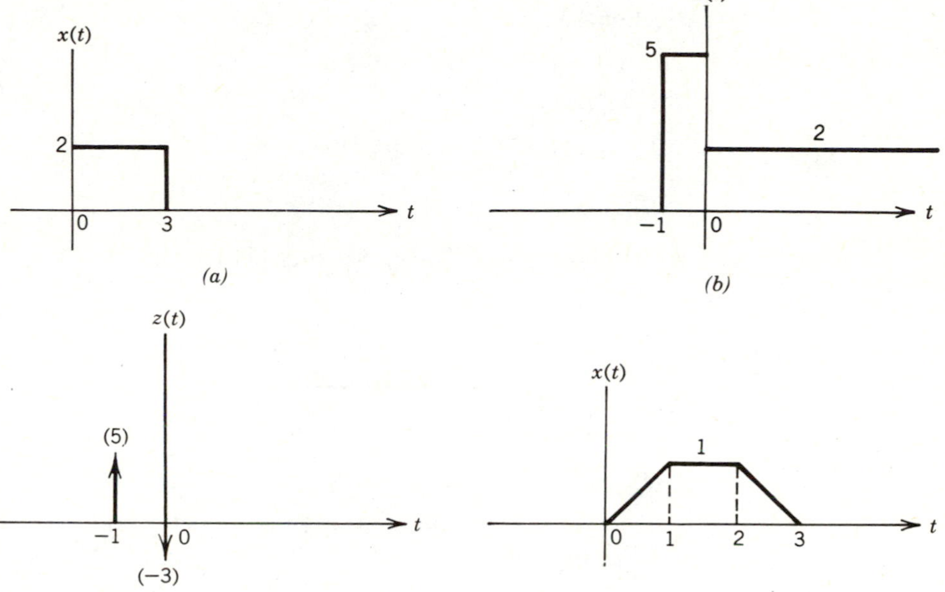

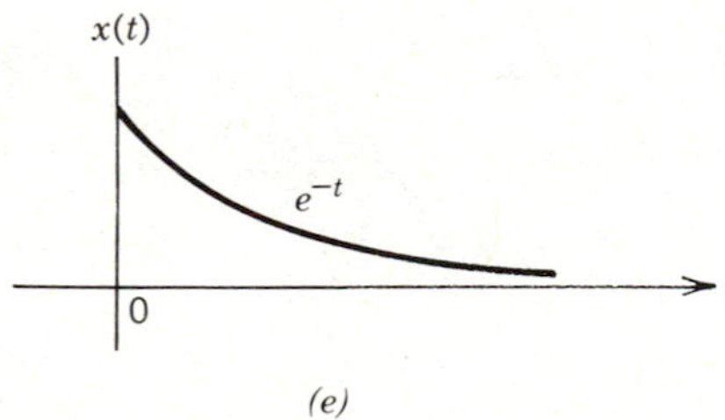

Figure 4.66

2. a. $v_1(t) = 0.5r(t-1) - 0.5r(t-3)$
b. $v_2(t) = 0.5r(t-1) - r(t-3) + 0.5r(t-5)$
c. $v_3(t) = \delta(t-1) + 2\delta(t-2) - \delta(t-3) + 3\delta(t-4)$

3. See Fig. 4.67.

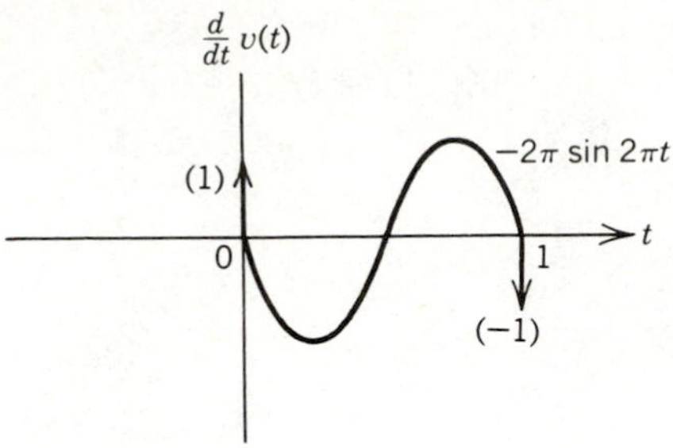

Figure 4.67

Section 4.3

$v(0^-) = v(0^+) = 10$ V

$i(0^-) = 0$

$i(0^+) = 2$ A up through the capacitor

Section 4.4

1. $i_L(t) = \frac{1}{3}(1 - e^{-3t})u(t)$ (See Fig. 4.68)

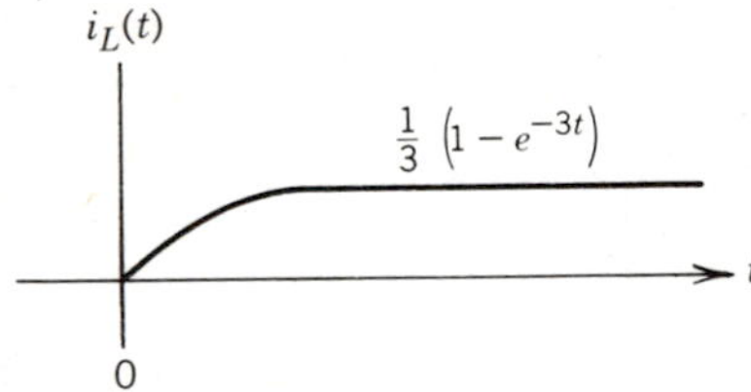

Figure 4.68

2. $v(t) = 15u(t) - 30u(t - 0.002) + 15u(t - 0.004)$
(See Fig. 4.69)

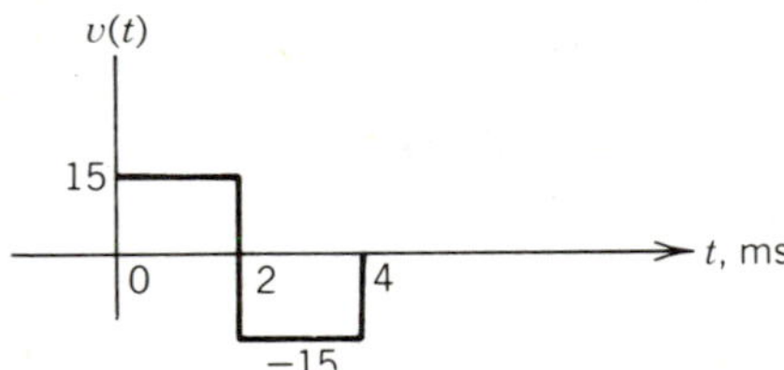

Figure 4.69

3. $w(t) = \frac{2}{9}(10)^{-3}(1 - 2e^{-3t} + e^{-6t})$ J
$p(t) = \frac{4}{3}(10)^{-3}(e^{-3t} - e^{-6t})$ W

4. 2.5 H

Section 4.5

$i_L(0^-) = 1$ A

$i_L(0^+) = 1$ A

$v_L(0^-) = 0$

$v_L(0^+) = -10$ V

CHAPTER 5 FIRST ORDER CIRCUITS

We have progressed to the point where we can have up to three different elements in addition to energy sources in our circuits, but we have not discussed ways of analyzing such circuits. If the circuit contains only energy sources (forcing functions) and resistors, we are experienced in applying our analysis tools to find the response of the circuit (voltage across and current through each element) to these forcing functions. Of the tools in our bag, only Ohm's law applies exclusively to resistors. The remaining tools apply to circuits containing any combination of resistors, inductors, and capacitors because such circuits are linear. Also, in Chapter 4 we developed the voltage–current relationships across capacitors and inductors, which are equivalent to Ohm's law for resistors.

In this chapter we will investigate circuits that contain a resistor and capacitor or a resistor and inductor. We leave until Chapter 6 the analysis problem for circuits containing all three elements. Since the v–i relationship across inductors and capacitors involve integral and differential relationships, the equations resulting from the application of KCL and KVL to circuits containing these elements will be integro-differential equations. These equations are much more difficult to solve than the algebraic equations associated with resistive circuits and considerable time and effort in the next several chapters will be devoted to the solution and interpretation of such equations.

For the equations encountered in the next two chapters, we will present only one method of solution for each type of equation. Many different methods exist and are covered in courses in calculus and differential equations. In these two chapters, our interest will primarily be in interpretation of the solution and not in the methods of generating the solution itself.

5.1 First Order Differential Equations

LEARNING OBJECTIVES

After completing this section you should be able to do the following:

1. Determine the order and characterize a given differential equation as
   a. Ordinary or partial.
   b. Linear or nonlinear.
   c. Time variant or time invariant.
   d. Homogeneous or nonhomogeneous.
2. Find the characteristic equation and homogeneous solution of a first order equation.
3. Use the method of undetermined coefficients to find the forced response of a first order equation.

The calculus was developed independently by Sir Issac Newton (1642–1727) and Gottfried Wilhelm Leibnitz (1646–1716). Both men were interested in describing the relationships that existed between position, velocity, and acceleration in the motion of objects, as well as a multitude of other similar relationships in applied science. It is not surprising that two people should have independently developed the calculus at that time, for it was "in the air," so to speak. Many people were aware of the limitations of algebra for the purpose of describing relationships between continuous quantities, and it was just a matter of time (and genius) before someone rectified the situation. Today we make a distinction between calculus and differential equations in most math courses, but there was no such distinction in the minds of Newton and Leibnitz. The study of relationships between continuous quantities is, today, usually termed "integral equations" and "differential equations." Integral equations contain functions and their integrals. Differential equations contain functions and their derivatives. Either type may be used in circuit analysis, but we will stick to differential equations.

The general form of the differential equation that we will be concerned with in this chapter is given by

$$a_1 \frac{d}{dt} x(t) + a_0 x(t) = f(t) \tag{5.1.1}$$

This equation is:

- Ordinary because it contains ordinary derivatives instead of partial derivatives.
- First order because the highest derivative term is of first order.
- Linear because there are no products or powers of the variable $x(t)$ or any derivatives of $x(t)$.
- Time invariant (or constant coefficient) because the coefficients a_0 and a_1 are not functions of time.
- Homogeneous if $f(t) = 0$. Otherwise it is nonhomogeneous. The term $f(t)$ is called the forcing function.

In solving any problem in mathematics it is essential to know the nature of the problem and its solution. Equation 5.1.1 is a relationship between *functions*. The variable $x(t)$ is a function of time, the derivative $(d/dt)x(t)$ is another function of time, and the forcing function $f(t)$ is a third function of time. Equation 5.1.1 states that the sum of the two functions on the left, when multiplied by the coefficients a_0 and a_1, must equal $f(t)$ at every value of time.

We assume that a_0, a_1, and $f(t)$ are known. The *solution* of Eq. 5.1.1 is therefore that function $x(t)$ which satisfies the equation. That is, we are not looking for a number, or a geometric vector, or a matrix when we try to solve a differential equation. We are looking for a function.

The art of solving differential equations is essentially a guessing game. We choose a function $x(t)$ and see if it satisfies the equation for all time. If not, we try another guess. A number of procedures have been developed since the time of Newton and Leibnitz for making intelligent guesses, and we will use the simplest of these procedures in our applications. But in order to illustrate the problem, assume for a moment that we have had no previous experience, and consider the simplest possible equation.

$$\frac{d}{dt}x(t)+x(t)=0 \tag{5.1.2}$$

We would naturally begin by selecting some function to try, such as the triangular function $x_1(t)$ shown in Fig. 5.1*a*. The derivative is the rectangular function shown in Fig. 5.1*b*. Equation 5.1.2 says that the sum of these two functions should be zero, and furthermore, this sum should be zero at every instant of time. A glance at Fig. 5.1 shows that this sum is not zero except at isolated points in the interval $0<t<2$, so $x_1(t)$ is certainly not a solution.

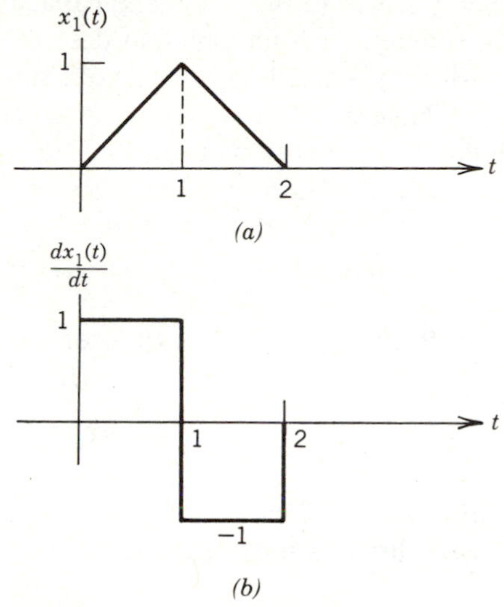

Figure 5.1. Trial solution for a differential equation.

At this point we might begin to ask some important questions. Does a function exist that satisfies the equation? If one such function exists, is it unique, or might there be more, possibly an infinite number of such functions? The answer is that there are an infinite number of functions that satisfy Eq. 5.1.2 as it is stated, but they are all of exponential form. If we further restrict Eq. 5.1.2 by specifying an initial condition, then there is exactly one solution. Knowing this, let us try another guess. We might try $x_2(t)=e^t$ and discover that this does not work. Other specific exponential forms will not work either, unless we get lucky and by chance hit upon the correct solution. A more intelligent approach is to try a general exponential form and see if we can determine the specific solution later. Thus, we try

$$x_3(t)=Ae^{st} \tag{5.1.3}$$

where A and s are numbers to be determined later. Substituting Eq. 5.1.3 into Eq. 5.1.2 gives

$$Ase^{st}+Ae^{st}=0 \tag{5.1.4}$$

Dividing each term by Ae^{st} gives

$$s+1=0 \tag{5.1.5}$$

This is called the *characteristic equation* and the value $s=-1$ is called the *natural frequency* of the differential equation. Therefore the solution to Eq. 5.1.2 is given by

$$x(t)=Ae^{-t} \tag{5.1.6}$$

Notice that we may use any value of A in the solution, giving rise to an infinite number of functions that satisfy the differential equation. However, if the value of $x(t)$ is known at any time (say, $t=0$), then the solution is unique. The initial energy stored in inductors and capacitors will be used to specify initial conditions for differential equations that describe circuits.

All first order, linear, ordinary, time-invariant homogeneous differential equations have a solution that is exponential. Therefore, our procedure will be to assume the solution given by Eq. 5.1.3, substitute this into the differential equation to obtain the characteristic equation, and find the natural frequency to use in the solution. The constant A in Eq. 5.1.3 will then be determined from initial conditions.

EXAMPLE 5.1.1 Find the solution to the differential equation given by

$$\frac{d}{dt}x(t)+2x(t)=0$$
$$x(0)=5 \tag{5.1.7}$$

Solution

Substitute the general solution given by Eq. 5.1.3 into Eq. 5.1.7 to obtain

$$Ase^{st}+2Ae^{st}=0$$

or

$$s+2=0$$

Therefore, the natural frequency is $s=-2$, giving

$$x(t)=Ae^{-2t} \tag{5.1.8}$$

The value of $x(t)$ at $t=0$ is given as 5. Setting $t=0$ in Eq. 5.1.8 gives

$$x(0)=5=Ae^{0}=A$$

so

$$x(t)=5e^{-2t}$$

To show that this is, indeed, the solution of Eq. 5.1.7, we have plotted $2x(t)$ and $(d/dt)x(t)$ in Fig. 5.2. The sum of these two functions is obviously zero for every t.

There are a number of methods for finding the particular solution, but we will present only one of these, the method of undetermined coefficients. When the forcing function is a function for which repeated differentiation yields only a finite number of independent derivatives, then a particular solution of Eq. 5.1.1 can be found by assuming $x(t)$ to be an arbitrary linear combination of $f(t)$ and

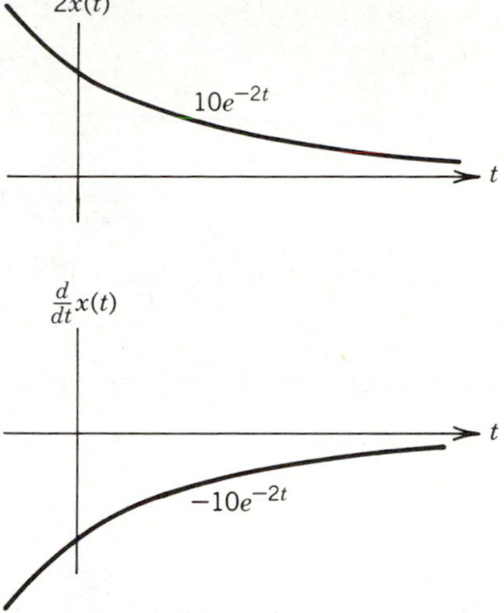

Figure 5.2. The solution in Example 5.1.1.

all its independent derivatives. The assumed solution is substituted into Eq. 5.1.1 and the constants (undetermined coefficients) are determined so that the resulting equation is satisfied. The class of forcing functions having only a finite number of linear independent derivatives consists of

Constant
t^n (n is a positive integer)
e^{at}
$\cos at$
$\sin bt$

and any others obtainable from these by a finite number of additions, subtractions, and multiplications.

In order to obtain an intuitive understanding of the method of undetermined coefficients, let us consider a simple example. Suppose the equation to be solved is given by

$$\frac{d}{dt}x(t)+x(t)=f(t) \tag{5.1.9}$$

where $f(t)$ is the ramp function $f(t)=t$ shown in Fig. 5.3. Equation 5.1.9 states that the solution $x(t)$ must be some function such that the sum $(d/dt)x(t)+x(t)$ will equal the function in Fig. 5.3. As a first guess, we might try $x(t)=t$, that is,

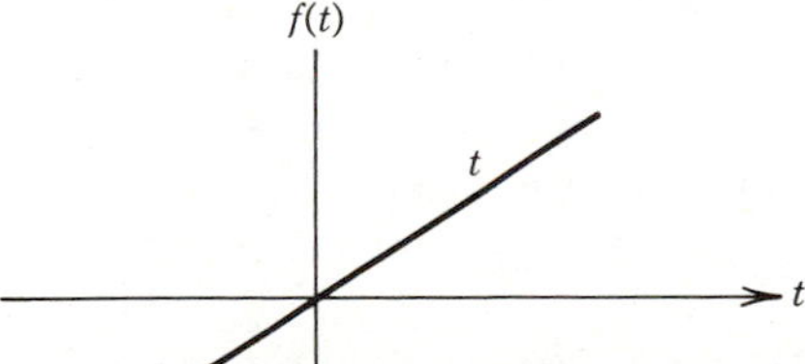

Figure 5.3. The forcing function $f(t)=t$.

let $x(t)=f(t)$. This would be the solution if the derivative was zero, which it is not. Instead, when

$$x(t)=t$$

$$\frac{d}{dt}x(t)+x(t)=t+1$$

we are close, but not quite correct. A better guess is

$$x(t)=t-1 \tag{5.1.10}$$

This is the correct solution because now

$$\frac{d}{dt}x(t)+x(t)=t$$

Instead of having to guess the correct solution by trial and error, it is better to try a general solution and solve for the coefficients. This is exactly what we did for the homogeneous solution, so there is great similarity between the two procedures. The general solution is given by

$$x(t)=At+B$$

Then

$$\frac{D}{dt}x(t)=A$$

Substituting into Eq. 5.1.9 gives

$$A+At+B=t$$

Equating like coefficients on each side of the equation gives

$$A+B=0$$
$$A=1$$

Upon solving these two independent equations we obtain the solution given in Eq. 5.1.10.

The method of undetermined coefficients may be formally stated as follows:

1. Assume a general solution of the form of the forcing function plus all its derivatives. (This is possible when the forcing function is one of the forms specified previously.)
2. Substitute the general solution into the given differential equation and solve for the undetermined coefficients.

The story is not complete at this point, for consider Eq. 5.1.1 again. If we add to the particular solution $x_p(t)$ the homogeneous solution $x_h(t)$, the sum remains equal to the forcing function $f(t)$. This is so because the left side of Eq. 5.1.1 equals zero when $x(t)=x_h(t)$. Therefore, the complete solution of Eq. 5.1.1 is found by the sum

$$x(t)=x_p(t)+x_h(t) \tag{5.1.11}$$

The arbitrary coefficients in the homogeneous solution are found from initial conditions after summing.

EXAMPLE 5.1.2 Find the complete solution to the differential equation given by

$$\frac{d}{dt}x(t) + 2x(t) = e^{-t}$$
$$x(0) = 5 \tag{5.1.12}$$

Solution

The homogeneous solution was found in Eq. 5.1.8.

$$x_h(t) = Ae^{-2t}$$

(Notice that we do not solve for the constant A at this point. We must wait until we sum the homogeneous and particular solutions.) The forced response is of the form

$$x_p(t) = Be^{-t}$$

Substituting into Eq. 5.1.12 gives

$$-Be^{-t} + 2Be^{-t} = e^{-t}$$

or $B = 1$. Therefore, the complete solution is given by

$$x(t) = x_h(t) + x_p(t) = Ae^{-2t} + e^{-t}$$

Only at this point may we solve for the arbitrary constant A from the given initial condition. Setting $t = 0$ in the complete solution gives

$$5 = A + 1$$

or

$$A = 4$$

Therefore, the final solution is given by

$$x(t) = 4e^{-2t} + e^{-t}$$

One more troublesome point must be investigated before completing our discussion of first order differential equations. Our procedure for finding the particular solution must be modified when the forcing function is of the form of the homogeneous solution. To illustrate, suppose that in Eq. 5.1.2 the forcing function had been

$$f(t) = e^{-2t}$$

Proceeding as before we would try

$$x_p(t) = Be^{-2t} \tag{5.1.13}$$

But upon substituting this into Eq. 5.1.12 we obtain

$$-2Be^{-2t} + 2Be^{-2t} = e^{-2t}$$

or

$$0 = 1$$

This is obviously an impossibility, and it is important that we be able to recognize and handle such cases. The source of the difficulty is found by noting that the assumed particular solution is also the homogeneous solution. Since the homogeneous solution gives zero for the left side of the differential equation (by definition), this situation will always be encountered in this case. Therefore, the assumed particular solution must be modified.

If $f(t)$ duplicates a term in the homogeneous solution, then a particular solution can always be found by assuming for $x_p(t)$, not the usual choice, but this choice multiplied by the lowest power of t that will eliminate the duplication. Thus, to solve our example we should use instead of Eq. 5.1.13,

$$x_p(t) = Bte^{-2t}$$

Proceeding as before, we then find the complete solution [with $x(0) = 5$] given by

$$x(t) = 5e^{-2t} + te^{-2t}$$

LEARNING EVALUATIONS

1. Find the characteristic equation and homogeneous solution of the differential equation given by

$$\frac{d}{dt}x(t) + 0.5x(t) = f(t)$$

2. Find the particular solution for the preceding equation when
   a. $f(t) = 1$
   b. $f(t) = t^2$
   c. $f(t) = e^{-0.5t}$
3. Find the complete solution for each case in problem 2 for the initial condition $x(0) = 1$.

5.2 The *RC* circuit

LEARNING OBJECTIVES

After completing this section you should be able to do the following:

1. Define the time constant for an *RC* circuit and graphically show the effect of the time constant on the voltages and currents in the circuit.
2. Find the current, resistor voltage, and capacitor voltage as a function of time in a first order *RC* circuit.

The circuit shown in Fig. 5.4 contains a resistor, capacitor, switch, and constant-voltage source V_s.

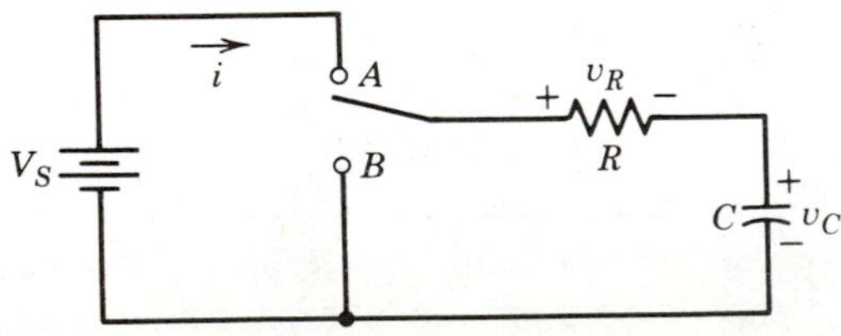

Figure 5.4. Basic *RC* circuit.

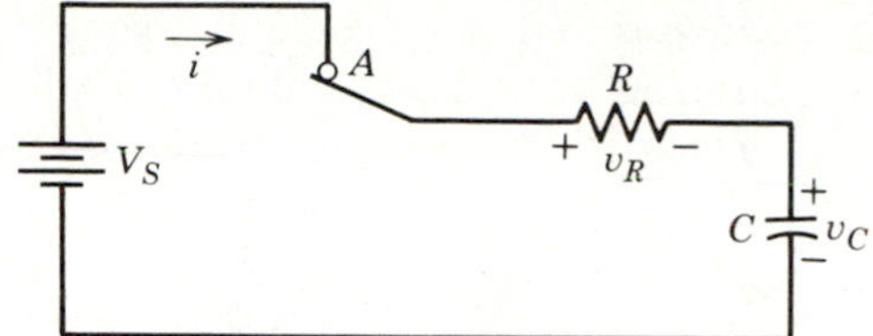

Figure 5.5. Circuit of Fig. 5.4 with switch placed in position *A* at $t = 0$.

We assume that there is no energy stored in the capacitor. Consequently, if the switch is placed in position *B*, nothing happens because the circuit does not contain a source of energy. When the switch is moved to position *A*, however, some very enlightening things occur. Let us assume that the switch is placed in position *A* at time $t = 0$. The resulting circuit in Fig. 5.5 will be used as a model to describe the physical phenomena that occur after $t = 0$. Writing KVL around the circuit yields

$$V_s = v_R + v_C, \qquad t \geq 0 \tag{5.2.1}$$

We know that at $t < 0$, v_C is zero because we have defined the capacitor to be initially uncharged and v_C is linearly related to the charge on the capacitor. Also, since the voltage across the capacitor cannot change instantaneously,

$$v_C = 0 \quad \text{at} \quad \text{t} = 0 \tag{5.2.2}$$

Therefore,

$$V_S = v_R \quad \text{at} \quad t = 0 \tag{5.2.3}$$

And from Ohm's law,

$$i(0) = \frac{v_R}{R} \tag{5.2.4}$$

On the other hand, we know that after the circuit has been in the position shown for a long time, the capacitor will act as an open circuit. This requires that

$$i = v_R = 0, \qquad t \gg 0 \tag{5.2.5}$$

and

$$v_C = V_S, \qquad t \gg 0 \tag{5.2.6}$$

The period from $v_C = 0$ to $v_C = V_S$ is called the "transient" period of the circuit and is that time from the initial application of the energy source until an equilibrium state is reached.

This transient period can be described qualitatively by recalling that current is the movement of charge and that charge can collect on the plates of a capacitor. At $t = 0$, a current i is established as given by Eq. 5.2.4. This current results in charge q being transferred to the plates of the capacitor. Since

$$v_C(t) = \frac{q(t)}{C}$$

v_C increases from zero to some positive value. From Eq. 5.2.1, if v_C increases, then v_R must decrease since V_S is a constant. As v_R decreases, the current i

decreases and the change in charge across the capacitor decreases. This process continues until the equilibrium condition of Eqs. 5.2.5 and 5.2.6 is reached.

This period can also be described mathematically by writing KVL to obtain

$$V_S = Ri + \frac{1}{C}\int_0^t i\,d\lambda, \qquad t \geq 0$$

Differentiating this equation gives

$$R\frac{di}{dt} + \frac{1}{C}i = 0 \qquad t \geq 0 \tag{5.2.7}$$

This is a homogeneous equation of the type discussed in the previous section. Therefore, assume a solution to Eq. 5.2.7 to be

$$i(t) = Ae^{st} \tag{5.2.8}$$

Placing Eq. 5.2.8 into Eq. 5.2.7 gives the characteristic equation

$$Rs + \frac{1}{C} = 0 \tag{5.2.9}$$

Solving for s yields the natural frequency given by

$$s = -\frac{1}{RC} \tag{5.2.10}$$

In order to evaluate the constant A, use Eqs. 5.2.3, 5.2.4, and 5.2.8 to show

$$i(0) = A = \frac{V_S}{R} \tag{5.2.11}$$

So our complete solution is

$$i(t) = \frac{V_S}{R}e^{-t/RC}, \qquad t \geq 0 \tag{5.2.12}$$

Further

$$v_R(t) = Ri(t) = V_S e^{-t/RC}, \qquad t \geq 0 \tag{5.2.13}$$

and from Eq. 5.2.1,

$$v_C(t) = V_S - V_S e^{-t/RC}, \qquad t \geq 0$$

or

$$v_C(t) = V_S(1 - e^{-t/RC}), \qquad t \geq 0 \tag{5.2.14}$$

Examination of Eq. 5.2.14 reveals that our qualitative discussion was correct. At $t = 0$, $v_C(t) = 0$ and at $t = \infty$, $v_C(t) = V_S$. A plot of $v_C(t)$ is shown in Fig. 5.6. A

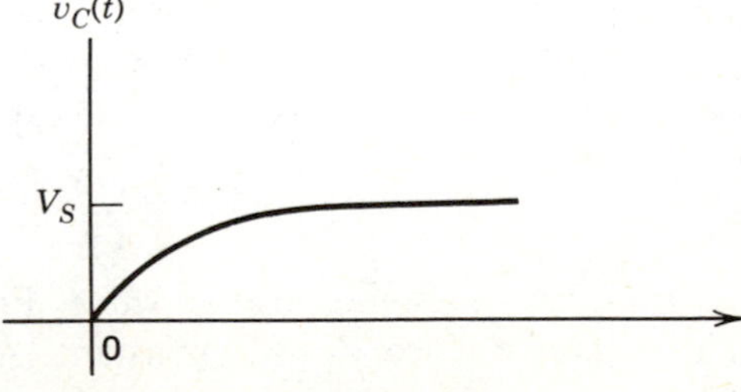

Figure 5.6. A plot of capacitor voltage $v_C(t)$ for circuit shown in Fig. 5.5.

discussion of the characteristics of this plot will be postponed temporarily.

Equation 5.1.14 can also be broken into two parts as

$$v_C(t) = v_{ss}(t) + v_{tr}(t), \qquad t \geq 0$$

where

$$v_{ss}(t) = V_S, \qquad t \geq 0$$

and

$$v_{tr}(t) = V_S e^{-t/RC}, \qquad t \geq 0$$

$v_{ss}(t)$ is called the "steady-state" portion of the response $v_C(t)$ because of its lasting, or long-term effects. $v_{tr}(t)$ is called the "transient" response because its effect is negligible after a period of time.

We used different terminology in Section 5.1, but with the same meaning. The following terms are equivalent.

Steady state
Particular
Forced response

Also, the following are equivalent.

Transient
Homogeneous
Source-free response

Referring again to Fig. 5.4, assume that the switch has been in position A for a long time. From our discussion, we know that the capacitor voltage will equal the battery voltage and that the voltage across the resistor is zero (no current flow). If we instantaneously move the switch to position B at a time we defined as $t = 0$, then the circuit in Fig. 5.7 results. Since the voltage across the capacitor cannot change instantaneously, we have that

$$v_C(0) = V_S \tag{5.2.15a}$$

$$v_R(0) = -V_S \tag{5.2.15b}$$

$$i(0) = \frac{v_R(0)}{R} = \frac{-V_S}{R} \tag{5.2.15c}$$

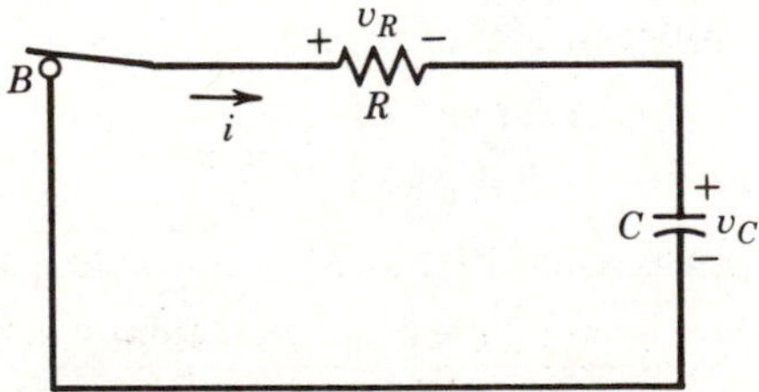

Figure 5.7. Circuit of Fig. 5.4 with switch in position B.

And for $t \geq 0$, the current through the capacitor must equal the current through the resistor since we only have one mesh in our circuit.

$$i = C\frac{dv_C}{dt} = \frac{v_R}{R} \tag{5.2.16}$$

Also, by observation,

$$v_R(t) = -v_C(t)$$

so Eq. 5.2.16 becomes

$$\frac{dv_C}{dt} = \frac{-v_C}{RC} = 0 \tag{5.2.17}$$

Equation 5.2.17 is exactly the same form as Eq. 5.2.7, so the solutions will also be of the same form, and using Eq. 5.2.12 we can write directly that

$$v_C(t) = V_S e^{-t/RC}, \qquad t \geq 0 \tag{5.2.18}$$

A plot of $v_C(t)$ is found in Fig. 5.8.

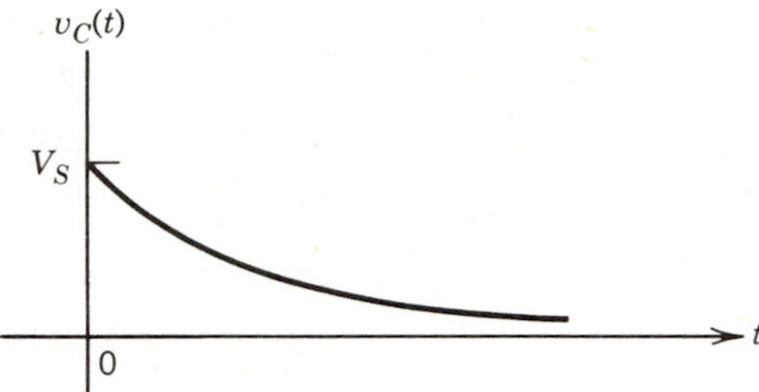

Figure 5.8. A plot of $v_C(t)$ for the circuit shown in Fig. 5.7.

It is seen that the voltage has the initial value of V_S and decays exponentially to zero as t increases. Equation 5.2.18 is seen to be the transient response of the circuit. Although always exponential, the exact shape of the curve will depend on the values of R and C. Also, it can be observed that the shapes of the curves in Figs. 5.6 and 5.8 are identical.

In an effort to characterize the shape of the curve in terms of the circuit values R and C, the concept of a "time constant" was developed.

> The time constant of an RC circuit is the time it would take the voltage across the capacitor to change from its initial value to its final value if the rate of change of the capacitor voltage would remain constant and equal to the initial rate of change.

In Fig. 5.8 the initial rate of change of $v_C(t)$ is

$$\left.\frac{dv_C(t)}{dt}\right|_{t=0} = \left.\frac{-V_S}{RC}e^{-t/RC}\right|_{t=0} = \frac{-V_S}{RC} \tag{5.2.19}$$

The time constant is usually denoted by the Greek letter tau. From our definition

$$\tau = \frac{\text{total change}}{\text{initial rate of change}} = \frac{\text{final value} - \text{initial value}}{\text{initial rate of change}}$$

or

$$\tau = \frac{0 - V_S}{-V_S/RC}$$

so

$$\tau = RC \tag{5.2.20}$$

Equation 5.2.18 can be expressed in terms of the time constant.

$$v_C(t) = V_S e^{-t/\tau}, \qquad t \geq 0 \tag{5.2.21}$$

Since the capacitor voltage does not, in fact, decrease at a constant rate, the question arising in your mind should be: How many time constants does it take for the capacitor voltage to reach its final value? Table 5.1 evaluates $v_C(t)$ for t equal to various multiples of the time constant using Eq. 5.2.21. For this calculation assume that $V_S = 1$. The values in Table 5.1 indicate that $v_C(t)$ decreases by 63.2% in one time constant but theoretically never reaches its final value. The final value is an asymptote that is reached only at $t =$ infinity. From a practical standpoint, however, we can say that $v_C(t)$ reaches its final value in about five time constants.

TABLE 5.1
VALUE OF *V*(*t*) FOR VARIOUS MULTIPLES OF THE TIME CONSTANT

t	$v_C(t)$	% of Initial Value
τ	0.368	36.8
2τ	0.135	13.5
3τ	0.0498	4.98
4τ	0.0183	1.83
5τ	0.00674	0.674
6τ	0.000912	0.0912
7τ	0.0000454	0.00454
8τ	3.72×10^{-44}	3.72×10^{-42}

In our developments so far in this chapter, we have assumed that the switches in the circuits of Figs. 5.4 and 5.7 were moved instantaneously at time $t = 0$. If we had instead chosen an arbitrary closing time of t_0, our results would have been slightly more complex. For instance, Eq. 5.2.14 would become

$$v_C(t) = V_S(1 - e^{-(t-t_0)/RC}), \qquad t \geq t_0 \tag{5.2.22}$$

and Eq. 5.2.18 would be

$$v_C(t) = V_S e^{-(t-t_0)/RC}, \qquad t \geq t_0 \tag{5.2.23}$$

In these equations, V_S is the value of $v_C(t_0)$. Obviously, Eqs. 5.2.22 and 5.2.23 reduce to Eqs. 5.2.14 and 5.2.18, respectively, for $t_0 = 0$. The proof of the validity of Eqs. 5.2.22 and 5.2.23 is left as an exercise.

EXAMPLE 5.2.1 A simple circuit contains a resistor, capacitor, and switch connected in series. The switch is closed at $t=0$ and the value of capacitor voltage at $t=0$ is 40 V. Find $v_C(t)$, the time constant, the value of $v_C(t)$ at $t=2$ s, and the length of time it takes $v_C(t)$ to reach its final value if $R=20\ \text{k}\Omega$ and $C=100\ \mu\text{F}$.

Solution

The circuit may be drawn as shown in Fig. 5.9.

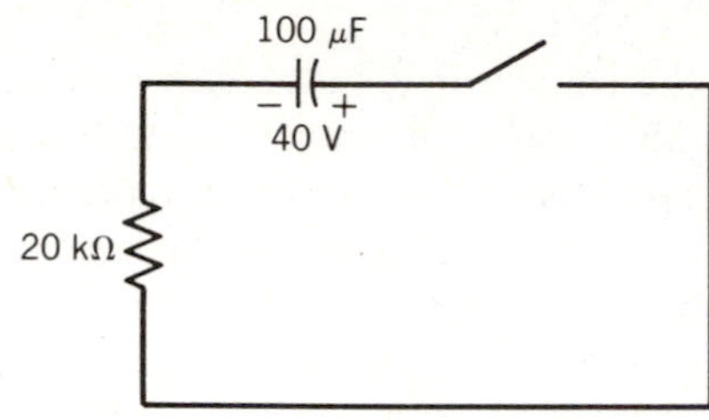

Figure 5.9

From Eq. 5.2.18, we can write

$$v_C(t) = v_C(0)e^{-t/RC}, \qquad t \geq 0$$

Using the values given in this problem

$$v_C(t) = 40e^{-t/(20\times10^3\times100\times10^{-6})}$$
$$= 40e^{-0.5t}, \qquad t \geq 0$$

The time constant is calculated as

$$\tau = RC = (20\times10^3)(100\times10^{-6})$$
$$= 2$$

The value of $v_C(t)$ at $t=2$ s is

$$v_C(2) = 40e^{-0.5(2)} = 40e^{-1} = 40(0.368)$$
$$= 14.72\ \text{V}$$

The time required to reach its final value is assumed to be five time constants:

$$5\tau = 5(2) = 10\ \text{s}$$

EXAMPLE 5.2.2 Consider the circuit shown in Fig. 5.10. The capacitor is initially uncharged and the switch is moved to position A at $t=0$. The switch is moved instantaneously to position B at $t=20$ ms. Find and plot $v_C(t)$ for $t \geq 0$.

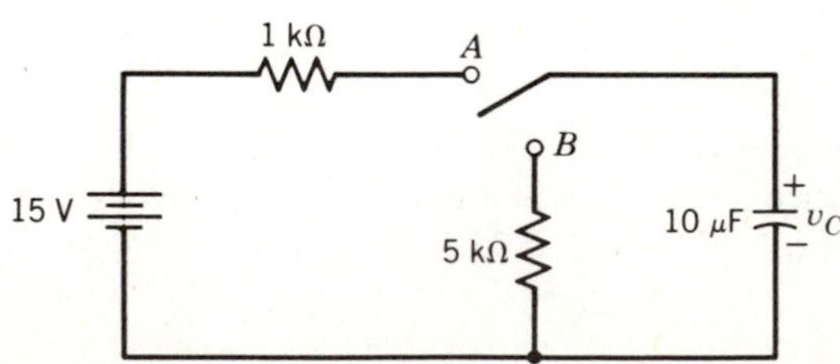

Figure 5.10

Solution

With the switch in position A, the solution to $v_C(t)$ is given by Eq. 5.2.14 as

$$v_C(t) = 15\left[1 - e^{-t/(1000\times 10\times 10^{-6})}\right]$$
$$= 15(1 - e^{-100t}), \qquad 0 \le t \le 0.02 \tag{5.2.24}$$

At $t = 20$ ms, we have

$$v_C(0.02) = 15(1 - e^{-2})$$
$$= 12.97 \text{ V}$$

This is the value of the voltage across the capacitor when the switch is moved to position B, so this becomes the *initial* value of $v_C(T)$ for the new series RC circuit. The capacitor voltage for the new circuit is given by Eq. 5.2.23 where

$$V_S = v_C(0.02) \quad \text{and} \quad t_0 = 0.02$$

so

$$v_C(t) = 12.97e^{-(t-0.02)/(5000\times 10\times 10^{-6})}$$
$$= 12.97e^{-20(t-0.02)}, \qquad t \ge 0.02 \tag{5.2.25}$$

Here we note that for circuit B, the time constant is 0.05 s so $v_C(t)$ will decay to zero in about $5(0.05) = 250$ ms. $v_C(t)$ is plotted in Fig. 5.11 using Eq. 5.2.24 for the first 20 ms, then using Eq. 5.2.25 for $t > 20$ ms.

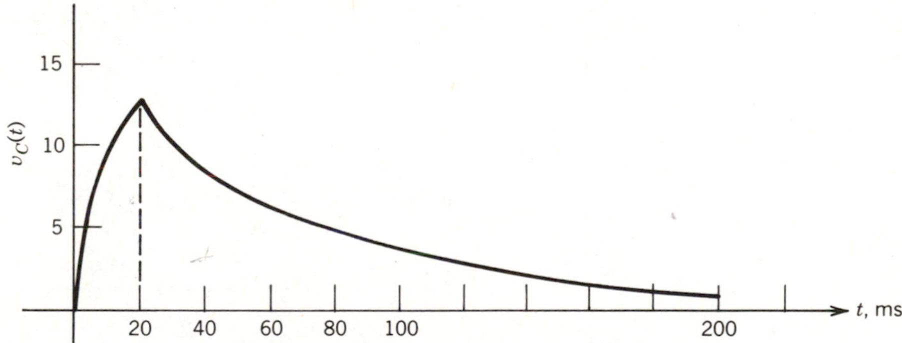

Figure 5.11. Plot of $v_C(t)$ from Example 5.2.2.

To complete our discussion of the simple RC circuit, we will examine the energy relationship that exists in the circuit using the circuit model shown in Fig. 5.7. We are specifically interested in the total energy dissipated in the resistor. The power dissipated in the resistor is

$$p(t) = \frac{v^2(t)}{R}$$

Using Eq. 5.2.18,

$$p(t) = \frac{V_S^2}{R} e^{-(2t/RC)} \tag{5.2.26}$$

The total energy expended in the resistor in the form of heat is

$$w(t)=\int_0^{\infty} p(t)\,dt=\int_0^{\infty}\frac{V_S^2}{R}e^{-(2t/RC)}\,dt$$

$$=\frac{V_S^2}{R}\left(-\frac{RC}{2}\right)e^{-(2t/RC)}\Bigg|_0^{\infty}$$

$$=\frac{1}{2}CV_S^2 \tag{5.2.27}$$

Since V_S is the initial value of the capacitor voltage, we see that the energy expended in the resistor is exactly equal to the amount of energy initially stored in the capacitor. At $t=\infty$, the capacitor voltage is zero so the energy stored in the capacitor is zero at this time. This complies with the law of conservation of energy.

LEARNING EVALUATIONS

1. Determine the time constant for each circuit shown in Fig. 5.12.

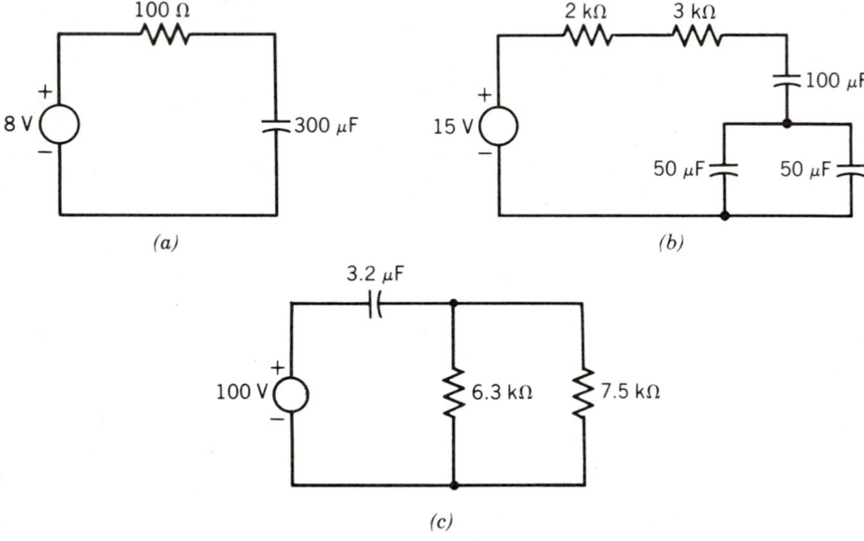

Figure 5.12

2. In Fig. 5.13 the switch is moved from position A to position B at $t=0$. Find and sketch the voltage $v_R(t)$.

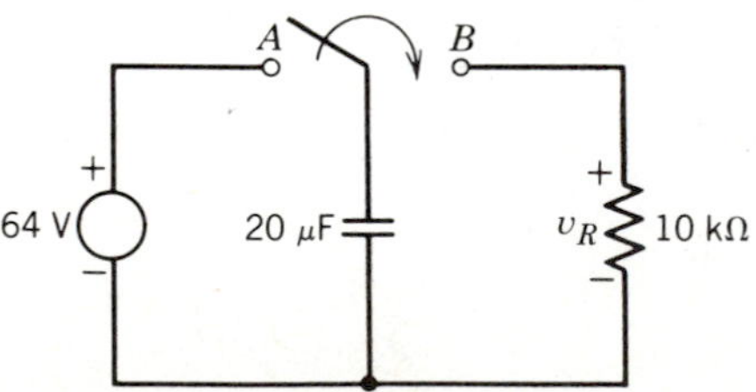

Figure 5.13

3. In Fig. 5.14 the switch is closed at $t = 0$. Find the current $i(t)$ if the capacitor is initially uncharged.

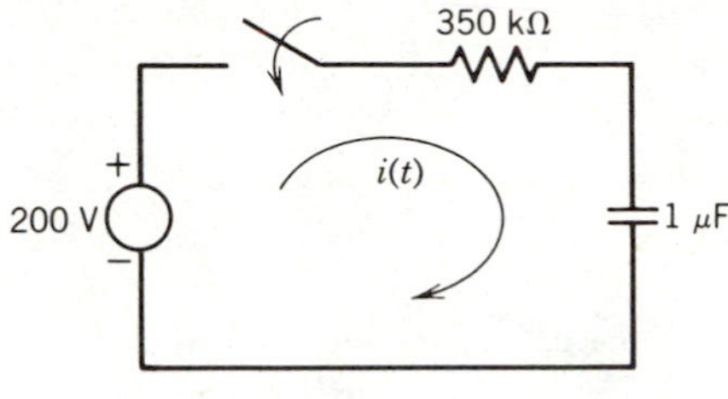

Figure 5.14

5.3 The *RL* Circuit

LEARNING OBJECTIVES

After completing this section you should be able to do the following:

1. Define the time constant for an *RL* circuit and graphically show the effect of the time constant on the voltages and currents in the circuit.
2. Find all currents and voltages in a first order *RL* circuit.

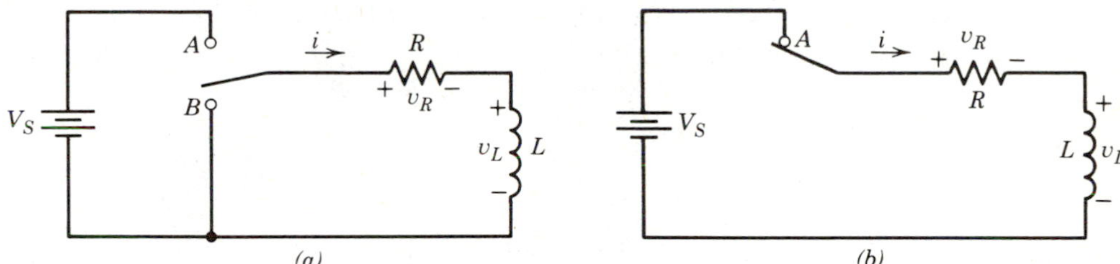

Figure 5.15. (*a*) Basic *RL* circuit. (*b*) Circuit of Fig. 5.15*a* with switch placed in position *A* at $t = 0$.

The circuit in Fig. 5.15*a* is similar to that of Fig. 5.4; the capacitor has been replaced with an inductor. Our investigation of this circuit will proceed in exactly the same manner that was successful in the previous section. We assume that there is no energy stored in the inductor and that the switch is moved to position *A* at some arbitrary time, say $t = 0$. The resulting circuit is shown in Fig. 5.15*b*. Writing KVL around the circuit gives

$$V_S = v_R + v_L, \qquad t \geq 0 \tag{5.3.1}$$

Since the energy in an inductor is related to the current through the inductor, the current i through the inductor prior to $t = 0$ must be zero because we assumed the energy stored was zero. Since the current through the inductor cannot change instantaneously,

$$i(t) = 0 \quad \text{at} \quad t = 0 \tag{5.3.2}$$

Therefore

$$v_R = Ri(t) = 0 \quad \text{at} \quad t = 0 \tag{5.3.3}$$

so

$$v_L = V_S \quad \text{at} \quad t = 0 \tag{5.3.4}$$

But we have also previously established that after a long period, the inductor will act as a short circuit to a constant voltage; this requires that

$$v_R = V_S, \qquad t \gg 0 \tag{5.3.5}$$

$$i = \frac{v_R}{R} = \frac{V_s}{R}, \qquad t \gg 0 \tag{5.3.6}$$

As before, the time from the conditions existing in Eqs. 5.3.2–5.3.4 to the conditions existing in Eqs. 5.3.5–5.3.6 is called the transient period. This is the time interval during which the current changes from its initial value to its final value.

We have at first glance an apparent paradox at $t = 0$. We say that the current through the inductor is zero but that the voltage across the inductor is V_S. Yet the inductor voltage can only exist if there is a rate of change of current. The answer to this paradox is that the current can be instantaneously zero and yet be changing value. Certainly all rapidly changing signals have an instantaneous value. We are just forcing the instantaneous value of the current to be zero at the instant $t = 0$. Consider the Indianapolis 500 automobile race. All cars traverse the course one lap to gain speed before the race actually begins. When a car crosses the starting line, the distance traveled in the race is zero, but the distance traveled is simultaneously changing at a finite rate.

If we apply Ohm's law and use Eq. 5.3.1 and 4.4.4, we obtain

$$V_S = Ri + L\frac{di}{dt} \tag{5.3.7}$$

This is a first order equation of the type discussed in Section 5.1. The homogeneous or transient solution is exponential:

$$i_{\text{tr}}(t) = Ae^{st} \tag{5.3.8}$$

The characteristic equation is found by substituting this solution into the homogeneous equation

$$0 = R + Ls$$

or

$$s = -\frac{R}{L}$$

Therefore, the transient solution is given by

$$i_{\text{tr}}(t) = Ae^{-(R/L)t} \tag{5.3.9}$$

The particular or forced response is of the form of a constant plus all its derivatives. Since the input voltage is constant,

$$i_{\text{ss}} = B$$

Substituting this into Eq. 5.3.7 gives

$$V_S = RB + 0$$

or

$$B = \frac{V_S}{R}$$

The sum of the transient and steadystate solution gives

$$i(t) = \frac{V_S}{R} + Ae^{-(R/L)t} \tag{5.3.10}$$

Since $i(0-) = 0$, and knowing that the inductor current cannot change instantaneously, we have at $t = 0$,

$$0 = \frac{V_S}{R} + A$$

Solving for A and substituting into Eq. 5.3.10 gives the final solution

$$i(t) = \frac{V_S}{R} - \frac{V_S}{R}e^{-(R/L)t} \tag{5.3.11}$$

A plot of Eq. 5.3.11 is shown in Fig. 5.16. Comparing Figs. 5.16 and 5.6 we see that current through an inductor and the voltage across a capacitor have the same exponential shape.

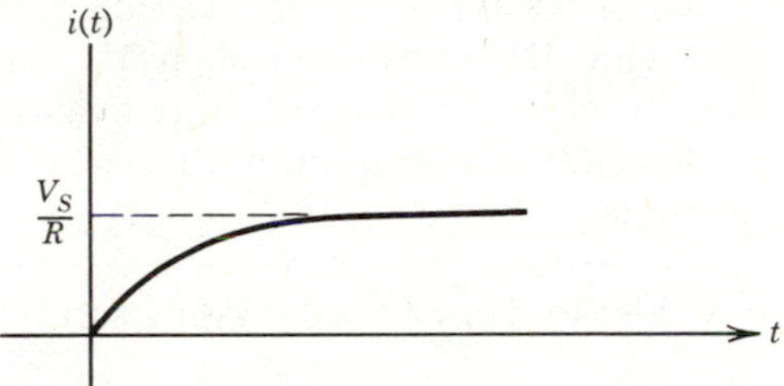

Figure 5.16. A plot of current, $i(t)$, for circuit in Fig. 5.15.

If we now instantaneously move the switch in Fig. 5.14 to position B after it has been in position A long enough for the transient portion of $i(t)$ to become negligible, Fig. 5.17 results. We now conveniently assume that $t = 0$ is the time when the switch is changed from position A to position B. Writing KVL, we obtain

$$Ri + L\frac{di}{dt} = 0, \qquad t \geq 0$$

or

$$\frac{di}{dt} + \frac{R_i}{L} = 0, \qquad t \geq 0 \tag{5.3.12}$$

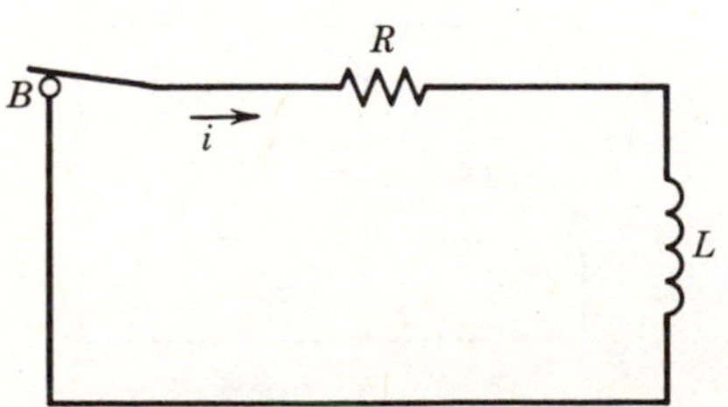

Figure 5.17. Circuit of Fig. 5.14 with switch in position B.

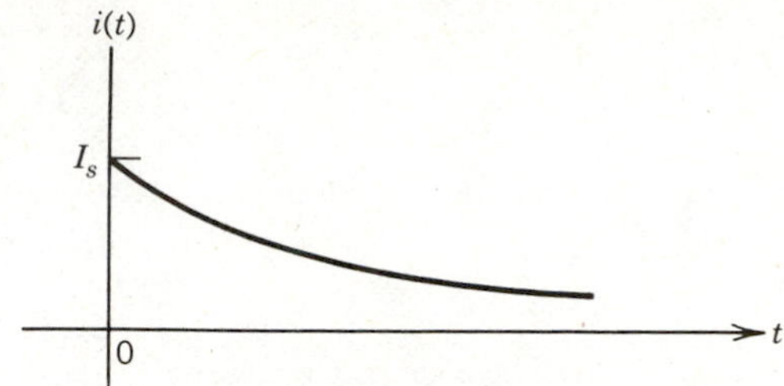

Figure 5.18. A Plot of $i(t)$ for the circuit in Fig. 5.17.

This equation is homogeneous, so the solution will be of the form

$$i(t) = I_s e^{-(R/L)t}, \qquad t \geq 0 \tag{5.3.13}$$

where I_s is the value of $i(t)$ at $t = 0$, $i(t)$ is plotted in Fig. 5.18, and, as we would expect, is seen to have the same shape as the curve in Fig. 5.8. The shape of the above curve is a function of the values of R and L. As before, we can characterize the shape of the curve by the concept of the "time constant" which for an RL circuit is defined as follows:

> The time constant of an RL circuit is the time it would take the current through the inductor to change from its initial value to its final value if the rate of change of the current were constant and equal to its initial rate of change.

Using an analysis similar to that of the preceding section, we can show that

$$\tau = \frac{L}{R}$$

A table identical to Table 5.1 could be developed for the RL circuit with $i(t)$ replacing $v_C(t)$ and the entries would not change.

EXAMPLE 5.3.1 A 5-mH inductor is connected in series with a resistor and at $t = 0$ the current in the inductor is 14 A. It is necessary for the current to be reduced to 60% of its initial value at $t = 20$ ms. Determine the necessary value of R and find the time constant of the circuit.

Solution

First we draw a model of our circuit as shown in Fig. 5.19.

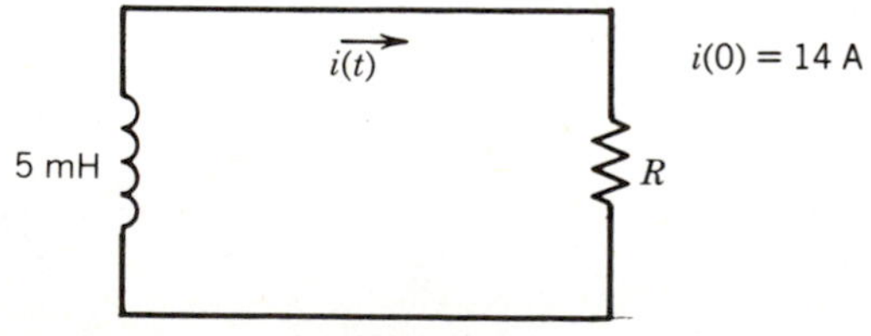

Figure 5.19

From Eq. 5.3.13, we can write

$$i(t) = i(0)e^{-(Rt/0.005)}$$

At $t = 20$ ms, we have

$$i(0.02) = 14e^{-R(0.02/0.005)} = 0.6(14)$$

or

$$e^{-4R} = 0.6$$
$$-4R = \ln(0.6) = -0.5108$$
$$R = 0.128\ \Omega$$

and

$$\tau = \frac{L}{R} = \frac{0.005}{0.128} = 0.039\ \text{s}$$

Up to this point we have assumed the switch in Fig. 5.14 is always moved instantaneously at time $t = 0$. Again, this was done for ease of discussion and mathematical development. A more general approach is to assume that the switch is moved at an arbitrary time t_0. With regard to the *RL* circuit, the transient response given by Eq. 5.3.13 becomes

$$i(t) = I_s e^{-R(t-t_0)/L}, \qquad t \geq t_0 \tag{5.3.14}$$

Here we realize that I_s is the value of $i(t_0)$. The proof of Eq. 5.3.14 is left to the student. If $t_0 = 0$, then the above equation degenerates to our original results.

EXAMPLE 5.3.2 The switch in the circuit shown in Fig. 5.20 has been in position *A* for at least 100 time constants. At $t = 12$ s, the switch is moved to position *B*. Find $i(t)$ and the current at $t = 12.25$ s.

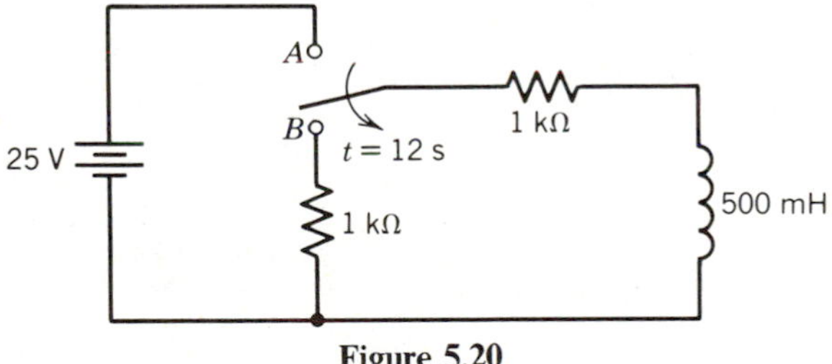

Figure 5.20

Solution

Since the switch has been in position *A* well past the transient period of the circuit, the inductor acts like a short circuit and the current is

$$i(t) = \frac{25}{1000} = 25\ \text{mA}$$

immediately prior to $t = 12$ s. Current cannot change instantaneously so

$$i(12) = 25\ \text{mA}$$

and we recognize that $t_0 = 12$ s. Using Eq. 5.3.14 we can write by inspection that

$$i(t) = 0.025e^{-[2(t-12)/0.50]}, \qquad t \geq 12$$

and

$$\begin{aligned} i(12.25) &= 0.025e^{-4(12.25-12)} \\ &= 0.025e^{-1} \\ &= 9.20 \text{ mA} \end{aligned}$$

LEARNING EVALUATIONS

1. Determine the time constant for each circuit shown in Fig. 5.21.
2. In Fig. 5.21*a* the switch is moved from position A to B at $t = 0$. Find and sketch the current $i(t)$.
3. In Fig. 5.21*b* the switch is closed at $t = 0$. Find and sketch the current $i(t)$.

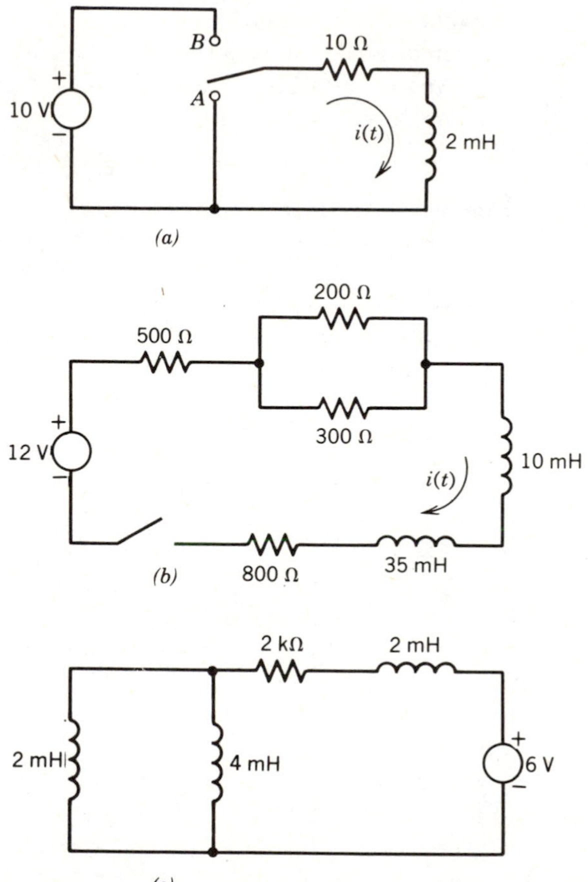

Figure 5.21

5.4 The Operational Amplifier Revisited

LEARNING OBJECTIVES

After completing this section you should be able to do the following:

1. Analyze op amp circuits containing resistors, inductors, and capacitors.
2. Simulate differential equations with analog computer circuits.
3. Design an integrate-and-dump binary signal detector.

The op amp was introduced in Section 3.6 and it may be helpful for you to review that section before continuing here. Resistors were the only circuit elements connected to the input and used as feedback elements. We now investigate the possibility of using capacitors and inductors, in addition to resistors, in op amp circuits.

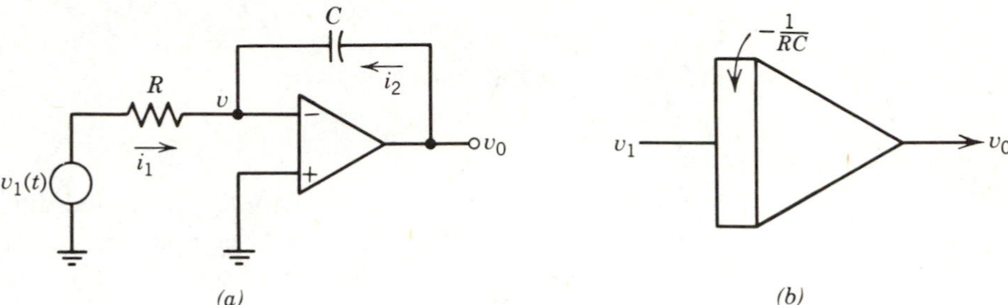

Figure 5.22. An integrator.

In Fig. 5.22*a* a capacitor is connected in the feedback circuit. Since the current into the amplifier is zero, we have

$$i_1 + i_2 = 0$$

$$\frac{v_1(t) - v(t)}{R} + C\frac{d}{dt}[v_0(t) - v(t)] = 0$$

Since $v(t)$ is approximately zero, this gives

$$\frac{d}{dt}v_0(t) = -\frac{1}{RC}v_1(t)$$

Integrating both sides of this equation gives

$$v_0(t) = -\frac{1}{RC}\int_{-\infty}^{t} v_1(\lambda)\,d\lambda \tag{5.4.1}$$

This is an integrator. The output voltage is negatively proportional to the integral of the input voltage. The symbol in Fig. 5.22*b* is used in place of the more complex circuit in Fig. 5.22*a*.

Since a capacitor in the feedback loop combined with an input resistor to the op amp performs integration, it seems likely that either circuit in Fig. 5.23 will

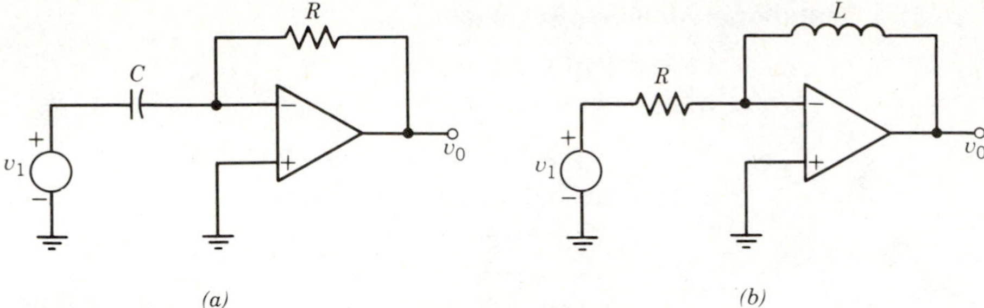

Figure 5.23. A differentiator.

perform differentiation, and this is true. Let us analyze the circuit in Fig. 5.23*b* to illustrate this fact. Summing currents at the inverting input gives

$$\frac{v_1(t)}{R} + \frac{1}{L}\int_{-\infty}^{t} v_0(\lambda)\, d\lambda = 0$$

Differentiating both sides gives

$$v_0(t) = -\frac{L}{R}\frac{d}{dt}v_1(t) \tag{5.4.2}$$

We now have methods of summing, integrating, and differentiating signals using op amp circuits. As it happens, the differentiator is almost never used in practice, but the summer and integrator are used extensively. Noise is the reason for the sparse use of differentiators. Noise can be defined as a waveform whose past values provide no insight into its future value. An equation cannot be written to describe the noise waveform for all time or for any discrete values of time. The noise waveform is called a "random signal" and is the nemesis of engineers. Noise in electric circuits can be explained as a form of Brownian motion, which we now describe.

In 1827 the renowned Scottish botanist Robert Brown was observing wheat pollen under a microscope. The pollen grains, which were suspended in water, moved about under the microscope in lifelike fashion. Naturally Brown became curious about this motion (now called "Brownian motion") and tried to discover its origin. He found that any small particle, when suspended in liquid, would exhibit this same form of lifelike motion. He even obtained a piece from the Sphinx, ground it into fine particles, and observed the same motion.

Brown concluded that he had discovered a fundamental form of life that was "present in all animate and inanimate objects." By the time he died in 1858 no one had discovered a better explanation, though many tried. It was not until 1905 that Albert Einstein correctly analyzed Brownian motion and explained that it was caused by thermal energy agitating the water molecules. He went on to predict that any system in thermal equilibrium would exhibit this same Brownian motion, and he correctly derived the current in a conductor caused by heat.[1]

[1]For a full account of this history, see D. K. C. MacDonald, *Noise and Fluctuations: An Introduction*, Wiley, New York, 1962.

Since an electric circuit element, such as a resistor, is a system in thermal equilibrium, we can expect random noise to be present as a voltage across its terminals. When this resistor is connected in a circuit, noise current will flow. This noise is called "Johnson noise," after the scientist J. B. Johnson, who measured this noise at Bell Labs in 1928. In an ordinary resistor at room temperature, the noise voltage across the terminals is on the order of microvolts.

A microvolt does not seem like much in view of the fact that v_1 and v_0 are about 1 to 10 V in a differentiator circuit. The problem arises from the fact that the derivative of a waveform depends not on its magnitude, but rather on its rate of change. The faster the rate of change, the larger the derivative of a noise waveform will be. Johnson noise contains all rates of change from very slow to extremely fast changes. (It is called white noise, in analogy with white light, which contains all visible wavelengths.) The rate of change is directly related to a term called the "frequency" of the waveform which we will discuss in Chapter 9. Therefore, the output of the circuit in Fig. 5.23 will contain considerable noise unless special precautions are taken to limit the frequency content of the noise input. Fortunately, most applications require only summers and integrators, as the following discussion illustrates.

Here is a second order differential equation.

$$\frac{d^2y}{dt^2} + a_1\frac{dy}{dt} + a_0y = x(t)$$

Solving for the highest derivative gives

$$\frac{d^2y}{dt} = x(t) - a_1\frac{dy}{dt} - a_0y(t) \tag{5.4.3}$$

This equation says that the second derivative of y is equal to the sum of three terms, the input $x(t)$, a first derivative term, and a term proportional to $y(t)$. A circuit to generate the second derivative may be obtained by drawing two integrator circuits as shown in Fig. 5.24*a*. If we arbitrarily assume that the input to the first integrator is d^2y/dt^2, then its output must be $-(dy/dt)$. (The gain is negative for each integrator; see Eq. 5.4.1.) If this signal is then supplied to the second integrator, its output must be $y(t)$. This situation is shown in Fig. 5.24*b*.

Refer once again to Eq. 5.4.3. The input signal $x(t)$ must be available for use, otherwise there is no need for a system to process this signal. The other two terms on the right, $-a_1(dy/dt)$ and $-a_0y(t)$, can be obtained from the output of the two integrator circuits in Fig. 5.24*b*, with a little additional circuitry. The sum of these three signals is the desired input to the first integrator. This leads to the simulation diagram in Fig. 5.24*c*. [This is called a simulation diagram because the system simulates a circuit that is described by the differential equation. Not only that, but the voltage output $y(t)$ is equal to the corresponding variable in the simulated circuit.]

The complete circuit diagram showing the connections to the op amps and the method of obtaining gains a_1 and a_0 is shown in Fig. 5.25. Here we assume that both a_1 and a_0 are less than 1, and that $1/RC = 1$. If not, then different values of R and C would have to be used to change the gain of one or more op amps from -1 to some larger value.

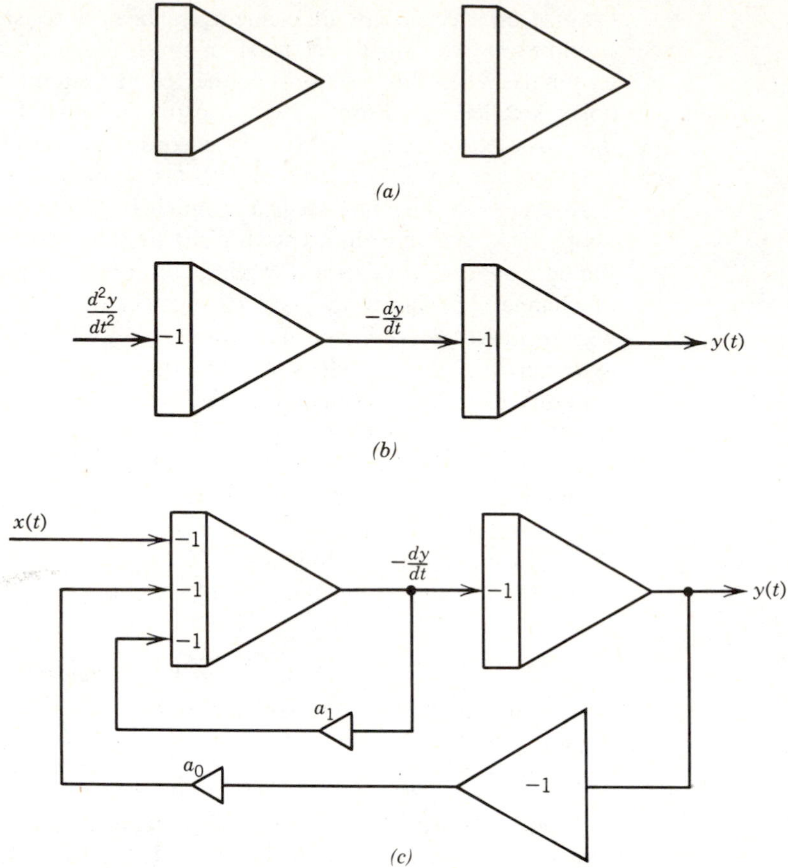

Figure 5.24. (*a*) Two integrators. (*b*) The integrators connected, each with gain $= -1$. (*c*) The simulation diagram for Eq. 5.4.3.

We have said nothing about initial conditions. These must be specified before the differential equation can be solved. For this equation, this means that dy/dt and $y(t)$ at $t = 0$ must be specified. These initial conditions may be stored as capacitor voltages on each integrator circuit. This is one reason that capacitors are used instead of the alternate *RL* combination that could be used to construct an integrator.

EXAMPLE 5.4.1 Draw a simulation diagram to solve the first order equation

$$\frac{dy}{dt} + 3y(t) = x(t) \tag{5.4.4}$$

with initial condition $y(0) = 5$.

Solution

Solving Eq. 5.4.4 for the highest derivative gives

$$\frac{dy}{dt} = x(t) - 3y(t)$$

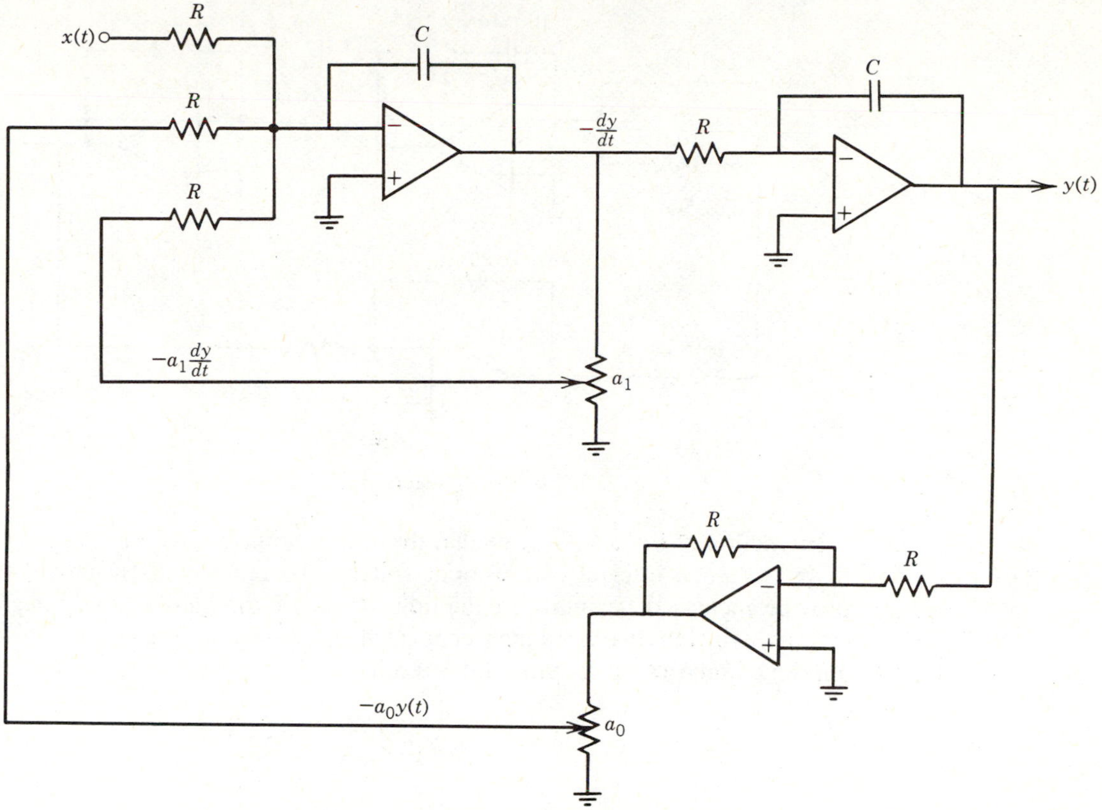

Figure 5.25

To draw the simulation diagram, begin with an integrator whose input is the highest derivative, as shown in Fig. 5.26*a*. If the input is dy/dt, then the output must be $-y(t)$. However, we need $-3y(t)$, plus $x(t)$, to supply to the integrator input. This can be accomplished by the circuit in Fig. 5.26*b*. Since the gain of the integrator circuit is $-1/RC$, the gain of the input signal $x(t)$ is -1, while the gain for the $y(t)$ term is -3. A circuit implementation of this is shown in Fig. 5.27.

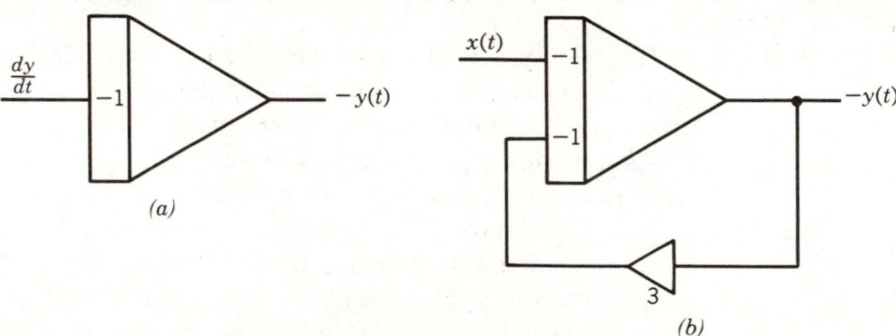

Figure 5.26

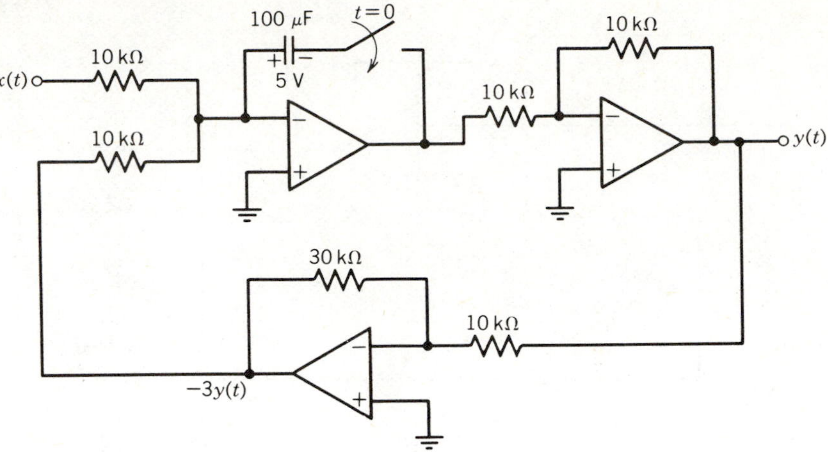

Figure 5.27

Inspection of Eq. 5.4.1 reveals that the output voltage of an integrator is related to the negative integral of the input voltage. To remove the negative sign and provide increased flexibility in the adjustment of the gain of our integrator, a circuit involving two op amps connected as shown in Fig. 5.28 is used. An analysis similar to earlier procedures results in

$$v_{out} = \frac{R_2}{R_1 R_3 C} \int_{-\infty}^{t} v_{in}(\lambda)\, d\lambda \tag{5.4.5}$$

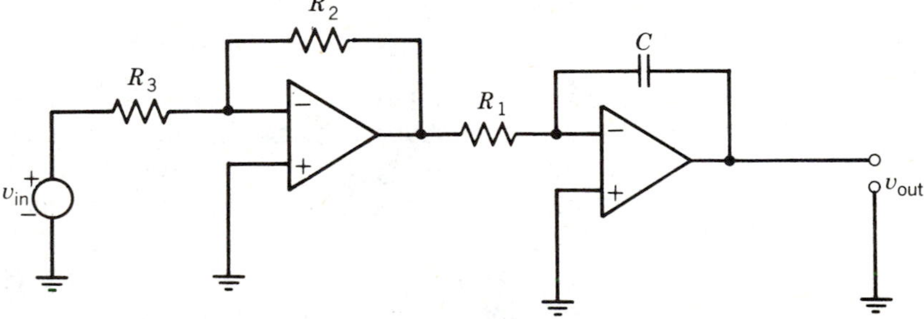

Figure 5.28. An inverter–integrator combination.

One very practical use of the circuit in Fig. 5.28 is in the detection of binary signals. The circuit is modified slightly by adding a switch to make an "integrate-and-dump" signal detector as shown in Fig. 5.29.

Assume that v_{in} is either $+V$ V or 0 V (binary or two-level signal) and that a new value of v_{in} is present every 10 ms. Also assume that noise is present in v_{in}. Figure 5.30*a* shows a waveform of $v_{in} = V$ while Fig. 5.30*b* shows the possible effects of noise. Our problem is to determine if v_{in} is $+V$ or zero in any given 10-ms interval. Inspection of Fig. 5.30*b* shows that we cannot just select a time (say, 5 ms) and look at the value of v_{in}, because it may be near zero or negative at that moment. If, however, we integrate v_{in} over the 10-ms period, our output will

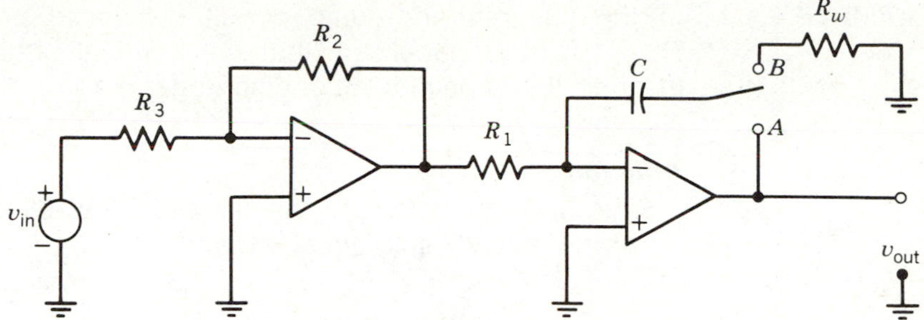

Figure 5.29. An integrate-and-dump signal detector.

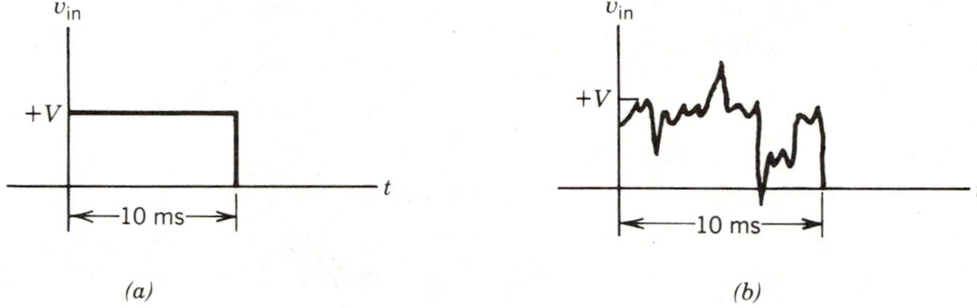

Figure 5.30. (*a*) $v_{in} = +V$. (*b*) $v_{in} = +V +$ noise.

be 0.01 V or zero for the noiseless case. When noise is present we see that it sometimes adds to the integral (area under the curve) and sometimes subtracts from it. Consequently the effect of noise in our signal is diminished by integrating and our decision concerning the proper value of v_{in} will be more accurate.

Operation of our integrate-and-dump detector is as follows:

1. The switch is placed in position B to guarantee that the capacitor is uncharged.
2. The switch is moved to position A at the start of v_{in} and left there 10 ms.
3. If at $t = 10$ ms,

$$v_{out} \geq \frac{1}{2}\left[0.01\left(\frac{R_2}{R_1 R_3 C}\right)V\right]$$

   we decide that $v_{in} = V$; otherwise, assume $V_{in} = 0.0$.
4. At $t = 10$ ms, we place the switch at B again to discharge the capacitor. [This requires at least five time constants, but since the only resistance present is that of the wire (R_w) in the circuit, this occurs very rapidly.]
5. Now move the switch to position A to determine the next value of v_{in}.
6. Repeat steps 2–5 until all sequential values of v_{in} are determined.

There are a few practical problems to overcome in our discussion (such as the time required to move the switch) but this should give you a basic understanding of an integrate-and-dump circuit as an example of the uses of RC circuits.

EXAMPLE 5.4.2 In the integrate-and-dump circuit discussed previously, $V=1$ mV and $R_w=1\ \Omega$. It is necessary that the nominal output be 5 V when $v_{in}=V$. Further, the capacitor must discharge in 1 μs. Choose values for R_1, R_2, R_3, and C.

Solution

First, we can calculate the largest allowable value of C. During discharge, the RC time constant is

$$\tau = R_w C = C$$

Further

$$5\tau \geq 1\ \mu\text{s}$$

so

$$5C \leq 1 \times 10^{-6}$$
$$C \leq 200\ \text{nF} \tag{5.4.6}$$

Next we must be careful not to saturate our first op amp. If its output is nominally 5 V, we will be safe, so

$$5 \geq \frac{R_2}{R_3} v_{in}$$

or

$$\frac{5}{0.001} > \frac{R_2}{R_3}$$

so

$$R_2 \leq 5000\ R_3 \tag{5.4.7}$$

Finally, from Eq. 5.4.2,

$$v_{out} = \frac{R_2}{R_1 R_3 C}[0.01(0.001)]$$

and

$$v_{out} = 5$$

so

$$R_2 = (5 \times 10^5)(R_1 R_3 C) \tag{5.4.8}$$

We have two inequalities and one equation to satisfy with four variables, so we are allowed an "engineering choice" as to the values of the four variables. Many choices are possible and only one of several equally correct solutions will be given here. Let us choose (Eq. 5.4.6)

$$C = 100\ \text{nF}$$

to give us a margin of safety on our discharge time. Next choose

$$R_2 = 1\ \text{k}\Omega$$
$$R_3 = 200\ \Omega$$

rather arbitrarily. This satisfies Eq. 5.4.7. Finally, using Eq. 5.4.8, we calculate

$$R_1 = 50\ \Omega$$

LEARNING EVALUATIONS

1. Construct an integrator using an op amp, a resistor, and an inductor.
2. Draw a simulation diagram to solve the equation

$$\frac{d^2y}{dt^2} + 0.5\frac{dy}{dt} + 0.6y(t) = x(t)$$

$$y(0) = 1$$

$$\frac{dy}{dt}(0) = 5$$

PROBLEMS

SECTION 5.1

1. Find and plot the solution $x(t)$ of the following differential equation

$$\frac{dx}{dt} + ax(t) = 0, \qquad x(0) = 1$$

   a. Let $a = 1$.
   b. Let $a = -1$.

2. Find the characteristic equation, natural frequency, and homogeneous solution for the equation

$$2\frac{dx}{dt} + 4x(t) = f(t)$$

3. Find the particular solution of the equation in Problem 2 if $f(t)$ is given by
   a. $f(t) = 2$
   b. $f(t) = 2t$
   c. $f(t) = 2e^{-2t}$
4. Find and plot the complete solution for each case in Problem 3 if $x(0) = 2$. Plot the solution for $0 < t < 5$.
5. Find and plot the complete solution for the differential equation

$$\frac{dx}{dt} + 0.2x = \cos t$$

$$x(0) = 1$$

   Plot your solution for $0 < t < 5$.

6. Find and plot the complete solution for the differential equation

$$\frac{dx}{dt} - 0.2x = 0.1$$

$$x(0) = 0$$

Plot your solution for $0 < t < 5$.

7. Find and plot the complete solution for the differential equation

$$\frac{dx}{dt} + 0.2x = e^{-0.2t}$$

$$x(0) = 1$$

Plot your solution for $0 < t < 5$.

SECTION 5.2

8. Find the time constant for the circuit in Fig. 5.31 when the switch is

a. In position A.
b. In position B.

9. Suppose the switch in Fig. 5.31 has been in position A for a long time. At $t = 0$ it is switched to position B. Find and plot:

a. The voltage across the capacitor.
b. The current in the circuit.

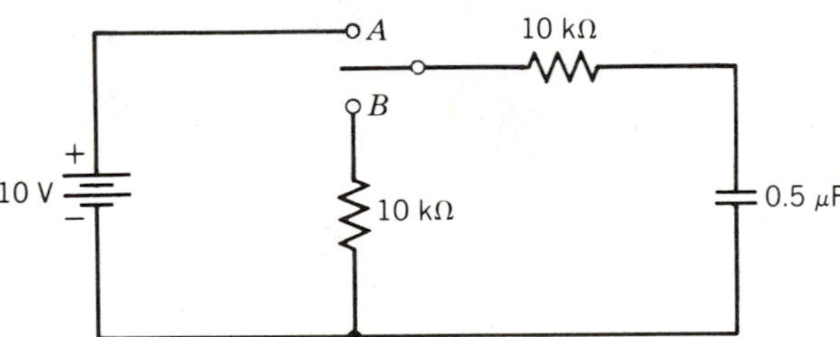

Figure 5.31

10. Suppose the switch in Fig. 5.31 has been in position B for a long time. At $t = 0$ it is switched to position A. Find and plot:

a. The voltage across the capacitor.
b. The current in the circuit.

11. The switch in Fig. 5.31 is changed from position B to position A at $t = 0$. Then at $t = 0.1$ s, it is switched back to position B. Find and plot the current in the circuit for $0 < t < 0.5$ s if the capacitor was uncharged before $t = 0$.

12. The switch in Fig. 5.32 is closed at $t = 0$. If the capacitor is initially uncharged, find the current in the circuit for $t > 0$.

13. Repeat Problem 12 if the switch is closed at $t = 2$ s instead of at $t = 0$.

14. Repeat Problem 12 if the source is $v(t) = 10 \cos 377t$.

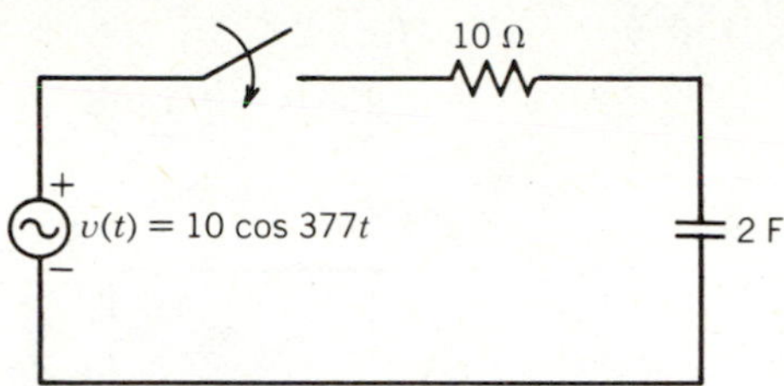

Figure 5.32

SECTION 5.3

15. Find the time constant for the circuit in Fig. 5.33 when the switch is:

a. In position A.
b. In position B.

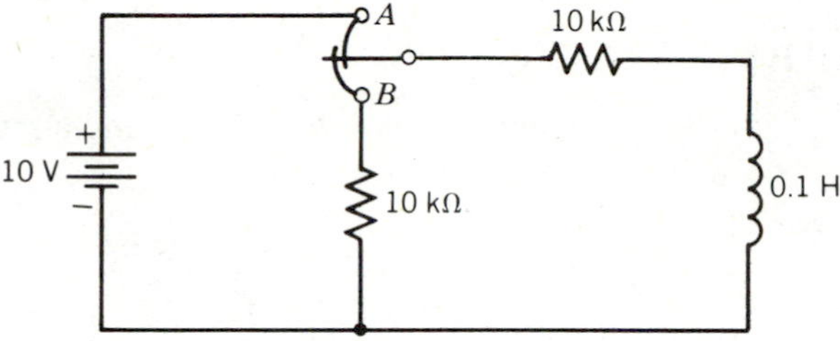

Figure 5.33

16. Suppose the switch in Fig. 5.33 has been in position A for a long time. At $t = 0$ it is switched to position B. Find and plot:

a. The current through the inductor.
b. The voltage across the inductor.

17. Suppose the switch in Fig. 5.33 has been in position B for a long time. At $t = 0$ it is switched to position A. Find and plot:

a. The current through the inductor.
b. The voltage across the inductor.

18. The switch in Fig. 5.33 is changed from position B to position A at $t = 0$. Then at $t = 20\ \mu$s it is switched back to position B. Find and plot the current in the inductor for $0 < t < 50\ \mu$s if the current was zero before $t = 0$.

19. The switch in Fig. 5.34 is closed at $t = 0$. If the initial inductor current is zero, find the current for $t > 0$.

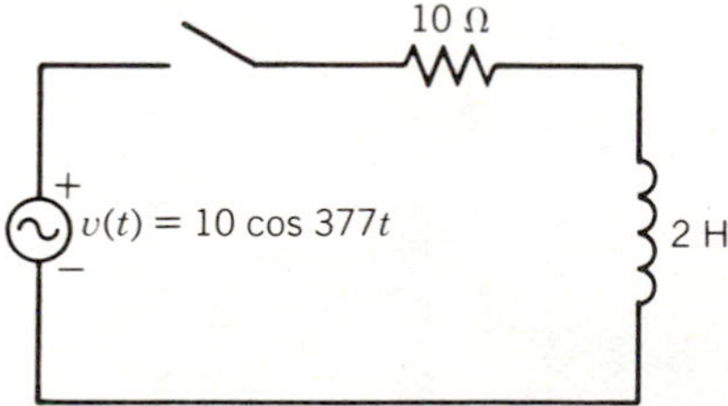

Figure 5.34

20. Repeat Problem 19 if $v(t) = 10e^{-5t}$.

21. The switch in Fig. 5.35 is opened at $t = 0$. Find the inductor current for $t > 0$.

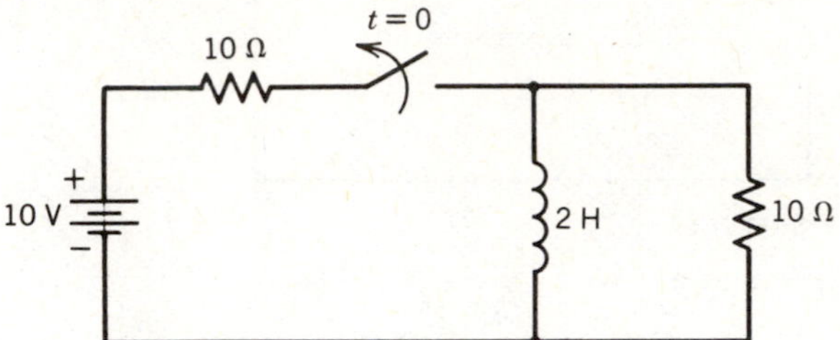

Figure 5.35

22. The switch in Fig. 5.35 has been open for a long time when it is closed at $t = 0$. Find the inductor current for $t > 0$.

SECTION 5.4

23. a. Find the unit step response for the op amp circuit in Fig. 5.36 if the initial capacitor voltage is zero.

b. Repeat if $v_c(0) = 2$ V.

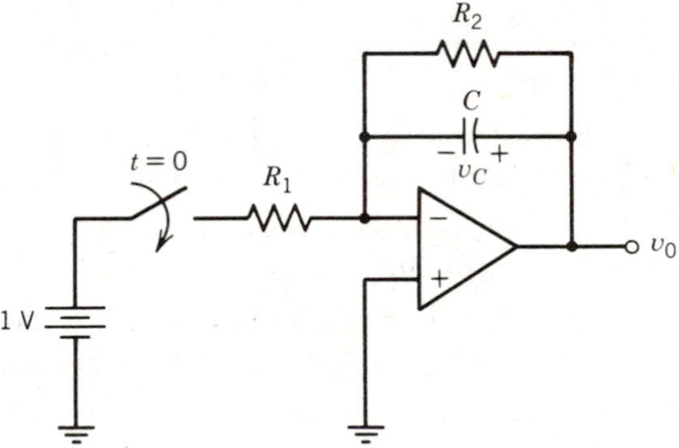

Figure 5.36

24. Find the unit step response for the op amp circuit in Fig. 5.37 if the initial inductor current is zero.

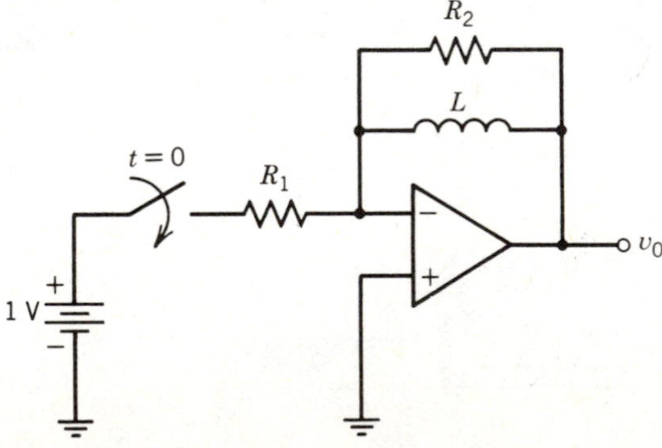

Figure 5.37

25. Draw a simulation diagram to solve the first order equation

$$\frac{dy}{dt} + 0.5y(t) = x(t)$$

with initial condition $y(0) = 2$. Find and plot the solution if $x(t) = u(t)$.

26. Draw a simulation diagram to solve the first order equation

$$\frac{dy}{dt} + 2y(t) = x(t)$$

with initial condition $y(0) = 2$.

27. Draw a simulation diagram to solve the second order equation

$$\frac{d^2y}{dt^2} + 2\frac{dy}{dt} + y(t) = x(t)$$

with initial conditions $y(0) = 1, (dy/dt)(0) = 0$.

28. Draw a simulation diagram to solve the third order equation

$$\frac{d^3y}{dt^3} + 2\frac{d^2y}{dt} + 3\frac{dy}{dt} + y(t) = x(t)$$

with initial conditions $y(0) = 1, (dy/dt)(0) = 0, (d^2y/dt^2)(0) = 2$.

29. Sketch an integrate-and-dump signal detector. Use resistors and an inductor.

LEARNING EVALUATION ANSWERS

Section 5.1

1. $s + 0.5 = 0$
$x_h(t) = Ae^{-0.5t}$

2. a. $x_p(t) = 2$
b. $x_p(t) = 2t^2 - 8t + 16$
c. $x_p(t) = te^{-0.5t}$

3. a. $x(t) = 2 - e^{-0.5t}$
b. $x(t) = 2t^2 - 8t + 16 - 15e^{-0.5t}$
c. $x(t) = te^{-0.5t} + e^{-0.5t}$

Section 5.2

1. a. 0.030 s
b. 0.250 s
c. 0.011 s

2. $v(t) = 64e^{-5t}V, t > 0$ (see Fig. 5.38)

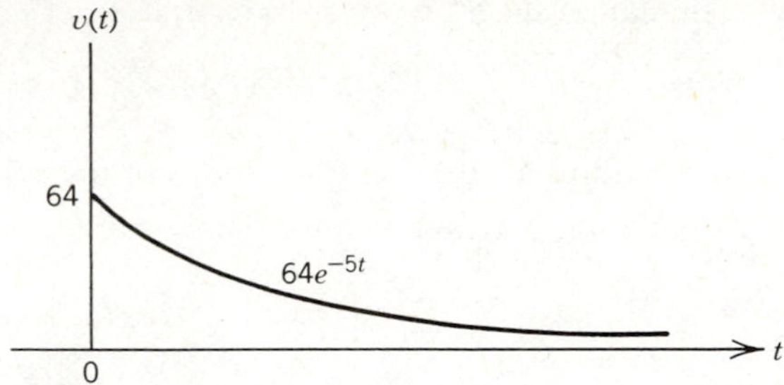

Figure 5.38

3. $i(t) = 0.571e^{-2.857t}$ mA, $t > 0$

Section 5.3

1. a. $T = 2(10)^{-4}$ s
   b. $T = 3.169(10)^{-5}$ s
   c. $T = 1.667(10)^{-6}$ s
2. $i(t) = 1 - e^{-5(10)^3 t}$ A, $t > 0$
3. $i(t) = 8.45 - 8.45e^{-31556t}$ mA, $t > 0$ (see Fig. 5.39)

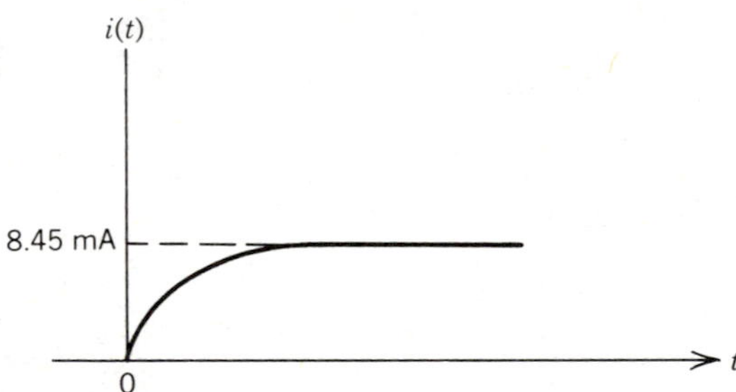

Figure 5.39

Section 5.4

1. See Fig. 5.40.

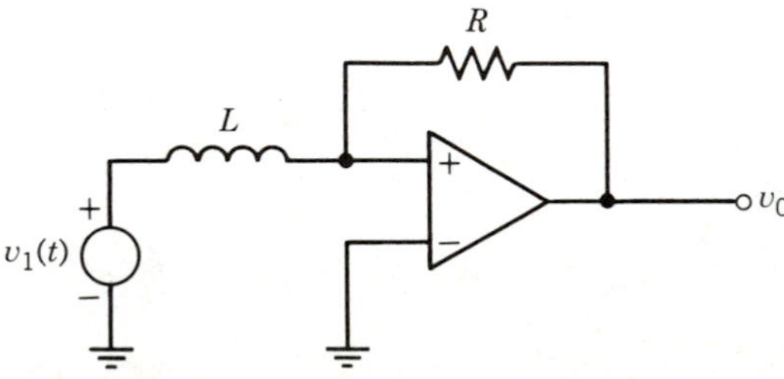

Figure 5.40

2. See Fig. 5.41.

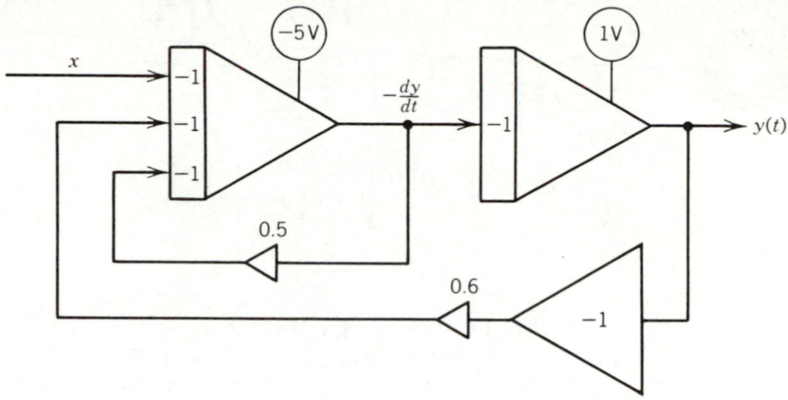

Figure 5.41

CHAPTER 6 SECOND ORDER CIRCUITS

We have previously investigated circuits containing single storage elements and found that the analysis of these circuits required the solution of a first order differential equation. In this chapter we allow two energy storage elements (inductors and/or capacitors) to be included in our circuits. Analysis of this circuit will require the solution of a second order differential equation (or two simultaneous first order equations). For this reason, circuits containing two energy storage elements are called "second order" circuits.

Although the analysis of second order circuits is more complex than the analysis of first order circuits, we will use the same approach taken in the previous chapter. Recognizing that the complete response of a circuit is made up of the forced response and transient or source-free response, we will determine the transient response of the source-free circuit first and later find the complete response. The elements in a circuit containing a resistor, inductor, and capacitor may be connected in series, in parallel, or series-parallel, and each case will be studied separately.

Second order circuits find wide use in engineering applications. Any physical system that can be described by a second order differential equation can be modeled by an *RLC* circuit. Observing the circuit response by altering values of one or more parameters allows us to predict how altering the value of the corresponding component in the physical system will affect its output. Also electric circuits are usually inexpensive to build compared to the physical system being modeled.

6.1 Second Order Differential Equations

LEARNING OBJECTIVES

After completing this section you should be able to do the following:

1. Find the characteristic equation, natural frequencies, and homogeneous solution of a second order equation.
2. Use the method of undetermined coefficients to find the forced response of a second order equation.
3. Find the complete solution by adding the homogeneous and forced solutions and then solving for arbitrary constants from given initial conditions.

A second order linear time invariant (LTI) ordinary differential equation has the form

$$a_2 \frac{d^2}{dt^2} x(t) + a_1 \frac{d}{dt} x(t) + a_0 x(t) = f(t) \tag{6.1.1}$$

where $f(t)$ is the forcing function and $x(t)$ is the unknown function that we seek as the solution. The coefficients a_0, a_1, and a_2 are constant (not functions of time). As in Section 5.1, we consider first the simplest case, where $f(t) = 0$ and all coefficients a_0, a_1 and a_2 equal 1.

$$\frac{d^2}{dt^2} x(t) + \frac{d}{dt} x(t) + x(t) = 0 \tag{6.1.2}$$

What is the solution to this homogeneous equation? Is there a solution, and if so is it unique? Since we had success with the exponential function in searching for the solution to a first order equation, it seems prudent to try the exponential form here. Assume

$$x(t) = Ae^{st} \tag{6.1.3}$$

Substituting Eq. 6.1.3 into Eq. 6.1.2 gives

$$s^2 Ae^{st} + sAe^{st} + ae^{st} = 0$$

Dividing each term by Ae^{st} gives the characteristic equation

$$s^2 + s + 1 = 0 \tag{6.1.4}$$

Solving for the two roots of this equation gives the natural frequencies

$$s_1 = -\frac{1}{2} + j\frac{\sqrt{3}}{2}, \qquad s_2 = -1 - j\frac{\sqrt{3}}{2}$$

Therefore, there must be two solutions, one using s_1 in Eq. 6.1.3, and the other using s_2 in Eq. 6.1.3. Since either solution will give zero for the sum of the three terms in Eq. 6.1.2, their sum must also be a solution. Therefore, the homogeneous solution is given by

$$x_h(t) = A_1 e^{s_1 t} + A_2 e^{s_2 t} \tag{6.1.5}$$

This is also called the transient solution $x_{\text{tr}}(t)$.

We should pause to note several features of the solution in Eq. 6.1.5. First, the roots of the characteristic equation (the natural frequencies) are complex numbers. There are three possible cases for the roots; they may be real and unequal, real and equal, or complex. When they are complex, they must be conjugate to each other. Second, there are two functions that satisfy a second order homogeneous equation. In general, there are n solutions to an nth order equation, although some solutions may be repeated. (Carl Gauss used the proof that every rational polynomial has a solution as his doctoral thesis. This is now known as the fundamental theorem of algebra.) Third, there are two constants A_1 and A_2 in the solution. This means that it will take two independent linear equations to solve for these, and the two equations are usually found from initial conditions. Therefore we will need two initial conditions, usually $x(0)$ and $(d/dt)x(0)$.

EXAMPLE 6.1.1 Find the solution of the homogeneous equation given by

$$\frac{d^2}{dt}x(t) + 5\frac{d}{dt}x(t) + 4x(t) = 0 \tag{6.1.6}$$

where

$$x(0) = 0$$
$$\frac{d}{dt}x(0) = 1$$

Solution

The characteristic equation is found by substituting Eq. 6.1.3 into Eq. 6.1.6 to get

$$s^2 + 5s + 4 = 0$$

The natural frequencies are therefore given by

$$s_1 = -4, \qquad s_2 = -1$$

Therefore, the homogeneous solution is given by

$$x_h(t) = A_1 e^{-4t} + A_2 e^{-t} \tag{6.1.7}$$

Notice that the natural frequencies are real and unequal. We next substitute the initial conditions into Eq. 6.1.7 to get

$$x(0) = 0 = A_1 + A_2$$
$$\frac{d}{dt}x(0) = 1 = -4A_1 - A_2$$

which gives $A_1 = \frac{1}{3}$, $A_2 = -\frac{1}{3}$. Therefore, the complete solution is given by

$$x_h(t) = \tfrac{1}{3}e^{-4t} - \tfrac{1}{3}e^{-t} \tag{6.1.8}$$

The particular solution of higher order differential equations by the method of undetermined coefficients is identical in principle to the solution of first order equations. If the forcing function is any of the forms listed in Section 5.1, or any combination of sums and products of these forms, then the solution may be found by the method of undetermined coefficients. We simply assume a solution of the form of the forcing function plus all its derivatives, substitute into the equation, and then solve for the undetermined coefficients. A review of Section 5.1 at this point would be appropriate if you are not familiar with this procedure. Here is an example involving a second order differential equation.

EXAMPLE 6.1.2 Find the particular solution for the equation given by

$$\frac{d^2}{dt^2}x(t) + 5\frac{d}{dt}x(t) + 4x(t) = 10e^{-2t} \tag{6.1.9}$$

Solution

The natural frequencies are -4 and -1. (See Eq. 6.1.8.) Therefore, the assumed particular (also called steady-state) solution is given by

$$x_p(t) = Be^{-2t}$$

Substituting this into Eq. 6.1.9 gives

$$4Be^{-2t} - 10Be^{-2t} + 4Be^{-2t} = 10e^{-2t}$$

Dividing by e^{-2t} and collecting terms gives $B = -5$. Therefore, the particular solution is given by

$$x_p(t) = -5e^{-2t} \tag{6.1.10}$$

The complete solution is found by adding the homogeneous and particular solutions. Only after the two solutions are added do we solve for the constants in the homogeneous solution from initial conditions. Here is an example.

EXAMPLE 6.1.3 Find the complete solution for the equation given by

$$\frac{d^2}{dt^2}x(t) + 5\frac{d}{dt}x(t) + 4x(t) = 10e^{-2t}$$

where

$$x(0) = 0$$

$$\frac{d}{dt}x(0) = 1$$

Solution

Combining the solutions of the previous two examples we have

$$x(t) = x_p(t) + x_h(t) = -5e^{-2t} + A_1e^{-4t} + A_2e^{-t}$$

Substituting initial conditions gives

$$x(0) = 0 = -5 + A_1 + A_2$$

$$\frac{d}{dt}x(0) = 1 = 10 - 4A_1 - A_2$$

From these two equations we obtain $A_1 = \frac{4}{3}$ and $A_2 = \frac{11}{3}$. Therefore, the complete solution is given by

$$x(t) = -5e^{-2t} + \tfrac{4}{3}e^{-4t} + \tfrac{11}{3}e^{-t}$$

LEARNING EVALUATION

Find the complete solution for the equation given by

$$\frac{d^2}{dt^2}x(t) + 6\frac{d}{dt}x(t) + 5x(t) = e^{-t}$$

where

$$x(0) = 1$$

$$\frac{d}{dt}x(0) = 0$$

6.2 The Source-Free Series *RLC* Circuit

LEARNING OBJECTIVES

After completing this section you should be able to do the following:

1. Derive the second order differential equation for the voltage across the capacitor in a source-free series *RLC* circuit.
2. Define the characteristic equation and find the roots of this equation.
3. State the solution to the second order differential equation for the voltage across the capacitor in a series *RLC* source-free circuit in terms of the roots of the characteristic equation.
4. Graph a circuit variable composed of two exponential functions by using the concept of the time constant.

A circuit containing a single resistor, inductor, and capacitor in series is shown in Fig. 6.1. We assume that an initial current $i(0) = I_0$ and an initial voltage $v_C(0) = V_0$ is present. An arbitrary beginning time of t_1 could have been chosen with the initial values of $i(t)$ and $v_C(t)$ specified as $i(t_1)$ and $v_C(t_1)$. This only complicates the development and blocks insight into the results, so t_1 is chosen to be zero without loss of generality.

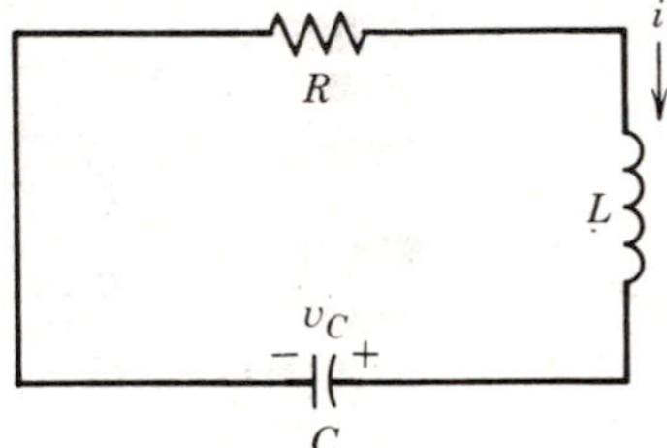

Figure 6.1. A series *RLC* circuit.

Using pre-established sign conventions, we can write KVL around the mesh as

$$v_R + v_L + v_C = 0 \tag{6.2.1}$$

or

$$Ri + L\frac{di}{dt} + \frac{1}{C}\int_{-\infty}^{t} i\,d\tau = 0 \tag{6.2.2}$$

If we differentiate both sides and divide by L, we obtain

$$\frac{d^2 i}{dt^2} + \frac{R}{L}\frac{di}{dt} + \frac{1}{LC}i = 0, \qquad t \geq 0 \tag{6.2.3}$$

If we had chosen to write Eq. 6.2.2 in terms of the voltage across the capacitor, v_C, instead of the current, i, we would have obtained from Eq. 6.2.1

$$R\left(C\frac{dv_C}{dt}\right) + L\frac{d}{dt}\left(C\frac{dv_C}{dt}\right) + v_C = 0, \qquad t \geq 0$$

or rearranging terms,

$$\frac{d^2v_C}{dt^2}+\frac{R}{L}\frac{dv_C}{dt}+\frac{1}{LC}v_C=0, \qquad t\geq 0 \tag{6.2.4}$$

Comparing Eqs. 6.2.3 and 6.2.4, we note that they are of the same form. The solution of either equation immediately indicates the solution of the other. This can be shown to be true of any variable (e.g., v_R) of a source-free *RLC* circuit. We will proceed using Eq. 6.2.4.

We must have two initial conditions specified for our second order differential equation in order to find a solution and both conditions must be in terms of the variable of the equation (in this case, v_C). We have one initial condition specified as

$$v_C(0)=V_0 \tag{6.2.5}$$

But the other is $i(0)$. In order to relate I_0 to our equation variable, we use the relation

$$i(t)=C\frac{dv(t)}{dt}$$

at $t=0$:

$$\left(\frac{dv_C}{dt}\right)\bigg|_{t=0}=\frac{1}{C}I_0 \tag{6.2.6}$$

Equation 6.2.4 must be satisfied at every instant of time and this is possible only if the waveforms of $v_C(t)$ and its first and second derivatives have the same shape. Also, at least one of the waveforms must be the negative of the others at all times so that the sum can be zero. Just as we found in the case of the *RL* and *RC* circuits, the exponential function meets these conditions. We wisely assume a solution of Eq. 6.2.4 to be of the form

$$v_C(t)=Ae^{st} \tag{6.2.7a}$$

then

$$\frac{d}{dt}v_C(t)=Ase^{st}=sv_C(t) \tag{6.2.7b}$$

and

$$\frac{d^2}{dt^2}v_C(t)=As^2e^{st}=s^2v_C(t) \tag{6.2.7c}$$

where it is assumed that s is not a function of t. Substituting Eqs. 6.2.7 into Eq. 6.2.4 results in

$$s^2v_C+\frac{R}{L}sv_C+\frac{1}{LC}v_C=0, \qquad t\geq 0$$

or

$$\left(s^2+\frac{R}{L}s+\frac{1}{LC}\right)v_C=0, \qquad t\geq 0 \tag{6.2.8}$$

Since Eq. 6.2.8 must be zero for all $t \geq 0$ either

$$v_C(t) = 0, \qquad t \geq 0 \tag{6.2.9}$$

or

$$\left(s^2 + \frac{R}{L}s + \frac{1}{LC}\right) = 0, \qquad t \geq 0 \tag{6.2.10}$$

Acceptance of Eq. 6.2.9 is precluded by initial conditions so Eq. 6.2.10 must hold.

Equation 6.2.10 is the *characteristic equation* for a second order linear homogeneous equation, such as that given by Eq. 6.2.4. Solution of this characteristic equation for s will give us the natural frequencies. Applying the quadratic formula to the characteristic equation shows that it has two roots:

$$s_1 = -\frac{R}{2L} + \sqrt{\left(\frac{R}{2L}\right)^2 - \frac{1}{LC}} \tag{6.2.11a}$$

$$s_2 = -\frac{R}{2L} - \sqrt{\left(\frac{R}{2L}\right)^2 - \frac{1}{LC}} \tag{6.2.11b}$$

Either the term $A_1e^{s_1t}$ or the term $A_2e^{s_2t}$ can satisfy Eq. 6.2.4. But since we have specified two initial conditions for Eq. 6.2.4, we must have two arbitrary constants in our solution in order to accommodate these initial conditions. Since either of the preceding terms satisfies our equation, their sum must also satisfy the equation, since it is linear:

$$v_C(t) = A_1e^{s_1t} + A_2e^{s_2t} \tag{6.2.12}$$

where s_1 and s_2 are given by Eq. 6.2.11 and A_1 and A_2 are determined by the initial conditions.

All astute students have already noticed that Eq. 6.2.12 breaks down if $s_1 = s_2$, since this would result in

$$v_C(t) = (A_1 + A_2)e^{s_1t} = Be^{s_1t}$$

and we would again have only one arbitrary constant with two initial conditions. Alteration of the form of the solution for this special case will be postponed temporarily.

For the case where $s_1 \neq s_2$, we find the initial conditions as

$$v_C(0) = V_0 = A_1 + A_2$$

$$I_0 = i(0) = C\left(\frac{dv_C}{dt}\right)\bigg|_{t=0} = C(s_1A_1 + s_2A_2)$$

Solving these two equations yields

$$A_1 = \frac{(I_0/C) - s_2V_0}{s_1 - s_2}, \qquad s_1 \neq s_2 \tag{6.2.13}$$

$$A_2 = \frac{s_1V_0 - (I_0/C)}{s_1 - s_2}, \qquad s_1 \neq s_2 \tag{6.2.14}$$

Substitution of Eqs. 6.2.11, 6.2.13, and 6.2.14 into Eq. 6.2.12 then results in a complete solution for $v_C(t)$.

EXAMPLE 6.2.1 Assume that the series *RLC* circuit in Fig. 6.1 has the following values:

$$R = 150\ \Omega \qquad L = 50\text{ mH} \qquad C = 10\ \mu\text{F} \qquad V_0 = 1\text{ V} \qquad I_0 = 10\text{ mA}$$

Find $v_C(t)$ and $i(t)$.

Solution

Using Eq. 6.2.11, we obtain

$$s_1 = -(1500) + \sqrt{2.25 \times 10^6 - (2 \times 10^6)}$$
$$= -1000$$

and

$$s_2 = -2000$$

By Eqs. 6.2.13 and 6.2.14,

$$A_1 = \frac{(10^{-2}/10^{-5}) + 2000(1)}{-1000 + 2000} = 3.0$$

$$A_2 = \frac{-1000(1) - (10^{-2}/10^{-5})}{1000} = -2.0$$

so

$$v_C(t) = 3.0e^{-1000t} - 2.0e^{-2000t}\text{ V}, \qquad t \geq 0$$

Note that we can check our results by letting $t = 0$ in our equation for $v_C(t)$ and see if $v_C(0)$ equals its given initial condition.

The current can be found by taking the derivative of $v_C(t)$ and multiplying by the capacitance.

$$i(t) = C\frac{dv_C}{dt} = 10^{-5}\left[(3.0)(-1000)e^{-1000t} - (2.0)(-2000)e^{-2000t}\right]$$
$$= 30e^{-1000t} + 40e^{-2000t}\text{ mA}$$

Again calculating $i(0)$ we get

$$i(0) = -30 + 40 = 10\text{ mA}$$

which checks with our initial assumptions. It should also be noted here that roundoff error can affect your ability to meet the initial conditions exactly. Also erroneous calculations of s_1 and s_2 will still allow you to meet initial conditions. For instance, if s_1 had been found to be -1500 instead of its true value of -1000, all subsequent calculations, though in error, would show that initial conditions are met.

The result, $v_C(t)$, can be graphed in several ways. By use of computer or calculator, values of v_C for increments of time can be calculated and then plotted directly. This results in an accurate graph. For an approximate graph of the signal one can use the following procedure:

1. Calculate the value of the first exponential for the first five multiples of its time constant and plot these. The multiplying factors are given in Table 6.1.
2. Repeat step 1 for the second exponential. See Table 6.2.

TABLE 6.1
FOR $3e^{-1000t}$, $\tau = 1$ ms

(t/τ)	t, ms	Multiplying Factor	$3e^{-t/\tau}$
0	0	1	3
1	1	0.368	1.104
2	2	0.135	0.405
3	3	0.0498	0.149
4	4	0.0183	0.055
5	5	0.0067	0.02

TABLE 6.2
FOR $-2e^{-2000t}$, $\tau = 0.5$ ms

(t/τ)	t, ms	Multiplying Factor	$-2e^{-t/\tau}$
0	0	1	−2.0
1	0.5	0.368	−0.736
2	1.0	0.135	−0.27
3	1.5	0.0498	−0.1
4	2.0	0.0183	−0.037
5	2.5	0.0067	−0.013

3. Plot each of the exponentials on the same graph and add the curves to obtain the resultant $v_C(t)$. This is illustrated in Fig. 6.2.

LEARNING EVALUATION

Solve the differential equation for current $i(t)$ in Fig. 6.1 if $i(0) = 0$, $v_C(0) = 1$. The circuit parameters are $R = 5\ \Omega$, $L = 4$ mH, and $C = 1000\ \mu$F.

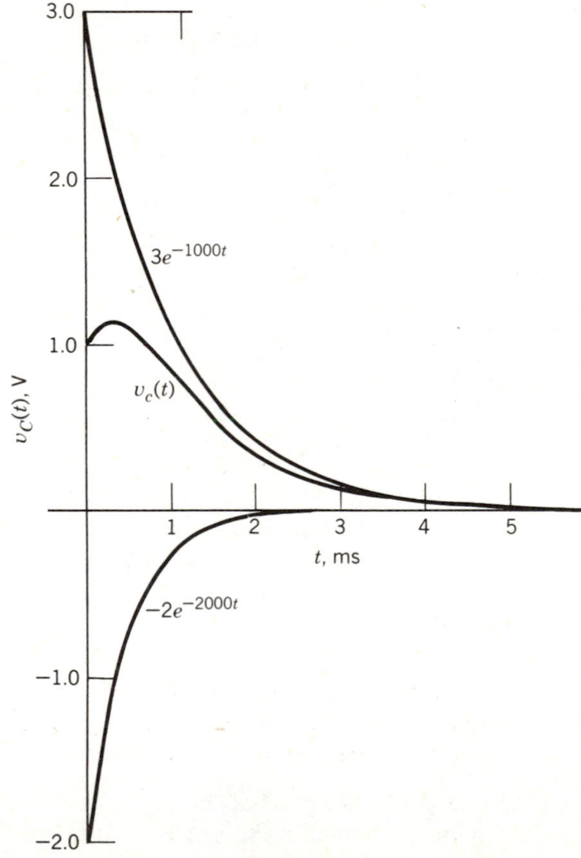

Figure 6.2

6.3 Classification of Responses

LEARNING OBJECTIVES

After completing this section you should be able to do the following:

1. Define and find the following parameters:
   Damping coefficient
   Natural frequency
   Damped natural frequency
2. Define whether a circuit is overdamped, critically damped, or underdamped based upon the foregoing three parameters.
3. Define and find the settling time for each of the three types of circuits mentioned above.
4. Solve for any voltage or current in a source-free series *RLC* circuit.

We found that the voltage across the capacitor of a series *RLC* circuit can be expressed by a second order differential equation (Eq. 6.2.4). Indeed, any circuit variable of a second order circuit can be described by a second order differential equation (e.g., the current i, see Eq. 6.2.3). Let us choose the general form of this second order differential equation to be

$$\frac{d^2x}{dt^2} + 2\alpha\frac{dx}{dt} + \omega_n^2 x = 0 \tag{6.3.1}$$

where x is our circuit variable. In particular, for $x = v_C$, by comparing Eq. 6.3.1 with Eq. 6.2.4, we see that

$$\alpha = \frac{R}{2L} \tag{6.3.2a}$$

$$\omega_n^2 = \frac{1}{LC} \tag{6.3.2b}$$

Then the roots of Eq. 6.3.1 can be expressed as

$$s_1 = -\alpha + \sqrt{\alpha^2 - \omega_n^2} \tag{6.3.3a}$$

$$s_2 = -\alpha - \sqrt{\alpha^2 - \omega_n^2} \tag{6.3.3b}$$

The relative values of α and ω_n determine the nature of the roots s_1 and s_2.

The Overdamped Case

If $\alpha > \omega_n$, then $\alpha^2 > \omega_n^2$ and $(\alpha^2 - \omega_n^2)$ is a positive number; s_1 and s_2 are real numbers. This condition is called the overdamped case. Since α and ω_n must be positive numbers for all physically realizable values of R, L, and C, we have that if $\alpha > \omega_n$, then

$$\alpha > \left(\alpha^2 - \omega_n^2\right)^{1/2}$$

This means that the roots s_1 and s_2 will always be *negative* real numbers and the circuit response will always be the algebraic sum of two negative exponential terms.

The example in the previous section will serve to help us understand the reason the case where $\alpha > \omega_n$ is called the "overdamped" case. In most applications, the transient or source-free responses of a circuit are undesirable and we want these responses to disappear as rapidly as possible. In Example 6.2.1, the transient response rose to some maximum value and eventually settled out at a final value of zero. As we discussed previously, the value actually reaches zero only at $t = \infty$, but at some point we say the response is negligible. Although there is no "industry standard" saying when this point is reached, in *RLC* circuits the following definition finds general acceptance.

> The *settling time* for a source-free *RLC* circuit is the time required for the response to reduce to 1% of its maximum value.

The procedure for calculating the settling time of a variable in a source-free *RLC* circuit is:

1. Differentiate the equation specifying the variable and set the result equal to zero.
2. Solve for $t = t_m$. This is the time when the maximum value occurs.
3. Put the value of t_m in the original equation and solve for the maximum value of the variable.
4. Find 1% of this maximum value and set the variable equation equal to this value.
5. Using the equation found in step 4, solve for t, which is the settling time, t_s.

In Example 6.2.1 we found that

$$v_C(t) = 3.0e^{-1000t} - 2.0e^{-2000t} \tag{6.3.4}$$

Performing step 1 yields

$$-1000(3)e^{-1000t_m} + 2(2000)e^{-2000t_m} = 0$$

or

$$-3000e^{-1000t_m} + 4000e^{-2000t_m} = 0 \tag{6.3.5}$$

To accomplish step 2, divide Eq. 6.3.5 by e^{-1000t_m},

$$4000e^{-1000t_m} = 3000$$

or

$$e^{-1000t_m} = 0.75$$

Taking the natural log of both sides gives

$$-1000t_m = -0.2876$$

so

$$t_m = 0.288 \text{ ms}$$

In step 3, we place this value of t_m in Eq. 6.3.4 and solve for $v_C(t_m) = v_{\max}$.

$$v_{\max} = 3e^{-1000(2.88\times10^{-4})} - 2e^{-2000(2.88\times10^{-4})} = 1.125$$

Step 4 is

$$(1.125)(0.01) = 3e^{-1000t_s} - 2e^{-2000t_s}$$

We know the last term can be ignored since it has long since decayed to near zero, so

$$0.01125 = 3.0e^{-1000t_s}$$

or

$$\ln(e^{-1000t_s}) = \ln(0.00375)$$

Hence,

$$t_s = 5.59 \text{ ms}$$

As we will demonstrate shortly, this is a relatively long settling time. The term "overdamped" means that the circuit values are such that the response is "sluggish", and consequently the transient response does not become negligible very rapidly.

To show that the circuit in Example 6.2.1 is in fact overdamped, we can calculate α and ω_n as

$$\alpha = \frac{R}{2L} = \frac{150}{2(50)(10^{-3})} = 1500.0$$

$$\omega_n = \sqrt{\frac{1}{LC}} = \sqrt{\frac{1}{(50)(10^{-3})(10)(10^{-6})}} = 1414.21$$

α is greater than ω_n so the circuit satisfies the overdamped criterion.

The Underdamped Case

If $\alpha < \omega_n$, the response is called "underdamped." For this case, $\alpha^2 < \omega_n^2$ and Eqs. 6.3.3 show that s_1 and s_2 are complex numbers. To facilitate further discussion, let us define another term, ω_d, as

$$\omega_d = \sqrt{\omega_n^2 - \alpha^2} \tag{6.3.6}$$

Then for $\alpha < \omega_n$, we have

$$\begin{aligned}\sqrt{\alpha^2 - \omega_n^2} &= \sqrt{-(\omega_n^2 - \alpha^2)} \\ &= \sqrt{-\omega_d^2} \\ &= j\omega_d\end{aligned}$$

So for the underdamped case Eqs. 6.3.3 become

$$s_1 = -\alpha + j\omega_d \tag{6.3.7}$$

$$s_2 = -\alpha - j\omega_d \tag{6.3.8}$$

The response for $v_C(t)$, using Eq. 6.2.12, is

$$\begin{aligned}v_C(t) &= A_1 e^{(-\alpha + j\omega_d)t} + A_2 e^{(-\alpha - j\omega_d)t} \\ &= e^{-\alpha t}\left(A_1 e^{j\omega_d t} + A_2 e^{-j\omega_d t}\right)\end{aligned}$$

Recall that Euler's formula states

$$e^{j\beta} = \cos\beta + j\sin\beta$$

so $v_C(t)$ can be expressed as

$$v_C(t) = e^{-\alpha t}\{A_1(\cos\omega_d t + j\sin\omega_d t) + A_2[\cos(-\omega_d t) + j\sin(-\omega_d)t]\}$$

But

$$\sin(-\beta) = -\sin\beta$$
$$\cos(-\beta) = \cos\beta$$

so

$$v_C(t) = e^{-\alpha t}[(A_1 + A_2)\cos\omega_d t + j(A_1 - A_2)\sin\omega_d t]$$

or

$$v_C(t) = e^{-\alpha t}(B_1\cos\omega_d t + B_2\sin\omega_d t) \tag{6.3.9}$$

where B_1 and B_2 must be real numbers since $v_C(t)$ is real and α and ω_d are real numbers.

Another trigonometric identity allows us to present Eq. 6.3.9 in an alternative form:

$$B_1\cos\beta + B_2\sin\beta = D\cos(\beta - \theta)$$

where

$$D = \sqrt{B_1^2 + B_2^2} \qquad \text{and} \qquad \theta = \tan^{-1}\left(\frac{B_2}{B_1}\right)$$

Hence,

$$v_C(t) = De^{-\alpha t}\cos(\omega_d t - \theta) \tag{6.3.10}$$

Expressing $v_C(t)$ as in Eq. 6.3.10 allows us to clearly see that $v_C(t)$ has an oscillatory component and an exponential component. The exponential component causes the response to eventually approach zero and the rate of approach is governed by the value of α. For this reason, α is called the *damping coefficient*.

To understand ω_d, first consider the case where $R = 0$. In this circumstance, $\alpha = 0$, $\omega_d = \omega_n$, and $v_C(t)$ would be a sinusoid oscillating at a radian frequency ω_d. Therefore, ω_n is the radian frequency at which the circuit response would oscillate if the resistance of the circuit were zero. Consequently, ω_n is often called the *natural frequency* of the circuit. ω_d is always less than ω_n if the resistance is greater than zero so ω_d is called the *damped natural frequency*.

For illustration, change the value of R in Example 6.2.1 to 100 Ω. Then

$$\alpha = 1000 \qquad \text{and} \qquad \omega_n = 1414.21$$

Since $\alpha < \omega_n$, $v_C(t)$ is given by Eq. 6.3.9 or Eq. 6.3.10.

$$\begin{aligned}\omega_d &= \sqrt{\omega_n^2 - \alpha^2}\\ &= \sqrt{(1414)^2 - (1000)^2}\\ &= 1000\end{aligned}$$

so

$$v_C(t) = e^{-1000t}(B_1 \cos 1000t + B_2 \sin 1000t) \qquad (6.3.11)$$

To evaluate B_1 and B_2 we must use the initial conditions.

$$v_C(0) = 1 = B_1$$

$$\begin{aligned} i(0) = C\left(\frac{dv_C}{dt}\right)\Bigg|_{t=0} &= (10^{-5})\big[(-10^3)e^{-1000t}B_1 \cos 1000t \\ &\quad + (-10^3)e^{-1000t}B_1 \sin 1000t \\ &\quad + (-10^3)e^{-1000t}B_2 \sin 1000t \\ &\quad + (10^3)B_2 e^{-1000t} \cos 1000t\big]\Big|_{t=0} \\ &= (10^{-5})(10^3 B_1 + 10^3 B_2) \end{aligned}$$

so

$$i(0) = 0.01 = -0.01 + 0.01B_2$$

or

$$B_2 = 2$$

Consequently,

$$v_C(t) = e^{-1000t}(\cos 1000t + 2\sin 1000t)\ \text{V}, \qquad t \geq 0$$

The alternative form for $v_C(t)$ (Eq. 6.3.10) yields

$$v_C(t) = \sqrt{5}\, e^{-1000t} \cos(1000t - 63.4^0)\ \text{V}, \qquad t \geq 0 \qquad (6.3.12)$$

Plotting $v_C(t)$ can be accomplished by using Eq. 6.3.12 and plotting the two functions separately, then plotting their product on the same graph.

Alternatively, a simple computer or calculator program can be written to calculate v_C for certain values of time. Table 6.3 lists values of v_C for specified times that were calculated using the BASIC program shown at the right of the table. The term $\cos 1000t$ has a period of approximately 6 ms, and this cyclic term can be seen to go through two cycles by observing the alternating sign of v_C, since the exponential component is always positive. The graph of $v_C(t)$ is shown in Fig. 6.3. Only the first 5 ms are plotted, since values subsequent to this could not be distinguished from zero on the graph.

To find the settling time for this underdamped example, we could proceed as before, but the mathematics becomes rather messy. Instead, let us again try finding a computer solution. Our approach will be to find the maximum value of v_C, then 1% of that value, and finally that time where the absolute value of $v_C(t)$ never exceeds this 1% value. Table 6.4 contains a BASIC program that calculates the settling time for our example to be 4.98 ms. This program is not optimized for minimum length but rather represents a straightforward computer solution. Many programs exist for finding a maximum in the shortest possible time, but are rather intricate and the time savings do not warrant the increased complexity and programming time associated with their use for "one-shot" programs such as the one in this example.

TABLE 6.3
DATA FOR FIG. 6.3 AND BASIC PROGRAM THAT GENERATED THE DATA

T = 0	V = 1.0013	100 T = 0
T = 2.00000E-04	V = 1.12866	110 V = SQR(5.0) * EXP(− 1000 * T) * cos(1000 T − 1.1065)
T = 4.00000E-04	V = 1.14011	120 PRINT "T = "; T, "V = "; V
T = 6.00000E-04	V = 1.0731	130 T = T + 0.0005
T = 8.00000E-04	V = .957905	140 if T < = 0.014 THEN 110
T = 9.99999E-04	V = .817943	150 END
T = 1.20000E-03	V = .67055	
T = 1.40000E-03	V = .527829	
T = 1.60000E-03	V = .397588	
T = 1.80000E-03	V = .284243	
T = 2.00000E-03	V = .189648	
T = 2.20000E-03	V = .113818	
T = 2.40000E-03	V = 5.55321E-02	
T = 2.60000E-03	V = 1.28249E-02	
T = 2.80000E-03	V = −1.66427E-02	
T = 3.00000E-03	V = −3.53053E-02	
T = 3.20000E-03	V = −4.55028E-02	
T = 3.40000E-03	V = −4.93580E-02	
T = 3.60000E-03	V = −4.87094E-02	
T = 3.80000E-03	V = −4.50842E-02	
T = 4.00000E-03	V = −3.97012E-02	
T = 4.20000E-03	V = −3.34925E-02	
T = 4.40000E-03	V = −2.71370E-02	
T = 4.60000E-03	V = −2.10993E-02	
T = 4.80000E-03	V = −1.56701E-02	
T = 5.0000E-03	V = −1.10045E-02	

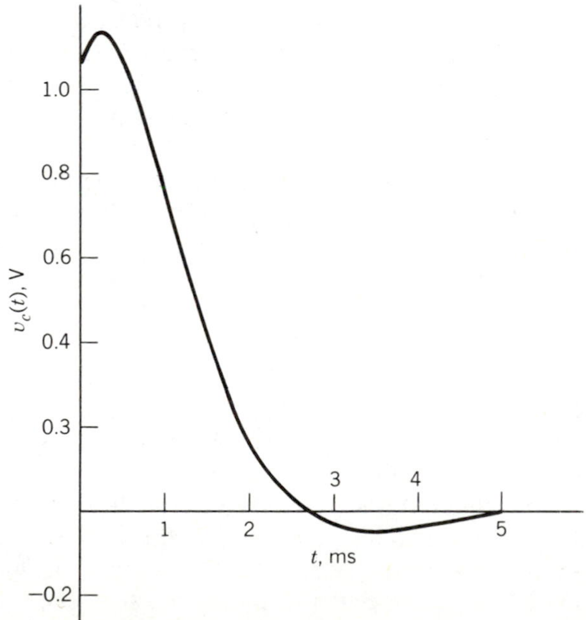

Figure 6.3. Plot of $v_C(t)$ for the underdamped case.

TABLE 6.4
A BASIC PROGRAM FOR CALCULATING SETTLING TIME

```
100 DEF FNV(T) = SQR(5.0) *EXP( - 1000 *T)*COS(1000*T - 1.1065)
110 T = 0
120 VO = FNV(T)                                        'CALCULATE V(0)
130 V2 = V0                                            'V2 = V(MAXIMUM)
140 T = T + 0.00001
150 V1 = FNV(T)
160 IF V1 < V2 THEN 190
170 V2 = V1
180 T2 = T                                             'T2 = T(MAXIMUM)
190 IF T < = 0.01 THEN 140                             'PROGRAM STOPS AT T = 10 MSEC
200 V3 = 0.01*V2                                       'V3 = SETTLING VOLTAGE
210 J = 0                                              'J CONTROLS SETTING VALUE OF T3
220 T = T2
230 T = T + 0.00001
240 V1 = FNV(T)
250 IF ABS (V1) > V3 THEN 320
260 IF J = 1 THEN 290
270 J = 1
280 T3 = T                                             'T3 = SETTLING TIME
290 IF T < = 0.01 THEN 230
300 PRINT "V MAX = "; V2, "T MAX = "; T2, "SETTLING TIME = "; T3
310 GO TO 340
320 J = 0
330 GO TO 290
340 STOP
350 END
```

The Critically Damped Case

We have shown that if $\alpha > \omega_n$, the series *RLC* circuit response is overdamped, and if $\alpha < \omega_n$, the response is underdamped. Only one additional situation can exist; where $\alpha = \omega_n$. This case is called the *critically damped* case. With values of inductance and capacitance constant, the overdamped case can be changed to the critically damped case by adjusting the value of resistance to make $\alpha = \omega_n$. In this situation, the settling time for the $\alpha = \omega_n$ case will always be less than that of the overdamped case; hence the name "critically damped."

Obviously α can also be made equal to ω_n by changing the values of L or C or both. However, changing L or C changes the natural frequency of the circuit. Direct comparisons of settling times of two circuits with different natural frequencies is not useful in illustrating the meanings of the overdamped and critically damped cases.

Examination of Eqs. 6.3.3 for $\alpha = \omega_n$ shows that

$$s_1 = s_2 = -\alpha \tag{6.3.13}$$

and the solution for the voltage across the capacitor of a series *RLC* circuit as shown in Fig. 6.1 is given by Eq. 6.2.12 as

$$v_C(t) = A_1 e^{-\alpha t} + A_2 e^{-\alpha t} = A e^{-\alpha t} \tag{6.3.14}$$

As pointed out in Section 6.2, this form is not sufficient since we have two initial conditions to satisfy and only one arbitrary constant in our equation. Our conclusion then must be that the exponential response is *not* all of the solution to the differential equation of Eq. 6.2.4 for $s_1 = s_2$.

To find the true solution to Eq. 6.2.4 when $\alpha = \omega_n$, we can use any one of several approaches:

1. We could try to guess at a new form of the solution and see if it satisfies the differential equations.
2. We could take the limit of Eq. 6.2.12 as $s_1 \to s_2$.
3. We could take the solution for the underdamped case (Eq. 6.3.9) and allow ω_d to approach zero.

Let us proceed by using the third method listed. We have from (Eq. 6.3.9)

$$v_C(t) = e^{-\alpha t}(B_1 \cos \omega_d t + B_2 \sin \omega_d t) \tag{6.3.15}$$

and we want to take the limit as $\omega_d \to 0$. Since B_1 and B_2 may be functions of ω_d, we must solve for these constants in terms of initial conditions before going further. To be as general as possible, assume that

$$v_C(0) = V_0 \qquad \text{and} \qquad i(0) = I_0$$

then from Eq. 6.3.15,

$$v_C(0) = V_0 = B_1$$

and

$$\frac{d}{dt} v_C(0) = \frac{I_0}{C} = \left[-\alpha e^{-\alpha t}(V_0 \cos \omega_d t + B_2 \sin \omega_d t) \right.$$
$$\left. + e^{-\alpha t}(-\omega_d V_0 \sin \omega_d t + \omega_d B_2 \cos \omega_d t) \right] \Big|_{t=0}$$

$$\frac{I_0}{C} = -\alpha V_0 + \omega_d B_2$$

or

$$B_2 = \frac{(I_0/C) + V_0}{\omega_d}$$

so Eq. 6.3.15 becomes

$$v_C(t) = e^{-\alpha t}\left[V_0 \cos \omega_d t + \left(\frac{I_0}{C} + \alpha V_0 \right) \frac{\sin \omega_d t}{\omega_d} \right] \tag{6.3.16}$$

The limit of $v_C(t)$ as $\omega_d \to 0$ is the sum of the limits of the two terms inside the brackets. It is helpful to evaluate these two terms independently.

$$\lim_{\omega_d \to 0} V_0 \cos \omega_d t = V_0$$

But

$$\lim_{\omega_d \to 0} \left(\frac{I_0}{C} + \alpha V_0 \right) \frac{\sin \omega_d t}{\omega_d}$$

is indeterminate. However, if we multiply by t/t, we find

$$\lim_{\omega_d \to 0} \left(\frac{I_0}{C} + \alpha V_0 \right) t \frac{\sin \omega_d t}{\omega_d t} = \left(\frac{I_0}{C} + V_0 \right) t$$

since

$$\lim_{x \to 0} \frac{\sin x}{x} = 1$$

Using these results we can state that

$$\lim_{\omega_d \to 0} v_C(t) = e^{-\alpha t} \left[V_0 + \left(\frac{I_0}{C} + \alpha V_0 \right) t \right]$$

Hence, the general solution for the case where $\alpha = \omega_n$ is

$$v_C(t) = (A + Bt) e^{-\alpha t} \tag{6.3.17}$$

where A and B are arbitrary constants evaluated by using the given initial conditions.

Returning once again to Example 6.2.1, adjust the value of R so $\alpha = \omega_n$.

$$\begin{aligned} R &= \frac{2L}{\sqrt{LC}} = \sqrt{\frac{4L}{C}} \\ &= \sqrt{\frac{4(50 \times 10^{-3})}{10^{-5}}} \\ &= 100\sqrt{2} \end{aligned}$$

Consequently,

$$\begin{aligned} \alpha &= \frac{R}{2L} = \frac{100\sqrt{2}}{2(50)(10^{-3})} \\ &= 1000\sqrt{2} \approx 1414.21 \end{aligned}$$

so

$$\begin{aligned} v_C(t) &= (A + Bt) e^{-1414.21t} \\ v_C(0) &= A = V_0 = 1 \end{aligned}$$

$$\frac{i(0)}{C} = \left. \frac{dv_C}{dt} \right|_{t=0} = \left. \left[(-1414.21A - 1414.21Bt) e^{-1414.21t} + Be^{-1414.21t} \right] \right|_{t=0}$$

Therefore,

$$\frac{i(0)}{C} = \frac{10 \times 10^{-3}}{10^{-5}} = -1414.21 + B$$

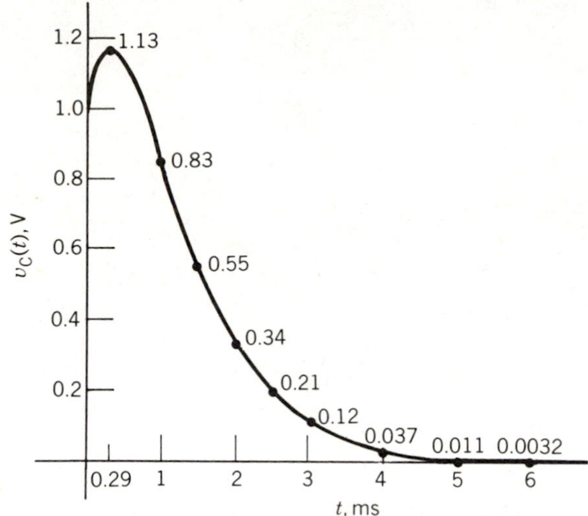

Figure 6.4. Plot of $v_C(t)$ for the critically damped case.

or

$$\begin{aligned} B &= 1000 + 1414.21 \\ &= 2414.21 \end{aligned}$$

Finally, then,

$$v_C(t) = (1 + 2424.21t)e^{-1414.21t}, \qquad t \geq 0 \tag{6.3.18}$$

$v_C(t)$ is plotted in Fig. 6.4. To find the settling time, we proceed as before by setting the derivative of Eq. 6.3.18 equal to zero and solving for t_m.

$$\frac{dv_C(t)}{dt} = \left(1000 - 3.4 \times 10^6 t_m\right)e^{-1414.21t_m} = 0$$

Since e^{-x} is monitonically decreasing and always greater than or equal to zero for positive x, we can set

$$1000 - 3.4 \times 10^6 t_m = 0$$

and find

$$t_m = 0.29 \text{ ms}$$

The value of $v_C(t_m)$ can be calculated to be 1.13 V. One percent of this is 0.0113 and occurs at

$$t_s = 5.0 \text{ ms}$$

As we predicted, this is less than the settling time found for the overdamped case. The underdamped case will often have a settling time equal to or slightly less than the critically damped case, as was demonstrated here.

It is important to realize that Eq. 6.3.1 represents the response of a second order circuit for *any* circuit variable. We have applied this only to the series *RLC* circuit where the circuit variable of interest was the capacitor voltage. *For this case*, α and ω_n were defined by Eqs. 6.3.2. For other cases, α and ω_n *may* be

TABLE 6.5
TRANSIENT RESPONSE SUMMARY FOR A SECOND ORDER CIRCUIT.
FORM OF THE DIFFERENTIAL EQUATION: $\frac{d^2x}{dt^2} + 2\alpha\frac{dx}{dt} + \omega_n^2 x = 0$, $x =$ circuit variable

Type of Response	Necessary Condition	Roots of the Characteristic Equation (s_1 s_2)	Circuit Response
Overdamped	$\alpha > \omega_n$	Negative, real, unequal	$x(t) = A_1e^{s_1t} + A_2e^{s_2t}$
Critically damped	$\alpha = \omega_n$	Negative, real, equal	$x(t) = (A + Bt)e^{st}$
Underdamped	$\alpha < \omega_n$	Complex conjugates; negative real parts	$x(t) = e^{-\alpha t}(B_1 \cos \omega_d t + b_2 \sin \omega_d t)$ or $x(t) = De^{-\alpha t}(\cos \omega_{dt} - \theta)$
$s_{1,2} = -\alpha \pm \sqrt{\alpha^2 - \omega_n^2}$		$\omega_d = \sqrt{\omega_n^2 - \alpha^2}$	$D = B_1^2 + B_2^2$ $\theta = \tan^{-1}\frac{B_2}{B_1}$

different functions of the circuit parameters. Two other cases will be examined in the next sections. Table 6.5 summarizes the salient features of the second order circuit transient response.

LEARNING EVALUATION

If in Fig. 6.1 $L = 4$ mH and $C = 1000\ \mu$F, find the values of R so that the circuit is

a. Underdamped.
b. Critically damped.
c. Overdamped.

6.4 The Source-Free Parallel *RLC* Circuit

LEARNING OBJECTIVES

After completing this section you should be able to do the following:

1. Solve for any voltage or current in a source-free parallel *RLC* circuit.
2. Find the expressions for the damping coefficient, natural frequency, and damped natural frequency in a source-free parallel *RLC* circuit.

A circuit containing a resistor, inductor, and capacitor connected in parallel is shown in Fig. 6.5. The necessary initial conditions are $v(0)$ and $i_L(0)$. Note that the same voltage, $v(t)$, appears across all three elements of the circuit. Let us try

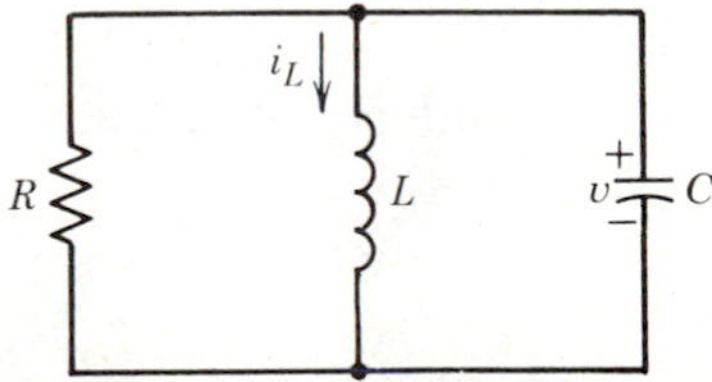

Figure 6.5. A parallel *RLC* circuit.

to find the current, $i_L(t)$, using the techniques of the previous sections. By KCL,

$$\frac{v}{R} + i_L + C\frac{dv}{dt} = 0$$

and

$$v = L\frac{di_L}{dt}$$

Combining these two equations and rearranging terms yields

$$\frac{d^2 i_L}{dt^2} + \frac{1}{RC}\frac{di_L}{dt} + \frac{1}{LC}i_L = 0 \qquad (6.4.1)$$

Comparing Eq. 6.4.1 with our general form of

$$\frac{d^2x}{dt^2} + 2\alpha\frac{dx}{dt} + \omega_n^2 x = 0$$

where x is the circuit variable under consideration, yields

$$\alpha = \frac{1}{2RC}, \qquad \omega_n = \frac{1}{\sqrt{LC}}$$

So for the parallel RLC circuit, ω_n is defined as before but α is a function of R and C instead of R and L. We solve for $i_L(t)$ by finding values of α and ω_n and determining if our solution is overdamped, critically damped, or underdamped. We then choose the proper form of the solution from Table 6.5.

EXAMPLE 6.4.1 For the circuit in Fig. 6.5, $R = 1\ \text{k}\Omega$, $L = 500$ mH, $C = 5\ \mu$F. Find the expression for the current $i_L(t)$ if $i_L(0) = 5$ mA and $v(0) = 1$ V.

Solution

First calculate α and ω_n:

$$\alpha = \frac{1}{2RC} = \frac{1}{2(1000)(5\times10^{-6})} = 100$$

$$\omega_n = \frac{1}{\sqrt{LC}} = 632.46$$

Consequently, the circuit is underdamped so

$$\omega_d = \sqrt{\omega_n^2 - \alpha^2} = 624.5$$

Choosing the form of the solution from Table 6.5, we have

$$i_L(t) = e^{-100t}(B_1 \cos 624.5t + B_2 \sin 624.5t)$$

Next evaluate B_1 and B_2 using initial conditions

$$i_L(0) = 5\text{ mA} = B_1$$

$$\left.\frac{di_L}{dt}\right|_{t=0} = \frac{v(0)}{L} = \frac{1}{0.5} = 2$$

$$\left.\frac{di_L}{dt}\right|_{t=0} = [e^{-100t}(-624.5B_1 \sin 624.5t + 624.5B_2 \cos 624.5t) - 100e^{-100t}(B_1 \cos 624.5t + B_2 \sin 624.5t)]|_{t=0}$$

or

$$2 = 624.5B_2 - 100B_1$$

Since

$$B_1 = 0.005$$

we find

$$B_2 = 0.004$$

Hence,

$$i_L(t) = e^{-100t}(5 \cos 624.5t + 4 \sin 624.5t) \text{ mA}, \qquad t \geq 0$$

A plot of the inductor current is shown in Fig. 6.6. The oscillatory nature of the underdamped response is clearly seen in this case. Physically, the energy in the circuit is being alternately stored in the inductor and capacitor. Every time the energy is transferred from one storage element to the other, some of the energy is dissipated (lost) in the resistor.

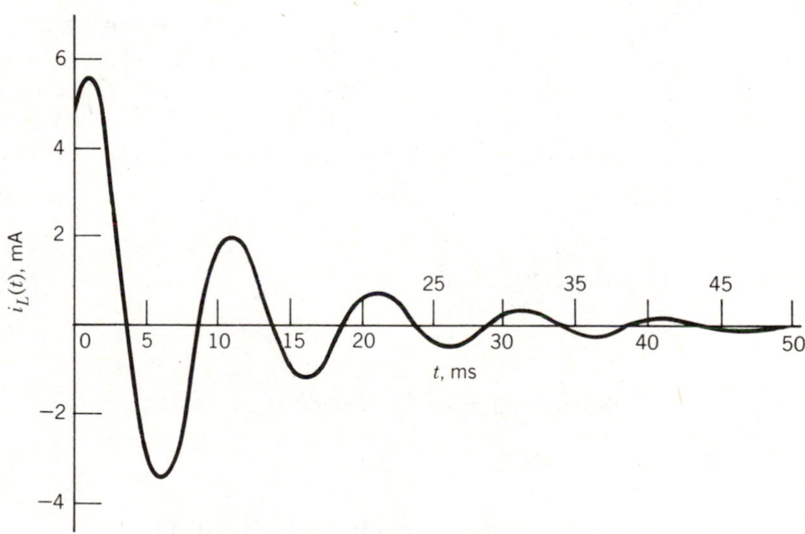

Figure 6.6. Plot of $i_L(t)$ for Example 6.4.1.

LEARNING EVALUATION

Solve for the voltage $v(t)$ in Example 6.4.1.

6.5 The Source-Free Series-Parallel Circuit

LEARNING OBJECTIVES

After completing this section you should be able to do the following:

1. Find the expressions for the damping coefficient, the natural frequency, and the damped natural frequency for any second order source-free series-parallel circuit.
2. Find the source-free response of any circuit variable of a series-parallel second order circuit similar to the circuit used as an example in this section.

Many different series-parallel circuits containing two energy storage elements can be constructed. One such circuit is shown in Fig. 6.7.

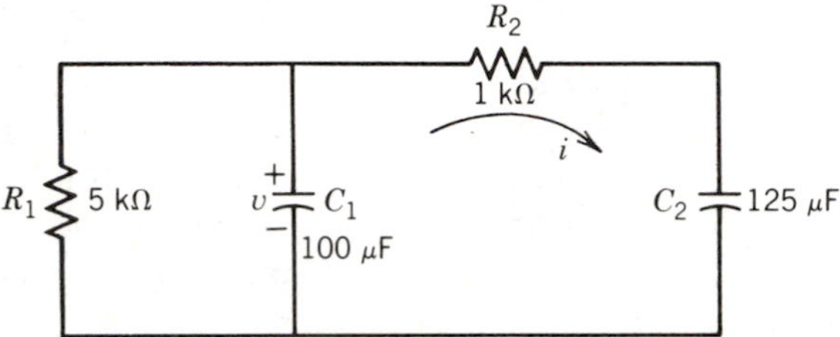

Figure 6.7. A series-parallel second order circuit.

This circuit contains two capacitors as our energy storage elements instead of one capacitor and one inductor but this does not violate any premise on which our efforts to date are based. We want to find the voltage, v, across capacitor C_1. To do this, let us apply KCL to the node connecting R_1, C_1, and R_2, and then apply KVL around the right-hand mesh. The nodal equation is

$$\frac{v}{R_1} + C_1\frac{dv}{dt} + i = 0 \tag{6.5.1}$$

The mesh equation is

$$R_2 i + \frac{1}{C_2}\int_{-\infty}^{t} i\,d\lambda - v = 0 \tag{6.5.2}$$

Solving Eq. 6.5.1 for i gives

$$i = -C_1\frac{dv}{dt} - \frac{v}{R_1}$$

Substituting this into Eq. 6.5.2 results in

$$R_2\left(-C_1\frac{dv}{dt} - \frac{v}{R_1}\right) + \frac{1}{C_2}\int_{-\infty}^{t}\left(-C_1\frac{dv}{dt} - \frac{v}{R_1}\right)d\lambda - v = 0 \tag{6.5.3}$$

Take the derivative of Eq. 6.5.3 in order to remove the integral and collect terms:

$$\frac{d^2v}{dt^2} + \frac{1}{C_1R_2}\left(\frac{R_2}{R_1} + \frac{C_1}{C_2} + 1\right)\frac{dv}{dt} + \frac{1}{C_1C_2R_1R_2}v = 0 \tag{6.5.4}$$

Examination of Eq. 6.5.4 reveals that

$$\alpha = \left(\frac{1}{2C_1R_1} + \frac{1}{2C_2R_2} + \frac{1}{2C_1R_2} \right)$$

$$\omega_n = \frac{1}{\sqrt{C_1C_2R_1R_2}}$$

Using the values shown in Fig. 6.7 yields

$$\alpha = 10$$

$$\omega_n = 4$$

Since $\alpha > \omega_n$, the system is overdamped and

$$v(t) = A_1e^{s_1t} + A_2e^{s_2t} \text{ V}, \qquad t \geq 0$$

We find s_1 and s_2 to be

$$s_1 = -\alpha + \sqrt{\alpha^2 - \omega_n^2} = -0.835$$

$$s_2 = -\alpha - \sqrt{\alpha^2 - \omega_n^2} = -19.165$$

So

$$v(t) = A_1e^{-0.835t} + A_2e^{-19.165t} \text{ V}, \qquad t \geq 0$$

Evaluating the constants A_1 and A_2 requires initial conditions be stated. Then we would proceed as before.

You should now feel comfortable finding the transient response to any second order circuit. We should bare our soul, however, and admit that we chose a relatively easy series-parallel circuit to investigate. Some other circuits of this type present difficulties in finding one second order differential equation in terms of the variable of interest. Take heart though; additional analysis techniques will be introduced later to make this task easier.

LEARNING EVALUATION
Solve for the current $i(t)$ in Fig. 6.7.

6.6 Complete Response for Second Order Circuits

LEARNING OBJECTIVES
After completing this section you should be able to do the following:

Find the complete response of a circuit variable in a series or parallel *RLC* circuit which contains the forcing function.

We now have all the information necessary to find the complete response of a second order circuit. Recall that the complete response is composed of a steady-state response and a transient response. That is, a voltage $v(t)$ can be expressed as

$$v(t) = v_{ss}(t) + v_{tr}(t) \qquad (6.6.1)$$

where $v_{ss}(t)$ is the steady-state response and $v_{tr}(t)$ is the transient response. In this section we will consider only those forcing functions that are dc sources. Usually other types of forcing functions are more easily dealt with using techniques to be introduced in later chapters.

Let us examine the process involved by modifying the circuit of Fig. 6.1 to include a forcing function, $Ku(t)$. The resulting circuit is given in Fig. 6.8 where we assume the values used in Example 6.2.1.

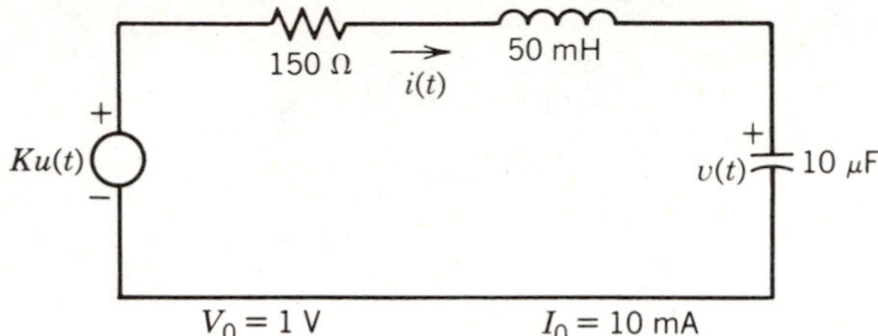

Figure 6.8. Series *RLC* circuit with forcing function.

The steady-state response is found by replacing the capacitor with an open circuit and the inductor with a short circuit. You will observe that $i(t)$ will be zero and

$$v_{ss}(t) = K \tag{6.6.2}$$

The transient response was found in Example 6.2.1 to be

$$v_{tr}(t) = Ae^{-1000t} + Be^{-2000t}, \qquad t \geq 0 \tag{6.6.3}$$

The key point in finding the complete response is realizing that we must evaluate the constants A and B *after* the form of the *complete* response is found. Combining Eqs. 6.6.2 and 6.6.3, we have

$$v(t) = K + Ae^{-1000t} + Be^{-2000t}, \qquad t \geq 0 \tag{6.6.4}$$

In this case, the initial conditions cannot change when the voltage source becomes active because the current in an inductor and the voltage across a capacitor cannot change instantaneously. At $t = 0$,

$$v(0) = K + A + B = 1 \tag{6.6.5}$$

and

$$\left.\frac{dv}{dt}\right|_{t=0} = \frac{10 \times 10^{-3}}{C} = 1000A = 2000B \tag{6.6.6}$$

Solving Eqs. 6.6.5 and 6.6.6. for A and B yields

$$A = 3 - 2K$$
$$B = K - 2$$

So Eq. 6.6.4 becomes

$$v(t) = K + (3 - 2K)e^{-1000t} + (K - 2)e^{-2000t}, \qquad t \geq 0 \tag{6.6.7}$$

If our forcing function is zero, Eq. 6.6.7 degenerates to the transient response found in Example 6.2.1. Equation 6.6.7 is plotted in Fig. 6.9 for $K = 10$ V.

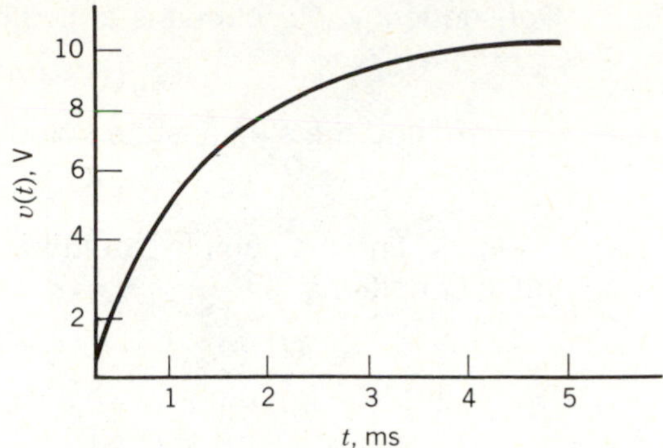

Figure 6.9. Response of circuit in Fig. 6.8.

EXAMPLE 6.6.1 For the circuit in Fig. 6.10, find the current $i(t)$. Assume the switch has been connected to the 15-V source for a long time and is instantaneously switched to the 10-V source at $t = 0$.

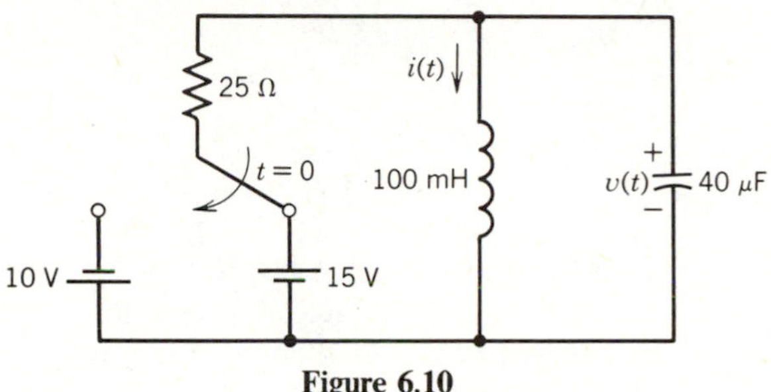

Figure 6.10

Solution

First evaluate the initial conditions. At $t = 0^-$, the inductor acts like a short circuit and the capacitor an open circuit, so

$$i(0^-) = \tfrac{15}{25} = 0.6 \text{ A}$$

$$v(0^-) = 0$$

Since the above voltage and current can not change instantaneously, our initial conditions become

$$i(0^+) = 0.6 \text{ A}$$

$$v(0^+) = 0$$

Second, find the transient response. For the parallel RLC circuit,

$$\alpha = \frac{1}{2RC} = 500$$

$$\omega_n = \frac{1}{\sqrt{LC}} = 500$$

Consequently, our circuit is critically damped and $i_{tr}(t)$ is

$$i_{tr}(t) = (A + Bt)e^{-500t}$$

Third, find the steady-state solution. By observation

$$i_{ss}(t) = -\tfrac{10}{25} = -0.4 \text{ A}$$

Finally, find the complete solution and evaluate the constants using the initial conditions

$$i(t) = i_{ss}(t) + i_{tr}(t) = -0.4 + (A + Bt)e^{-500t}$$
$$i(0) = 0.6 = -0.4 + A$$

So

$$A = 1.0$$

Now

$$\left.\frac{di}{dt}\right|_{t=0} = \frac{v(0)}{L} = 0 = \left.(-500Ae^{-500t} + Be^{-500t} - 500Bte^{-500t})\right|_{t=0}$$
$$0 = -500A + B$$

or

$$B = 500$$

Hence,

$$i(t) = -0.4 + (1 + 500t)e^{-500t} \text{ A}, \qquad t \geq 0$$

LEARNING EVALUATION

For the circuit in Fig. 6.10, find the current $i(t)$. Assume the switch has been connected to the 10-V source for a long time and is instantaneously switched to the 15-V source at $t = 0$.

PROBLEMS

SECTION 6.1

1. For each of the four differential equations that follow: (1) Find the natural frequencies (roots of the characteristic equation) and display them on an Argand diagram (the complex plane). (2) Find and plot the homogeneous solution, assuming that the initial conditions are $x(0) = 1, (dx/dt)(0) = 0$. Can you draw any conclusion about the relationship between the location of the natural frequencies in the complex plane and the nature of the natural response?

a. $\dfrac{d^2x}{dt^2} + 3\dfrac{dx}{dt} + 2x = 0$

b. $\dfrac{d^2x}{dt^2} - 3\dfrac{dx}{dt} + 2x = 0$

c. $\frac{d^2x}{dt^2} + 2\frac{dx}{dt} + 2x = 0$

d. $\frac{d^2x}{dt^2} - 2\frac{dx}{dt} + 2x = 0$

2. Write the homogeneous differential equation that corresponds to each pair of natural frequencies given below.

a. $s_1 = 2, s_2 = 3$
b. $s_1 = -2, s_2 = 3$
c. $s_1 = 3 + j4, s_2 = 3 - j4$
d. $s_1 = -3 + j4, s_2 = -3 - j4$
e. $s_1 = 0, s_2 = 1$

3. Sketch the natural response that corresponds to each pair of natural frequencies in Problem 2. Assume initial conditions $x(0) = 1, (dx/dt)(0) = 0$.

4. Find the forced response for each differential equation in Problem 1 if:

a. The forcing function is a unit step.
b. The forcing function is $f(t) = e^{-t}$.

5. Find the complete response for each differential equation in Problem 1 if $x(0) = 1, (dx/dt)(0) = 0$. The forcing function is a unit step.

6. Repeat Problem 5 if the forcing function is $f(t) = e^{-t}$.

SECTION 6.2

7. Find and plot the voltage across the capacitor in Fig. 6.11 for $t > 0$ if the switch is changed from position A to B at $t = 0$. Assume the switch was in position A for a long time before $t = 0$. The parameter values are $R = 6\ \Omega$, $L = 1$ H, $C = 0.2$ F.

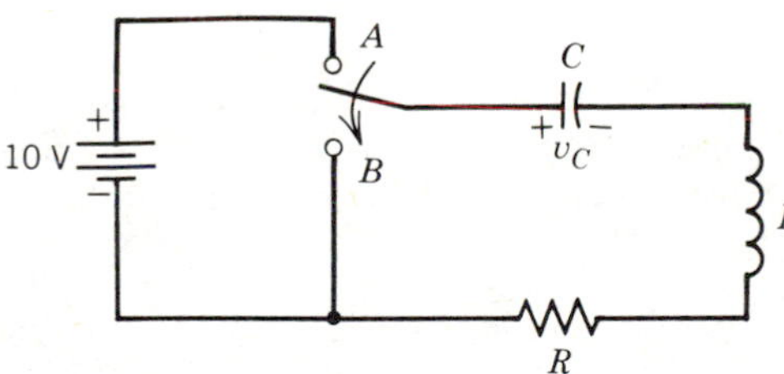

Figure 6.11

8. Draw a series RLC circuit and label the values of R and C that correspond to each pair of natural frequencies given below if $L = 1$ H.

a. $s_1 = -1, s_2 = -5$
b. $s_1 = -2, s_2 = -3$
c. $s_1 = -1 + j1, s_2 = -1 - j1$

9. Find and plot the voltage across the inductor in Problem 7.

10. Repeat Problem 7 if $R = 2\ \Omega$, $L = 1$ H, $C = 0.5$ F.

SECTION 6.3

11. The characteristic equations for four systems are given below. For each system find: (1) the natural frequencies, (2) the settling time, and (3) determine whether the circuit is overdamped, critically damped, or underdamped.

a. $s^2 + 3s + 2 = 0$
b. $s^2 + 2s + 2 = 0$
c. $s^2 + 4s + 4 = 0$
d. $s^2 + 5s + 6 = 0$

12. Repeat Problem 11 for systems described by the following differential equations.

a. $\dfrac{d^2x}{dt^2} + 3\dfrac{dx}{dt} + 2x = f(t)$
b. $\dfrac{d^2x}{dt^2} + 20\dfrac{dx}{dt} + 400x = f(t)$
c. $\dfrac{d^2x}{dt^2} + 4(10)^6\dfrac{dx}{dt} + 3(10)^{12}x = f(t)$
d. $\dfrac{d^2x}{dt^2} + 6(10)^6\dfrac{dx}{dt} + 9(10)^{12}x = f(t)$

13. Repeat Problem 11 for a series *RLC* circuit (such as in Fig. 6.11) if the parameter values are given by

a. $R = 6\ \Omega$, $L = 1\text{H}$, $C = 0.2$ F
b. $R = 5\ \Omega$, $L = 1$ H, $C = \frac{1}{6}$ F
c. $R = 2\ \Omega$, $L = 1$ H, $C = 0.5$ F
d. $R = 4\ \Omega$, $L = 1$ H, $C = 0.25$ F

SECTION 6.4

14. Find and plot the current through the inductor in Fig. 6.12 for $t > 0$ if the switch is changed from position A to B at $t = 0$. Assume the switch was in position A for a long time before $t = 0$. The parameter values are $R = \frac{1}{6}\Omega$, $L = 0.2$ H, $C = 1$ F.

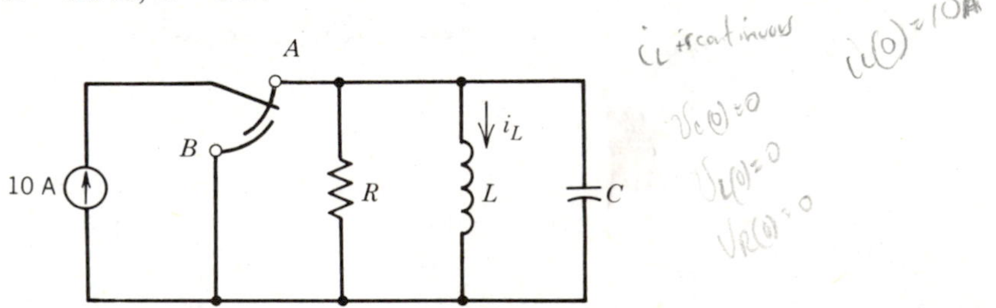

Figure 6.12

15. Draw a parallel *RLC* circuit and label the values of R and L that correspond to each pair of natural frequencies given below if $C = 1$ F.

a. $s_1 = -1,\ s_2 = -5$
b. $s_1 = -2,\ s_2 = -3$
c. $s_1 = -1 + j1,\ s_2 = -1 - j1$

16. Find and plot the current through the capacitor in problem 14.
17. Repeat Problem 14 if $R = 0.5\ \Omega$, $L = 0.5$ H, and $C = 1$ F.

SECTION 6.5

18. Find the voltage $v(t)$ and the current $i(t)$ in Fig. 6.13 if $v(0) = 0$ and $i(0) = 1$ A.

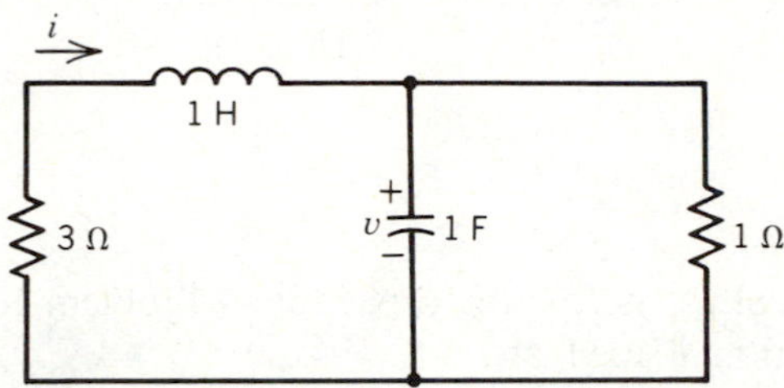

Figure 6.13

19. Find the voltage v in Fig. 6.14 if the switch is changed from position A to B at $t = 0$.

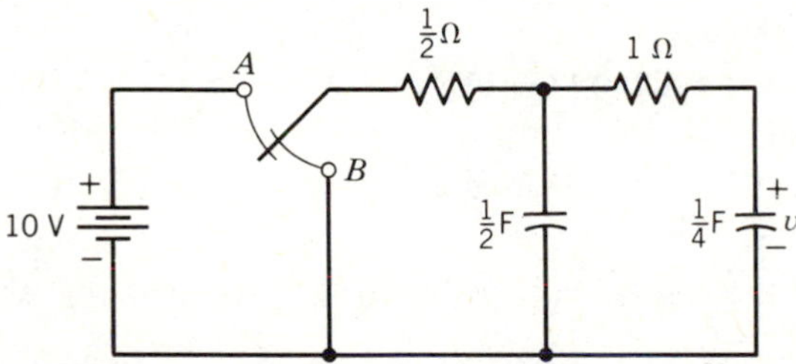

Figure 6.14

20. Find the voltage $v(t)$ for all time in Fig. 6.15.

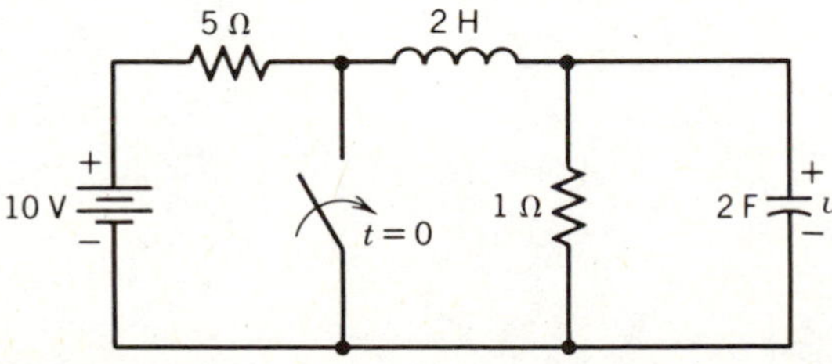

Figure 6.15

21. Find the voltage v in fig. 6.16 if the switch is changed from position A to B at $t = 0$.

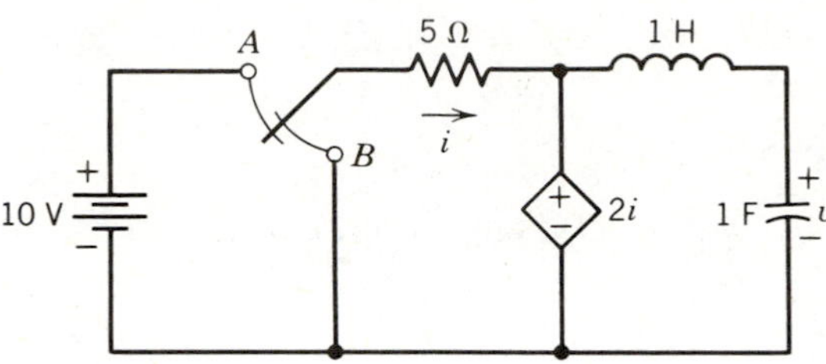

Figure 6.16

SECTION 6.6

22. Find the voltage across the capacitor in Problem 10 if the switch is changed from position B to A at $t = 0$.
23. Find the current through the inductor in Problem 14 if the switch is changed from position B to A at $t = 0$.
24. Find the voltage $v(t)$ in Fig. 6.14 if the switch is changed from position B to A at $t = 0$.
25. Repeat Problem 24 if the forcing function is $10e^{-t}$ instead of the 10-V battery.
26. Find the voltage $v(t)$ in Fig. 6.15 if the switch is opened at $t = 0$ after being closed for a long time.
27. Repeat Problem 26 if the forcing function is $10 \cos 2\pi t$ instead of the 10-V battery.
28. Find the voltage $v(t)$ in Fig. 6.16 if the switch is changed from position B to A at $t = 0$.

LEARNING EVALUATION ANSWERS

Section 6.1

$$x(t) = \tfrac{1}{4}te^{-t} + \tfrac{19}{16}e^{-t} - \tfrac{3}{16}e^{-5t}, \qquad t > 0$$

Section 6.2

$$i(t) = \tfrac{1}{3}e^{-1000t} - \tfrac{1}{3}e^{-250t}\ \text{A}, \qquad t > 0$$

Section 6.3

a. $R < 4\ \Omega$
b. $R = 4\ \Omega$
c. $R > 4\ \Omega$

Section 6.4

$$v(t) = -e^{-1000t}(1.251\cos 624.5t + 3.561\sin 624.5t)\ \text{V}, \qquad t > 0$$

Section 6.5

$$i(t) = A_1 e^{-0.835t} + A_2 e^{-19.165t}\ \text{V}, \qquad t > 0$$

Section 6.6

$$i(t) = -0.6 + (1 + 500t)e^{-500t}\ \text{A}, \qquad t > 0$$

CHAPTER

7 DIGITAL SYSTEMS

Our efforts in circuit analysis in the previous chapters have been directed at circuits that were "analog" in nature; that is, the circuit variables could take on an infinite number of values. Circuits that are "digital" in nature have circuit variables that can take on only a finite number of values (often only two values are allowed: zero and one). A well-known subset of digital circuits is digital computers, which now affect nearly every area of our daily lives, from controlling the traffic flow in our town to providing protection from enemy nuclear attack.

No one person is responsible for the invention of the modern digital computer, but the most important advance in its development is probably due to John von Neumann (1903–1957). Von Neumann was a Hungarian–American mathematician who contributed important work in many branches of advanced mathematics, including quantum mechanics, game theory, and computer programming. His major contribution to computer development was the invention of the stored-program concept. Before his invention, digital computers were wired to solve a specific problem. In order to use the computer to solve a different problem, it had to be rewired to the specific configuration required for the new problem. Von Neumann conceived the concept of controlling the computer by means of a program stored in its own memory, so that changing programs was no more difficult than changing the contents of memory. It is this invention that has led to the proliferation of the digital computer in our daily lives.

No less a revolution has occurred in signal processing. Because of the low cost and high accuracy of digital circuits, and because of the widespread knowledge of digital circuits caused by the ubiquitous computer, most signal processing is (or soon will be) done by computer circuits. A basic understanding of the theory as well as the methods of digital-signal processing, therefore, becomes as important as the corresponding understanding of analog-signal processing. In this text we will introduce the theory of digital-signal processing, leaving the methods to texts on logic circuits.

This chapter first introduces the concept of difference equations and models of elements of a digital circuit. Then a method of solving difference equations is explained. A mathematical treatment of digital signals is provided. Complete solutions to first and second order digital systems are also developed.

7.1 Difference Equations and Digital Circuits

LEARNING OBJECTIVES

After completing this section you should be able to do the following:

1. Write a difference equation that corresponds to a given system diagram.

2. Draw the direct form I realization that corresponds to a given linear difference equation.

A discrete-time signal is a series of numbers where each number represents that value of the signal at some unit of time. Usually each number in the series is separated from its immediate neighbors by a constant unit of time. A discrete-time system often can be represented by a mathematical expression called a *difference equation*. A difference equation is a function of a variable where the variable is restricted in value to a subset of the set of integers.

As an example, consider the analog equation

$$x(t) = 10t - 15, \qquad 0 \le t \le 4$$

and the difference equation

$$x(n) = 10n - 15, \qquad n = 0, 1, 2, 3, 4$$

$x(t)$ is an analog signal and can take on an infinite number of values between $t = 0$ and $t = 4$ but $x(n)$ can assume only five values, one for each n specified. We say that $x(n)$ is a discrete-time signal represented by the series of numbers (-15, -5, 5, 15, 25). This can be written as

$$x(0) = -15$$
$$x(1) = -5$$
$$x(2) = 5$$
$$x(3) = 15$$
$$x(4) = 25$$

Note that in this case the discrete signal equals the analog signal for those values of t equal to n. This is not always the case.

A discrete-time system (also called a digital system) is constructed of elements that operate on a discrete-time signal to produce another discrete-time signal according to some well-defined rule. In analog systems, our elements were the resistor, inductor, capacitor, and energy sources. For each element, we developed the operation it performed on an input signal in order to generate an output signal. For instance, if the input signal to a capacitor was $v(t)$, we established that its output signal was $i(t) = C(dv/dt)$.

The elements of our digital system are going to be the unit delay, the constant multiplier, the adder, the multiplier, and the branch. The basic operations performed by these elements are shown in Fig. 7.1. The unit delay operation shown in Fig. 7.1*a* produces the output equal to the input but delayed by one time unit. This is expressed as

$$y(n) = x(n-1) \tag{7.1.1}$$

The use of the symbol z^{-1} to denote delay will be explained in Chapter 14.

The constant multiplier shown in Fig. 7.1*b* multiplies the input signal by a constant according to the equation

$$y(n) = Ax(n) \tag{7.1.2}$$

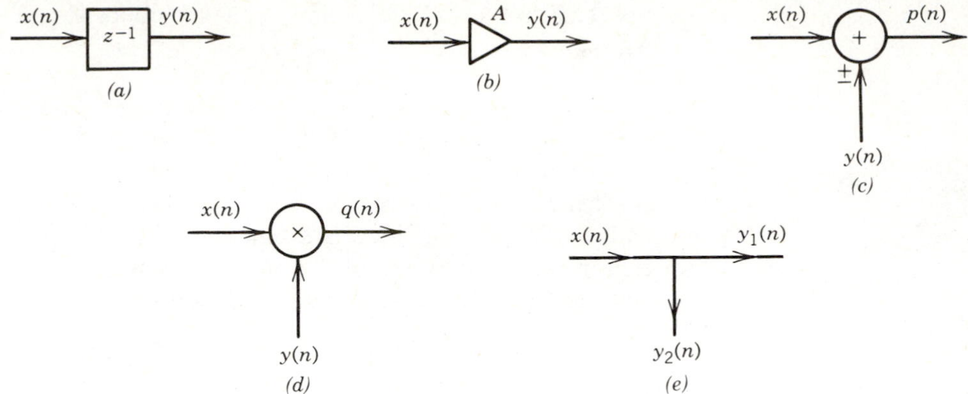

Figure 7.1. Discrete-time signal operations. (*a*) Unit delay, $y(n) = x(n-1)$. (*b*) Constant multiplier, $y(n) = Ax(n)$. (*c*) Adder, $p(n) = x(n) \pm y(n)$. (*d*) Multiplier, $q(n) = x(n)y(n)$. (*e*) Branching, $y_1(n) = x(n)$, $y_2(n) = x(n)$.

The adder operation shown in Fig. 7.1*c* adds two input signals to produce the output given by

$$p(n) = x(n) \pm y(n) \tag{7.1.3}$$

The multiplier operation in Fig. 7.1*d* produces the product of the two input signals.

$$q(n) = x(n) \cdot y(n) \tag{7.1.4}$$

The branching operation in Fig. 7.1*e* refers to the process of connecting a signal to two or more points in the system. Thus, both y_1 and y_2 are equal to the input x,

$$\begin{aligned} y_1(n) &= x(n) \\ y_2(n) &= x(n) \end{aligned} \tag{7.1.5}$$

These elements may be combined to form a system diagram that describes the operation of a discrete-time system. Mathematically, a discrete-time system is described by a difference equation relating input to output. For each difference equation, there is a corresponding system diagram consisting of combinations of the various elements shown in Fig. 7.1. There may be several forms for this diagram, but they are all equivalent. To illustrate, let us begin with the difference equation given by

$$y(n+1) = a_0 x(n) - b_0 y(n) \tag{7.1.6}$$

Draw a unit delay block as shown in Fig. 7.2*a*, and assume that the input is $y(n+1)$. Then the output of the unit delay must be $y(n)$. Next, Eq. 7.1.6 tells us that if we add $a_0 x(n)$ to $-b_0 y(n)$ the result will be $y(n+1)$. This is shown in Fig. 7.2*b*. Now $x(n)$ is the system input (which we assume to be available), and $y(n)$ is available as the unit delay output in Fig. 7.2*a*. Hence, if Figs. 7.2*a* and 7.2*b* are combined by the appropriate connections as in Fig. 7.2*c*, the result is a system diagram for Eq. 7.1.6.

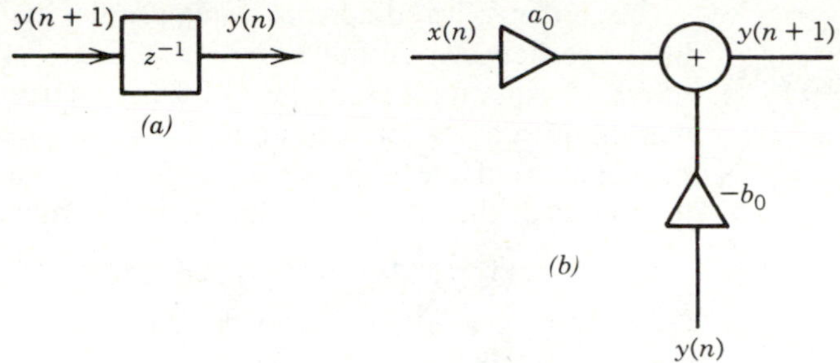

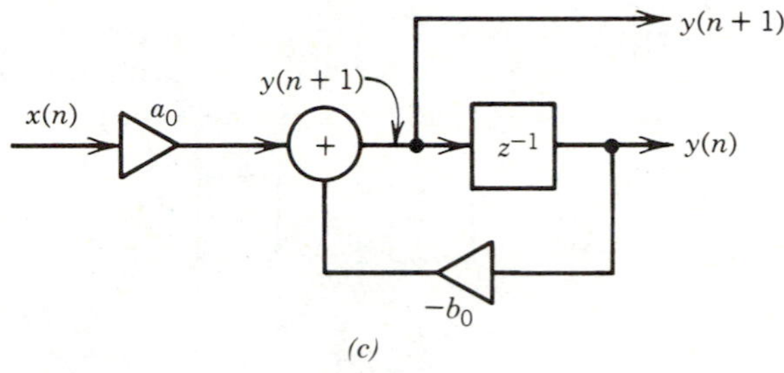

Figure 7.2. Realization of the Equation $y(n+1) = a_0 x(n) - b_0 y(n)$.

EXAMPLE 7.1.1 Find the difference equation that describes the system shown in Fig. 7.3

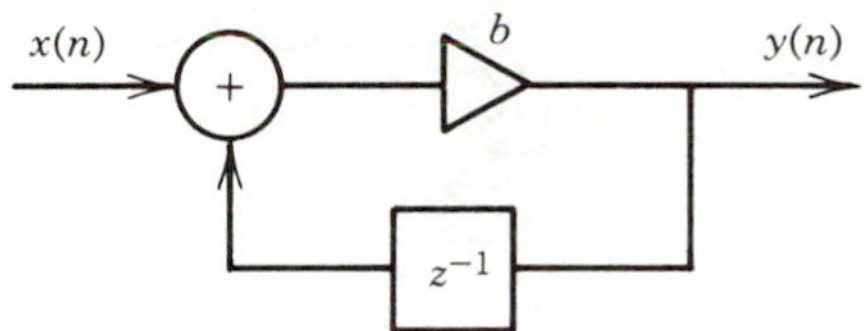

Figure 7.3. Example 7.1.1.

Solution

The input to the unit delay is $y(n)$, so the output must be $y(n-1)$. Therefore, the summer output is $x(n) + y(n-1)$. Since $y(n)$ is b times the summer output, we have

$$y(n) = bx(n) + by(n-1) \tag{7.1.7}$$

EXAMPLE 7.1.2 Draw a system diagram for the discrete-time system described by

$$y(n) = ax(n) - ax(n-1) \tag{7.1.8}$$

Solution

See Fig. 7.4.

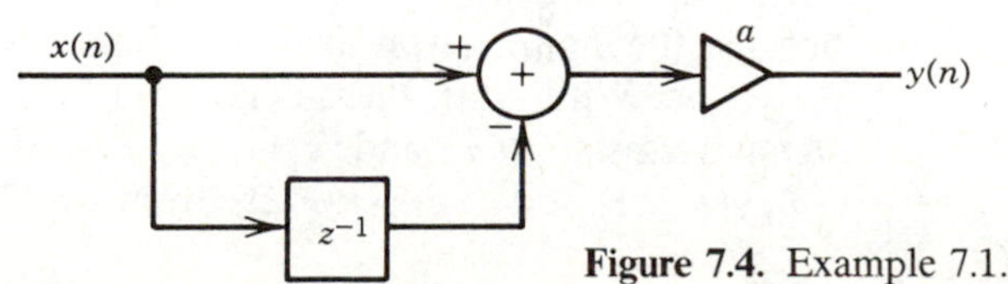

Figure 7.4. Example 7.1.2.

The order of a difference equation is defined to be the maximum time difference between output terms. Equations 7.1.6 and 7.1.7 describe first order difference equations, while Eq. 7.1.8 is of zero order. The number of unit delay operations in a system diagram is at least as large as the order of the corresponding equation. Here is a second order equation, so the number of unit delay operations in the system diagram is two or more.

$$y(n) = a_0 x(n) + a_1 x(n-1) - b_1 y(n-1) - b_2 y(n-2) \tag{7.1.9}$$

To realize a system diagram for this equation, we begin by drawing two unit delay operators as shown in Fig. 7.5*a*. Two delays are chosen because the system is second order. Next, the output is assumed to be available. (This is an essential feature of all simulation diagrams, whether digital or analog. We just arbitrarily assume that somehow the system output will be available when it is needed. We are never disappointed.) If $y(n)$ is available, then when it is supplied to the first delay operator and a connection is made to the second operator, the two terms $y(n-1)$ and $y(n-2)$ become available, as shown in Fig. 7.5*b*.

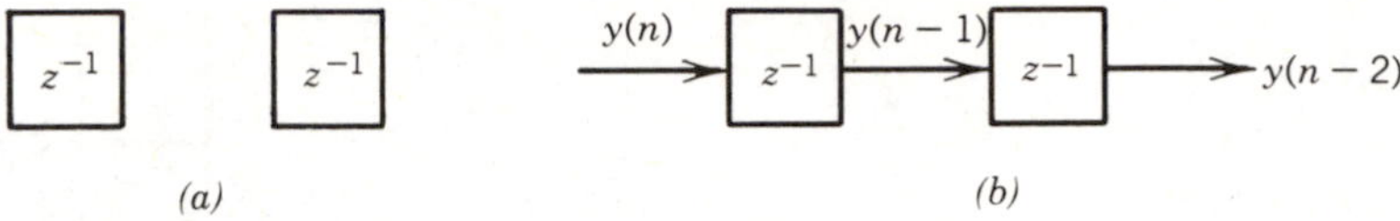

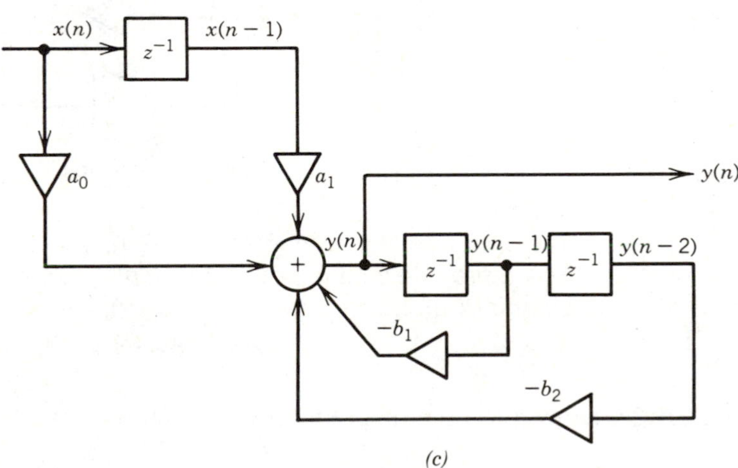

Figure 7.5. Realization of a second order difference equation.

Now the right side of Eq. 7.1.9 can be constructed by using another unit delay, because the input $x(n)$ is always available. After all, $x(n)$ is the input signal to be processed. Without it, there is no need for a system. Figure 7.5*c* shows the input signal terms $a_0 x(n)$ and $a_1 x(n-1)$ summed together with the output terms $-b_1 y(n-1)$ and $-b_2 y(n-2)$. From Eq. 7.1.9, this sum is recognized as $y(n)$, so the summer output is supplied to the first unit delay operator to complete the system.

Diagrams similar to that in Fig. 7.5 are called the "direct form I realization" of linear difference equations. Notice that all the input terms are generated on the left side of the adder while all of the output variable values are generated on the right side of the adder, both by use of unit delays and constant multipliers. In this form, the output, $y(n)$, is always found at the output of the adder.

We can express the general form of a linear difference equation containing $k+1$ input terms and $k+1$ output terms as follows:

$$y(n) = \sum_{i=0}^{k} a_i x(n-i) - \sum_{i=1}^{k} b_i y(n-i) \tag{7.1.10}$$

This equation is said to represent a kth order system. Notice that the variable of summation, i, starts at zero for the input terms but must start at 1 for the output terms since the zero term, $y(n)$, appears on the left side of the equation.

The direct form I realization for Eq. 7.1.10 is shown in Fig. 7.6. If any given term does not appear in the difference equation, its multiplier, b, is simply set to zero. For instance, consider the difference equation.

$$5x(n-4) + 0.77x(n-2) - 2x(n) = y(n) - 0.5y(n-1) + 4y(n-3) \tag{7.1.11}$$

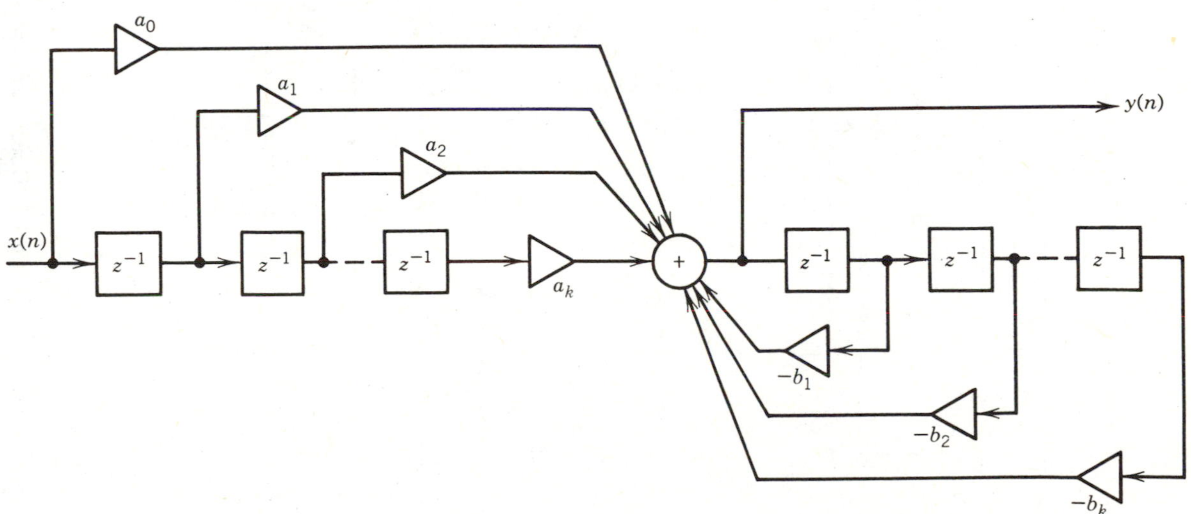

Figure 7.6. Direct form I realization.

The direct form I realization of this equation is shown in Fig. 7.7*a*. The shortened (and preferred) version is shown in Fig. 7.7*b* where the zero multipliers have simply been omitted.

The advantage of the direct form I realization is its ease of construction from the difference equation. Its chief disadvantage is the need for up to $2k$ time delay operations for a kth order system. There is another realization, called direct form II, that removes this disadvantage.

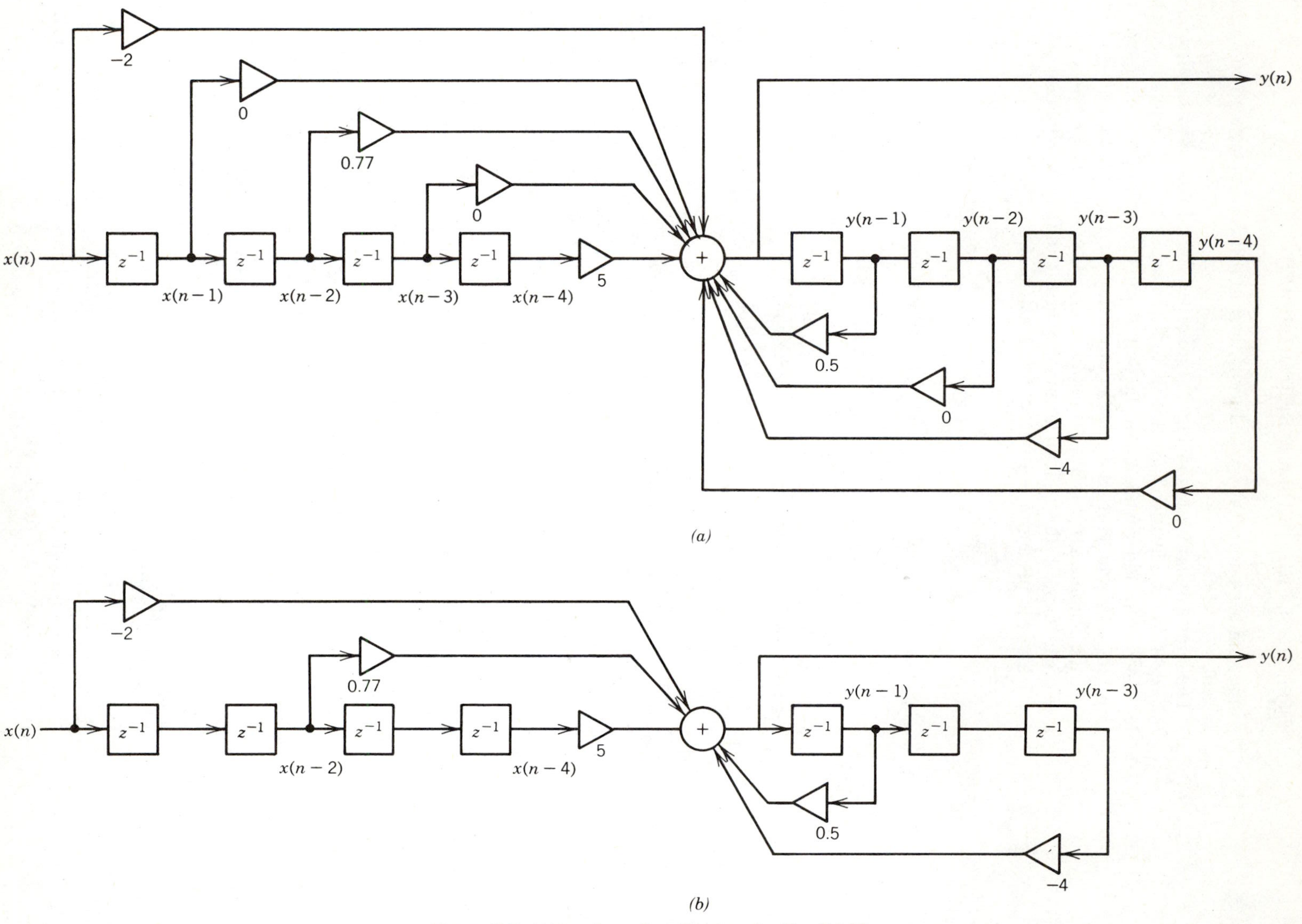

Figure 7.7. Direct form I realization for Eq. 7.1.11.

The minus sign on the second summation in Eq. 7.1.10 is there for two reasons. First, a more natural way to write Eq. 7.1.10 is to have all the output terms on one side of the equal sign, and all input terms on the other side. That is,

$$\sum_{i=0}^{k} b_i y(n-i) = \sum_{i=0}^{k} a_i x(n-i) \tag{7.1.12}$$

This form explains the necessity for the minus sign in Eq. 7.1.10, but another reason is for ease in discussing transfer functions and z transforms later in the text.

EXAMPLE 7.1.3 Draw the direct form I realization for the following difference equation

$$y(n) = 2x(n) + 0.6x(n-1) - 0.5y(n-1) + 0.2y(n-2)$$

Solution

This is a second order difference equation, so two delay units are required for the output terms $[y(n-i)]$. The input terms are $x(n)$ and $x(n-1)$, so only one delay unit is required for the input terms. The realization is shown in Fig. 7.8.

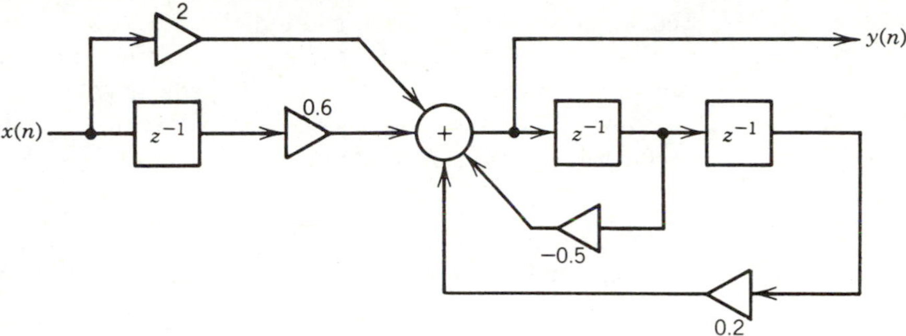

Figure 7.8

LEARNING EVALUATIONS

1. Draw the direct form I realization for the following difference equation.

$$y(n) = 2x(n-1) - 0.2y(n-1) + 0.3y(n-3)$$

2. Write the difference equation corresponding to the system diagram shown in Fig. 7.9.

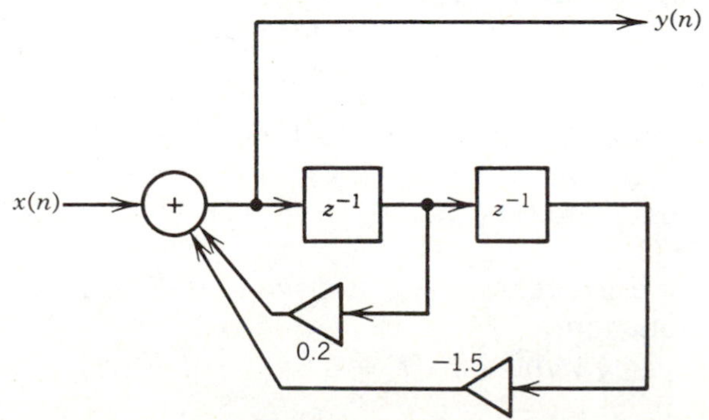

Figure 7.9

7.2 Iterative Solution of Difference Equations

LEARNING OBJECTIVES

After completing this section you should be able to do the following:

1. Define what is meant by the solution of a difference equation.
2. Find the iterative solution of a given difference equation.

In dealing with *RLC* circuits, we obtained a differential equation relating the output variable to the input function. We then solved this equation to find an expression for the output variable. The same approach is used with digital systems. The output is related to the input by a difference equation and we must develop a means of solving this equation in order to find an expression for the output variable.

The term "the solution of a difference equation" means to find that sequence which satisfies the equation. In other words, the solution is a discrete-time function, not a number or some other mathematical entity. A step-by-step (iterative) procedure may be used to find this function for a given difference equation with known input and initial conditions.

It is customary to apply the input beginning at discrete time $n=0$, although there is nothing sacred about $n=0$. Remember in solving *RLC* equations, we assumed a starting point of $t=0$ for convenience. This implies that the input sequence has the form given by

$$\ldots, 0, 0, 0, x(0), x(1), x(2), \ldots$$

Let us apply such a sequence to the discrete-time system described by the equation

$$y(n) = a_0 x(n) + a_1 x(n-1) - b_1 y(n-1) \tag{7.2.1}$$

Solving for $y(0)$ gives

$$y(0) = a_0 x(0) - b_1 y(-1)$$

since $x(-1)=0$. Here we see that it is necessary to know the value $y(-1)$ in order to calculate $y(0)$. [We assume $x(0)$ is known.] The term $y(-1)$ is called the initial condition. In general, for an nth order difference equation it is necessary to know n initial conditions, usually in the form of $y(-1), y(-2), \ldots, y(-n)$.

The next step is to solve for $y(1)$.

$$y(1) = a_0 x(1) + a_1 x(0) - b_1 y(0)$$

Notice that the previously calculated value of $y(0)$ is used, along with the known input values $x(1)$ and $x(0)$, to calculate $y(1)$. Knowing $y(1)$ it now becomes possible to calculate $y(2)$, and so on. This process may be continued indefinitely.

EXAMPLE 7.2.1 Find the response to a discrete unit step of the system described by the equation

$$y(n) = 0.5x(n) - 0.5x(n-1) + 0.9y(n-1) \quad \text{with} \quad y(-1) = 0$$

Solution

By a discrete unit step we mean the input sequence $x(n)$ given by

$$\ldots, 0, 0, 0, 1, 1, 1, 1, \ldots$$

where the first "1" occurs at $n = 0$. Knowing $y(-1)$ and all values of $x(n)$, the first value of the response can be calculated.

$$y(0) = 0.5x(0) - 0.5x(-1) + 0.9y(-1) = 0.5$$

Then

$$y(1) = 0.5x(1) - 0.5x(0) + 0.9y(0) = 0.5(1) - 0.5(1) + 0.9(0.5) = 0.45$$

$$y(2) = 0.5x(2) - 0.5x(1) + 0.9y(1) = 0.5(1) - 0.5(1) + 0.9(0.45) = 0.405$$

and so on. It is easy to see that a general rule for calculating $y(n)$ can be derived from the above result.

$$y(n) = 0.5(0.9)^n, \qquad n > 0$$

This is a closed-form expression for the system response, and obviously of much greater value than the set of numbers found by the step-by-step method of solution. Our purpose in this chapter is to introduce methods for finding closed-form solutions to difference equations.

The step-by-step method of solution may be applied to any difference equation, but the difficulty lies in the fact that no general expression for the response is derived by this method. If the value of the response for large n is desired, then the number of steps required to find this value is n. Here is one more example before beginning our search for methods of finding closed-form solutions.

EXAMPLE 7.2.2 Find the step-by-step solution to a unit step input for the system pictured in Fig. 7.10 if $y(-1) = 0$ and $y(-2) = 0$.

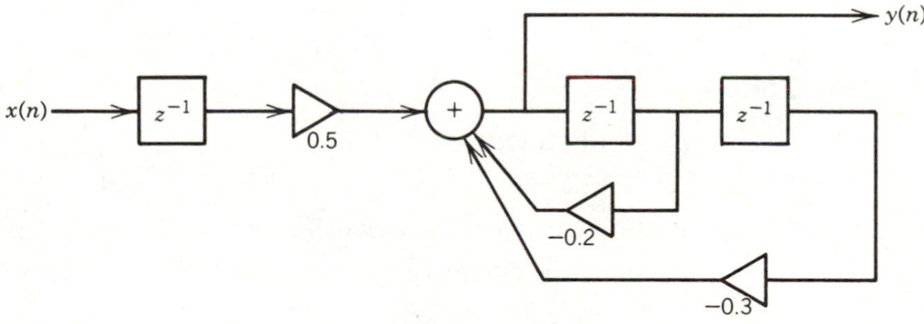

Figure 7.10

Solution

The equation describing the system is given by

$$y(n) = 0.5x(n-1) - 0.2y(n-1) - 0.3y(n-2)$$

Starting at $n = 0$,

$$y(0) = 0.5x(-1) - 0.2y(-1) - 0.3y(-2) = 0$$

$$y(1) = 0.5x(0) - 0.2y(0) - 0.3y(-1) = 0.5$$

$$y(2) = 0.5x(1) - 0.2y(1) - 0.3y(0) = 0.5(1) - 0.2(0.5) = 0.4$$
$$y(3) = 0.5x(2) - 0.2y(2) - 0.3y(1) = 0.5(1) - 0.2(0.4) - 0.3(0.5) = 0.27$$
$$y(4) = 0.5x(3) - 0.2y(3) - 0.3y(2) = 0.5(1) - 0.2(0.27) - 0.3(0.4) = 0.326$$

and so on. This process may be continued indefinitely.

We observe that it is not easy to discover a general rule for finding the response here, as it was in Example 7.2.1. This illustrates the need for a systematic method of solution.

LEARNING EVALUATIONS

1. Solve the following difference equation for the first five values of $y(n)$.

$$y(n) = 0.5y(n-1) + \frac{1}{y(n-1)}$$
$$y(-1) = 1$$

2. Find the first five values of the response to a unit step for the system pictured in Fig. 7.11 if $y(-1) = 0$ and $y(-2) = 0$.

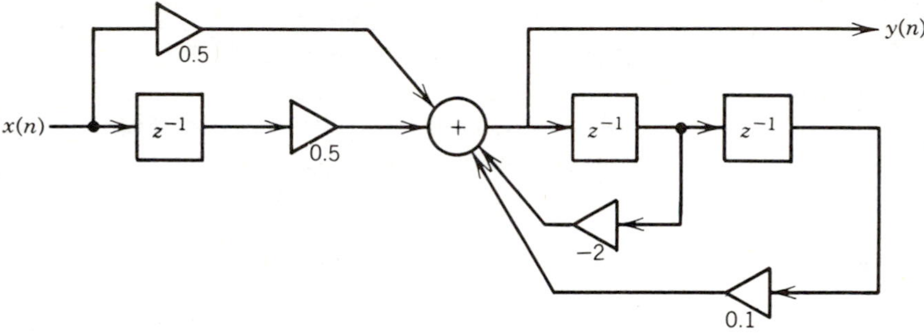

Figure 7.11

7.3 Digital Signals

LEARNING OBJECTIVES

After completing this section you should be able to do the following:

1. Write equations for signals in terms of impulse, step, and ramp signals.
2. Apply the difference and summation operators to signals.
3. Describe the relationship between the impulse, step, and ramp signals in terms of the difference and sum operators.

A discrete-time signal is a sequence of numbers. Perhaps it would be worthwhile to review the meaning of the terms sequence and series, since we will use both terms frequently. A sequence is a function whose domain is a set of integers. A discrete-time signal (also called a digital signal) is shown in Fig. 7.12 as the function $x(n)$. This is a sequence. The domain consists of all values of n shown scaled along the abscissa.

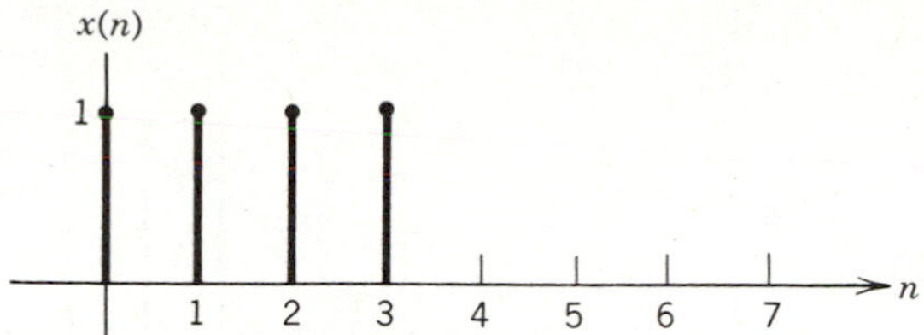

Figure 7.12. A typical digital signal.

A series of n terms is an expression of the form

$$r_1 + r_2 + r_3 + \cdots + r_n$$

where each term is formed by some definite rule.

A series then is also a sum. Various series can be derived by summing terms in the sequence shown in Fig. 7.12. Some of these are listed below.

$$S_1 = \sum_{i=0}^{1} x(i) = 2$$

$$S_2 = \sum_{i=0}^{3} x(i) = 4$$

$$S_3 = \sum_{i=-\infty}^{+\infty} x(i) = 4$$

Thus the two terms, sequence and series, have quite distinct meanings. A sequence is a function, while a series (also called a sum of a sequence) is a single number, the sum of terms in a sequence.

There are several digital signals that will occur frequently in the future, and some special notation is commonly used to describe these special signals. You will hopefully recognize these as digital counterparts of some basic analog signals introduced in previous chapters; namely, the unit step, delta, and ramp functions. To distinguish between the analog and digital versions, we could call the latter digital unit step, digital delta function, and so on. However, the more common method of differentiation is by the independent variable of the function. In other words, the analog unit step is designated by $u(t)$ while the digital unit step will be designated by $u(n)$.

The digital unit step function is shown in Fig. 7.13a and is defined by the equation

$$u(n) = \begin{matrix} 1 & n \geq 0 \\ 0 & n < 0 \end{matrix} \tag{7.3.1}$$

Notice that the value of this function changes from zero to 1 at a value of time $n = 0$. That is, when the argument is zero the value of the function changes. This same observation applies to $u(n-a)$ and $u(n+a)$ shown in Figs. 7.13b and 7.13c. The function $u(n-a)$ changes value when $n - a = 0$, or when $n = a$. Likewise, $u(n+a)$ changes value when $n + a = 0$, or $n = -a$.

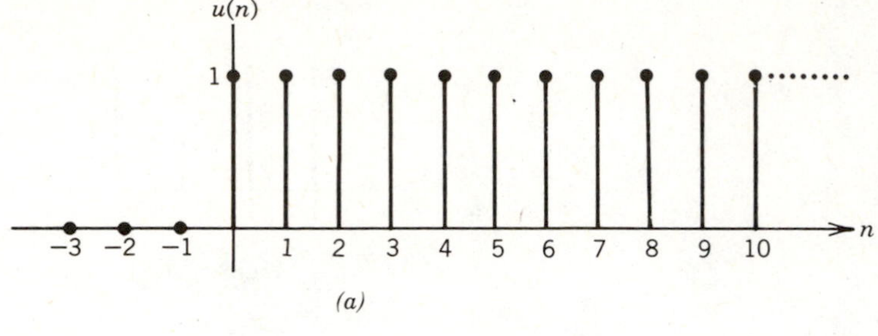

(a)

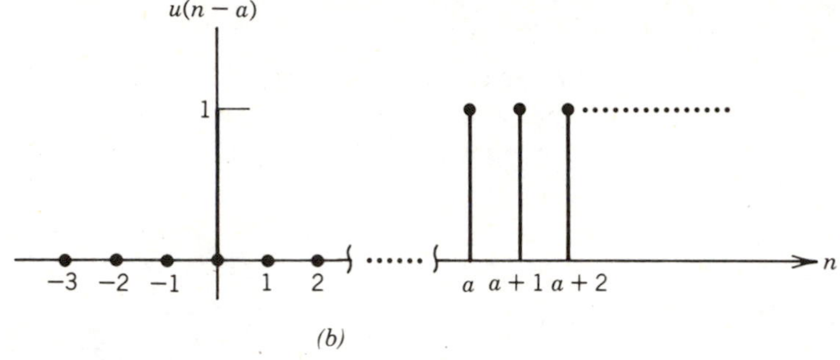

(b)

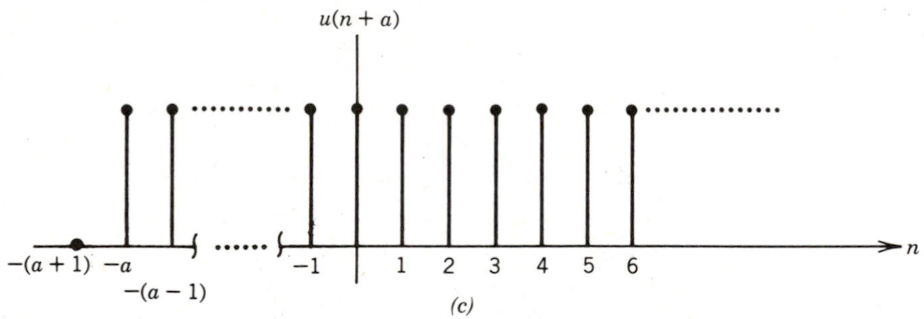

(c)

Figure 7.13. The digital unit step function.

Another important function is the digital delta function $\delta(n)$ shown in Fig. 7.14*a*. This function consists of a single nonzero value at $n = 0$. When the δ function is displaced along the time axis by time a, as shown in Fig. 7.14*b*, it is denoted by the symbol $\delta(n - a)$. Thus the notation used in specifying the δ function is similar to that used for the unit step function. Other names for the

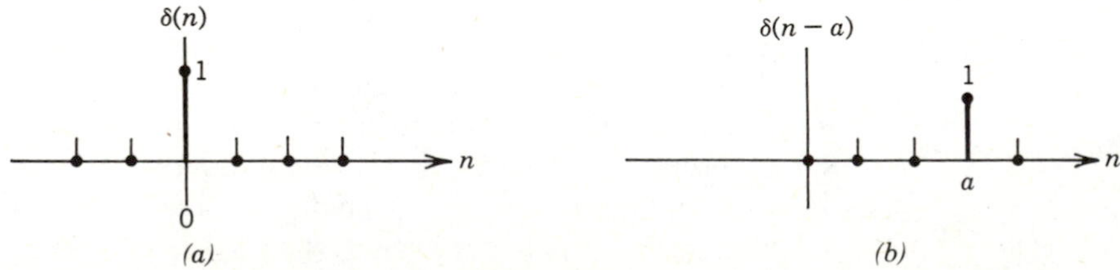

Figure 7.14. The digital δ function.

delta function are the dirac function and the impulse function. The term dirac is used in honor of the English physicist P. A. M. Dirac, who first used the δ function to model point charges. The δ function is specified by

$$\delta(n) = \begin{cases} 1 & n = 0 \\ 0 & \text{otherwise} \end{cases} \tag{7.3.2}$$

A third signal to receive special treatment and a special name is the digital unit ramp, $r(n)$. It is specified by the formula

$$r(n) = \begin{cases} n+1 & n \geq 0 \\ 0 & n < 0 \end{cases} \tag{7.3.3}$$

This is illustrated in Fig. 7.15*a*. The unit ramp is a special case of the general ramp function. Recall that for analog signals a ramp function was given by Eq. 4.2.3.

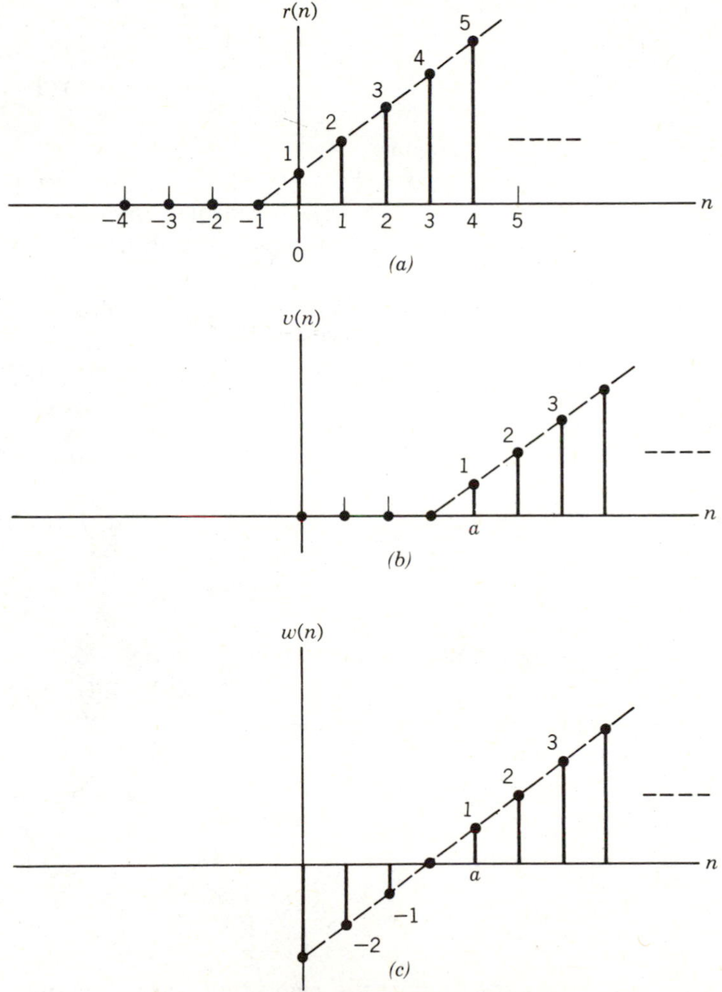

Figure 7.15. The digital ramp function.

$$x(t) = \begin{cases} 0 & t < t_1 \\ A(t - t_1) & t \geq t_1 \end{cases}$$

For digital signals, a ramp function is given by

$$x(n) = \begin{cases} 0 & n < a \\ A(n - a + 1) & n \geq a \end{cases}$$

If $A = 1$ and $a = 0$, then this equation reduces to the unit ramp of Eq. 7.3.3. The reason for defining the unit ramp such that its value is 1 when n is 0 is one of mathematics, and will become clear shortly. Another way to write eq. 7.3.3 is to use the step function to "cut off" the ramp at $n = 0$, that is,

$$r(n) = (n + 1)u(n) \tag{7.3.4}$$

This same procedure may be used to specify many signals, for example, $v(n)$ and $w(n)$ in Figs. 7.15*b* and 7.15*c* are given by

$$v(n) = (n - a + 1)u(n - a)$$
$$w(n) = (n - a + 1)u(n)$$

Composite Signals

Equations 7.3.3 and 7.3.4 illustrate two ways to specify signals, either directly, as in Eq. 7.3.3, or in terms of elementary signals such as the step and ramp as in Eq. 7.3.4. Here are some examples that illustrate these procedures.

EXAMPLE 7.3.1 Refer again to Fig. 7.12. The signal there can be expressed in two ways.

$$x(n) = \begin{cases} 0 & n < 0 \\ 1 & 0 \leq n \leq 3 \\ 0 & n > 3 \end{cases}$$

or

$$x(n) = u(n) - u(n - 4)$$

EXAMPLE 7.3.2 The truncated ramp $x(n)$ shown in Fig. 7.16*a* may be expressed as

$$x(n) = \begin{cases} 0 & n < 0 \\ 2(n + 1) & 0 \leq n \leq 3 \\ 8 & n > 3 \end{cases}$$

or as the sum of two ramp functions

$$x(n) = 2(n + 1)u(n) - 2[(n + 1) - 4]u(n - 4)$$

as illustrated in Fig. 7.16*b*.

EXAMPLE 7.3.3 The signal $x(n)$ shown in Fig. 7.17*a* may be expressed either by the formula

$$x(n) = \begin{cases} 0 & n < 0 \\ n + 1 & 0 \leq n \leq 3 \\ 4 & 4 \leq n \leq 7 \\ [12 - (n + 1)] & 7 < n \leq 11 \\ 0 & n > 11 \end{cases}$$

or as the sum of the ramp functions shown in Fig. 7.17*b*.

$$x(n) = r(n)u(n) - r(n - 4)u(n - 4) - r(n - 8)u(n - 8) + r(n - 12)u(n - 12)$$

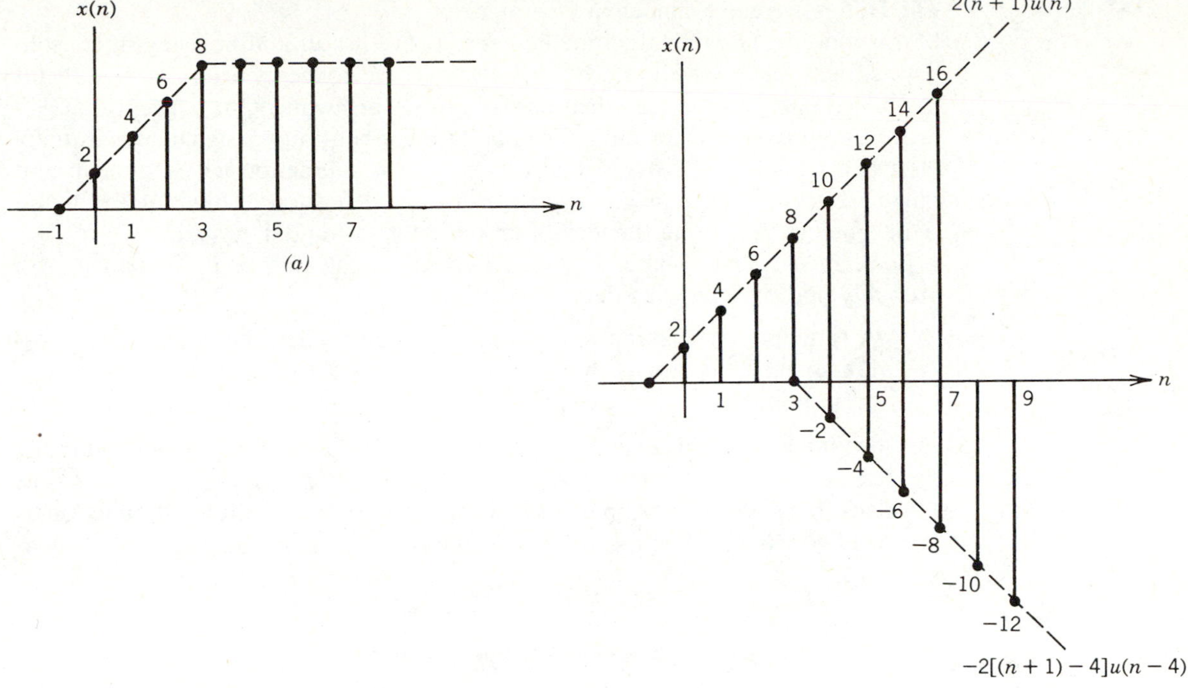

Figure 7.16. Figures for Example 7.3.2.

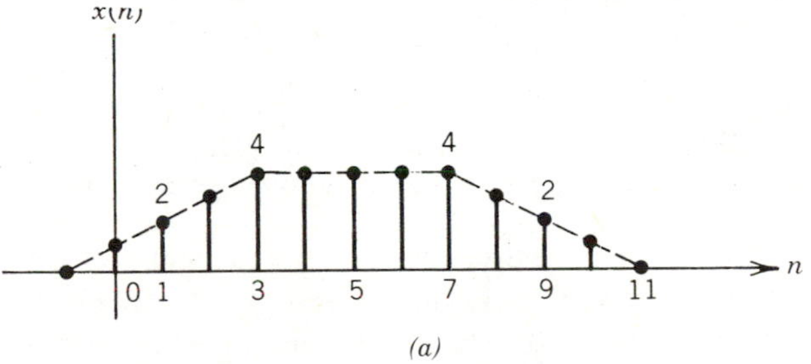

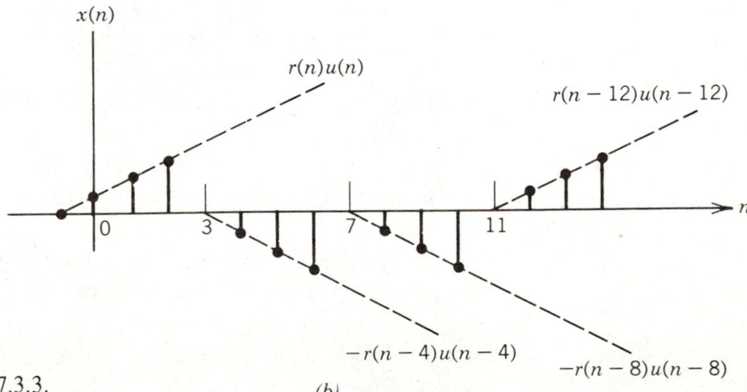

Figure 7.17. Functions used in Example 7.3.3.

The Difference and Summation Operators

Many concepts from abstract mathematics prove useful and unifying in consolidating ideas related to the study of systems. These concepts need not be difficult to understand, for otherwise their use is hardly worthwhile. One concept that will prove useful is that of operator. The operator for continuous signals was initially discussed in Chapter 5. Recall that an operator is a function whose domain and range are sets of functions. Let us first review the definition of function for digital signals before presenting the idea of an operator for digital systems.

Look in the first chapter of your algebra or calculus text. There you will probably find the following definition of function.

> A function is a relationship between two sets (called the domain and range) such that for each element in the first set (the domain) there corresponds exactly one element in the second set (the range).

The idea inherent in this definition is pictured in Fig. 7.18. The domain and range are pictured as baskets containing x's and y's, and the function f is pictured as a "black box" with one input and one output. The definition of function is satisfied by this black box if each input produces exactly one output.

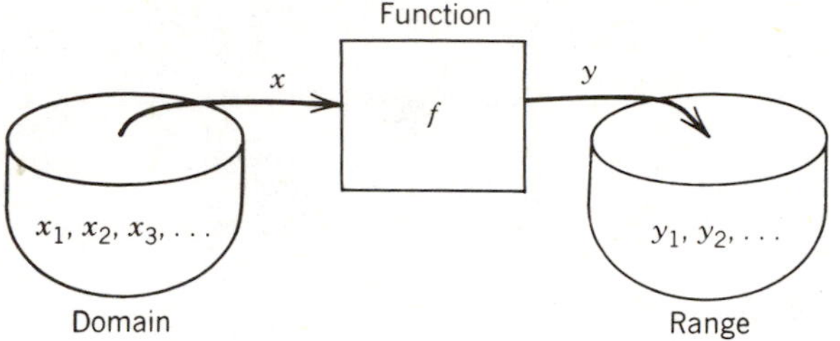

Figure 7.18. A function $y = f(x)$ is a black box with one input and one output.

Originally the concept of function applied only to numbers and both the domain and range were sets of numbers. This is the most common situation, and the one to which the graph of a function is applied. The domain is represented by valves along the abscissa (the x axis) and the range is represented by values along the ordinate (the y axis). In modern times the scope of applications has been extended until there is no restriction on the domain and range; they may contain anything. Here is an example of a function that has nothing to do with numbers.

Suppose I have a basket full of cards. On each card is written an instruction. You select a card, read the instruction, and perform the indicated act.

This is a function. The domain consists of the set of instructions. The range consists of the set of actions. Notice that nowhere is there a number present in all this.

Since an operator is a function whose domain and range consist of functions, this is another example of a function whose domain and range are not sets of numbers. They are sets of functions. In this chapter we will deal with discrete-time functions $x(n)$, $y(n)$, and so on. Thus, a digital system may be modeled as an operator with domain $\{x(n)\}$ and range $\{y(n)\}$. Our purpose for presenting the above material is to introduce the difference and summation operators.

The *difference operator* Δ is defined by

$$\Delta x(n) = x(n) - x(n-1) \tag{7.3.5}$$

[Formally, this is called the backward difference operator. The forward difference operator is equal to $x(n+1) - x(n)$. We will use only the backward difference operator and simply call it the difference operator.] A digital system that performs the difference operation is shown in Fig. 7.19*a*, where the response is $y(n) = \Delta x(n)$.

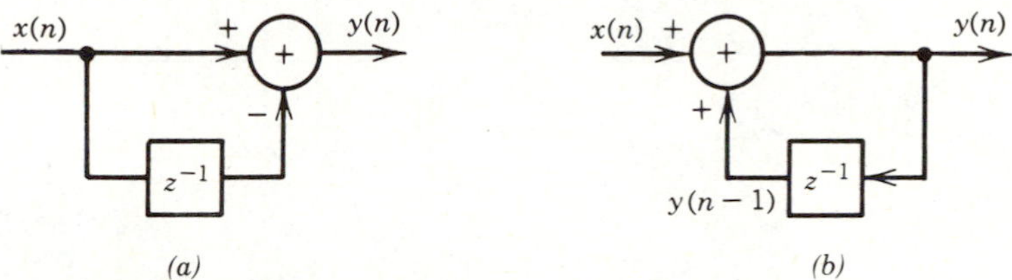

Figure 7.19. Digital systems that perform the Δ and σ operations. (*a*) The difference operator $y(n) = x(n) - x(n-1)$. (*b*) The summation operator $y(n) = \sum_{k=-\infty}^{n} x(k)$, also called the σ operator.

A sleight of hand is present in Fig. 7.19*a*, which may be overlooked by the casual reader. The inputs to the summer have all previously been positive signals. In this case the output of the unit delay is shown as negative so the summer is really acting as a "subtractor." Formally the output of the unit delay could be passed through a constant multiplier with a gain of -1. Practically, a summer is the operational amplifier introduced in Section 3.6., and can find either the sums or differences of signals. Hence, the constant multiplier is in fact not needed and the system as shown in Fig. 7.19*a* is appropriate. The mental picture that one should obtain from this is that of a function (sequence) $x(n)$ being supplied to the system. As a result, the function (sequence) $y(n)$ emerges.

The *summation operator* is defined by

$$\sigma x(n) = \sum_{k=-\infty}^{n} x(k) = \cdots + x(n-2) + x(n-1) + x(n) \tag{7.3.6}$$

This operator sums all values of the function x over the infinite past. A digital system that performs the operation defined by Eq. 7.3.6 is shown in Fig. 7.19*b*.

Note that the upper limit on the sum in Eq. 7.3.6 is n. Consequently, summation is a function of n. Therefore, the mental picture of this system is again that of an input sequence producing an output sequence.

In using the σ operator, the assumption is made that $x(n) = 0$ for all $n < n_0$, where n_0 is any finite value of discrete time. An equivalent way to specify σ is to start the index of summation at $k = n_0$ instead of $k = -\infty$. In either case, the purpose is to make sure that $\sigma x(n)$ is finite for all finite valued sequences $x(n)$.

EXAMPLE 7.3.4 Apply the difference and summation operators to the signal $x(n)$ shown in Fig. 7.20*a*.

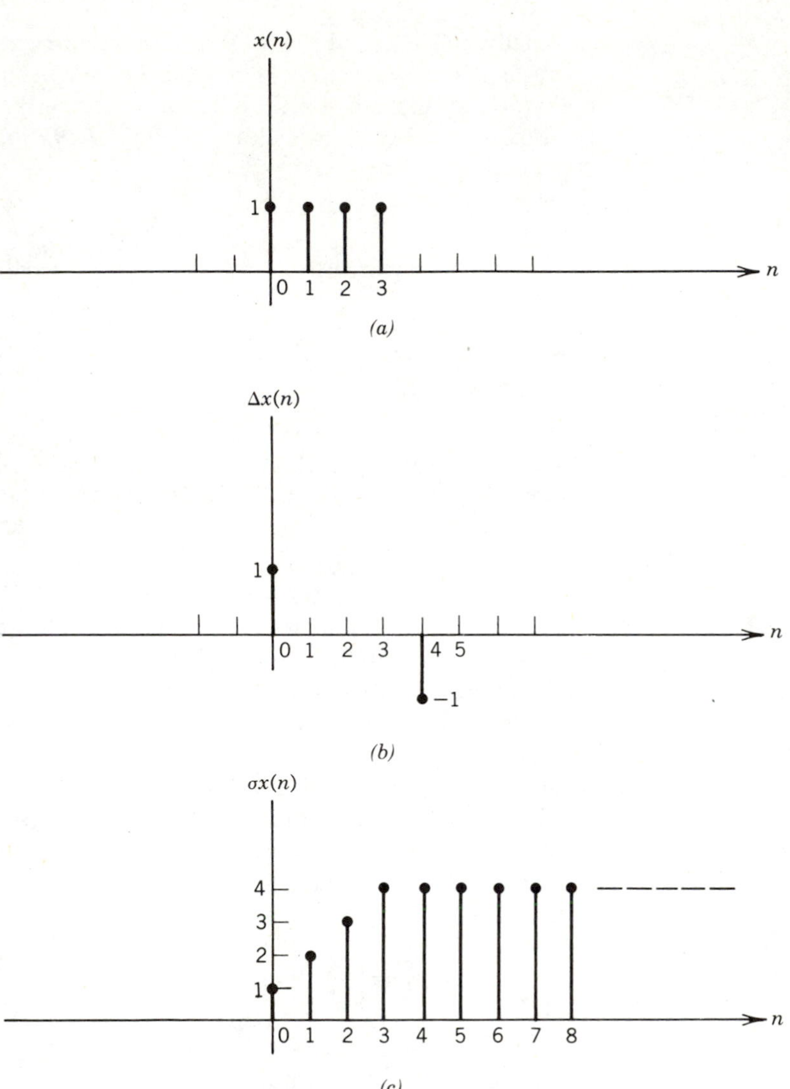

Figure 7.20. Example 7.3.4.

Solution

First apply the Δ operator. For any $n < 0$, $x(n) = x(n-1) = 0$. Hence, $\Delta x(n) = 1$ for $n = 0$. This reasoning continues for all n to produce the result shown in Fig. 7.20*b*. Notice that the negative pulse occurs at $n = 4$, not $n = 3$, in $\Delta x(n)$.

Next apply the σ operator. For any $n < 0$, there is nothing to sum. All values of x are zero, so their sum is zero. For $n > 0$ however, there are nonzero values to sum. But here some confusion may arise over the meaning of k and n in Eq. 7.3.6. The summation operator is more complex than the difference operator because it involves two steps. First the independent variable is changed from n to k in Fig. 7.20*a*. Then a value of n is chosen and all values of x are summed from $k = -\infty$ to $k = n$. This gives

$$\sigma x(n) = \begin{matrix} 0 & n < 0 \\ (n+1) & 0 \leq n \leq 3 \\ 4 & n > 3 \end{matrix}$$

as shown in Fig. 7.20*c*.

The Δ and σ operations are inverses of one another. That is, if Δ is first applied to a signal $x(n)$, followed by σ, the overall result is $x(n)$.

Likewise, if σ is first applied, followed by Δ, the result is again $x(n)$.

$$\Delta\sigma x(n) = \sigma\Delta x(n) = x(n)$$

One purpose for introducing the Δ and σ at this time is to illustrate the relationships that exist between the impulse, step, and ramp functions. First consider the application of the Δ operator to the unit ramp:

$$\begin{aligned} \Delta r(n-a) &= \Delta[(n+1-a)u(n-a)] \\ &= (n+1-a)u(n-a) - (n+1-a-1)u(n-a-1) \\ &= \begin{matrix} 0 & n < a \\ 1 & n \geq a \end{matrix} \\ &= u(n-a) \end{aligned}$$

Hence, we see that applying the Δ operator to the unit ramp results in the unit step. If we had defined the unit ramp as $r(n) = n$ for all $n > 0$, then the unit step would have been displaced one discrete-time unit to the right when derived from the unit ramp by using the Δ operator. Hence, using our definition of the unit ramp, Eq. 7.3.3, allows us to remain mathematically correct in our definition of the unit step function.

If we apply the Δ operator to the unit step we obtain

$$\Delta u(n-a) = u(n-a) - u(n-a-1)$$

$$= \begin{matrix} 0 & n < a \\ 1 & n = a \\ 0 & n > a \end{matrix}$$

or

$$\Delta u(n-a) = \delta(n-a)$$

The difference operator applied to the unit step results in the delta function. Based on our finding of taking the derivatives of ramp and step functions in Chapter 4, we conclude that the difference operator in digital systems is analogous to the differential operator in continuous systems.

Now let us apply the summation operator to the delta function.

$$\sigma\delta(n) = \sum_{k=-\infty}^{n} \delta(n)$$

$$= \begin{matrix} 0 & n < 0 \\ 1 & n \geq 0 \end{matrix}$$

So the unit step results from the application of the summation operator to the delta function. Similarly we can show that using the σ operator on the unit step will result in the unit ramp. Therefore, the summation operator in digital systems is analogous to the integral operator in analog systems.

LEARNING EVALUATIONS

1. Express the signal shown in Fig. 7.21 in two ways:
   a. By a formula.
   b. In terms of impulse, step, and ramp signals.

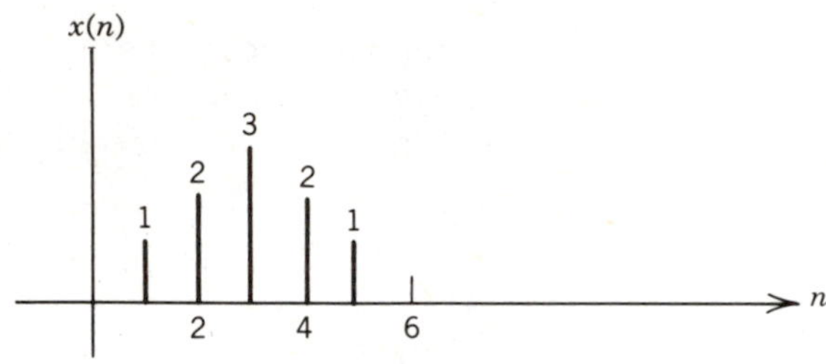

Figure 7.21

2. Apply the difference operator to the signal $x(n)$ shown in Fig. 7.21.
3. Apply the summation operator to the answer found in Problem 2. Check that $x(n)$ is the result.
4. Apply the summation operator to the unit ramp function to obtain the unit parabola, $p(n)$. Write the equation for $p(n)$.

7.4 First Order Digital Systems

LEARNING OBJECTIVES

After completing this section you should be able to do the following:

1. Find the characteristic equation for a first order system.
2. Find the natural frequency of a given first order system.
3. Find the source free response of a first order system.
4. Find the forced response of a first order system.
5. Find the complete solution by summing the source-free and forced solutions, and then evaluate the constant in the solution from initial conditions.

With $x(n)$ as the system input and $y(n)$ as the output, we have shown (Eq. 7.1.12) that the general form of a difference equation is given by

$$\sum_{i=0}^{k} b_i y(n-i) = \sum_{i=0}^{k} a_i x(n-i) \tag{7.4.1}$$

where $b_0 = 1$ and the a_i and b_i are arbitrary constants. Equation 7.4.1 is said to be of order k if $b_k \neq 0$. In this section we will study first order systems only. For first order systems, Eq. 7.4.1 reduces to

$$y(n) + b_1 y(n-1) = a_0 x(n) + a_1 x(n-1)$$

In Chapter 5 we encountered the first order analog system which resulted in a first order differential equation relating a dependent variable to an independent variable of the system. The solution of this equation was found to consist of two parts; the source-free (transient) component and the forced or steady-state component. The complete solution to the equation was found by summing the two components and evaluating the constants using initial conditions.

For first order digital systems, the same situation exists. A difference equation relates a dependent variable (which we have been calling the system output) to an independent variable (which we have been calling the system input). The complete solution to the difference equation will again be found to be made up of two components; the transient response and the forced response. Just as in analog systems, the complete solution will be determined by summing the two components and evaluating the unknown constants using the given initial condition.

The Source-Free Solution

The source-free (transient) solution is found by setting the forcing function equal to zero. Equation 7.4.1 becomes

$$\sum_{i=0}^{k} b_i y(n-i) = 0, \qquad b_0 = 1 \tag{7.4.2}$$

For a first order system, Eq. 7.4.2 becomes

$$y(n) + by(n-1) = 0 \tag{7.4.3}$$

Our approach to solving this equation will be the one we found successful in Chapters 5 and 6. We will guess the form of the solution and insert it into the equation to see if it indeed satisfies the equation.

Since Eq. 7.4.3 involves one unit delay, the operator z must be present. Since we experienced success with the exponential form earlier we will try it again. So assume the solution to Eq. 7.4.3 is

$$y_{\text{tr}}(n) = Az^n \tag{7.4.4}$$

where A is a constant and z is in general a complex number. The proof that z^{-1} is in fact the unit delay will be delayed until Chapter 14. The subscript "tr" indicates this is the transient solution.

To find a solution to Eq. 7.4.3, insert Eq. 7.4.4 into Eq. 7.4.3 and evaluate:

$$Az^n + bAz^{n-1} = 0$$

Factoring yields

$$Az^n(1+bz^{-1})=0$$

or

$$1+bz^{-1}=0 \tag{7.4.5}$$

Finally

$$z^{-1}=-\frac{1}{b}$$

or

$$z=-b$$

Substituting this result into Eq. 7.4.4 gives

$$y_{\text{tr}}(n)=A(-b)^n \tag{7.4.6}$$

Equation 7.4.6 is the general transient solution for systems described by first order difference equations.

To illustrate, assume that a digital system is described by the first order difference equation

$$y(n)+0.5y(n-1)=0$$

Using Eq. 7.4.5 the transient solution can be immediately written as

$$y_{\text{tr}}(n)=A(-0.5)^n$$

Equation 7.4.5 is called the characteristic equation for the system and $z=-b$ is called a root of the characteristic equation. The number $z=-b$ is also called a natural frequency of the system.

For a first order system, the natural frequency will always be a real number. For a second or higher order system, there will be as many natural frequencies as the order of the system (though some roots may be repeated) and these roots may be complex numbers.

EXAMPLE 7.4.1 Find the characteristic equation, the natural frequency, and the transient solution for a system described by the following difference equation.

$$y(n)=x(n)+0.9y(n-1)$$

Solution

The transient equation is given by

$$y(n)-0.9y(n-1)=0$$

Substitution of the assumed solution $y(n)=Az^n$ gives

$$Az^n-0.9z^{-1}Az^n=0$$

or

$$1-0.9z^{-1}=0$$

This is the characteristic equation. Solving for z gives $z=0.9$. Therefore, the transient solution is

$$y(n) = A(0.9)^n$$

This solution is plotted in Fig. 7.22. The exponential nature of the solution is clearly seen.

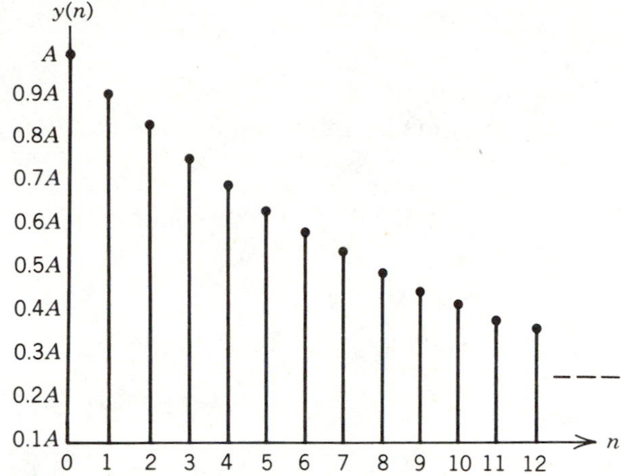

Figure 7.22. Plot for Example 7.4.1.

The Forced Response

Now we must attack the problem of finding the forced (steady-state) solution. When the forcing function is one of the forms a^n or e^{bn}, n^p, $\sin cn$, $\cos cn$, or composed of sums or products of these forms, we may use the method of undetermined coefficients. This is true because when these functions are shifted they do not change form. To show this, consider first the exponential. If $y(n) = a^n$, then $y(n+1) = a^{n+1} = a \cdot a^n$ and $y(n+m) = a^m a^n$. Thus the form a^n is retained.

Next, consider $y(n) = n^p$, where p is a positive integer such as $1, 2, 3, \ldots$. Shifting, we obtain

$$y(n+m) = (n+m)^p = n^p + pn^{p-1}m + \cdots + pnm^{p-1} + m^p$$

where the expression on the right is the binomial expansion. Each term on the right contains a power of n.

Finally, for the sine and cosine terms, if $y(n) = \cos cn$, then

$$y(n+m) = \cos c(n+m) = (\cos cm)\cos cn - (\sin cm)\sin cn$$

and if $y(n) = \sin cn$, then

$$y(n+m) = \sin c(n+m) = (\sin cm)\cos cn + (\cos cm)\sin cn$$

Thus, the $\sin cn$ and $\cos cn$ forms are retained.

We may speak of families of terms a^n, n^p, $\cos cn$, $\sin cn$. If the forcing function is a member of one of these families, then the steady-state solution will also be a member of this family. Knowing this, it remains only to solve for the undetermined coefficients in the solution. We now illustrate this procedure by continuing Example 7.4.1.

EXAMPLE 7.4.2 Find the steady-state solution for the system described by

$$y(n) = x(n) + 0.9y(n-1) \tag{7.4.7}$$

where

$$x(n) = (0.4)^n u(n)$$

and

$$y(-1) = 0$$

Solution

Assume that the steady-state solution is given by

$$y_{ss}(n) = B(0.4)^n$$

where B is the undetermined coefficient. Then

$$y_{ss}(n-1) = B(0.4)^{n-1} = \frac{B}{0.4}(0.4)^n$$

Substituting into Eq. 7.4.7 gives

$$B(0.4)^n = (0.4)^n + \frac{0.9}{0.4}B(0.4)^n$$

or

$$B = 1 + 2.25B$$

so

$$B = -0.8$$

Therefore, the forced or steady-state solution is

$$y_{ss}(n) = -0.8(0.4)^n$$

The Complete Solution

The complete solution is the sum of the source-free and forced solutions. The source-free solution contains an arbitrary constant at this stage, and it is found by substituting initial conditions into the complete solution. Let us complete the solution of the problem presented in the previous two examples in order to illustrate.

EXAMPLE 7.4.3 Find the complete solution for the system described by

$$y(n) = x(n) + 0.9y(n-1)$$

where

$$x(n) = 0.4u(n)$$
$$y(-1) = 0$$

Solution

From the previous two examples, the source-free and steady-state solutions are given by

$$y_{tr}(n) = A(0.9)^n$$
$$y_{ss}(n) = -0.8(0.4)^n$$

The complete solution is the sum

$$\begin{aligned} y(n) &= y_{tr}(n) + y_{ss}(n) \\ &= A(0.9)^n - 0.8(0.4)^n \end{aligned} \tag{7.4.8}$$

The arbitrary constant A can be determined if $y(n)$ is known for any value of $n \geq 0$. The given initial condition $y(-1) = 0$ is not sufficient, because the input $x(n)$ is applied at $n = 0$. That is, the solution given in Eq. 7.4.8 is valid for $n \geq 0$, so substitution of values of y for $n < 0$ is not appropriate. The step-by-step method may be used to solve for $y(0)$, however, and this may be substituted into Eq. 7.4.8. From Eq. 7.4.7,

$$y(0) = x(0) + 0.9y(-1)$$
$$= 1 + 0 = 1$$

Substituting into Eq. 7.4.8 gives

$$y(0) = A(0.9)^0 - 0.8(0.4)^0$$

or

$$1 = A - 0.8$$
$$A = 1.8$$

Therefore, the complete solution is given by

$$y(n) = 1.8(0.9)^n - 0.8(0.4)^n$$

There are several points to note about the source-free, forced, and complete solutions of first order digital systems. First, consider the source-free solution and its relation to the characteristic equation. Figure 7.23 illustrates three values of the natural frequency (a root of the characteristic equation) plotted on the complex plane. We use the complex plane because the roots of the characteristic equation may be real or complex values for higher order systems. Of course, there is only one root for a first order equation, and it must be real. A unit circle is inscribed in the plane to illustrate the magnitude of the natural frequency.

For the situation pictured in Fig. 7.23a, the natural frequency z_1 is real, positive, and has magnitude less than 1. For this case $y_{tr}(n)$ decays with time since the form $A(z_1)^n$ approaches zero as n increases.

In Fig. 7.23b, the natural response again decays with time because $|z_2| < 1$, but the values of the response alternate between negative and positive values, depending on whether n is odd or even. In Fig. 7.23c, the natural response grows without bound, because $A(z_3)^n$ increases with increasing n whenever $|z_3| > 1$. A system whose natural response increases with time is called *unstable*. A *stable* system has $|z_i| < 1$ for all natural frequencies z_i, and, hence, the natural response decays with time.

For a linear system, the forced response is related to the input signal in a simple way. Figure 7.24b illustrates the relationship between the input $x(n)$ and the forced response $y_{ss}(n)$ for the system in Example 7.4.2. Figure 7.24a illustrates the natural response, and Fig. 7.24c graphs the complete response. Notice that the form of the forced response is the same as the form of the input. In general, the forced response always has the same form as the forcing function plus all its differences. The only reason for restricting the type of input signal that could be handled by the method of undetermined coefficients in our earlier discussion was so we could find these differences.

The complete response is found after adding the source-free and forced solutions. Only then can any constants in the natural response be found by

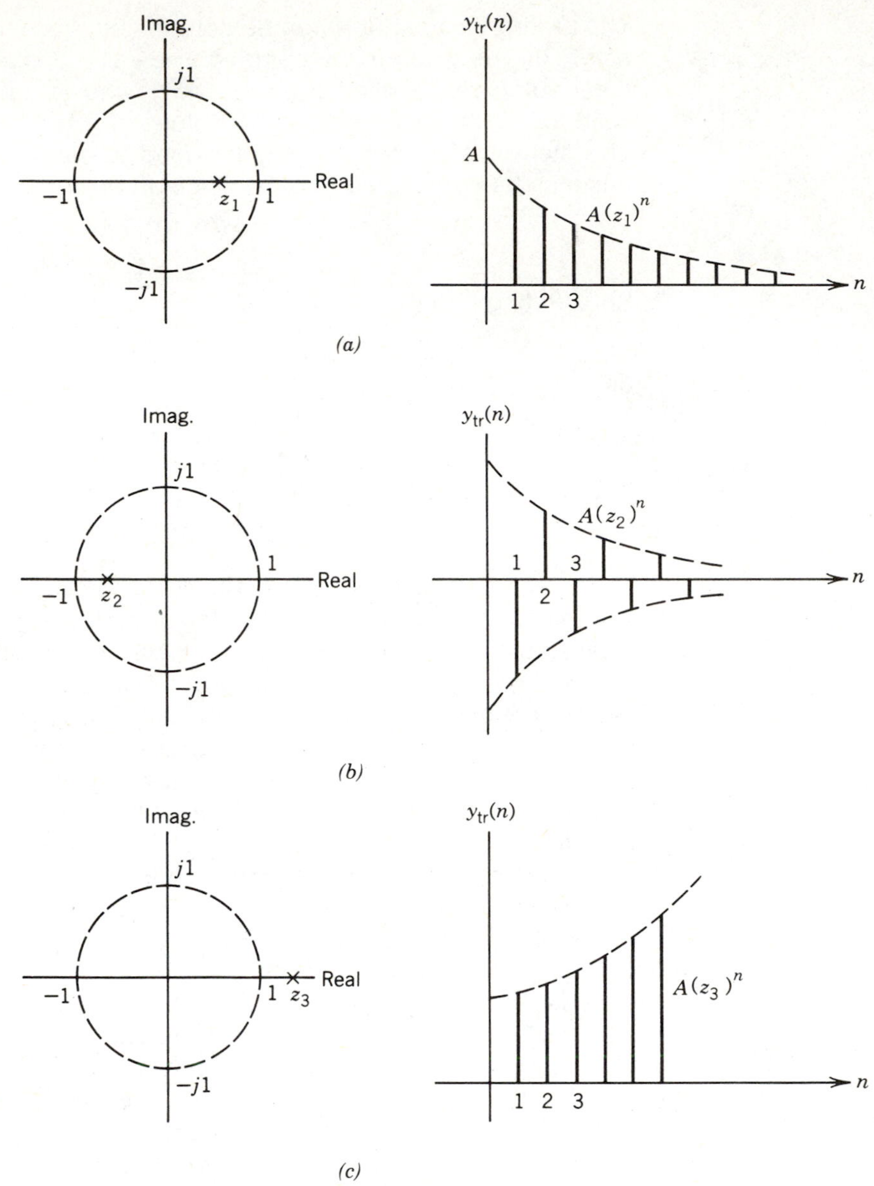

Figure 7.23. Illustrating the relationship between the natural frequency and the natural response of a system.

substituting initial conditions. Here is one more example worked in its entirety to illustrate the complete procedure.

EXAMPLE 7.4.4 Find the complete response for the system described by

$$y(n) = x(n) - 0.8y(n-1) \tag{7.4.9}$$

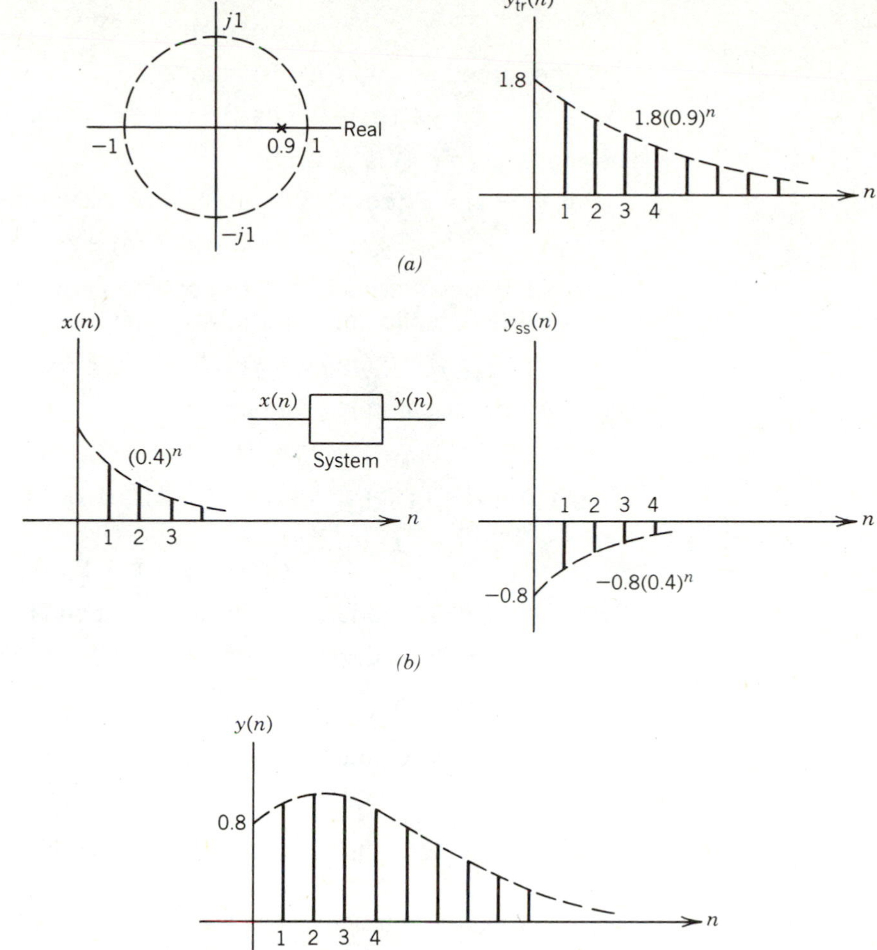

Figure 7.24. Plot of solution of example 7.4.2.

where

$$x(n) = \cos(0.3\pi n)u(n)$$

and

$$y(-1) = 0$$

Solution

The source-free response is found by substituting $y_{tr}(n) = A(z)^n$ into the source-free equation.

$$y(n) + 0.8y(n-1) = 0$$

The characteristic equation is given by

$$z + 0.8 = 0$$

so the source-free response is

$$y_{tr}(n) = A(-0.8)^n$$

Next, the forced solution is assumed to be

$$y_{ss}(n) = B_1 \cos(0.3\pi n) + B_2 \sin(0.3\pi n)$$

Therefore, $y_{ss}(n-1)$ is given by

$$y_{ss}(n-1) = B_1[\cos(0.3\pi)\cos(0.3\pi n) + \sin(0.3\pi)\sin(0.3\pi n)] + B_2[\cos(0.3\pi)\sin(0.3\pi n) - \sin(0.3\pi)\cos(0.3\pi n)]$$

Substituting these terms into the describing equation and solving for the undetermined coefficients B_1 and B_2 gives

$$y_{ss}(n) = 0.56976\cos(0.3\pi n) - 0.25081\sin(0.3\pi n)$$

The complete solution is given by

$$y(n) = A(-0.8)^n + y_{ss}(n)$$

Using the given initial condition $y(-1) = 0$ we find from Eq. 7.4.9 that $y(0)$ is given by

$$y(0) = x(0) + 0 = 1$$

Hence, $A = 0.43024$, and the complete solution is given by

$$y(n) = 0.43024(-0.8)^n + 0.56976\cos(0.3\pi n) - 0.25081\sin(0.3\pi n)$$

LEARNING EVALUATIONS

1. Find the transient response of the system described by
$$y(n) = x(n) + 0.2y(n-1)$$
2. Find the steady-state solution of the above system if $x(n) = u(n)$, the unit step function.
3. Find the complete solution if $y(-1) = 0$.

7.5 Second Order Digital Systems

LEARNING OBJECTIVES

After completing this section you should be able to do the following:

1. Find the source-free solution of a second order system.
2. Find the forced solution of a second order system.
3. Sum the source-free and forced solutions to find the complete solution, and then evaluate the constants in the solution from initial conditions.

For a second order system, Eq. 7.4.1 reduces to

$$b_0 y(n) + b_1 y(n-1) + b_2 y(n-2) = a_0 x(n) + a_1 x(n-1) + a_2 x(n-2) \tag{7.5.1}$$

As in the previous section, we will apply the same methods used in analog systems to solve Eq. 7.5.1 for $y(n)$. The results will be identifiably similar to those found in Chapter 6 for second order analog systems.

The Source-Free Response

In the previous section we saw that the transient response of a first order system was exponential. That is, for a system described by the difference equation,

$$y(n) + b_1 y(n-1) = 0 \tag{7.5.2}$$

The solution is of the form

$$y(n) = Az^n = A(-b_1)^n \tag{7.5.3}$$

Let us examine the meaning of this result a bit further before discussing the second order equation.

Equation 7.5.2 states that the discrete-time function $y(n)$ is related to a shifted version of $y(n)$ as illustrated in Fig. 7.25. The value of b_1 in the diagram is equal to -0.9, but our discussion applies to any real value for b_1. Figure 7.25a shows

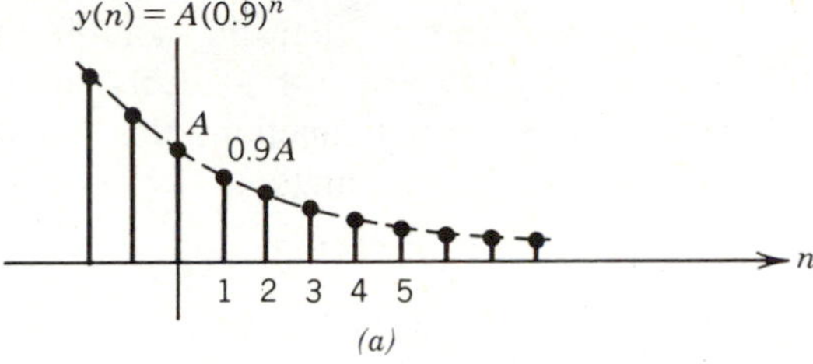

(a)

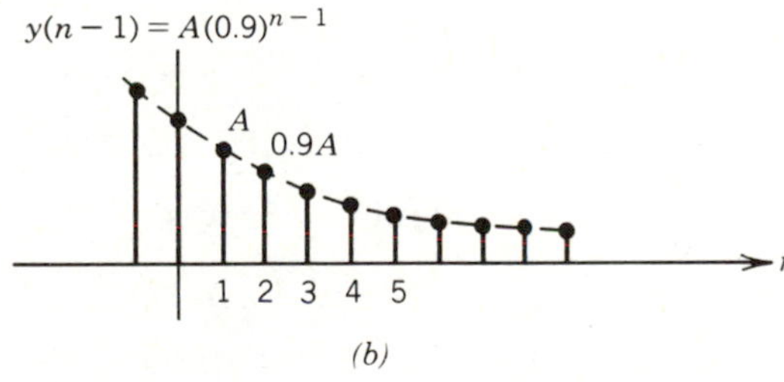

(b)

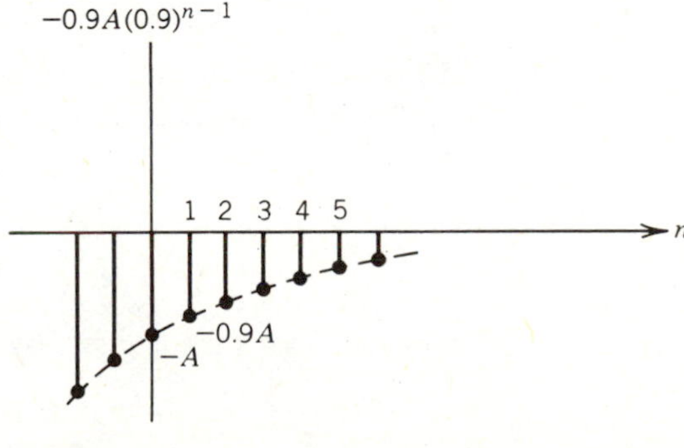

(c)

Figure 7.25. Results of the shifting operation.

the solution $y(n) = A(0.9)^n$. Figure 7.25b shows the shifted version of the solution, $y(n) = A(0.9)^{n-1}$, and Fig. 7.25c shows this shifted version multiplied by b_1. According to Eq. 7.5.2, the sum of the two functions in Figs. 7.25a and 7.25c must be zero for every value of n. It is easy to see that this sum is zero for every n. [As an exercise, the reader is encouraged to pick some other value of b_1 and plot the three functions $y(n)$, $y(n-1)$, and $b_1 y(n-1)$, where $y(n) = A(-b_1)^n$. Add $y(n)$ to $b_1 y(n-1)$ and see that the sum is again zero.]

Is there any other function besides the exponential that satisfies Eq. 7.5.2? As it happens, the answer is no. Search as you will, you will be unable to discover another type of function that satisfies the general first order, source-free equation.

Now consider the source-free second order difference equation. From Eq. 7.5.1 this is seen to be

$$y(n) + b_1 y(n-1) + b_2 y(n-2) = 0 \tag{7.5.4}$$

This equation is somewhat different from the first order equation, because here the sum of three functions must be zero. The problem is to find a function $y(n)$ so that the shifted version $b_1 y(n-1)$ plus the again shifted version $b_2 y(n-2)$ is zero. It so happens that the solution to any order source-free equation is exponential. However, for a second order equation, there may be one or two different exponential functions that satisfy the equation. Another complicating factor for second and higher order equations is the possibility that these solutions may be complex valued.

The procedure used to find the solution to Eq. 7.5.1 is, as before, to assume a solution

$$y(n) = Az^n$$

and to set $a_0 = a_1 = a_2 = 0$. Placing our assumed solution into Eq. 7.5.1 yields

$$b_0 Az^n + b_1 Az^{n-1} + b_2 Az^{n-2} = 0$$

Dividing both sides by Az^n gives

$$b_0 + b_1 z^{-1} + b_2 z^{-2} = 0$$

Multiplying both sides by z^2 results in

$$b_0 z^2 + b_1 z + b_2 = 0 \tag{7.5.5}$$

Equation 7.5.5 is the characteristic equation for the transient response of a second order system. The roots of the equation can be found by use of the quadratic equation:

$$z_{1,2} = \frac{-b_1 \pm \sqrt{b_1^2 - 4b_0 b_2}}{2b_0} \tag{7.5.6}$$

The solution of Eq. 7.5.4 then is given by

$$y_{\text{tr}}(n) = A_1 z_1^n + A_2 z_2^n \tag{7.5.7}$$

The nature of the roots (natural frequencies) z_1 and z_2 is determined by the discriminant

$$b_1^2 - 4b_0b_2 \tag{7.5.8}$$

If Eq. 7.5.8 is greater than zero, the roots are real and unequal; if less than zero, the roots are complex; if equal to zero, the roots are equal and as in Chapter 6, further measures must be taken.

In fact, this all should look very familiar to you because the situation is identical to that found in the last chapter when we discussed overdamped, underdamped, and critically damped transient responses to second order continuous systems.

As an example of the first case (real, unequal roots), consider the second order system described by the equation

$$y(n) - y(n-1) + 0.1875y(n-2) = 0$$

The characteristic equation is

$$z^2 - z + 0.1875 = 0$$

Solving this yields

$$z_1 = 0.75$$
$$z_2 = 0.25$$

so

$$y_{tr}(n) = A_1(0.75)^n + A_2(0.25)^n \tag{7.5.9}$$

In the second case, when the roots are real and equal, we can rely on our previous experience summarized in Table 6.5, and assume that the form of the transient solution is given by

$$y_{tr}(n) = A_1 z^n + A_2 n z^n \tag{7.5.10}$$

where $z = z_1 = z_2$. The fact that this form is valid may be demonstrated as follows. For real, equal roots, the source-free equation has the form

$$y(n) - 2ay(n-1) + a^2 y(n-2) = 0 \tag{7.5.11}$$

The natural frequency is $z = a$. Now assume a general form for the solution given by

$$y_{tr}(n) = f(n)a^n$$

Upon substituting into Eq. 7.5.11 we find that the function $f(n)$ must satisfy

$$f(n) - 2f(n-1) + f(n-2) = 0.$$

Aside from the trivial solution $f(n) = 0$, there are only two functions that satisfy this condition.

$$f(n) = A_1$$
$$f(n) = A_2 n$$

Hence, Eq. 7.5.10 is the general form for the solution with real and equal roots.

The third case, where roots are complex, is in principle identical to the case of real and unequal roots. The only difference is that the roots are complex rather than real numbers.

EXAMPLE 7.5.1 Find the solution to the homogeneous equation given by

$$y(n) - 0.5y(n-1) + 0.0625y(n-2) = 0$$

Solution

The characteristic equation is given by

$$z^2 - 0.5z + 0.0625 = 0$$

which has repeated roots $z_{1,2} = 0.25$. Therefore, the transient solution is

$$y_{tr}(n) = A_1(0.25)^n + A_2 n(0.25)^n$$

For the third case of complex valued[1] roots, one can see from Eq. 7.5.6 that the roots form a complex conjugate pair. That is, $z_1 = z_2^*$. The form of the solution is given by Eq. 7.5.7, but this form contains complex numbers and is somewhat inconvenient. A form containing only real numbers will be derived after the next example.

EXAMPLE 7.5.2 Find the solution to the source-free equation given by

$$y(n) - 0.5y(n-1) + 0.125y(n-2) = 0$$

Solution

The characteristic equation is given by

$$z^2 - 0.5z + 0.125 = 0$$

The natural frequencies of the system (the roots of the characteristic equation) are the complex numbers z_1 and z_2, given by

$$z_1 = 0.25 + j0.25$$

$$z_2 = 0.25 - j0.25$$

Therefore the transient response is given by

$$y_{tr}(n) = A_1(0.25 + j0.25)^n + A_2(0.25 - j0.25)^n$$

Both roots z_1 and z_2 are complex numbers in the above example, and the constants A_1 and A_2 may also be complex, depending on the initial values of $y(n)$. The values of $y(n)$ must always be real, however, since there is no mechanism for generating complex values in the difference equation. That is, if the step-by-step method is used to find succeeding values of $y(n)$ from any given initial conditions, these values will always be real numbers.

To put the above response in a better form, convert each frequency from rectangular to polar form. Let

$$z_1 = |z_1|e^{j\Theta}$$

$$z_2 = |z_2|e^{-j\Theta}$$

where $|z_1| = |z_2|$ because $z_1 = z_2^*$. Next, rewrite Eq. 7.5.7 as

[1]See Appendix 2 for a review of complex numbers. The appendixes are no less important than any other material, and should not be skipped unless the student is familiar with their content. The material in the appendixes is there because it applies to more than one section of the text.

$$\begin{aligned} y(n) &= A_1\left(|z_1|e^{j\Theta}\right)^n + A_2\left(|z_2|e^{-j\Theta}\right)^n \\ &= |z_1|^n\left(A_1 e^{jn\Theta} + A_2 e^{-jn\Theta}\right) \\ &= |z_1|^n\left[A_1(\cos n\Theta + j\sin n\Theta) + A_2(\cos n\Theta - j\sin n\Theta)\right] \end{aligned}$$

where we used Euler's formula in the last step. Collecting terms, we have

$$y(n) = |z_1|^n\left[(A_1 + A_2)\cos(n\Theta) + (jA_1 - jA_2)\sin(n\Theta)\right]$$

Since $z_1 = z_2^*$, it must also be true that $A_1 = A_2^*$ if $y(n)$ is real valued. Hence, $A_1 + A_2$ is a real number, and $jA_1 - jA_2$ is also a real number. Therefore, we may rename the constants as

$$C_1 = A_1 + A_2$$
$$C_2 = jA_1 - jA_2$$

Therefore,

$$y(n) = |z_1|^n\left[C_1\cos(n\Theta) + C_2\sin(n\Theta)\right] \tag{7.5.12}$$

EXAMPLE 7.5.3 Use Eq. 7.5.12 to solve the transient equation

$$y(n) - 0.5y(n-1) + 0.125y(n-2) = 0$$

Solution

The roots of the characteristic equation are

$$z_1 = (0.25 + j0.25) = \frac{\sqrt{2}}{4}e^{j\pi/4}$$

$$z_2 = (0.25 - j0.25) = \frac{\sqrt{2}}{4}e^{-j\pi/4}$$

Substituting into Eq. 7.5.12 gives

$$y(n) = \left(\frac{\sqrt{2}}{4}\right)^n\left\{C_1\cos\left(\frac{n\pi}{4}\right) + C_2\sin\left(\frac{n\pi}{4}\right)\right\}$$

Note: The angle must be measured in radians, not in degrees, when converting to polar form.

The Forced Response

The method of solution used in this chapter applies to *any* order equation, just so long as the forcing function is one of the forms listed in Section 7.4. Recall that this implies that $x(n)$ is of the form

$$a^n \quad \text{or} \quad e^{bn}, n^p, \sin(cn), \cos(cn)$$

or sums or products of such forms. Therefore, the procedure for finding the forced response of a second order system is the same as that of a first order system. Here are some examples to illustrate this procedure.

EXAMPLE 7.5.4 Find the steady-state (forced) solution of the following difference equation.

$$y(n) - y(n-1) + 0.1875y(n-2) = 0.5x(n) + 0.5x(n-1) \tag{7.5.13}$$

where $x(n) = u(n)$.

Solution

The forcing function is a unit step, and hence a member of the family of functions for which our approach is appropriate. [For example, $u(n) = n^0$, $n \geq 0$.] Therefore, the forced response is a constant (the same form as the input) plus all shifted versions of a constant (which is still a constant). The assumed solution is given by

$$y_{ss}(n) = C$$

so

$$y_{ss}(n-1) = C$$

and

$$y_{ss}(n-2) = C$$

Substituting into Eq. 7.5.13 gives

$$C - C + 0.1875C = 0.5 + 0.5$$

or

$$C = \tfrac{16}{3}$$

Hence, the forced solution is $y_{ss}(n) = \frac{16}{3}$. The general solution of Eq. 7.5.13 is found by adding the transient and steady-state solutions to obtain

$$y(n) = A_1(0.75)^n + A_2(0.25)^n + 5.333 \tag{7.5.14}$$

where the transient solution is given by Eq. 7.5.9. Initial conditions can now be used to solve for the constants A_1 and A_2.

A problem in semantics may be encountered in the preceding example. The forcing function is a unit step, not a constant, and hence, the shifted versions $u(n)$ are $u(n-1)$ and $u(n-2)$. Yet in the solution given above, we ignored the fact that the forcing function changed value at $n = 0$, and simply treated the system input as a constant for all time. This is an important feature of all classical solutions of difference (and differential) equations. The problem is first solved as if the forcing function was present for all time. After finding the solution, initial conditions are then substituted to account for any discontinuities in the input.

EXAMPLE 7.5.5 Find the forced response for the system in Example 7.5.3 if the forcing function is given by

$$x(n) = (0.9)^n$$

Solution

The system equation to be solved is given by

$$y(n) - 0.5y(n-1) + 0.125y(n-2) = (0.9)^n \tag{7.5.15}$$

The steady-state solution is assumed to be

$$y_{ss}(n) = B(0.9)^n$$

so

$$y_{ss}(n-1) = B(0.9)^{n-1} = \left(\frac{B}{0.9}\right)(0.9)^n$$

$$y_{ss}(n-2) = B(0.9)^{n-2} = \left(\frac{B}{0.81}\right)(0.9)^n$$

Substituting these values into Eq. 7.5.15 gives

$$B(0.9)^n - \left(\frac{1}{2}\right)\left(\frac{B}{0.9}\right)(0.9)^n + 0.125\left(\frac{B}{0.81}\right)(0.9)^n = (0.9)^n$$

or $B = 1.67$. The particular solution is therefore given by

$$y_{ss}(n) = 1.67(0.9)^n$$

The general solution is

$$y(n) = \left(\frac{\sqrt{2}}{4}\right)^n \left[C_1 \cos\left(\frac{n\pi}{4}\right) + C_2 \sin\left(\frac{n\pi}{4}\right)\right] + 1.67(0.9)^n$$

The constants C_1 and C_2 may now be found from initial conditions.

Excitation at a Natural Frequency

Suppose the system equation is given by

$$y(n) - (a+b)y(n-1) + aby(n-2) = x(n) \tag{7.5.16}$$

If $a = b$, then the roots of the characteristic equation are repeated. In that case, the transient solution is not given by the standard form expressed by Eq. 7.5.7, but instead is given by Eq. 7.5.10. That is, one of the terms in the solution is multiplied by time n.

A similar situation occurs if the forcing function contains a frequency that is equal to one of the natural frequencies. Let $a \neq b$ in Eq. 7.5.16. The characteristic equation is

$$z^2 - (a+b)z + ab = 0$$

from which the natural frequencies are

$$z_1 = a, \qquad z_2 = b$$

The transient response is given by

$$y_{tr}(n) = A_1(a)^n + A_2(b)^n$$

Now let us suppose that the forcing function contains a natural frequency, say

$$x(n) = (a)^n$$

If the usual procedure is followed, where the steady-state solution is assumed to be

$$y_{ss}(n) = B(a)^n$$

things just do not work out. If we determine the "undetermined coefficient" B so that $y(n)$ satisfies Eq. 7.5.16, the solution reduces to the nonsensical situation

$$0 = (a)^n$$

To avoid this difficulty, the steady-state solution should be

$$y_{ss}(n) = Bn(a)^n \tag{7.5.17}$$

Now substitution of this assumed solution into the system equation provides the proper answer. In terms of the general form given by Eq. 7.5.16, the solution proceeds as follows.

$$y_{ss}(n-1) = B(n-1)a^{n-1} = \left(\frac{B}{a}\right)(n-1)a^n$$

$$y_{ss}(n-2) = B(n-2)a^{n-2} = \left(\frac{B}{a^2}\right)(n-2)a^n$$

Substituting into Eq. 7.5.16 gives

$$Ba^n\left(\frac{a-b}{a}\right) = a^n$$

Since this must hold for all n,

$$B = \frac{a}{a-b} \tag{7.5.18}$$

The general solution of Eq. 7.5.16 is therefore

$$y(n) = A_1(a)^n + A_2(b)^n + \frac{a}{a-b}(a)^n \tag{7.5.19}$$

EXAMPLE 7.5.6 Find the general solution of the following difference equation.

$$y(n) - y(n-1) + \tfrac{3}{16}y(n-2) = (0.75)^n \tag{7.5.20}$$

Solution

This system has the source-free solution given by Eq. 7.5.9

$$y_{tr}(n) = A_1(0.75)^n + A_2(0.25)^n$$

The forcing function contains one of the natural frequencies, so we must assume

$$y_{ss}(n) = Bn(0.75)^n$$

With $a = 0.75$ and $b = 0.25$ in Eq. 7.5.19, the general solution is given by

$$y(n) = A_1(0.75)^n + A_2(0.25)^n + 1.5n(0.75)^n$$

We did not discuss excitation at a natural frequency for first order systems, but our motive was to simplify that introduction to the subject. These same principles apply to any order system. If the forcing function contains a system frequency, the steady-state solution contains a term given by Eq. 7.5.17.

In the rare case that the system has repeated roots, and the forcing function also contains that frequency, Eq. 7.5.17 must be modified by raising n to the second power.

The Complete Response

The source-free response contains arbitrary constants. The complete response of a discrete-time system is the sum of the source-free and forced response, and therefore also contains arbitrary constants. In Section 7.4 we noted that the correct procedure is to wait until adding the two responses before solving for the one arbitrary constant present in the first order solution. Since we are now dealing with second order systems, there are two constants to solve for, so we need two initial conditions. Again, the proper time to solve for the arbitrary constants is after adding the transient and forced response to obtain the general solution.

EXAMPLE 7.5.7 Find the complete response for the system described by

$$y(n) - y(n-1) + 0.1875y(n-2) = (0.75)^n, \qquad n \geq 0 \qquad (7.5.21)$$
$$y(-1) = 0$$
$$y(-2) = 0$$

Solution

The general solution was found in Example 7.5.6, given by

$$y(n) = A_1(0.75)^n + A_2(0.25)^n + 1.5n(0.75)^n \qquad (7.5.22)$$

But this solution is valid only for $n \geq 0$. Hence, substitution of the given initial conditions in this solution in order to solve for A_1 and A_2 is inappropriate. We must apply the step-by-step method to Eq. 7.5.21 to find values of $y(0)$ and $y(1)$.

$$y(0) = y(-1) - 0.1875y(-2) + (0.75)^0 = 1$$
$$y(1) = y(0) - 0.1875y(-1) + (0.75)^1 = 1.75$$

Substituting these values into Eq. 7.5.22 results in two equations, given by

$$A_1 + A_2 = 1$$
$$0.75A_1 + 0.25A_2 = 0.625$$

Solving gives $A_1 = 0.75$, $A_2 = 0.25$, so the complete response is given by

$$y(n) = 0.75(0.75)^n + 0.25(0.25)^n + 1.5n(0.75)^n$$

EXAMPLE 7.5.8 Find the solution for the system given by

$$y(n) - 0.5y(n-1) + 0.125y(n-2) = u(n)$$
$$y(-1) = 1$$
$$y(-2) = 0 \qquad (7.5.23)$$

Solution

The transient solution was found in Example 7.5.3 and is given by

$$y_{tr}(n) = \left(\frac{\sqrt{2}}{4}\right)^n \left[C_1 \cos\left(\frac{n\pi}{4}\right) + C_2 \sin\left(\frac{n\pi}{4}\right)\right]$$

To find the steady-state solution, assume

$$y_{ss}(n) = A$$

Substituting this constant into Eq. 7.5.23 gives

$$A - 0.5A + 0.125A = 1$$

or

$$A = 1.60$$

Therefore, the general solution is given by

$$y(n) = \left(\frac{\sqrt{2}}{4}\right)^n \left[C_1 \cos\left(\frac{n\pi}{4}\right) + C_2 \sin\left(\frac{n\pi}{4}\right)\right] + 1.6$$

Using the step-by-step method on Eq. 7.5.23 gives

$$y(0) = 0.5y(-1) + 0.125y(-2) + 1 = 1.5$$
$$y(1) = 0.5y(0) + 0.125y(-1) + 1 = 1.875$$

Substituting these conditions into the general solution, we have

$$1.5 = C_1 + 1.6$$
$$1.875 = \frac{\sqrt{2}}{4}\left[C_1 \cos\left(\frac{\pi}{4}\right) + C_2 \sin\left(\frac{\pi}{4}\right)\right] + 1.6$$

or

$$C_1 = -0.1$$
$$C_2 = 1.2$$

Therefore, the complete solution is

$$y(n) = \left(\frac{\sqrt{2}}{4}\right)^n \left[-0.1 \cos\left(\frac{n\pi}{4}\right) + 1.2 \sin\left(\frac{n\pi}{4}\right)\right] + 1.6$$

LEARNING EVALUATIONS

1. Find the transient response of the system described by

$$y(n) - y(n-1) + 0.5y(n-2) = x(n)$$

2. List the general form of the forced response for each forcing function in the following chart.

	$x(n)$	$y(n)$
1	$3(0.2)^n$	
2	$5\cos(0.2n)$	
3	$2n$	
4	$u(n)$	
5	$2\sin(0.5n)$	

3. The general solution of the difference equation

$$y(n)+2y(n-1)+y(n-2)=u(n)$$
$$y(-1)=1$$
$$y(-2)=1$$

is given by

$$y(n)=A_1(-1)^n+A_2n(-1)^n+\tfrac{1}{4}$$

Solve for the constants A_1 and A_2.

PROBLEMS

SECTION 7.1

1. Draw the direct form I realization for the following difference equations.

a. $y(n)=0.3x(n)-0.2y(n-1)$

b. $y(n)=0.3x(n)+0.2x(n-1)-0.2y(n-1)$

c. $y(n)=x(n)+0.5y(n-1)+0.5y(n-2)$

d. $y(n)=0.5x(n)+0.2x(n-1)-0.4y(n-1)-0.3y(n-2)-0.3y(n-3)$

2. Write the corresponding difference equation for each diagram in Fig. 7.26.

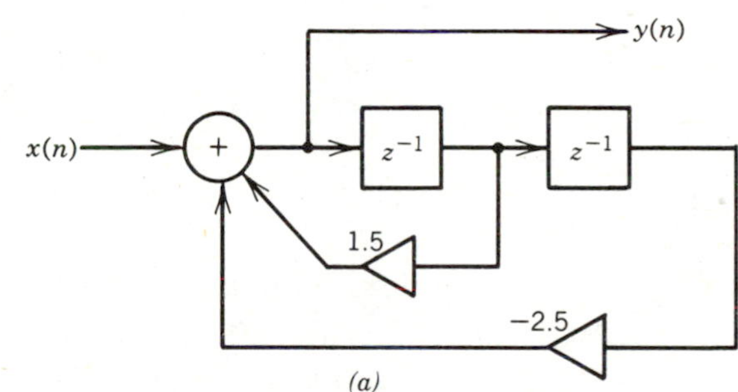

(a)

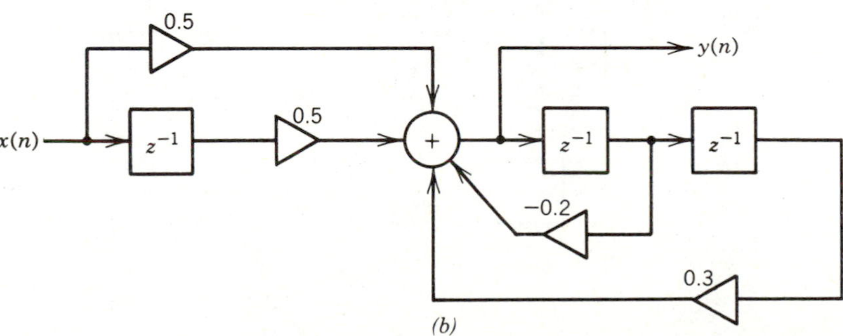

(b)

Figure 7.26

3. Draw the direct form I realization for the following difference equations.
   a. $y(n) = 4x(n) + 16x(n-1) + y(n-2)$
   b. $y(n) = x(n) - 12.5x(n-1) + 2.7y(n-1) + 0.3y(n-2)$
   c. $y(n) = x(n-1) - y(n-2)$
   d. $y(n) = 0.4x(n) - 0.5x(n-1) + 2y(n-1)$
4. Write the corresponding difference equation for each diagram in Fig. 7.27.

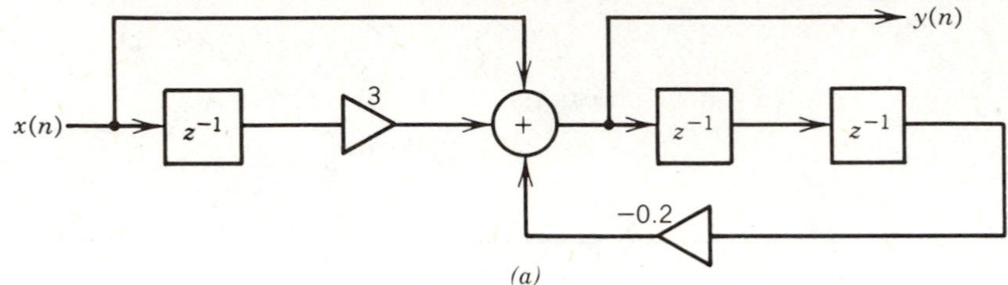

(a)

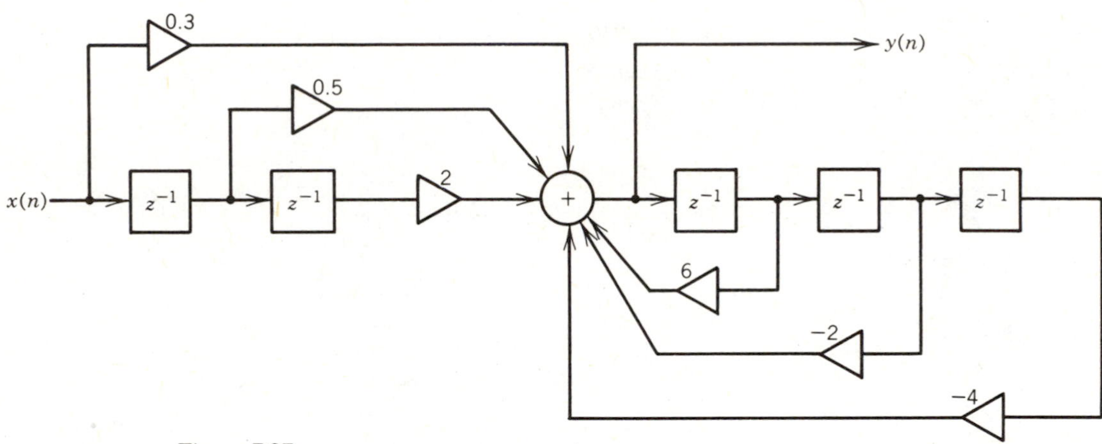

Figure 7.27 (b)

5. Draw the direct form I realization for the following difference equations.
   a. $y(n) = 0.2x(n) + 0.3x(n-1) + 4y(n-1) - y(n-2)$
   b. $y(n) = 3x(n) + 5x(n-1) - 3y(n-1) + 2y(n-2) + y(n-3)$
6. Write the corresponding difference equation for the diagram in Fig. 7.28.

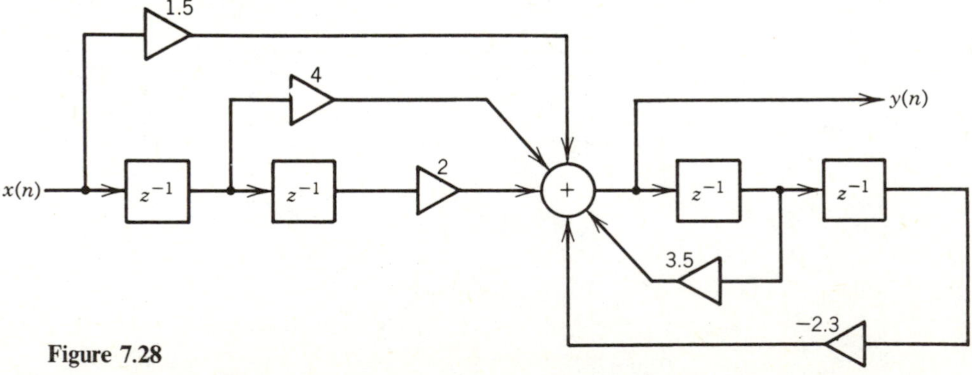

Figure 7.28

SECTION 7.2

7. Many useful algorithms can be cast in the form of difference equations. One of the better known algorithms is the following, used to find the square root of a number x.

$$y(k) = \frac{0.5x(k)}{y(k-1)} + 0.5y(k-1)$$

where $x(k)$ is a step of height equal to the number whose square root we wish to calculate. Use initial condition $y(-1) = 1$. Find $\sqrt{3}$, $\sqrt{10}$, and $\sqrt{20}$.

8. Pascal's triangle displays the binomial coefficients in recursive form, given by

$$\begin{array}{ccccccccc} & & & 1 & & 1 & & & \\ & & 1 & & 2 & & 1 & & \\ & 1 & & 3 & & 3 & & 1 & \\ 1 & & 4 & & 6 & & 4 & & 1 \end{array}$$

This process may be continued indefinitely. The next row after the one shown (the fifth row) is found by summing terms in the fourth row. The fifth row is 1 5 10 10 5 1, where $5 = 1 + 4$ and $10 = 4 + 6$.

The terms in the nth row may also be generated by a difference equation, given by

$$y(k) = \frac{(n-k+1)}{k} y(k-1), \qquad k = 1, 2, \ldots, n$$

$$y(0) = 1$$

a. Generate the fifth row ($n = 5$) from this equation,

b. Generate the sixth row and check by completing this row in the triangle.

9. Find the first three values of the response to a unit step by the system in Fig. 7.29. Assume $y(-1) = 1$ and $y(-2) = 2$.

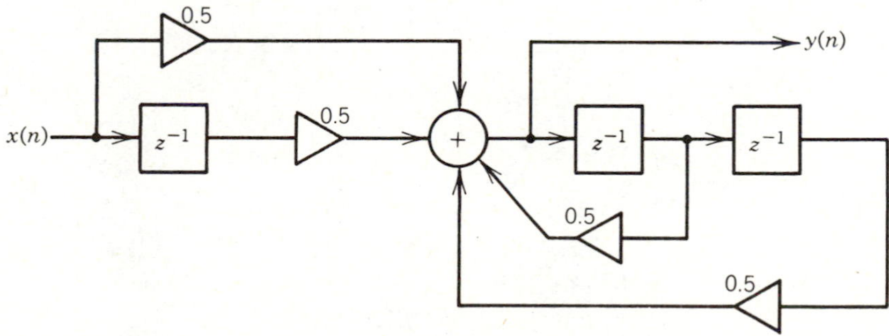

Figure 7.29

10. For the system in Fig. 7.30, calculate the first four values of the response to $x(n)$. Assume $y(-1) = y(-2) = 0$.

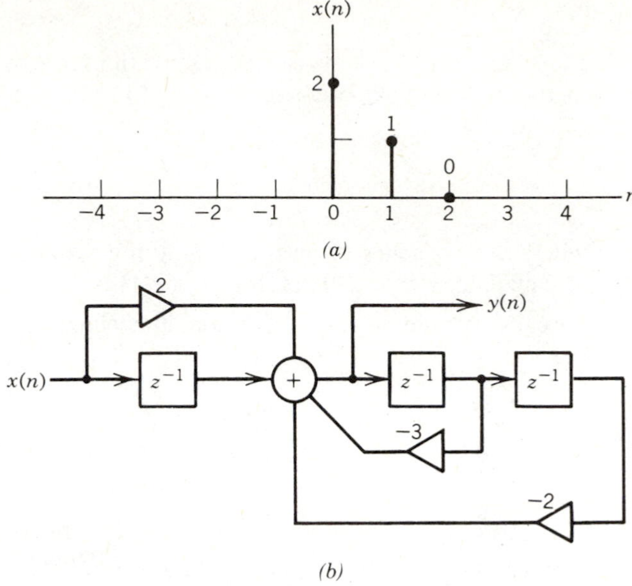

Figure 7.30

11. A system is described by the following difference equation. Find the first five values of the response of the system to an input $x(n)$.

$$y(n) = x(n) - 2x(n-1) + y(n-1)$$
$$y(-1) = 0$$

Assume that $x(n)$ is described by Fig. 7.31.

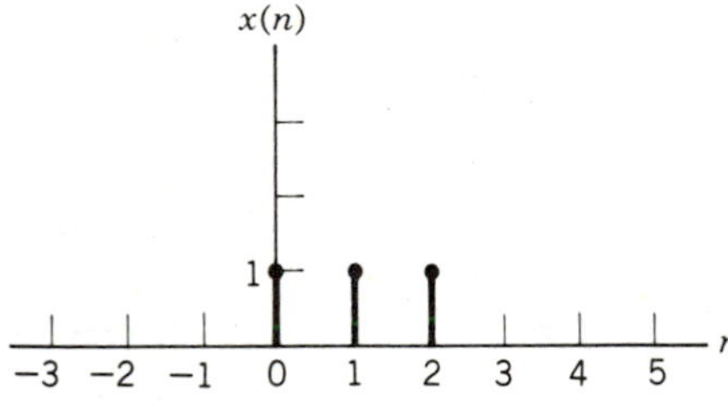

Figure 7.31

SECTION 7.3

12. Sketch the following functions.

a. $\delta(n-3)$
b. $u(n+3)$
c. $nu(n-3)$
d. $(n-3)u(n-3)$
e. $\delta(n) - \delta(n-3)$
f. $u(n) - u(n-3)$
g. $nu(n) - (n-3)u(n-3)$
h. $3nu(n) - (n-3)u(n-3)$

13. Describe each signal shown in Fig. 7.32 in each of two ways, by a direct formula specifying $x(n)$ for each n, and in terms of elementary signals $\delta(n)$, $u(n)$, and so on.

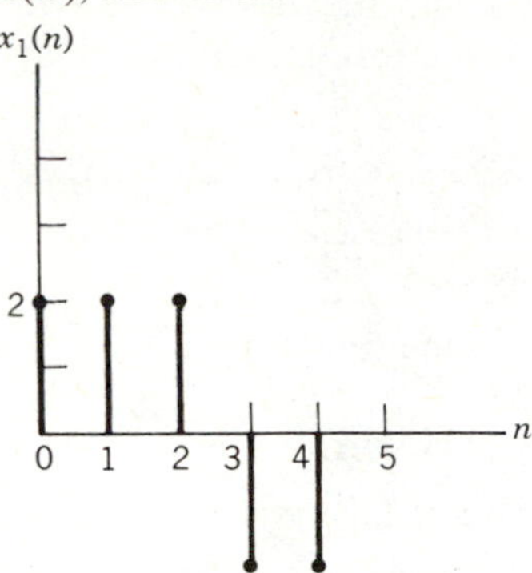

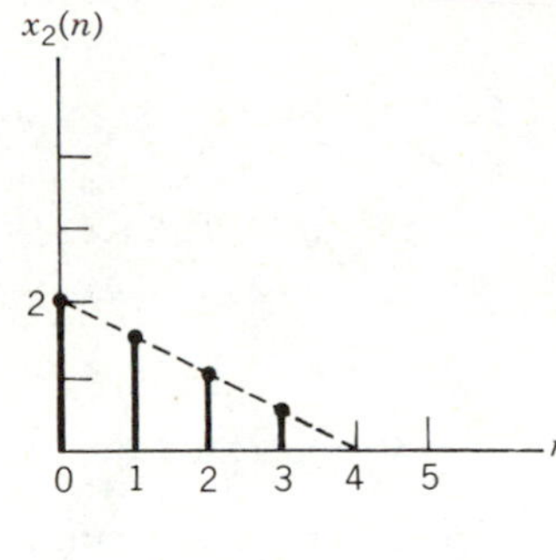

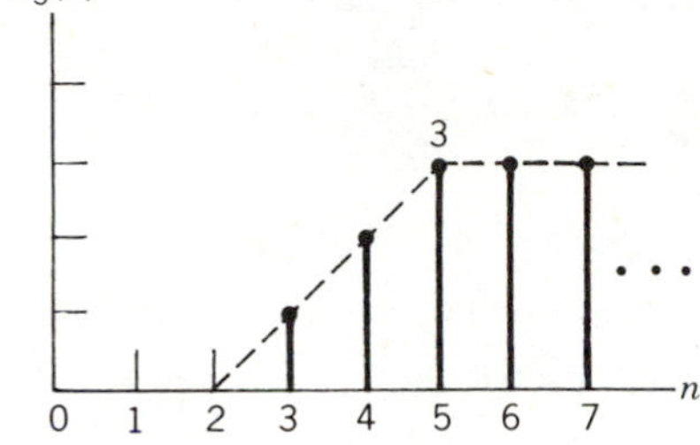

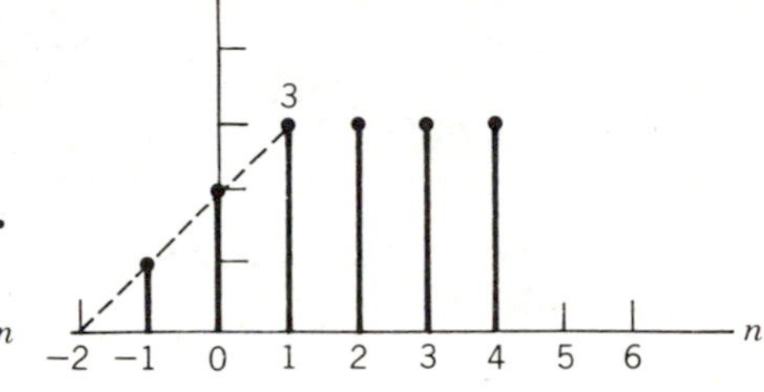

Figure 7.32

14. Show that the digital system in Fig. 7.19*b* will perform the σ operation on the input signal. Hint: Write down the system output at times n, $n-1$, $n-2$, and so on, and then substitute.

15. Apply the Δ operation to each signal in Problem 12. Sketch the result.

16. Apply the Δ operation to each signal in Problem 13. Sketch the result.

17. Apply the σ operation to each signal in Problem 12. Sketch the result.

18. Apply the σ operation to each signal in Problem 13. Sketch the result.

19. Describe the following signal (Fig. 7.33) in each of two ways, by a direct formula specifying $x(n)$ for each n, and in terms of elementary signals $\delta(n)$, $u(n)$, and so on.

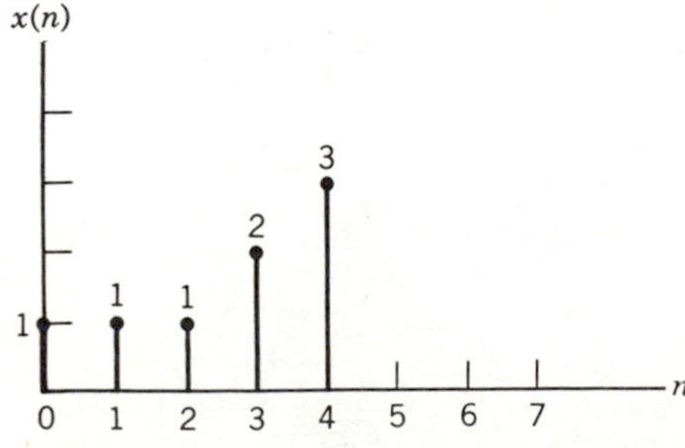

Figure 7.33

20. Apply the Δ operator to the signal given in Problem 19. Sketch the result. Apply the σ operator to the result, and show that this results in $x(n)$.

21. Apply the σ operator to the signal given in Problem 19. Sketch the result. Apply the Δ operator to the result, and show that this results in $x(n)$.

22. Sketch the following functions.
    a. $4\delta(n+6)+nu(n+1)-(n+3)u(n-4)+3u(n-6)$
    b. $2\delta(n)+4u(n)+nu(n-1)-(n+4)u(n-4)$
    c. $\delta(n)+u(n)-u(n-5)$

23. Apply the Δ operation to the signal given in Problem 22a.

SECTION 7.4

24. Find the natural response of the system described by

$$y(n)=x(n)+0.7y(n-1)$$

25. Find the particular solution for the system in Problem 24 if $x(n)$ is given by
    a. $x(n)=u(n)$
    b. $x(n)=(0.3)^n u(n)$
    c. $x(n)=n^2 u(n)$
    d. $x(n)=\cos(0.1\pi n)u(n)$

26. Find the complete response for each part of Problem 25 if $y(-1)=0$.

27. Find the natural response of the system described by

$$y(n)=x(n)+0.2y(n-1)$$

28. A system is described by

$$y(n)=0.5x(n)-0.1y(n-1) \qquad [y(-1)=0]$$

    a. Find the natural response of the system.
    b. If $x(n)=(0.4)^n u(n)$, find the complete response.

29. For the system described by

$$y(n)=1.5x(n)+0.6y(n-1)$$

find the complete response if $x(n)=6u(n)$ and $y(-1)=1$.

30. Find the complete response of the system described by

$$y(n)=x(n)+y(n-1)$$

if $x(n)=\cos(0.2\pi n)u(n)$ and if $y(-1)=0$.

31. Find the forced response of the system described by

$$y(n)=0.3x(n)+0.3y(n-1)$$

to the input $x(n)=e^{0.7n}$.

32. Is the system described in Problem 28 stable? Explain.

SECTION 7.5

33. Find the natural response of the system described by

$$y(n) - 0.64y(n-2) = x(n)$$

34. Find the forced response of the system

$$y(n) + 2y(n-1) + y(n-2) = x(n)$$

if $x(n)$ equals:

a. $4n$
b. $4u(n)$
c. $6(0.4)^n$

35. Find the complete response of the system

$$y(n) + 0.6y(n-1) + 0.09y(n-2) = 2n$$

if $y(-1) = 2$ and $y(-2) = 1$.

36. For the system

$$y(n) - 1.6y(n-1) + 0.64y(n-2) = x(n)$$

a. Find the natural response.
b. Find the forced response if $x(n) = u(n)$.
c. If $y(-1) = 0$ and $y(-2) = 0$, find the complete solution.

37. Find the natural response of the system described by

$$y(n) + 0.25y(n-1) + 0.015y(n-2) = 0$$

Is this system stable? Explain.

LEARNING EVALUATION ANSWERS

Section 7.1

1. See Fig. 7.34

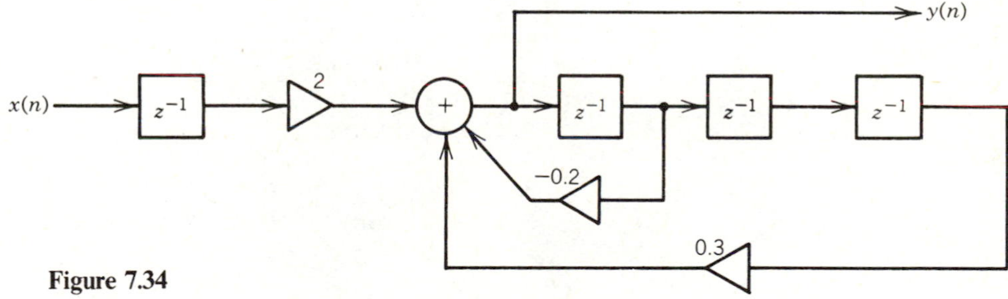

Figure 7.34

2. $y(n) = x(n) + 0.2y(n-1) - 1.5y(n-2)$

Section 7.2

1. $y(0) = 1.5$
$y(1) = 1.41667$
$y(2) = 1.41422$
$y(3) = 1.41421$
$y(4) = 1.41421$

2. $y(n) = 0.5x(n) + 0.5x(n-1) - 2y(n-1) + 0.1y(n-2)$
$y(0) = 0.5$
$y(1) = 0.0$
$y(2) = 1.05$
$y(3) = 1.10$
$y(4) = 3.305$

Section 7.3

1. a. $x(n) = \begin{cases} 0 & n < 0 \\ n & 0 < n < 3 \\ (6-n) & 3 < n < 6 \\ 0 & n > 6 \end{cases}$

b. $x(n) = r(n)u(n) - 2r(n-3)u(n-3) + r(n-6)u(n-6)$

2. $\Delta x(n) = u(n-1) - 2u(n-4) + u(n-7)$

3. $\sigma[\Delta x(n)] = x(n)$

4. $\sigma r(n) = \frac{1}{2}n(n+1)u(n)$

Section 7.4

1. $y_{tr}(n) = A(0.2)^n$

2. $y_{ss}(n) = 1.25$

3. $y(n) = 1.25 - 0.25(0.2)^n,\ n \geq 0$

Section 7.5

1. $y_{tr}(n) = A_1(0.5 + j0.5)^n + A_2(0.5 - j0.5)^n$

2.

	$x(n)$	$y(n)$
1	$3(0.2)^n$	$B(0.2)^n$
2	$5\cos(0.2n)$	$B_1\cos(0.2n) + B_2\sin(0.2n)$
3	$2n$	$Bn + C$
4	$u(n)$	B
5	$2\sin(0.5n)$	$B_1\cos(0.5n) + B_2\sin(0.5n)$

3. $A_1 = -\frac{9}{4},\ A_2 = -\frac{3}{2}.$

CHAPTER

8 LINEAR TIME-INVARIANT (LTI) SYSTEMS

Up to this point in the text, we have tacitly assumed that all circuits and systems to which we applied our analysis tools met two constraints: linearity and time invariance. In this chapter we will define both linearity and time invariance and demonstrate how to test a system for these properties. Further we will show that the properties become additional tools we can use in finding the response of systems to various inputs.

These properties will also lead us to another very important analysis technique called convolution. The chapter is concluded by finding system responses using the convolution technique.

Keep in mind that from the very first of the text, our goal has been to develop more and more sophisticated tools that will allow us to find the output (response) across any element of increasing complex systems, given some initial conditions on the system and/or some input to the system. We started with an elementary element (the resistor) and an elementary tool (Ohm's law). Combinations of elements we called circuits and more analysis tools were developed (for example, Thevenin's theorem). As we encountered discrete-time components, we began referring to our collection of elements as a system rather than a circuit and continued to develop more analysis tools.

This chapter will add even more tools to our bag, but it is still the function of the engineer to select the most appropriate tool for any given analysis task.

8.1 Linearity

LEARNING OBJECTIVES

After completing this section you should be able to do the following:

1. Define linearity.
2. Be able to test a given system for linearity.

Consider the system model shown in Fig. 8.1. The input is x, the response is y, and the letter L is used for the mathematical operation relating input to output.

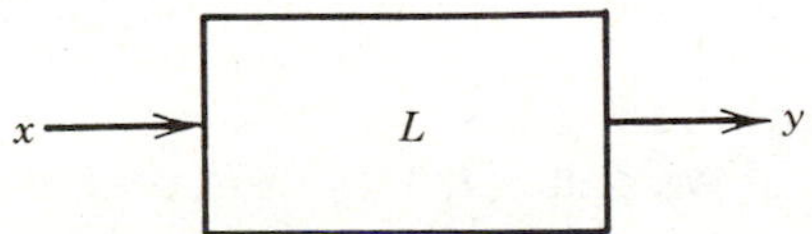

Figure 8.1. The transformation $y = L(x)$.

This is written as

$$y = L(x) \tag{8.1.1}$$

Definition 8.1.1. Linearity. The transformation in Eq. 8.1.1 is linear if

$$L[a_1x_1(t) + a_2x_2(t)] = a_1L[x_1(t)] + a_2L[x_2(t)] \tag{8.1.2}$$

for all $a_1, a_2, x_1(t), x_2(t)$.

This definition embodies two criteria. If $a_1 = a_2 = 1$, then

$$L[x_1(t) + x_2(t)] = L[x_1(t)] + L[x_2(t)] \tag{8.1.3}$$

This property is called *additivity*. If $x_2(t) = 0$, then

$$L[a_1x_1(t)] = a_1L[x_1(t)] \tag{8.1.4}$$

This property is called *homogeneity*. Therefore, a system is linear if it is both additive and homogeneous. Otherwise it is nonlinear.

The additivity and homogeneity properties are closely related. If $x_1 = x_2$ in Eq. 8.1.3 (additivity), then this is the same as $a_1 = 2$ in Eq. 8.1.4 (homogeneity). In practical systems, additivity implies homogeneity. The converse is not true, however, as we show in Example 8.1.2. *Superposition* is another name for additivity. Therefore, in practical systems superposition implies linearity.

The concept of linearity applies to much more than systems. In fact, linearity permeates all of mathematics and the physical sciences. There are no linear systems if we adhere strictly to Definition 8.1.1. Any physical system will have limits such as saturation, maximum power, and so on, in continuous-time systems, and limited word length in discrete-time systems. We assume our mathematical models represent systems that are linear over a defined range of operations.

EXAMPLE 8.1.1 The current–voltage characteristic of a diode can often be approximated by $i(t) = v^2(t)$ for positive voltages as shown in Fig. 8.2. Determine whether or not this system is linear.

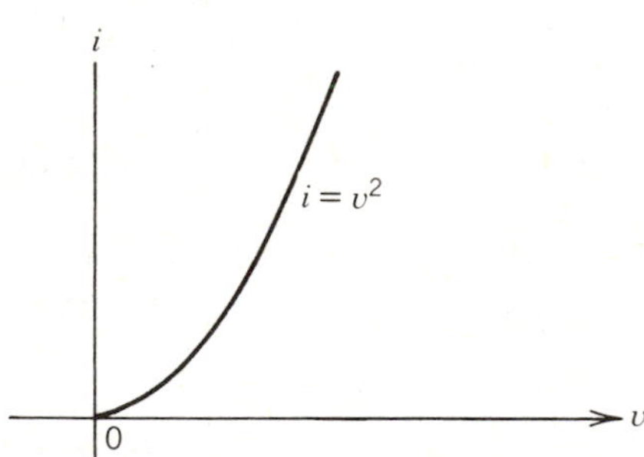

Figure 8.2. Nonlinear diode characteristic.

Solution

If we call $v(t)$ the input and $i(t)$ the output, then the transformation L of Eq. 8.1.1 is given by

$$i(t) = L\{v(t)\} = v^2(t), \qquad v > 0$$

First assume that $v(t) = v_1(t) + v_2(t)$, where both voltages are positive. Then

$$\begin{aligned} L[v(t)] &= L[v_1(t) + v_2(t)] \\ &= v_1^2(t) + 2v_1(t)v_2(t) + v_2^2(t) \\ &= L[v_1(t)] + L[v_2(t)] + 2v_1(t)v_2(t) \end{aligned}$$

The system is not additive because of the term $2v_1(t)v_2(t)$.

Next assume that $v(t) = av_1(t)$. Then

$$L[v(t)] = L[av_1(t)] = a^2L[v_1(t)]$$

so that the system is not homogeneous either. Since the system is neither additive nor homogeneous, it is certainly not linear.

EXAMPLE 8.1.2 Suppose a system consists of the two cascaded units shown in Fig. 8.3. The first unit sums $x^3 + \dot{x}^3$, and the second unit takes the cube root of the sum. Determine if this system is linear.

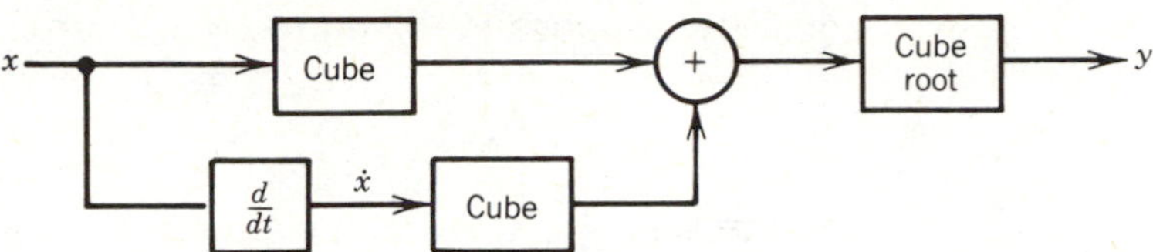

Figure 8.3. A homogeneous but non-additive system.

Solution

The response to input x_1 is

$$y_1 = \sqrt[3]{x_1^3 + \dot{x}_1^3}$$

This system is homogeneous, because if we multiply the input by any constant a, the output is given by

$$y = \sqrt[3]{(ax_1)^3 + (a\dot{x}_1)^3} = a\sqrt[3]{x_1^3 + \dot{x}_1^3} = ay_1$$

This system is not additive, however, for the response to $x_1 + x_2$ is given by

$$y = \sqrt[3]{(x_1 + x_2)^3 + (\dot{x}_1 + \dot{x}_2)^3} \neq y_1 + y_2$$

where y_1 is the response to x_1 and y_2 is the response to x_2. Therefore the system is not linear.

This is an example of a system that is homogeneous but not additive. Thus, a homogeneous system is not necessarily linear. It is true, however, that any practical additive system is also homogeneous.

EXAMPLE 8.1.3 The relationship between input x and output y is shown in Fig. 8.4. Is the system linear?

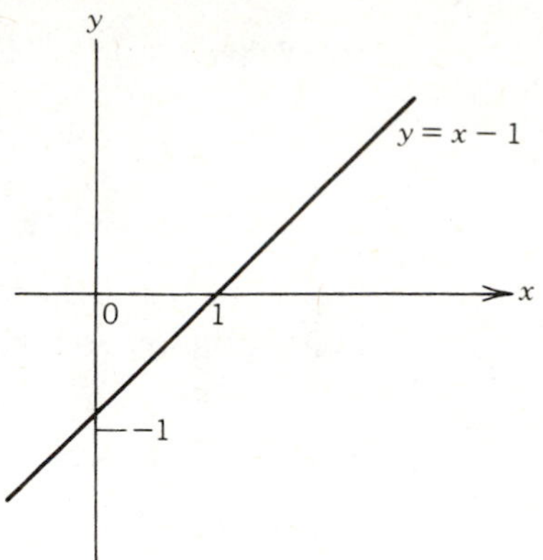

Figure 8.4. A straight-line relationship is not linear unless the line passes through the origin.

Solution

Despite the straight-line relationship, the system is not linear. Testing for additivity, suppose $x_1 = 2$ and $x_2 = 3$. Then

$$y_1 = x_1 - 1 = 1$$

$$y_2 = x_2 - 1 = 2$$

and the sum $y_1 + y_2 = 3$. But suppose $x = x_1 + x_2 = 5$. Then

$$y = x - 1 = 5 - 1 = 4$$

which is not 3 as required by the additivity property. For the system to be linear, the straight line in Fig. 8.4 must pass through the origin.

The concept of linearity applies equally to continuous-time and discrete-time systems. Instead of dealing with transformations of continuous-time signals as in the previous examples, we may apply Definition 8.1.1 to transformations of discrete-time signals. Here are a few examples to illustrate the procedure used to test for linearity in discrete-time systems.

EXAMPLE 8.1.4 A first order system is described by

$$y(k) = 0.2x(k) + 0.8y(k-1), \qquad k > 0$$
$$y(-1) = 0 \tag{8.1.5}$$

The input is x and the output is y. Is the system linear?

Solution

For an input x_1 the response is y_1, and for input x_2, the response is y_2, each found from Eq. 8.1.5 and given by

$$y_1(k) = 0.2x_1(k) + 0.8y_1(k-1), \qquad y_1(-1) = 0$$

$$y_2(k) = 0.2x_2(k) + 0.8y_2(k-1), \qquad y_2(-1) = 0$$

Now suppose that the input is the sum $x_1(k) + x_2(k)$ for all $k > 0$. If we add the above two equations we obtain

$$y_1(k) + y_2(k) = 0.2[x_1(k) + x_2(k)] + 0.8[y_1(k-1) + y_2(k-1)]$$

which indicates that the system is additive and therefore linear.

As a further example, consider a root extraction algorithm that can be modeled as a discrete-time system. We can find the square root of a positive number x by forming a guess y_1 and calculating

$$y_2 = \frac{(x/y_1) + y_1}{2} \tag{8.1.6}$$

The number y_2 will be closer to $\sqrt{x}$ than was our original guess y_1. This process may be repeated, using the previously calculated y_2 as a new guess, resulting in another answer that is even closer to the true value $\sqrt{x}$.

This algorithm works because of the following arguments: (1) Suppose $y_1 = \sqrt{x}$. Then each term on the right side of Eq. 8.1.6 is equal to $\sqrt{x}$, so this formula correctly computes $\sqrt{x}$ if our guess is correct. (2) If $y_1 \neq \sqrt{x}$, then either

$$\frac{x}{y_1} < \sqrt{x} < y_1$$

which occurs if y_1 is too large, or

$$y_1 < \sqrt{x} < \frac{x}{y_1}$$

which occurs if y_1 is too small. In either case, the formula of Eq. 8.1.6 sets y_2 equal to the average of these two terms, and thus y_2 must lie between y_1 and x/y_1.

This algorithm may be extended to the computation of the nth root. For example, the cube root iteration is given by

$$y_2 = \frac{(x/y_1^2) + 2y_1}{3}$$

and for the nth root, by

$$y = \frac{\{[x/y_1^{(n-1)}] + (n-1)y_1\}}{n}$$

This algorithm can be recast in the form of a difference equation. For the square root we have

$$y(k) = \frac{0.5x(k)}{y(k-1)} + 0.5y(k-1) \tag{8.1.7}$$

where $x(k)$ is a step function of height equal to the number whose square root we wish to calculate.

This system is not linear. Suppose $x_1(k) = 2$ for all $k > 0$. Then $y_1(k)$ converges to 1.414 for large k. Now double the input so that $x_2(k) = 4$. The response now converges to 2, that is, $2x_1$ does not produce $2y_1$.

LEARNING EVALUATIONS

1. Define linearity.

2. Is the circuit shown in Fig. 8.5 linear? The input is $v_1(t)$ and the output is $v_2(t)$.

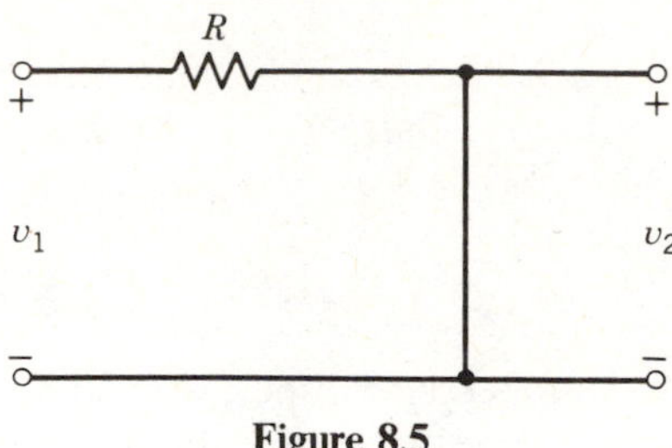

Figure 8.5

3. Test the following systems for linearity. The input is $x(t)$ and the output is $y(t)$ in each case.

a. $y(t) = 2tx(t)$
b. $y(t) = x(0) + 3x(t), t > 0$
c. $y(t) = x(t-1)$
d. $y(t) = x^2(t)\cos t$

8.2 Time Invariance

LEARNING OBJECTIVES

After completing this section you should be able to do the following:

1. Define time invariance.

2. Test a given system for time invariance.

Analog and discrete-time systems are classified either as time varying or as time invariant. All of the systems we have considered in this text so far would be classified as time invariant. This means that the system output is not a function of when the input is applied. A flashlight is an example of a time-invariant system. Whether I turn the switch of the flashlight on now or two days from now the result will be the same. On the other hand, taking a test is a time-varying system where the output (test score) is a function of length of time between exposure to the material covered on the test and the time of the test. Based on this understanding, we will now formalize the definition of time invariance.

Definition 8.2.1. ***Time Invariance.*** The transformation 8.1.1 is time invariant if

$$y(t+\varepsilon) = L[x(t+\varepsilon)] \qquad \text{for all} \quad \varepsilon,\ x \tag{8.2.1}$$

Time invariance means that the response to the input is independent of the time at which the input is applied. If the input is delayed (or advanced) in time, then the output is delayed (or advanced) by the same amount. Otherwise there is no change in the output. For continuous-time systems, the parameter t in Eq. 8.2.1 is a particular value of the continuous variable t. For discrete-time systems we must set $t = t_k$, a particular value of the discrete variable t_k. Examples of time-invariant system models are

1. $y(t) = 2x(t)$

2. $y(t_k)=x^2(t_k)$
3. $y(t)=\int_{-\infty}^{t} x(\lambda)\,d\lambda$
4. $y(k)=\sum_{n=-\infty}^{k} x(n)$

Also, here are some examples of time-varying system models.

5. $y(t)=y(t)\cos\omega_0 t$
6. $y(t)=\int_0^t x(\lambda)\,d\lambda$
7. $y(k)=\sum_{n=0}^{k} x(n)$

That the systems in examples 6 and 7 are time varying, while those in 3 and 4 are not, can be seen by shifting the input in time and computing the output. For example, let us compare the continuous-time systems in 3 and 6. Suppose the system input is the square pulse $x_1(t)$ shown in Fig. 8.6. The response of both

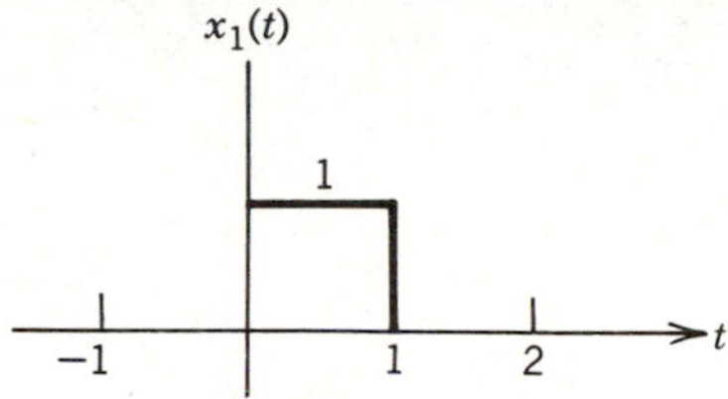

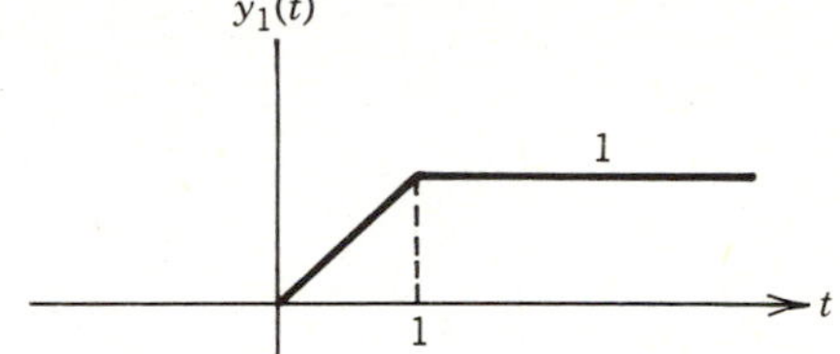

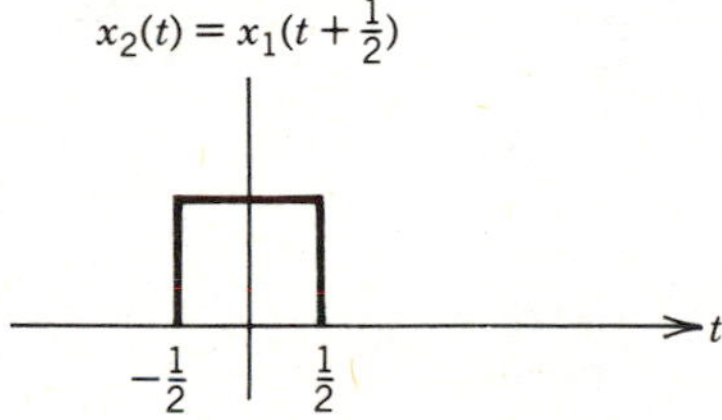

Response of system 3 to $x_2(t)$

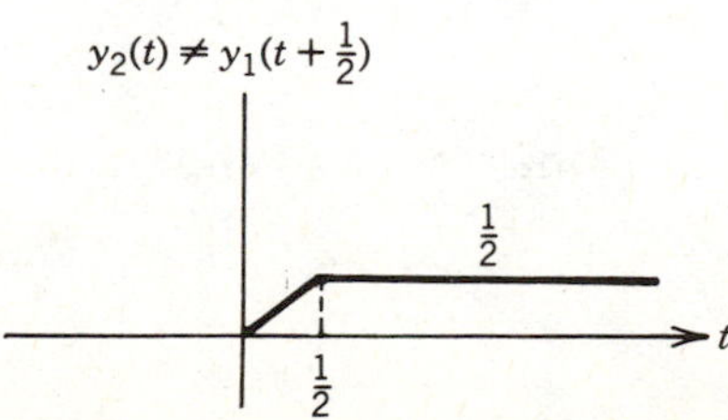

Response of system 6 to $x_2(t)$

Figure 8.6. System 3 is time invariant while system 6 is not.

systems 3 and 6 is $y_1(t)$. But now let us shift the input to the left so that $x_2(t)$ is the input, where $x_2(t) = x_1(t + \frac{1}{2})$. Now the response of system 3 differs from that of system 6. The lower limit of zero on the integral in 6 causes the difference in the response, as shown in the diagram.

In a similar manner, the response of system 7 is affected by the lower limit of zero on the summation. Any portion of the input signal $x(k)$ that occurs before $k = 0$ does not affect the output. Hence, a shift in the input signal around $k = 0$ causes more than a corresponding shift in the response.

The system in example 5 is time varying because the input is multiplied by a function of t. To illustrate this point let us suppose that $\omega_0 = 2\pi$. Then $\cos 2\pi t$ is shown plotted in Fig. 8.7*a*. If the input is $x_1(t)$ shown in Fig. 8.7*b*, then the response is $y_1(t)$ as shown. In Fig. 8.7*c*, the input is shifted 0.5 s to the left. The response $y_2(t)$ is given by $x_2(t) \cos 2\pi t$ which is not $y_1(t + \frac{1}{2})$, as it would be for a time-invariant system.

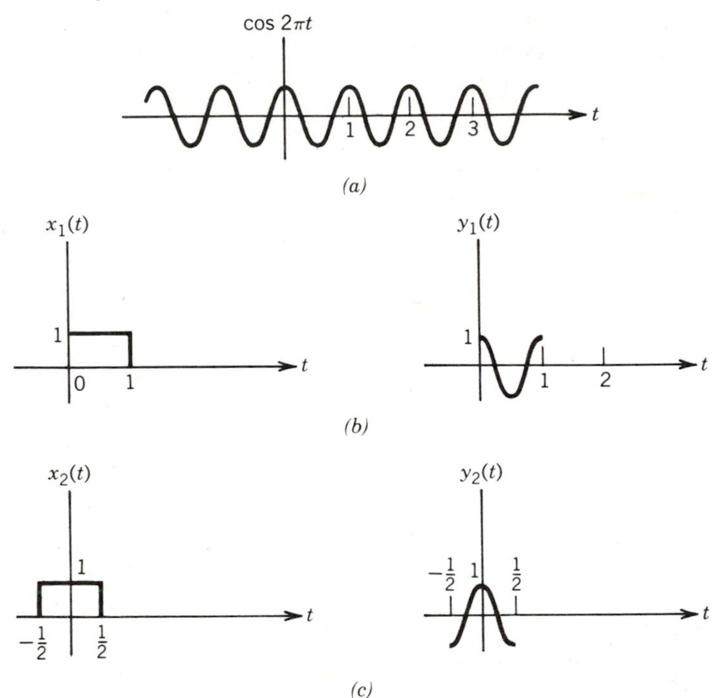

Figure 8.7. System 5 is time varying.

EXAMPLE 8.2.1 Test the system modeled by Eq. 8.2.2 for time invariance. The input is x and the output is y.

$$y(t) = x(0) + 2x(t) \tag{8.2.2}$$

Solution

Suppose that the input signal is $x_1(t)$ shown in Fig. 8.8. Then the response is given by

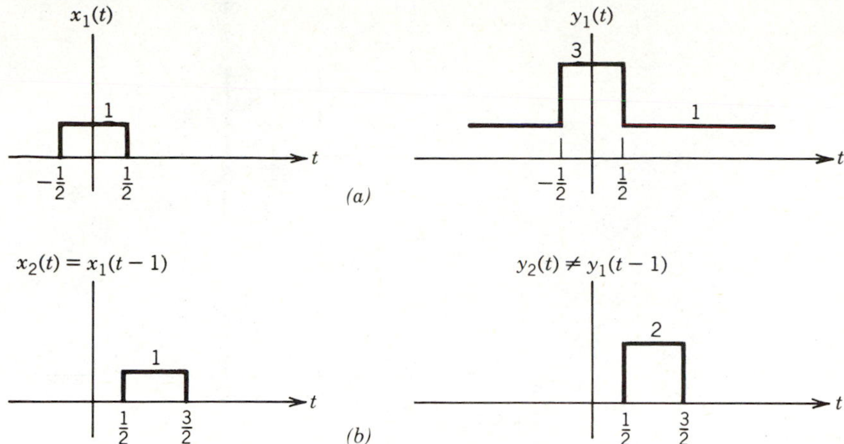

Figure 8.8. Input–output signals for a time-varying system.

$$y_1(t) = 1 + 2x_1(t)$$

as plotted in the figure. In Fig. 8.8*b* the input is shifted to the right by one time unit, but since $x(0)$ is now zero, instead of 1, the response is given by

$$y_2(t) = 2x_2(t)$$

Thus $y_2(t) \neq y_1(t-1)$ and the system is time varying.

EXAMPLE 8.2.2 The square root algorithm of Eq. 8.1.7 is nonlinear. Is it time invariant?

Solution

Yes, for our definition of time invariance implies that any system, when started in its initial state, should be independent of the time at which the input is applied. For the algorithm of Eq. 8.1.7, let us agree to always start the system with initial guess $y = 1.0$. This is pictured in Fig. 8.9. If we wish to find the square root of 2, then $x(k) = 2$, $k = 1, 2, 3, \ldots$. The input and response are pictured in Figs 8.10*a* and 8.10*b*. If we delay the input so that $x(k) = 2$, $k = 3, 4, 5, \ldots$, then the response is also shifted as pictured in Figs. 8.10*c* and 8.10*d*.

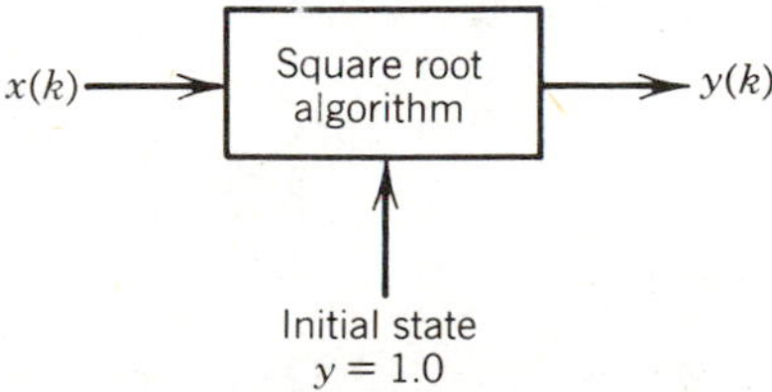

Figure 8.9. The square root algorithm with initial state $y = 1$.

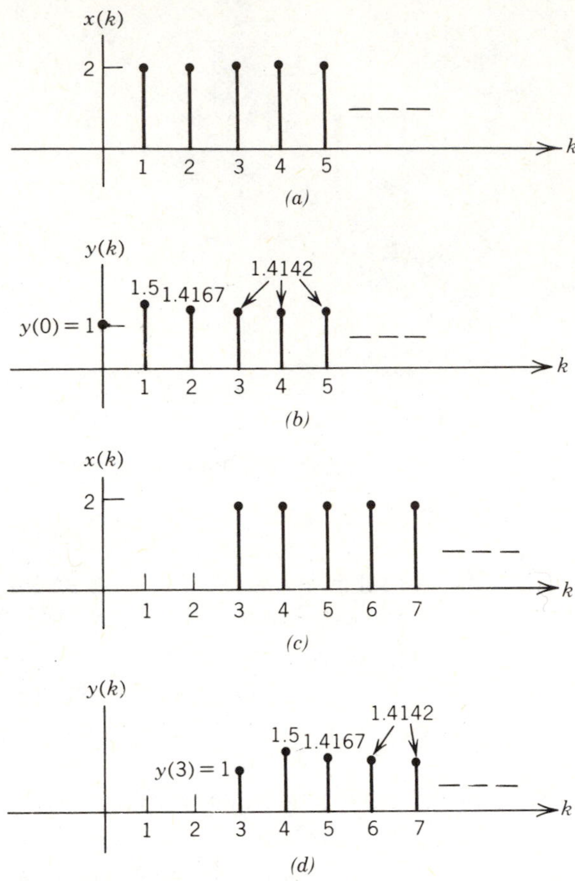

Figure 8.10. Time invariance of the square root algorithm.

LEARNING EVALUATIONS

1. Define time invariance.
2. Test the following systems for time invariance. The input is $x(t)$ and the output is $y(t)$ in each case.
   a. $y(t) = 2tx(t)$
   b. $y(t) = x(0) + 3x(t),\ t > 0$
   c. $y(t) = x(t-1)$
   d. $y(t) = x^2(t)\cos t$

8.3 Use of the LTI Properties

LEARNING OBJECTIVES

After completing this section you should be able to do the following:

1. State the LTI properties.

2. Given an input–output pair, be able to use the LTI properties to find the response to a second input, where the second input is simply related to the first input.

In the previous two sections of this chapter we have shown that there are three properties embodied in our definitions for linearity and time invariance. Any LTI system possesses these three properties. These are the following:

Property 1. Additivity. The sum of two inputs produces the sum of two outputs, that is, $x_1 + x_2$ gives $y_1 + y_2$.

Property 2. Homogeneity. A multiple of the input produces that multiple of the output, that is, ax gives ay.

Property 3. Time invariance. A time shift of the input produces a time shift of the output, that is, $x(t-a)$ gives $y(t-a)$.

We will refer to these as the LTI properties.

The LTI constraint is exceedingly strong, for knowledge of a single input–output pair will (usually) provide enough information so that we can calculate the response to an arbitrary input. Here is an example.

Suppose we are given the input–output pair $x_1(t)$, $y_1(t)$ shown in Fig. 8.11*a*. If we know that the system is LTI we can calculate the response to $x_2(t)$, $x_3(t)$, and $x_4(t)$ by applying the LTI properties.

In Fig. 8.11*b* we see that

$$x_2(t) = 2x_1(t)$$

Therefore, we immediately know that

$$y_2(t) = 2y_1(t)$$

Examination of Fig. 8.11*c* shows that

$$x_3(t) = -x_1(t-1)$$

Hence

$$y_3(t) = -y_1(t-1)$$

Finally in Fig. 8.11*d* we have

$$\begin{aligned} x_4(t) &= 2x_1(t) - x_1(t-1) \\ &= x_2(t) + x_3(t) \end{aligned}$$

so

$$y_4(t) = y_2(t) + y_3(t)$$

as shown.

In this example, we expressed the input $x_4(t)$ as the sum of signals for which we knew the outputs. That is, we were able to find the response to $x_4(t)$ from knowledge of the $x_1(t)$, $y_1(t)$ pair by the LTI properties. Thus, knowledge of a single input–output pair, say $x_1(t)$, $y_1(t)$, will allow us to find the response to an

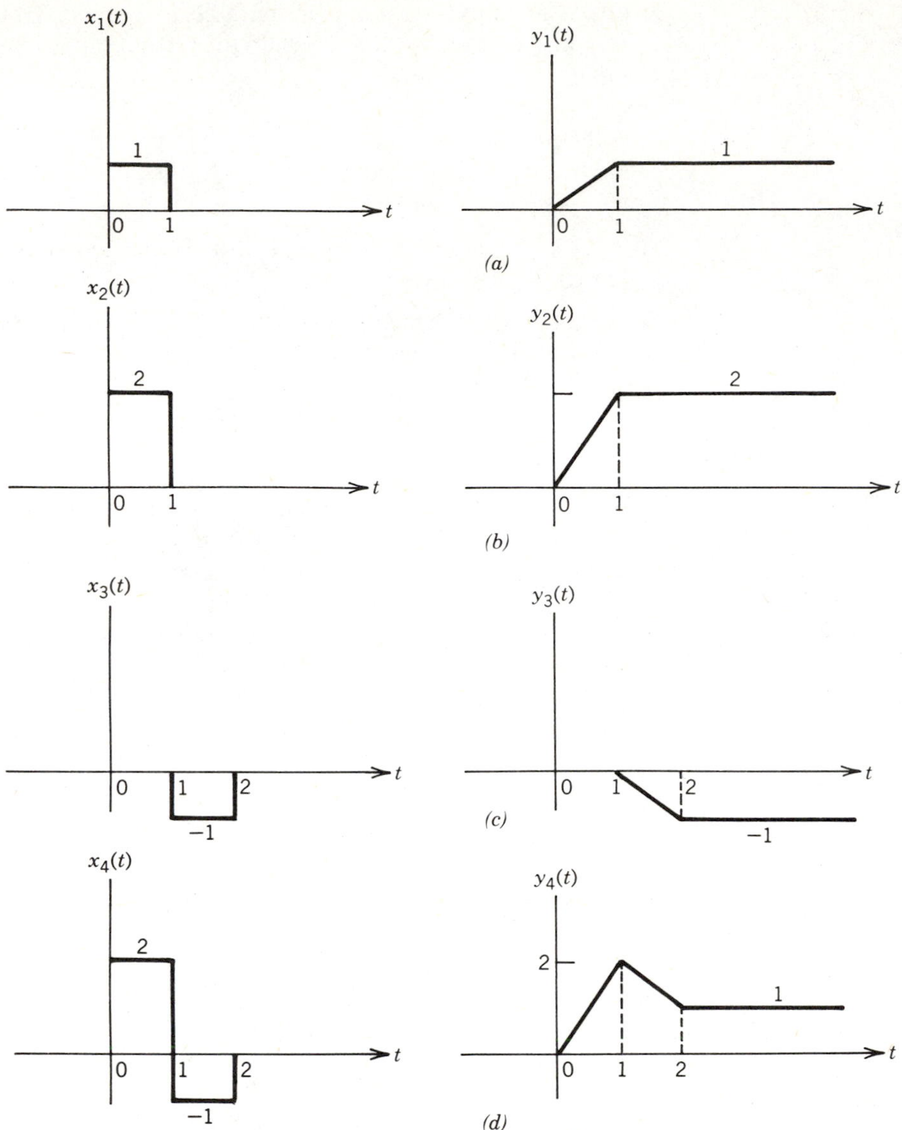

Figure 8.11. Input–output pairs for a continuous-time LTI system.

arbitrary signal, say $x_4(t)$, if we can express $x_4(t)$ as the sum of terms that are related to $x_1(t)$ in some elementary way.

Suppose we wished to characterize our LTI system by an input–output pair. We should choose an input–output pair that is "general" in the sense that most other input–output pairs can be determined from this one. This is possible. In fact, if we know the response to a unit step input, then we can determine the response to any other input. Since the unit impulse is linearly related to the unit step, knowledge of the response of a system to the unit impulse, called the impulse response, will also allow us to determine the response to arbitrary inputs.

For LTI systems, knowledge of almost any input–output pair will allow us to determine the response to an arbitrary input. There are exceptions. Knowledge of the response to a single sinusoidal signal, for example, will be sufficient only for finding the response to any other sinusoid at the same frequency.

There are three closely related methods for finding the response of continuous-time LTI systems to arbitrary input signals. These are convolution, direct solution of differential equations, and transform methods. In essence these three methods are nothing more than convenient ways to sum the response to simple signals so as to find the response to a complicated signal, as we did for the $x_4(t)$, $y_4(t)$ pair in Fig. 8.11.

We have already addressed the problem of finding the system response using the direct solution method. This approach is difficult and somewhat tedious for complicated systems. The convolution approach will be introduced and used shortly in this chapter. The transform method will be used in the last chapters of the text.

It is important to keep all of this in proper perspective. Given a system (e.g., electrical circuit) with some input and/or initial energy storage, we want to be able to find a system output (perhaps the voltage across a capacitor). We have introduced many elements now that can be used to build our system. Some elements are analog and some are discrete. Knowing the input–output relationship across the element allows us to write an equation relating the system input to the system output. Our task, then, is to develop methods that will allow us to solve this describing equation for the system output. All of the systems considered in Chapters 1–7 of the text were LTI systems so the methods now being introduced apply to those systems.

Before we move on to convolution, let us consider some more examples of applying LTI properties to both continuous- and discrete-time systems.

EXAMPLE 8.3.1 An LTI system has the response $y_1(t)$ to the input $x_1(t)$ shown in Fig. 8.12.

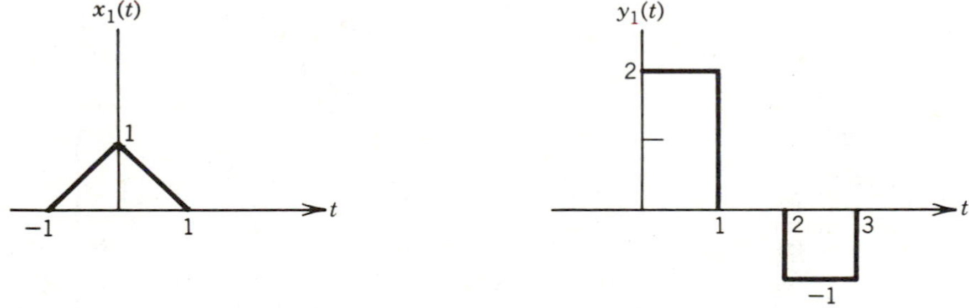

Figure 8.12. The input–output pair x_1, y_1.

a. Sketch the input $2x_1(t-1)$ and the corresponding response.

b. Express the input $x_2(t)$ shown in Fig. 8.13 as the sum

$$x_2(t) = \sum_n a_n x_1(t-n), \qquad n = \cdots -2, -1, 0, 1, 2 \cdots$$

That is, find each value of a_n.

c. Find and sketch $y_2(t)$, the response to $x_2(t)$.

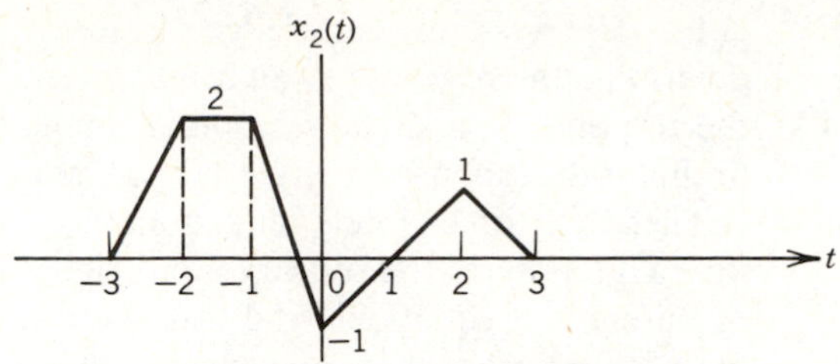

Figure 8.13. The signal $x_2(t)$.

Solution

a. See Fig. 8.14*a*.

b. See Fig. 8.14*b*. This shows $x_2(t)$, which is expressed mathematically by

$$x_2(t) = 2x_1(t+2) + 2x_1(t+1) - x_1(t) + x_1(t-2)$$

Thus $a_{-2} = 2$, $a_{-1} = 2$, $a_0 = -1$, $a_1 = 0$, $a_2 = 1$. All others are zero.

c.

$$\begin{aligned} y_2(t) &= \sum_n a_n y_1(t-n) \\ &= 2y_1(t+2) + 2y_1(t+1) - y_1(t) + y_1(t-2) \end{aligned}$$

as in Fig. 8.14*c*.

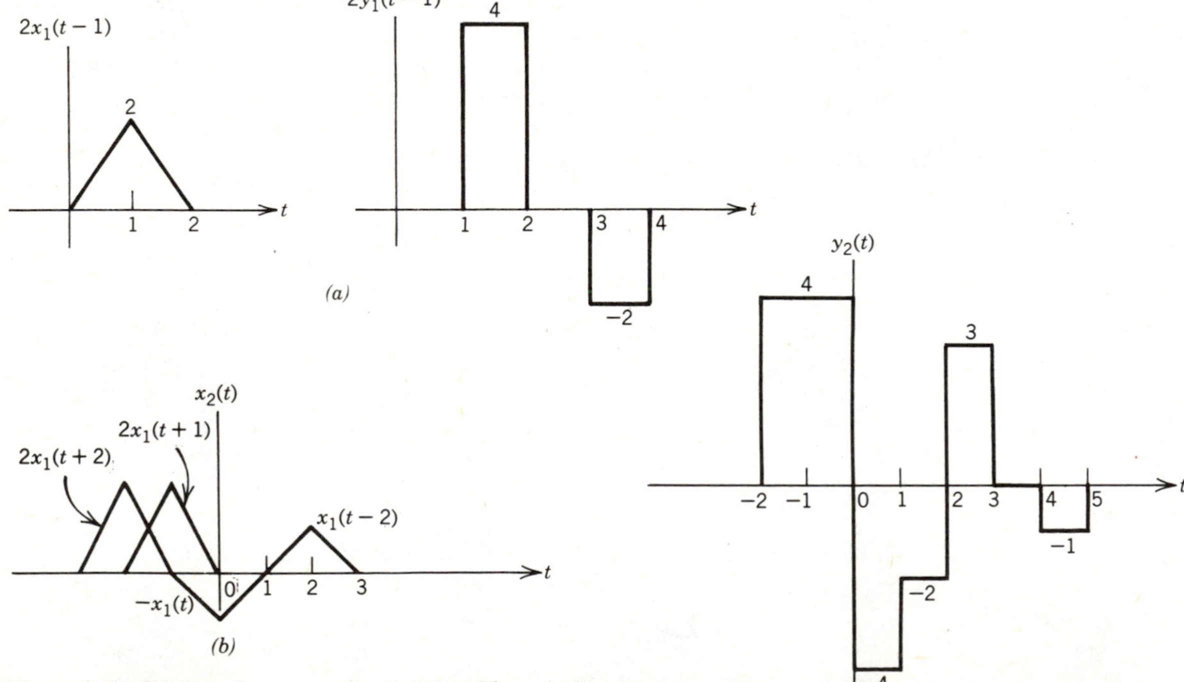

Figure 8.14. Finding the response $y_2(t)$. (a) The solution to a. (b) The solution to b. (c) The solution to c.

The LTI properties play the same role for discrete-time systems as they do for continuous-time systems. That is, knowledge of a single input–output pair will usually allow us to calculate the response to an arbitrary input. As with continuous-time systems, there are three closely related methods for the analysis of discrete-time LTI systems: convolution, difference equations, and transforms. Each of these methods is based on the ability to sum the responses to simple signals so as to find the response to a complicated signal.

In Fig. 8.15 the input–output pair $x_1 - y_1$ is shown. We can find the response to x_4 by applying the LTI properties. That is, since

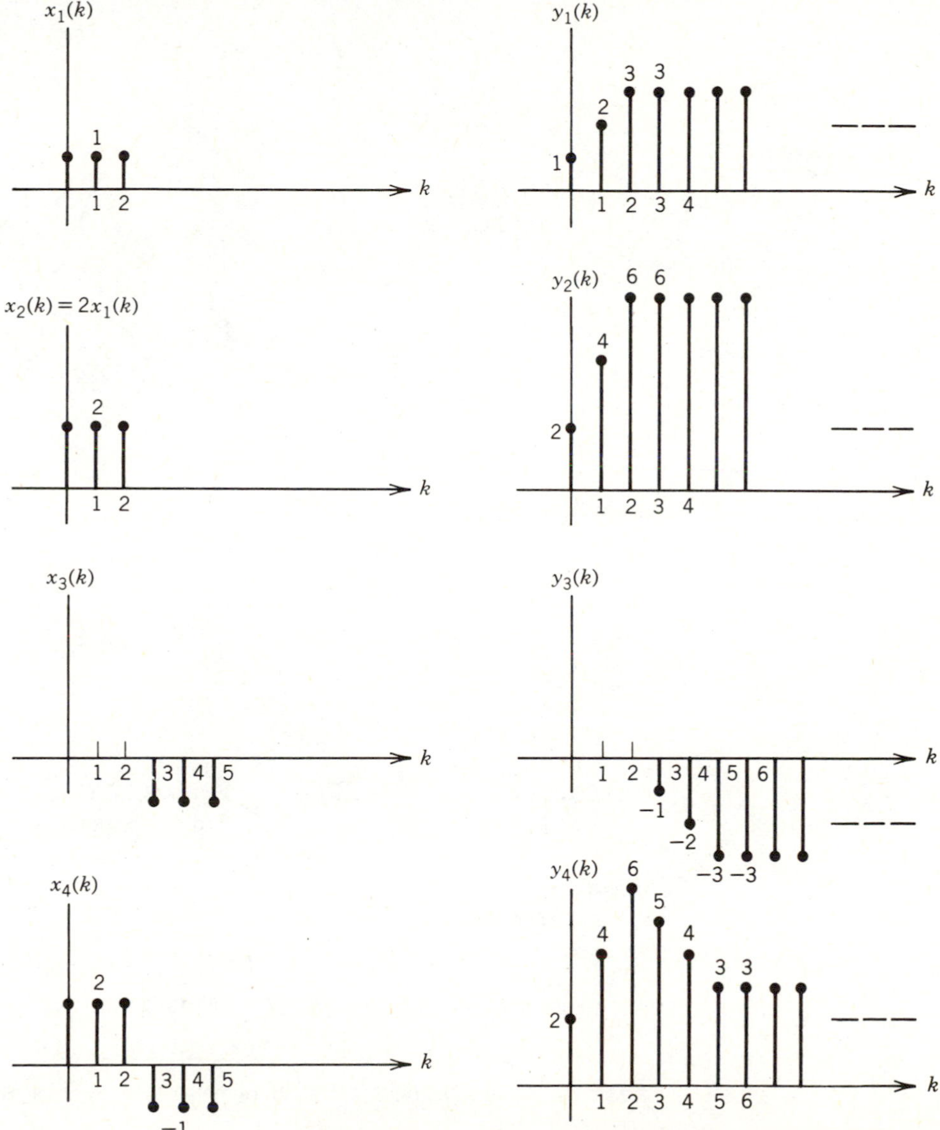

Figure 8.15. Input–output pairs for a discrete-time LTI system.

$$x_4(k) = 2x_1(k) + x_3(k)$$

where x_3 is $-x_1$ shifted to the right by three units, then

$$y_4(k) = 2y_1(k) + y_3(k)$$

This is the same procedure that we followed in Fig. 8.11.

EXAMPLE 8.3.2 An LTI discrete-time system has the response $y_1(k)$ to the input $x_1(k)$ shown in Fig. 8.16. Find the response to the unit alternating sequence $x_2(k)$.

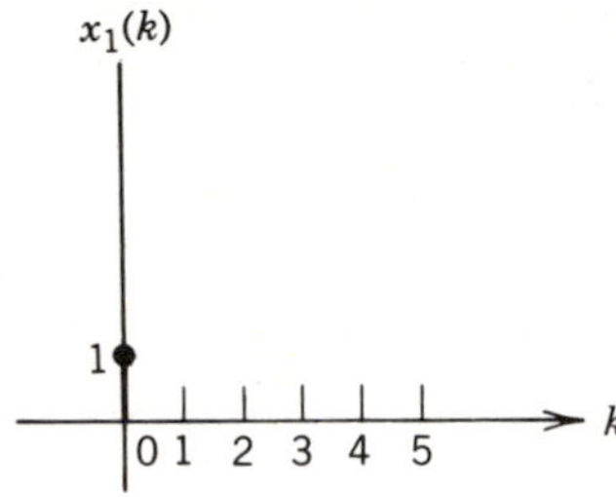

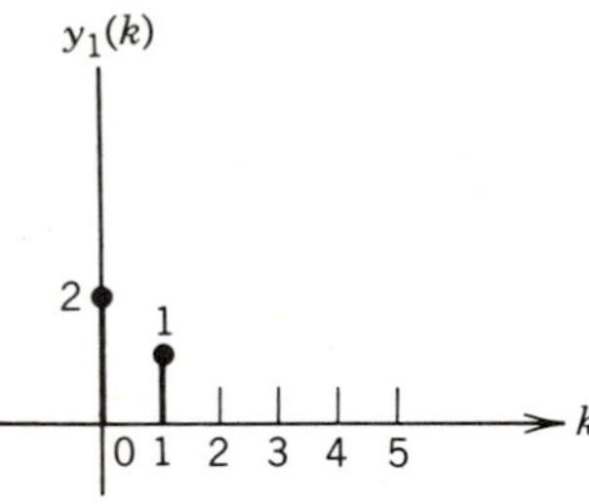

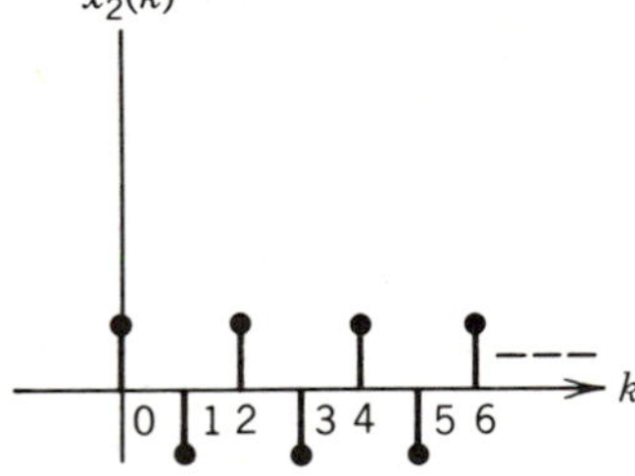

Figure 8.16. Given the x_1, y_1 pair, find the response to x_2.

Solution

We can express $x_2(k)$ in terms of $x_1(k)$ as

$$x_2(k) = x_1(k) - x_1(k-1) + x_1(k-2) - x_1(k-3) + \cdots$$

Therefore, by the LTI properties, $y_2(k)$ is given by

$$y_2(k) = y_1(k) - y_1(k-1) + y_1(k-2) - y_1(k-3) + \cdots$$

These terms are plotted in Fig. 8.17 followed by their sum $y_2(k)$.

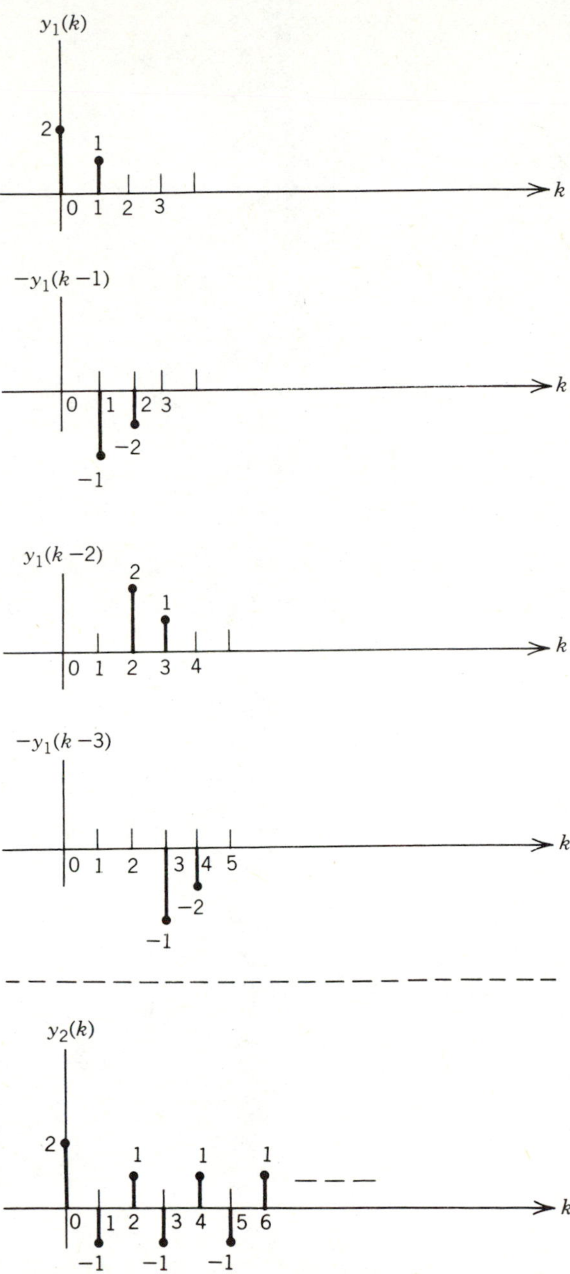

Figure 8.17. Application of the LTI properties to find y_2.

EXAMPLE 8.3.3 The input–output pair for an LTI system is shown in Fig. 8.18 as $x_1(t), y_1(t)$. Find the response to $x_2(t)$.

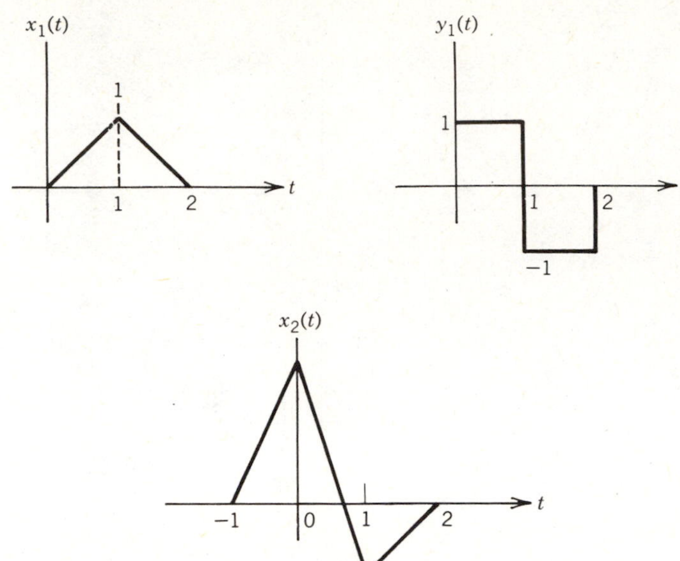

Figure 8.18. Given the x_1, y_1 pair, find the response to x_2.

Solution

See Fig. 8.19. The functions $x_2(t)$ and $y_2(t)$ are given by

$$x_2(t) = 2x_1(t+1) - x_1(t)$$

$$y_2(t) = 2y_1(t+1) - y_1(t)$$

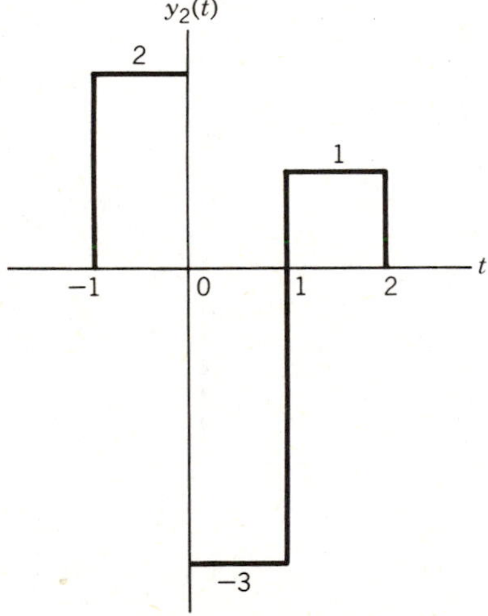

Figure 8.19. The response y_2.

EXAMPLE 8.3.4 The output of an LTI system is $y_1(t)$ if the input is $x_1(t)$ shown in Fig. 8.20. Find the response to $x_2(t)$.

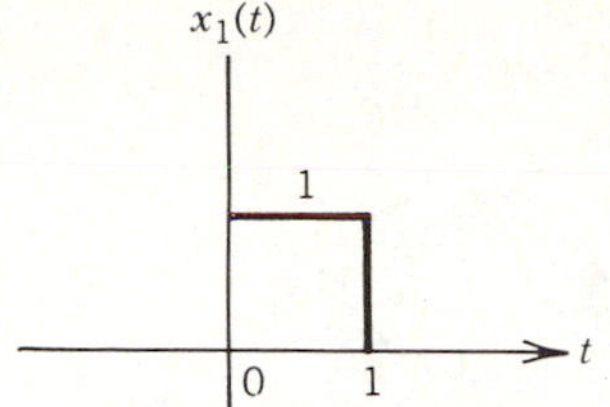

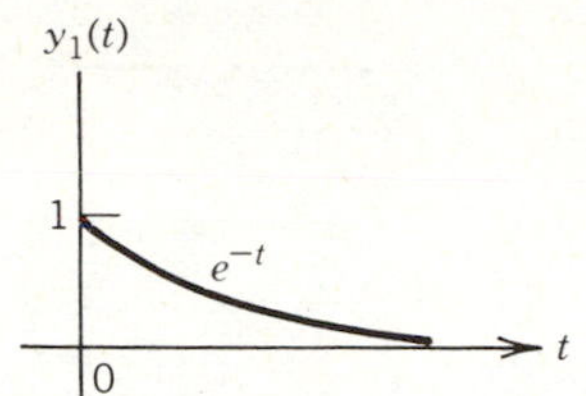

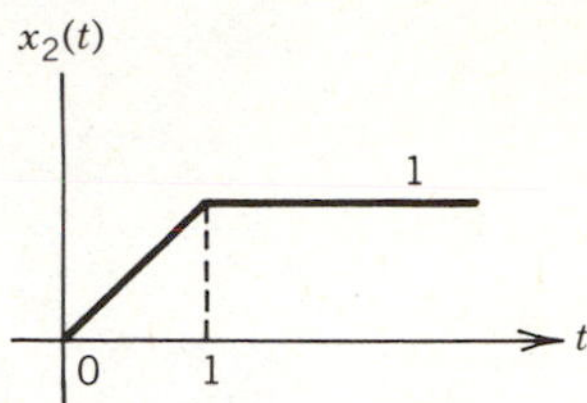

Figure 8.20. Given the x_1, y_1 pair, find the response to x_2.

Solution

To use the method we have developed we must be able to express $x_2(t)$ as the sum of signals simply related to $x_1(t)$. Obviously this cannot be done as in the previous examples. But notice that $x_2(t)$ is the integral of $x_1(t)$.

$$x_2(t) = \int_{-\infty}^{t} x_1(\lambda)\, d\lambda$$

Therefore $y_2(t)$, the response to $x_2(t)$, must be the integral of $y_1(t)$.

$$y_2(t) = \int_{-\infty}^{t} y_1(\lambda)\, d\lambda = 1 - e^{-t}, \qquad t > 0$$

as shown in Fig. 8.21.

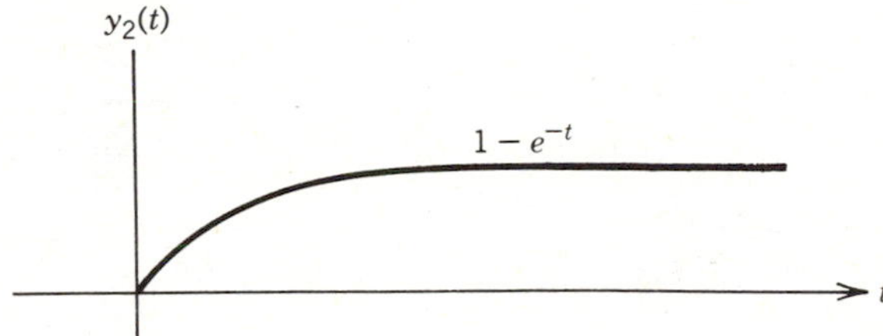

Figure 8.21. The response y_2.

Integration and differentiation are linear operations. It therefore follows that for an LTI system, integration (differentiation) of the input produces an output that is the integral (derivative) of the original output. That is, if x_1, y_1 is an input–output pair, then $x_2 = \int_{-\infty}^{t} x_1\, d\lambda$ and $y_2 = \int_{-\infty}^{t} y_1\, d\lambda$ form an input–output pair. Also $x_3 = (d/dt)x_1(t)$ and $y_3 = (d/dt)y_1(t)$ form an input–output pair.

The above examples are, for the most part, impractical, for it would be difficult to construct actual LTI systems that are characterized by the input–output pairs given. Here are some practical examples.

EXAMPLE 8.3.5 When a unit step voltage $x_1(t) = u(t)$ is applied to the LTI circuit in Fig. 8.22, the response is

$$y_1(t) = \left\{\tfrac{1}{2} - \tfrac{1}{2}e^{-10t}\right\} u(t)$$

a. Find the response to a unit ramp, $x_2(t) = tu(t)$.

b. Find the response to a unit impulse, $x_3(t) = \delta(t)$.

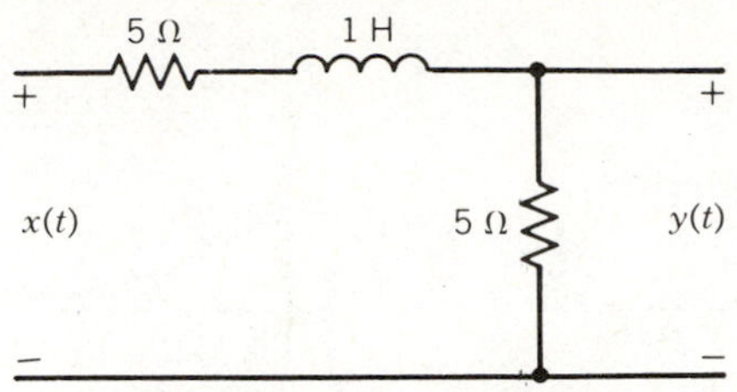

Figure 8.22. An LTI circuit.

Solution

a. Since $x_2(t) = \int_{-\infty}^{t} x_1(\lambda)\, d\lambda$, the response is given by

$$y_2(t) = \int_{-\infty}^{t} y_1(\lambda)\, d\lambda = \int_0^t \left(\frac{1}{2} - \frac{1}{2} e^{-10\lambda}\right) d\lambda$$

$$= \left(\frac{1}{2}\lambda + \frac{1}{20} e^{-10\lambda}\right)\Big|_0^t$$

$$= \left[\frac{t}{2} - \frac{1}{20}(1 - e^{-10t})\right] u(t)$$

b. With $x_3(t)$ the derivative of $x_1(t)$ we differentiate $y_1(t)$ to obtain

$$y_3(t) = \frac{d}{dt} y_1(t) = 5e^{-10t} u(t)$$

EXAMPLE 8.3.6 The response of the first order LTI system of Eq. 8.1.5 to a unit step $x_1(k) = u(k)$ is $y_1(k) = (1 - 0.8^k)u(k)$. Find the response to $x_2(k) = ku(k)$.

Solution

Since

$$x_2(k) = \sum_{j=-\infty}^{k} x_1(j)$$

the response is given by

$$y_2 = \sum_{j=-\infty}^{k} y_1(j) = \sum_{j=0}^{k} [1 - (0.8)^j] u(k)$$

$$= \left(k - \sum_{j=0}^{k} 0.8^k\right) u(k)$$

LEARNING EVALUATIONS

1. State the LTI properties
2. One input–output pair is $p(t) - q(t)$ shown in Fig. 8.23. Given that the system is LTI, find the output if $x(t)$ is the input.

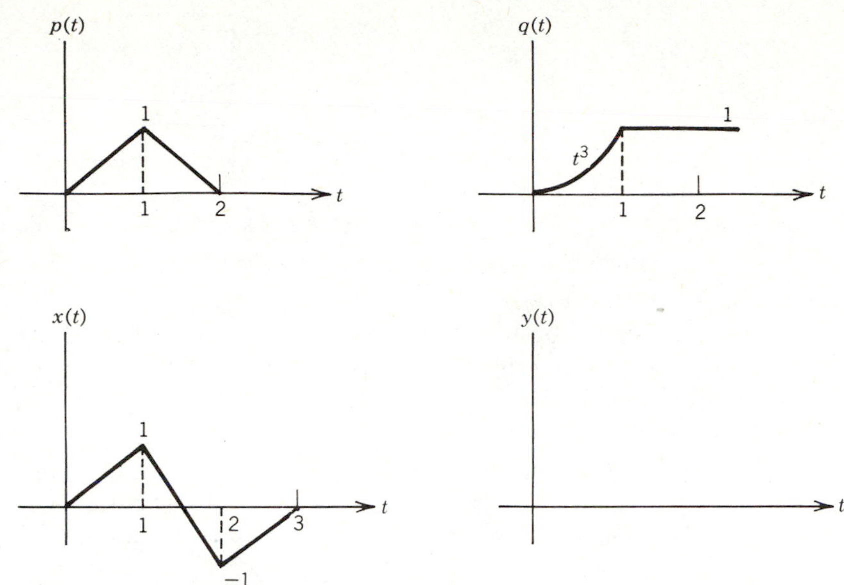

Figure 8.23

8.4 Convolution

LEARNING OBJECTIVES

After completing this section you should be able to do the following:

1. Evaluate convolution integrals.
2. Evaluate convolution summations.

One method used to determine the output of a system (either analog or digital) when the input is known is convolution. This method of solution has existed for a long time but was not popular until the advent of the digital computer. Convolution is the easiest of all methods of solution to implement on a computer. Consequently it is very important that engineers be familiar with this technique.

We will first introduce convolution by use of continuous systems, then show that this method applies also to discrete systems.

Convolution is a binary operation. A binary operation maps an ordered pair of elements from a set into a single element of the set. Given a set S, we select two elements from S (an ordered pair) and these two elements are used to produce a third element in the set S. For example, if S is the set of real numbers, then addition is a binary operation on S. Two numbers, say 2 and 4, are selected. Their sum produces a third element of S, the number 6.

A black box with two inputs and one output is a good illustration of this concept. The binary operation "addition" is illustrated in Fig. 8.24a where $c = a + b$, and "multiplication" is illustrated in Fig. 8.24b where $c = a \times b$. There is no reason to restrict the inputs to be numbers. Convolution is a binary

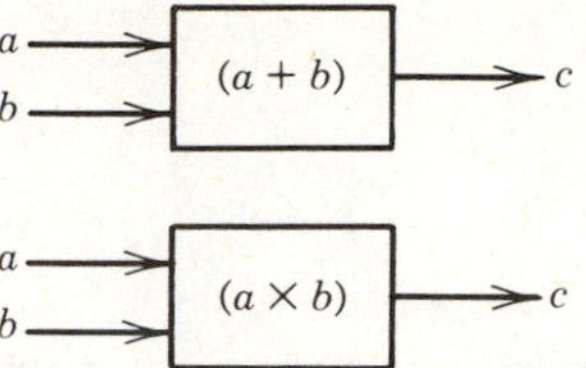

Figure 8.24. Addition and multiplication are binary operations.

operation where the two inputs are functions and the output is a function. The symbol $*$ is used for convolution, just as $+$ and $\times$ are used for addition and multiplication.

Continuous-Time Convolution

Convolution is defined in Eq. 8.4.1. Thus the two inputs to the black box are $f_1(t)$ and $f_2(t)$, and the output is $f_3(t)$ as shown in Fig. 8.25.

$$f_3(t) = f_1(t) * f_2(t) \tag{8.4.1a}$$

$$= \int_{-\infty}^{\infty} f_1(\lambda) f_2(t-\lambda)\, d\lambda \tag{8.4.1b}$$

$$= \int_{-\infty}^{\infty} f_1(t-\lambda) f_2(\lambda)\, d\lambda \tag{8.4.1c}$$

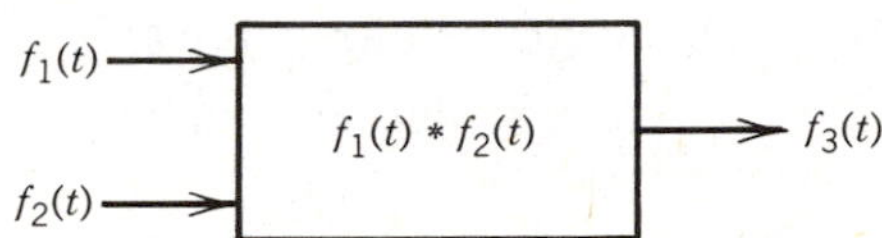

Figure 8.25. Convolution is a binary operation.

Equation 8.4.1b states that in order to find $f_3(t)$, we must perform the following steps:

1. Find $f_1(\lambda)$. This is accomplished by simply substituting λ for t in the expression of $f_1(t)$.
2. Find $f_2(t-\lambda)$. Again simply substitute $(t-\lambda)$ for t in the expression of $f_2(t)$.
3. Multiply $f_1(\lambda)$ and $f_2(t-\lambda)$ together and integrate over all λ.
4. Repeat steps 1–3 for all possible values of t.

If we want to use Eq. 8.4.1c, we simply interchange f_1 and f_2 in the foregoing steps. The choice between Eqs. 8.4.1b and 8.4.1c is based on the functions $f_1(t)$ and $f_2(t)$. If $f_2(t)$ is a less complicated function than $f_1(t)$, then Eq. 8.4.1b is most often used. If the reverse is true, it is usually to your advantage to use Eq. 8.4.1c.

We now illustrate the evaluation of Eq. 8.4.1 with several examples, beginning with the simplest cases and progressing to more difficult situations.

EXAMPLE 8.4.1 Convolve the two step functions shown in Fig. 8.26.

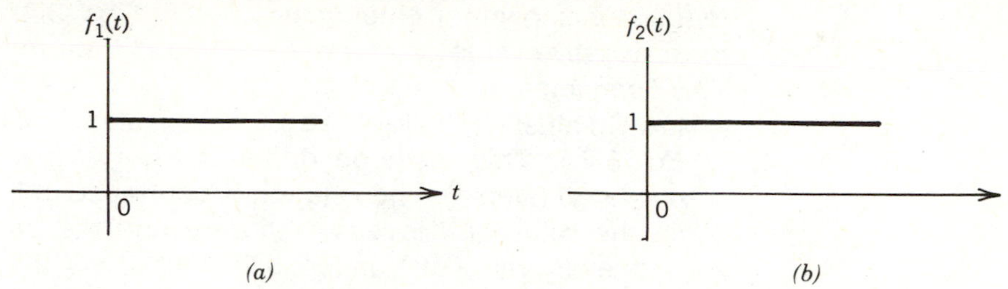

Figure 8.26. Two step functions to be convolved.

Solution

Since $f_1(t)$ and $f_2(t)$ are identical, we will arbitrarily choose to use Eq. 8.4.1b. Therefore $f_1(\lambda)$ and $f_2(t-\lambda)$ must be found. To find $f_1(\lambda)$ is easy; we simply replace t by λ in Fig. 8.26*a*.

To find $f_2(t-\lambda)$, we can write

$$f_2(t) = \begin{cases} 1 & t>0 \\ 0 & t<0 \end{cases}$$

Then substitute $(t-\lambda)$ for t to get

$$f_2(t-\lambda) = \begin{cases} 1 & t-\lambda>0 \\ 0 & t-\lambda<0 \end{cases}$$

A preferred form is

$$f_2(t-\lambda) = \begin{cases} 1 & t>\lambda \\ 0 & t<\lambda \end{cases}$$

This same result can be obtained graphically. We begin in Fig. 8.27*a* by plotting $f_2(\lambda)$ versus λ. This function is "flipped" in Fig. 8.27*b* to obtain

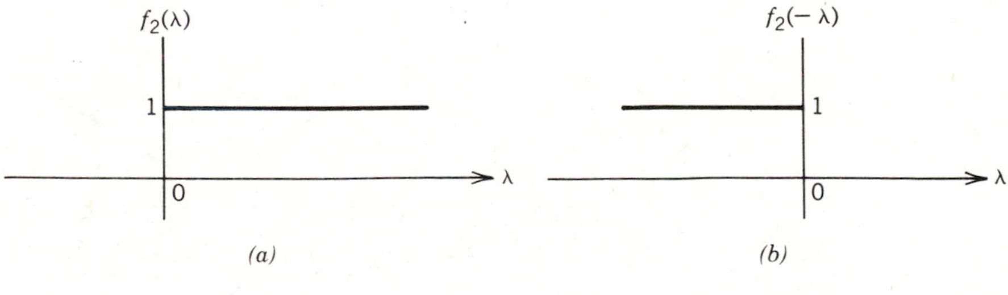

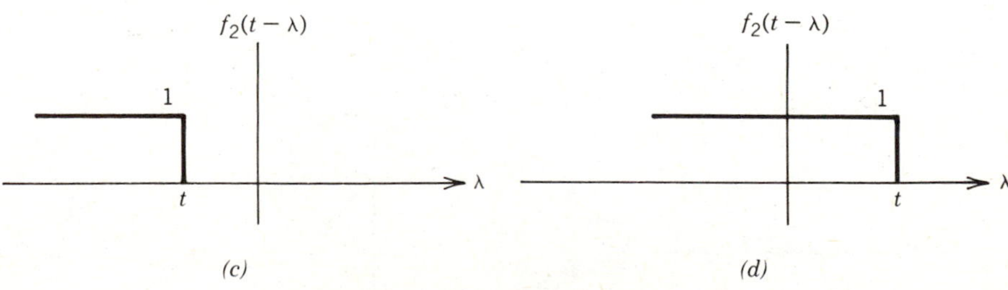

Figure 8.27. Flipping and slipping f_2.

$f_2(-\lambda)$. Note that $f_2(-\lambda) = f_2(0-\lambda)$, or this is $f_2(t-\lambda)$ if $t=0$. This same function is shown for other values of t in Figs. 8.27*c* and Fig. 8.27*d*, first for a negative value of t (say, $t=-1$) and then for a positive value of t. This completes steps 1 and 2.

Now we must multiply $f_1(\lambda)$ by $f_2(t-\lambda)$, step 3, and integrate according to Eq. 8.4.1. This must be done for every value of t in the interval $-\infty < t < \infty$ (step 4). The solution is continued in Fig. 8.28, where in Fig. 8.28*a* the value of t is less than zero and the product $f_1(\lambda)f_2(t-\lambda)$ is zero for every value of λ. In Fig. 8.28*b* with $t>0$, the product $f_1(\lambda)f_2(t-\lambda)$ is equal to 1 for $0<\lambda<t$ and zero elsewhere. Thus $f_3(t)$ is given by

$$f_3(t) = \int_{-\infty}^{0} 0\,d\lambda + \int_0^t 1\,d\lambda + \int_t^{\infty} 0\,d\lambda = t, \qquad t>0$$

This is plotted in Fig. 8.28*c* and we see that the convolution of two unit steps results in the unit ramp.

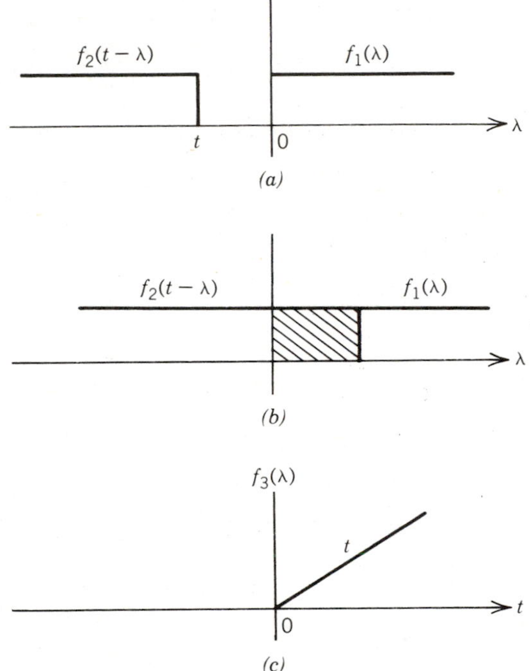

Figure 8.28. The process of convolving two step functions.

EXAMPLE 8.4.2 Convolve the two functions shown in Fig. 8.29.

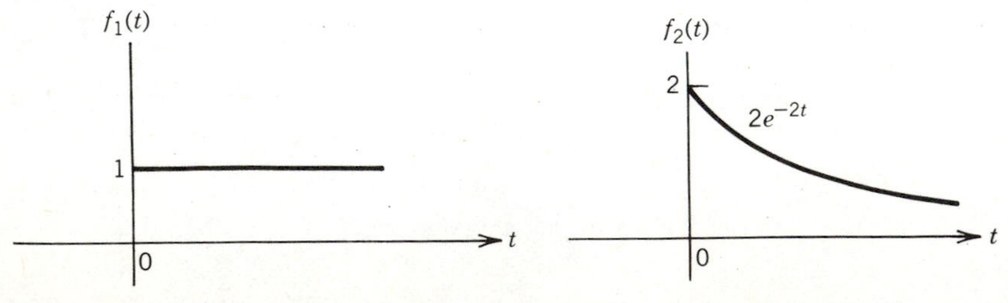

Figure 8.29. A step and an exponential function to be convolved.

Solution

We will graphically flip and slip $f_2(t)$ to illustrate how this is done. [It would be easier to flip and slip $f_1(t)$ since it is a simpler function than $f_2(t)$ and we would normally choose the easier approach.] Therefore, the following formula will be used.

$$f_3(t) = \int_{-\infty}^{\infty} f_1(\lambda) f_2(t-\lambda)\, d\lambda$$

Study Fig. 8.30. We proceed from $f_2(\lambda)$ in Fig. 8.30*a* to $f_2(-\lambda)$ in Fig. 8.30*b* to $f_2(t-\lambda)$ in Figs. 8.30*c* and 8.30*d*. To evaluate $f_3(t)$ we note that $f_1(\lambda) f_2(t-\lambda)$ is zero for all $t<0$. Therefore, $f_3(t)$ is zero for all $t<0$. Now for $t>0$ we have (Fig. 8.30*d*)

$$f_3(t) = \int_0^t 2e^{-2(t-\lambda)}\, d\lambda = 2e^{-2t} \int_0^t e^{-2\lambda}\, d\lambda = 1 - e^{-2t}, \qquad t>0$$

$f_3(t)$ is plotted in Fig. 8.30*e*.

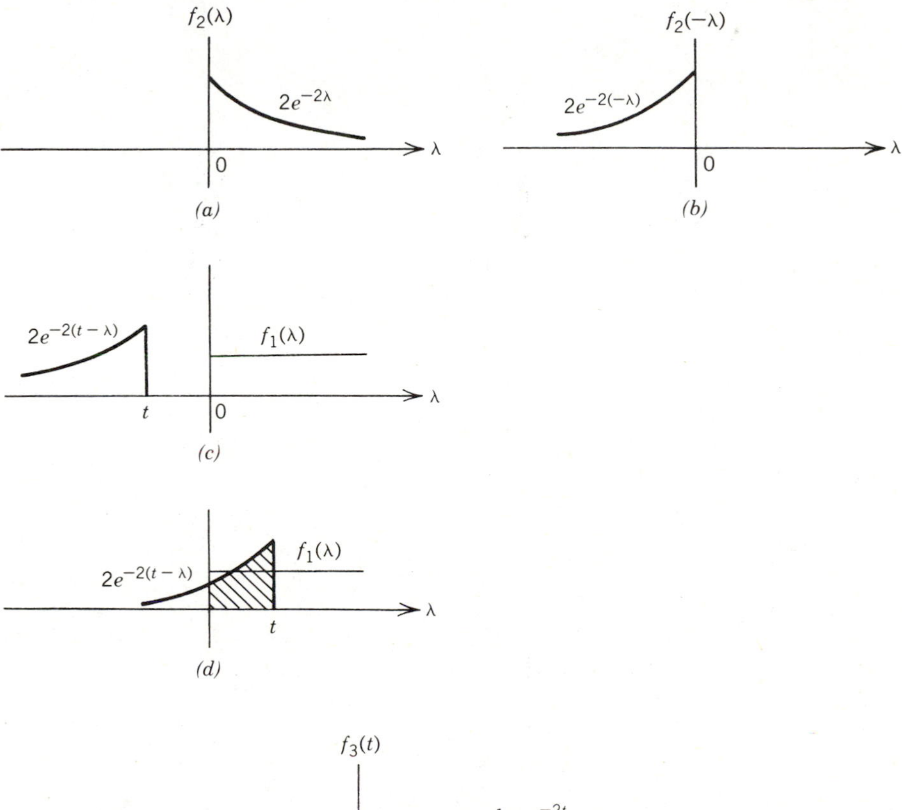

Figure 8.30. The convolution of a step and exponential function.

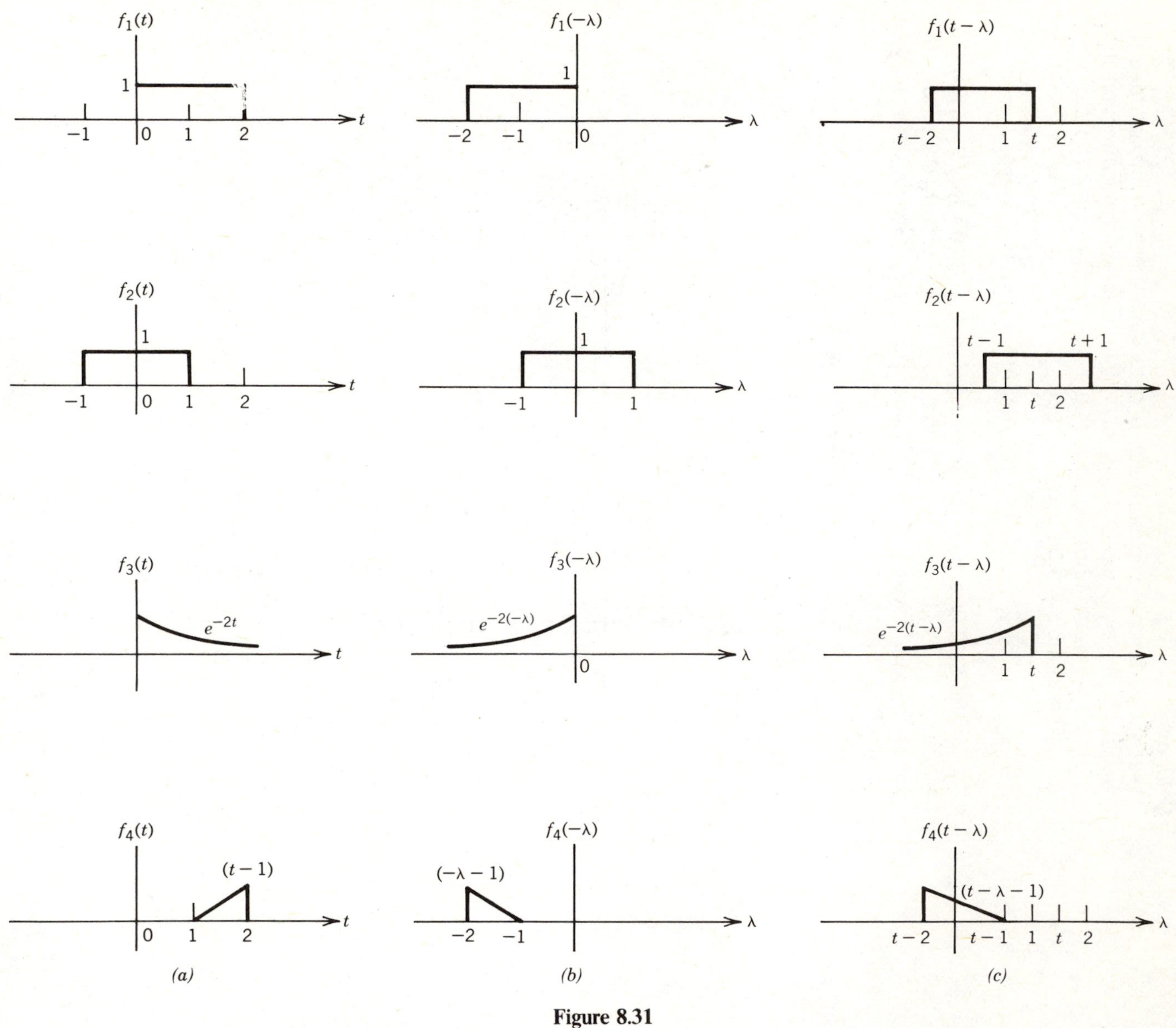

f1(t)
f1(−λ)
f1(t − λ)
f2(t)
f2(−λ)
f2(t − λ)
t − 1
t + 1
f3(t)
f3(−λ)
f3(t − λ)
e^{−2t}
e^{−2(−λ)}
e^{−2(t−λ)}
f4(t)
f4(−λ)
f4(t − λ)
(t − 1)
(−λ − 1)
(t − λ − 1)
t − 2
(a)
(b)
(c)

Figure 8.31

You must be able to both plot and write the expression for $f(t-\lambda)$ so that the limits of integration can be established and the integral evaluated.

EXAMPLE 8.4.3 For each function $f(t)$ shown in Fig. 8.31*a* plot $f(t-\lambda)$ versus λ for a value of t given by $t=1.5$. Also label the figures with the correct equation.

Solution

In each graph of Fig. 8.31*a* we plot $f(-\lambda)$ in Fig. 8.31*b* and then $f(t-\lambda)$ for $t=1.5$ in Fig. 8.31*c*.

EXAMPLE 8.4.4 Convolve the functions f_1 and f_2 shown in Figs. 8.32*a* and 8.32*b*.

Solution

Our first task is to choose either Eq. 8.4.1b or 8.4.1c. Since $f_2(t)$ is less complex than $f_1(t)$ we choose Eq. 8.4.1b. Now we must find $f_2(t-\lambda)$. From Fig. 8.32*b*, we can write

$$f_2(t) = -3, \qquad 1 < t < 2$$

so

$$f_2(t-\lambda) = -3, \qquad 1 < t-\lambda < 2$$

or

$$f_2(t-\lambda) = -3, \qquad -2 < \lambda - t < -1$$

Finally

$$f_2(t-\lambda) = -3, \qquad t-2 < \lambda < t-1$$

Figure 8.32*c* plots $f_2(\lambda)$ and $f_2(t-\lambda)$ for $t<1$. The functions do not overlap so their product, and consequently the integral of their product, is zero.

For $1 < t < 2$, the situation shown in Fig. 8.32*d* exists. From observation we can write

$$\begin{aligned} f_3(t) &= \int_0^{t-1} (-3)2e^{-2\lambda}\, d\lambda \qquad 1 < t < 2 \\ &= 3[e^{2(1-t)} - 1] \qquad 1 < t < 2 \end{aligned}$$

For the case where $t > 2$, we use Fig. 8.32*e* and write

$$\begin{aligned} f_3(t) &= \int_{t-2}^{t-1} (-3)2e^{-2\lambda}\, d\lambda, \qquad t > 2 \\ &= 3\{e^2 - e^4\}e^{-2t}, \qquad t > 2 \end{aligned}$$

The complete function $f_3(t)$ is plotted in Fig. 8.32*f*.

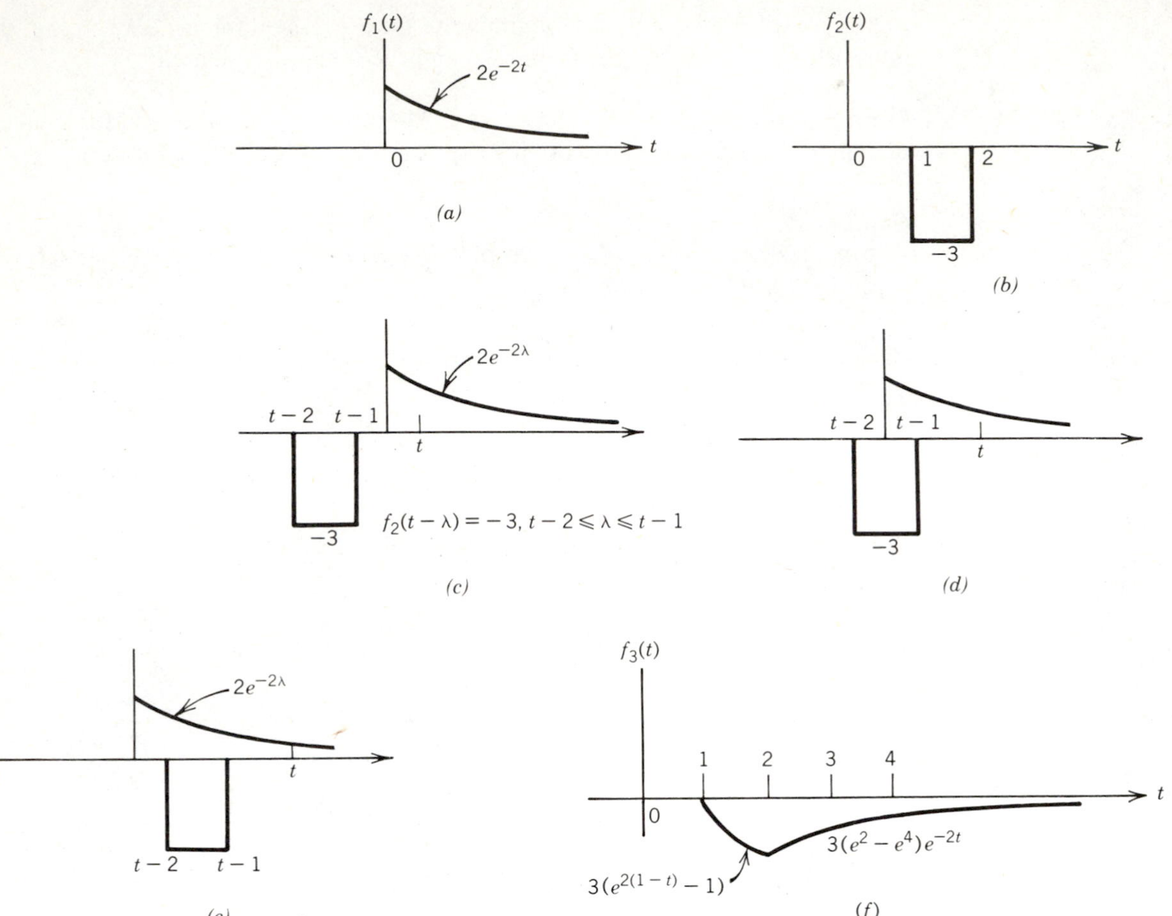

Figure 8.32. Convolution where one of the functions assumes negative values.

In progressing from simple to more complex cases of convolution we have had to evaluate more and more integrals. In Examples 8.4.1 and 8.4.2 there were only two integrals, and one of those was zero. In this last example, there were three separate integrals to evaluate, with one equal to zero. If we convolve f_2 with f_4 in Fig. 8.31, for example, five integrals must be evaluated, with two of them equal to zero. The point we are making here is that the number of regions of t for which different integrals must be evaluated can be any number, from 1 on up. This is understandable since the convolution integral must be evaluated for each value of t for $-\infty < t < \infty$.

A very important concept involves convolution where one function is the impulse function. This is an intermediate step from convolution of continuous signals to convolution of discrete-time signals. An impulse function may be considered either a continuous signal that is zero for all values of time except one, or it may be considered a discrete-time signal where all values in the sequence are zero, save one. We will consider two examples involving impulse functions.

EXAMPLE 8.4.5 Convolve the two signals shown in Fig. 8.33*a* and Fig. 8.33*b*.

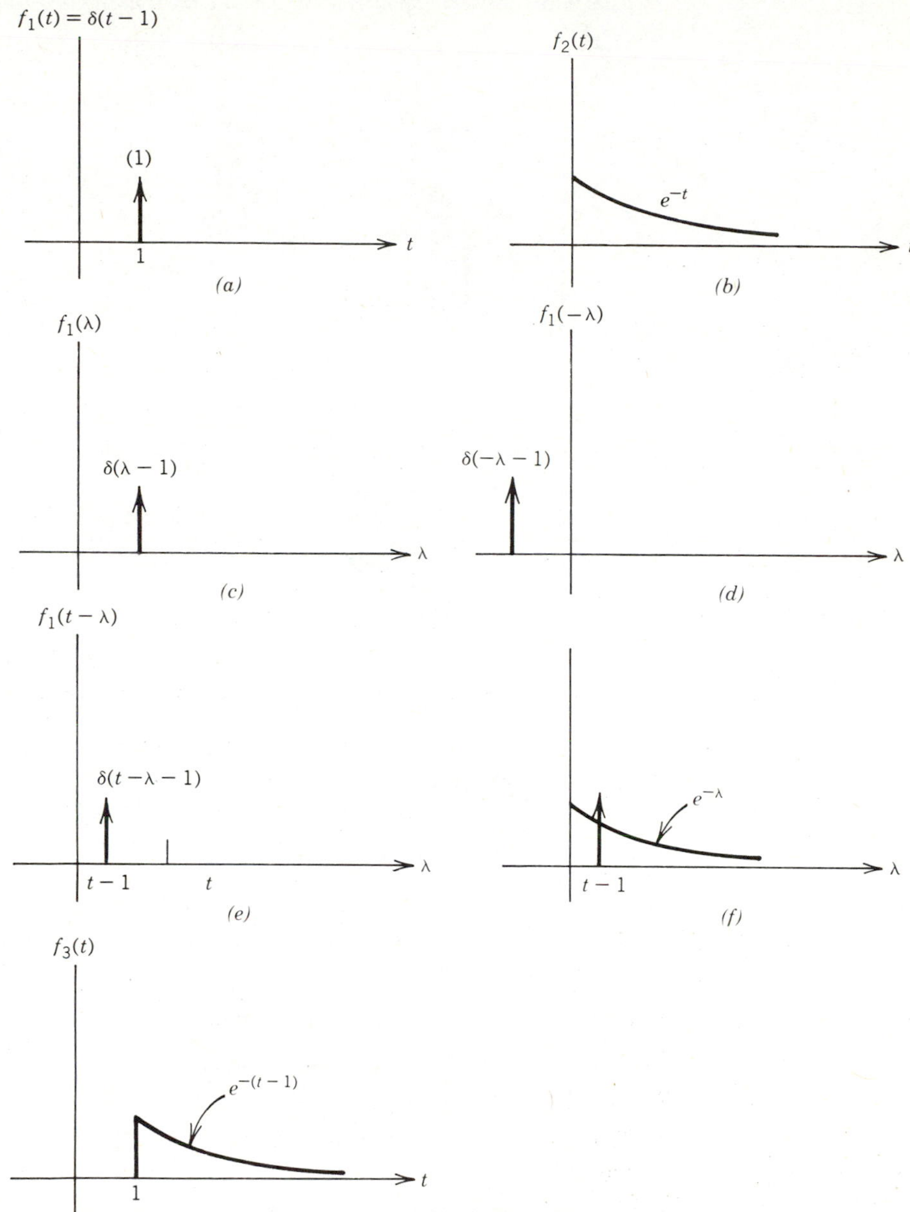

Figure 8.33. Convolution with an impulse.

Solution

We will find $f_1(t-\lambda)$. The process of flipping and slipping the δ function is illustrated in Figs. 8.33*c*, 8.33*d*, and 8.33*e*. For $t<1$, the product is zero. Then in Fig. 8.33*f* the product is shown for $t>1$. In this case $f_3(t) = \int_1^\infty \delta(t-2-1)e^{-\lambda}\,d\lambda = e^{-(t-1)} t > 1$. The result is plotted in Fig. 8.33*g*.

EXAMPLE 8.4.6 Here is one more example of the impulse in the convolution process. Figure 8.34 shows $f_3(t)$ as the result of the convolution operation on $f_1(t)$ and $f_2(t)$.

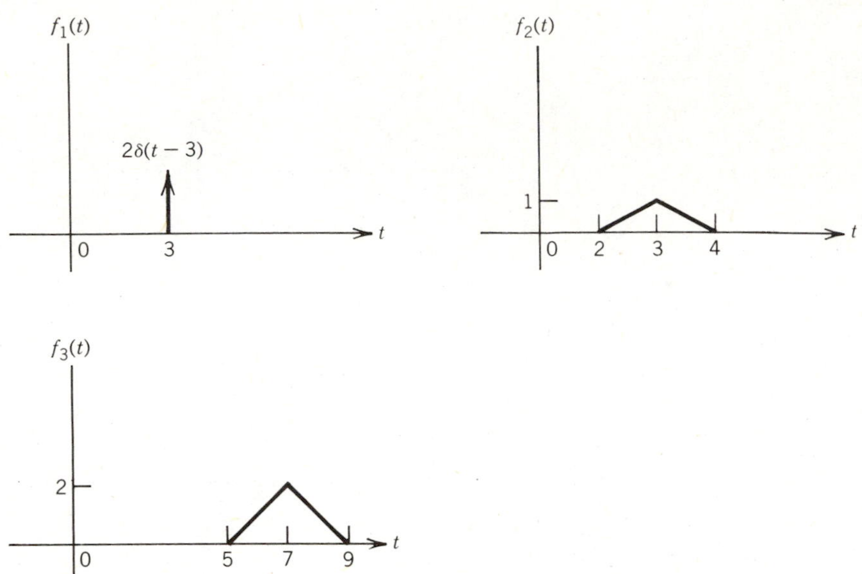

Figure 8.34. Another example of convolution with an impulse.

From these two examples we see that convolution with an impulse is quite simple, for it amounts to a time translation of the function that is convolved with the impulse. Notice that in Fig. 8.33, $f_2(t)$ is shifted to the right by one time unit to produce $f_3(t)$. In Fig. 8.34, $f_2(t)$ is shifted to the right three time units and multiplied by the area under the impulse. In general, to convolve $f(t)$ with $A\delta(t-t_0)$ we simply multiply $f(t)$ by A and shift it by t_0 units.

Let us now restate a step-by-step procedure for evaluating convolution integrals based on our experience above:

1. Plot f_1 and f_2 as functions of λ rather than t.
2. Select one function to flip and slip, say f_2.
3. Flip $f_2(\lambda)$ to obtain $f_2(-\lambda)$.
4. Slip f_2 to left or right until the point originally at the origin coincides with the present value of t.
5. The area under the product $f_1(\lambda)f_2(t-\lambda)$ is the value of $f_3(t)$ for that one value of t.
6. Vary t from $-\infty$ to $+\infty$.

Discrete-Time Convolution

Given two discrete-time signals $v_1(n)$ and $v_2(n)$, their convolution is the binary operation given by

$$v_3(n) = v_1(n) * v_2(n) \tag{8.4.2a}$$

$$= \sum_{k=-\infty}^{+\infty} v_1(k)v_2(n-k) \tag{8.4.2b}$$

$$= \sum_{k=-\infty}^{+\infty} v_1(n-k)v_2(k) \tag{8.4.2c}$$

The evaluation of this summation is identical in principle to the evaluation of the continuous-time integral. Here are some examples.

EXAMPLE 8.4.7 In Fig. 8.35, $v_1(n)$ consists of three pulses occurring at $n=2,3,4$. The function $v_2(n)$ is a unit step beginning at $n=0$. Find $v_3(n) = v_1(n) * v_2(n)$.

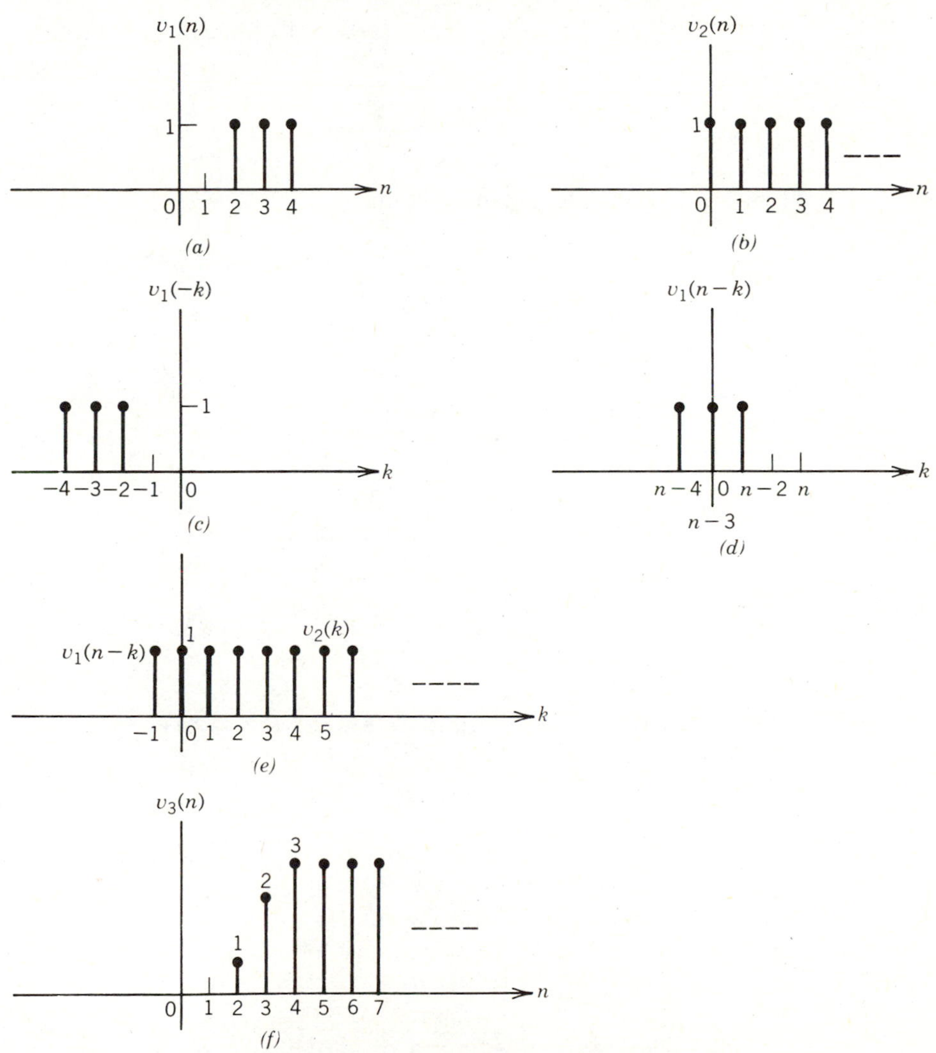

Figure 8.35. The convolution $v_3(n) = v_1(n) * v_2(n)$.

Solution

To convolve these functions we choose to flip and slip $v_1(n)$ as in Figs. 8.35*c* and 8.35*d*, with $n=3$ in Fig. 8.35*d*. In Fig. 8.35*e* the two functions are shown together, so that at $n=3$ the convolution summation is 2. The function $v_3(n)$ is given by

$$v_3(n) = \begin{matrix} 0, & n<2 \\ 1, & n=2 \\ 2, & n=3 \\ 3, & n>4 \end{matrix}$$

as you can see by evaluating the convolution summation at $n=2$, 3, and 4.

There are several closed-form identities that are useful in evaluating convolution summations. The most frequently used is the finite geometric series, given by

$$\sum_{k=0}^{n} a^k = \frac{1-a^{n+1}}{1-a}, \qquad a \neq 1 \tag{8.4.3}$$

This can be shown by writing

$$S = \sum_{k=0}^{n} a^k = 1 + a + a^2 + \cdots + a^n$$

Now multiply both sides by a to obtain

$$aS = a + a^2 + a^3 + \cdots + a^{n+1}$$

Next subtract aS from S to obtain

$$S - aS = 1 - a^{n+1}$$

Solving for S gives Eq. 8.4.3.

Here is another frequently used identity, which can be derived by similar procedures.

$$\sum_{k=0}^{n} ka^k = \frac{a}{(1-a)^2}\left[1-(n+1)a^n + na^{n+1}\right], \qquad a \neq 1 \tag{8.4.4}$$

EXAMPLE 8.4.8 Convolve the two functions $v_1(n)$ and $v_2(n)$ in Figs. 8.36*a* and 8.36*b*.

Solution

We flip and slip v_1 in Figs. 8.36*c* and 8.36*d*. In Fig. 8.36*e*, $n<0$ so $v_3(n)=0$. In Fig. 8.36*f* we illustrate the situation for n somewhere between 0 and 5, and v_3 is given by

$$v_3(n) = \sum_{k=0}^{n} v_1(n-k)v_2(k) = \sum_{k=0}^{n} 0.9^{(n-k)}$$

Now make a change of variable $j = n-k$ to put this expression in the form of Eq. 8.4.3.

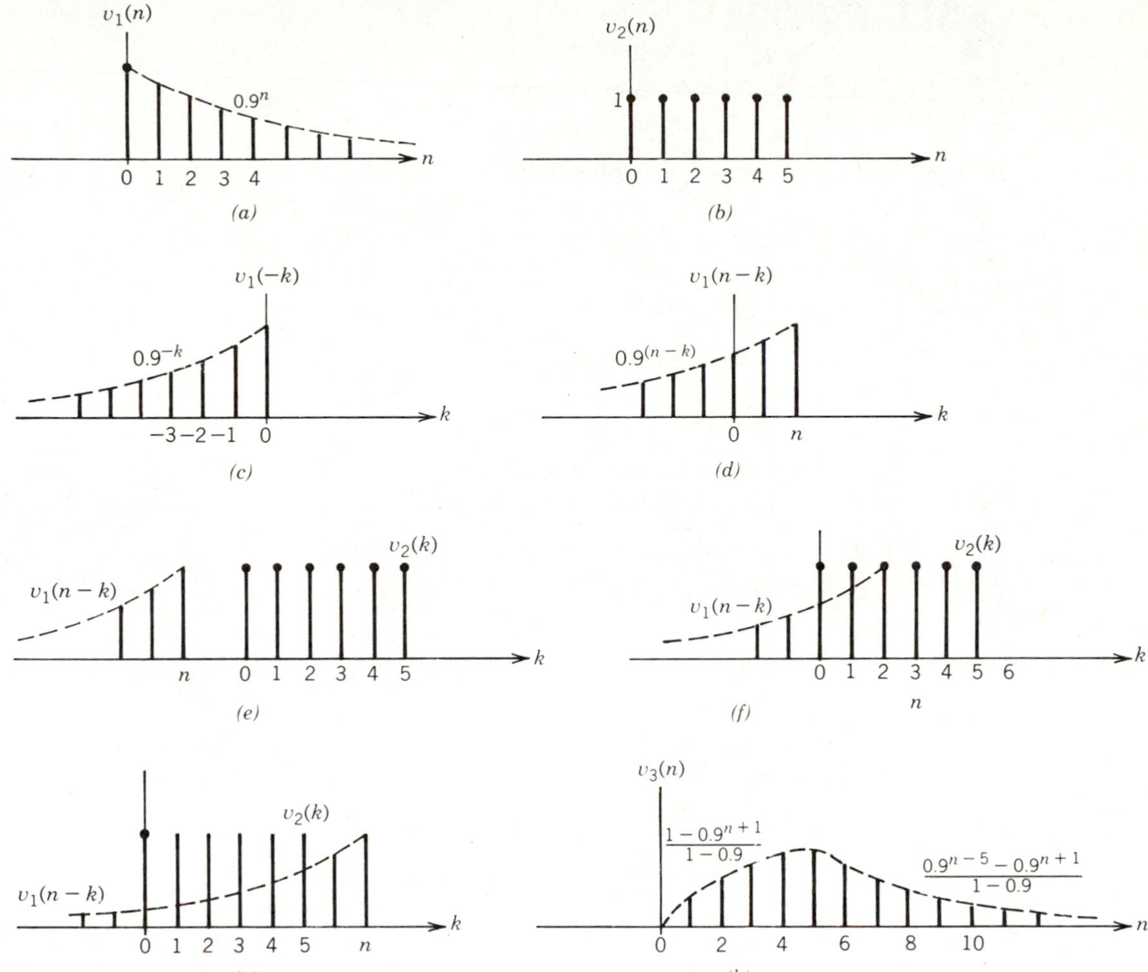

Figure 8.36. The convolution of v_1 with v_2.

$$v_3(n) = \sum_{j=0}^{n} 0.9^j = \frac{1 - 0.9^{(n+1)}}{1 - 0.9}, \qquad 0 < n < 5$$

In Fig. 8.36*g* we have $n > 5$ so $v_3(n)$ is

$$v_3(n) = \sum_{k=0}^{5} 0.9^{(n-k)} = \sum_{j=n-5}^{n} 0.9^j$$

where the last expression is obtained by again setting $j = n - k$.

In order to obtain a closed-form expression for $v_3(n)$ let S be defined by

$$S = \sum_{j=n-5}^{n} 0.9^j = 0.9^{n-5} + 0.9^{n-4} + \cdots + 0.9^n$$

Multiply both sides by 0.9.

$$0.9S = 0.9^{n-4} + 0.9^{n-3} + \cdots + 0.9^{n} + 0.9^{n+1}$$

Subtract $0.9S$ from S to obtain

$$S - 0.9S = 0.9^{n-5} - 0.9^{n+1}$$

Finally, replace the left side by $(1 - 0.9)S$ and divide by $(1 - 0.9)$ to obtain

$$S = \sum_{j=n-5}^{n} 0.9^{j} = \frac{0.9^{n-5} - 0.9^{n+1}}{1 - 0.9}$$

The general form of the above equation (replace 0.9 with a) is

$$S = \sum_{j=n-5}^{n} a^{j} = \frac{a^{n-5} - a^{n+1}}{1 - a}, \qquad a \neq 1 \tag{8.4.5}$$

The procedure for evaluating convolution summations is identical in principle to the procedure for evaluating convolution integrals. Here is the procedure.

1. Plot v_1 and v_2 as functions of k rather than n.
2. Select one function to flip and slip, say v_2.
3. Flip $v_2(k)$ to obtain $v_2(-k)$.
4. Slip v_2 to left or right until the point originally at the origin coincides with the present value of n.
5. The summation of the product $v_1(k)v_2(n-k)$ is the value of $v_3(n)$ for that one value of n.

LEARNING EVALUATIONS

1. Convolve the two continuous-time functions shown in Fig. 8.37.

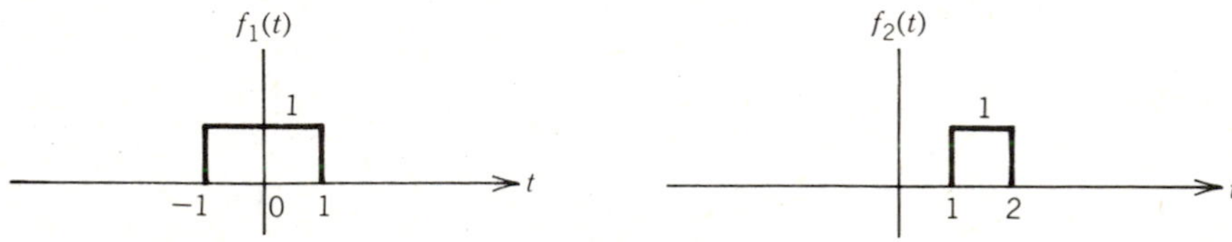

Figure 8.37

2. Convolve the two discrete-time functions shown in Fig. 8.38.

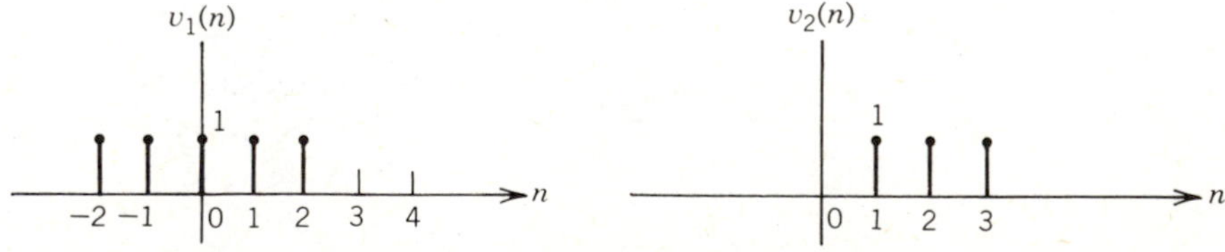

Figure 8.38

8.5 System Response by Convolution

LEARNING OBJECTIVES

After completing this section you should be able to do the following:

1. Derive the convolution summation by application of the LTI properties.
2. Find the response of LTI systems to arbitrary input signals by convolution.

Now that we have a basic understanding of how to convolve two functions, either analog or digital, we are in a position to use this knowledge to find a system output given an arbitrary system input. We have shown earlier that if we know the system output (response) to a given input, and if our system is linear and time invariant (LTI), then we can use the LTI properties to find the output for a new input that is linearly related to the original input. You should review Examples 8.3.1 and 8.3.2 to make sure you understand this principle.

In order to provide maximum flexibility when using this approach to finding the system output, we would like to find an input signal with the following properties:

1. The system response to this input signal can be easily determined.
2. Most other inputs can be expressed as the sum of terms related to this input signal.

Two signals meet this criteria. One is the unit impulse [$\delta(t)$ for continuous systems, $\delta(n)$ for discrete systems]. The other is the exponential function (e^{st} for continuous systems, z^n for discrete systems.) We will use the unit impulse in the remainder of this chapter but will employ the exponential function often in the remainder of the text.

The fact that convolution can be used to find the response of an LTI system may be demonstrated by a derivation based on the LTI properties.

Consider the discrete system shown in Fig. 8.39*a* where the input is the unit impulse. Recall that the output $h(n)$ is called the impulse response of the system. Since the impulse is applied at time $n = 0$, the output is zero for all values of $n < 0$. This situation is shown in Fig. 8.39*b* where some arbitrary shape for $h(n)$ is assumed. The response at time $n = a$ to an impulse applied at $n = 0$ is $y(a) = h(a)$. Now suppose the input is delayed by j units. Because of LTI, the output is also delayed j units as shown in Fig. 8.39*c*. So the output now at time $n = a$ is

$$y(a) = h(a - j)$$

If the impulse occurring at time $n = j$ is no longer the unit impulse but one of value

$$x(j)\delta(n - j)$$

the output at time $n = a$ will be

$$y(a) = x(j)h(a - j)$$

This is due to the homogeneity principle of LTI systems.

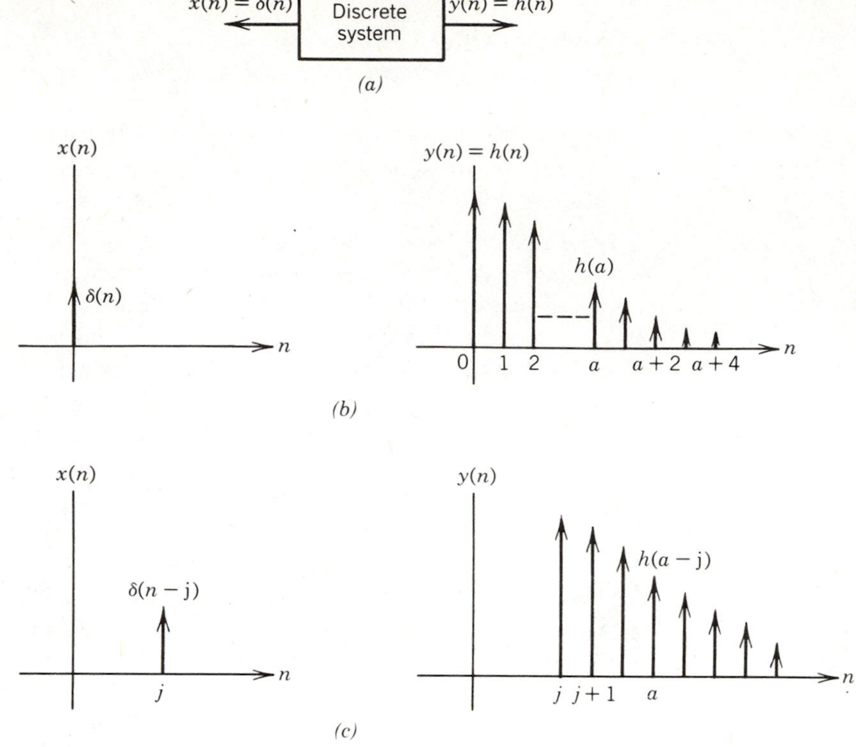

Figure 8.39. Development of the discrete convolution equation.

Now let us add a second impulse to our input function which occurs at time $n-k$ and assume the impulse is

$$x(k)\delta(n-k)$$

By the LTI properties the output at time $n=a$ to the input signal

$$x(n)=x(k)\delta(n-k)+x(j)\delta(n-j)$$

will be

$$y(a)=x(k)h(a-k)+x(j)h(a-j)$$

This also could be written as

$$y(n)=\sum_{i=-\infty}^{\infty} x(i)h(n-i) \tag{8.5.1}$$

where all the $x(i)$ terms are zero except for $i=j$ and $i=k$ and a has been replaced by n.

Equation 8.5.1 will be recognized as identical to Eq. 8.4.2b. Hence, Eq. 8.5.1 is the convolution summation for digital systems where $x(i)$ is the magnitude of the discrete signal at time $n=i$ and $h(n-i)$ is the impulse response of the system.

The derivation of the convolution integral is similar in principle to the derivation of the summation. The continuous nature of the mathematics causes some difficulty, and since no additional insight would be gained, we will bypass this derivation. But it is simply a matter of selecting an appropriate input–output pair and showing that the response of an LTI continuous-time system is the convolution of the input $x(t)$ and the impulse response $h(t)$.

From this discussion we see that convolution is a method for finding the response of an LTI system to an arbitrary input. The impulse response, $h(n)$ or $h(t)$, may be convolved with the input to obtain the output. (The easiest method to find the impulse response is usually by Laplace or Z transforms, which are yet to be discussed. In this chapter we will usually simply state the impulse response.)

EXAMPLE 8.5.1 A discrete-time system is described by the difference equation

$$y(k+1) = 0.8y(k) + x(k)$$

Find the response by convolution to the input $x(k)$ shown in Fig. 8.40*a*.

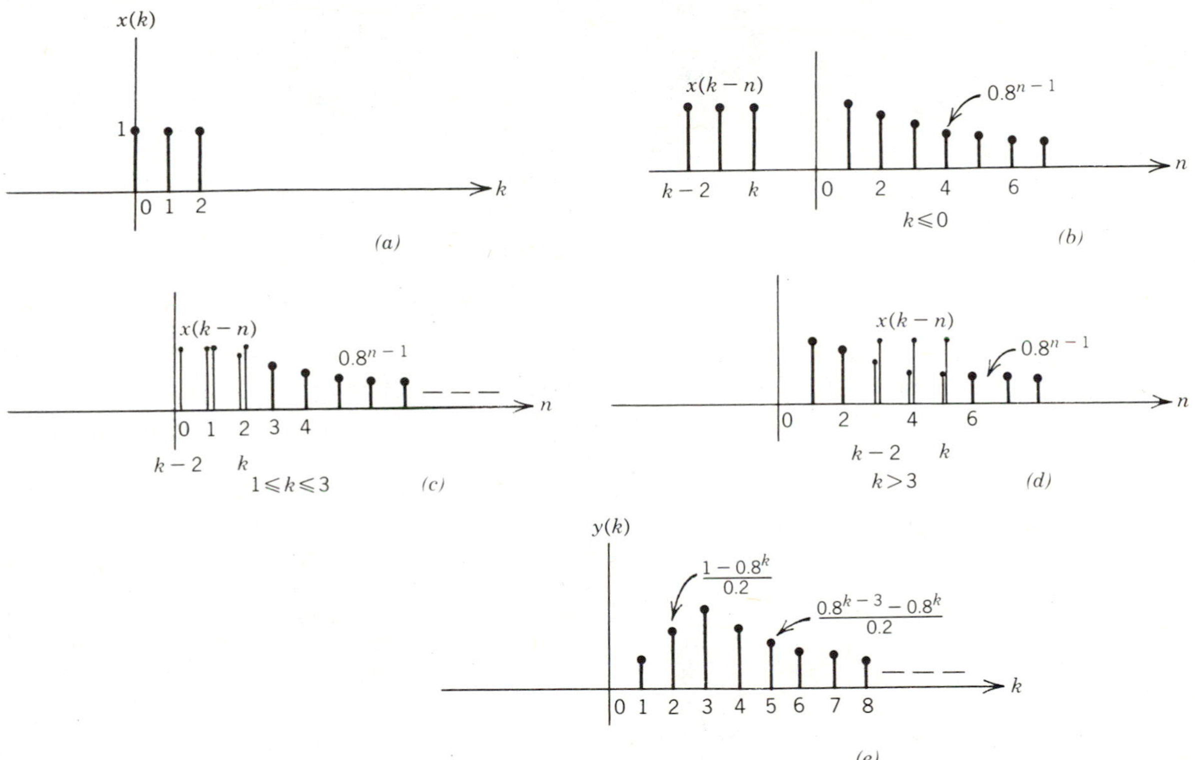

Figure 8.40

Solution

The impulse response must first be found. Applying the input $x(0) = 1$, $x(k) = 0$, for all other k, with initial value $y(0) = 0$, we find

$$y(1) = 0.8(0) + 1 = 1$$

$$y(2) = 0.8(1) + 0 = 0.8$$

$$y(3) = 0.8(0.8) + 0 = 0.8^2$$

$$y(4) = 0.8(0.8)^2 + 0 = 0.8^3$$

The general rule is evidently given by

$$h(k) = 0.8^{k-1}, \qquad k > 0$$

Now this impulse response is convolved with the input as shown in Figs. 8.40*b*, 8.40*c*, and 8.40*d* to obtain

$$y(k) = 0, \qquad k \leq 0$$

$$y(k) = \sum_{n=1}^{k} 0.8^{(n-1)}, \qquad 1 < k \leq 3$$

Let $j = n - 1$ in this summation to obtain the form given in Eq. 8.4.3.

$$y(k) = \sum_{j=0}^{k-1} 0.8^j = \frac{1 - 0.8^k}{0.2}, \qquad 1 < k \leq 3$$

For $k > 3$ we have

$$y(k) = \sum_{n=k-2}^{k} 0.8^{(n-1)} = \sum_{j=k-3}^{k-1} 0.8^j$$

$$= \frac{0.8^{(k-3)} - 0.8^k}{0.2}, \qquad k > 3$$

This is placed in the form of Eq. 8.4.5 by the index change $j = n - 1$. The response $y(k)$ is shown in Fig. 8.40*e*.

EXAMPLE 8.5.2 The impulse response for the *RC* low-pass filter in Fig. 8.41 can be found by methods introduced in Chapter 5 to be

$$h(t) = \frac{1}{RC} e^{-t/RC}, \qquad t > 0$$

Find the response to the square pulse input $x(t)$ shown in the figure if $RC = 2$.

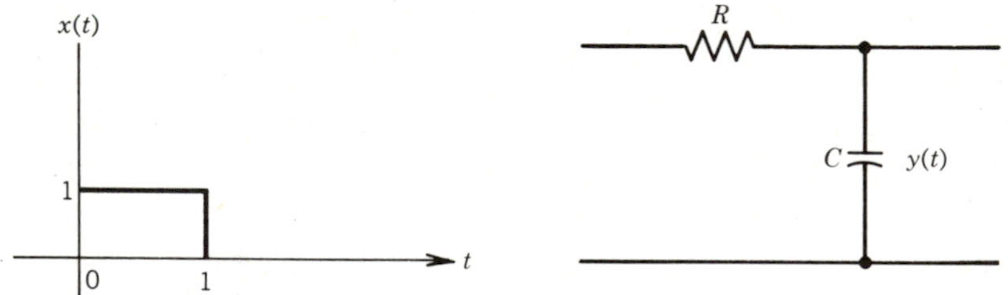

Figure 8.41

Solution

For $t < 0$, the product of the input and the impulse response is zero, as shown in Fig. 8.42*a*. Hence,

$$y(t) = 0, \qquad t < 0$$

The situation for other ranges of t is pictured in Fig. 8.42b and 8.42c. The resulting convolution integral is given by

$$y(t) = \int_0^t \frac{1}{2} e^{-\lambda/2}\, d\lambda = 1 - e^{-t/2}, \qquad 0 < t < 1$$

$$y(t) = \int_{t-1}^t \frac{1}{2} e^{-\lambda/2}\, d\lambda = e^{-(t-1)/2} - e^{-t/2}, \qquad 1 < t$$

The complete response $y(t)$ is shown in Fig. 8.42d.

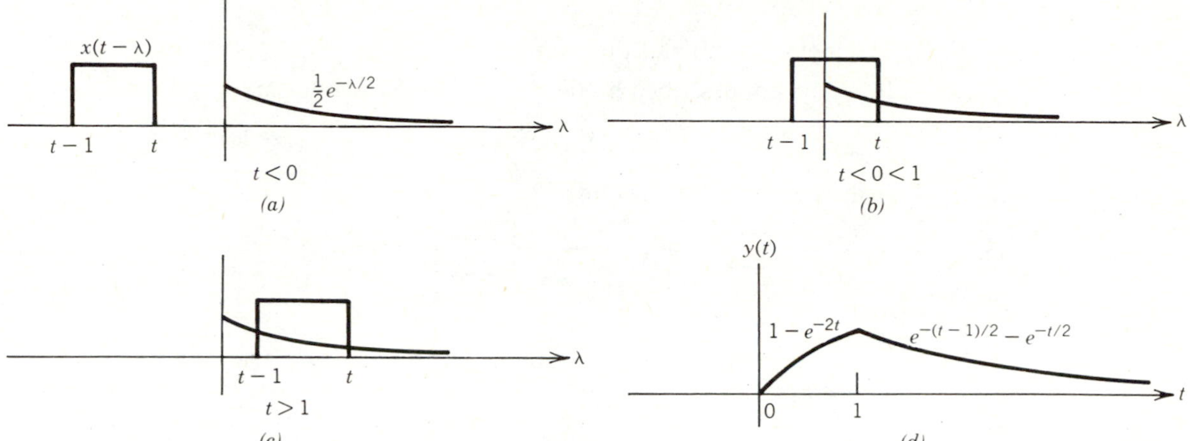

Figure 8.42. Response of circuit in Fig. 8.11.

LEARNING EVALUATIONS

1. Derive the convolution summation by application of the LTI properties.
2. Find the response of the first order digital system described by the equation

$$y(k) = 0.5y(k-1) + 0.5x(k-1)$$

to the input signal

$$x(k) = u(k) - u(k-4)$$

PROBLEMS

SECTION 8.1

1. Test the following systems for linearity.
   - **a.** $y(t) = x(t)\sin t$
   - **b.** $y(t) = x(t) + 1$
   - **c.** $y(t) = 64tx(t)$
   - **d.** $y(t) = x(t) + x(t-2)$
   - **e.** $y(t) = 20x(t)$
   - **f.** $y(t) = x^4(t)$

2. Is the circuit shown in Fig. 8.43 linear?

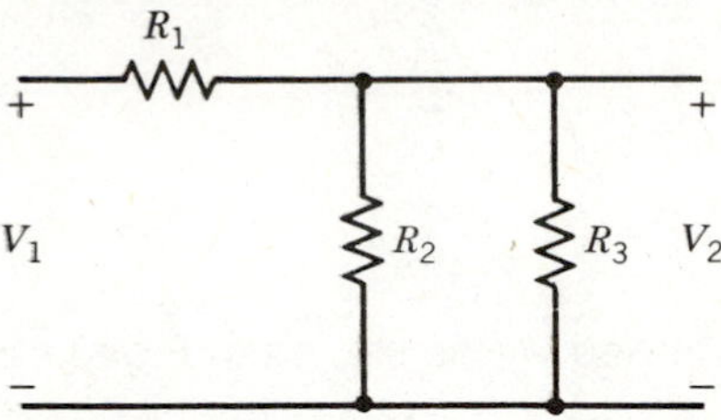

Figure 8.43

The input is $v_1(t)$ and the output is $v_2(t)$.

3. Is the system described by the following equation linear?

$$y(t)=y(t-1)-(1+t)x(t)+[x(t-1)]^2$$

4. Is the system in Fig. 8.44 linear?

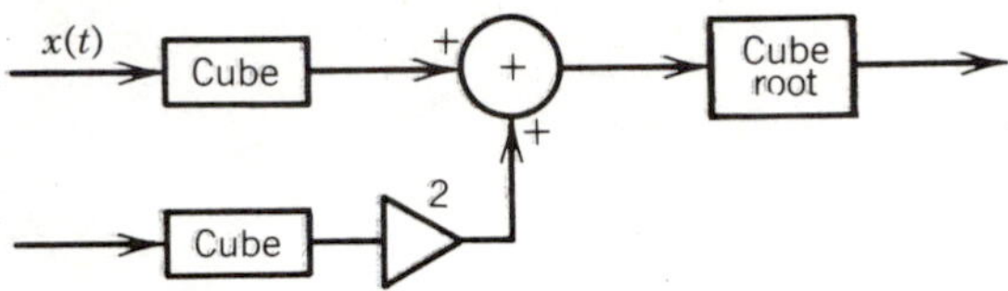

Figure 8.44

5. Is this a linear relationship for all x?

$$y(t)=x(0)+x(t)$$

SECTION 8.2

6. Test the following systems for time invariance. $x(t)$ is the input, and $y(t)$ is the output.
   a. $y(t)=2x(t)+3x(t-1)-4y(t-1)$
   b. $y(t)=t^2x(t)+2tx(t-1)+y(t-1)$
   c. $y(t)=x^2(t)$
   d. $y(t)=\dfrac{1+t}{2}x(t)-x(t-1)-y(t-1)$
   e. $y(t)=x(t)\sin t$
   f. $y(t)=3x(t)+16y(t-1)+4(t+1)x(t-1)$
   g. $y(t)=x(t-2)$
7. Is the circuit in Fig. 8.45 time invariant?

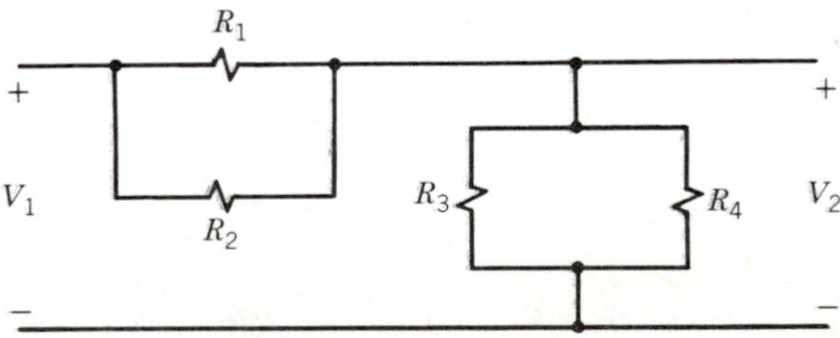

Figure 8.45

8. Test the system in Fig. 8.46 for time invariance.

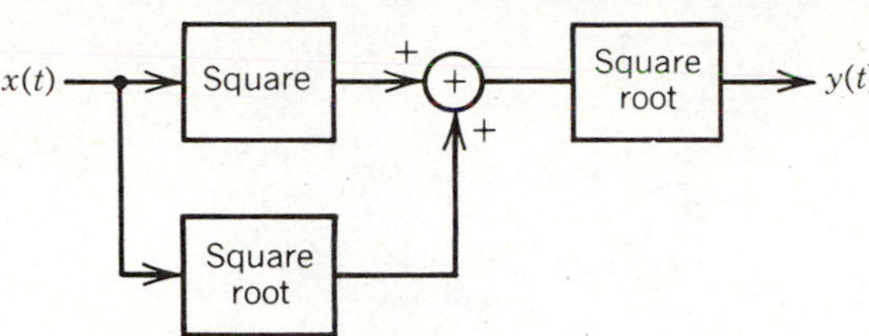

Figure 8.46

SECTION 8.3

9. Given the $x_1 - y_1$ pair, find $y_2(t)$ using the LTI properties if $x_2(t)$ is as shown in Fig. 8.47.

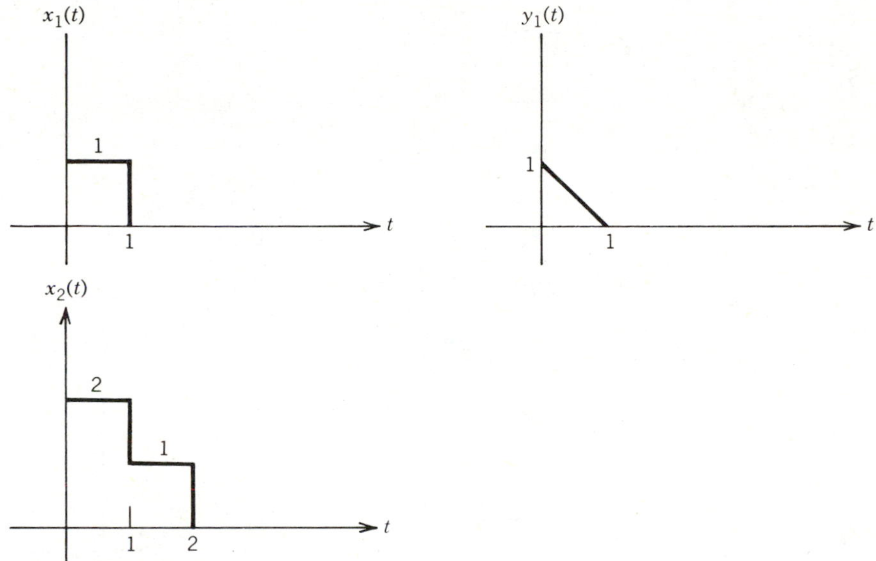

Figure 8.47

10. An input–output pair of a system is shown in Fig. 8.48. Given that the system is LTI, find the output of the system corresponding to $x_2(t)$.

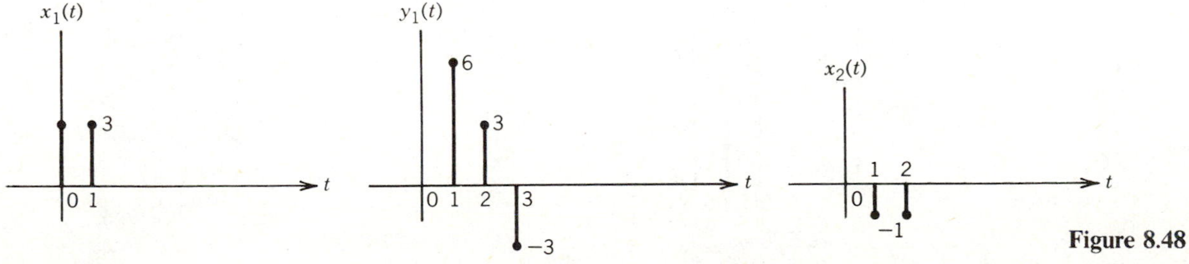

Figure 8.48

11. An input–output pair to a system is shown in Fig. 8.49. Given that the system is LTI, find the response to the input $x_2(t)$.

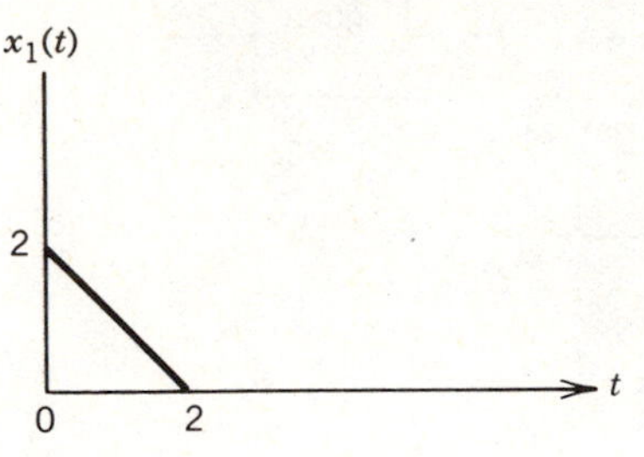

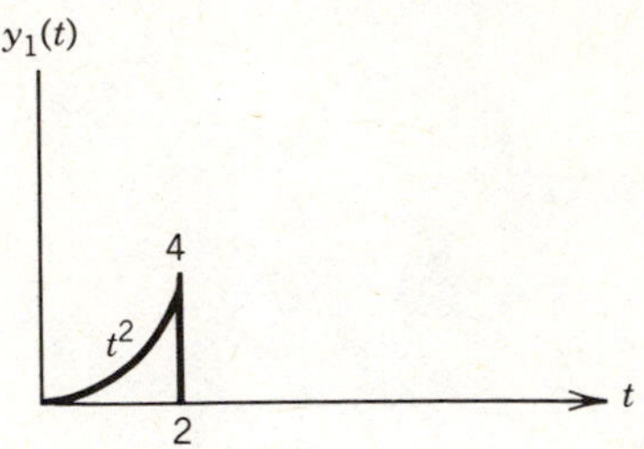

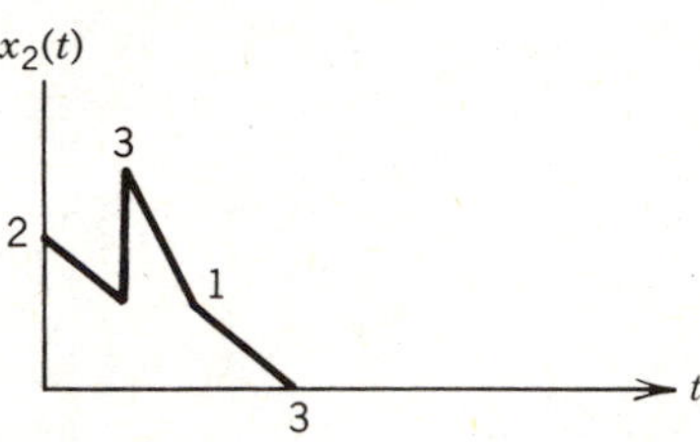

Figure 8.49

12. An input–output pair to a system is shown in Fig. 8.50. Given that the system is LTI, find the response to the input $x_2(t)$.

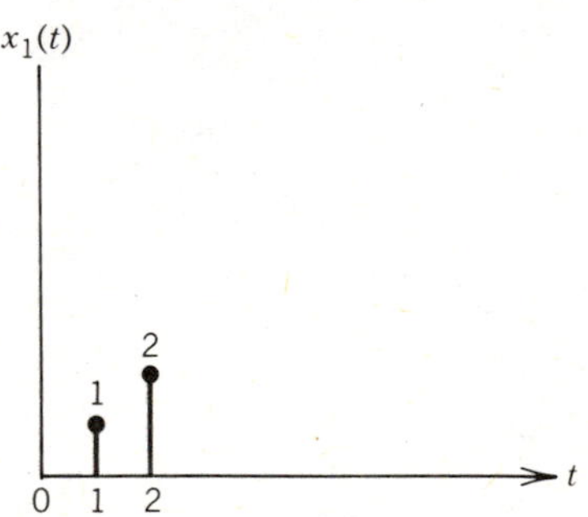

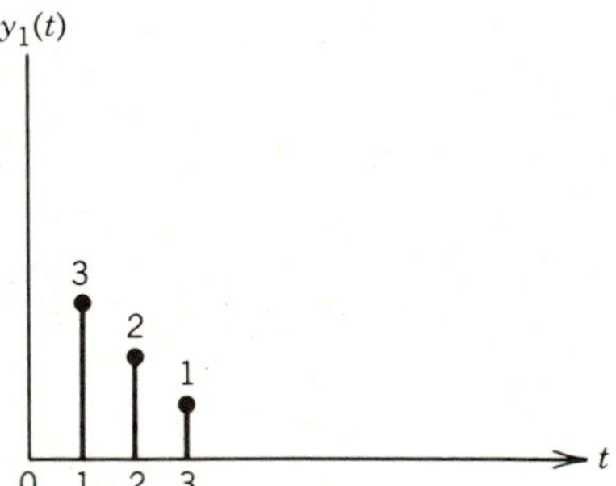

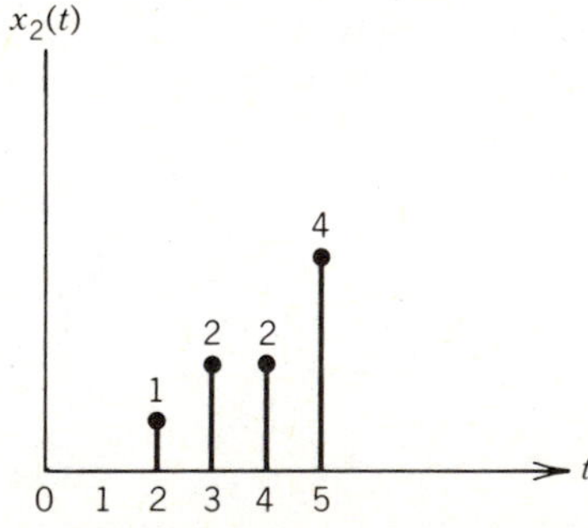

Figure 8.50

13. State the LTI properties.

SECTION 8.4

14. In Fig. 8.51, convolve the following functions:

a. f_1 and f_2
b. f_2 and f_3
c. f_1 and f_3

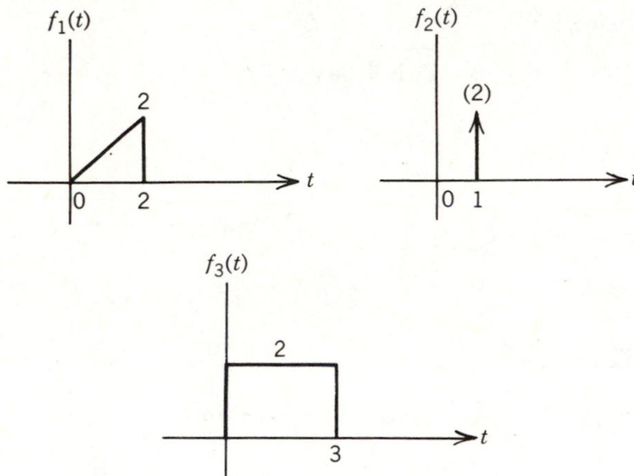

Figure 8.51

15. In Fig. 8.52, convolve the following functions:

a. v_1 and v_2
b. v_2 and v_3
c. v_1 and v_3

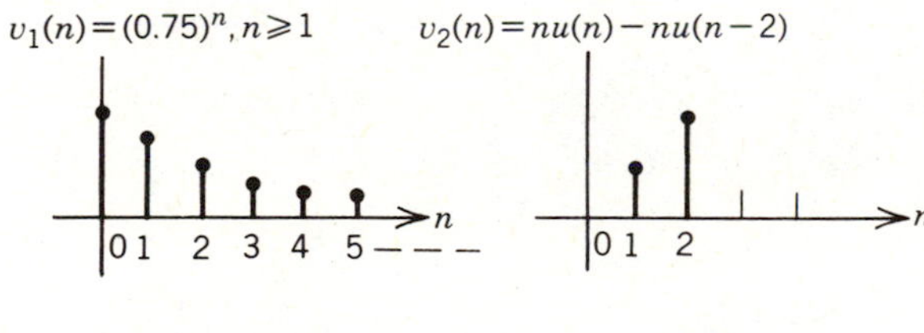

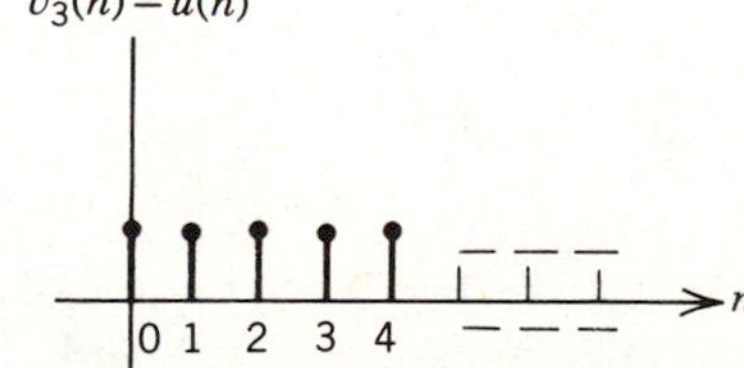

Figure 8.52

16. Convolve $f_1(t)$ with $f_2(t)$ in Fig. 8.53.

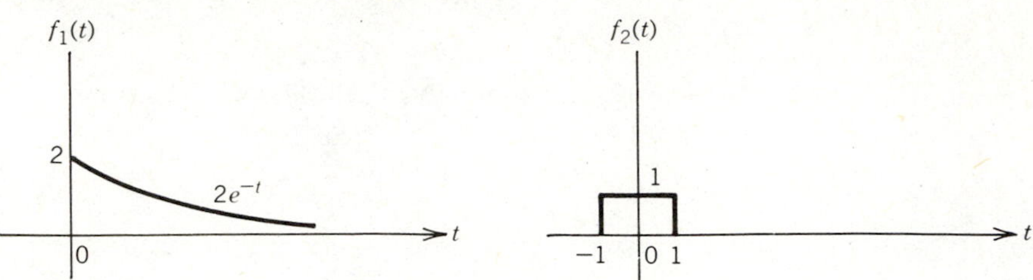

Figure 8.53

17. Convolve $v_1(n)$ with $v_2(n)$ in Fig. 8.54.

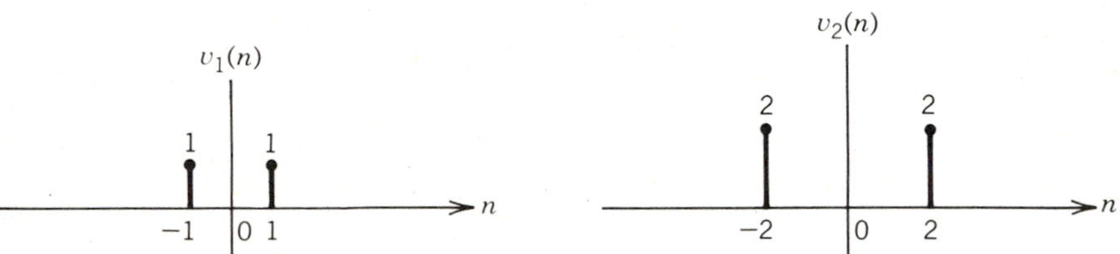

Figure 8.54

18. In Fig. 8.55, convolve the following functions:

a. f_1 and f_2
b. f_1 and f_3
c. f_1 and f_4
d. f_2 and f_3
e. f_2 and f_4
f. f_3 and f_4

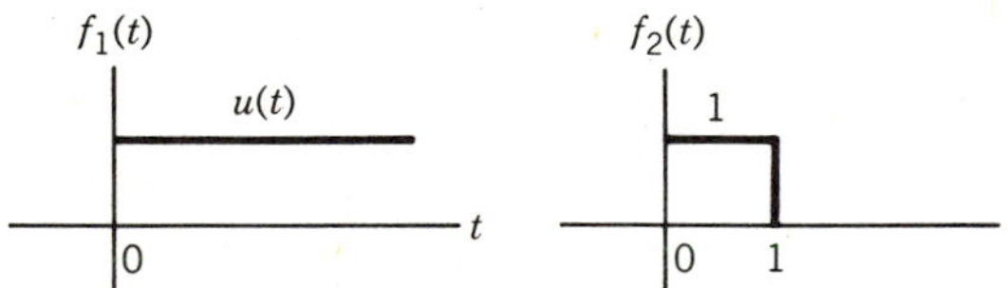

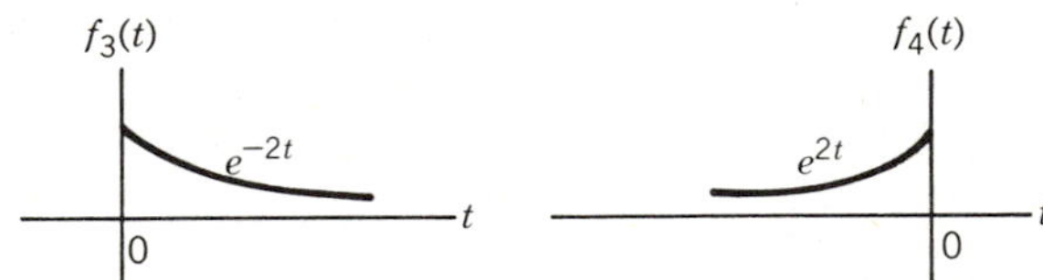

Figure 8.55

19. Discover the relationship between A_1, A_2, and A_3, where A_1 and A_2 are finite areas under functions $f_1(t)$ and $f_2(t)$, and A_3 is the area under $f_3(t) = f_1(t) * f_2(t)$.

20. In Fig. 8.56, convolve the following functions:

a. v_1 and v_2
b. v_1 and v_3
c. v_1 and v_4
d. v_2 and v_3
e. v_2 and v_4
f. v_3 and v_4

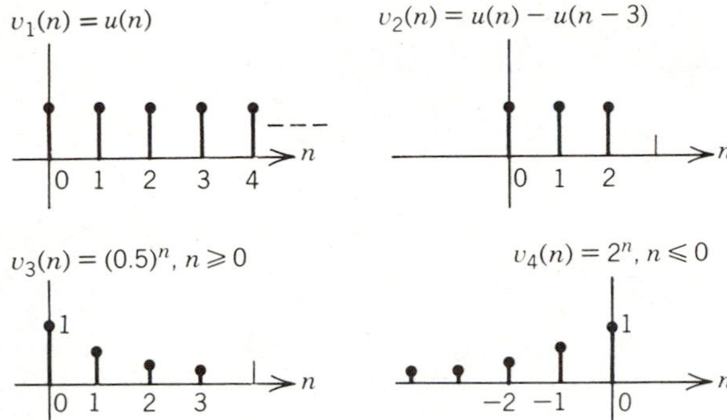

Figure 8.56

SECTION 8.5

21. Find the response of a low-pass filter whose impulse response is $h(t) = e^{-t}u(t)$ to the input $x(t) = u(t) - u(t-2)$. Compare to Problem 22.

22. Find the response of a digital system described by the difference equation

$$y(k) = 0.2y(k-1) + 0.8x(k-1)$$

to the input $x(k) = u(k) - u(k-5)$.

23. The impulse response of a discrete-time system is given by

$$h(k) = 0.9^k - 0.2^k$$

Find the response of this system to a unit step $x(k) = u(k)$.

24. For the circuit in Fig. 8.57; $R = 3$ MΩ, $C = 2$ μF. For an input, $x(t) = u(t) - u(t-2)$, determine $y(t)$.

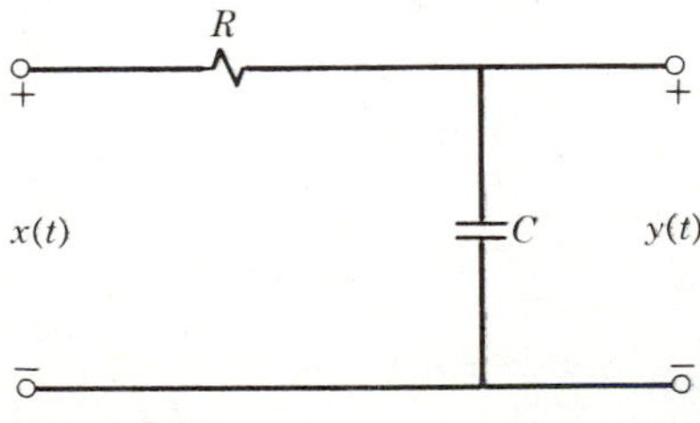

Figure 8.57

25. A digital system is described by the difference equation

$$y(n) = x(n) - y(n-1)$$

a. Find the impulse response if $y(-1) = 0$.
b. Find the response to $x(n) = 4u(n)$.

26. If the impulse response to a discrete-time system is

$$h(n) = 0.6^n, \qquad n > 0$$

find the response to $x(n) = u(n)$.

27. If the impulse response to a system is as shown in Fig. 8.58, determine the response to an input $x(t) = u(t-1)$.

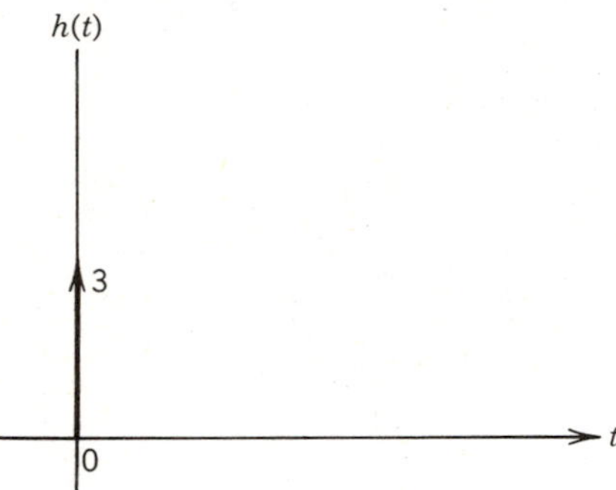

Figure 8.58

LEARNING EVALUATION ANSWERS

Section 8.1

1. See Definition 8.1.1.
2. Yes.
3. **a** and **c** are linear. **b** is linear only if $x(0) = 0$. **d** is nonlinear.

Section 8.2

1. See Definition 8.2.1.
2. Systems **a**, **b**, and **d** are time varying. Only **c** is time invariant.

Section 8.3

1. Additivity: $y_1 + y_2 = L(x_1 + x_2)$
   Homogeneity: $ay = L(ax)$
   Time Invariance: $y(t+\epsilon) = L[x(t+\epsilon)]$
2. Since $x(t) = p(t) - p(t-1)$, then $y(t) = q(t) - q(t-1)$ as shown in Fig. 8.59.

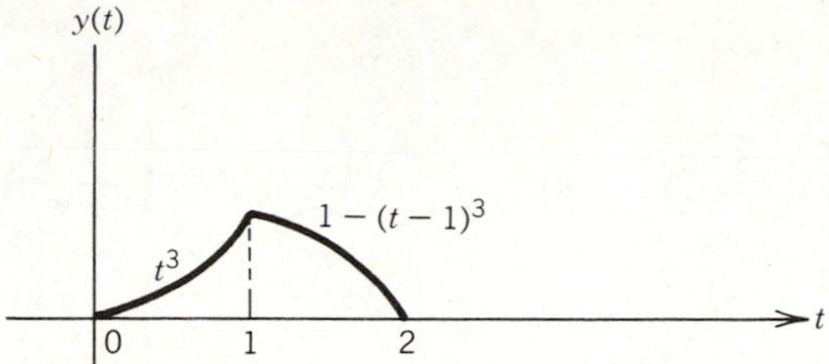

Figure 8.59

Section 8.4

1. See Fig. 8.60.

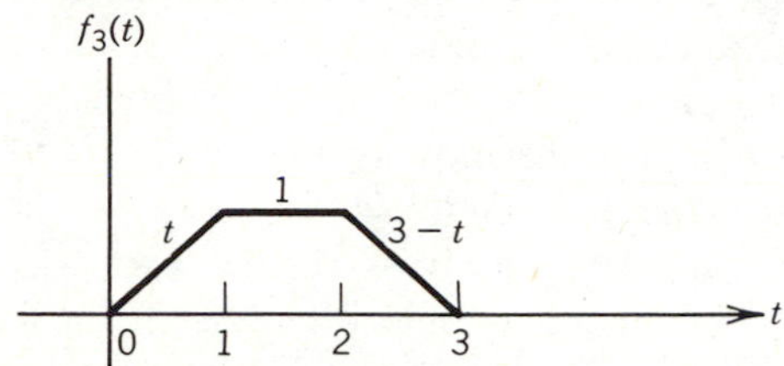

Figure 8.60

2. See Fig. 8.61.

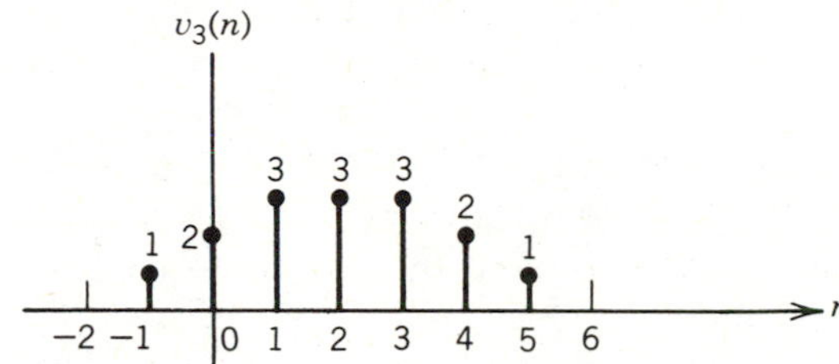

Figure 8.61

Section 8.5

1. See Fig. 8.39 and the accompanying discussion.

2.

$$
\begin{aligned}
h(k) &= 0.5^k, k > 0 \\
&= 0, k < 0
\end{aligned}
$$

Convolving $h(k)$ with $x(k)$ gives the response shown in Fig. 8.62.

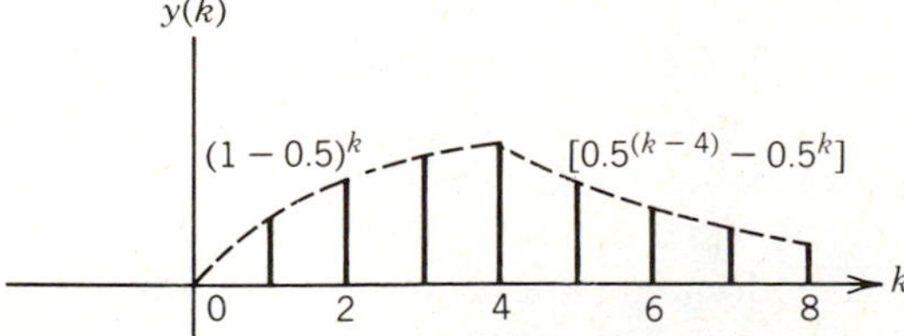

Figure 8.62

CHAPTER

SINUSOIDAL ANALYSIS

We have so far dealt comprehensively only with the constant (dc) and exponential forcing functions. We are now in a position to consider a third (and perhaps the most useful) forcing function: the sinusoid. We have encountered this function briefly in the past but have postponed its serious consideration until we could introduce another analysis technique called transform analysis.

Sinusoidal functions appear spontaneously in nature. Vibrations of a piano string are sinusoidal as is the motion of a spring fixed at one end with a force suddenly applied to the other end. We also show in a later chapter that functions that periodically repeat themselves can be expressed as sums of sinusoids. Therefore, if we can find the output of a circuit for a sinusoidal forcing function, we can find the output for any periodic function by using superposition.

Transform analysis involves changing (transforming) a circuit and its forcing function to a different but equivalent form for analysis and then changing the result back (inverse transforming) to the original form to find the answer. We are not unfamiliar with this procedure in everyday life. When faced with the problem of moving massive amounts of high-quality coal from the Wyoming area to Texas, Arkansas, and other remote locations, engineers developed one solution by transforming the coal to a slurry of powder and water at the mining site. This transformed material is moved by pipeline to its destination where the water is removed and the coal converted to its original form. You have also used transform techniques previously in mathematics. The logarithm is a transform method that allows us to convert multiplication–division problems into addition–subtraction problems.

An analogy for transform analysis is the famous story of *Through the Looking Glass* by Lewis Carroll. In passing through the looking glass, Alice encounters a topsy-turvy world in which time runs backward, animals talk, and, in general, things are just not normal. There are two entirely different worlds on either side of the looking glass, and so it is with transform analysis. Up until now we have dealt only with the time domain (the real world). All of the functions that we have encountered, such as the current in a resistor, or the voltage applied to a circuit, have been functions of time. But it is possible to take the transform of these quantities (pass through the looking glass) and represent them as functions of frequency. Now the frequency domain is nothing like the topsy-turvy world that Alice encountered, but things are different. For example, we will find that instead of convolving two time functions (in the real world), the corresponding operation in the frequency domain (wonderland) is multiplication. Since it is much simpler to multiply two functions than it is to convolve two functions, it will often simplify our analysis of LTI systems to do just that. After multiplying the two functions, the inverse transform must be evaluated to return to the time domain, so this procedure may or may not simplify matters. Sinusoidal analysis is one case where matters are greatly simplified by this procedure, and we will amply illustrate this fact in this and the next chapter.

9.1 Sinusoidal Functions

LEARNING OBJECTIVES

After completing this section you should be able to do the following:

1. Write the equation of a sinusoidal function from the graph.
2. Draw the graph of a sinusoidal function from the equation.
3. Determine the phase relationship between two sinusoids.
4. Find the mean square and root mean square value of a given waveform.

We will use the term "sinusoidal function" to represent sine and cosine functions. The function

$$f(t) = A \sin \omega t \tag{9.1.1}$$

is illustrated in Fig. 9.1. Notice that there are two scales marked along the abscissa, one scale in units of time t and the other scale in units of angle ϕ, where $\phi = \omega t$. The units of angle may be more familiar to you, but our interest is in sinusoids as a function of time. From the diagram we see that one period T is equal to 2π radians (rad). That is, when $\omega t = 2\pi$, the value of t is T. Hence,

$$T = \frac{2\pi}{\omega} \tag{9.1.2}$$

The time T represents the time required to complete one cycle of the waveform. The *frequency* of a sinusoidal function is the number of cycles completed by the function in 1 s and is represented by the symbol f. By observation,

$$f = \frac{1}{T} \tag{9.1.3}$$

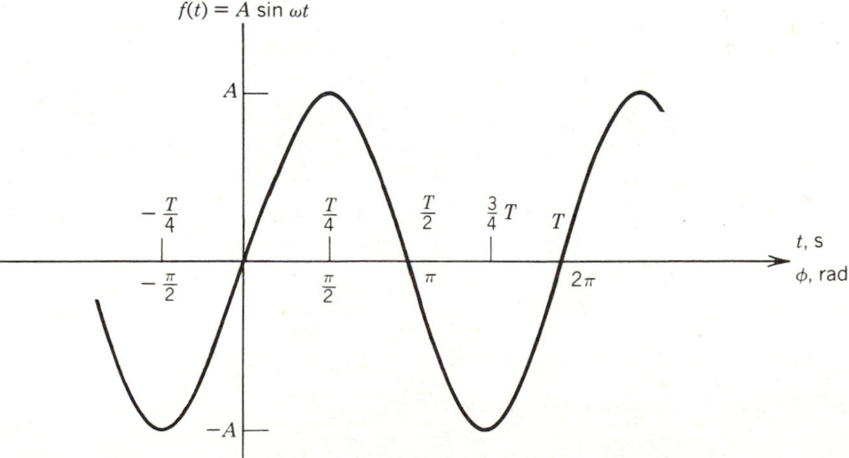

Figure 9.1. The sinusoid $A \sin \omega t$.

the unit for frequency is seen, from Eq. 9.1.3, to be cycles per second (cps) and this name was used for many years. However, the name of the SI unit for frequency is the hertz and this has found almost universal acceptance. One hertz is equal to 1 cps and is abbreviated Hz. Combining Eqs. 9.1.2 and 9.1.3, we find that

$$\omega = 2\pi f \tag{9.1.4}$$

ω is called the "radian" or "angular" frequency of the sinusoid and has the units of radians per second.

The power delivered to our homes in the United States is sinusoidal and has a frequency of 60 Hz. This waveform then (from Eq. 9.1.3) completes one cycle every $\frac{1}{60}$ s. Most other countries have power generated at different frequencies, the most common being 50 Hz.

The sinusoid is periodic, meaning that

$$f(t) = f(t+T) \qquad \text{for all } t \tag{9.1.5}$$

If we pick any value of time t, the value of $f(t)$ is the same at that time as it is one period later, at $t + T$. Periodic functions exist for all time, meaning that they were present before the world was formed, and they will be present after we cease to exist. If there is a beginning or an end to a function, then Eq. 9.1.5 is not true for all time, and the function cannot be periodic. This is important to us because when we speak of steady-state sinusoidal analysis, we necessarily imply that there is no transient response. Any transient response must have died out ages ago.

A general sinusoid is pictured in Fig. 9.2*a*. Notice that a sinusoid is completely characterized by three parameters, amplitude A, frequency ω (or f), and phase θ. Once these three parameters are specified, the equation for the sinusoid may be written immediately, and the graph may be constructed.

In Fig. 9.2*b*, two sinusoids are shown that can be expressed mathematically as

$$x(t) = A\sin(\omega t + \theta_1)$$
$$y(t) = B\sin(\omega t - \theta_2)$$

What is the phase relationship between $x(t)$ and $y(t)$? If we pick any point on the $x(t)$ curve, we observe that the same point occurs on the $y(t)$ curve $\theta_1 + \theta_2$ rad later. Therefore, $x(t)$ is said to *lead* $y(t)$ by $\theta_1 + \theta_2$ rad. It is equally correct to state that $y(t)$ *lags* $x(t)$ by $\theta_1 + \theta_2$ rad. It only makes sense to discuss the phase relationship between two sinusoids when they are of the same frequency. Sinusoids of different frequencies do not have phase relations.

EXAMPLE 9.1.1 Draw the graph of the function

$$v(t) = 2\sin\left(628t + \frac{\pi}{2}\right)$$

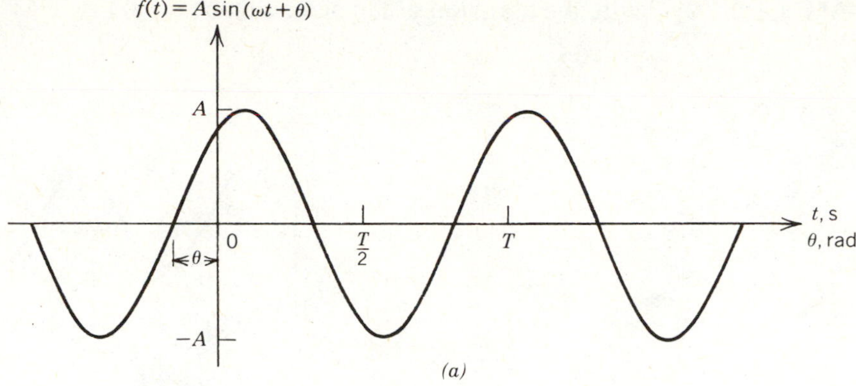

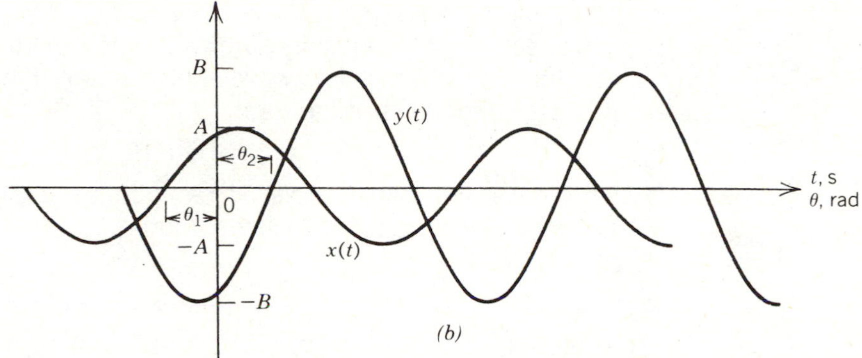

Figure 9.2. Sinusoidal functions. (*a*) A sinusoid $A \sin(\omega t + \theta)$. (*b*) Two sinusoids with different phase angles.

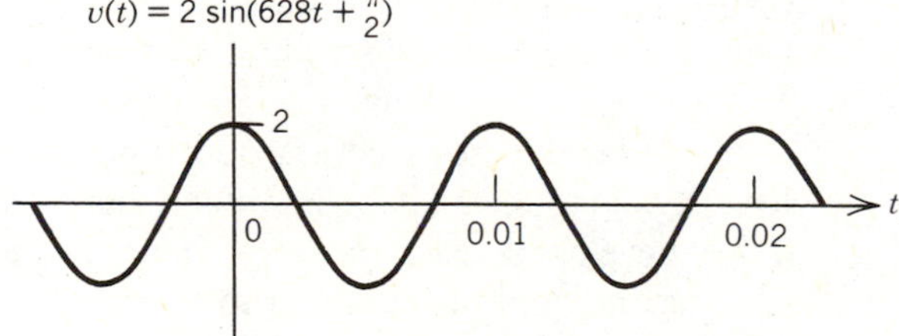

Figure 9.3

Solution

The period is $T = 2\pi/\omega = 2\pi/628 = \frac{1}{100}$ s. The graph is shown in Fig. 9.3. Recall that $\cos\omega t = \sin(\omega t + \pi/2)$ so this is a cosine wave, and it could be represented by

$$v(t) = 2\cos(628t)$$

EXAMPLE 9.1.2 Write the equation of the sinusoid shown in Fig. 9.4.

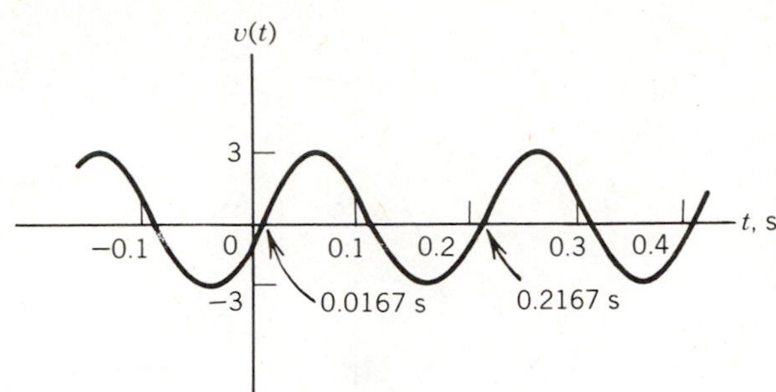

Figure 9.4

Solution

The period is $T = 0.2$ s. Hence, $\omega = 2\pi/T = 10\pi$ rad/s. The amplitude is $A = 3$, so the only remaining problem is to determine the phase angle θ. The function $v(t)$ is a sine wave shifted to the right by 0.0167 s, meaning that the equation is of the form

$$v(t) = 3\sin(10\pi t - \theta)$$

where θ is the phase angle corresponding to 0.0167 s. Since a phase angle of 2π corresponds to 0.2 s, then

$$\theta = \frac{2\pi(0.0167)}{0.2} = \frac{\pi}{6}$$

Thus, $v(t) = 3\sin(10\pi t - \pi/6)$.

The rms Value of a Function

In the early days of electric power generation a controversy arose over whether ac or dc power was best. Thomas Edison was an advocate for direct current distribution. It was one of the few times that he was wrong, and what must have been particularly galling to him was that a former employee and bitter enemy, Nikola Tesla, was the leading spokesman for alternating current distribution.

The difficulty involved transporting electricity over wires without too much loss. It was found that high voltage increased the efficiency of transmission, and Tesla developed transformers that could convert ac to higher voltage at the sending end, and then convert to lower voltage at the receiving end. Since Edison could not match this performance with dc, his side eventually lost out; but not without a bitter fight. He even went so far as to lobby for ac in the new electric chair, first used in New York State. He then pointed with horror at the danger of ac. Ironically, technology has just recently progressed to the point where dc offers some advantages under certain conditions for long-distance power transmission.

This history is related to our present topic because some basis has to be used for comparing ac with dc. Since the biggest concern was over heat loss in transmission, average power was used. Recall that instantaneous power in a resistor is given by

$$p(t)=v(t)i(t)=i^2(t)R=\frac{v^2(t)}{R} \tag{9.1.6}$$

where $v(t)$ and $i(t)$ are the voltage across and the current through the resistor R, respectively. The mean or average value of any function is the ratio of area to base. Find the area, divide by the base, and this gives the average value. Thus the *average power* is given by

$$P_{av}=\frac{1}{T}\int_0^T p(t)\,dt \tag{9.1.7}$$

where T is the period for an ac waveform. Of course, any of the terms in Eq. 9.1.6 may be substituted for $p(t)$ in Eq. 9.1.7 to obtain average power. The average power is called the *mean square value* if $R=1$. Thus the mean square value for a current $i(t)$ is given by

$$I_{ms}=\frac{1}{T}\int_0^T i^2(t)\,dt$$

The *root mean square* (rms) value is given by

$$I_{rms}=\sqrt{\frac{1}{T}\int_0^T i^2(t)\,dt} \tag{9.1.8}$$

The same reasoning applies to voltage, so that the rms value of $v(t)$ is given by

$$V_{rms}=\sqrt{\frac{1}{T}\int_0^T v^2(t)\,dt} \tag{9.1.9}$$

We must keep in mind that we are dealing with periodic functions here. When the function is not periodic, the mean square value is defined by

$$V_{ms}=\lim_{T\to\infty}\frac{1}{2T}\int_{-T}^{T} v^2(t)\,dt \tag{9.1.10}$$

The rms value is then the square root of this average. Equation 9.1.10 is a general formula that applies to both periodic and nonperiodic functions.

Consider the following question. Given a dc voltage of constant value V_{dc}, what ac voltage will produce the same average power in a resistor R? Two such waveforms are shown in Fig. 9.5. The dc power is given by

$$P_{dc}=\frac{V_{dc}^2}{R}$$

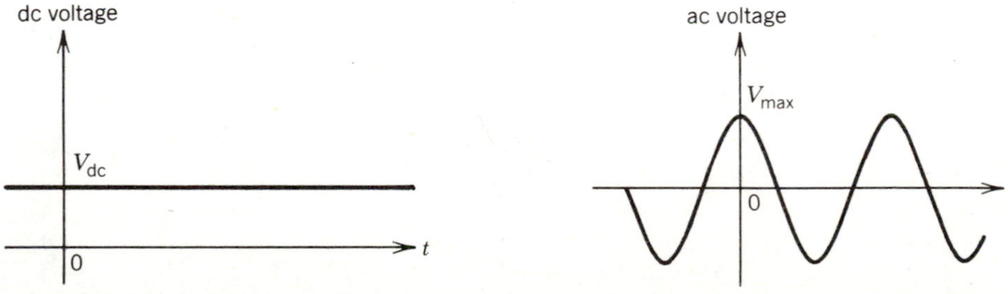

Figure 9.5. dc and ac waveforms that produce the same average power in a resistor R.

The ac power, from Eqs. 9.1.6 and 9.1.7, is given by

$$P_{ac} = \frac{1}{RT}\int_0^T V_{max}^2 \cos^2(\omega t)\, dt = \frac{V_{max}^2}{2R}$$

Comparing, we get

$$V_{dc} = \frac{V_{max}}{\sqrt{2}}$$

This is the effective or rms value of the ac waveform, as you can see by using Eq. 9.1.9.

EXAMPLE 9.1.3 Find the rms value of each waveform in Fig. 9.6.

Solution

The following steps are used in Eq. 9.1.9: (1) Square $v(t)$, (2) find the average value of $v^2(t)$, and (3) take the square root to obtain the rms value. The square of each voltage waveform is shown in Fig. 9.7. The rms value for Fig. 9.7*a* is

$$V_{1,\text{rms}} = \sqrt{\frac{1}{T}\int_0^T 4\, dt} = 2\text{ V}$$

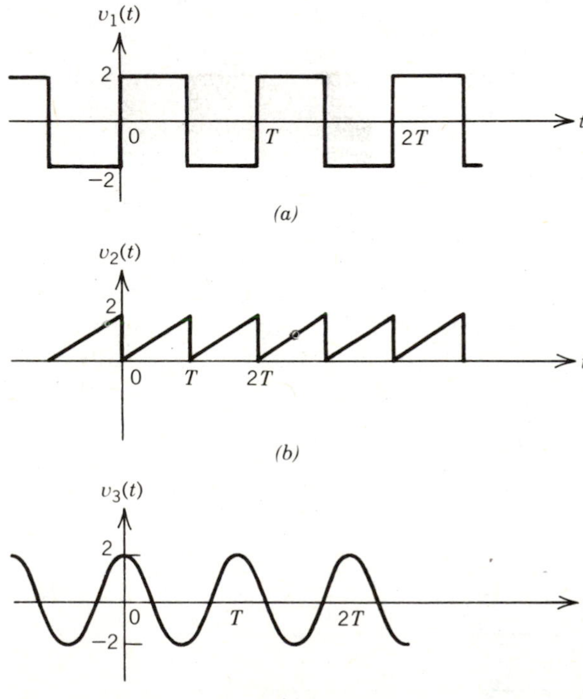

Figure 9.6

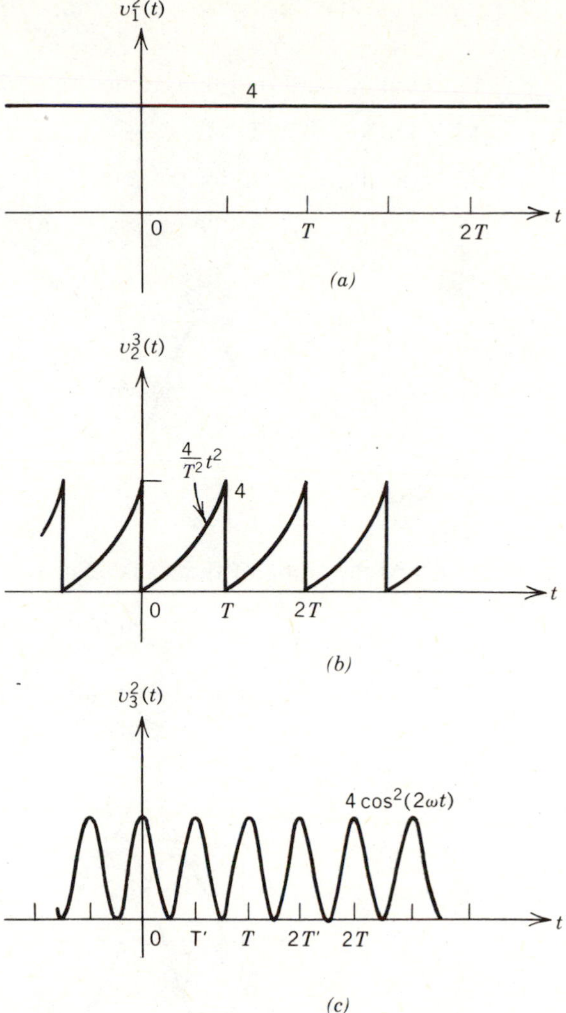

Figure 9.7. Square of the waveforms in Fig. 9.6.

For Fig. 9.7*b*,

$$V_{2,\text{rms}} = \sqrt{\frac{1}{T}\int_0^T \frac{4}{T^2}t^2\,dt} = \frac{2}{\sqrt{3}}\ \text{V}$$

For Fig. 9.7*c*,

$$V_{3,\text{rms}} = \sqrt{\frac{1}{T}\int_0^T 4\cos^2(2\omega t)\,\text{dt}} = \sqrt{2}\ \text{V}$$

One of the most widely used formulas in all of electrical engineering is the relationship

$$V_{rms} = \frac{V_{max}}{\sqrt{2}} \tag{9.1.11}$$

but keep in mind that this applies only to sinusoidal waveforms.

LEARNING EVALUATIONS

1. Write the equation for the sinusoid shown in Fig. 9.8.

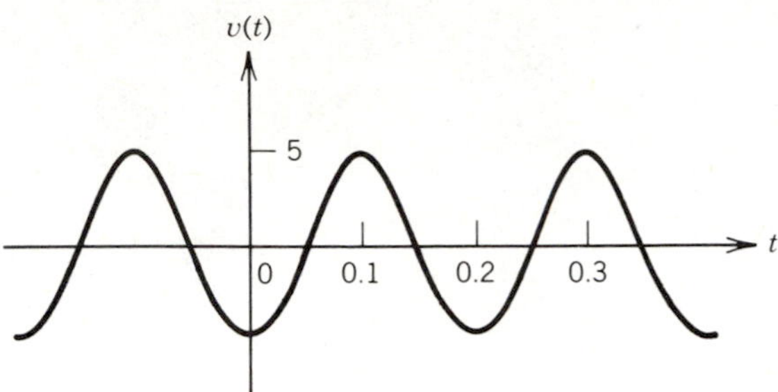

Figure 9.8

2. Draw the graph of the sinusoid given by

$$v(t) = 10\cos\left(25\pi t + \frac{\pi}{4}\right)$$

3. Determine the phase relationship between $x(t)$ and $y(t)$ in Fig. 9.9.

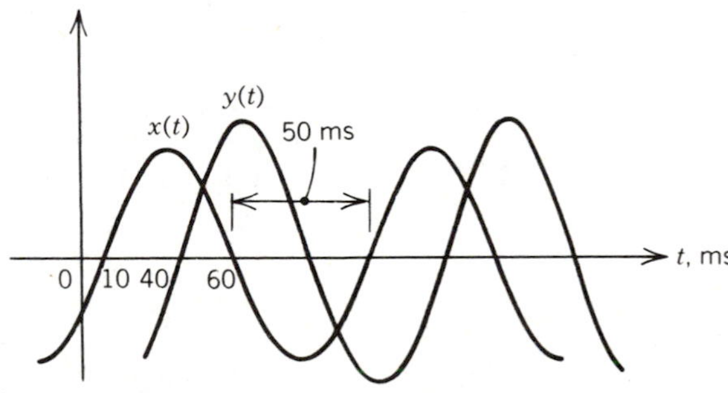

Figure 9.9

4. Find the rms value of the waveform in Fig. 9.10.

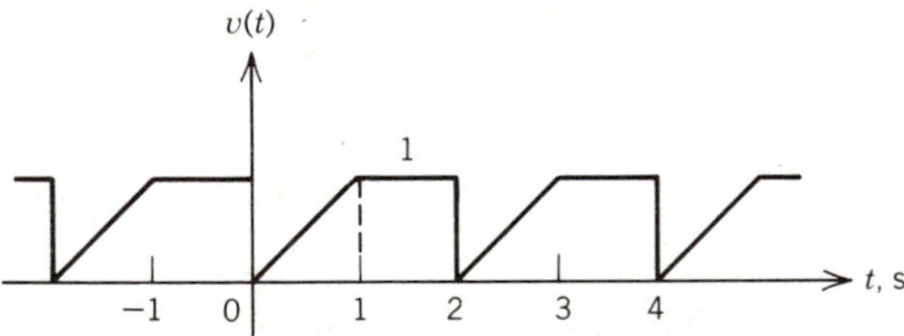

Figure 9.10

9.2 The Phasor Transform

LEARNING OBJECTIVES

After completing this section you should be able to do the following:

1. Find the phasor transform of a sinusoid.
2. Express the phasor transform in its exponential, polar, and rectangular forms.
3. Evaluate expressions requiring the addition, subtraction, multiplication, and division of phasors.

We are now capable of analyzing circuits of the type shown in Fig. 9.11 where the forcing function is sinusoidal. We could write mesh or node equations and use integral and differential calculus to solve for voltages and currents. However, this method is quite tedious. Charles Proteus Steinmetz (1865–1923), a German-born American electrical engineer, wishing to avoid this effort, succeeded in developing a method that reduced the difficulty of analysis of these circuits to that approaching (but not quite reaching) the simplicity of resistive circuit analysis. This method involves the *phasor transform*. In this approach integration and differentiation of sinusoids is replaced by the multiplication and division of complex numbers.

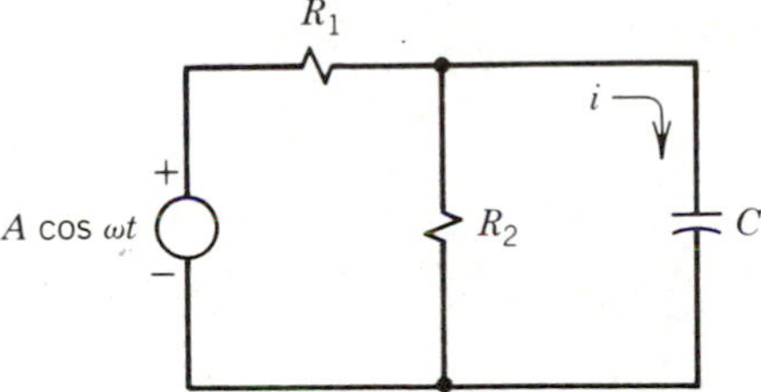

Figure 9.11. An *RC* circuit with a sinusoidal forcing function.

To understand the basis for what we will define as the phasor transform, we must first realize that any linear operation (in particular, integration and differentiation) on a sinusoid of a given frequency results in another sinusoid of the *same frequency*. This is shown in Fig. 9.12. The amplitude may change and the phase may be shifted but the frequency is the same.

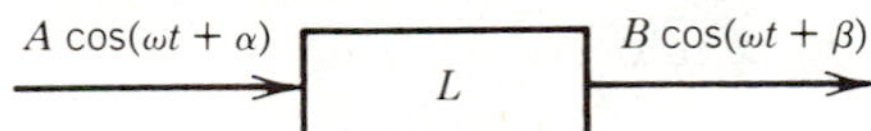

Figure 9.12. Linear transformation of a sinusoid.

Consider the sinusoid

$$y(t) = A \cos(\omega t + \theta) \tag{9.2.1}$$

Recalling that

$$e^{jx} = \cos x + j \sin x$$

Eq. 9.2.1 can be written as

$$y(t) = \mathrm{Re}[Ae^{j(\omega t+\theta)}]$$
$$= \mathrm{Re}[Ae^{j\theta}e^{j\omega t}]$$

All of the information contained in $y(t)$ that is subject to change by a linear system is included in the term

$$\mathbf{A} = Ae^{j\theta} \tag{9.2.2}$$

A is defined as the exponential form of the *phasor transform* of $A\cos(\omega t+\theta)$ and is represented in boldface type. The phasor transform (usually referred to simply as a "phasor") is most often used in another form

$$\mathbf{A} = A\underline{/\theta} \tag{9.2.3}$$

where $\underline{/\theta}$ is shorthand notation for the mathematically correct term $e^{j\theta}$. That is,

$$\underline{/\theta} = e^{j\theta} \tag{9.2.4}$$

Equation 9.2.3 is the polar form of the phasor transform and may be new to you. Consider the operation of multiplication of two phasors

$$(A\underline{/\alpha})(B\underline{/\beta}) \tag{9.2.5}$$

Using Eq. 9.2.4 in Eq. 9.2.5 we have

$$Ae^{j\alpha}Be^{j\beta} = ABe^{j\alpha}e^{j\beta} = ABe^{j(\alpha+\beta)}$$

or

$$(A\underline{/\alpha})(B\underline{/\beta}) = AB\underline{/\alpha+\beta} \tag{9.2.6}$$

Similarly for division,

$$\frac{A\underline{/\alpha}}{B\underline{/\beta}} = \frac{A}{B}\underline{/\alpha-\beta} \tag{9.2.7}$$

A third form of the phasor transform is

$$\mathbf{A} = A(\cos\theta + j\sin\theta) \tag{9.2.8}$$

This, of course, is called the rectangular form.

In summary, a phasor is a complex number that contains the amplitude and phase information associated with a *cosine* function at a given frequency.

EXAMPLE 9.2.1 Find the phasor transform of each of the following sinusoids:

a. $x(t) = 4\cos 200t$
b. $y(t) = -7.33\cos(100t - 15°)$
c. $z(t) = 56\sin(377t + 30°)$

Solution

a). The amplitude is 4 and the phase is 0° so the phasor is

$$\mathbf{X} = 4\underline{/0°}$$

b. The amplitude is -7.33 and the phase is $-15°$ so the phasor is

$$\mathbf{Y} = -7.33\angle -15°$$

c. This function is a sine wave and must be converted to a cosine by the trigonometric relation

$$\sin x = \cos\left(x - \frac{\pi}{2}\right)$$

so

$$\begin{aligned} 56\sin(377t + 30°) &= 56\cos(377t + 30° - 90°) \\ &= 56\cos(377t - 60°) \end{aligned}$$

Therefore the phasor is

$$\mathbf{Z} = 56\angle -60°$$

EXAMPLE 9.2.2 If

$$\mathbf{X} = -3.3\angle 22°$$

and

$$\mathbf{Y} = 14.9\angle -86°$$

find

$$\mathbf{R}_1 = \mathbf{XY}$$

and

$$\mathbf{R}_2 = \frac{\mathbf{X}}{\mathbf{Y}}$$

Solution

Using Eq. 9.2.6, we have

$$\begin{aligned} \mathbf{R}_1 = \mathbf{XY} &= (-3.3\angle 22°)(14.9\angle -86°) \\ &= -49.17\angle -64° = 49.17\angle -64 + 180 = 49.17\angle 116° \end{aligned}$$

Then using Eq. 9.2.7, we find

$$\mathbf{R}_2 = \frac{\mathbf{X}}{\mathbf{Y}} = \frac{-3.3\angle 22°}{14.9\angle -86°} = -0.22\angle 108° = 0.22\angle -72°$$

Let us emphasize again that operations with phasors apply only when all functions are of the same frequency.

We have now defined multiplication and division with phasors but what about addition and subtraction? In Example 9.2.2, if we want to find a phasor **S** where

$$\begin{aligned} \mathbf{S} &= \mathbf{X} + \mathbf{Y} \\ &= -3.3\angle 22° + 14.9\angle -86° \end{aligned}$$

we are confronted with the inability to add phasors in polar form. We must convert the phasors to rectangular form using Eq. 9.2.8. In this case

$$\begin{aligned}\mathbf{X} &= -3.3(\cos 22° + j\sin 22°)\\ &= -3.3(0.927 + j0.375)\\ &= = -3.059 - j1.236\end{aligned}$$

and

$$\begin{aligned}\mathbf{Y} &= 14.9[\cos(-86°) + j\sin(-86°)]\\ &= 1.039 - j14.864\end{aligned}$$

Then

$$\begin{aligned}\mathbf{S} &= \mathbf{X} + \mathbf{Y}\\ &= -3.059 - j1.236 + 1.039 - j14.864\\ &= -2.020 - j16.100\end{aligned}$$

Recall that a complex number of rectangular form

$$a + jb \tag{9.2.9}$$

can be converted to polar form

$$c\underline{/\phi}$$

by the following identities:

$$c = \sqrt{a^2 + b^2}\,; \qquad \phi = \tan^{-1}\frac{|b|}{|a|} \tag{9.2.10}$$

So

$$\mathbf{S} = 16.226\underline{/262.849°}$$

Notice that we must be extremely careful to evaluate the angle θ properly. Figure 9.13a shows $\theta = \phi$ as defined in Eq. 9.2.10 where a and b are positive. The angle used in the polar form of the phasor transform is always with reference to the positive real axis. In Fig. 9.13a there is no problem since Eq. 9.2.10 identifies the same angle. Figure 9.13b depicts the case where a is positive and b is negative. Calculation of θ must be related to ϕ depicted in the figure. From observation we can write $\theta = 360° - \phi$. In Fig. 9.13c the real component is negative so the angle defined by Eq. 9.2.10 is as shown in the figure. The angle θ is related to ϕ by

$$\theta = 180° - \phi$$

Finally, in Fig. 9.13d both real and imaginary components are negative and

$$\theta = 180° + \phi$$

In most sinusoidal steady-state analysis you will be required to convert between rectangular and polar form repeatedly. Many calculators have rectangular ↔ polar conversion functions and this eases the problem considerably.

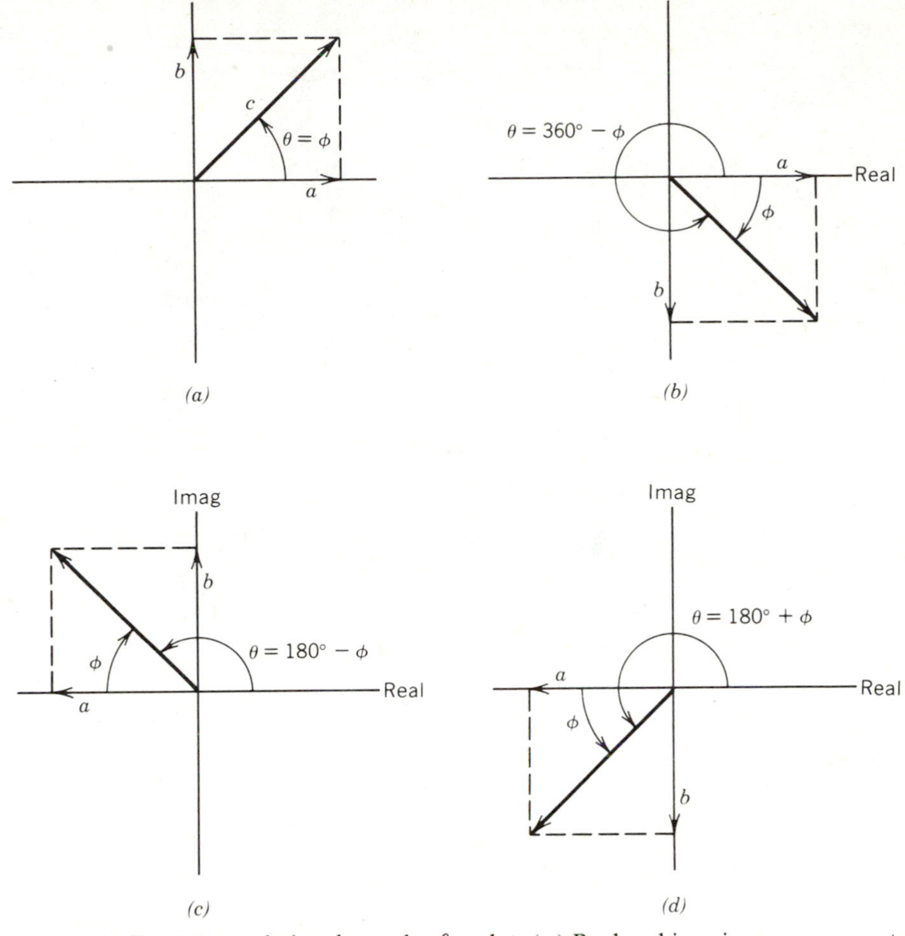

Figure 9.13. Four cases relating the angles θ and ϕ. (a) Real and imaginary components are positive. (b) Real component is positive, imaginary is negative. (c) Real component is negative and imaginary component is positive. (d) Real and imaginary components are negative.

EXAMPLE 9.2.3 If

$$\mathbf{V} = \frac{2 + j7}{8\angle 30^\circ + 4\angle -238^\circ} + 3\angle -57^\circ$$

find **V** in polar form.

Solution

The solution is found by completing the following steps:

1. Convert each phasor in the denominator to rectangular form.
2. Add the values found in step 1 and convert back to polar form.
3. Convert the numerator to polar form.

4. Find the quotient.
5. Convert the quotient and the remaining term to rectangular form.
6. Add the results of step 5 and convert to polar form.

Now carry out each step:

1. $8\angle 30^\circ = 6.93 + j4$

   $4\angle -238^\circ = -2.12 + j3.39$

2. $$\begin{aligned} 8\angle 30^\circ + 4\angle -238^\circ &= 6.93 + j4 - 2.12 + j3.39 \\ &= 4.81 + j7.39 \\ &= 8.82\angle 56.94^\circ \end{aligned}$$

3. $2 + j7 = 7.28\angle 74.05^\circ$

4. $\dfrac{7.28\angle 74.05}{8.82\angle 56.94} = 0.826\angle 17.10^\circ$

5. $0.826\angle 17.10^\circ = 0.79 + j0.24$

   $3\angle -57^\circ = 1.63 - j2.52$

6. $0.79 + j0.24 + 1.63 - j2.52 = 2.422 - j2.273$, so

$$\mathbf{V} = 3.322\angle -43.17^\circ$$

We must be careful not to lose sight of reality. We know that current is the movement of electrons and that we can convey information by controlling this movement in some predetermined fashion or by monitoring the movement caused by some external event. In order to be able to discuss, interpret, or alter this movement of electrons, we attempt to describe the movement by mathematical expressions (models). One such expression is the sinusoidal function which simply means that the electrons are moving in a manner described by the sinusoid. In this section we have established an alternate but equivalent way of describing the motion of the electrons—by the phasor transform. This means of modeling the behavior of the electrons in our circuit offers us some mathematical advantages in analyzing our circuit, which we will now make use of.

LEARNING EVALUATIONS

1. Find the phasor transform of each of the following functions.
   - a. $33.6\cos(722t + 13^\circ)$
   - b. $-14.8\sin\left(377t - \dfrac{\pi}{6}\right)$
   - c. $10\sin(1000t)$
2. For each form of the phasor below find the other two forms.
   - a. $14e^{-j\pi/2}$

b. $394.2\angle 48^\circ$

c. $-10 + j18$

3. Evaluate the following expressions.

a. $$\frac{(10\angle 30^\circ)(14 + j5)}{6e^{j45^\circ} - 8\angle 230^\circ}$$

b. $$\frac{(-6^{-j10})(30e^{-j\pi/10})}{(3\angle 65^\circ)(4\angle -65^\circ)}$$

9.3 Impedance and Admittance

LEARNING OBJECTIVES

After completing this section you should be able to do the following:

1. Derive the relationship between voltage and current phasors across circuit elements R, L, C.
2. Define impedance and admittance.
3. Transform a circuit from the time domain to the frequency domain.
4. Calculate the impedance or admittance between any two terminals of a network.
5. Combine impedances and admittances to achieve circuit reduction.

We have shown that sinusoidal sources can mathematically be represented by phasors but the question that now must be addressed is what effect this has on the voltage–current relationships for resistors, inductors, and capacitors. To answer this question, consider the circuit in Fig. 9.14 where v is the complex function

$$v(t) = Ae^{j(\omega t + \theta)}$$

and its phasor is

$$\mathbf{V} = A\angle\theta$$

If the element shown is a resistor of resistance R, then

$$i(t) = \frac{v}{R} = \frac{A}{R}e^{j(\omega t + \theta)}$$

or in phasor form

$$\mathbf{I} = \frac{A\angle\theta}{R} = \frac{\mathbf{V}}{R}$$

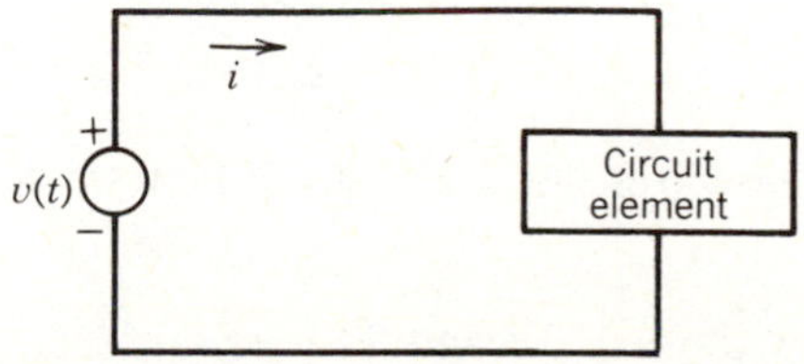

Figure 9.14. A circuit with a general circuit element.

Rearranging,

$$\mathbf{V} = R\mathbf{I} \tag{9.3.1}$$

that is, given a phasor voltage **V** applied to a resistive circuit, the resulting phasor current **I** is found by dividing the phasor voltage by the resistance R. Big deal, you say! This is the rule that we have always used. You are right for resistive elements but you are in for a surprise with the inductor and capacitor.

Assume now that the element shown is a capacitor. We know that

$$i(t) = C\frac{dv(t)}{dt} = C\frac{dAe^{j(\omega t+\theta)}}{dt}$$
$$= j\omega CAe^{j(\omega t+\theta)}$$

In phasor form,

$$\mathbf{I} = j\omega C\mathbf{V} \tag{9.3.2}$$

Equation 9.3.2 says that given a phasor voltage **V** across a capacitor, we can find the phasor current, **I**, through the capacitor simply by multiplying **V** by $j\omega C$! Differentiation is no longer necessary.

To illustrate using Fig. 9.14, assume

$$v(t) = 5\cos(10t + 30°)$$

and

$$C = 50\ \mu\text{F}$$

First let us find $i(t)$ the "conventional" way using our proven time domain technique.

$$i(t) = C\frac{dv}{dt} = (50 \times 10^{-6})(5)\frac{d}{dt}\cos(10t + 30°)$$
$$= -(10)(50 \times 10^{-6})(5)\sin(10t + 30°)$$
$$= -2.5\sin(10t + 30°)\ \text{mA} \tag{9.3.3}$$

Now transform the voltage and capacitance to the frequency domain and try our new technique.

$$\mathbf{V} = 5\angle 30°$$
$$j\omega C = j(10)(50 \times 10^{-6}) = (50 \times 10^{-5})\,j = 50 \times 10^{-5}\angle 90°$$

Using Eq. 9.3.2,

$$\mathbf{I} = j\omega C\mathbf{V} = (50 \times 10^{-5}\angle 90°)(5\angle 30°)$$
$$= 2.5 \times 10^{-3}\angle 120°$$

Now transform **I** back to the time domain:

$$i(t) = 2.5\cos(10t + 120°)\ \text{mA} \tag{9.3.4}$$

Are Eqs. 9.3.3 and 9.3.4 equivalent? If they are not, we have wasted a lot of time and effort in this chapter. Remember that

$$-\sin x = \sin(x + 180°)$$

and

$$\sin x = \cos(x - 90°)$$

Using these, Eq. 9.3.3 becomes

$$\begin{aligned} -2.5\sin(10t + 30°) &= 2.5\sin(10t + 30° + 180°) \\ &= 2.5\cos(10t + 30° + 180° - 90°) \\ &= 2.5\cos(10t + 120°) \end{aligned}$$

which is identical with Eq. 9.3.4.

Equation 9.3.2 can be rewritten as

$$\frac{\mathbf{V}}{\mathbf{I}} = \frac{1}{j\omega C} \tag{9.3.5}$$

The ratio of phasor voltage across a capacitor to phasor current through the capacitor is defined as *impedance* of the capacitor. This impedance must have the units of ohms since it is the ratio of voltage to current. Consequently, while in the time domain a capacitor has a capacitance, C, and voltage and current are related by

$$i = C\frac{dv}{dt}$$

in the frequency domain it has impedance, $(1/j\omega C)$, and voltage and current phasors are related by

$$\mathbf{V} = \frac{1}{j\omega C}\mathbf{I}$$

It will be useful in the future to recognize the following equivalent forms of the impedance of a capacitor:

$$\frac{1}{j\omega C} = \frac{-j}{\omega C} = \frac{1}{\omega C}\angle -90° = \frac{1}{\omega C\angle 90°} \tag{9.3.6}$$

Finally, assume the element in Fig. 9.14 is an inductor. We can write

$$\begin{aligned} i &= \frac{1}{L}\int_{-\infty}^{t} v\,dt = \frac{1}{L}\int_{-\infty}^{t} Ae^{j(\omega t+\theta)}\,dt \\ &= \frac{1}{j\omega L}Ae^{j(\omega t+\theta)} \end{aligned}$$

In phasor form,

$$\mathbf{I} = \frac{1}{j\omega L}\mathbf{V}$$

or

$$\mathbf{V} = j\omega L\mathbf{I} \tag{9.3.7}$$

Using our previous definition, we see that the impedance of an inductor is $j\omega L$.

To summarize, if we have a *circuit* with sinusoidal forcing functions and R, L, C components, we can *transform the circuit* from the time domain to the frequency domain by:

1. Replacing all forcing functions with their phasor equivalents.
2. Leaving all resistors unchanged.
3. Replacing all capacitors with their impedances $1/j\omega C$.
4. Replacing all inductors with their impedances $j\omega L$.

The voltage–current relationship for each element is then given by

$$\mathbf{V} = R\mathbf{I}$$

and

$$\mathbf{V} = j\omega L\mathbf{I}$$

and

$$\mathbf{V} = \frac{1}{j\omega C}\mathbf{I}$$

The impedance of a *network* is defined as the ratio of transformed voltage to transformed current. Thus impedance is a complex number. Figure 9.15 pictures a network with one terminal pair (port) where we can measure voltage and current. If the network is LTI, and if the voltage and current are sinusoidal, then the impedance $\mathbf{Z}(j\omega)$ is given by

$$\mathbf{Z}(j\omega) = \frac{\mathbf{V}(j\omega)}{\mathbf{I}(j\omega)} \tag{9.3.8}$$

Do not confuse the capital $\mathbf{Z}$ used here for impedance with the lower case z used in connection with digital systems.

The *admittance* $\mathbf{Y}(j\omega)$ is the inverse of $\mathbf{Z}(j\omega)$.

$$\mathbf{Y}(j\omega) = \frac{1}{\mathbf{Z}(j\omega)} = \frac{\mathbf{I}(j\omega)}{\mathbf{V}(j\omega)} \tag{9.3.9}$$

Admittance is a measure of how much current will flow for a given voltage. The

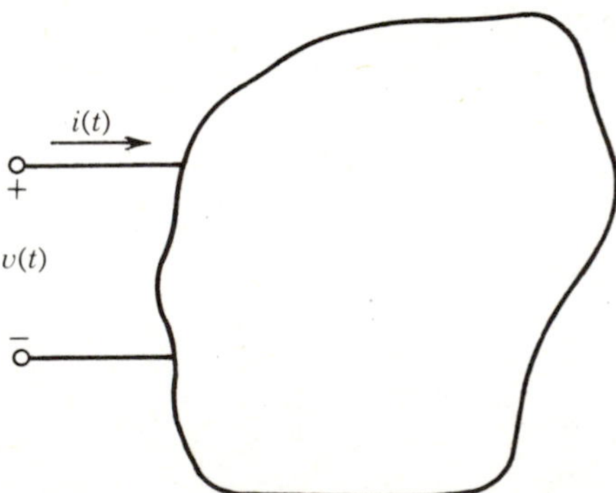

Figure 9.15. A one-terminal pair network.

larger the admittance, the larger is the current. Both admittance and impedance are complex numbers.

When we wish to talk about either impedance or admittance without specifying which is meant, we use the term immittence. Since immittence is a complex quantity, it specifies not only the relation between the magnitude of voltage and current, but also the relation between the angles. The immittence of a network is calculated by the same methods that were used for resistive circuits. The only difference is that we are now using complex numbers instead of real numbers to characterize each circuit element.

Impedance

The impedance of a series network is found by adding the impedance of each element. The impedance of the series RL circuit in Fig. 9.16a is given by

$$\mathbf{Z}_1(j\omega) = R + j\omega L \tag{9.3.10}$$

The impedance of the series RC circuit in Fig. 9.16b is given by

$$\begin{aligned}\mathbf{Z}_2(j\omega) &= R + \frac{1}{j\omega C} \\ &= R - j\frac{1}{\omega C}\end{aligned} \tag{9.3.11}$$

The impedance of the series RLC circuit in Fig. 9.16c is given by

$$\mathbf{Z}_3(j\omega) = R + j\left(\omega L - \frac{1}{\omega C}\right) \tag{9.3.12}$$

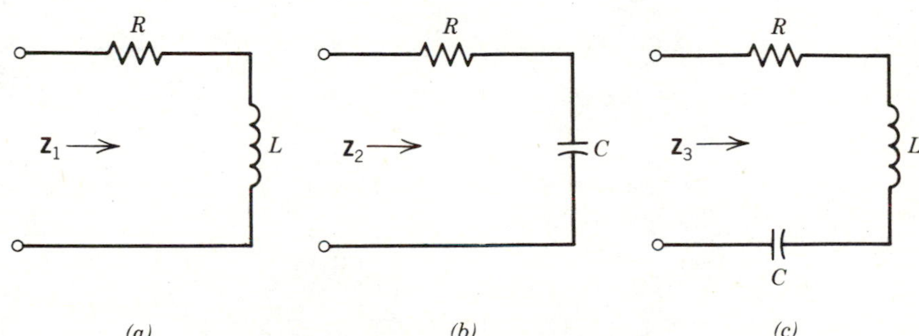

Figure 9.16. Network impedance.

The impedance of a parallel network is found by the same procedure used for parallel resistors. The impedance of the parallel RL network in Fig. 9.17a is given by

$$\mathbf{Z}_1(j\omega) = \frac{1}{(1/R) + (1/j\omega L)} = \frac{j\omega RL}{R + j\omega L} \tag{9.3.13}$$

The impedance of the parallel RC network in Fig. 9.17b is given by

$$\mathbf{Z}_2(j\omega) = \frac{1}{(1/R) + j\omega C} = \frac{R}{1 + j\omega RC} \tag{9.3.14}$$

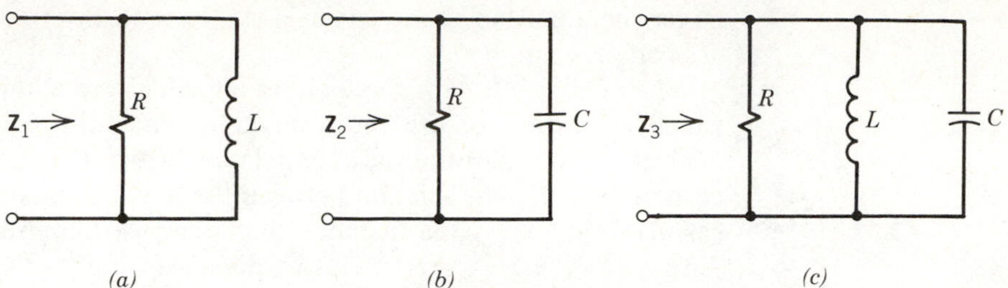

Figure 9.17. Networks with parallel elements.

The impedance of the parallel *RLC* network in Fig. 9.17*c* is given by

$$\mathbf{Z}_3(j\omega) = \frac{1}{(1/R)+(1/j\omega L)+j\omega C} = -\frac{j\omega RL}{j\omega L + R - \omega^2 RLC} \tag{9.3.15}$$

EXAMPLE 9.3.1 Find the impedance of the network in Fig. 9.18*a* if the sinusoidal forcing function has frequency $f = 60$ Hz.

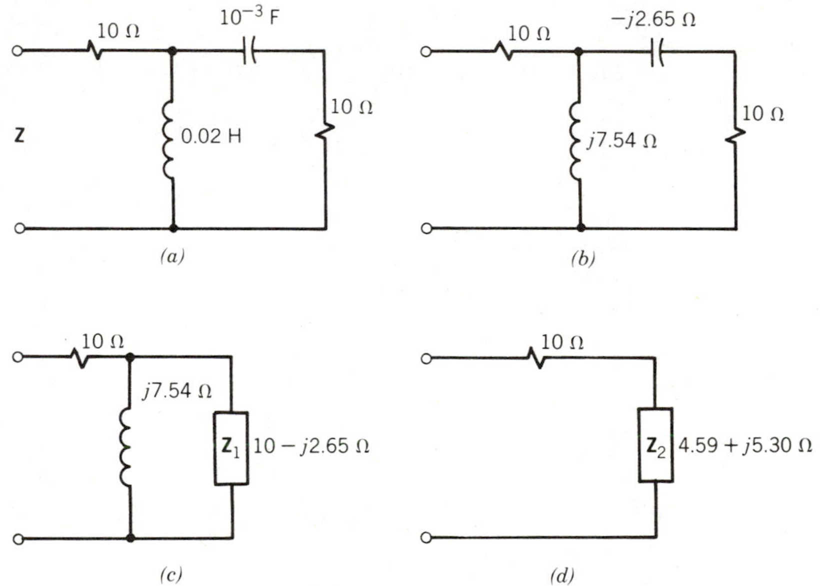

Figure 9.18

Solution

The first step is to replace each element by its impedance at $f = 60$ Hz (or $\omega = 377$ rad/s.) This is shown in Fig. 9.18*b*. Starting at the opposite end of the network from the input port, we combine the resistor and capacitor impedance to obtain

$$\mathbf{Z}_1 = 10 - j2.65$$

This impedance is now in parallel with the inductor impedance as shown in Fig. 9.18*c*. Combining, we obtain

$$\mathbf{Z}_2 = \frac{1}{\dfrac{1}{j7.54} + \dfrac{1}{10 - j2.65}} = \frac{j7.54(10 - j2.65)}{10 + j(7.54 - 2.65)} = 4.59 + j5.30$$

This impedance is in series with the 10-Ω resistor as shown in Fig. 9.18*d*. Adding, we obtain the final answer.

$$\mathbf{Z}(j\omega) = 10 + (4.59 + j5.30) = 14.59 + j5.30 = 15.52\angle 19.56^\circ$$

Some special terminology has arisen in connection with impedance since its inception by Steinmetz about the turn of the century. The real part of the impedance is called resistance, although there may be no apparent connection to the resistor values in the circuit. Notice that in Example 9.3.1 there are two 10-Ω resistors, yet there is no simple way to arrive at a value of 14.59 for the real part of $\mathbf{Z}$ in the final answer. The imaginary part of the impedance is termed the *reactance*, and it is labeled X. Thus $\mathbf{Z} = R + jX$ is the impedance of any network.

Admittance

Since admittance is the reciprocal of impedance, we can calculate it by first finding $\mathbf{Z}$ and then using $\mathbf{Y} = 1/\mathbf{Z}$. However, it will pay later dividends to be equally fluent in calculating complex admittance. The admittance of a parallel network is found by adding. The admittance of the parallel RL network in Fig. 9.19*a* is found by first relabeling the elements by their admittance, Fig. 9.19*b*, and then adding.

$$\mathbf{Y}_1(j\omega) = \frac{1}{R} - \frac{j}{\omega L} \tag{9.3.16}$$

The admittances for Figs. 9.19*c* and 9.19*e* are found in the same manner.

$$\mathbf{Y}_2(j\omega) = \frac{1}{R} + j\omega C \tag{9.3.17}$$

$$\mathbf{Y}_3(j\omega) = \frac{1}{R} + j\left(\omega C - \frac{1}{\omega L}\right) \tag{9.3.18}$$

You should check to see that $\mathbf{Y}_1$, $\mathbf{Y}_2$, and $\mathbf{Y}_3$ are the reciprocals of $\mathbf{Z}_1$, $\mathbf{Z}_2$, and $\mathbf{Z}_3$ given in Eqs. 9.3.13, 9.3.14, and 9.3.15, respectively.

Since admittance is the reciprocal of impedance, the simplest procedure for finding the admittance of series elements is usually to find the impedance first, then invert. The admittance of the series RL circuit in Fig. 9.20*a* is given by

$$\mathbf{Y}_1(j\omega) = \frac{1}{R + j\omega L} \tag{9.3.19}$$

$\mathbf{Y}_2$ is given by

$$\mathbf{Y}_2(j\omega) = \frac{1}{R + (1/j\omega C)} = \frac{j\omega C}{1 + j\omega RC} \tag{9.3.20}$$

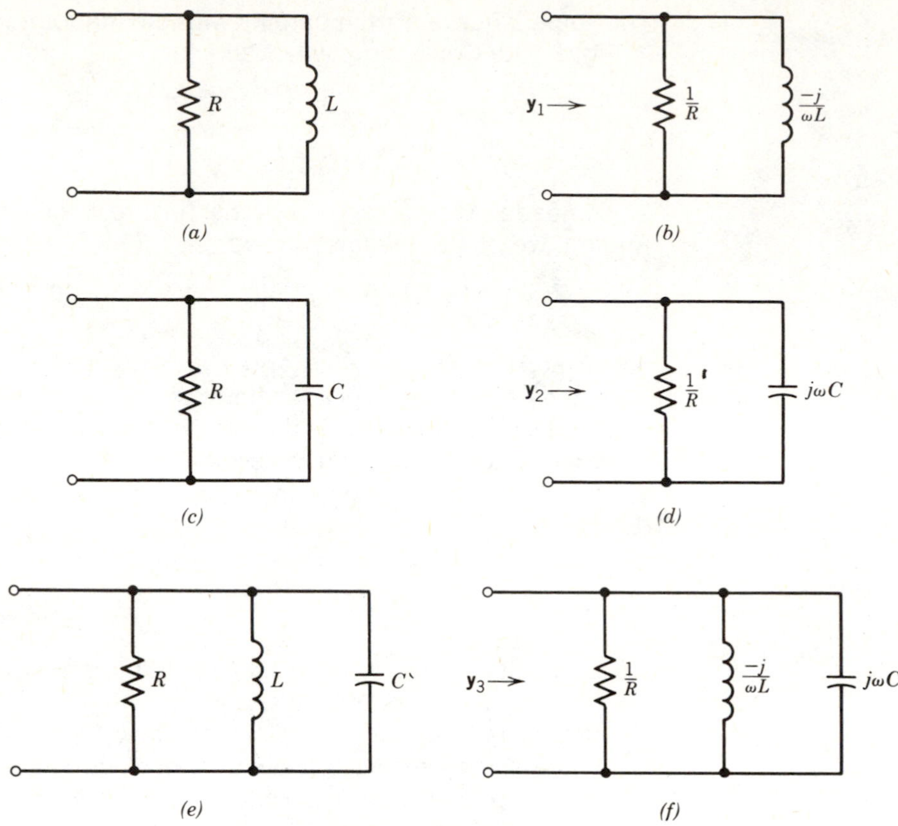

Figure 9.19. Admittance functions.

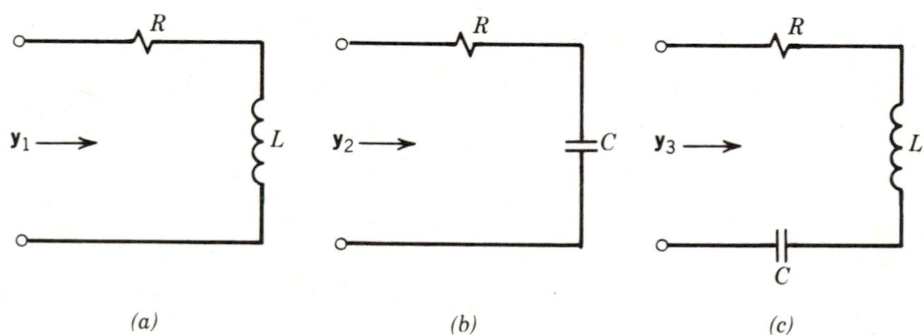

Figure 9.20. Admittance of series circuit.

And $\mathbf{Y}_3$ is given by

$$\mathbf{Y}_3(j\omega) = \frac{1}{R + j\omega L + (1/j\omega C)} = \frac{j\omega C}{(1 - \omega^2 LC) + j\omega RC} \tag{9.3.21}$$

EXAMPLE 9.3.2 Calculate the admittance of the network in Example 9.3.1.

Solution

The obvious method of solution is to invert the answer to Example 9.3.1.

$$\mathbf{Y}(j\omega) = \frac{1}{\mathbf{Z}(j\omega)} = \frac{1}{14.59 + j5.30} = 0.061 - j0.022$$

Let us try to arrive at this same answer from scratch. The circuit of Fig. 9.18*a* is represented in the frequency domain in terms of admittance in Fig. 9.21*a*. Combining the series capacitor and resistor gives

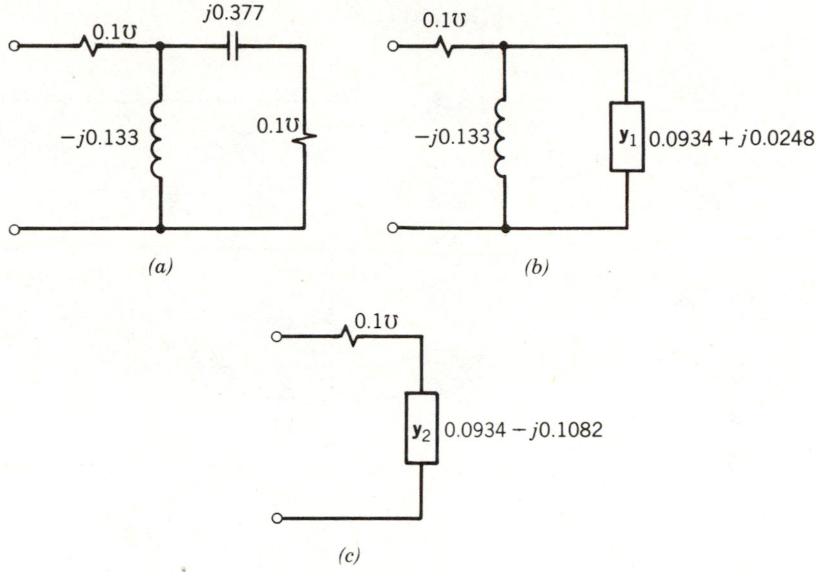

Figure 9.21

$$\mathbf{Y}_1 = \frac{j0.0377}{0.1 + j0.377} = 0.0934 + j0.0248$$

This is in parallel with the inductor as shown in Fig. 9.21*b*. Adding $\mathbf{Y}_1$ to the inductor admittance gives

$$\mathbf{Y}_2 = 0.0934 - j0.1082$$

as shown in Fig. 9.21*c*. Finally, combining $\mathbf{Y}_2$ with the 0.1-℧ conductor gives

$$\mathbf{Y} = \frac{0.1\mathbf{Y}_2}{0.1 + \mathbf{Y}_2} = 0.0616 - j0.022$$

which compares favorably with $1/\mathbf{Z}$.

Here is some more special terminology. The real part of the admittance is called *conductance*, and it is labeled G. The imaginary part is called *susceptance*, and it is labeled B. Therefore, we write

$$\mathbf{Y} = G + jB$$

The susceptance is positive for a capacitor, and negative for an inductor.

LEARNING EVALUATIONS

1. Transform the circuits in Figure 9.22 to the frequency domain.

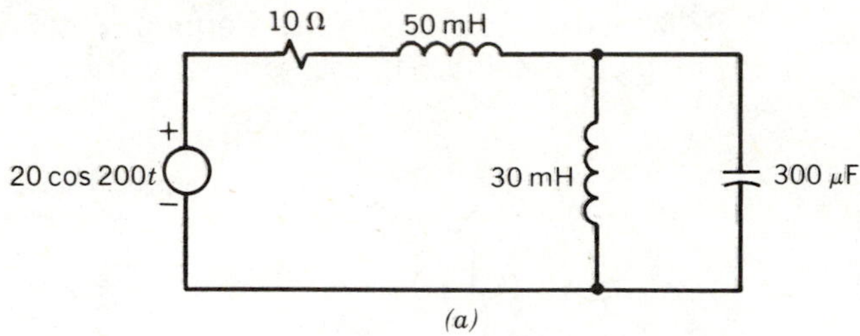

(a)

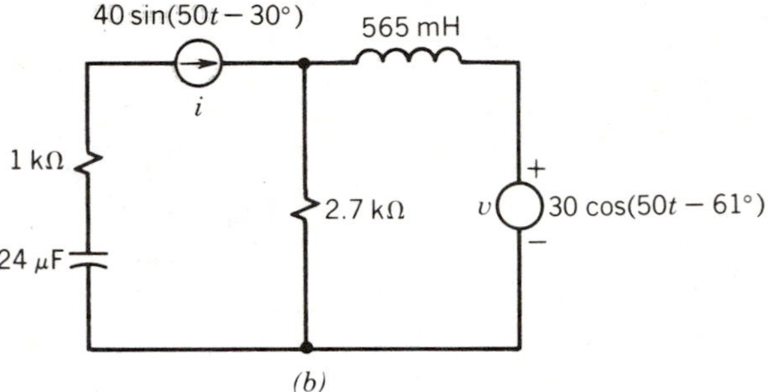

(b)

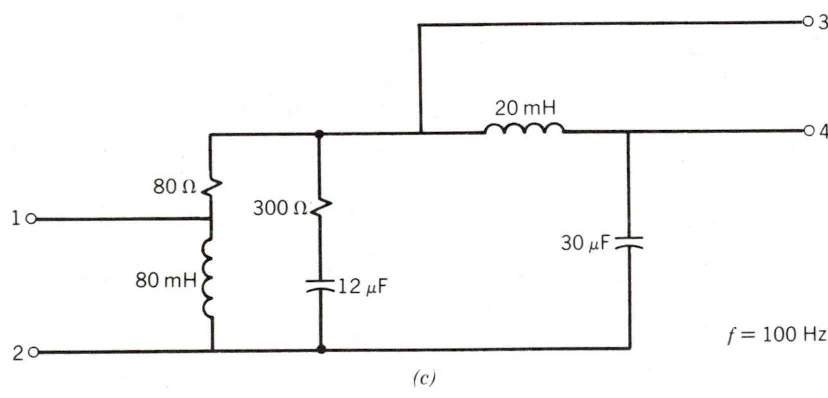

(c)

Figure 9.22

2. Calculate the impedance of the network in Fig. 9.22*c* between
   a. Terminals 1–2
   b. Terminals 3–4
3. Repeat Learning Evaluation 2, except find admittance instead of impedance.
4. Calculate the impedance of the network in Fig. 9.23 if the frequency of the sinusoidal forcing function is 60 Hz.
5. Find the admittance of the network in Fig. 9.23. The frequency is 60 Hz.

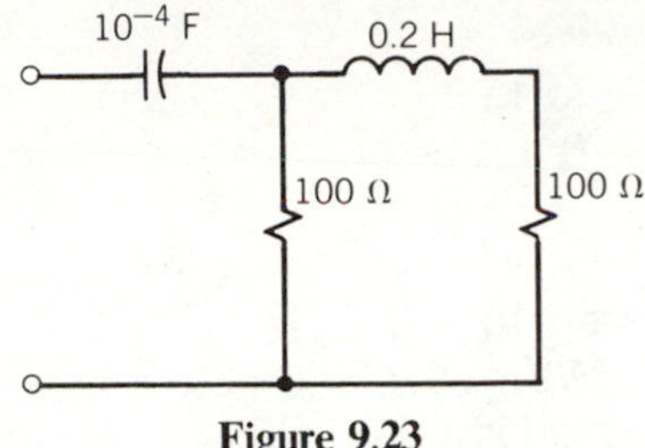

Figure 9.23

9.4 Steady-State Sinusoidal Analysis

LEARNING OBJECTIVES

After completing this section you should be able to do the following:

1. State and use the three-step procedure for analyzing circuits with sinusoidal forcing functions.
2. Use any of the methods introduced in Chapters 1–3 to accomplish objective 1.

The phrase "steady-state sinusoidal analysis" implies that all transients in our circuit have died out (steady state) and that all forcing functions in our circuit are sinusoidal in nature. As we have repeated throughout the text, analysis of a circuit means finding all voltages and currents in the circuit that are of interest to us. This still holds true for sinusoidal circuits. The only difference is that we will use complex numbers in sinusoidal circuit analysis.

The approach, based on previous sections of this chapter, is quite straightforward. The procedure involves the following steps:

1. Transform the circuit to the frequency domain.
   a. Replace all sinusoidal functions by their phasor transforms.
   b. Replace all circuit elements with their equivalent impedance or admittance.
2. Use any of the techniques developed previously to solve for the desired phasor variables.
3. Transform these phasor variables back to the time domain.

EXAMPLE 9.4.1 Find $i(t)$ in Fig. 9.24a if $R = 1$ kΩ, $C = 0.2$ μF, and $v(t) = 50\cos(2000\pi t)$. Also find the voltage, $v_c(t)$, across the capacitor.

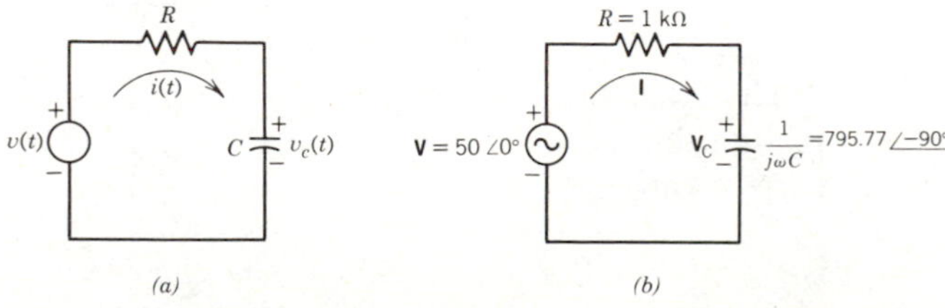

Figure 9.24. A series RC circuit with sinusoidal forcing functions. (a) Time domain circuit. (b) Frequency domain circuit.

Solution

Step 1a

The phasor transform of $v(t)$ is

$$\mathbf{V} = 50\angle 0^\circ$$

Step 1b

We observe that $\omega = 2000\pi$. The resistance R is unaffected by the transform and is 1 kΩ. The impedance of the capacitor is

$$\frac{1}{j\omega C} = \frac{1\angle -90^\circ}{(2000)(0.2)(10^{-6})} = 795.77\angle -90^\circ$$

The transformed circuit is shown in Fig. 9.24b.

Step 2

By Ohm's law,

$$\mathbf{I} = \frac{\mathbf{V}}{1000 + 795.77\angle -90^\circ} = \frac{50\angle 0^\circ}{1000 - j795.77}$$

$$= \frac{50\angle 0^\circ}{1277.99\angle -38.51^\circ}$$

$$= 0.0391\angle 38.51^\circ$$

By voltage division,

$$\mathbf{V}_c = \frac{795.77\angle -90^\circ}{1000 + 795.77\angle -90^\circ} 50\angle 0^\circ$$

$$= \frac{39788.50\angle -90^\circ}{1277.99\angle -38.51^\circ}$$

$$= 31.13\angle -51.49^\circ$$

Step 3

Transforming back to the time domain, we have

$$i(t) = 0.0391 \cos(2000\pi t + 38.51^\circ)\ \text{A}$$

and

$$v_c(t) = 31.12 \cos(2000\pi t - 51.49^\circ)\ \text{V}$$

The Example 9.4.1 points out a very important fact concerning the phase relationship between the sinusoidal voltage across a capacitor and the sinusoidal current through the capacitor:

> For sinusoidal circuits, the current through a capacitor leads the voltage across the capacitor by 90°.

EXAMPLE 9.4.2 Refer to Fig. 9.25*a*. If $v(t) = 10\sin 377t$, find $v_L(t)$ and $i_L(t)$ and determine the phase relationship between these two variables.

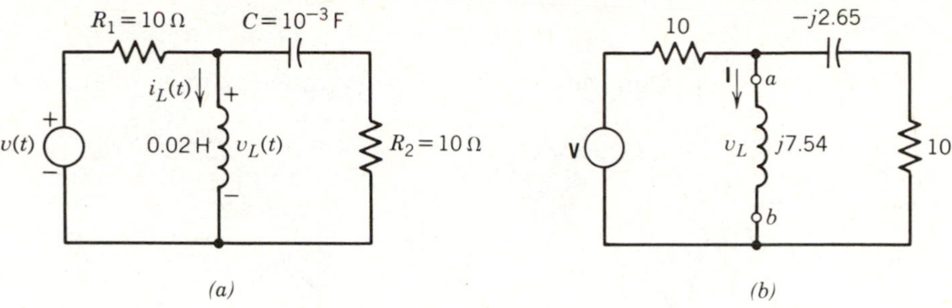

Figure 9.25. Circuit for Example 9.4.2.

Solution

Step 1a

The phasor transform of $v(t)$ is

$$\mathbf{V} = 10\angle -90^\circ$$

where we note that $v(t)$ must be expressed in "cosine" form.

Step 1b

We observe that $\omega = 377$. The resistor values remain unchanged. The impedance of the capacitor is

$$\frac{1}{j\omega C} = \frac{1}{(377)(10^{-3})}\angle -90^\circ = 2.65\angle -90^\circ = -j2.65$$

The transformed circuit is shown in Fig. 9.25*b*.

Step 2

There are numerous approaches to solving the problem. For practice, let us use Thévenin's theorem with the inductor representing the load. First we must find the Thévenin voltage which is the voltage across terminals a–b when the inductor is removed from the circuit. By voltage division, this is

$$\mathbf{V}_{TH} = \frac{10 - j2.65}{10 + (10 - j2.65)}\mathbf{V} = \frac{10.35\angle -14.86^\circ}{20.17\angle -7.55^\circ}10\angle -90^\circ$$

$$= 5.13\angle -97.30^\circ$$

To use Thévenin's theorem with phasor analysis we must replace the term "Thévenin resistance" with the more general term "Thévenin impedance," which is found in exactly the same manner as we did previously. Examining our circuit, we observe that with respect to terminals a–b, the 10-Ω impedance on the left is in parallel with the series combination of the 10-Ω and $-j2.65$-Ω impedances on the right. Therefore,

$$\mathbf{Z}_{TH} = \frac{(10)(10 - j2.65)}{10 + 10 - j2.65} = \frac{103.5\angle -14.84^\circ}{20.17\angle -7.55^\circ}$$

$$= 5.13\angle -7.30^\circ\ \Omega$$

Our Thévenin equivalent circuit is shown in Fig. 9.26. By inspection,

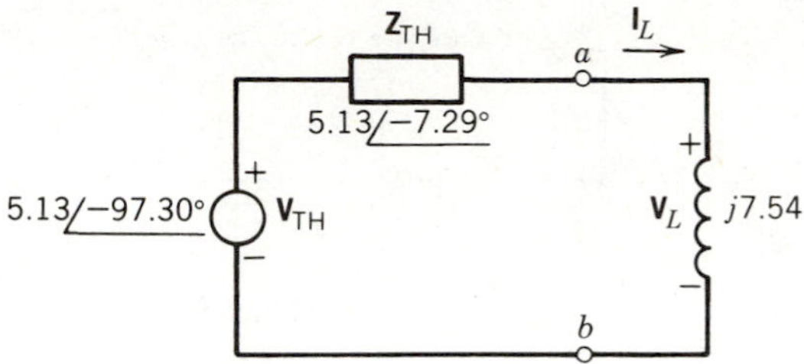

Figure 9.26. Thévenin equivalent of Fig. 9.25*b*.

$$\mathbf{I}_L = \frac{5.13\angle -97.30^\circ}{5.13\angle -7.30^\circ + j7.54}$$

$$= \frac{5.13\angle -97.30^\circ}{8.56\angle 53.55}$$

$$= 0.60\angle -150.86^\circ$$

and

$$\mathbf{V}_L = (\mathbf{I}_L)(j7.54)$$

$$= 4.52\angle -60.86^\circ$$

Step 3

Translating back to the time domain results in

$$v_L(t) = 4.52\cos(377t - 60.86^\circ)$$

and

$$i_L(t) = 0.60\cos(377t - 150.86^\circ)$$

From this we see that the voltage leads the current by 90°. This important concept can be more formally stated as:

> For sinusoidal circuits, the voltage across an inductor always leads the current through the inductor by 90°.

Mesh and Nodal Analysis

For more complicated circuits, step 2 in our procedure will usually involve use of mesh or nodal analysis techniques which were introduced in Chapter 3 for resistive circuits. We will review these methods here as they apply to transformed circuits. Consider the circuit in Fig. 9.27. Writing mesh equations gives

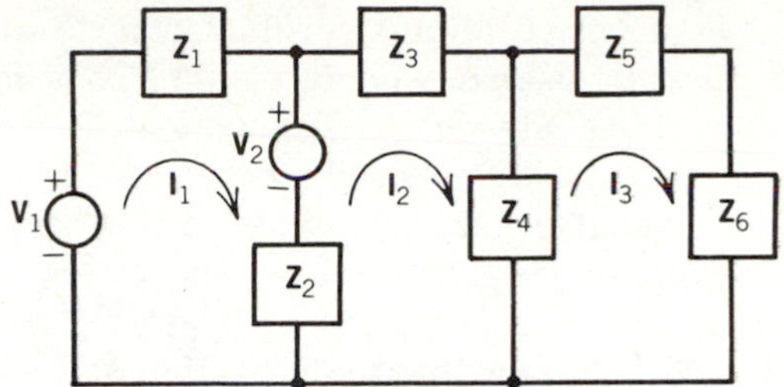

Figure 9.27. Transformed circuit with three meshes.

$$\mathbf{V}_1 - \mathbf{V}_2 = (\mathbf{Z}_1 + \mathbf{Z}_2)\mathbf{I}_1 - \mathbf{Z}_2\mathbf{I}_2$$
$$\mathbf{V}_2 = -\mathbf{Z}_2\mathbf{I}_1 + (\mathbf{Z}_2 + \mathbf{Z}_3 + \mathbf{Z}_4)\mathbf{I}_2 - \mathbf{Z}_4\mathbf{I}_3$$
$$0 = -\mathbf{Z}_4\mathbf{I}_2 + (\mathbf{Z}_4 + \mathbf{Z}_5 + \mathbf{Z}_6)\mathbf{I}_3$$

These equations may be written in matrix form as

$$\begin{bmatrix} \mathbf{V}_1 - \mathbf{V}_2 \\ \mathbf{V}_2 \\ 0 \end{bmatrix} = \begin{bmatrix} (\mathbf{Z}_1 + \mathbf{Z}_2) & -\mathbf{Z}_2 & 0 \\ -\mathbf{Z}_2 & (\mathbf{Z}_2 + \mathbf{Z}_3 + \mathbf{Z}_4) & -\mathbf{Z}_4 \\ 0 & -\mathbf{Z}_4 & (\mathbf{Z}_4 + \mathbf{Z}_5 + \mathbf{Z}_6) \end{bmatrix} \begin{bmatrix} \mathbf{I}_1 \\ \mathbf{I}_2 \\ \mathbf{I}_3 \end{bmatrix} \tag{9.4.1}$$

Note that the elements of the square matrix are impedances. This is a matrix equation of the form

$$\mathbf{V} = \mathbf{Z}\mathbf{I} \tag{9.4.2}$$

The matrix **V** is $n \times 1$, **Z** is $n \times n$, and **I** is $n \times 1$, where there are n meshes in the network.

Here are some important observations about Eq. 9.4.1. The three entries in the **V** matrix represent the forcing function for each mesh. $\mathbf{V}_1 - \mathbf{V}_2$ is the sum of all voltage sources (forcing functions) around mesh 1. $\mathbf{V}_2$ is the forcing function in mesh 2, and there are no sources in mesh 3. A source is positive in Eq. 9.4.1, if it will force current to flow in the assumed direction. Thus V_2 is negative with regard to mesh 1, but positive in mesh 2.

The main diagonal of **Z** consists of the self-impedances in each mesh, and all terms are positive. The off-diagonal terms are all negative and they represent between-mesh impedances. The impedance matrix is of the form

$$\mathbf{Z} = \begin{bmatrix} \mathbf{Z}_{11} & \mathbf{Z}_{12} & \mathbf{Z}_{13} \\ \mathbf{Z}_{21} & \mathbf{Z}_{22} & \mathbf{Z}_{23} \\ \mathbf{Z}_{31} & \mathbf{Z}_{32} & \mathbf{Z}_{33} \end{bmatrix} \tag{9.4.3}$$

$\mathbf{Z}_{11}$, $\mathbf{Z}_{22}$, and $\mathbf{Z}_{33}$ represent the sum of all impedances in each mesh, respectively. The matrix is symmetrical, meaning that $\mathbf{Z}_{12} = \mathbf{Z}_{21}$, $\mathbf{Z}_{13} = \mathbf{Z}_{31}$, and $\mathbf{Z}_{23} = \mathbf{Z}_{32}$. This happens in part because of the lack of controlled or dependent sources. With reference to Fig. 9.27, $\mathbf{Z}_{12} = -\mathbf{Z}_2$, $\mathbf{Z}_{13} = 0$ because there are no impedances common to mesh 1 and 3, and $\mathbf{Z}_{23} = -\mathbf{Z}_4$.

These observations will allow us to write the matrix equation directly from the circuit diagram without the time consuming and error-prone intermediate step of writing individual mesh equations. Remember an important point about the mesh currents. We assumed all currents to be in the clockwise direction. It is essential that all be in the same direction, either cw or ccw, before our procedure is applicable.

EXAMPLE 9.4.3 Write the mesh equations in matrix form for the circuit in Fig. 9.28.

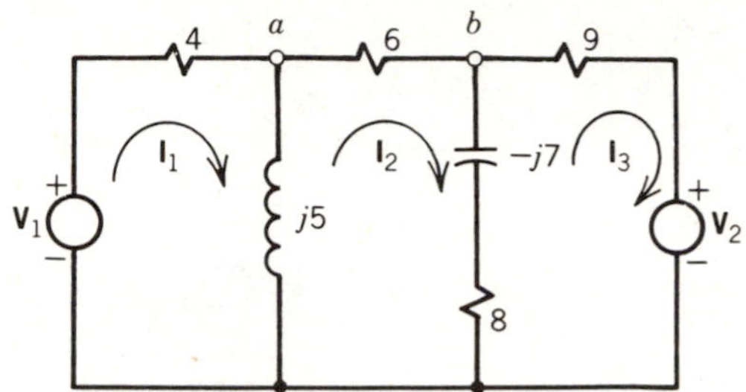

Figure 9.28. Circuit for Example 9.4.3.

Solution

$$\begin{bmatrix} \mathbf{V}_1 \\ 0 \\ -\mathbf{V}_2 \end{bmatrix} = \begin{bmatrix} (4+j5) & -j5 & 0 \\ -j5 & (14-j2) & (-8+j7) \\ 0 & (-8+j7) & (17-j7) \end{bmatrix} \begin{bmatrix} \mathbf{I}_1 \\ \mathbf{I}_2 \\ \mathbf{I}_3 \end{bmatrix}$$

Notice that $\mathbf{V}_1$ is positive in mesh 1, and $\mathbf{V}_2$ is negative in mesh 3. The impedance $\mathbf{Z}_{22}$ is

$$\mathbf{Z}_{22} = j5 + 6 - j7 + 8 = 14 - j2$$

The other terms should be self-explanatory.

EXAMPLE 9.4.4 Write the mesh equations in matrix form for the circuit in Fig. 9.29.

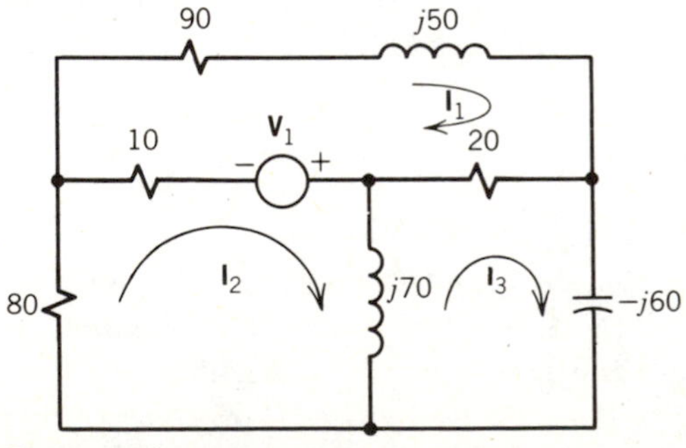

Figure 9.29. Circuit for Example 9.4.4.

Solution

$$\begin{bmatrix} -\mathbf{V}_1 \\ \mathbf{V}_1 \\ 0 \end{bmatrix} = \begin{bmatrix} (120+j50) & -10 & -20 \\ -10 & (90+j70) & -j70 \\ -20 & -j70 & (20+j10) \end{bmatrix} \begin{bmatrix} \mathbf{I}_1 \\ \mathbf{I}_2 \\ \mathbf{I}_3 \end{bmatrix}$$

Cramer's rule may now be used to solve for any mesh current. The only difference between the material presented here and that in Chapter 3 is the use of complex numbers. Voltages, currents, and impedances are generally complex valued. Keep in mind that the complex valued voltages and currents actually represent sinusoidal time functions.

The application of Cramer's rule to Eq. 9.4.1 results in the following solutions for $\mathbf{I}_1$, $\mathbf{I}_2$, and $\mathbf{I}_3$.

$$\mathbf{I}_1 = \frac{\begin{vmatrix} \mathbf{V}_1 - \mathbf{V}_2 & -\mathbf{Z}_2 & 0 \\ \mathbf{V}_2 & (\mathbf{Z}_2 + \mathbf{Z}_3 + \mathbf{Z}_4) & -\mathbf{Z}_4 \\ 0 & -\mathbf{Z}_4 & (\mathbf{Z}_4 + \mathbf{Z}_5 + \mathbf{Z}_6) \end{vmatrix}}{\begin{vmatrix} (\mathbf{Z}_1 + \mathbf{Z}_2) & -\mathbf{Z}_2 & 0 \\ -\mathbf{Z}_2 & (\mathbf{Z}_2 + \mathbf{Z}_3 + \mathbf{Z}_4) & -\mathbf{Z}_4 \\ 0 & -\mathbf{Z}_4 & (\mathbf{Z}_4 + \mathbf{Z}_5 + \mathbf{Z}_6) \end{vmatrix}}$$

$$\mathbf{I}_2 = \frac{\begin{vmatrix} (\mathbf{Z}_1 + \mathbf{Z}_2) & \mathbf{V}_1 - \mathbf{V}_2 & 0 \\ -\mathbf{Z}_2 & \mathbf{V}_2 & -\mathbf{Z}_4 \\ 0 & 0 & (\mathbf{Z}_4 + \mathbf{Z}_5 + \mathbf{Z}_6) \end{vmatrix}}{|\mathbf{Z}|}$$

$$\mathbf{I}_3 = \frac{\begin{vmatrix} (\mathbf{Z}_1 + \mathbf{Z}_2) & -\mathbf{Z}_2 & \mathbf{V}_1 - \mathbf{V}_2 \\ -\mathbf{Z}_2 & (\mathbf{Z}_2 + \mathbf{Z}_3 + \mathbf{Z}_4) & \mathbf{V}_2 \\ 0 & -\mathbf{Z}_4 & 0 \end{vmatrix}}{|\mathbf{Z}|}$$

The denominator is always the same, $|\mathbf{Z}|$. The numerator is found by replacing the appropriate column in $\mathbf{Z}$ by the forcing function. That is, replace column 1 by $\mathbf{V}$ in the numerator to solve for $\mathbf{I}_1$, and so on.

EXAMPLE 9.4.5 Use mesh analysis to solve for $\mathbf{V}_b$ in Fig. 9.30.

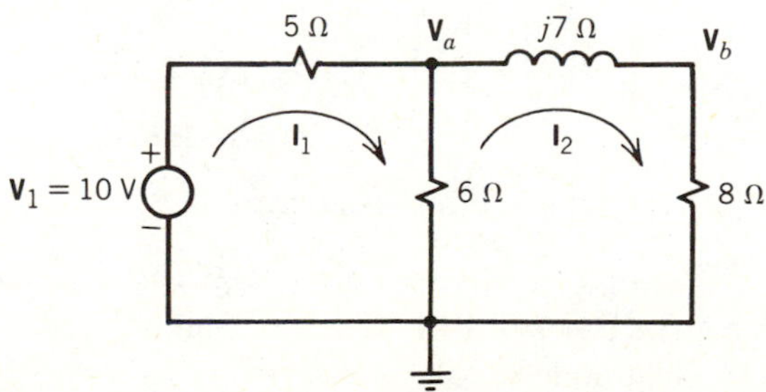

Figure 9.30. Circuit for Example 9.4.5.

Solution

Since $\mathbf{V}_b$ is the voltage across the 8-Ω resistor, we will find $\mathbf{I}_2$ first.

$$\mathbf{I}_2 = \frac{\begin{vmatrix} 11 & 10 \\ -6 & 0 \end{vmatrix}}{\begin{vmatrix} 11 & -6 \\ -6 & (14 + j7) \end{vmatrix}} = \frac{60}{140.9\angle 33.1^\circ} = 0.426\angle -33.1^\circ$$

Then $\mathbf{V}_b = 8\mathbf{I}_2 = 3.41\angle -33.1^\circ$.

Nodal analysis is also a very useful technique in sinusoidal analysis. Consider the network in Fig. 9.31. There are three nodes labeled 1, 2, and 3. The voltage at each node with respect to the ground node is $\mathbf{V}_1$, $\mathbf{V}_2$, and $\mathbf{V}_3$, respectively. There are two sinusoidal current sources, and the circuit elements are represented by their admittances. Writing nodal equations gives

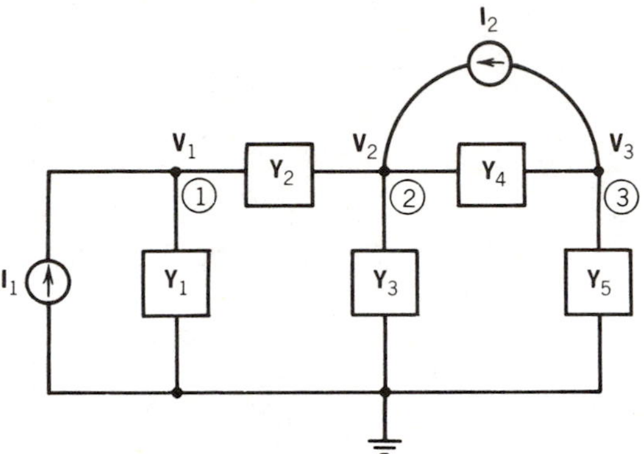

Figure 9.31. A three-node circuit.

$$\begin{aligned} \mathbf{I}_1 &= (\mathbf{Y}_1 + \mathbf{Y}_2)\mathbf{V}_1 - \mathbf{Y}_2\mathbf{V}_2 \\ \mathbf{I}_2 &= -\mathbf{Y}_2\mathbf{V}_1 + (\mathbf{Y}_2 + \mathbf{Y}_3 + \mathbf{Y}_4)\mathbf{V}_2 - \mathbf{Y}_4\mathbf{V}_3 \\ -\mathbf{I}_2 &= -\mathbf{Y}_4\mathbf{V}_2 + (\mathbf{Y}_4 + \mathbf{Y}_5)\mathbf{V}_3 \end{aligned}$$

These equations may be written in matrix form as

$$\begin{bmatrix} \mathbf{I}_1 \\ \mathbf{I}_2 \\ -\mathbf{I}_2 \end{bmatrix} = \begin{bmatrix} (\mathbf{Y}_1 + \mathbf{Y}_2) & -\mathbf{Y}_2 & 0 \\ -\mathbf{Y}_2 & (\mathbf{Y}_2 + \mathbf{Y}_3 + \mathbf{Y}_4) & -\mathbf{Y}_4 \\ 0 & -\mathbf{Y}_4 & (\mathbf{Y}_4 + \mathbf{Y}_5) \end{bmatrix} \begin{bmatrix} \mathbf{V}_1 \\ \mathbf{V}_2 \\ \mathbf{V}_3 \end{bmatrix} \tag{9.4.4}$$

This is a matrix equation of the form

$$\mathbf{I} = \mathbf{Y}\mathbf{V} \tag{9.4.5}$$

The matrix $\mathbf{I}$ is $n \times 1$, $\mathbf{Y}$ is $n \times n$, and $\mathbf{V}$ is $n \times 1$, where n is the number of independent nodes.

Notice that the **I** matrix represents the current forcing function at each node. The sources $\mathbf{I}_1$ and $\mathbf{I}_2$ are positive at nodes 1 and 2, but $\mathbf{I}_2$ is negative at node 3. If the source forces current into the node it is positive, otherwise it is negative.

The main diagonal of the **Y** matrix consists of the admittances common to each node, and all terms are positive. The off-diagonal terms represent the between-node admittances, and they are all negative. The admittance matrix has the form

$$\mathbf{Y} = \begin{bmatrix} \mathbf{Y}_{11} & \mathbf{Y}_{12} & \mathbf{Y}_{13} \\ \mathbf{Y}_{21} & \mathbf{Y}_{22} & \mathbf{Y}_{23} \\ \mathbf{Y}_{31} & \mathbf{Y}_{32} & \mathbf{Y}_{33} \end{bmatrix} \tag{9.4.6}$$

The admittance matrix for a passive network is symmetrical because $\mathbf{Y}_{ij} = \mathbf{Y}_{ji}$ for all i, j.

We may write equations directly in matrix form according to the following algorithm.

1. The forcing function consists of all current sources into each node.

$$\mathbf{I} = \begin{bmatrix} \mathbf{I}_{\mathrm{I}} \\ \mathbf{I}_{\mathrm{II}} \\ \mathbf{I}_{\mathrm{III}} \end{bmatrix}$$

   where $\mathbf{I}_{\mathrm{I}}$ is the algebraic sum of all sources into node 1, $\mathbf{I}_{\mathrm{II}}$ is the algebraic sum of all sources into node 2, and $\mathbf{I}_{\mathrm{III}}$ is the algebraic sum of all sources into node 3.

2. The admittance matrix consists of a main diagonal $\mathbf{Y}_{11}, \mathbf{Y}_{22}, \mathbf{Y}_{33}$, where $\mathbf{Y}_{ii}$ is the sum of all admittances common to node i. The off-diagonal terms are all negative, and they consist of admittances between nodes.

EXAMPLE 9.4.6 Write the nodal equations in matrix form for the circuit in Fig. 9.32.

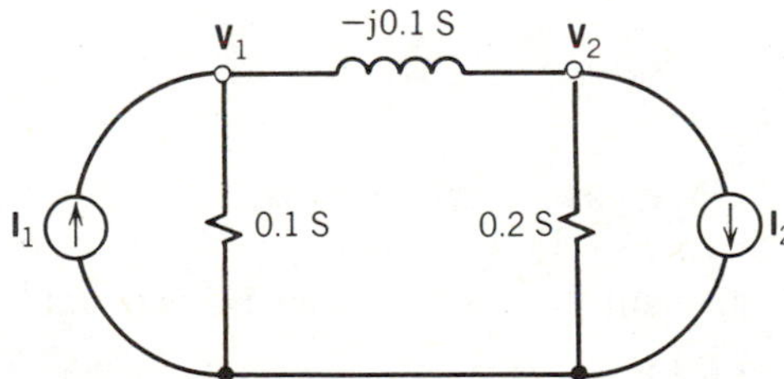

Figure 9.32. Circuit for Example 9.4.6.

Solution

$$\begin{bmatrix} \mathbf{I}_1 \\ -\mathbf{I}_2 \end{bmatrix} = \begin{bmatrix} (0.1 - j0.1) & j0.1 \\ j0.1 & (0.2 - j0.1) \end{bmatrix} \begin{bmatrix} \mathbf{V}_1 \\ \mathbf{V}_2 \end{bmatrix}$$

So far, all of our examples for both mesh analysis and nodal analysis have contained the "correct" form of sources. voltage sources in series with impedances were used in mesh analysis, and current sources in parallel with admittances were used in nodal analysis. Let us illustrate the procedure when this is not the case by using nodal analysis to solve Example 9.4.5.

EXAMPLE 9.4.7 Use nodal analysis to find $\mathbf{V}_b$ in Fig. 9.30.

Solution

Use source transformation to replace the voltage source in series with an impedance by the equivalent current source in parallel with an impedance, as shown in Fig. 9.33. The circuit in Fig. 9.30 is now redrawn in Fig. 9.34 with this equivalent source. All impedances are replaced by their admittances. Now we may write the matrix equation using Cramer's rule directly to solve for $\mathbf{V}_b$.

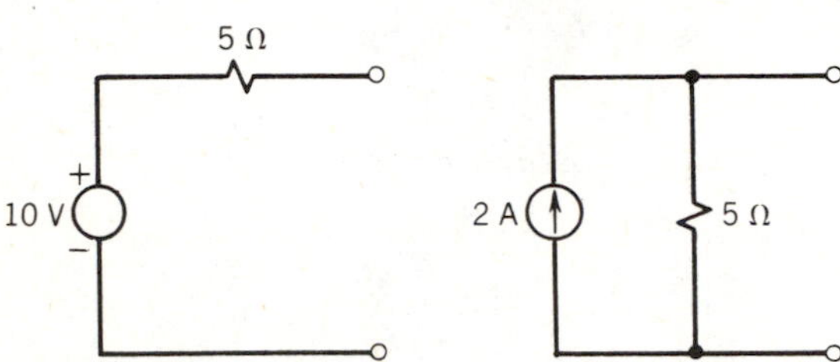

Figure 9.33. Source transformation for the 10-V source in series with the 5-Ω resistor.

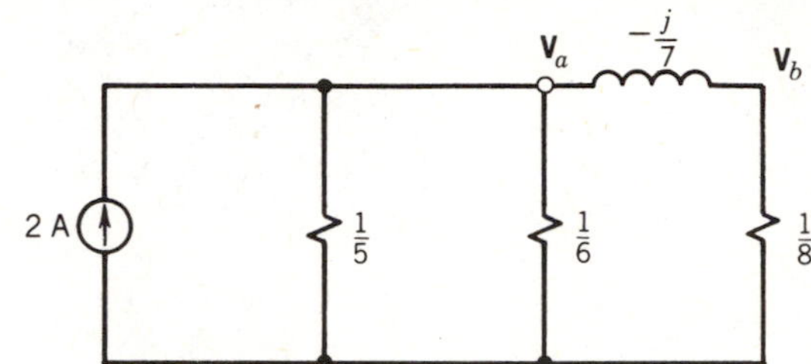

Figure 9.34. Equivalent circuit for circuit in Fig. 9.30.

$$\mathbf{V}_b = \frac{\begin{vmatrix} \frac{1}{5}+\frac{1}{6}-\frac{j}{7} & 2 \\ \frac{j}{7} & 0 \end{vmatrix}}{\begin{vmatrix} \frac{1}{5}+\frac{1}{6}-\frac{j}{7} & \frac{j}{7} \\ \frac{j}{7} & \frac{1}{8}-\frac{j}{7} \end{vmatrix}} = \frac{0.286\angle -90^\circ}{0.084\angle -56.87^\circ} = 3.41\angle -33.1^\circ$$

We have seen in this section that all the techniques for circuit analysis that we have previously employed are applicable to sinusoidal analysis when the circuit elements are in their transformed state.

LEARNING EVALUATIONS

1. Find $i(t)$ in Fig. 9.35 if $v(t) = 32.4\cos(400t + 28^\circ)$

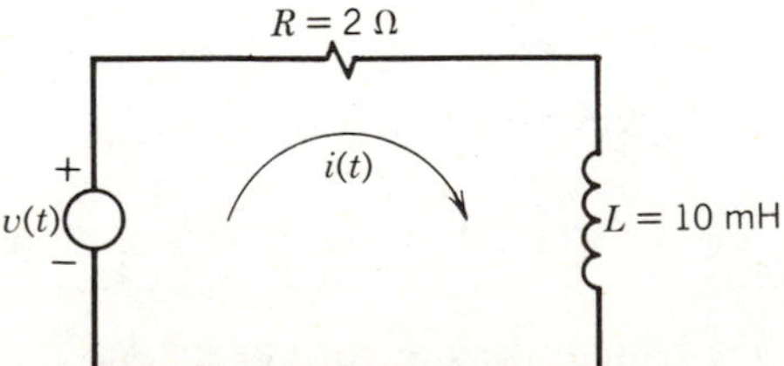

Figure 9.35

2. For Example 9.4.2, assume the frequency of the input, $v(t)$, has changed to 500 rad/s. Find the new values of $v_L(t)$ and $i_L(t)$.
3. Solve for the current $i_L(t)$ in Fig. 9.36 if $v_1(t) = 45\cos(2\pi t + 36°)$. Do this in two ways.
   a. Write mesh equations and solve for $i_L(t)$ directly.
   b. Write nodal equations, find the voltage across the inductor, then use Ohm's law.

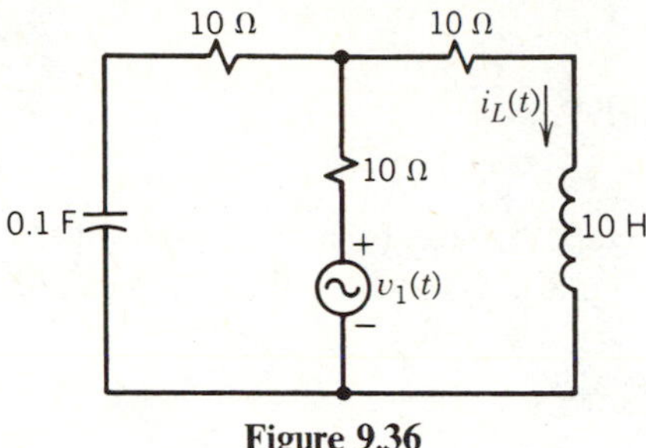

Figure 9.36

9.5 Transfer Functions

LEARNING OBJECTIVES

After completing this section you should be able to do the following:

1. Define continuous-time transfer function.
2. Define discrete-time transfer function.
3. Find the transfer function for a given circuit containing R, L, and C components.
4. Find the transfer function from the difference equation for a discrete-time system.

In the previous chapter we mentioned that the exponential signal could play the same type of role as the δ function in the analysis of LTI systems. This was because the signal e^{st} for continuous-time systems, and z^k for discrete-time systems, had the following two desirable properties.

1. The system response to this input signal can easily be determined.
2. Most other input signals can be expressed as the sum of terms related to this input signal.

This section is devoted to illustrating property 1, and most of the remainder of this text is devoted to frequency analysis, which is the use of property 2 with the exponential forcing function.

Complex Frequency

The terms s in e^{st} and z in z^k are complex terms called the complex frequency. For our discussion, consider the term s where

$$s = \sigma + j\omega \tag{9.5.1}$$

where σ is called the neper frequency and ω the radian frequency. σ is measured in units called nepers per second, while ω is measured in radians per second. When no confusion will result, either term is referred to simply as frequency.

We can relate our standard continuous-time signals to the complex exponential. If

$$v(t) = Ae^{st} \tag{9.5.2}$$

when

$$\sigma = \omega = 0$$

then

$$v(t) = A$$

Here A must be a real constant if $v(t)$ is to be a real signal. Hence, a constant signal is one where the complex frequency is zero. However, if

$$\sigma = a$$

and

$$\omega = 0$$

where a is an arbitrary real constant, then Eq. 9.5.2 becomes

$$v(t) = Ae^{at}$$

which of course is our standard exponential function. Again A must be real if $v(t)$ is to be real. Now if

$$\sigma = 0$$

and

$$\omega = \omega_1$$

where ω_1 is a real constant, Eq. 9.5.2 is

$$v(t) = Ae^{j\omega_1 t}$$

In this case A is, in general, complex. If $v(t)$ is to be a real function, then we must relate it to the real part of $Ae^{j\omega_1 t}$

$$v(t) = \text{Re}(Ae^{j\omega_1 t})$$

but this is the cosine function we defined earlier in this chapter.

Finally, if

$$\sigma = \sigma$$

and

$$\omega = \omega$$

then

$$\begin{aligned} v(t) &= Ae^{(\sigma + j\omega)t} \\ &= Ae^{\sigma t}e^{j\omega t} \end{aligned}$$

Again for $v(t)$ to be a real function we must insist that

$$\begin{aligned} v(t) &= \text{Re}\{Ae^{\sigma t}e^{j\omega t}\} \\ &= A_1 e^{\sigma t}\cos(\omega t + \theta) \end{aligned}$$

So we see that we can represent an exponentially damped sinusoid with complex frequency having nonzero real and imaginary parts.

We will develop the transfer function concept using the complex frequency s but realizing that for sinusoidal circuits

$$s = j\omega \tag{9.5.3}$$

Similar arguments hold true for the discrete-time systems. For discrete-time sinusoidal circuits,

$$z = e^{j\omega} \tag{9.5.4}$$

Continuous-Time Transfer Function $H(s)$

It is a relatively simple matter to characterize a continuous-time LTI system by the input–output pair

$$e^{st} - \text{response to } e^{st}$$

where s is a complex number. Here is an example.

Find the output voltage $v_2(t)$ if the input to the RC low-pass filter in Fig. 9.37 is an exponential $v_1(t) = e^{st}$.

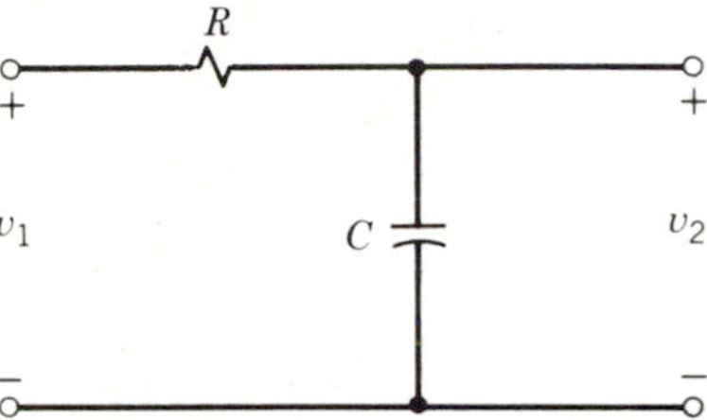

Figure 9.37. A low-pass RC filter.

The differential equation relating input to output may be found by using Kirchhoff's law, combined with the component equations. The result is

$$\frac{d}{dt}v_2(t) = -\frac{v_2(t)}{RC} + \frac{v_1(t)}{RC}$$

The steady-state solution, with $v_1(t) = e^{st}$, is given by

$$v_{2\mathrm{ss}}(t) = ke^{st} = \frac{1/RC}{s + 1/RC}e^{st}$$

and the transient solution is given by

$$v_{2t}(t) = Ae^{-t/RC}$$

The complete solution is

$$v_2(t) = \frac{1/RC}{s + 1/RC}e^{st} + Ae^{-t/RC} \tag{9.5.5}$$

where the constant A is determined from initial conditions.

With $H(s)$ defined by

$$H(s) = \frac{1/RC}{s + 1/RC}$$

the voltage $v_2(t)$ is composed of two parts. The forced or steady-state response is $H(s)e^{st}$, and the source-free or transient response is $Ae^{-t/RC}$. The term $H(s)$ is called the transfer function, system function, or filter characteristic of the network. Note that it is defined by

$$\frac{\text{steady-state response to } e^{st}}{e^{st}} = \frac{H(s)e^{st}}{e^{st}} = H(s) \tag{9.5.6}$$

This definition is valid so long as two conditions are met: (1) the system is asymptotically stable, and (2) the system is physically realizable. A system is asymptotically stable if it always returns to the same state after the input is removed. A system is physically realizable if the output does not occur before the input is applied. For future reference we formalize the definition of the transfer function.

> For an LTI, asymptotically stable, physically realizable continuous-time system, the function
>
> $$H(s) = \frac{\text{steady-state response to } e^{st}}{e^{st}}$$
>
> is said to be the transfer function of the system.

The transfer function can also be defined in terms of the impulse response $h(t)$ as

$$H(s) = \int_{-\infty}^{+\infty} h(t)e^{-st}\,dt \tag{9.5.7}$$

This is the more general definition. Equation 9.5.6 is equivalent to Eq. 9.5.7 under the conditions of asymptotic stability and physical realizability.

Let us return now to the complete solution given by Eq. 9.5.5. If the input is an exponential signal, then we can find the response by using Eq. 9.5.5 except when $s = 1/RC$. But suppose the input signal has the same frequency as the natural response, that is, $s = 1/RC$. Equation 9.5.5 does not provide the correct expression for the response in this case, and it is easy to see why. The transfer function is infinite if $s = 1/RC$. We call this a *pole* of the transfer function. This is an important fact to remember. The poles of the transfer function are the natural frequencies of the system.

Calculating $H(s)$ for Circuits

We said at the beginning of this section that it is a relatively simple matter to characterize an LTI system by the system function. Here is a procedure to use with electric circuits. This procedure can be justified by going the longer route of solving the system differential equation with an e^{st} forcing function.

1. Replace each capacitor by the equivalent impedance $Z_c(s) = 1/sC$. (For sinusoidal circuit, remember that $s = j\omega$ so this is our familiar impedance term.)

2. Replace each inductor by the equivalent impedance $Z_L(s) = sL$.
3. Resistors remain unchanged. That is, $Z_R(s) = R$.
4. Find the ratio of response/input, as in dc circuit theory. This ratio is $H(s)$.

For example, the RC low-pass filter in Fig. 9.38a is replaced by the equivalent circuit in Fig. 9.38b. The ratio $V_2(s)/V_1(s)$ is the transfer function, given by

$$H(s) = \frac{V_2(s)}{V_1(s)} = \frac{1/sC}{R + 1/sC} = \frac{1/RC}{s + 1/RC}$$

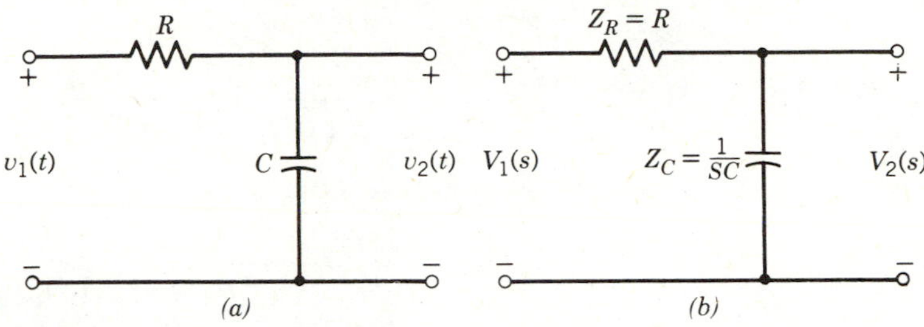

Figure 9.38. The RC circuit in the time domain and the frequency domain.

EXAMPLE 9.5.1 Find the transfer function $I_L(s)/V(s)$ for the circuit in Fig. 9.39a.

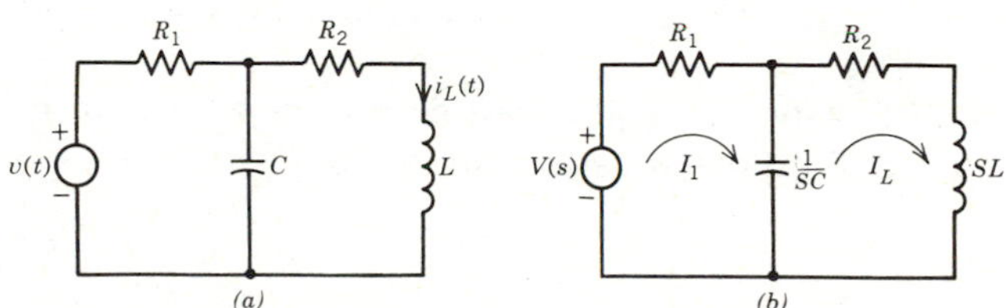

Figure 9.39. The time and frequency domain representations of a circuit.

Solution

Replace the elements by their impedance to obtain the circuit shown in Fig. 9.39b. The procedure for solution is now identical to that used in dc circuit theory. There are several equivalent methods available to us. Probably the most common method is to write loop equations, from which we obtain

$$V(s) = I_1\left(R_1 + \frac{1}{sC}\right) - I_L\left(\frac{1}{sC}\right)$$

$$0 = -I_1\left(\frac{1}{sC}\right) + I_L\left(R_2 + sL + \frac{1}{sC}\right)$$

or

$$I_L(s) = \frac{\begin{vmatrix} \left(R_1 + \frac{1}{sC}\right) & V(s) \\ -\frac{1}{sC} & 0 \end{vmatrix}}{\begin{vmatrix} \left(R_1 + \frac{1}{sC}\right) & -\frac{1}{sC} \\ -\frac{1}{sC} & \left(R_2 + sL + \frac{1}{sC}\right) \end{vmatrix}}$$

$$= \frac{V(s)}{s^2LCR_1 + s(CR_1R_2 + L) + (R_1 + R_2)}$$

Therefore, $H(s)$ is given by

$$H(s) = \frac{I_L(s)}{V(s)} = \frac{1}{s^2LCR_1 + s(CR_1R_2 + L) + (R_1 + R_2)}$$

The transfer function is a relation between the transform of the system input and the transform of the system output, where the input and output can be any measurable parameters. In Example 9.5.1, the transfer function has the units of admittance since the input excitation is a voltage and the response is a current. The point is that $H(s)$ can be the ratio of transforms of any current or voltage to any other current or voltage in the circuit.

EXAMPLE 9.5.2 A typical equivalent circuit for an electronic circuit is shown in Fig. 9.40*a*. Find the voltage gain V_2/V_1.

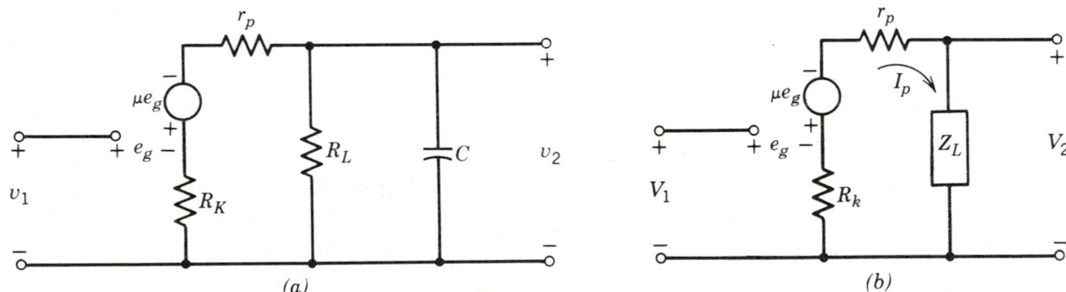

Figure 9.40. The equivalent circuit for an electronic circuit.

Solution

Replace the parallel R_LC circuit by an equivalent impedance as shown in Fig. 9.40*b*. Now write two loop equations.

$$V_1 - e_g + I_pR_k = 0$$

$$I_pR_k + \mu e_g + I_pr_p + I_pZ_L = 0$$

Solve for e_g from the first equation and substitute into the second equation to obtain

$$0 = I_p[(\mu + 1)R_k + r_p + Z_L] + \mu V_1$$

and with $I_p = V_2/Z_L$ we obtain

$$0 = \frac{V_2}{Z_L}[(\mu + 1)R_k + r_p + Z_L] + \mu V_1$$

or finally,

$$H(s) = \frac{V_2}{V_1} = \frac{-\mu Z_L}{(\mu + 1)R_k + r_p + Z_L}$$

$$= \frac{-\mu R_L}{s[CR_L R_k(\mu + 1) + CR_L r_p] + (\mu + 1)R_k + r_p + R_L}$$

EXAMPLE 9.5.3 Find the transfer function $H(s)$ for the circuit in Fig. 9.41*a*. The input is $v_1(t)$ and the output is $i_2(t)$.

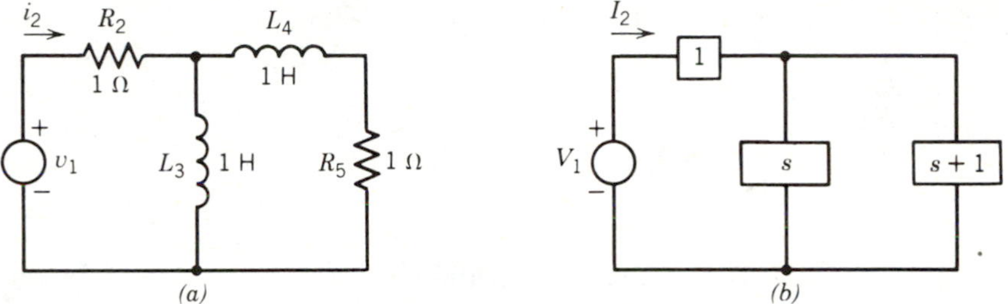

Figure 9.41. The time domain representation of a circuit.

Solution

Transform the circuit as shown in Fig. 9.41*b*. Direct circuit reduction then results in

$$\frac{I_2}{V_1} = H(s) = \frac{2s + 1}{s^2 + 3s + 1} \tag{9.5.8}$$

EXAMPLE 9.5.4 In Example 9.5.3, assume that

$$v_1(t) = 10\cos(2t + 30°)$$

Find $i_2(t)$ using the transfer function.

Solution

We observe that

$$\mathbf{V}_1 = 10\angle 30°$$

and

$$s = j\omega = j2$$

Using these values in Eq. 9.5.8 yields

$$\mathbf{I}_2 = 10\underline{/30^\circ}\ \frac{2(j2)+1}{(j2)^2+3(j2)+1}$$

$$= 10\underline{/30^\circ}\ \left(\frac{1+j4}{-3+j6}\right)$$

$$= 6.1\underline{/-10.60^\circ}$$

So

$$i_2(t) = 6.1\cos(2t - 10.60^\circ)$$

Discrete-Time Transfer Functions $H(z)$

In direct analogy with continuous time systems, we will use the input–output pair

$$z^k - \text{response to } z^k$$

where z is a real or complex number called the frequency. If z^k is the system input, then the response will consist of two terms, the transient solution and the steady-state solution. The coefficient of the steady-state solution will be the transfer function. Here is an example.

Find a closed form expression for $y(k)$ if $x(k) = z^k$ in the following equation.

$$y(k+1) + 0.8y(k) = x(k) \tag{9.5.9}$$

The transient solution is given by

$$y_t(k) = a(-0.8)^k$$

The steady-state solution is found by assuming an exponential form as follows.

$$y_s(k) = bz^k$$

$$y_s(k+1) = bz(z)^k$$

Substitution into Eq. 9.5.9 gives

$$bz(z)^k + 0.8b(z)^k = (z)^k$$

$$b = \frac{1}{z+0.8}$$

Therefore, the complete solution is given by

$$y(k) = a(-0.8)^k + \frac{1}{z+0.8}(z)^k$$

where the constant a is found from initial conditions. Define $H(z)$ by

$$H(z) = \frac{1}{z+0.8}$$

so that the solution is given by

$$y(k) = a(-0.8)^k + H(z)\cdot(z)^k$$

That is, $H(z)$ is the steady-state or forced response. In direct analogy with the continuous-time case, the transfer function is defined as follows.

> For an LTI, asymptotically stable, physically realizable discrete-time system, the function
>
> $$H(z) = \frac{\text{steady-state response to } z^k}{z^k} \tag{9.5.10}$$
>
> is said to be the transfer function of the system.

Another definition in terms of the impulse response $h(k)$ is given by

$$H(z) = \sum_{k=-\infty}^{\infty} h(k) z^{-k} \tag{9.5.11}$$

As with the continuous-time definition, Eq. 9.5.10 is equivalent to Eq. 9.5.11 under the conditions of asymptotic stability and physical realizability.

Calculating $H(z)$ from Difference Equations

Here is an algorithm to use for calculating the transfer function of a discrete-time system from the difference equation. This algorithm allows us to avoid solving for the steady-state solution with an exponential forcing function. The procedure is presented without justification, for it depends on the properties of the z transform. This topic is the subject of Chapter 14, and justification for the procedure will be given there (see Section 14.3).

With x denoting input and y denoting output, the model has the form

$$\begin{aligned} b_k y(n-k) + b_{k-1} y(n-k-1) + \cdots + b_0 y(n) \\ = a_l x(n-l) + a_{l-1} x(n-l-1) + \cdots + a_0 x(n) \end{aligned}$$

or

$$\sum_{i=0}^{k} b_i y(n-i) = \sum_{i=0}^{l} a_i x(n-i) \tag{9.5.12}$$

Here is an algorithm for finding the transfer function of this digital system. Replace each $y(n-i)$ term by $z^{-i}Y(z)$, and replace each $x(n-i)$ term by $z^{-i}X(z)$. Then solve for the ratio $Y(z)/X(z)$. This is the transfer function. Therefore, we have

$$H(z) = \frac{Y(z)}{X(z)} = \frac{a_l z^{-l} + a_{l-1} z^{-l+1} + \cdots + a_0}{b_k z^{-k} + b_{k-1} z^{-k+1} + \cdots + b_0} = \frac{\sum_{i=0}^{l} a_i z^{-i}}{\sum_{i=0}^{k} b_i z^{-i}}$$

This procedure is justified by the delay property of z transforms. It is this property that allows us to replace terms such as $y(n-i)$ by $z^{-i}Y(z)$. See Eq. 14.3.3.

EXAMPLE 9.5.5 Find the transfer function for the following first order system.

$$y(k) = 0.7x(k-1) + 0.3y(k-1) \tag{9.5.13}$$

Solution

$$Y(z) = 0.7z^{-1}X(z) + 0.3z^{-1}Y(z)$$

$$H(z) = \frac{Y(z)}{X(z)} = \frac{0.7z^{-1}}{1 - 0.3z^{-1}} = \frac{0.7}{z - 0.3}$$

EXAMPLE 9.5.6 Find the transfer function for the following second order system.

$$y(n) = 0.01x(n-1) + 0.02x(n-2) + 1.9y(n-1) + 0.9y(n-2)$$

Solution

$$Y(z) = 0.01z^{-1}X(z) + 0.02z^{-2}X(z) + 1.9z^{-1}Y(z) - 0.9z^{-2}Y(z)$$

$$H(z) = \frac{Y(z)}{X(z)} = \frac{0.01z^{-1} + 0.2z^{-2}}{1 - 1.9z^{-1} + 0.9z^{-2}}$$

$$= \frac{0.01z + 0.02}{z^2 - 1.9z + 0.9}$$

LEARNING EVALUATIONS

1. If the input signal is $v_1(t)$ and the output is $i_1(t)$, find the transfer function for the circuit in Fig. 9.42.

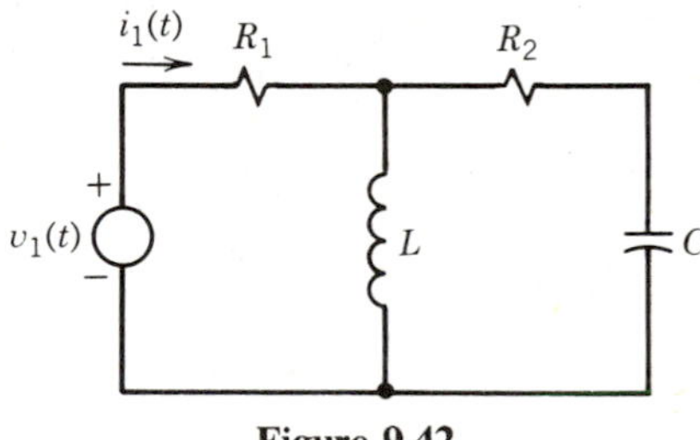

Figure 9.42

2. Find the discrete-time transfer function from the following difference equation.

$$y(k) = 0.5x(k-1) + 0.8y(k-1)$$

9.6 Sinusoidal Analysis of Digital Systems

LEARNING OBJECTIVES

After completing this section you should be able to do the following:

1. Express a sinusoidal discrete-time signal in the frequency domain.
2. Find the sinusoidal transfer function of a given digital system.
3. Find the steady-state sinusoidal response of a digital system by the transform method.

The procedure for finding the sinusoidal response of a digital system is analogous to the procedure for an analog system. There are several differences caused by the discrete nature of the signal, but the same principles apply to both cases. These principles are based on the LTI properties.

The input signal is assumed to be samples of a continuous sinusoid,

$$x(t) = A\cos(\omega t + \theta) \tag{9.6.1}$$

as shown in Fig. 9.43. If this signal is sampled every T_s s, the resulting discrete-time signal is equal to $x(t)$ at the sampling instants, $t = nT_s$. Therefore,

$$x(nT_s) = A\cos(\omega nT_s + \theta),$$

or

$$x(n) = A\cos(n\omega T_s + \theta) \tag{9.6.2}$$

This is our digital input signal. It has angular frequency ω and sampling period T_s. Notice that $\omega \neq 2\pi/T_s$.

The sampling frequency f_s is related to T_s by

$$f_s = \frac{1}{T_s} \tag{9.6.3}$$

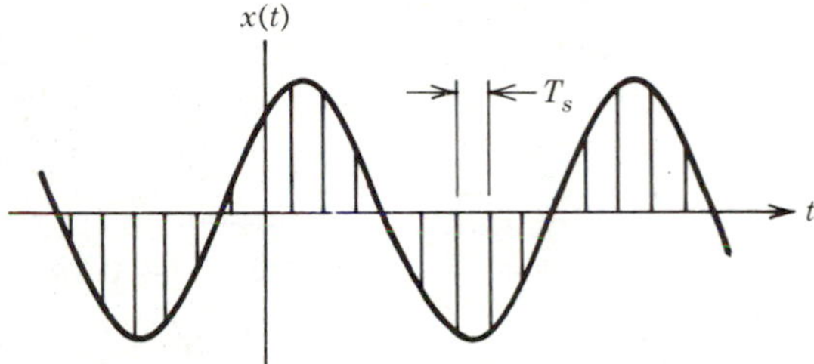

Figure 9.43. A sinusoidal waveform sampled every T_s s.

This is more properly called the sampling rate, for it represents the rate in samples per second at which the signal is sampled.

Let us define the *folding frequency* f_0 as $f_s/2$.

$$f_0 = \frac{f_s}{2} = \frac{1}{2T_s} \tag{9.6.4}$$

This parameter is important because it represents the highest unambiguous frequency possible in a digital system. By this we mean that if $x(t)$ is a sinusoid of frequency $f_2 > f_0$, then the samples of this signal could also be caused by a sinusoid of frequency $f_1 < f_0$. An example to illustrate this point is pictured in Fig. 9.44.

The *normalized frequency* ν is defined as

$$\nu = \frac{f}{f_0} = 2T_s f \tag{9.6.5}$$

This parameter is convenient when studying the sinusoidal response of digital systems because $\nu = 1$ represents the following frequency, and $\nu = 2$ represents

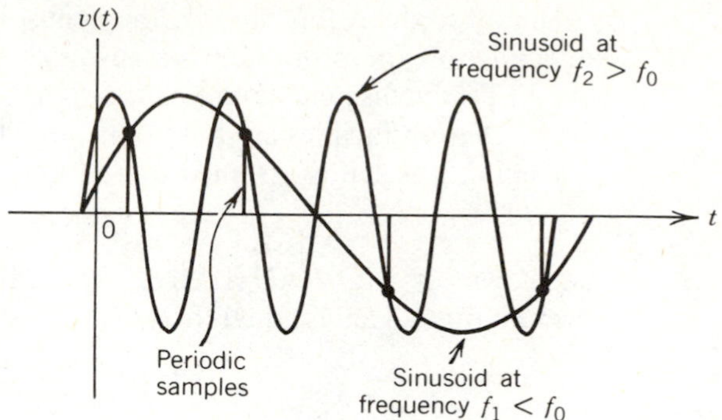

Figure 9.44. Two sinusoids about the folding frequency f_0.

the sampling frequency. Now we can express the input sinusoid $x(n)$ in terms of ν as

$$x(n) = A\cos(n\pi\nu + \theta) \tag{9.6.6}$$

EXAMPLE 9.6.1 The sinusoidal signal $x(t) = 10\cos(377t)$ is sampled at a rate $f_s = 500$ Hz. Express this signal in terms of normalized frequency.

Solution

$x(n) = 10\cos(2\pi 60\ nT_s)$, where $T_s = 1/f_s = 0.002$. Therefore,

$$x(n) = 10\cos(0.24\pi n)$$

Another method of arriving at the same expression is to use Eq. 9.6.5 where $\nu = 2T_s f = 2(0.002)60 = 0.24$. Then Eq. 9.6.6 gives

$$x(n) = 10\cos(0.24\pi n)$$

The phasor transform of a sampled sinusoid is defined as a complex number, where the cosine waveform is used as the standard.

$$A\cos(n\pi\nu + \theta) \leftrightarrow Ae^{j\theta} \tag{9.6.7}$$

Therefore the phasor transform of a sine function is rotated 90°.

$$A\sin(n\pi\nu + \theta) \leftrightarrow Ae^{j(\theta - \pi/2)} \tag{9.6.8}$$

EXAMPLE 9.6.2 The phasor transforms of several sinusoidal functions are given below.

Solution

$$10\cos(n\pi\nu) \leftrightarrow 10e^{j0} = 10 = 10\underline{/0^\circ}$$

$$10\sin(n\pi\nu) \leftrightarrow 10e^{-j\pi/2} = -j10 = 10\underline{/-90^\circ}$$

$$10\cos\left(n\pi\nu + \frac{\pi}{6}\right) \leftrightarrow 10e^{j\pi/6} = 10\underline{/30^\circ} = 5 + j8.66$$

$$2\cos(n\pi\nu) + 3\sin(n\pi\nu) \leftrightarrow 2 + 3e^{-j\pi/2} = 3.6055e^{-j0.98279}$$

We gave an algorithm in the previous section for calculating the transfer function for a discrete-time system from the describing difference equation. With

x denoting input, and y denoting output, the difference equation is of the form

$$\sum_{i=0}^{k} b_i y(n-i) = \sum_{i=0}^{l} a_i x(n-i) \tag{9.6.9}$$

Our algorithm replaced each $y(n-i)$ term by $z^{-i}Y(z)$, and replaced each $x(n-i)$ term by $z^{-i}X(z)$. Then the ratio $Y(z)/X(z)$ is the transfer function $H(z)$.

For steady-state sinusoidal analysis the value of z is always on the unit circle in the z plane (this is explained in Chapter 14). That is,

$$z = e^{j\pi\nu} \tag{9.6.10}$$

Therefore the procedure for calculating the transfer function for sinusoidal analysis is identical to that in Section 9.5 with z given by Eq. 9.6.10.

EXAMPLE 9.6.3 Find the sinusoidal transfer function for the following first order system.

$$y(k) = 0.7x(k-1) + 0.3y(k-1)$$

Solution

$$Y(z) = 0.7z^{-1}X(z) + 0.3z^{-1}Y(z)$$

$$H(z) = \frac{0.7z^{-1}}{1 - 0.3z^{-1}} = \frac{0.7}{z - 0.3}$$

Now substitute Eq. 9.6.10.

$$H(\nu) = \frac{0.7}{e^{j\pi\nu} - 0.3}$$

EXAMPLE 9.6.4 Suppose that the signal applied to the system in the previous example is the sampled sinusoid of Example 9.6.1. That is, the angular frequency is $\omega = 377$ rad/s, and the sampling rate is $f_s = 500$ Hz. Evaluate $H(\nu)$.

Solution

The normalized frequency is given by Eq. 9.6.5.

$$\nu = 2T_s f = 2(0.002)(60) = 0.24$$

Hence,

$$H(\nu) = H(0.24) = \frac{0.7}{e^{j\pi 0.24} - 0.3} = 0.867e^{-j1.011} = 0.867\angle -57.93^\circ$$

The final step in finding the sinusoidal steady-state response of digital systems is to multiply $X(\nu)$ by $H(\nu)$ and find the inverse transform of the product.

EXAMPLE 9.6.5 If the signal in Example 9.6.1 is applied to the system in Example 9.6.4, the output signal is

$$Y(\nu) = H(\nu)X(\nu) = 8.665e^{-j57.93^\circ}$$

Therefore,

$$y(n) = 8.665\cos(0.24\pi n - 57.93^\circ)$$

EXAMPLE 9.6.6 Determine the response of a digital system described by the equation

$$y(n) = 0.1x(n-1) + 0.2x(n-2) + 1.9y(n-1) - 0.9y(n-2)$$

if the input is a cosine wave of amplitude 10 and frequency 100 Hz. The sampling frequency is 1 kHz.

Solution

The normalized frequency is

$$\nu = 2T_s f = 2(10)^{-3}(100) = 0.2$$

The transform of the input signal is

$$10\cos(n\pi\nu) \leftrightarrow 10$$

The transfer function is given by

$$H(z) = \frac{0.1z + 0.2}{z^2 - 1.9z + 0.9}$$

or

$$H(\nu) = \frac{0.1e^{j\pi\nu} + 0.2}{e^{j2\pi\nu} - 1.9e^{j\pi\nu} + 0.9} = \frac{0.1e^{j0.2\pi} + 0.2}{e^{j0.4\pi} - 1.9e^{j0.2\pi} + 0.9}$$

$$= \frac{0.2809 + j0.0588}{(0.3090 + j0.9510) - (1.5371 + j1.1168) + 0.9}$$

$$= 0.13707e^{-j1.5218}$$

The response is given by

$$Y(\nu) = X(\nu)H(\nu) = 1.3707e^{-j1.5218}$$

or

$$y(n) = 1.3707\cos(0.2\pi n - 87.19°)$$

LEARNING EVALUATIONS

1. A sinusoidal signal $x(t) = 5\cos(\omega t + 60°)$ has frequency $\omega = 377$ rad/s. If this signal is sampled at 800 Hz, find the normalized frequency and the transform.
2. Find the value of the transfer function of the system described by

$$y(n) = 0.5x(n-1) + 0.1y(n-1)$$

The input signal is the sinusoid given in the above problem.

3. Find the response $y(n)$ for the above input signal and system.

9.7 A Glance Back

Whew! This has been a long and involved chapter. It may be well to reflect a minute on the topics covered. We were primarily interested in developing an ability to analyze circuits with sinusoidal forcing functions. To do this required the introduction of several new concepts.

The idea of transform analysis was introduced and resulted in the phasor transform, which allows us to use complex algebra and tools from previous chapters to solve sinusoidal circuit analysis problems. Also the effect of frequency on our standard circuit elements required that we define and use the terms impedance and admittance. At this point we found sinusoidal circuit analysis for continuous systems to be rather straightforward although often tedious.

The concept of transfer functions was introduced for several reasons. It allowed us to briefly explore the idea of complex frequency and see that sinusoids form a subset of complex frequency signals. It also allowed us to see the use of transfer functions in continuous-time sinusoidal analysis. Finally, it provided a mechanism whereby we could perform sinusoidal analysis of digital systems without theoretical principles that will not be encountered until Chapter 14.

PROBLEMS

SECTION 9.1

1. a. Write the equations for $v_1(t)$ and $v_2(t)$ in Fig. 9.45. First express each function as a cosine function, then as a sine function. Notice that $v_2(t)$ is simply $v_1(t)$ plus a constant.
   b. Can you express $v_2(t)$ as the square of a sinusoidal term?

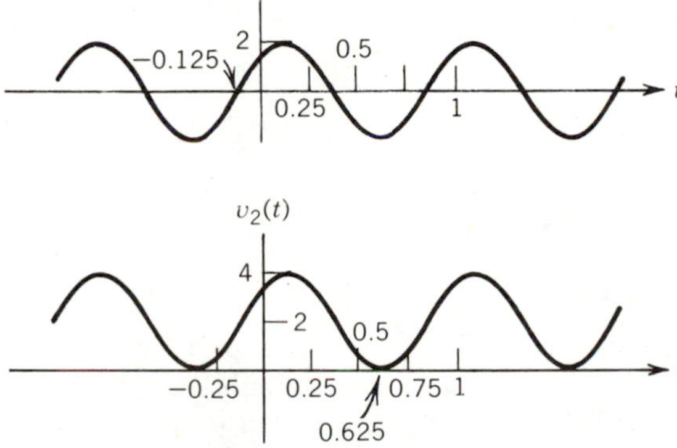

Figure 9.45

2. The frequency of a sinusoidal signal is often estimated in practice by counting the zero crossings, that is, how many times the signal has a zero value in a specified time period. If there are 245 zero crossings of a signal in 10 ms, what is the frequency?
3. Sketch a few cycles of each sinusoid.
   a. $v_1(t) = 10\cos\left(2\pi t - \dfrac{\pi}{3}\right)$
   b. $v_2(t) = 5\cos\left(4\pi t + \dfrac{\pi}{3}\right)$
   c. $v_3(t) = 163\cos(377t - 30°)$
4. Find the rms value of each waveform in Fig. 9.46.

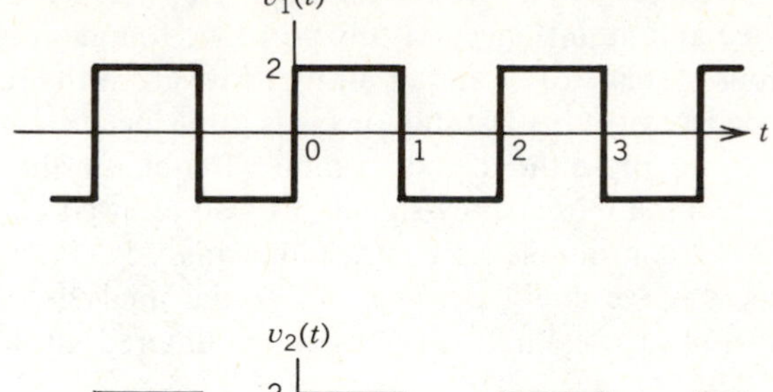

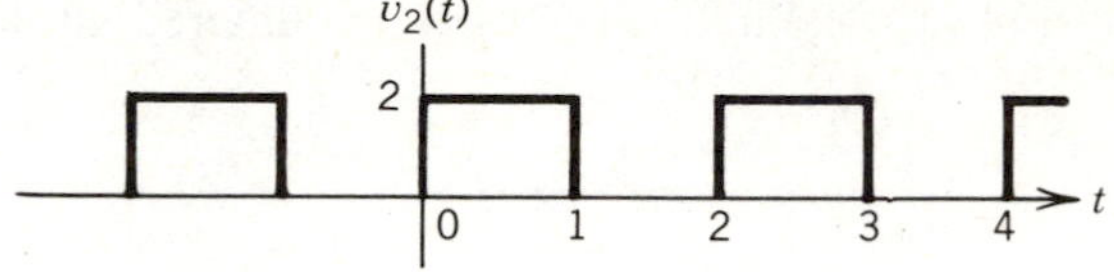

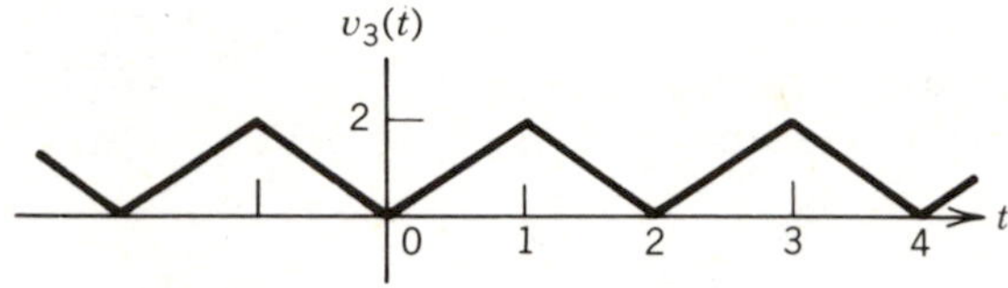

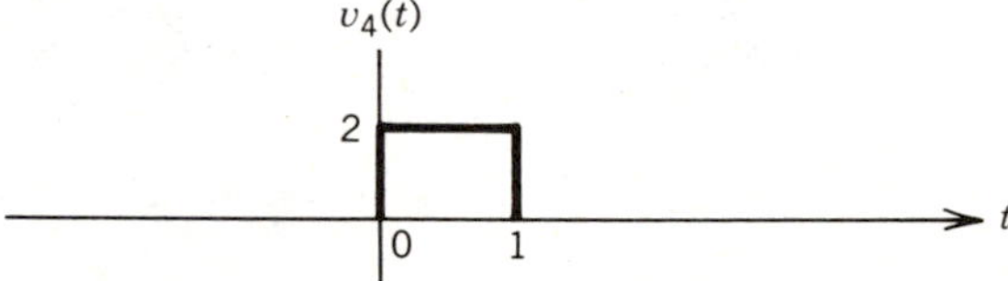

Figure 9.46

5. In many practical situations a periodic waveform may have no known analytical representation. A graph may be the only available way to determine the function. Suppose that the signal $v(t)$ in Fig. 9.47 has a period of 10 ms and that 10 values of $v(t)$ are recorded once each millisecond. These are given by the values in Table 9.1. Estimate the rms value of $v(t)$.

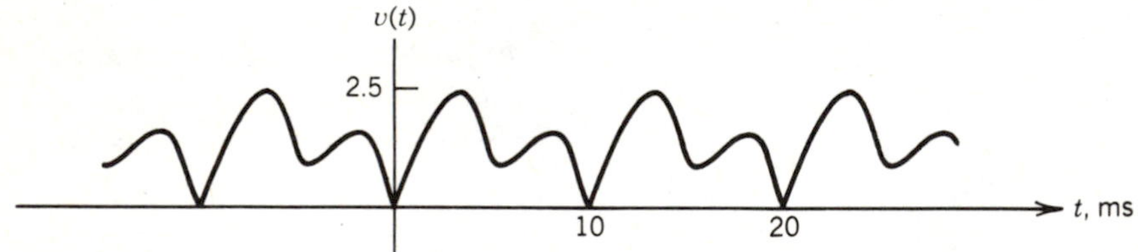

Figure 9.47

TABLE 9.1

t, ms	$v(t)$
1	1.0
2	2.5
3	3.0
4	2.5
5	1.0
6	1.4
7	1.7
8	2.0
9	1.0
10	0.0

6. Signals may be classified as either a power signal or an energy signal. Power signals have nonzero but finite average power. Energy signals have nonzero but finite total energy. The average power in a signal $v(t)$ is defined by

$$P = \lim_{T \to \infty} \frac{1}{2T} \int_{-T}^{T} v^2(t)\, dt$$

The total energy in a signal $v(t)$ is defined by

$$E = \int_{-\infty}^{\infty} v^2(t)\, dt$$

Classify each signal in Fig. 9.46 as either a power signal or an energy signal.

SECTION 9.2

7. Determine the phasor transform of each sinusoid below.

a. $v_1(t) = 10\cos(377t - \pi/6)$
b. $v_2(t) = 10\sin(377t - \pi/6)$
c. $v_3(t) = 10\cos(377t - \pi/6) + 10\sin(377t - \pi/6)$

8. Three sinusoidal functions $v_1(t)$, $v_2(t)$, and $v_3(t)$ are given by
$v_1(t) = 2\cos\omega t$
$v_2(t) = 3\cos(\omega t - \pi/3)$
$v_3(t) = 2\sin(\omega t - \pi/3)$
Write an expression in terms of a single sinusoid for each function below.

a. $v_4(t) = v_1(t) + v_2(t)$
b. $v_5(t) = v_1(t) - v_2(t)$
c. $v_6(t) = v_1(t) + v_3(t)$
d. $v_7(t) = v_2(t) + v_3(t)$

9. Express each phasor shown below as a sinusoidal time function if the frequency is known to be $f = 60$ Hz.

a. $\mathbf{V}_1 = 10e^{j\pi/2}$
b. $\mathbf{V}_2 = 5e^{-j\pi/3}$
c. $\mathbf{V}_3 = 3 + j4$

10. Express each phasor below as a sinusoidal time function if the frequency is known to be $\omega = 2\pi$ rad/s.

a. $\mathbf{V}_1 = \left(\dfrac{3 + j4}{2 - j3}\right)$

b. $\mathbf{V}_2 = \dfrac{5\angle 60^\circ (2 + j3)}{10\angle 30^\circ}$

SECTIONS 9.3 AND 9.4

11. Find the voltage between terminals a–b in Fig. 9.48. The source is $i(t) = 0.1\cos 2t$ A.

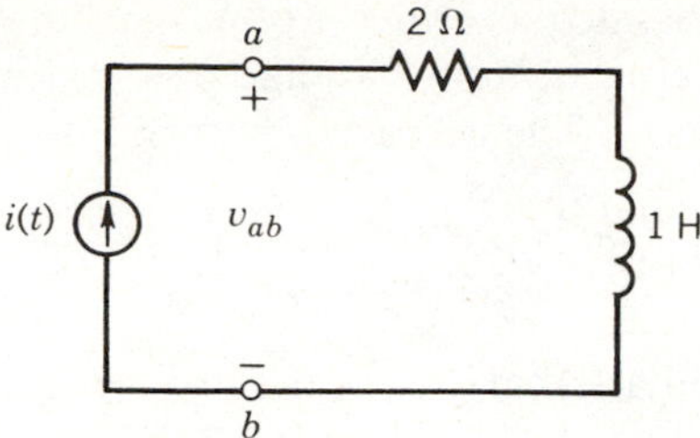

Figure 9.48

12. Find the current supplied by the voltage source in Fig. 9.49. The source is $v(t) = 10 \cos 2t$ V.

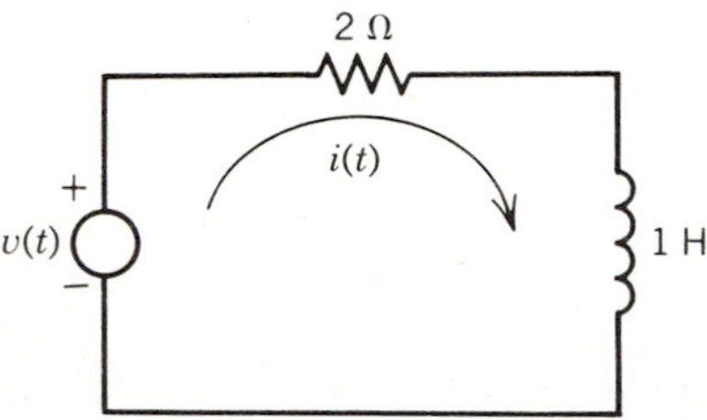

Figure 9.49

13. Use Thévenin's theorem to find the voltage across R_2 in Fig. 9.25. $v(t) = 10 \sin 377t$.
14. Use Norton's theorem to find the voltage across R_2 in Fig. 9.25. $v(t) = 10 \sin 377t$.
15. Use mesh analysis to find the current through R_2 in Fig. 9.25. $v(t) = 10 \sin 377t$.
16. Use nodal analysis to find the voltage across R_2 in Fig. 9.25. $v(t) = 10 \sin 377t$.
17. Write the mesh equations in matrix form for the circuit in Fig. 9.50. Solve for both mesh currents $\mathbf{I}_1$ and $\mathbf{I}_2$ in phasor form. Finally, find $i_1(t)$ and $i_2(t)$ if the forcing function is a cosine waveform with frequency $f = 10^4$ Hz.

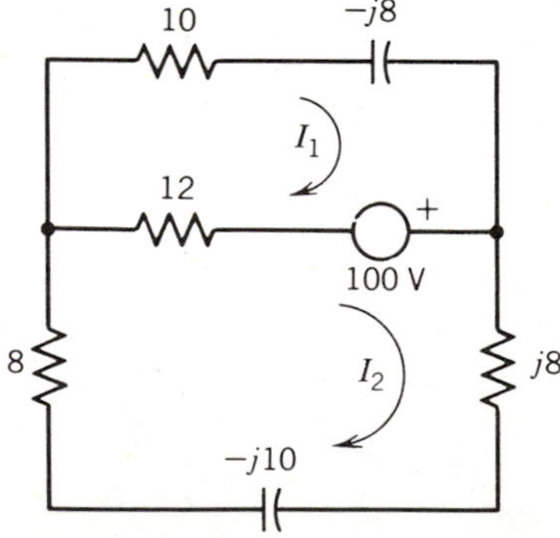

Figure 9.50

18. Write the mesh equations in matrix form and solve for mesh currents $\mathbf{I}_1$ and $\mathbf{I}_2$ in Fig. 9.51.

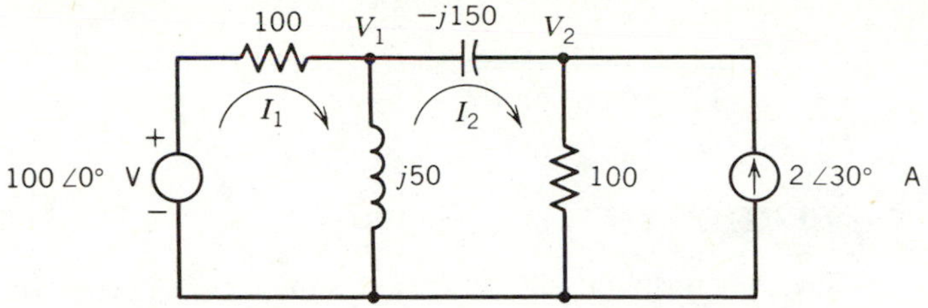

Figure 9.51

19. Write the nodal equations in matrix form and solve for the node voltages $\mathbf{V}_1$ and $\mathbf{V}_2$ in Fig. 9.51.

20. Find the steady-state response $v_2(t)$ in Fig. 9.52.

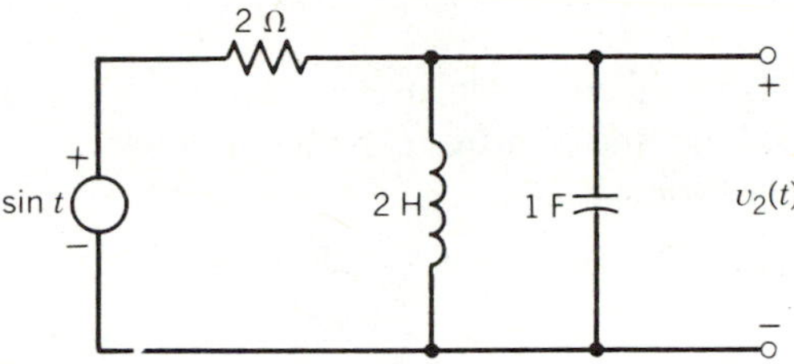

Figure 9.52

21. Find the steady-state response $v_2(t)$ in Fig. 9.53.

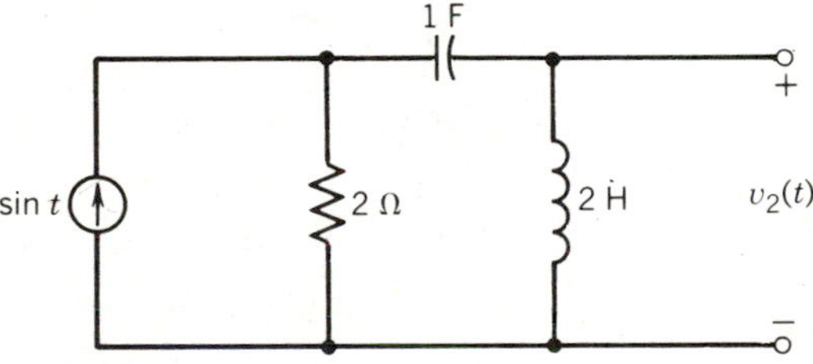

Figure 9.53

SECTION 9.5

22. Solve for $v_2(t)$ in Fig. 9.52 by finding the transfer function of the circuit $H(s)$. Then solve for $V_2(s)$ from $V_2(s) = H(s)V_1(s)$. Finally, convert $V_2(s)$ to the time domain to obtain $v_2(t)$.

23. Repeat Problem 22 for the circuit in Fig. 9.53.

24. Find the discrete-time transfer function $H(z)$ from the difference equation

$$y(k) = 0.6y(k-1) + 0.4y(k-2) + x(k-1)$$

25. Find the discrete-time transfer function for the system described by the difference equation

$$\sum_{i=0}^{N} b_1 y(k-i) = \sum_{i=0}^{N} a_i x(k-i)$$

That is, derive the general form of the transfer function in terms of the coefficients a_i and b_i.

SECTION 9.6

26. A sinusoidal signal $x(t)$ is sampled at a uniform rate of 1000 samples per second, where

$$x(t) = 5\cos(500\pi t)$$

Find the steady-state response to this signal for the system described by the difference equation

$$y(k) = 0.6y(k-1) + 0.4y(k-2) + x(k-1)$$

27. The transfer function, when evaluated at any one frequency ν, is a complex number. Find and plot this complex number as a function of ν for the digital system in Problem 26.

28. Determine the response of a digital system described by the equation

$$y(n) = 0.2y(n-1) + 0.8x(n-1)$$

if the input is a cosine wave of amplitude 10 and frequency 200 Hz. The sampling frequency is 1 kHz.

29. Determine the response of a digital system described by the equation

$$y(n) = 0.3y(n-1) + 0.4y(n-2) + 0.5x(n) + 0.5x(n-1)$$

if the input is a cosine wave of amplitude 10 and frequency 250 Hz. The sampling frequency is 1 kHz.

30. a. A simple first order low-pass filter is described by the difference equation

$$y(k) = 0.3x(k-1) + 0.7y(k-1)$$

Find and compare the response of this filter to two different sinusoidal signals.

$$x_1(t) = 10\cos 50\pi t$$

$$x_2(t) = 10\cos 100\pi t$$

The sampling frequency is 500 Hz.

b. A better low-pass filter is described by

$$y(n) = 1.14y(n-1) - 0.41y(n-2) + 0.067x(n) + 0.134x(n-1) + 0.067x(n-2)$$

Find the response of this filter to each sinusoid in **a**.

LEARNING EVALUATION ANSWERS

Section 9.1

1. $v(t) = -5\cos(10\pi t) = 5\cos(10\pi t + \pi)$
2. See Fig. 9.54.
3. $x(t)$ leads $y(t)$ by 0.4π rad.
4. $V_{\text{rms}} = \sqrt{2/3}$

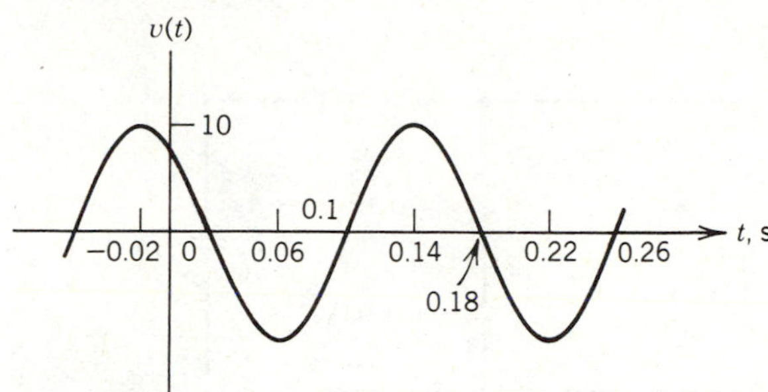

Figure 9.54

Section 9.2

1. a. $33.6\angle 13°$
   b. $14.8\angle \pi/3$
   c. $10\angle -90°$
2. a. $14\angle -90°$, $-j14$
   b. $263.77 + j292.95$, $394.2e^{j0.84}$
   c. $20.59\angle 119.05$, $20.59e^{j119.05°}$
3. a. $10.63\angle 1.78°$
   b. $29.15\angle -138.96°$

Section 9.3

1. See Fig. 9.55.
2. a. $54.84\angle 57.45°$
   b. $13.11\angle 80.92°$
3. a. $0.02\angle -57.45°$
   b. $0.08\angle -80.92°$
4. $56.48 + j13.86\ \Omega$
5. $0.0167 - j0.0041$ ℧

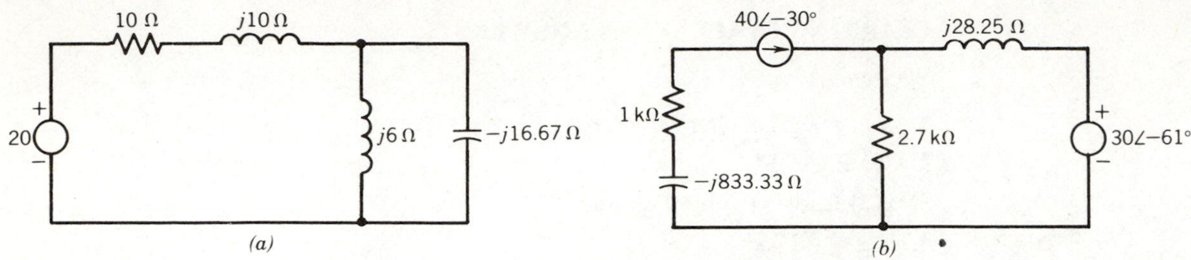

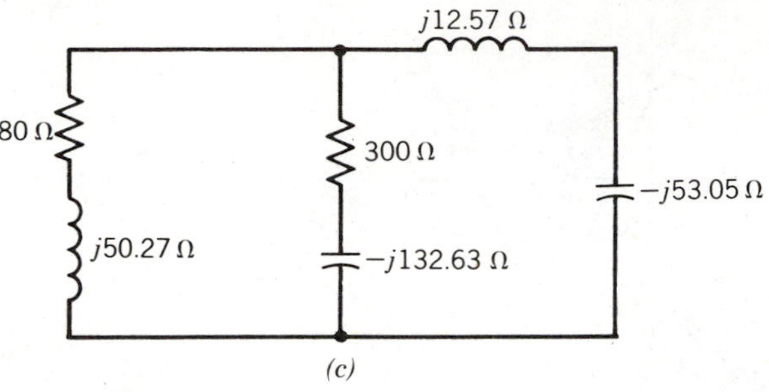

Figure 9.55

Section 9.4

1. $i(t) = 7.24\cos(400t - 35.4°)$
2. $i_L(t) = 0.47\cos(1000\pi t - 157.6°)$
   $v_L(t) = 4.71\cos(1000\pi t - 67.6°)$
3. $i_L(t) = 0.3535\cos(2\pi t - 45°)$

Section 9.5

1. $H(s) = \dfrac{s^2LC + sR_2C + 1}{s^2LC(R_1 + R_2) + s(L + R_1R_2C) + R_1}$
2. $H(z) = \dfrac{0.5}{z - 0.8}$

Section 9.6

1. $\nu = 0.15$
   $X(\nu) = 5e^{j60°} = 5e^{j1.0472 \text{ rad}}$
2. $H(\nu) = \dfrac{0.5}{e^{j\pi\nu} - 0.1}$
   $H(0.15) = 0.54823e^{-j0.52104}$
3. $Y(\nu) = H(\nu)X(\nu) = 2.74115e^{j0.52104}$
   $y(n) = 2.74115\cos(0.15\pi n + 0.52616)$

CHAPTER 10 POWER SYSTEM ANALYSIS

A study of sinusoidal circuits is not complete until the problem of delivering, measuring, and altering power is investigated. A significant number of electrical engineers spend their professional careers in this area generally described as power system analysis. A separate but obviously equally important area is that of power generation. Power generation, however, is not a circuit analysis problem and this topic is left for other courses.

In this chapter we will review and explore in greater detail the concepts of power and introduce a new term called the power factor. We will also discuss "three-phase" power and determine its advantages. One section will discuss the methods of measuring power and the last section will introduce the transformer as a circuit element for altering the component parts of electrical power.

While not spanning the entire area of power, these topics provide the fundamentals necessary for further study of this area.

10.1 ac Power

LEARNING OBJECTIVES

After completing this section you should be able to do the following:

1. Determine the instantaneous power (as a function of time) in a circuit with arbitrary forcing function.
2. Find the average power and the power factor in steady state ac circuits.
3. Correct power factor in ac circuits.

We have previously discussed or referred to many of the concepts in this section, but often in an incidental way. The concepts of average power, rms value, and power factor are so important to every practicing electrical engineer that we must investigate these concepts more throughly.

We have seen that the instantaneous power in any element is given by

$$p(t) = v(t)i(t) \tag{10.1.1}$$

where $v(t)$ is the voltage across the element, and $i(t)$ is the current through the element. The average of this quantity over an interval of time $t_1 < t < t_2$ is the average power, P.

$$P = \frac{1}{t_2 - t_1}\int_{t_1}^{t_2} p(t)\,dt \tag{10.1.2}$$

A difficulty arises with this definition in that the average power depends on the limits t_1 and t_2. This does not coincide with our usual experience, for average power in a resistor that is supplied by a constant or periodic source can be measured with a thermometer. The heat produced is proportional to the energy dissipated by the resistor.

To avoid this difficulty, the limits t_1 and t_2 are allowed to approach $-\infty$ and $+\infty$, respectively. That is, average power is usually defined by

$$P = \lim_{T\to\infty} \frac{1}{2T}\int_{-T}^{T} p(t)\,dt \tag{10.1.3}$$

If the function $p(t)$ is periodic with period T, this reduces to

$$P = \frac{1}{T}\int_{0}^{T} p(t)\,dt \tag{10.1.4}$$

Since we are concerned primarily with periodic functions in this chapter we will most often use Eq. 10.1.4.

We shall begin our analysis of power and energy in ac circuits with the simple case illustrated in Fig. 10.1 where a sinusoidal source is applied to a resistor. The voltage and current are given by

$$v(t) = V\cos\omega t$$

$$i(t) = \frac{V}{R}\cos\omega t = I_R\cos\omega t$$

The instantaneous power is the product

$$\begin{aligned} p(t) &= v(t)i(t) = VI_R\cos^2\omega t \\ &= \frac{VI_R}{2}(1+\cos 2\omega t) \end{aligned} \tag{10.1.5}$$

The average power is found from Eq. 10.1.4

$$P = \frac{1}{T}\int_0^T p(t)\,dt = \frac{VI_R}{2} = V_{\text{rms}}I_{R\,\text{rms}} \tag{10.1.6}$$

where we have applied Eq. 9.1.11 in order to express the average power in terms of rms values.

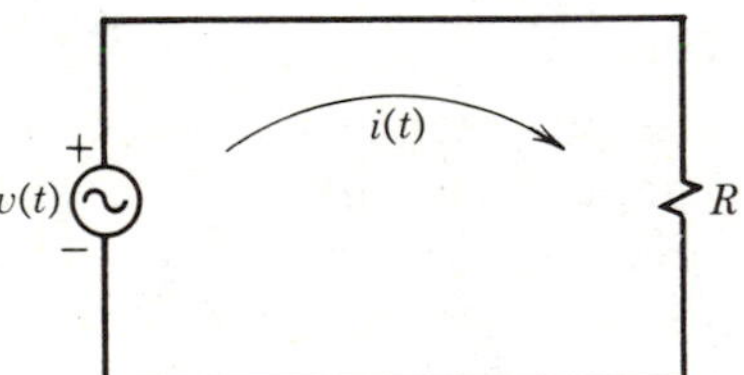

Figure 10.1. A sinusoidal source in series with a resistor.

An inductor driven by a sinusoidal voltage is shown in Fig. 10.2. The voltage and current are given by

$$v(t) = V\cos\omega t$$

$$i(t) = \frac{1}{L}\int_{-\infty}^{t} v(\lambda)\,d\lambda = \frac{V}{\omega L}\sin\omega t = I_L\sin\omega t$$

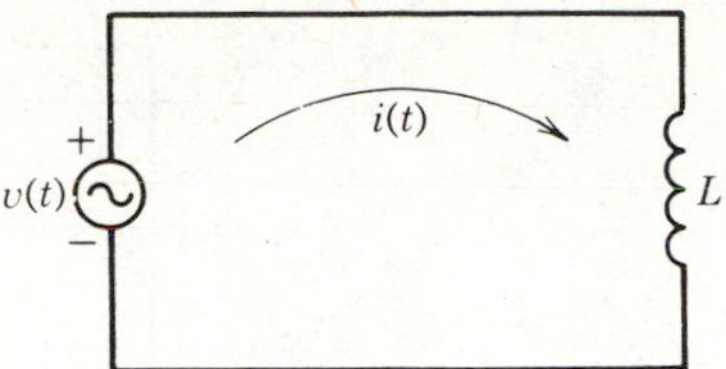

Figure 10.2. A sinusoidal source in series with an inductor.

The instantaneous power is given by

$$p(t) = v(t)i(t) = VI_L \cos \omega t \sin \omega t$$
$$= \frac{VI_L}{2} \sin 2\omega t$$

The integral of a sinusoid over one or more periods is zero, so the average power is zero.

$$P_L = \frac{1}{T}\int_0^T p(t)\,dt = 0 \tag{10.1.7}$$

A similar analysis for the capacitor (Fig. 10.3) shows that the instantaneous power is given by

$$p(t) = -\frac{VI_C}{2}\sin 2\omega t$$

Here, again, the average power is zero.

$$P_C = \frac{1}{T}\int_0^T p(t)\,dt = 0 \tag{10.1.8}$$

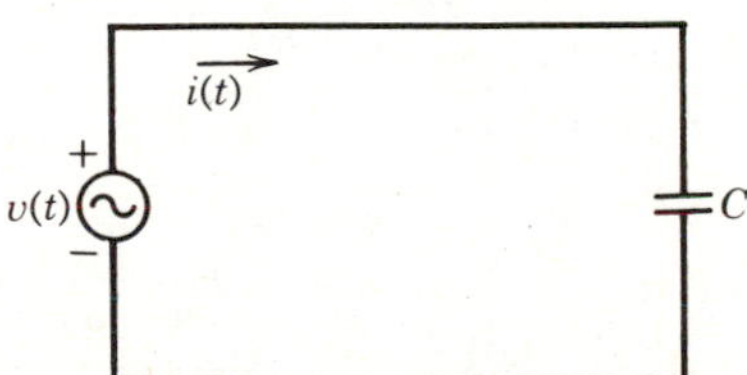

Figure 10.3. A sinusoidal source in series with a capacitor.

The calculation of average power in a circuit supplied by a steady-state sinusoidal source becomes complicated if we persist in using the above formulas. It becomes complicated in any case, but less so if we use some of the phasor techniques developed by Steinmetz, Tesla, and others. Let us assume a voltage source of the form

$$v(t) = V\cos \omega t \leftrightarrow Ve^{j0} = V\angle 0 \tag{10.1.9}$$

Then the current is of the form

$$i(t) = I\cos(\omega t + \theta) \leftrightarrow I\angle \theta \tag{10.1.10}$$

where θ is positive in a capacitive circuit, and negative in an inductive circuit.

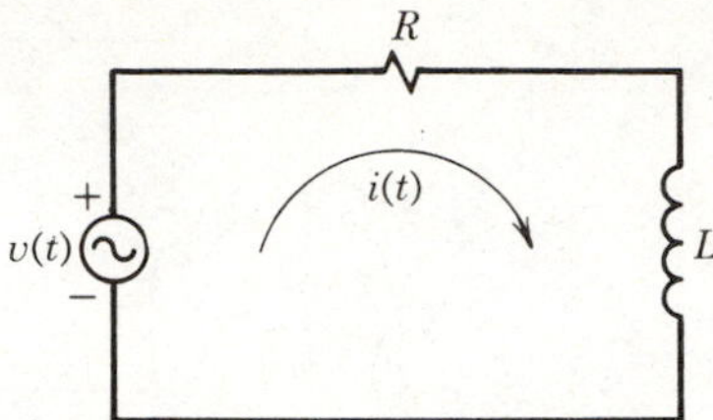

Figure 10.4. A series RL circuit with sinusoidal forcing function.

Consider the RL series circuit illustrated in Fig. 10.4. If the voltage is given by Eq. 10.1.9 then the current phasor is

$$\mathbf{I} = \frac{V}{R + j\omega L} = \frac{V}{\sqrt{R^2 + \omega^2 L^2}} e^{-j\tan^{-1}(\omega L/R)}$$

so that $i(t)$ is given by Eq. 10.1.10 with

$$I = \frac{V}{\sqrt{R^2 + \omega^2 L^2}} \quad \text{and} \quad \theta = -\tan^{-1}\left(\frac{\omega L}{R}\right).$$

The average power is

$$P = \frac{1}{T}\int_0^T VI\cos(\omega t)\cos(\omega t + \theta)\,dt$$

Substitution of the trigonometric identity

$$\cos\alpha\cos\beta = \tfrac{1}{2}\cos(\alpha - \beta) + \tfrac{1}{2}\cos(\alpha + \beta)$$

gives

$$P = \frac{1}{T}\int_0^T \frac{VI}{2}\cos\theta\,dt + \frac{1}{T}\int_0^T \frac{VI}{2}\cos(2\omega t + \theta)\,dt$$

The second term is zero, giving

$$\boxed{P = \frac{VI}{2}\cos\theta = V_{\text{rms}} I_{\text{rms}}\cos\theta} \tag{10.1.11}$$

Equation 10.1.11 tells us that we can find the average power simply by calculating the three terms V_{rms}, I_{rms}, and $\cos\theta$ and multiplying them together. This approach is normally much easier than using Eq. 10.1.4.

EXAMPLE 10.1.1 If ordinary household voltage is applied to the circuit in Fig. 10.4 with $R = 10\ \Omega$ and $L = 0.05$ H, find the average power dissipated in the circuit.

Solution

Let $\mathbf{V}_{\text{rms}} = 120\underline{/0°}$ V, $f = 60$ Hz.

$$\mathbf{I}_{\text{rms}} = \frac{120\underline{/0°}}{R + j\omega L} = \frac{120\underline{/0°}}{10 + j377(0.05)} = 5.62\underline{/-62.05°}$$

Thus, $P = 120(5.62)\cos(62.05°) = 316.3$ W.

All of this energy is dissipated in the resistor as heat. The inductor can only store energy.

The term $\cos\theta$ in Eq. 10.1.11 is called the *power factor*.

$$\boxed{pf = \cos\theta = \frac{P}{V_{\text{rms}} I_{\text{rms}}}} \qquad (10.1.12)$$

The term $V_{\text{rms}} I_{\text{rms}}$ is called the *apparent power S*.

$$\boxed{S = V_{\text{rms}} I_{\text{rms}}} \qquad (10.1.13)$$

These terms are displayed on the *power triangle* shown in Fig. 10.5. The third leg of this triangle is labeled Q and is called the *reactive power*. The reactive power has the same units as real power P, but to differentiate from real power it is given the artificial unit of vars. (Var is an abbreviation of volt–amperes–reactive.) Similarly, the apparent power is given units of volt–amperes.

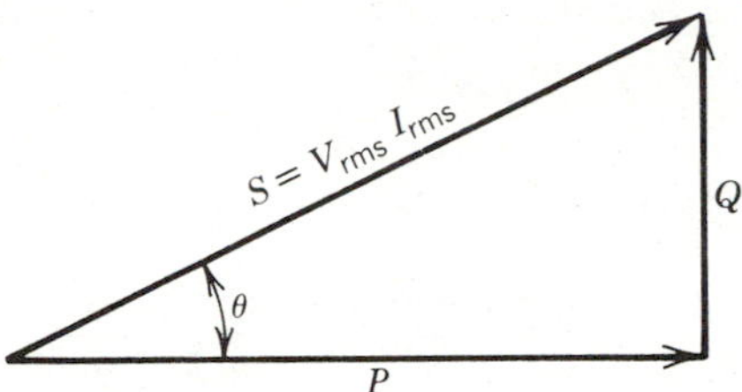

Figure 10.5. The power triangle.

The "real power," P, is the power actually dissipated where the voltage and current are sinusoidal in nature. If the load is purely resistive, then $\cos\theta = 1$ and the real power equals the apparent power. The "reactive power," Q, is an imaginary component forming the third side of the power triangle. We can now define a complex power **P** where

$$\mathbf{P} = P + jQ = S\angle\theta$$

This power triangle provides considerable simplification in sinusoidal power calculations. The angle θ is the angle between voltage and current, the apparent power S is the product $V_{\text{rms}} I_{\text{rms}}$, and from these two quantities P and Q are easily determined.

One of the more common but important problems in power transmission is that of power factor correction. The customer places a load, such as a motor or furnace, across the power line, so that all loads are in parallel. This is illustrated in Fig. 10.6, where $\mathbf{Z}_1$ and $\mathbf{Z}_2$ represent the various appliances across the power

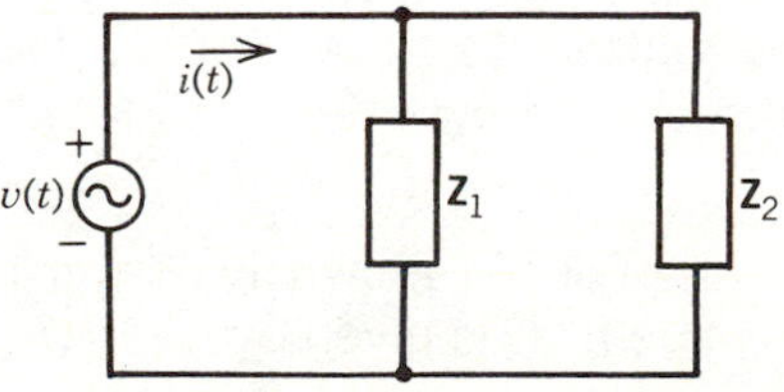

Figure 10.6. Parallel loads on a power line.

line. Loads are usually inductive; for example, an induction motor is inductive. The power factor is called lagging in this case, meaning that the current is lagging the voltage. The voltage is used as reference, since it is constant for all loads.

A plot of the current locus for a parallel RL circuit is shown in Fig. 10.7 to illustrate our discussion. If a capacitor of appropriate value is placed across the line (in parallel with R and L) the angle θ will be reduced to zero. This will have the effect of reducing the line current **I**, although it will have no effect on $\mathbf{I}_R$ and $\mathbf{I}_L$. There are several reasons for power factor correction. Only one (reduction of line current) will be discussed here.

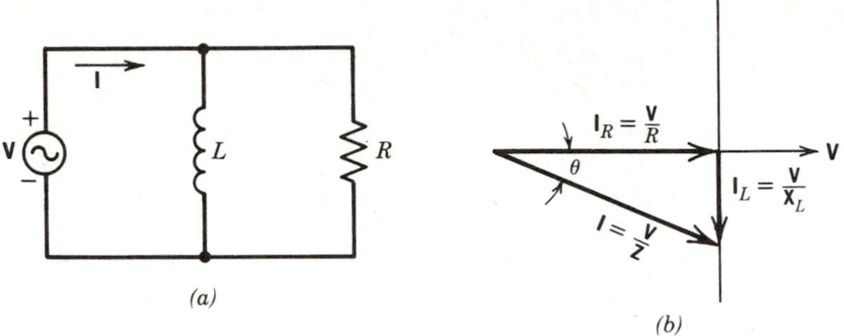

Figure 10.7. Current locus for parallel RL circuit.

The reduction of line current is important because of economic reasons, but this fact is not readily apparent. Let us postulate an ideal power generating system to generate the power supplied to the parallel RL circuit in Fig. 10.7*a*. The amount of fuel supplied to the system is directly proportional to the power in the resistor. Since the inductance dissipates no energy, except for a small amount radiated by the electromagnetic field around the inductor, the fuel supplied to the system need not increase no matter how large is the inductor current. Thus, in our ideal system, the power company need not worry about the increase in line current due to the inductor.

The difficulty arises because the system is not ideal; there is resistance in the power line. In this case, our model should include a resistance in series with the source. Although small, this resistance does dissipate energy, especially in a long line. Hence, the power company is vitally interested in reducing the power factor, which reduces the line current. There are two ways the power company can correct this economic situation: either charge the customer who has a low power factor a higher rate, or request that the customer connect a capacitor across the line to bring the power factor to within acceptable limits.

EXAMPLE 10.1.2 Suppose an industrial plant uses 450 kW of power at 0.85 lagging power factor. The line voltage is 15 kV at 60 Hz. Find the capacitor value necessary to correct the power factor to (a) 1, and (b) 0.95 lagging.

Solution

a. The power triangle is shown in Fig. 10.8 where $\Theta = \cos^{-1}(0.85) = 31.8°$. The value of Q is given by

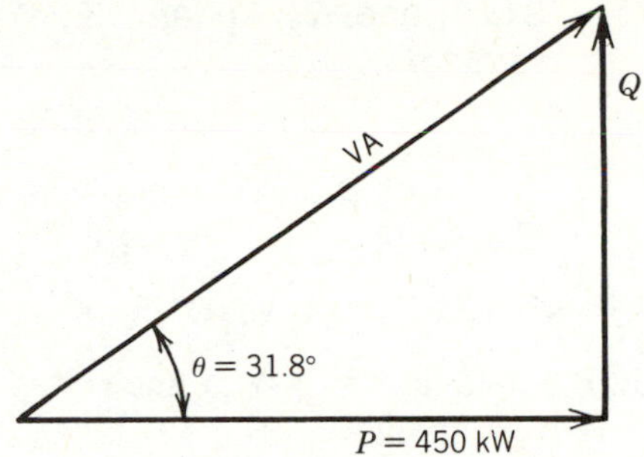

Figure 10.8. Power triangle for Example 10.1.2a.

$$Q = P \tan \Theta = 278.88 \text{ kvars}$$

Therefore to correct the power factor to 1, a capacitor that will draw 277.88 kvars leading should be used.

$$Q_C = \frac{V^2}{X_C} = 278.88(10)^3$$

or

$$X_C = \frac{V^2}{Q_C} = \frac{[15(10)^3]^2}{278.88(10)^3} = 8.068(10)^2$$

Since $X_C = 1/\omega C$, we have

$$C = \frac{1}{8.068(10)^2(377)} = 3.288(10)^{-6} \text{ F}$$

Although this value of capacitance is common, it is expensive to construct one that can withstand the 15-kV line voltage.

b. If the power is to be corrected to 0.95 lagging, this means that $\Theta = 18.2°$ is desired. As shown in Fig. 10.9, a value of Q_C that will reduce the angle from 31.8 to 18.2° must be used. Let

$$Q_C = Q - Q_2$$

where $Q = 278.88$ kvars is the original value of Q corresponding to

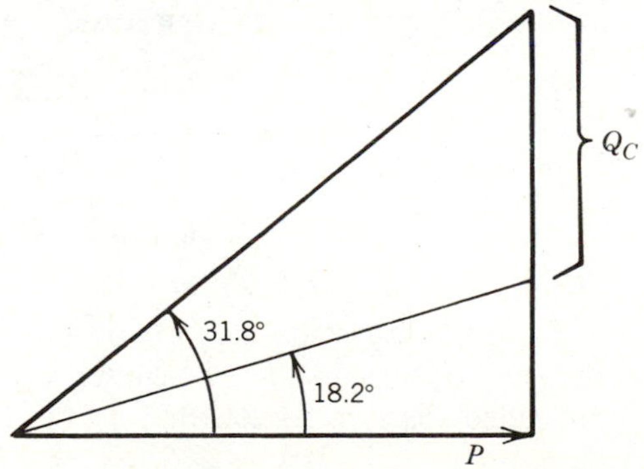

Figure 10.9. Power triangle for Example 10.1.2b.

31.8°, and $Q_2 = P\tan(18.2°) = 147.91$ kvars is the final value. Therefore,

$$Q_C = 278.88 - 147.91 = 130.97 \text{ kvars}$$

Following the above analysis we find the desired value of capacitance, $C = 1.544\ \mu\text{F}$.

LEARNING EVALUATIONS

1. If a load draws 1 kW at line voltage $V_{rms} = 115$ V, 60 Hz, find the power loss in a 10-Ω line if
   a. pf = 0.8
   b. pf = 0.9
2. What values of capacitor must be used to correct the above system power factors to 0.95?

10.2 Polyphase Circuits

LEARNING OBJECTIVES

After completing this section you should be able to do the following:

1. Find the line and phase currents, line and phase voltages, and power in a balanced three phase circuit.
2. Analyze three phase circuits with an unbalanced load.

Almost all power is generated as balanced three-phase voltages. This means three voltages of equal magnitude but separated in phase by 120°. There are several advantages to polyphase power, especially to three-phase power. Power in a three-phase circuit is constant, rather than pulsating as in a single-phase circuit. Three-phase motors start and run much better than single-phase motors. Also, the weight and size of the conductors in a polyphase system is less than that required for a single-phase system that delivers the same power.

We will need double subscript notation in this section. By V_{ab} we mean the voltage at a with respect to b, that is, the voltage at point a is positive as shown in Fig. 10.10 if $\mathbf{V}_{ab}$ is positive. The double subscript notation on current $\mathbf{I}_{ab}$ means that current flows from a to b.

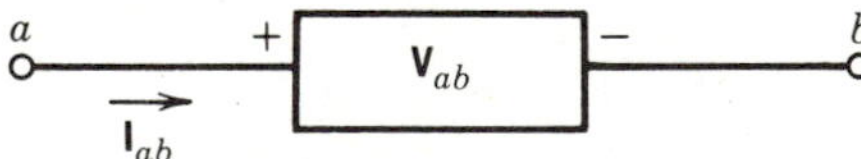

Figure 10.10. Subscript convention.

Three-phase systems are of two varieties, Y- and Δ-connected. A representation of a Y- connected source is shown in Fig. 10.11. The voltages $\mathbf{V}_{an}$, $\mathbf{V}_{bn}$, and $\mathbf{V}_{cn}$ are called phase voltages with

$$\mathbf{V}_{an} = V_p\angle 0°$$

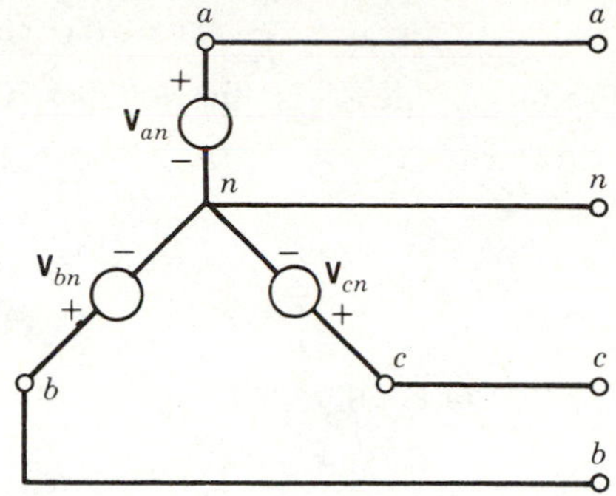

Figure 10.11. A Y-connected three-phase circuit.

$$\mathbf{V}_{bn} = V_p\angle -120^\circ \qquad (10.2.1)$$
$$\mathbf{V}_{cn} = V_p\angle 120^\circ$$

as shown in Fig. 10.12. This figure is called a "phasor diagram" and is simply a plot of the magnitude and phase of each of the phasors in our circuit. These directions are arbitrary, and we could just as well have taken them in the other direction with $\mathbf{V}_{an} = V_p\angle 0^\circ$, $\mathbf{V}_{bn} = V_p\angle 120^\circ$, and $\mathbf{V}_{cn} = V_p\angle -120^\circ$. We chose the directions in Eq. 10.2.1 so that if the phasors in Fig. 10.12 rotate in the ccw direction, the phasor tips pass a fixed point in the order a, b, c. This sequence is called the positive sequence.

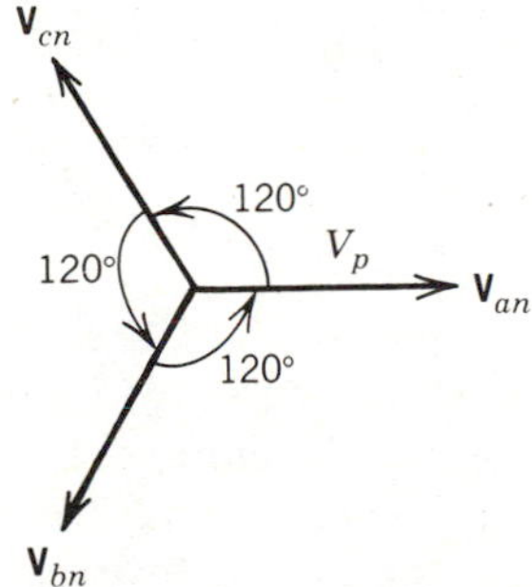

Figure 10.12. Phase voltages defined in Eq. 10.2.1.

The line-to-line voltages, or simply line voltages, are found by addition from the phase voltages. For example.

$$\begin{aligned}\mathbf{V}_{ab} &= \mathbf{V}_{an} + \mathbf{V}_{nb}\\ &= \mathbf{V}_{an} - \mathbf{V}_{bn}\\ &= V_p\angle 0^\circ - V_p\angle -120^\circ = \sqrt{3}\, V_p\angle 30^\circ\end{aligned}$$

In a similar manner we have

$$\mathbf{V}_{bc} = \sqrt{3}\, V_p\angle -90^\circ$$

$$\mathbf{V}_{ca} = \sqrt{3}\, V_p \angle 150^\circ$$

The magnitude of the line voltages is $V_L = \sqrt{3}\, V_p$, so

$$\begin{aligned} \mathbf{V}_{ab} &= V_L \angle 30^\circ \\ \mathbf{V}_{bc} &= V_L \angle -90^\circ \\ \mathbf{V}_{ca} &= V_L \angle 150^\circ \end{aligned} \qquad (10.2.2)$$

The phasor addition is illustrated in Fig. 10.13.

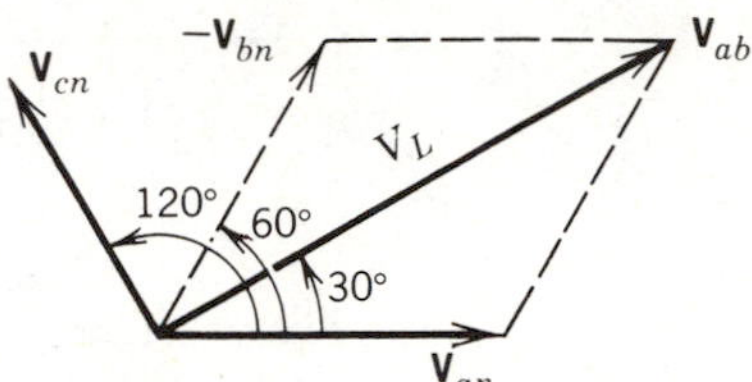

Figure 10.13. Phasor addition.

A balanced Y–Y system is shown in Fig. 10.14. Both the source and load are Y-connected, and the same impedance $\mathbf{Z}_p$ is present in each phase of the load. The line currents are therefore given by

$$\begin{aligned} \mathbf{I}_{aA} &= \frac{\mathbf{V}_{an}}{\mathbf{Z}_p} = \frac{V_p \angle 0^\circ}{\mathbf{Z}_p} \\ \mathbf{I}_{bB} &= \frac{\mathbf{V}_{bn}}{\mathbf{Z}_p} = \frac{V_p \angle -120^\circ}{\mathbf{Z}_p} \\ \mathbf{I}_{cC} &= \frac{\mathbf{V}_{cn}}{\mathbf{Z}_p} = \frac{V_p \angle 120^\circ}{\mathbf{Z}_p} \end{aligned} \qquad (10.2.3)$$

The current in each line is also the current in each phase for the balanced Y-connection. Therefore, the phase currents are balanced. Since their sum is the current in the neutral line (by KCL), this neutral current must be zero. Thus, the neutral line may be removed without changing the line currents.

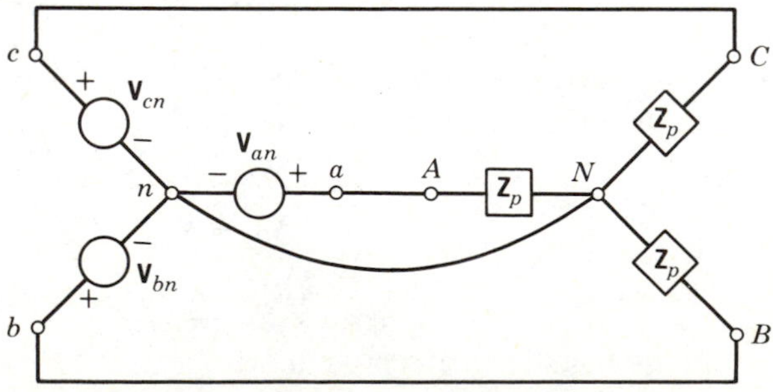

Figure 10.14. A Y-load connected to a Y-source.

If $\mathbf{Z}_p$ has the angle θ, and if $I_L = I_p$ is the line and phase current, respectively, we have

$$\begin{aligned} \mathbf{I}_{aA} &= I_L\angle{-\theta} = I_p\angle{-\theta} \\ \mathbf{I}_{bB} &= I_L\angle{-\theta - 120°} = I_p\angle{-\theta - 120°} \\ \mathbf{I}_{cC} &= I_L\angle{-\theta + 120°} = I_p\angle{-\theta + 120°} \end{aligned} \tag{10.2.4}$$

The average power in each phase is

$$P_p = (V_p)_{\text{rms}}(I_p)_{\text{rms}} \cos\theta \tag{10.2.5}$$

and the total power delivered to the load is

$$P = 3P_p \tag{10.2.6}$$

EXAMPLE 10.2.1 Find the line current, the power per phase, and the total power in the three-phase balanced circuit shown in Fig. 10.14 if $\mathbf{Z}_p = 5 + j3\ \Omega$. The phase voltage is 115 V rms.

Solution

The impedance expressed in polar coordinates is given by

$$\mathbf{Z}_p = 5 + j3 = 5.83\angle{30.96°}$$

Therefore, the line currents are given by

$$\mathbf{I}_{aA} = \frac{115}{5.83\angle{30.96°}} = 19.72\angle{-30.96°}$$

$$\mathbf{I}_{bB} = 19.72\angle{-150.96°}$$

$$\mathbf{I}_{cC} = 19.72\angle{89.04°}\ L$$

where these are rms values. The power per phase is therefore given by

$$P_p = (115)(19.72)\cos(30.96°) = 1944.85\ \text{W}$$

The total power is

$$P = 3P_p = 5834.56\ \text{W}$$

A delta-connected load (Δ-connected) is shown in Fig. 10.15. If the source is Y- or Δ-connected, then the circuit is Y–Δ or Δ–Δ. It is obvious that there can be no

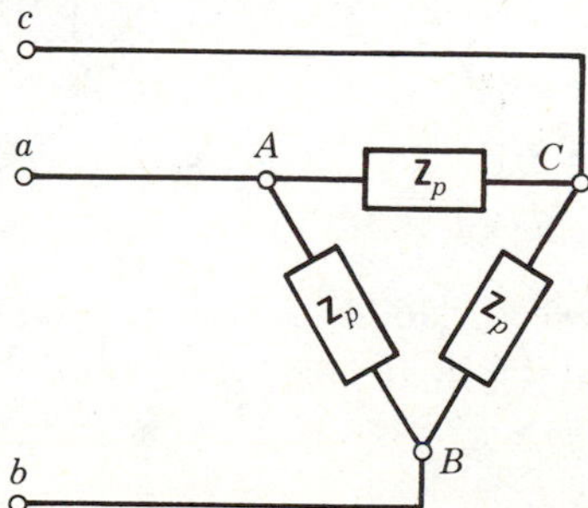

Figure 10.15. A delta-connected load.

neutral connection to a Δ-connected load. In practice, the source is always Y-connected, because if there is a slight imbalance in Δ-connected sources, a current will flow around the Δ. Since each source impedance is small, a small imbalance in the voltages can cause a large current to flow. This problem is not present in Y-connected sources.

The line voltages are equal to the phase voltages in a Δ-connected load. If the line-to-line voltages for the source are given by Eq. 10.2.2, then the load voltages are given by

$$\begin{aligned} \mathbf{V}_{AB} &= V_L\angle 30^\circ \\ \mathbf{V}_{BC} &= V_L\angle -90^\circ \\ \mathbf{V}_{CA} &= V_L\angle 150^\circ \end{aligned} \tag{10.2.7}$$

The current in each phase is found directly by Ohm's law.

$$\begin{aligned} \mathbf{I}_{AB} &= \frac{\mathbf{V}_{AB}}{\mathbf{Z}_p} = I_p\angle 30^\circ - \theta \\ \mathbf{I}_{BC} &= \frac{\mathbf{V}_{BC}}{\mathbf{Z}_p} = I_p\angle -90^\circ - \theta \\ \mathbf{I}_{CA} &= \frac{\mathbf{V}_{CA}}{\mathbf{Z}_p} = I_p\angle 150^\circ - \theta \end{aligned} \tag{10.2.8}$$

where $I_p = V_L/|\mathbf{Z}_p|$. The phase currents are not equal to the line currents, as they were in the Y-connected load. From Fig. 10.15 we see that

$$\begin{aligned} \mathbf{I}_{aA} &= \mathbf{I}_{AB} + \mathbf{I}_{AC} \\ &= \mathbf{I}_{AB} - \mathbf{I}_{CA} \end{aligned}$$

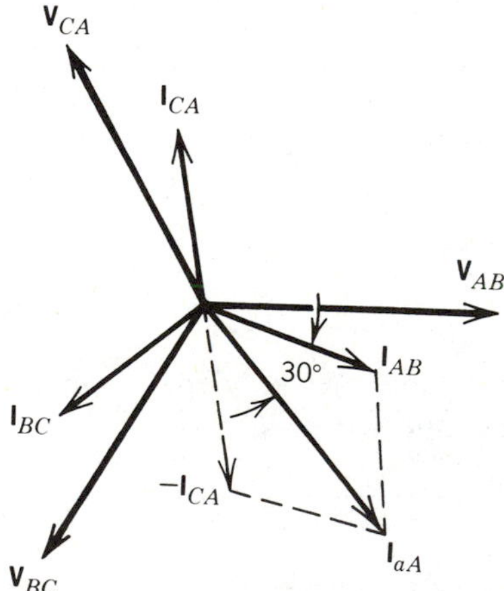

Figure 10.16. Current phasor addition.

The geometry of this vector addition is illustrated in Fig. 10.16. The current $\mathbf{I}_{aA}$ has magnitude $\sqrt{3}\,I_p$ and is at an angle 30° behind $\mathbf{I}_{AB}$. Therefore,

$$\begin{aligned} \mathbf{I}_{aA} &= \sqrt{3}\, I_p \angle -\theta \\ \mathbf{I}_{bB} &= \sqrt{3}\, I_p \angle -120^\circ - \theta \\ \mathbf{I}_{cC} &= \sqrt{3}\, I_p \angle 120^\circ - \theta \end{aligned} \tag{10.2.9}$$

EXAMPLE 10.2.2 Suppose the Δ-connected load of Fig. 10.15 with $\mathbf{Z}_p = 5 + j3\ \Omega$ is connected to a Y-connected source. The phase voltage in the source is 115 V rms. Find the phase currents, line currents, and power in the load.

Solution

The line voltage has magnitude $V_L = \sqrt{3}\, V_p = \sqrt{3}(115) = 199$-V rms. If we assume the voltages given by Eq. 10.2.7, then the load phase currents are given by

$$\mathbf{I}_{AB} = \frac{\mathbf{V}_{AB}}{\mathbf{Z}_p} = \frac{199\angle 30^\circ}{5.83\angle 30.96^\circ} = 34.16\angle -0.96^\circ$$

$$\mathbf{I}_{BC} = \frac{199\angle -90^\circ}{5.83\angle 30.96^\circ} = 34.16\angle -120.96^\circ$$

$$\mathbf{I}_{CA} = \frac{199\angle 150^\circ}{5.83\angle 30.96^\circ} = 34.16\angle 199.04^\circ$$

where these are rms values. The line currents are found using Eq. 12.2.9.

$$\mathbf{I}_{aA} = \sqrt{3}(34.13)\angle -30.96^\circ = 58.17\angle -30.96^\circ$$

$$\mathbf{I}_{bB} = 59.17\angle -150.96^\circ$$

$$\mathbf{I}_{cC} = 59.17\angle 89.04^\circ$$

The power per phase is

$$\begin{aligned} P_p &= I_p V_p \cos\theta = (34.13)(199)\cos(30.96^\circ) \\ &= 5834.6 \text{ W} \end{aligned}$$

The total power is

$$P = 3P_p = 17{,}504 \text{ W}$$

A common situation is to have a balanced Y-connected source, but an unbalanced load. The load is either Y-connected (with no neutral line) or Δ-connected. In either case, the line currents are unbalanced. This situation can be analyzed by using KVL and KCL, for these laws are fundamental and always apply to any circuit. Here is an example.

EXAMPLE 10.2.3 Suppose that a Y-connected source is connected to the unbalanced load shown in Fig. 10.17. If each source voltage is 115-V rms, and $\mathbf{Z}_1 = 3 + j4$, $\mathbf{Z}_2 = 4 + j3$, $\mathbf{Z}_3 = 5 + j0$, find each line current.

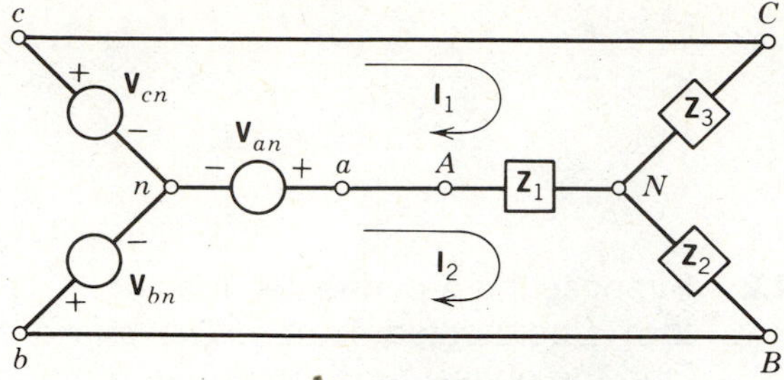

Figure 10.17. A Y–Y system with unbalanced loads.

Solution

Figure 10.12 illustrates the source voltages if we assume a positive phase sequence. Assuming two mesh currents $\mathbf{I}_1$ and $\mathbf{I}_2$ and writing KVL gives

$$0 = \mathbf{V}_{an} - \mathbf{V}_{cn} + \mathbf{I}_1\mathbf{Z}_3 + (\mathbf{I}_1 - \mathbf{I}_2)\mathbf{Z}_1$$

$$0 = \mathbf{V}_{bn} - \mathbf{V}_{an} + (\mathbf{I}_2 - \mathbf{I}_1)\mathbf{Z}_1 + \mathbf{I}_2\mathbf{Z}_2$$

Substituting values we have

$$0 = 115\underline{/0^\circ} - 115\underline{/120^\circ} + 5\mathbf{I}_1 + (3 + j4)\mathbf{I}_1 - (3 + j4)\mathbf{I}_2$$

$$0 = 115\underline{/-120^\circ} - 115\underline{/0^\circ} + (3 + j4)\mathbf{I}_2 - (3 + j4)\mathbf{I}_1 + (4 + j3)\mathbf{I}_2$$

Solving for $\mathbf{I}_1$ and $\mathbf{I}_2$ gives

$$\mathbf{I}_1 = 26.61\underline{/105.26}$$

$$\mathbf{I}_2 = 15.8\underline{/26.82}$$

Solving for the line currents in Fig. 10.17 gives

$$\mathbf{I}_{aA} = \mathbf{I}_2 - \mathbf{I}_1 = 28.09\underline{/-41.31^\circ}$$

$$\mathbf{I}_{bB} = -\mathbf{I}_2 = 15.8\underline{/-153.18^\circ}$$

$$\mathbf{I}_{cC} = \mathbf{I}_1 = 26.61\underline{/105.26^\circ}$$

LEARNING EVALUATION

Find the phase currents and line currents in the balanced Δ-connected load shown in Fig. 10.15 if the source is 115-V rms and Y-connected. Each impedance is $\mathbf{Z} = 8 + j6\ \Omega$.

10.3 Power Measurements

LEARNING OBJECTIVES

After completing this section you should be able to do the following:

1. Connect two wattmeters properly to measure three-phase power.
2. Find the impedance of a balanced load from the two wattmeter readings.

The measurement of the average real power delivered to a load is essential in many situations. Power measurements that vary significantly in time may indicate imminent failure of one or more components in our system. For instance, the resistance value of a power resistor may change dramatically as the resistor is stressed because of age, environment, and so on. This could be reflected in an altered power absorption by the load. Utility companies are obviously interested in measuring power delivered to a load so that they can properly bill the customer.

In order to measure average power we must measure the voltage across and current through the load. A wattmeter is an instrument constructed to do this. It consists of two coils; one to measure current and one to measure voltage. The coils are located and connected such that they produce an electromagnetic torque that is proportional to the product of the voltage and current. Usually the current coil is fixed and the voltage coil is able to rotate about a fixed point. The deflection of a pointer attached to the moving coil is proportional to the power being measured.

There is an inertia associated with this mechanical system that limits its response so that it averages out the instantaneous values of voltage and current and therefore produces a reading of average power. The frequency response of the wattmeter limits its use to dc and relatively low-frequency ac power measurements.

The average power absorbed by a single-phase load may be measured by a wattmeter, as shown in Fig. 10.18. The current coil is connected in series with the load, and the voltage coil is in parallel. If the voltage across the load, and the current into the load are both positive, the wattmeter reads upscale and indicates the average power absorbed by the load. If the voltage and current are both negative, the same upscale reading is obtained.

Incidentally, the "power" meter attached to our homes is a watt-hour meter, and thus measures energy, not power. It multiplies average power times the length of time that the power is used.

If three wattmeters are connected in a three-phase load in such a way that one measures the power in each phase, the sum of all the wattmeter readings will be the total power supplied to the load. There is no problem if the load is Y-connected with the neutral line available. But if the neutral line is inaccessible,

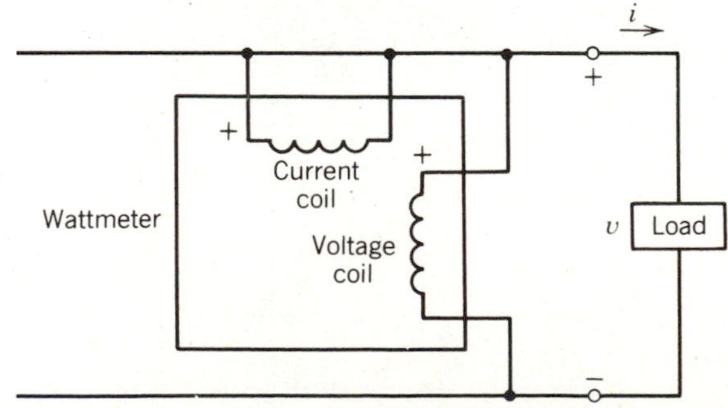

Figure 10.18. A wattmeter used with a single-phase load.

or if the load is Δ-connected, then it is not clear that three-phase power can be measured. The following analysis is presented to illustrate that three-phase power can always be measured, and in fact only two wattmeters need be used.

We will begin by analyzing the Δ-connected load with three wattmeters connected as shown in Fig. 10.19. The current coils of each wattmeter are connected in each line, and the voltage coils are connected between each line and a central point labeled n. The average power read by meter W_1 is

$$P_1 = \frac{1}{T}\int_0^T v_{an}(t)\, i_{aA}(t)\, dt \tag{10.3.1}$$

with similar expressions for P_2 and P_3. The total of all three wattmeter readings is

$$\begin{aligned} P_w &= P_1 + P_2 + P_3 \\ &= \frac{1}{T}\int_0^T v_{an} i_{aA}\, dt + \frac{1}{T}\int_0^T v_{bn} i_{bB}\, dt + \frac{1}{T}\int_0^T v_{cn} i_{cC}\, dt \\ &= \frac{1}{T}\int_0^T (v_{an} i_{aA} + v_{bn} i_{bB} + v_{cn} i_{cC})\, dt \end{aligned} \tag{10.3.2}$$

We wish to establish the fact that the sum of the three wattmeter readings in Fig. 10.19 equals the power in the load. The power in the load is

$$P_L = \frac{1}{T}\int_0^T (v_{ab} i_{AB} + v_{bc} i_{BC} + v_{ca} i_{CA})\, dt \tag{10.3.3}$$

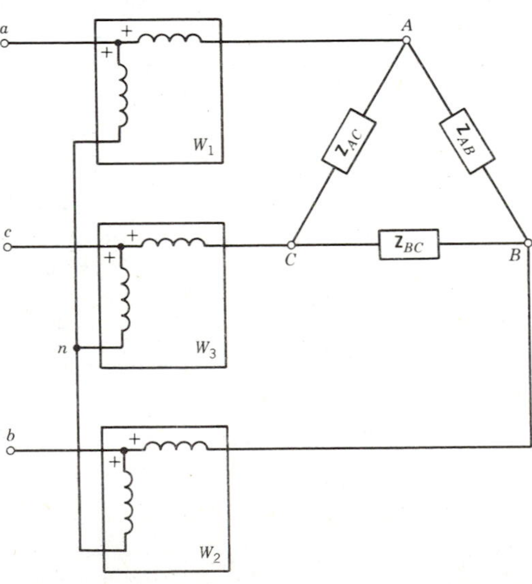

Figure 10.19. A wattmeter used with a Δ-connected load.

To show that P_w in Eq. 10.3.2 equals P_L in Eq. 10.3.3, express the line currents in terms of phase currents.

$$i_{aA} = i_{AB} - i_{CA}$$
$$i_{bB} = i_{BC} - i_{AB}$$
$$i_{cC} = i_{CA} - i_{BC}$$

Substitute into Eq. 10.3.2.

$$P = \frac{1}{T}\int_0^T [v_{an}(i_{AB} - i_{CA}) + v_{bn}(i_{BC} - i_{AB}) + v_{cn}(i_{CA} - i_{BC})]\, dt$$
$$= \frac{1}{T}\int_0^T [(v_{an} - v_{bn})i_{AB} + (v_{bn} - v_{cn})i_{BC} + (v_{cn} - v_{an})i_{CA}]\, dt$$

But

$$v_{an} - v_{bn} = v_{an} + v_{nb} = v_{ab}$$
$$v_{bn} - v_{cn} = v_{bc}$$
$$v_{cn} - v_{an} = v_{ca}$$

Therefore

$$P = \frac{1}{T}\int_0^T (v_{ab}i_{AB} + v_{bc}i_{BC} + v_{ca}i_{CA})\, dt$$

which is Eq. 10.3.3.

An additional bonus is readily apparent from our analysis. The point n is arbitrary. If we connect point n to point c, wattmeter W_3 will read zero, for its voltage coil will have zero voltage across it. Then the sum $W_1 + W_2$ will be the total power in the load. The load need not be balanced for the two wattmeter readings to equal the total power.

When the load is balanced, we can determine the load impedances from the two wattmeter readings. Consider the balanced load with two wattmeters connected as in Fig. 10.20. The impedance has angle θ, that is, $\mathbf{Z} = |Z|\angle\theta$. Wattmeter W_1 reads power given by

$$P_1 = |\mathbf{V}_{ac}|\,|\mathbf{I}_{aA}|\cos\phi_1$$

where ϕ_1 is the angle between $\mathbf{V}_{ac}$ and $\mathbf{I}_{aA}$. Also we are using rms values for $\mathbf{V}$ and $\mathbf{I}$ so we do not have to divide by 2. Likewise

$$P_2 = |\mathbf{V}_{bc}|\,|\mathbf{I}_{bB}|\cos\phi_2$$

where ϕ_2 is the angle between $\mathbf{V}_{bc}$ and $\mathbf{I}_{bB}$. Notice that $|\mathbf{V}_{ac}| = |\mathbf{V}_{bc}| = V_L$, and $|\mathbf{I}_{aA}| = |\mathbf{I}_{bB}| = I_L$. To determine the angles ϕ_1 and ϕ_2, assume a Y-connected source with $\mathbf{V}_{an}$, $\mathbf{V}_{bn}$, and $\mathbf{V}_{cn}$ given by Eq. 10.2.1. That is, the positive phase sequence balanced source with each phase voltage magnitude given by V_p. Then each line voltage is displaced 30° with magnitude $V_L = \sqrt{3}\,V_p$ as given by Eq. 10.2.2. These line voltages are illustrated in Fig. 10.21. The current through each leg of the Δ-connected load may now be determined.

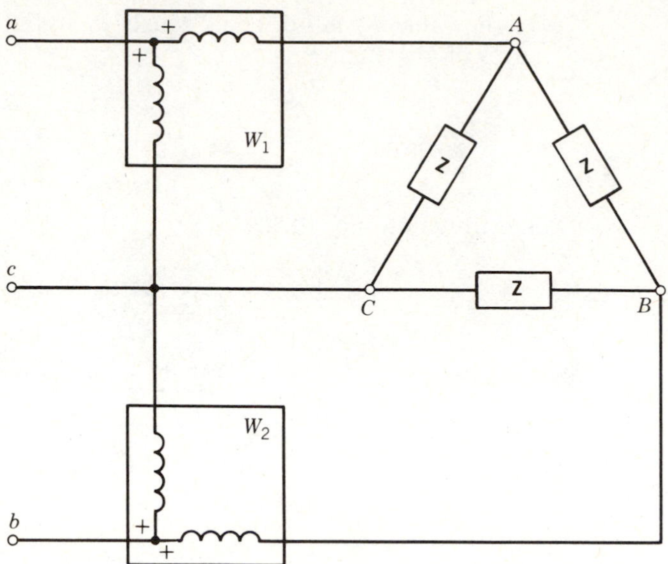

Figure 10.20. Two-wattmeter method for measuring three-phase power.

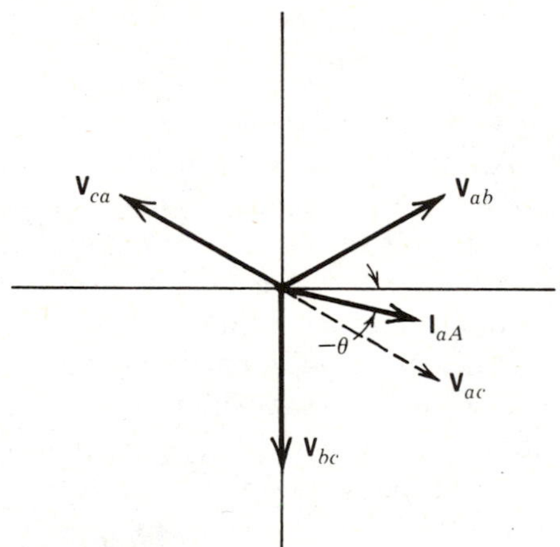

Figure 10.21. Phasor diagram for line voltages.

$$\mathbf{I}_{AB} = \frac{\mathbf{V}_{ab}}{\mathbf{Z}} = \frac{V_L \angle 30° - \theta}{|\mathbf{Z}|}$$

$$\mathbf{I}_{BC} = \frac{\mathbf{V}_{bc}}{\mathbf{Z}} = \frac{V_L \angle -90 - \theta}{|\mathbf{Z}|}$$

$$\mathbf{I}_{CA} = \frac{\mathbf{V}_{ca}}{\mathbf{Z}} = \frac{V_L \angle 150° - \theta}{|\mathbf{Z}|}$$

where θ is the impedance angle. The line currents are now found by phasor algebra.

$$\mathbf{I}_{aA} = \mathbf{I}_{AB} - \mathbf{I}_{CA} = I_L \angle -\theta$$
$$\mathbf{I}_{bB} = \mathbf{I}_{BC} - \mathbf{I}_{AB} = I_L \angle -120^\circ - \theta$$
$$\mathbf{I}_{cC} = \mathbf{I}_{CA} - \mathbf{I}_{BC} = I_L \angle 120 - \theta$$

We have illustrated the first of these in Fig. 10.21. From the diagram it is apparent that $\phi_1 = \theta - 30^\circ$. By a similar argument we find $\phi_2 = \theta + 30^\circ$.

Taking the ratio of P_1 to P_2 we get

$$\frac{P_1}{P_2} = \frac{\cos(30^\circ - \theta)}{\cos(30^\circ + \theta)}$$

Using the trigonometric identity

$$\cos(\alpha + \beta) = \cos\alpha\cos\beta - \sin\alpha\sin\beta$$

this ratio becomes

$$\frac{P_1}{P_2} = \frac{\cos 30^\circ \cos\theta + \sin 30^\circ \sin\theta}{\cos 30^\circ \cos\theta - \sin 30^\circ \sin\theta}$$
$$= \frac{\sqrt{\frac{3}{2}}\cos\theta + \frac{1}{2}\sin\theta}{\sqrt{\frac{3}{2}}\cos\theta - \frac{1}{2}\sin\theta}$$

Cross multiplying and collecting terms we obtain

$$\sqrt{3}(P_1 - P_2)\cos\theta = (P_1 + P_2)\sin\theta$$

or

$$\frac{\sin\theta}{\cos\theta} = \frac{\sqrt{3}(P_1 - P_2)}{P_1 + P_2} = \tan\theta$$

Therefore the power factor angle is

$$\theta = \tan^{-1}\frac{\sqrt{3}(P_1 - P_2)}{P_1 + P_2} \tag{10.3.4}$$

EXAMPLE 10.3.1 Suppose the balanced load in Fig. 10.20 has line voltage of 115-V rms and wattmeter readings $P_1 = 300$ W, $P_2 = 100$ W. Find the power factor and impedance **Z**.

Solution

From Eq. 10.3.4 the pf angle is

$$\theta = \tan^{-1}\frac{\sqrt{3}(300 - 100)}{300 + 100} = \tan^{-1}(0.866) = 41^\circ$$

The total power absorbed by the load is 400 W, so the per phase power is 400 / 3. Since

$$P_{AB} = |\mathbf{V}_{ab}|\,|\mathbf{I}_{AB}|\cos\theta$$

the phase current is

$$|\mathbf{I}_{AB}| = \frac{P_{AB}}{|\mathbf{V}_{ab}|\cos\theta}$$

Therefore

$$|\mathbf{Z}| = \frac{|\mathbf{V}_{ab}|}{|\mathbf{I}_{AB}|} = \frac{|\mathbf{V}_{ab}|^2\cos\theta}{P_{AB}}$$

$$= \frac{(115)^2\cos 41°}{133} = 75\ \Omega$$

Thus

$$\mathbf{Z} = 75\angle 41° = 56.6 + j49.2\ \Omega$$

LEARNING EVALUATION

The balanced Δ-connected load has line voltage of 120-V rms and wattmeter readings $P_1 = 500$ W, $P_2 = 150$ W. Find the pf angle θ and the impedance.

10.4 Transformers

LEARNING OBJECTIVES

After completing this section you should be able to do the following:

1. Write mesh equations for circuits with transformers, properly using the dot convention.
2. Analyze circuits containing ideal transformers.

Moving electric charge produces an electromagnetic field. If the field is time varying, then it will induce an electromotive force in a nearby circuit. This is the principle of the transformer, which serves to magnetically couple energy from one circuit to another. Consider the circuit shown in Fig. 10.22 consisting of two inductors, L_1 and L_2, in close proximity to one another. If current i_1 is changing with time, then the field about L_1 will also be changing with time. This will induce a voltage v_2 across L_2. The magnitude and direction of v_2 will depend on the time rate of change of i_1, and also on the physical construction of the transformer. We can express this relationship by

$$v_2(t) = M_{21}\frac{di_1}{dt} \tag{10.4.1}$$

The term M_{21} is termed the coefficient of mutual inductance, or simply mutual inductance. It is a measure of the voltage v_2 produced by the time rate of change in i_1.

If we change Fig. 10.22 by connecting the source to the right-hand pair of terminals instead of the left-hand pair and then measure the voltage v_1 we obtain

$$v_1(t) = M_{12}\frac{di_2}{dt}$$

The units of mutual inductance are henries, just as they are for self-inductance L.

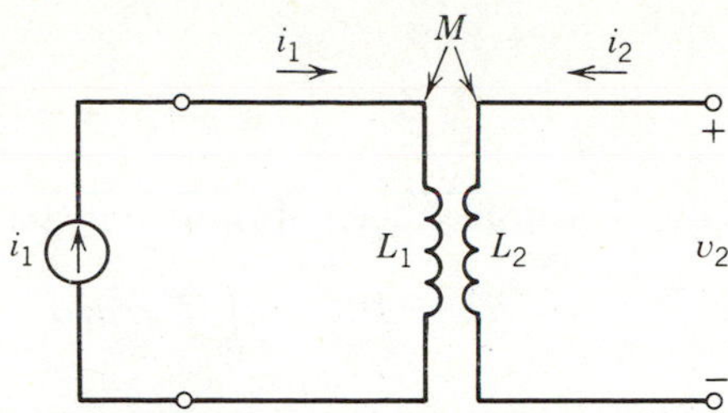

Figure 10.22. A circuit with mutual inductance.

Transformers may be wound on a core of either magnetic or nonmagnetic material. When the transformer is wound on magnetic material the relationship between ϕ and i is nonlinear. If the material is nonmagnetic then the relationship between current i and the magnetic flux ϕ is linear.

$$\phi = k_p N i$$

where k_p is the permeance of the space occupied by the field, and N is the number of turns in the coil. The permeance is a function of the physical dimensions and the permeability of the material containing the flux field.

When there is a linear relationship between flux and current the mutual inductance terms are equal.

$$M_{12} = M_{21} = M$$

There is more than one way to demonstrate this equality, but the most informative way is to calculate the energy stored in the two coils under the particular conditions illustrated in Fig. 10.23. We will also use this analysis to indicate some other facts about linear transformers.

Energy is the integral of power. Assume to $t = 0$ that there is no energy stored in the coils. The total energy stored in both coils by currents i_1 and i_2 at time t_2 is given by

$$W = \int_0^{t_2} v_1 i_1 \, dt + \int_0^{t_2} v_2 i_2 \, dt$$

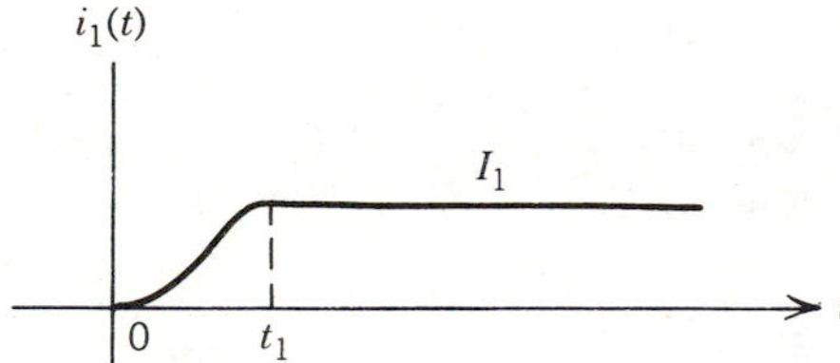

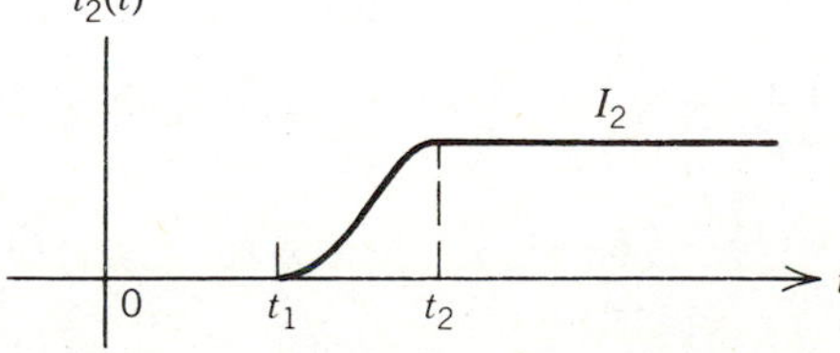

Figure 10.23. Illustrating the manner in which i_1 and i_2 vary.

or

$$W = \int_0^{t_1} v_1 i_1\,dt + \int_{t_1}^{t_2} v_1 i_1\,dt + \int_0^{t_1} v_2 i_2\,dt + \int_{t_1}^{t_2} v_2 i_2\,dt.$$

Let us consider each of these four terms separately. The first term is

$$W_1 = \int_0^{t_1} v_1 i_1\,dt = \int_0^{t_1}\left(L_1\frac{di_1}{dt} + M_{12}\frac{di_2}{dt}\right)i_1\,dt$$

Since current i_2 does not change during the interval $0 < t < t_1$, the value of this integral is

$$W_1 = \int_0^{t_1}\left(L_1\frac{di_1}{dt}\right)i_1\,dt = \int_0^{I_1} L_1 i_1\,di_1 = \frac{1}{2}L_1 I_1^2$$

The second term is

$$W_2 = \int_{t_1}^{t_2} v_1 i_1\,dt = \int_{t_1}^{t_2}\left[L_1\frac{di_1}{dt} + M_{12}\frac{di_2}{dt}\right]i_1\,dt$$

Current i_1 does not change during this interval, so

$$W_2 = \int_{t_1}^{t_2}\left(M_{12}\frac{di_2}{dt}\right)i_1\,dt = \int_{i_2(t_1)}^{i_2(t_2)} M_{12} I_1\,di_2$$

$$= M_{12} I_1 I_2$$

The third term is zero because i_2 is zero for $0 < t < t_1$. Finally, the last term is given by

$$W_4 = \int_{t_1}^{t_2} v_2 i_2\,dt = \int_{t_1}^{t_2}\left(M_{21}\frac{di_1}{dt} + L_2\frac{di_2}{dt}\right)i_2\,dt$$

Since i_1 is constant in this interval we have

$$W_4 = \int_{t_1}^{t_2}\left(L_2\frac{di_2}{dt}\right)i_2\,dt = \frac{1}{2}L_2 I_2^2$$

Summing all four terms gives

$$W = \tfrac{1}{2}L_1 I_1^2 + \tfrac{1}{2}L_2 I_2^2 + M_{12} I_1 I_2 \tag{10.4.2}$$

If the procedure of first increasing i_1 and then increasing i_2 is reversed, the total energy stored is given by

$$W = \tfrac{1}{2}L_1 I_1^2 + \tfrac{1}{2}L_2 I_2^2 + M_{21} I_1 I_2 \tag{10.4.3}$$

The only difference between Eqs. 10.4.2 and 10.4.3 is the mutual inductance term. Since the total energy stored is the same, we must have

$$M_{12} = M_{21} = M \tag{10.4.4}$$

Since Eqs. 10.4.2 and 10.4.3 are valid for arbitrary time t_2 we can express the energy as a function of time by

$$w(t) = \tfrac{1}{2}L_1 i_1^2(t) + \tfrac{1}{2}L_2 i_2^2(t) + M i_1(t) i_2(t) \tag{10.4.5}$$

If the coils are constructed so that M is positive, then a reversal of the winding of one coil will reverse the sign on M. In order to simplify matters we will use the dot convention to specify the proper sign on M. Given two coupled inductors, if the current entering the dotted end of one coil is increasing ($di/dt > 0$) then it induces a positive voltage at the dotted end of the other coil. This is illustrated in Fig. 10.24. With this convention we are now assuming the "passive sign" on M, that is, M is always positive and the dots specify whether + or − is attached to the relationship in Eq. 10.4.1.

One side of the transformer is called the primary winding, and the other is called the secondary winding. The primary is conventionally the side where the source is attached. Thus energy is transformed from the primary side to the secondary side. Notice that in our discussion so far the secondary has been open so that no current flows. If we now close the secondary by attaching some impedance, then current will flow in the secondary coil. This current will in turn induce a voltage in the primary, which must either add or subtract to the voltage that was present before the secondary path was completed.

The pair of equations that describe the transformer for the general case of $i_1 \neq 0$ and $i_2 \neq 0$ are given by

$$v_1 = L_1 \frac{di_1}{dt} + M \frac{di_2}{dt} \tag{10.4.6}$$

$$v_2 = M \frac{di_1}{dt} + L_2 \frac{di_2}{dt} \tag{10.4.7}$$

Here we have assumed the dot configuration as that in Figs. 10.24*a* or 10.24*d*. If the dots had been as in Figs. 10.24*b* or 10.24*c*, then the sign on the mutual inductance terms would have been negative. If the source is sinusoidal we can write Eqs. 10.4.6 and 10.4.7 in terms of phasors.

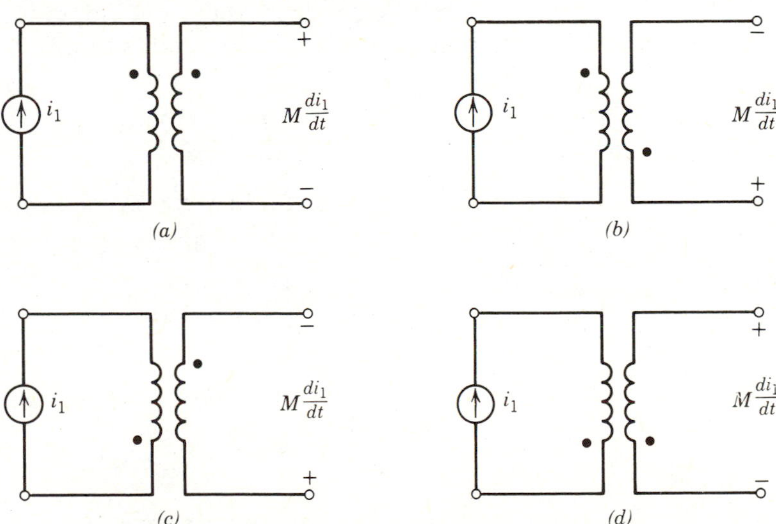

Figure 10.24. Dot convention for induced voltages.

$$\mathbf{V}_1 = j\omega L_1 \mathbf{I}_1 + j\omega M \mathbf{I}_2 \tag{10.4.8}$$

$$\mathbf{V}_2 = j\omega M \mathbf{I}_1 + j\omega L_2 \mathbf{I}_2 \tag{10.4.9}$$

EXAMPLE 10.4.1 Find phasor currents $\mathbf{I}_1$ and $\mathbf{I}_2$ and the voltage across the 100-Ω resistor in Fig. 10.25. The sinusoidal frequency is $\omega = 10$ rad/s.

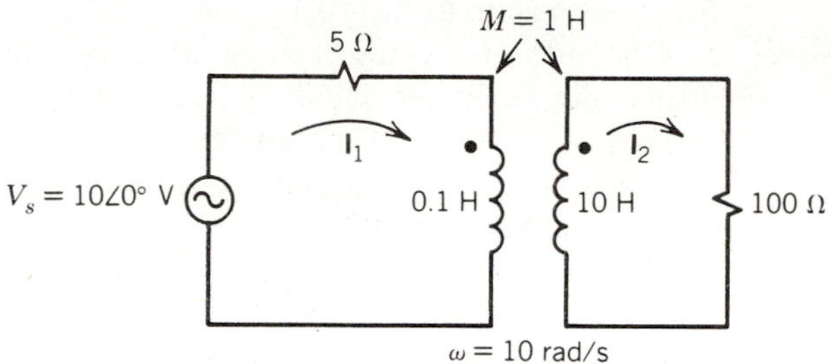

Figure 10.25. Circuit for Example 10.4.1.

Solution

Writing mesh equations we have

$$10 = (5 + j1)\mathbf{I}_1 - j10\mathbf{I}_2$$

$$0 = -j10\mathbf{I}_1 + (100 + j100)\mathbf{I}_2$$

The negative signs on the mutual inductance terms occur because of the dot configuration and the assumed current directions. In the first equation, since $\mathbf{I}_2$ leaves the dot in the secondary, the voltage induced in the primary is opposite in sign to the self inductance voltage term. The same reasoning applies to the second equation. Solving for $\mathbf{I}_1$ from the second equation we obtain

$$\mathbf{I}_1 = \left(\frac{100 + j100}{j10}\right)\mathbf{I}_2 = (10 - j10)\mathbf{I}_2$$

Upon substituting this into the first equation we obtain

$$\mathbf{I}_2 = 0.09836 + j0.08197$$

and

$$\mathbf{I}_1 = 1.80328 - j0.16393$$

Also

$$\mathbf{V}_2 = 100\mathbf{I}_2 = 9.836 + j8.197$$

$$= 12.80\angle 39.81^\circ$$

Notice that the output voltage $\mathbf{V}_2$ is larger in magnitude than the input voltage $\mathbf{V}_1$, and is proportionally larger than the voltage across the primary coil. Thus we may use a transformer to increase (or decrease) voltage.

Our analysis of energy in the primary and secondary of a transformer may be used to show that the value of M is constrained by

$$M \le \sqrt{L_1 L_2} \tag{10.4.10}$$

This result can be derived by noting that the total energy stored in a pair of coupled coils can never be negative. Complete the square in Eq. 10.4.5 to obtain

$$W(t) = \frac{1}{2}\left\{\left[\sqrt{L_1}\, i_1(t) + \frac{M}{\sqrt{L_1}} i_2(t)\right]^2 + \left(L_2 - \frac{M^2}{L_1}\right) i_2^2(t)\right\} \tag{10.4.11}$$

To ensure that $w(t)$ be positive it is sufficient to require that the term $(L_2 - M^2/L_1)$ be greater than or equal to zero. Thus,

$$|M| \le \sqrt{L_1 L_2}$$

The coefficient of coupling, k, is defined as the ratio of the actual value of M to its maximum value.

$$k = \frac{M}{\sqrt{L_1 L_2}} \tag{10.4.12}$$

It is evident that $0 \le k \le 1$.

The transformer in Example 10.4.1 has $M = \sqrt{L_1 L_2}$, or $k = 1$. This implies that all of the flux linking L_1 also links L_2. Although this ideal is impossible in practice, power transformers come close since they are wound on an iron core. The iron core provides a low reluctance path for the magnetic flux, thus linking both coils with substantially all of the flux. The assumption of an ideal transformer ($k = 1$) simplifies our analysis.

Eliminating $\mathbf{I}_2$ between Eqs. 10.4.8 and 10.4.9 gives

$$\mathbf{V}_1 = \frac{M}{L_2} V_2 + j\omega \frac{L_1 L_2 - M^2}{L_2} \mathbf{I}_1$$

Likewise, eliminating $\mathbf{I}_1$ gives

$$\mathbf{V}_2 = \frac{M}{L_1} \mathbf{V}_1 + j\omega \left(\frac{L_1 L_2 - M^2}{L_1}\right) \mathbf{I}_2$$

Setting $M = \sqrt{L_1 L_2}$ in these equations gives

$$\mathbf{V}_1 = \sqrt{\frac{L_1}{L_2}}\, \mathbf{V}_2$$

and

$$\mathbf{V}_2 = \sqrt{\frac{L_2}{L_1}}\, \mathbf{V}_1$$

The turns ratio N_2/N_1 is given approximately by

$$\frac{N_2}{N_1} = \sqrt{\frac{L_2}{L_1}}$$

so that we may rewrite the above equations as

$$\frac{\mathbf{V}_1}{\mathbf{V}_2} = \frac{N_1}{N_2} \qquad (10.4.13)$$

By a similar analysis we may show that when L_1 and L_2 are large

$$\frac{\mathbf{I}_1}{\mathbf{I}_2} = \frac{N_2}{N_1} \qquad (10.4.14)$$

This additional assumption (large L_1 and L_2) is satisfied by the transformers used in power transmission systems.

EXAMPLE 10.4.2 Compare the actual ratios $\mathbf{V}_2/\mathbf{V}_1$ and $\mathbf{I}_2/\mathbf{I}_1$ to those predicted by Eqs. 10.4.13 and 10.4.14 in Example 10.4.1.

Solution

The turns ratio is

$$\frac{N_2}{N_1} = \sqrt{\frac{L_2}{L_1}} = \sqrt{\frac{10}{0.1}} = 10$$

Since $|\mathbf{V}_2| = 12.5$ (from Example 10.4.1) then by Eq. 10.4.13,

$$|\mathbf{V}_1| = \frac{N_1}{N_2}|\mathbf{V}_2| = 0.1(12.5) = 1.25$$

In order to compare this value to that obtained by circuit analysis, find the voltage across the 5-Ω resistor and subtract this from the source to obtain V_1.

$$\mathbf{V}_R = 5\mathbf{I}_1 = 9.01 - j0.82$$

Hence

$$\mathbf{V}_1 = 10 - \mathbf{V}_R = 0.99 + j0.82 = 1.29\angle 39.63^\circ$$

which is in close agreement with that obtained by Eq. 10.4.13.

The agreement with Eq. 10.4.14 is not so close, however. Since $N_2/N_1 = 10$, we should have $\mathbf{I}_1/\mathbf{I}_2 = 10$. The ratio of $\mathbf{I}_1$ to $\mathbf{I}_2$ from Example 10.4.1, however, gives

$$\frac{\mathbf{I}_1}{\mathbf{I}_2} = \frac{1.81}{0.13} = 14.14$$

The disagreement arises because the impedances of L_1 and L_2 are not large when compared to the other impedances in the circuit.

LEARNING EVALUATIONS

1. Write mesh equations for the two mesh circuit shown in Fig. 10.26.

2. If $M = \sqrt{L_1 L_2}$ in Fig. 10.26, what would be the turns ratio N_2/N_1?

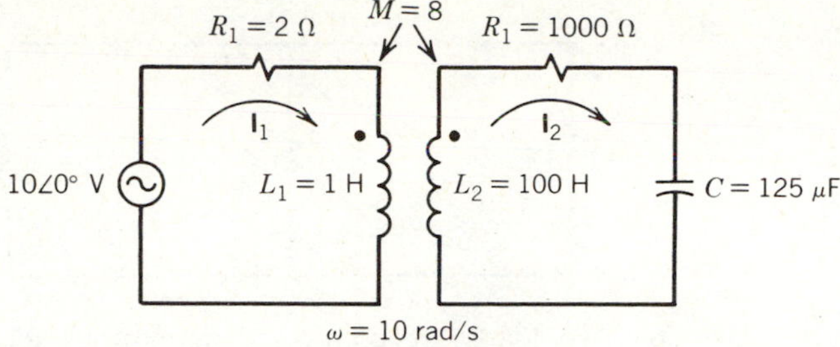

Figure 10.26

PROBLEMS

SECTION 10.1

1. The 100-V source is connected to the parallel RL circuit shown in Fig. 10.27 at $t = 0$. If the initial inductor current is zero, find and plot the instantaneous power $p(t)$ supplied by the source.

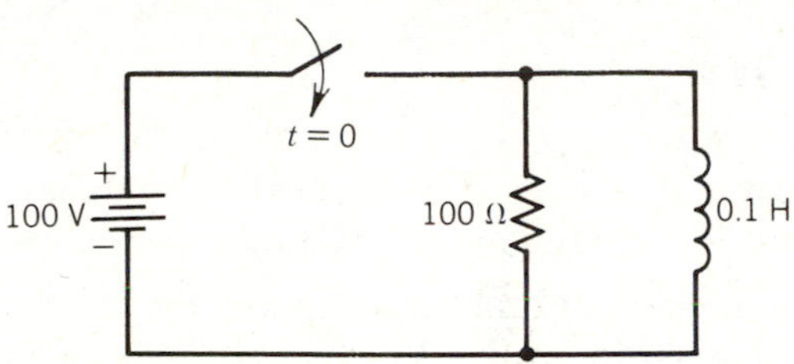

Figure 10.27

2. The battery and switch in Fig. 10.27 is replaced by a steady-state sinusoidal source

$$v(t) = 100\cos(1500t)$$

   a. Find and plot instantaneous power $p(t)$.
   b. Find the average power P.
   c. Find the power factor.

3. Calculate the value of the parallel capacitor necessary to increase the power factor to 0.95 in Problem 2.

4. Determine the power factor for a load:
   a. Consisting of a 100-μF capacitor in parallel with the series combination of 10-Ω resistance and 0.05-H inductance.
   b. That is inductive and draws 10-A rms and 0.9 kW at 120-V rms.

5. The power factor of the total load supplied by the current source in Fig. 10.28 is 0.85 lagging. What is the frequency of the source?

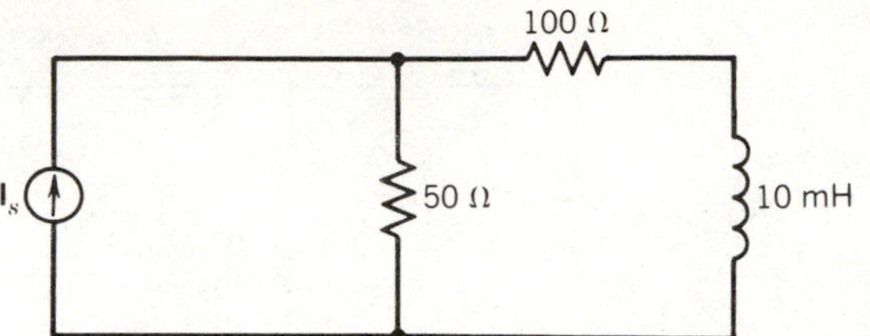

Figure 10.28

6. Two loads are connected in parallel across a 120-V line as shown in Fig. 10.29. Load 1 absorbs 10 kW and 12 kvars. Load 2 absorbs 6 kvars at 0.8 pf leading. Find the equivalent impedance presented to the source.

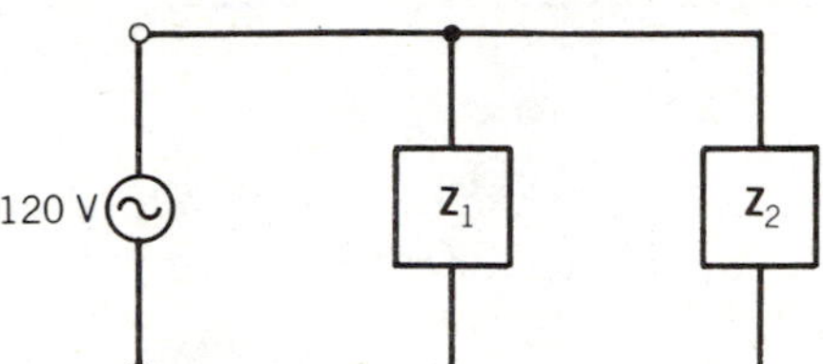

Figure 10.29

SECTION 10.2

7. A three phase source with line voltages of 120-V rms is used to supply a balanced Y-connected load consisting of three impedances $\mathbf{Z}_L = 3 + j4\ \Omega$. Find the line current and the average power absorbed by the load.
8. For the situation in Problem 7 suppose the line impedances $\mathbf{Z}_a$ are taken into account, as shown in Fig. 10.30. Let $\mathbf{Z}_a = 0.1 + j0.1\ \Omega$. Find the line current, the power absorbed by the load, and the total power delivered by the source.

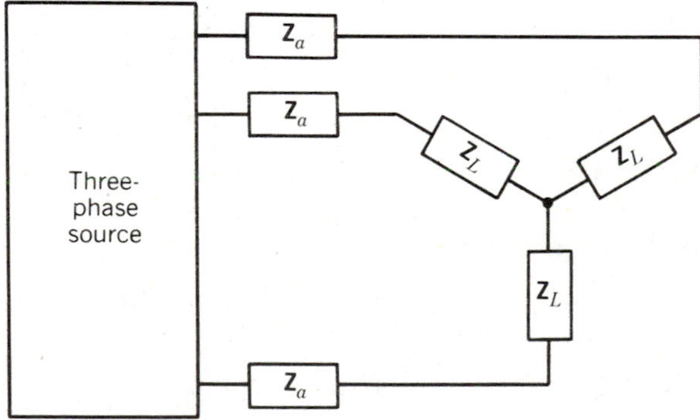

Figure 10.30

9. The power factor for the load in Problem 8 is 0.6. If the power factor is corrected to 0.8, find the savings in power delivered by the source. That is, repeat Problem 8 for the new load and determine the savings in power that must be delivered by the source.
10. Recall that we studied Δ-Y transformations in Chapter 2. Suppose that we wish to replace the load impedance in Problem 7 by an equivalent Δ-con-

nected load. Find the impedance in each leg of the equivalent Δ-connected load that will absorb the same power and require the same line currents.

11. Draw the phasor diagram for the line and phase voltages and the line and phase currents for the circuit in Fig. 10.31. The source voltages are 120-V rms positive phase sequence, and the balanced load impedance is $\mathbf{Z}_L = 4 + j3\ \Omega$.

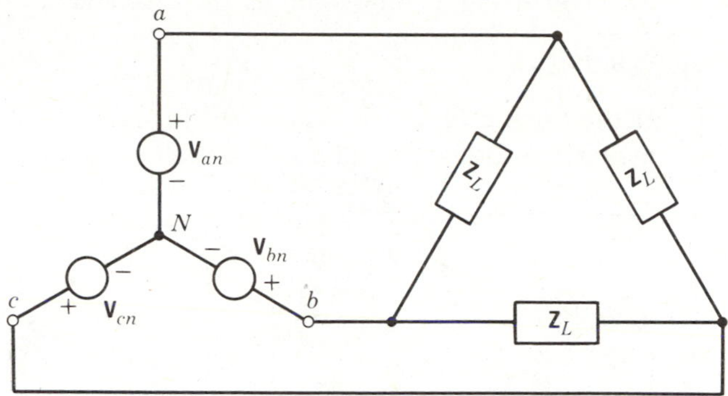

Figure 10.31

12. Find the total power delivered to the load in Problem 11.

13. A balanced three-phase load absorbs 10 kW at a lagging power factor of 0.9 when the line voltage is 120-V rms. Find four equivalent circuits to model the load.

SECTION 10.3

14. Suppose the balanced load in Fig. 10.20 has line voltage 115-V rms and that each load impedance is $Z_L = 50 + j50\ \Omega$. Find each wattmeter reading.

15. A single wattmeter is connected in a three-phase circuit as shown in Fig. 10.32. If the line voltage is 500 V and the balanced load impedance is $\mathbf{Z}_L = 100 + j100\ \Omega$, find the wattmeter reading. The phase sequence is positive.

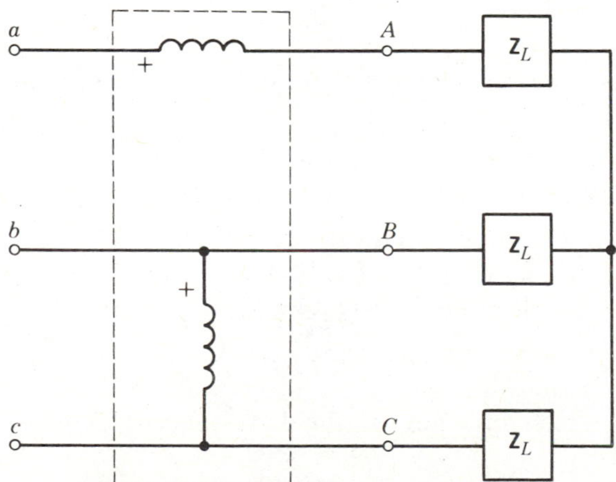

Figure 10.32

16. Suppose the load in Fig. 10.32 is Δ-connected. That is, the load has $\mathbf{Z}_L = 100 + j100\ \Omega$ in each leg, but it is Δ-connected instead of Y-connected. If the line voltage is 500-V rms, find the wattmeter reading.

17. The ratio of the reading of wattmeter no. 1 of Fig. 10.20 to the reading of wattmeter no. 2 is a function of the load power factor. Compute and plot a curve of this function (assuming the reactance of the load is inductive).

SECTION 10.4

18. If the forcing function in Fig. 10.33 is sinusoidal with frequency ω, find an expression for the impedance of the network.

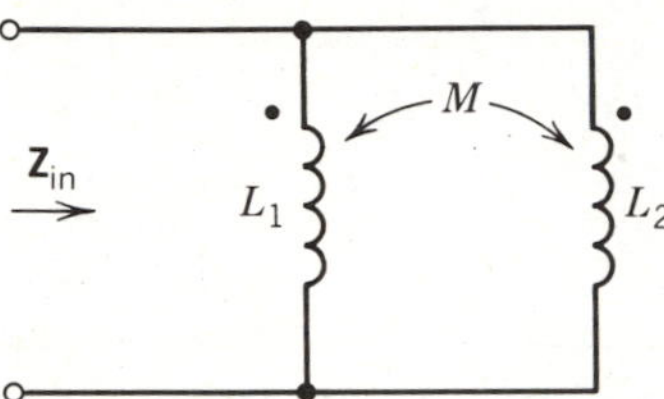

Figure 10.33

19. Repeat Problem 18 if the dot on the second coil is reversed.

20. Find the current in the secondary circuit in Fig. 10.34. The frequency of the sinusoidal source is $\omega = 100$ rad/s. The mutual inductance is $M = 0.02$ H.

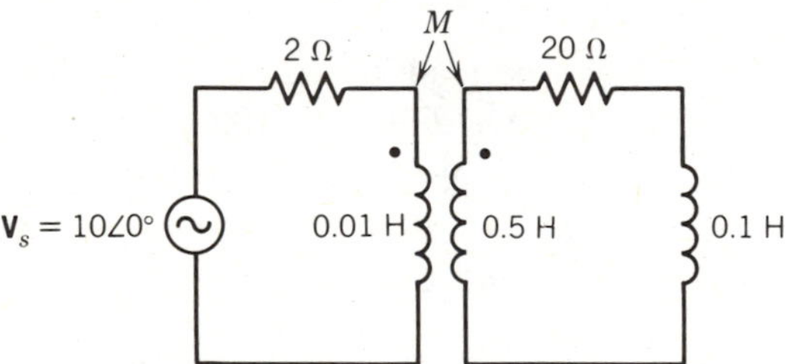

Figure 10.34

21. Two magnetically coupled coils have $L_1 = 8$ mH, $L_2 = 2$ mH, and $M = 3$ mH.
    a. What is the coefficient of coupling?
    b. What is the maximum value of M?

22. A transformer with the dot placed on the primary is shown in Fig. 10.35. The dot on the secondary has been obliterated, but the experimental setup can be used to determine where it belongs.
    a. If the dc voltmeter kicks upscale when the switch is closed, where does the missing dot belong?
    b. Which direction will the dc voltmeter kick when the switch is opened?

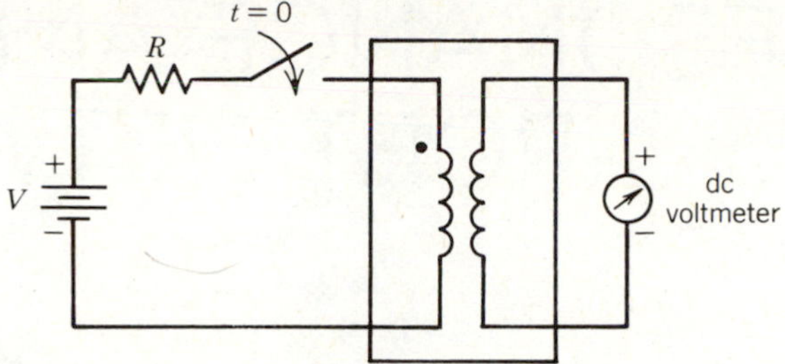

Figure 10.35

LEARNING EVALUATION ANSWERS

Section 10.1

1. a. 1181 W
   b. 934 W
2. a. 84 μF
   b. 31 μF

Section 10.2

$\mathbf{I}_{AB} = 19.92\angle -6.87°$

$\mathbf{I}_{BC} = 19.92\angle -126.87°$

$\mathbf{I}_{CA} = 19.92\angle 113.13°$

$\mathbf{I}_{aA} = 34.5\angle -36.87°$

$\mathbf{I}_{bB} = 34.5\angle -156.87°$

$\mathbf{I}_{cC} = 34.5\angle 83.13°$
(rms units)

Section 10.3

$\theta = 43°$
$\mathbf{Z} = 71.1 + j66.3$

Section 10.4

1. $10 = (2 + j10)I_1 - j80I_2$
   $0 = -j80I_1 + (1000 + j200)I_2$
2. $\dfrac{N_2}{N_1} = \dfrac{L_2}{L_1} = 10$

CHAPTER

11 RESONANCE AND TWO-PORT NETWORKS

11.1 Series Resonance

LEARNING OBJECTIVES

After completing this section you should be able to do the following:

1. Plot response versus frequency for *RL*, *RC*, and *RLC* series circuits.
2. Find the resonant frequency and 3-dB frequencies for a series *RLC* circuit.
3. Determine the bandwidth and quality factor for series *RLC* circuits.

Loci in circuit analysis are graphical representations of two general classes of parametric relations, current loci, and voltage loci. For current loci, the voltage is held constant while a circuit parameter or the frequency is varied. For voltage loci, it is the current that is held constant. We first consider voltage loci.

The *RL* series circuit in Fig. 11.1 has sinusoidal voltage applied. Writing KVL in terms of phasors gives

$$\mathbf{V} = \mathbf{I}R + j\mathbf{I}X_L = \mathbf{IZ} \qquad (11.1.1)$$

The complex numbers represented by each term in this equation are plotted in Fig. 11.2. Figure 11.2*a* represents the voltage diagram, with current I as reference, and Fig. 11.2*b* is called the impedance triangle. The phase angle θ is the angle between the applied current and the resulting voltage. Thus if $v(t) = V\cos(\omega t)$ then $i(t) = I\cos(\omega t - \theta)$.

The current lags the voltage since the reactance is inductive. Since counterclockwise is the positive direction of rotation, Fig. 11.2*a* illustrates the current $\theta°$ behind the voltage.

Now consider the effect of varying the frequency while holding the current constant in Fig. 11.1. At $f = 0$, $X_L = 0$, and the impedance is resistive. Therefore, $\mathbf{V} = \mathbf{I}R$, which is indicated in Fig. 11.2*c* as V_0. As frequency increases, $X_L = 2\pi fL$

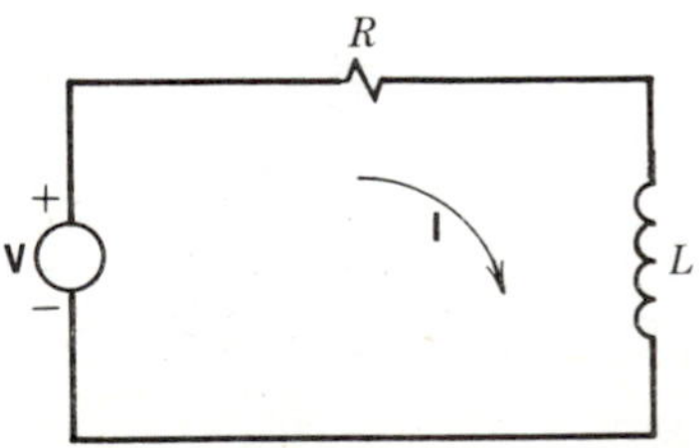

Figure 11.1. A series *RL* circuit.

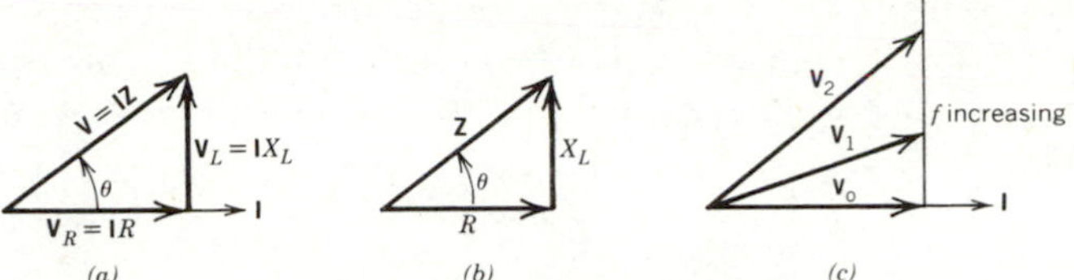

Figure 11.2. Voltage diagram and locus and impedance diagram for circuit in Fig. 11.1.

increases, giving the locus of voltage as shown in Fig. 11.2*c*. Note that the angle θ changes from 0 to 90° as f increases from 0 Hz to ∞.

Next let us consider the current locus. Equation 11.1.1 may be rearranged to give

$$\mathbf{I} = \frac{\mathbf{V}}{R + jX_L} = \frac{\mathbf{V}}{R + j(2\pi f L)} \tag{11.1.2}$$

The current **I** varies with frequency f if **V**, R, and L are held constant. The current locus is a semicircle as shown in Fig. 11.3*a*. At $f = 0$ the impedance is $Z = R$, and current and voltage are in phase. As f increases, the denominator of Eq. 11.1.2 becomes larger with increasing positive angle. Thus **I** becomes smaller with increasing negative angle. This is reflected in the current locus diagram. Figure 11.3*b* illustrates the admittance of the circuit as frequency increases. This admittance diagram is derived directly from Fig. 11.3*a* by dividing each term by **V**.

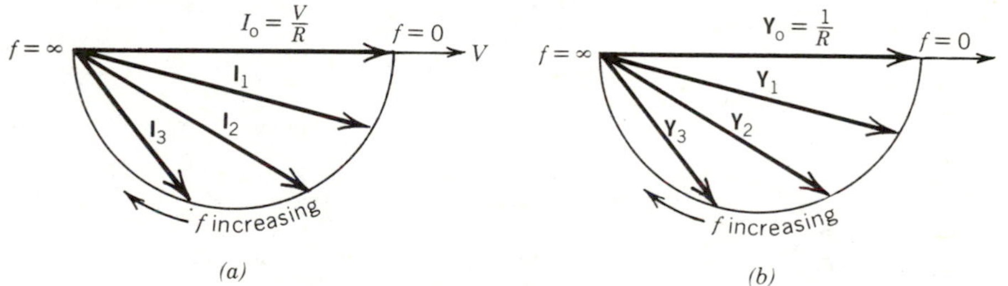

Figure 11.3. Loci for circuit in Fig. 11.1

The fact that the current locus is a semicircle may be proved as follows: From Eq. 11.1.2.

$$\mathbf{I} = \frac{\mathbf{V}}{\mathbf{Z}} \tag{11.1.3}$$

From Fig. 11.2*b*,

$$\mathbf{Z} = \frac{R}{\cos\theta} \tag{11.1.4}$$

Substituting Eq. 11.1.4 into Eq. 11.1.3 gives

$$\mathbf{I} = \frac{\mathbf{V}}{R}\cos\theta \tag{11.1.5}$$

which is the equation of a circle in polar coordinates, with diameter V/R and center at $(V/2R, 0)$.

Figure 11.3 can serve as the current locus when **V**, R, and f are held constant with L varying from zero to infinity. This is apparent from Eq. 11.1.2.

Next consider the RC circuit in Fig. 11.4. Writing KVL gives

$$\mathbf{V} = \mathbf{I}R - j\mathbf{I}X_c = \mathbf{I}\mathbf{Z} \tag{11.1.6}$$

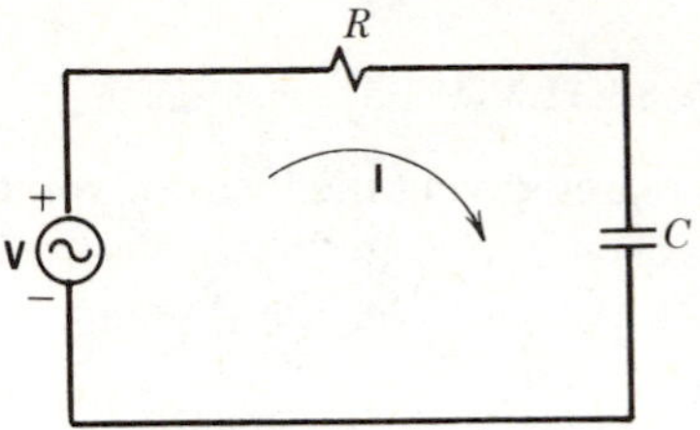

Figure 11.4. A series RC circuit.

The terms in this equation are illustrated in Fig. 11.5*a*, while the impedance diagram is shown in Fig. 11.5*b*. The voltage locus is shown in Fig. 11.5*c*, indicating that the phase angle is negative and changes from 0 to $-90°$ as f increases.

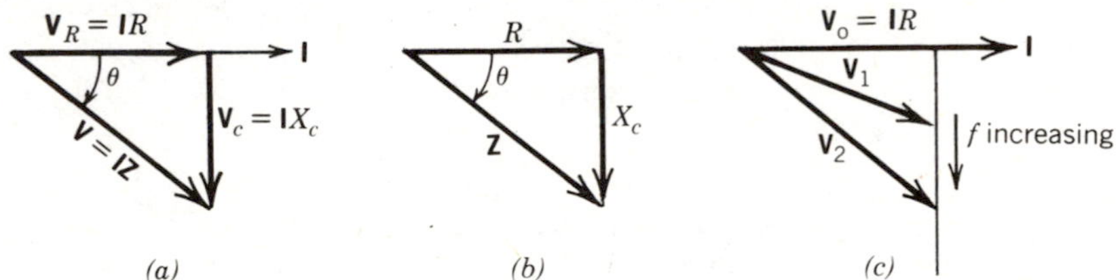

Figure 11.5. Voltage diagram and locus and impedance triangle for circuit in Fig. 11.4.

To plot the current locus we solve Eq. 11.1.6 for current.

$$\mathbf{I} = \frac{\mathbf{V}}{R - jX_c} = \frac{\mathbf{V}}{R - \dfrac{j}{2\pi fC}} \tag{11.1.7}$$

Capacitive reactance varies inversely with frequency. At $f = 0$, $X_c = \infty$, and the current is zero. As f increases, X_c becomes smaller, and current increases until $\mathbf{I} = \mathbf{V}/R$ at $f = \infty$. The locus of **I** from Eq. 11.1.7 is plotted in Fig. 11.6*a*. Figure 11.6*b* illustrates the admittance. Notice that current leads voltage in both Figs. 11.5 and 11.6.

Let us next consider the RLC circuit shown in Fig. 11.7. Writing KVL gives

$$\mathbf{V} = R\mathbf{I} + j\mathbf{I}(X_L - X_c) = \mathbf{I}\mathbf{Z} \tag{11.1.8}$$

The terms in this equation are illustrated in Fig. 11.8*a*, and the impedance diagram is shown in Fig. 11.8*b*. Since X_L and X_c both depend on frequency, these diagrams illustrate the situation for only one frequency. In this case

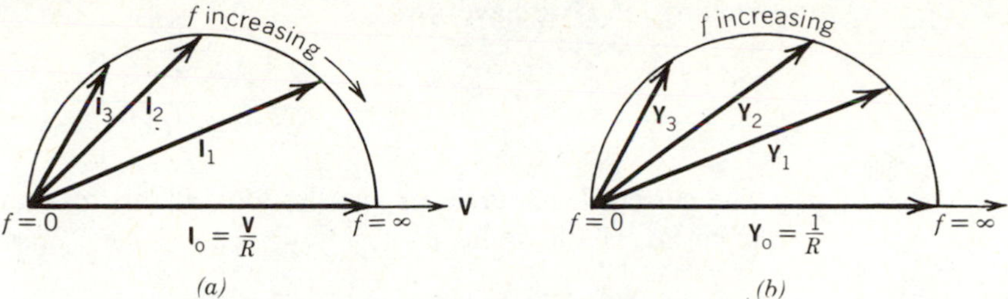

Figure 11.6. Loci for circuit in Fig. 11.4.

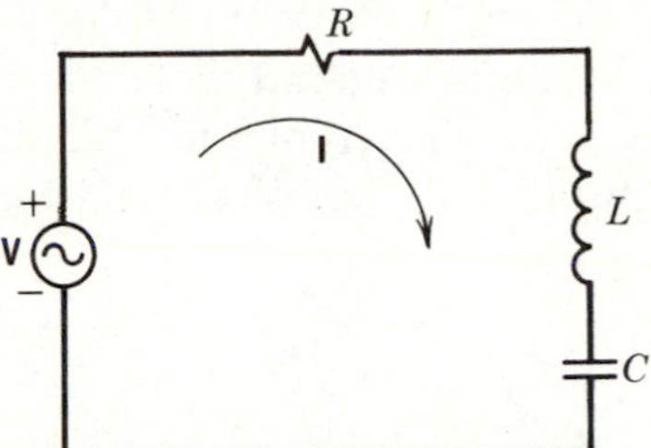

Figure 11.7. A series *RLC* circuit.

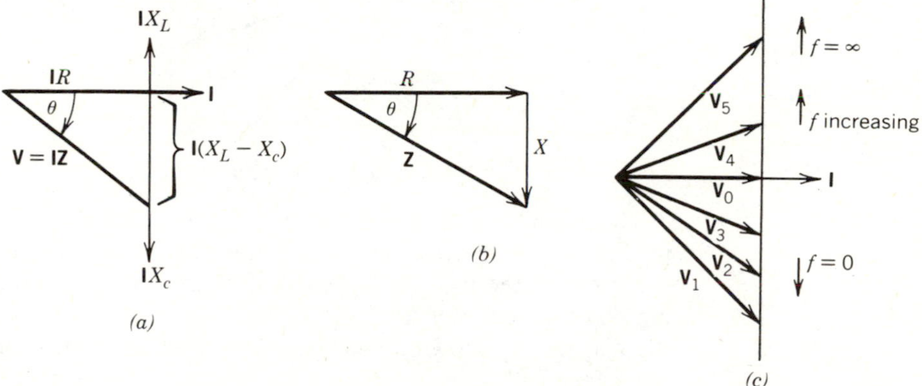

Figure 11.8. Voltage diagram and locus and impedance triangle for circuit in Fig. 11.7.

$X_c > X_L$, and the resulting diagram shows current leading the voltage. As f increases, X_L increases and X_c decreases. There is a point at which $X_L = X_c$ and $\mathbf{Z} = R$. The current and voltage are in phase, meaning that $\theta = 0$. As f increases still further, X_L becomes greater than X_c, and the current lags the voltage. This is illustrated by the voltage locus in Fig. 11.8*c*.

The frequency where $X_L = X_c$ is termed the *resonant frequency* f_r. It may be found by setting $X_L = X_c$ to give

$$2\pi fL = \frac{1}{2\pi fC} \tag{11.1.9}$$

Solving for f we have

$$\boxed{f_r = \frac{1}{2\pi\sqrt{LC}}} \tag{11.1.10}$$

The current locus gives us another look at this resonant condition. Solving for **I** from Eq. 11.1.8 we have

$$\mathbf{I} = \frac{\mathbf{V}}{R + j(X_L - X_c)} = \frac{\mathbf{V}}{R + j\left(2\pi f L - \dfrac{1}{2\pi f C}\right)} \tag{11.1.11}$$

The locus diagram in Fig. 11.9 is a plot of this equation for constant **V**, R, L, and C as f is varied from 0 to ∞. Notice that at resonance the current has maximum magnitude and the voltage and current are in phase.

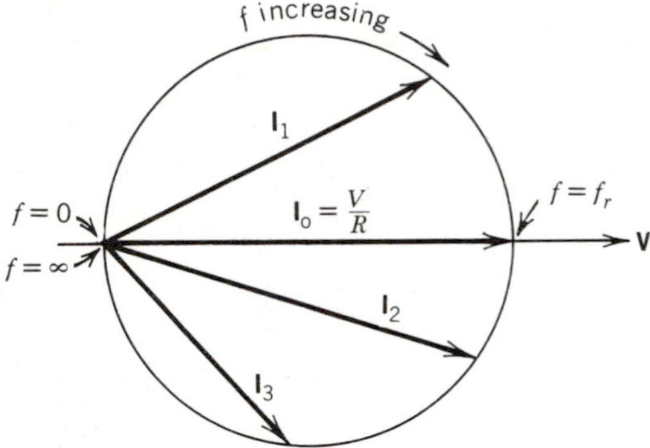

Figure 11.9. Current locus for circuit in Fig. 11.7.

The magnitude of current versus frequency for the series RLC circuit is shown in Fig. 11.10. Notice that since the current at resonance is $\mathbf{I}_r = \mathbf{V}/R$, the sharpness of the peak in this curve depends on R. For small R the peak will be high. For large R the curve will be flat with a broad low peak. This is indicated in the diagram by showing two frequency response curves, one for large R and the other for small R.

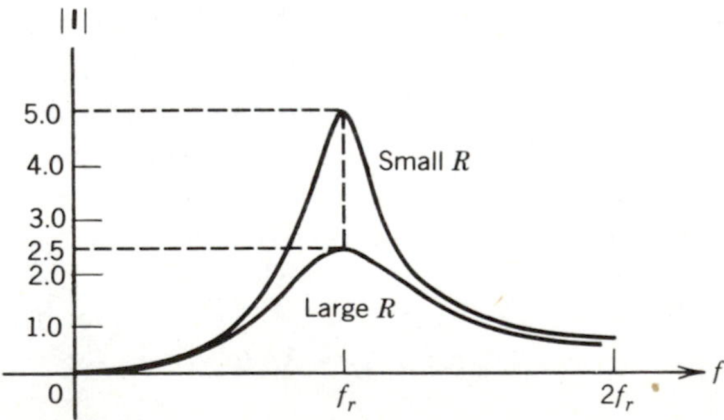

Figure 11.10. Current magnitude versus frequency for series RLC circuit.

The current phasor is a complex quantity. With applied voltage as reference, the current phasor has positive phase for low frequencies, as illustrated by current **I** in the locus diagram of Fig. 11.9. As frequency increases through resonance, the current phasor passes through zero phase to lagging phase, as illustrated by $\mathbf{I}_2$ and $\mathbf{I}_3$ in Fig. 11.9. The total phase change is from $+90°$ at $f=0$ to $-90°$ at $f=\infty$. Two phase angle plots are shown in Fig. 11.11 as a function of frequency, one for large R and the other for small R.

Bandwidth and Quality Factor

From Figs. 11.10 and 11.11 it is evident that a difference in sharpness of tuning, or peak width, is related to the value of R in the circuit. An arbitrary but universal measure of sharpness of tuning is found by selecting two frequencies, f_1 and f_2, with $f_1 < f_r < f_2$. These are the frequencies where the average power dissipated by the resistor is one-half the average power dissipated at resonance. The maximum current at frequency f_r is $\mathbf{I}_r = \mathbf{V}/R$. At the lower half-power frequency f_1, the current is $\mathbf{V}/(\sqrt{2}\,R)$, and at f_2 the current is also $\mathbf{V}/(\sqrt{2}\,R)$. Thus

$$|\mathbf{I}_1| = |\mathbf{I}_2| = \frac{I_r}{\sqrt{2}} \simeq 0.707 I_r \tag{11.1.12}$$

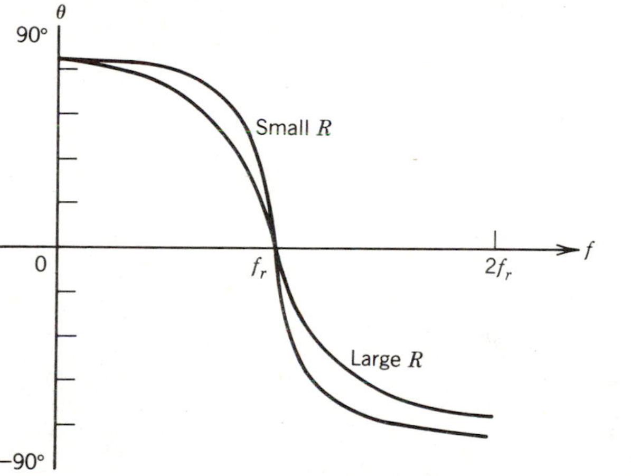

Figure 11.11. Phase angle plots for series *RLC* circuit.

The frequencies f_1 and f_2 are called the *half-power points*, and they are illustrated in Fig. 11.12.

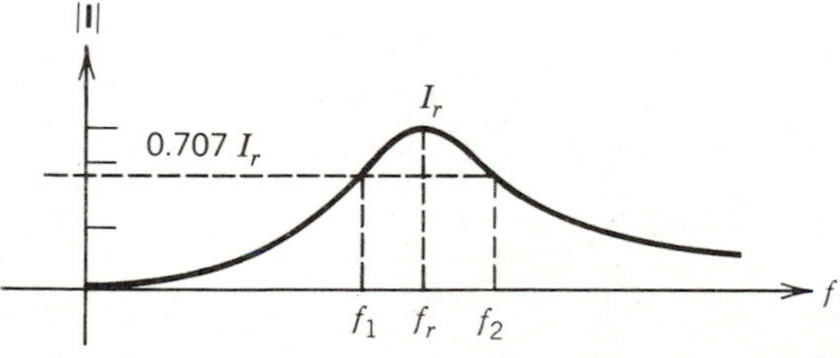

Figure 11.12. Half-power points.

Since the current is reduced by $\sqrt{2}$ at the half-power points, $|\mathbf{Z}| = \sqrt{2}\,R = \sqrt{R^2 + X^2}$, from which we find

$$X = R \tag{11.1.13}$$

At the lower frequency, f_1, the circuit is capacitive,

$$X = \frac{1}{2\pi f_1 C} - 2\pi f_1 L = R \tag{11.1.14}$$

and

$$\mathbf{Z} = \sqrt{2}\,R\underline{/-45^\circ} \tag{11.1.15}$$

At the higher frequency, f_2, the circuit is inductive,

$$X = 2\pi f_2 L - \frac{1}{2\pi f_2 C} = R \tag{11.1.16}$$

and

$$\mathbf{Z} = \sqrt{2}\,R\underline{/+45^\circ} \tag{11.1.17}$$

Figure 11.13 shows the impedance for these two cases. Notice particularly that the impedance angle, which is also the angle between voltage and current, is $\pm 45^\circ$ at the half-power points. Thus we may determine the frequencies f_1 and f_2 from either the magnitude plot (where $\mathbf{I}_1$ and $\mathbf{I}_2$ are $0.707 I_r$) or from the phase angle plot (where the angle is $+45^\circ$ or -45°).

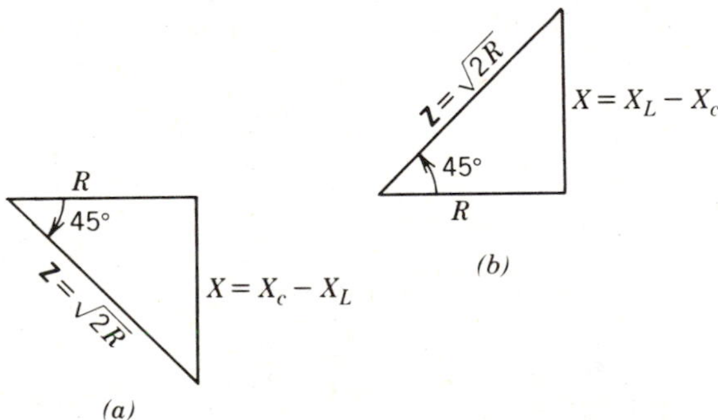

Figure 11.13. Impedance diagrams at half-power points.

We have previously found a formula for the resonant frequency in Eq. 11.1.10. The resonant frequency depends only on the product LC. Let us now determine the half-power points. The result will necessarily be more complex than Eq. 11.1.10 because we have determined that the values of f_1 and f_2 depend on R, in addition to depending on L and C. Start with Eq. 11.1.14 which involves only f_1.

$$\frac{1}{2\pi f_1 C} - 2\pi f_1 L = R$$

From this we obtain the quadratic form

$$f_1^2 + \frac{R}{2\pi L} f_1 - \frac{1}{4\pi^2 CL} = 0$$

Thus f_1 is given by

$$\boxed{f_1 = \frac{1}{4\pi L}\left(-R + \sqrt{R^2 + \frac{4L}{C}}\right)} \tag{11.1.18}$$

Similarly, starting with Eq. 11.1.16, we obtain

$$\boxed{f_2 = \frac{1}{4\pi L}\left(+R + \sqrt{R^2 + \frac{4L}{C}}\right)} \tag{11.1.19}$$

Since frequency as we are using it here represents sinusoidal frequency of oscillation, it cannot be negative. Therefore, only the positive sign before the radical is used.

The *bandwidth* of a circuit is the difference between the half-power frequencies.

$$\boxed{B = f_2 - f_1 \text{ Hz}} \tag{11.1.20}$$

Subtracting Eq. 11.1.18 from 11.1.19 gives

$$B = f_2 - f_1 = \frac{1}{4\pi L}\left(R + \sqrt{R^2 + \frac{4L}{C}} + R - \sqrt{R^2 + \frac{4L}{C}}\right) \tag{11.1.21}$$

or

$$\boxed{B = \frac{R}{2\pi L}}$$

The ratio of resonant frequency over bandwidth is a normalized measure of sharpness of tuning. It is called the *quality factor Q*.

$$Q = \frac{f_r}{B} = \frac{1}{2\pi\sqrt{LC}} \times \frac{2\pi L}{R} = \frac{\omega_r L}{R} \tag{11.1.22}$$

$$\boxed{Q = \frac{1}{R}\sqrt{\frac{L}{C}}}$$

The concept of quality factor, of which this is a special but important case, finds application to any system, electrical or not, where energy storage elements are present. It is defined by

$$Q = 2\pi \frac{\text{maximum energy stored}}{\text{total energy lost per cycle}} \tag{11.1.23}$$

For example, suppose a tennis ball rebounds to half its original height when dropped. The energy lost must be half the original potential energy, because after one bounce it regains half this energy. Thus, for our tennis ball

$$Q = 2\pi(2) = 4\pi$$

The resonant frequency is directly related to the half-power points. If we multiply Eqs. 11.1.18 and 11.1.19, we obtain

$$f_1 f_2 = \frac{1}{4\pi L}\left(-R + \sqrt{R^2 + \frac{4L}{C}}\right)\frac{1}{4\pi L}\left(R + \sqrt{R^2 + \frac{4L}{C}}\right)$$

$$= \frac{1}{4\pi^2 LC} = f_r^2$$

or

$$\boxed{f_r = \sqrt{f_1 f_2}} \qquad (11.1.24)$$

That is, f_r is the geometric mean of f_1 and f_2. Look at Fig. 11.12. You may have thought there was an error in drawing the diagram since $f_2 - f_r$ is larger than $f_r - f_1$, but f_r is not halfway between f_1 and f_2.

LEARNING EVALUATIONS

1. A series *RL* circuit (Fig. 11.7) has $R = 1\ \text{k}\Omega$ and $L = 10$ mH. Make a sketch of current amplitude versus frequency and a sketch of current phase versus frequency. Indicate the 3-dB point on each graph. Assume $V = 1$.
2. A series *RLC* circuit has $R = 10\ \Omega$, $C = 10\ \mu\text{F}$, and $L = 4$ mH. Find the resonant frequency, the quality factor, and the bandwidth.

11.2 Parallel Resonance

LEARNING OBJECTIVES

After completing this section you should be able to do the following:

1. Plot response versus frequency for *RL*, *RC*, and *RLC* parallel circuits.
2. Find the resonant and 3-dB frequencies for parallel circuits.
3. Determine the bandwidth and quality factor for parallel circuits.

The *RC* parallel circuit in Fig. 11.14 has sinusoidal current applied. Writing KCL at the top node gives

$$\mathbf{I} = \mathbf{V}G + j\mathbf{V}B_C = \mathbf{V}\mathbf{Y} \qquad (11.2.1)$$

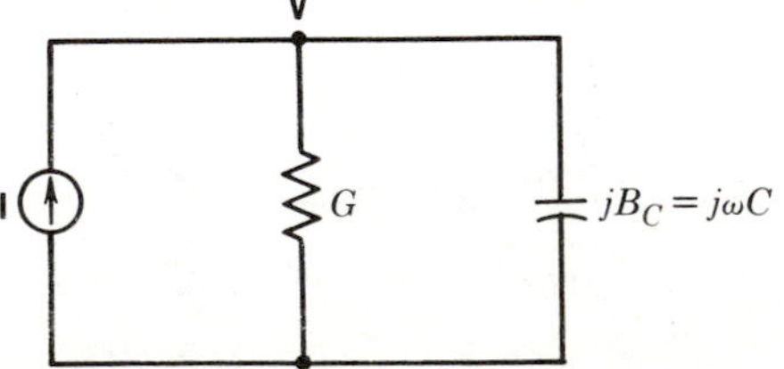

Figure 11.14. A parallel *RC* circuit.

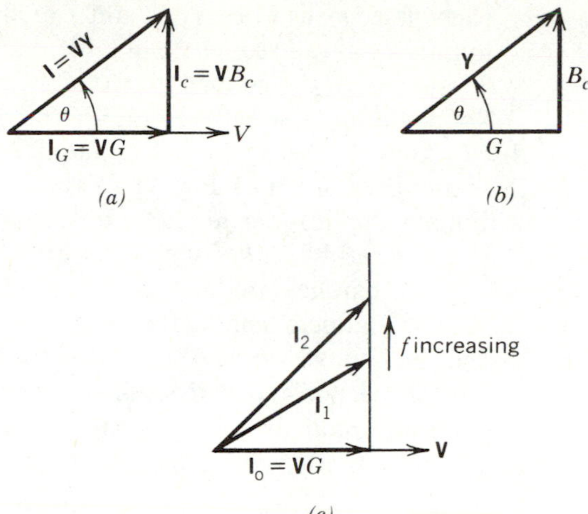

Figure 11.15. Diagrams relating to circuit in Fig. 11.14.

Figure 11.15 contains diagrams representing the terms in this equation. Figure 11.15*a* plots these terms for a particular frequency. Note that current leads voltage as it should for an *RC* circuit. Figure 11.15*b* is the admittance triangle, which is found from Fig. 11.15*a* by dividing each term by **V**. Figure 11.15*c* is the current locus found by varying frequency while holding the other parameters constant. Current $\mathbf{I}_0$ occurs at $f = 0$, and $\mathbf{I}_1$ and $\mathbf{I}_2$ are current phasors for increasing frequency. At $f = \infty$, $\mathbf{I} = \infty$ and the angle between **I** and **V** is 90°.

The voltage locus for the parallel *RC* circuit may be found by solving for **V** in Eq. 11.2.1.

$$\mathbf{V} = \frac{\mathbf{I}}{\mathbf{Y}} = \frac{\mathbf{I}}{G + jB_C} \tag{11.2.2}$$

The voltage **V** varies with frequency in a semicircle if **I**, G, and C are held constant. The voltage locus is shown in Fig. 11.16*a*. At $f = 0$ the admittance is $Y = G$, and the current and voltage are in phase. As f increases, the denominator of Eq. 11.2.2 becomes larger with increasing positive angle. Thus **V** becomes smaller with increasing negative angle. This is reflected in the voltage locus diagram. Figure 11.16*b* illustrates the impedance of the circuit as frequency increases.

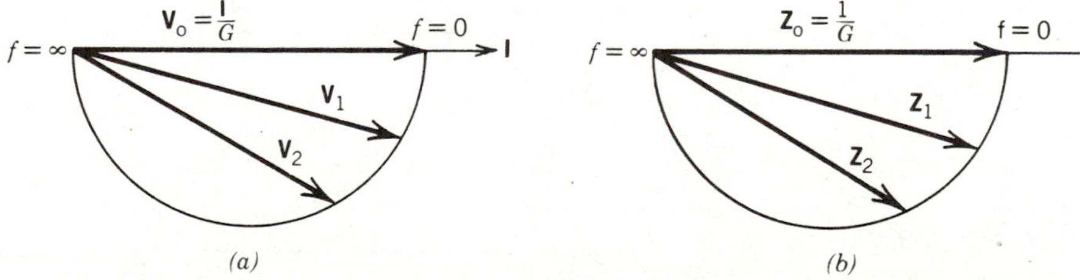

Figure 11.16. Voltage locus and frequency diagram for circuit in Fig. 11.14.

The voltage locus when **V**, G, and f are held constant with C varying from zero to infinity, is also a semicircle like the one in Fig. 11.16. At $C = 0$, the voltage is $\mathbf{I}/G$. At $C = \infty$, the voltage is zero.

Notice the analogy between the RC parallel circuit and the RL series circuit as discussed in the last section. The parallel RC circuit in Fig. 11.14 is the *dual* of the series RL circuit in Fig. 11.1. The principle of duality is an important one with many applications in circuit analysis.

The *principle of duality* states that every circuit has a dual. The dual of a series circuit is a parallel circuit. Resistors are replaced by conductors, capacitors are replaced by inductors, inductors are replaced by capacitors, nodes are replaced by meshes and conversely, and voltage sources are replaced by current sources and conversely. Hence, the analysis of series circuits in Section 11.1 will provide clues that allow us to analyze parallel circuits by invoking the principle of duality. Here is the procedure for deriving dual circuits.

1. In the center of each mesh of a given network place a node symbol. Number these nodes with the mesh numbers. Also place a reference node outside the circuit. Step 1 is illustrated in Fig. 11.17*a*, where a two-mesh circuit containing a resistor, inductor, and capacitor, along with two voltage sources, is illustrated.
2. Draw a dashed line between each interior node and the ground node passing through each element that is common to only one mesh. Also draw dashed lines between interior nodes that pass through elements common to two meshes. Step 2 is illustrated in Fig. 11.17*b*.

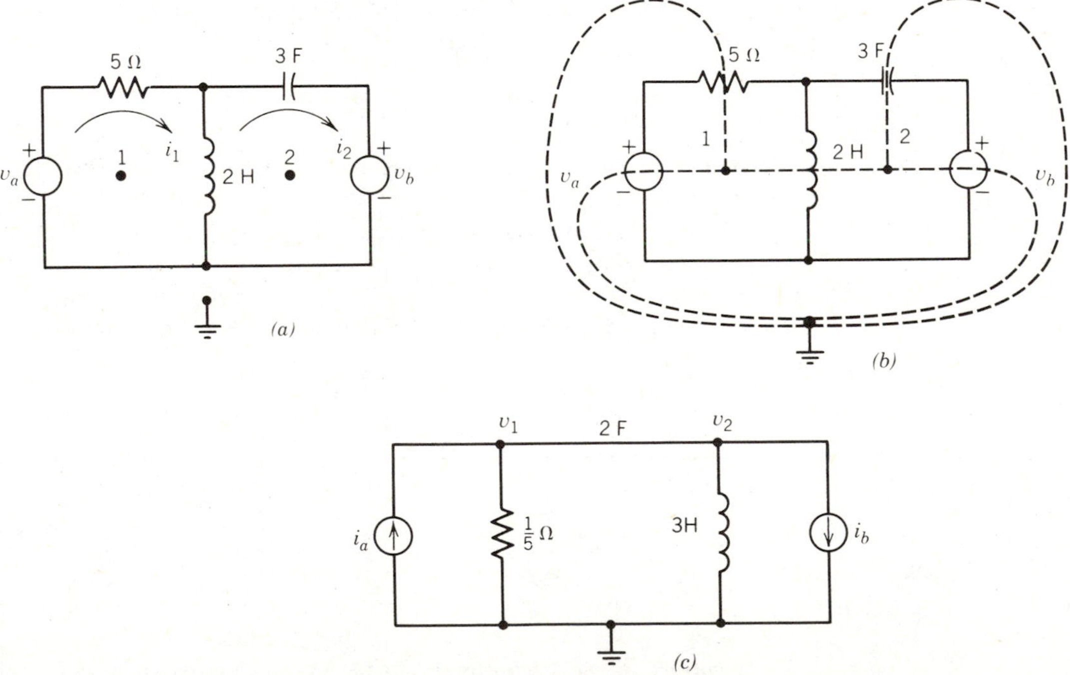

Figure 11.17. Construction of a dual circuit.

3. Connect the dual of each element between nodes. Voltage sources that force mesh currents in the assumed direction (v_a in Fig. 11.17*a*) are replaced by current sources into the corresponding node (i_a in Fig. 11.17*c*). Since v_b opposes mesh current i_2, source i_b in Fig. 11.17*c* is away from node 2.

Let us now illustrate the validity of the dual circuit by writing the mesh equations for the circuit in Fig. 11.17*a*.

$$v_a = 5i_1 + 2\frac{di_1}{dt} - 2\frac{di_2}{dt}$$

$$-v_b = -2\frac{di_1}{dt} + 2\frac{di_2}{dt} + \frac{1}{3}\int_{-\infty}^{t} i_2(\lambda)\, d\lambda$$

The node equations for Fig. 11.17*c* give

$$i_a = 5v_1 + 2\frac{dv_1}{dt} - 2\frac{dv_2}{dt}$$

$$-i_b = -2\frac{dv_1}{dt} + 2\frac{dv_2}{dt} + \frac{1}{3}\int_{-\infty}^{t} v_2(\lambda)\, d\lambda$$

Notice that the form of the first two equations is identical to the form of the second set. Only the names of the variables have been changed.

For ac circuits, we deal with impedance in series circuits and write KVL. The dual of these circuits is found by replacing each impedance by its admittance. Here is an example.

EXAMPLE 11.2.1 Find the dual circuit for the two-mesh circuit in Fig. 11.18*a*.

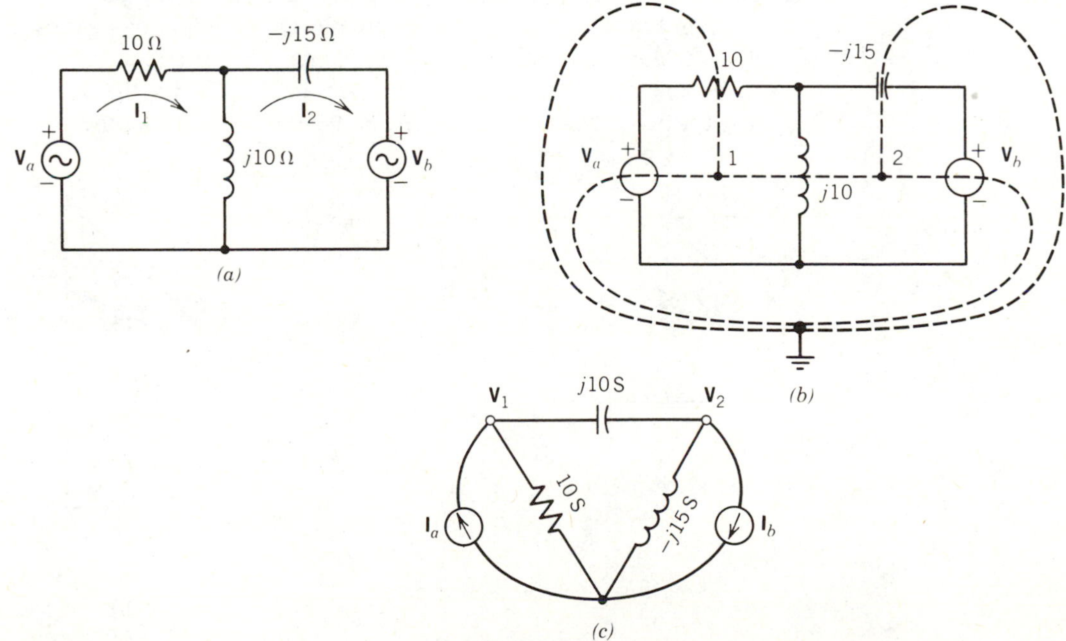

Figure 11.18

Solution

Steps 1 and 2 of our procedure are illustrated in Fig. 11.18*b*. To determine the admittance values in Fig. 11.18*c*, use the following chart.

$$R\ (\Omega) \rightarrow G\ (\text{S})$$
$$jX_L\ (\Omega) \rightarrow jB_c\ (\text{S})$$
$$-jX_c\ (\Omega) \rightarrow -jB_L\ (\text{S})$$

Here are the mesh equations for Fig. 11.18*a* in matrix form.

$$\begin{bmatrix} \mathbf{V}_a \\ -\mathbf{V}_b \end{bmatrix} = \begin{bmatrix} (10+j10) & -j10 \\ -j10 & -j5 \end{bmatrix} \begin{bmatrix} \mathbf{I}_1 \\ \mathbf{I}_2 \end{bmatrix}$$

The correspondence node equations for the circuit in Fig. 11.18*c* are given by

$$\begin{bmatrix} \mathbf{I}_a \\ -\mathbf{I}_b \end{bmatrix} = \begin{bmatrix} (10+j10) & -j10 \\ -j10 & -j5 \end{bmatrix} \begin{bmatrix} \mathbf{V}_1 \\ \mathbf{V}_2 \end{bmatrix}$$

In view of the principle of duality we may now complete our investigation of parallel networks by referring to the dual series network analysis of the previous section. The dual of a parallel *RL* circuit is the series *RC* circuit in Fig. 11.4. Hence the current locus is pictured in Fig. 11.5, with **I** replacing **V**, and **Y** replacing **Z**. Likewise the voltage locus of a parallel *RL* circuit is found from Fig. 11.16.

A parallel *RLC* circuit is shown in Fig. 11.19. The current locus shown in Fig. 11.20 is derived by duality from Fig. 11.8, and the voltage locus in Fig. 11.21 is derived from Fig. 11.9. Frequency response curves for the parallel *RLC* circuit are shown in Fig. 11.22. These curves are plotted by applying a sinusoidal current of constant amplitude but varying frequency to the circuit in Fig. 11.19 and measuring the phasor response **V**. Figure 11.22*a* shows the amplitude of the response for two different values of *R*, and Fig. 11.22*b* shows the phase angle.

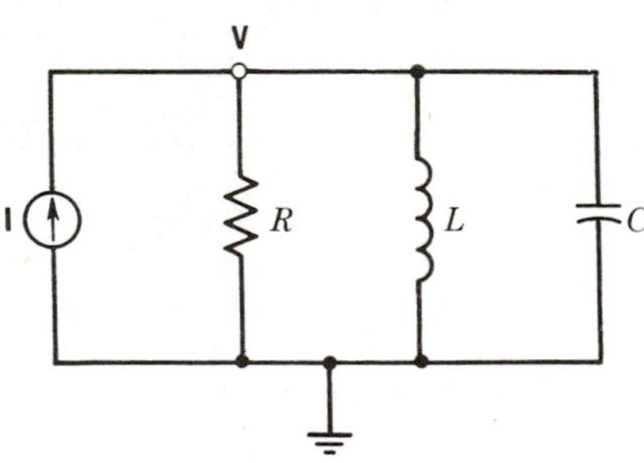

Figure 11.19. A parallel *RLC* circuit.

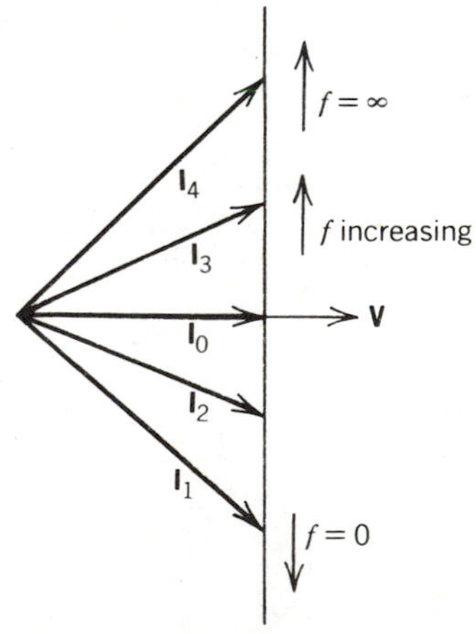

Figure 11.20. Current locus for parallel *RLC* circuit.

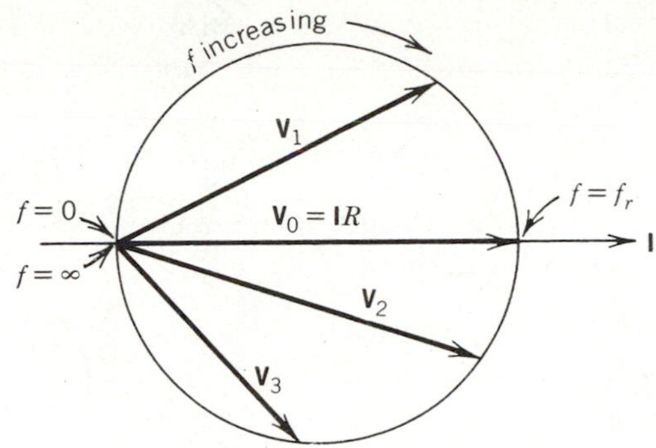

Figure 11.21. Voltage locus for parallel *RLC* circuit.

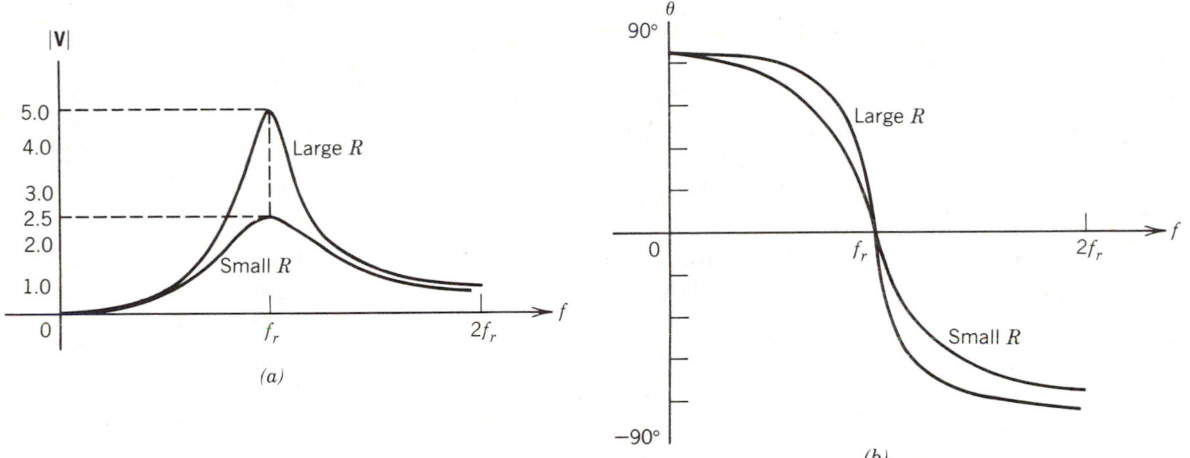

Figure 11.22. Frequency response curves for parallel *RLC* circuit.

Most practical variable frequency circuits use the parallel *RLC* network for tuning. For example, the common AM table radio has a variable capacitor that is controlled by the tuning knob. The bandwidth is controlled by the value of R in the circuit, which is chosen to give a bandwidth that passes the signal from one station while rejecting other stations that operate at a nearby frequency. The Federal Communications Commission (FCC) regulates the frequency of commercial stations in each area of the United States so that no two stations operate at similar frequencies. You may occasionally be able to detect two stations at the same frequency, but this is usually due to freak atmospheric conditions rather than to any laxity on the part of the FCC.

The half-power points f_1 and f_2 are the frequencies where the maximum voltage V_r is reduced in magnitude by $\sqrt{2}$.

$$\mathbf{V}_1 = \mathbf{V}_2 = \frac{\mathbf{V}_r}{\sqrt{2}} \simeq 0.707\mathbf{V}_r \tag{11.2.3}$$

The half-power points are also those frequencies where the angle between voltage and current is $\pm 45°$. The bandwidth is defined to be the difference between f_2 and f_1.

$$B = f_2 - f_1 \text{ Hz} \tag{11.2.4}$$

Applying duality to the analysis of series circuits we have from Eqs. 11.1.21 and 11.1.22 that

$$\boxed{B = f_2 - f_1 = \frac{G}{2\pi C}} \tag{11.2.5}$$

and

$$\boxed{Q = \frac{\omega_r C}{G} = \frac{R}{\omega_r L} = R\sqrt{\frac{C}{L}}} \tag{11.2.6}$$

Also,

$$\boxed{f_r = \sqrt{f_1 f_2}} \tag{11.2.7}$$

LEARNING EVALUATIONS

1. A parallel RC circuit (Fig. 11.14) has $R = 1$ kΩ and $C = 10$ μF. Make a sketch of voltage amplitude versus frequency and a sketch of voltage phase versus frequency. Indicate the 3-dB point on each graph. Assume $I = 1$.
2. A parallel RLC circuit has $R = 10$ kΩ, $C = 10$ μF, and $L = 4$ mH. Find the resonant frequency, the quality factor, and the bandwidth.

11.3 Two-Port Networks

LEARNING OBJECTIVE

After completing this section you should be able to define a port.

A network port is a pair of terminals with the property that the current entering one terminal is equal to the current leaving the other. The circuits shown in Fig. 11.23 are called one-port networks because each has one pair of terminals defined external to the network, and the current entering one terminal is equal to the current leaving the other if an external path is closed in some manner.

Figure 11.23*a* represents a very simple one-port network consisting of a single circuit element connected between the terminals *a–b*. The circuit in Fig. 11.23*b* is much more complex but still meets the definition of a one-port network. All one-port networks can be reduced to the form shown in Fig. 11.23*c* by use of Thévenin's theorem. We know that reducing a one-port circuit to its Thévenin's or Norton's equivalent often greatly reduces the analysis task when the circuit is connected to other devices.

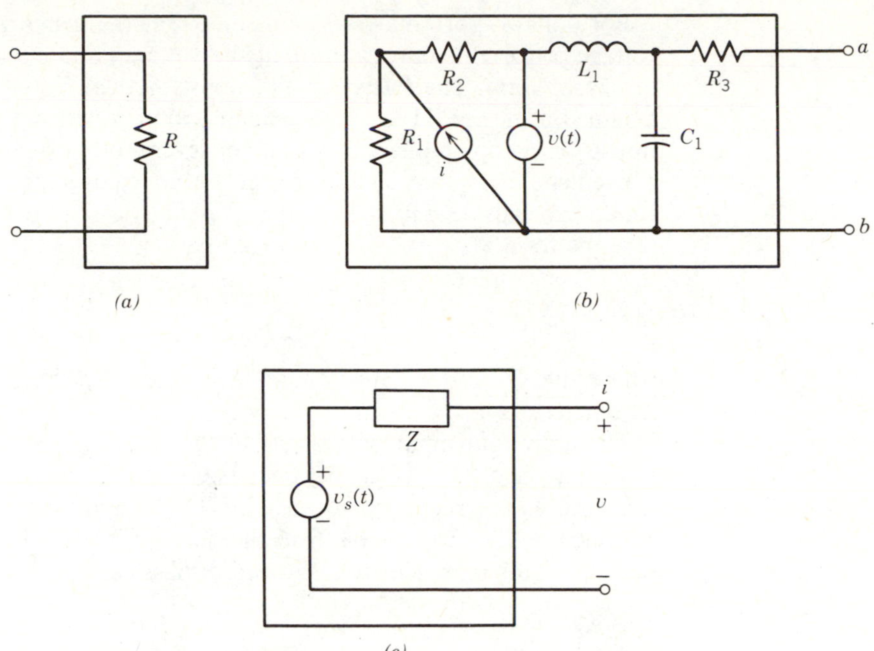

Figure 11.23. Several one-port networks.

A general two-port network is shown in Fig. 11.24. The box shown in the figure is often called a "black box" as mentioned in Chapter 3. The word "black" was presumably chosen to indicate something unknown or mysterious was inside the box. There are four variables associated with our two-port network and it is our goal to derive relationships between these four variables so that if any two are known, the other two can be calculated. These relationships will be stated in terms of network parameters. The parameters themselves will be related to whatever is inside the black box but once the parameters are measured or calculated or otherwise determined, the contents of the box can be ignored and all other relations established using the parameters. This is exactly what we did with Thévenin's theorem; once the Thévenin equivalent of a circuit was found, the actual circuit configuration was no longer needed for future calculations.

The one other restriction we must establish in defining a port is that the current entering one terminal of a port must equal the current leaving the other terminal. This is seen in Fig. 11.23c where the current, i, entering terminal a must certainly be the current exiting terminal b. In Fig. 11.24, $i_1 = i_1'$, $i_2 = i_2'$ is required by the definition.

Figure 11.24. A general two-port network.

For our two-port analysis, we assume that the network is free of independent voltage or current sources and that the voltages and currents can be represented in phasor form. The reference directions for these voltages and currents will be assumed as in Fig. 11.24. This general two-port network can represent a filter, a transistor, an op amp circuit, or any of several other such devices. It will often be convenient to replace such a device by an equivalent model consisting of the equations that we will derive to represent two-port networks. These equations take the form

$$\mathbf{Q}_1(\omega) = K_{11}(\omega)\mathbf{P}_1(\omega) + K_{12}(\omega)\mathbf{P}_2(\omega) \tag{11.3.1a}$$

$$\mathbf{Q}_2(\omega) = K_{21}(\omega)\mathbf{P}_1(\omega) + K_{22}(\omega)\mathbf{P}_2(\omega) \tag{11.3.1b}$$

where the $\mathbf{Q}_i$ and $\mathbf{P}_i$ are voltage or current. The K_{ij} terms are the network parameters.

There are four input parameters in Fig. 11.24; $\mathbf{I}_1$, $\mathbf{I}_2$, $\mathbf{V}_1$, and $\mathbf{V}_2$. Any two of these may be chosen as the parameters $\mathbf{Q}_1$ and $\mathbf{Q}_2$ in Eq. 11.3.1. This means that there are six different ways to write Eq. 11.3.1, and each way gives rise to different parameters K_{ij}. Table 11.1 lists the six possible combinations along with the corresponding parameter set. We will discuss each of these cases in turn.

TABLE 11.1
THE SIX-NETWORK PARAMETERS

Case	$\mathbf{Q}_1(\omega)$	$\mathbf{Q}_2(\omega)$	Parameters
1	$\mathbf{V}_1(\omega)$	$\mathbf{V}_2(\omega)$	Impedance (z) parameters
2	$\mathbf{I}_1(\omega)$	$\mathbf{I}_2(\omega)$	Admittance (y) parameters
3	$\mathbf{I}_1(\omega)$	$\mathbf{V}_2(\omega)$	Hybrid (g) parameters
4	$\mathbf{V}_1(\omega)$	$\mathbf{I}_2(\omega)$	Hybrid (h) parameters
5	$\mathbf{V}_1(\omega)$	$\mathbf{I}_1(\omega)$	*ABCD* parameters
6	$\mathbf{V}_2(\omega)$	$\mathbf{I}_2(\omega)$	$\mathcal{ABCD}$ parameters

LEARNING EVALUATIONS

1. Draw three distinct two-port networks.
2. Draw three distinct one-port networks.

11.4 Impedance Parameters

LEARNING OBJECTIVES

After completing this section you should be able to do the following:

1. Define the impedance parameters for any given two-port network.
2. Use the impedance parameters for a two-port network to find voltages and currents or voltage gains and current gains through components attached to the two-port network.
3. Synthesize simple networks to realize z-parameter models.

By our assumptions and by circuit restrictions we placed on ourselves in the text, we know that we are dealing with a linear, time-invariant network. If we assume that $\mathbf{I}_1$ and $\mathbf{I}_2$ in Fig. 11.24 are known, then $\mathbf{V}_1$ and $\mathbf{V}_2$ must be linearly related to these two currents and this general relationship can be expressed as

$$\mathbf{V}_1 = z_{11}\mathbf{I}_1 + z_{12}\mathbf{I}_2 \tag{11.4.1a}$$

$$\mathbf{V}_2 = z_{21}\mathbf{I}_1 + z_{22}\mathbf{I}_2 \tag{11.4.1b}$$

The terms z_{11}, z_{12}, z_{21}, and z_{22} are called impedance parameters because they must have the units of ohms.

In order to evaluate these parameters, we set the currents to zero alternately. In other words, by observation of Eq. 11.4.1,

$$z_{11} = \left.\frac{\mathbf{V}_1}{\mathbf{I}_1}\right|_{\mathbf{I}_2=0} \tag{11.4.2a}$$

$$z_{21} = \left.\frac{\mathbf{V}_2}{\mathbf{I}_1}\right|_{\mathbf{I}_2=0} \tag{11.4.2b}$$

$$z_{12} = \left.\frac{\mathbf{V}_1}{\mathbf{I}_2}\right|_{\mathbf{I}_1=0} \tag{11.4.2c}$$

$$z_{22} = \left.\frac{\mathbf{V}_2}{\mathbf{I}_2}\right|_{\mathbf{I}_1=0} \tag{11.4.2d}$$

Given a black box as in Fig. 11.24, we can determine the impedance parameters of z_{11} and z_{21} by attaching a known current source as shown in Fig. 11.25*a* and measuring $\mathbf{V}_1$ and $\mathbf{V}_2$, then using Eqs. 11.4.2a and 11.4.2b. Next reconnect the

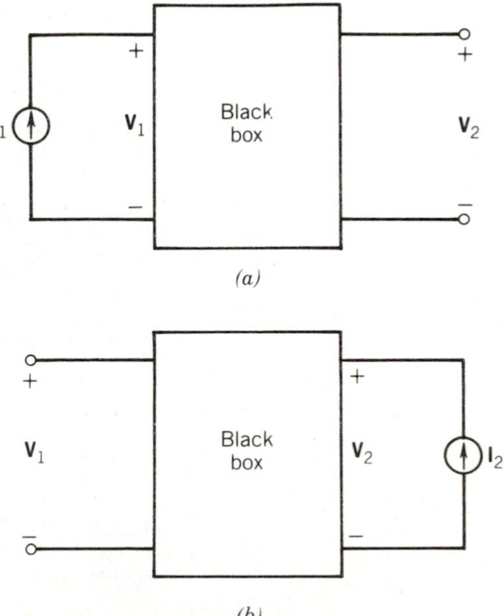

Figure 11.25. Circuits for measuring impedance parameters.

circuit as in Fig. 11.25*b* and again measure $\mathbf{V}_1$ and $\mathbf{V}_2$. Finally, using Eq. 11.4.2c and 11.4.2d, we calculate z_{12} and z_{22}. Because of the nature of this calculation these parameters are also called the open-circuit impedance parameters or sometimes simply the z parameters.

Assume that the contents of our black box is the T network shown in Fig. 11.26. If we open circuit $\mathbf{I}_2$ ($\mathbf{I}_2 = 0$) then

$$\mathbf{V}_1 = (\mathbf{Z}_a + \mathbf{Z}_c)\mathbf{I}_1$$

or

$$z_{11} = \left.\frac{\mathbf{V}_1}{\mathbf{I}_1}\right|_{\mathbf{I}_2=0} = \mathbf{Z}_a + \mathbf{Z}_c$$

and

$$\mathbf{V}_2 = \mathbf{Z}_c\mathbf{I}_1$$

so

$$z_{21} = \left.\frac{\mathbf{V}_2}{\mathbf{I}_1}\right|_{\mathbf{I}_2=0} = \mathbf{Z}_c$$

Next open circuit $\mathbf{I}_1$ ($\mathbf{I}_1 = 0$) and observe

$$\mathbf{V}_1 = \mathbf{Z}_c\mathbf{I}_2$$

or

$$z_{12} = \left.\frac{\mathbf{V}_1}{\mathbf{I}_2}\right|_{\mathbf{I}_1=0} = \mathbf{Z}_c$$

and

$$\mathbf{V}_2 = \mathbf{Z}_b\mathbf{I}_2 + \mathbf{Z}_c\mathbf{I}_2$$

so

$$z_{22} = \left.\frac{\mathbf{V}_2}{\mathbf{I}_2}\right|_{\mathbf{I}_1=0} = \mathbf{Z}_b + \mathbf{Z}_c$$

Once the z parameters have been determined, any two-port voltages or currents can be determined if the other two are known.

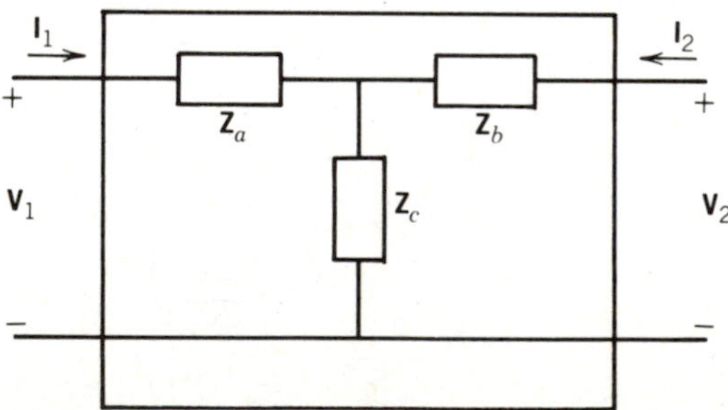

Figure 11.26. Black box containing a T impedance network.

The biggest advantage of z parameters lies in the fact that they can be determined experimentally by open circuiting one port at a time. This can rarely do any damage to the network inside the box.

EXAMPLE 11.4.1 Find the z parameters for the circuit in Fig. 11.27.

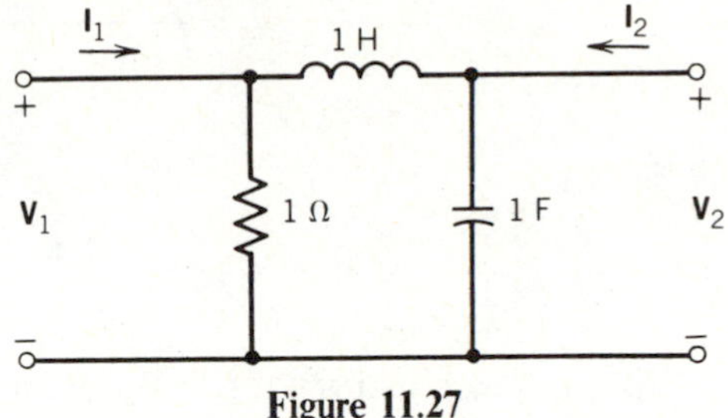

Figure 11.27

Solution

With $\mathbf{I}_2 = 0$ we may solve for z_{11} and z_{21}. the circuit in Fig. 11.28*a* is used for this by writing nodal equations to solve for $\mathbf{V}_1$ and $\mathbf{V}_2$ when $\mathbf{I}_1$ is the source. Notice that admittance values are used in the circuit.

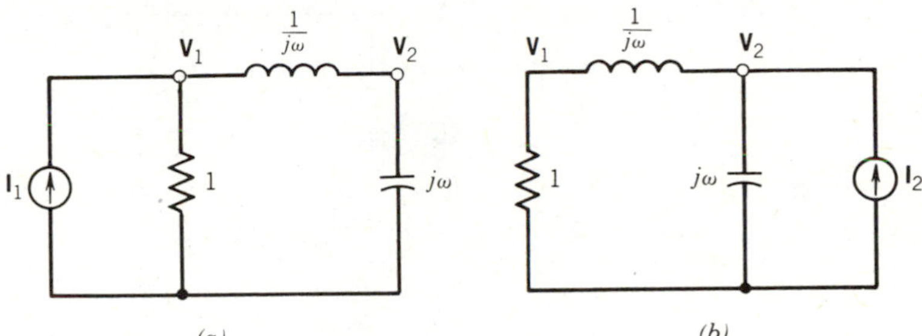

Figure 11.28

$$\mathbf{V}_1 = \frac{\begin{vmatrix} \mathbf{I}_1 & -\dfrac{1}{j\omega} \\ 0 & \left(j\omega + \dfrac{1}{j\omega}\right) \end{vmatrix}}{\begin{vmatrix} \left(1 + \dfrac{1}{j\omega}\right) & -\dfrac{1}{j\omega} \\ -\dfrac{1}{j\omega} & \left(j\omega + \dfrac{1}{j\omega}\right) \end{vmatrix}} = \frac{(1-\omega^2)\mathbf{I}_1}{1-\omega^2+j\omega}$$

So

$$z_{11} = \left.\frac{\mathbf{V}_1}{\mathbf{I}_1}\right|_{\mathbf{I}_2=0} = \frac{1-\omega^2}{1-\omega^2+j\omega}$$

Also from Fig. 11.28*a*,

$$\mathbf{V}_2 = \frac{\begin{vmatrix} 1 + \dfrac{1}{j\omega} & \mathbf{I}_1 \\ -\dfrac{1}{j\omega} & 0 \end{vmatrix}}{1 + j\omega + \dfrac{1}{j\omega}} = \frac{\mathbf{I}_1}{1 - \omega^2 + j\omega}$$

So

$$z_{11} = \left.\frac{\mathbf{V}_1}{I_1}\right|_{\mathbf{I}_2=0} = \frac{1 - \omega^2}{1 - \omega^2 + j\omega}$$

In Fig. 11.28*b*, current $\mathbf{I}_1 = 0$, so this diagram may be used to solve for $\mathbf{z}_{12}$ and $\mathbf{z}_{22}$. The results are

$$\mathbf{V}_1 = \frac{\begin{vmatrix} 0 & -\dfrac{1}{j\omega} \\ \mathbf{I}_2 & j\omega + \dfrac{1}{j\omega} \end{vmatrix}}{\begin{vmatrix} 1 + \dfrac{1}{j\omega} & -\dfrac{1}{j\omega} \\ -\dfrac{1}{j\omega} & j\omega + \dfrac{1}{j\omega} \end{vmatrix}} = \frac{\mathbf{I}_2}{1 - \omega^2 + j\omega}$$

So

$$z_{12} = \left.\frac{\mathbf{V}_1}{\mathbf{I}_2}\right|_{\mathbf{I}_1=0} = \frac{1}{1 - \omega^2 + j\omega}$$

Also from Fig. 11.28*b*,

$$\mathbf{V}_2 = \frac{\begin{vmatrix} 1 + \dfrac{1}{j\omega} & 0 \\ -\dfrac{1}{j\omega} & \mathbf{I}_2 \end{vmatrix}}{1 + j\omega + \dfrac{1}{j\omega}} = \frac{(1 + j\omega)\mathbf{I}_2}{1 - \omega^2 + j\omega}$$

So

$$z_{22} = \left.\frac{\mathbf{V}_2}{\mathbf{I}_2}\right|_{\mathbf{I}_1=0} = \frac{1 + j\omega}{1 - \omega^2 + j\omega}$$

The z parameter may be displayed in matrix form as

$$\begin{bmatrix} z_{11} & z_{12} \\ z_{21} & z_{22} \end{bmatrix} = \begin{bmatrix} \dfrac{1-\omega^2}{1-\omega^2+j\omega} & \dfrac{1}{1-\omega^2+j\omega} \\ \dfrac{1}{1-\omega^2+j\omega} & \dfrac{1+j\omega}{1-\omega^2+j\omega} \end{bmatrix}$$

Once the z parameters are known we may find any other desired network function from them. For example, suppose we wish to determine the voltage transfer function $\mathbf{V}_2/\mathbf{V}_1$ with $\mathbf{I}_2 = 0$. Rewriting Eqs. 11.4.1. with $\mathbf{I}_2 = 0$ gives

$$\mathbf{V}_1 = z_{11}\mathbf{I}_1$$
$$\mathbf{V}_2 = z_{21}\mathbf{I}_1$$

Then we find directly that $\mathbf{V}_2/\mathbf{V}_1$ is given by

$$\frac{\mathbf{V}_2}{\mathbf{V}_1} = \frac{z_{21}}{z_{22}} \tag{11.4.3}$$

A more complex case is illustrated in Fig. 11.29. If we know the z parameters for the circuit in the black box, then the voltage gain $\mathbf{V}_2/\mathbf{V}_1$ may be found by noting that

$$\mathbf{V}_2 = -\mathbf{I}_2 R_L$$

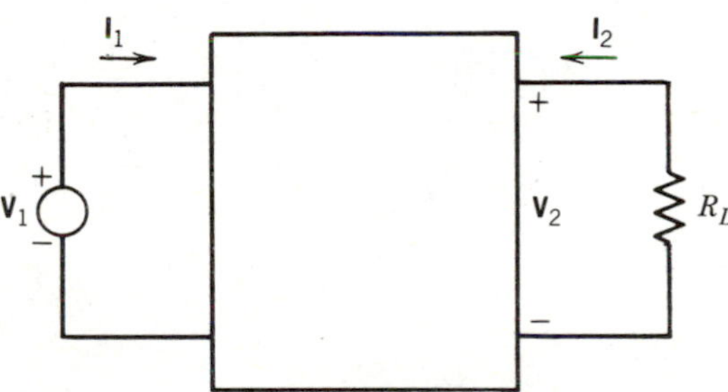

Figure 11.29

This equation may be combined with Eqs. 11.4.1 to give an overall voltage gain of

$$\frac{\mathbf{V}_2}{\mathbf{V}_1} = \frac{1}{\dfrac{z_{11}}{z_{21}}\left(1 + \dfrac{z_{22}}{R_L}\right) - \dfrac{z_{12}}{R_L}}$$

Notice that this reduces to Eq. 11.4.3 if $R_L = \infty$; that is, if $\mathbf{I}_2 = 0$.

EXAMPLE 11.4.2 Calculate the voltage $\mathbf{V}_2$ if the source and load shown in Fig. 11.30 are attached to the circuit in Example 11.4.1.

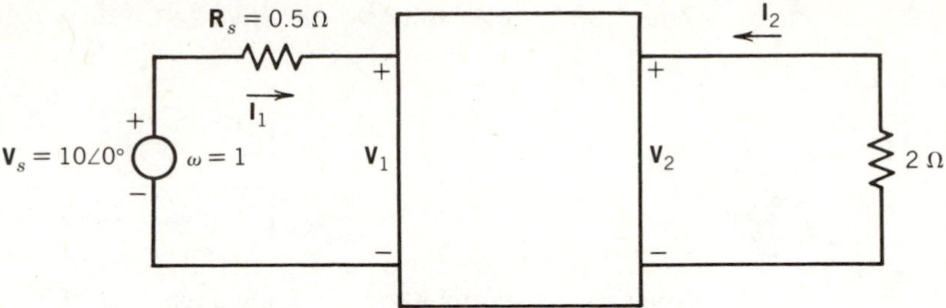

Figure 11.30. Circuit for example 11.4.2.

Solution

First observe that

$$\mathbf{V}_s = 0.5\mathbf{I}_1 + \mathbf{V}_1 \tag{11.4.4}$$

$$\mathbf{V}_2 = -2\mathbf{I}_2 \tag{11.4.5}$$

Next calculate values for the z parameters at the source frequency $\omega = 1$ rad/s.

$$z_{11} = \left.\frac{1-\omega^2}{1-\omega^2+j\omega}\right|_{\omega=1} = 0$$

$$z_{21} = z_{12} = \left.\frac{1}{1-\omega^2+j\omega}\right|_{\omega=1} = -j$$

$$z_{22} = \left.\frac{1+j\omega}{1-\omega^2+j\omega}\right|_{\omega=1} = 1-j$$

Equations 11.4.1a and 11.4.1b can be written using these values of the z parameters as

$$\mathbf{V}_1 = -j\mathbf{I}_2 \tag{11.4.6}$$

$$\mathbf{V}_2 = -j\mathbf{I}_1 + (1-j1)\mathbf{I}_2 \tag{11.4.7}$$

Equations 11.4.4–11.4.7 represent four equations in four unknowns since $\mathbf{V}_s$ is known. We need only to find $\mathbf{V}_2$. After some algebra we find $\mathbf{V}_2 = 1.54 - j7.69$ V.

To this point our discussion has centered on finding the z parameters for a given network. Let us now consider the inverse problem of finding a network representation for a given set of parameters. This is a synthesis problem, so there is no unique solution. We will describe a few simple forms for reciprocal and nonreciprocal networks. By a reciprocal network we mean one in which $z_{12} = z_{21}$, which occurs when the circuit contains only resistors, inductors, and capacitors.

The T network shown in Fig. 11.31*a* is applicable for the case when $z_{12} = z_{21}$. The z parameters for this network are easily shown to be given by

$$\begin{bmatrix} z_{11} & z_{12} \\ z_{21} & z_{22} \end{bmatrix} = \begin{bmatrix} \mathbf{Z}_a + \mathbf{Z}_c & \mathbf{Z}_c \\ \mathbf{Z}_c & \mathbf{Z}_b + \mathbf{Z}_c \end{bmatrix}$$

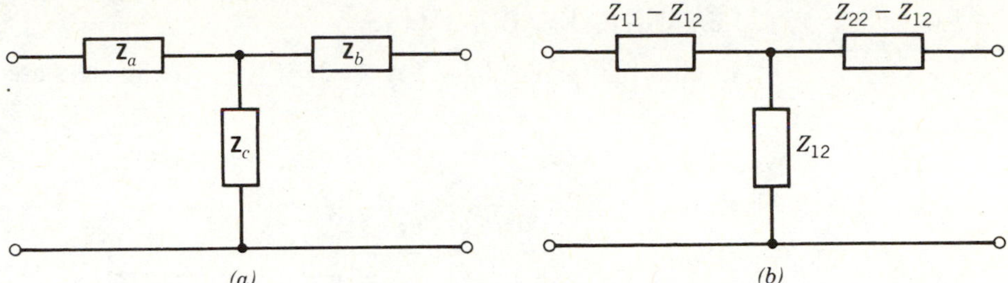

Figure 11.31

Equating like terms and solving for $\mathbf{Z}_a$, $\mathbf{Z}_b$, and $\mathbf{Z}_c$ gives

$$\mathbf{Z}_a = z_{11} - z_{12}$$
$$\mathbf{Z}_b = z_{22} - z_{12}$$
$$\mathbf{Z}_c = z_{12}$$

This gives the diagram in Fig. 11.31*b*.

As an example, suppose the z parameter matrix is given by

$$\begin{bmatrix} 5 & 1 \\ 1 & 4 \end{bmatrix}$$

Then $\mathbf{Z}_a = 5 - 1 = 4\ \Omega$, $\mathbf{Z}_b = 4 - 1 = 3\ \Omega$, and $\mathbf{Z}_c = 1\ \Omega$.

For nonreciprocal networks (those containing dependent sources and/or nonlinear devices) the two controlled-source realization shown in Fig. 11.32*a* is one of the most useful. The one controlled-source realization in Fig. 11.32*b* is also useful. Notice that this realization reduces to the T network above if $z_{21} = z_{12}$.

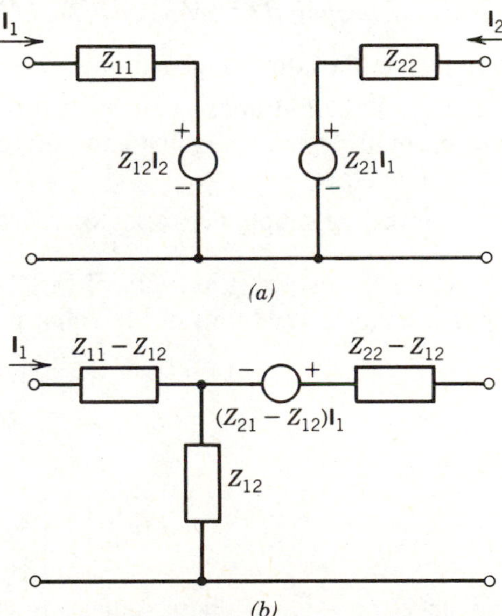

Figure 11.32. Realizations for nonreciprocal z parameters. (*a*) A two controlled-source realization. (*b*) A one controlled-source realization.

LEARNING EVALUATIONS

1. Find the current $\mathbf{I}_2$ in Fig. 11.33 if the z parameters are given by

$$z_{11} = 10$$
$$z_{12} = z_{21} = 5$$
$$z_{22} = 20$$

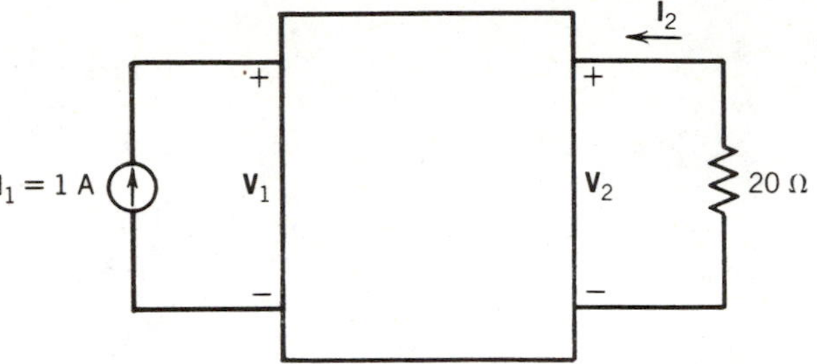

Figure 11.33. Circuit for Section 11.4 Learning Evaluation.

2. Find a T realization for z parameters given by

$$\begin{bmatrix} z_{11} & z_{12} \\ z_{21} & z_{22} \end{bmatrix} = \begin{bmatrix} 2 + j\omega & j\omega \\ j\omega & 3 + j2\omega \end{bmatrix}$$

11.5 Admittance Parameters

LEARNING OBJECTIVES

After completing this section you should be able to do the following:

1. Define the admittance parameters for any given two port network.
2. Use the admittance parameters for a two-port network to find voltages and currents or voltage gains and current gains through components attached to the two-port network.
3. Synthesize simple networks to realize y-parameter models.

If, for the two-port network of Fig. 11.24, we assume that the currents at each port are linear functions of the voltages across the ports we can write

$$\mathbf{I}_1 = y_{11}\mathbf{V}_1 + y_{12}\mathbf{V}_2 \tag{11.5.1a}$$

$$\mathbf{I}_2 = y_{21}\mathbf{V}_1 + y_{22}\mathbf{V}_2 \tag{11.5.1b}$$

or in matrix form

$$\begin{bmatrix} \mathbf{I}_1 \\ \mathbf{I}_2 \end{bmatrix} = \begin{bmatrix} y_{11} & y_{12} \\ y_{21} & y_{22} \end{bmatrix} \begin{bmatrix} \mathbf{V}_1 \\ \mathbf{V}_2 \end{bmatrix} \tag{11.5.1c}$$

Since the y terms relate voltages to currents, they must have the properties of admittance and consequently are called "admittance parameters" or often simply y parameters. Obviously Eq. 11.4.1 and 11.5.1 are related so a relationship can be established between the y parameters and the z parameters. We will not do it here, however, since you will be given that privilege in the problem set at the end of this chapter.

We can evaluate the y parameters in a way similar to the z parameters if we let $\mathbf{V}_1 = 0$, then $\mathbf{V}_2 = 0$. Under these conditions we obtain

$$y_{11} = \left. \frac{\mathbf{I}_1}{\mathbf{V}_1} \right|_{\mathbf{V}_2=0} \tag{11.5.2a}$$

$$y_{12} = \left. \frac{\mathbf{I}_1}{\mathbf{V}_2} \right|_{\mathbf{V}_1=0} \tag{11.5.2b}$$

$$y_{21} = \left. \frac{\mathbf{I}_2}{\mathbf{V}_1} \right|_{\mathbf{V}_2=0} \tag{11.5.2c}$$

$$y_{22} = \left. \frac{\mathbf{I}_2}{\mathbf{V}_2} \right|_{\mathbf{V}_1=0} \tag{11.5.2d}$$

As we observe, all y parameters can be measured or calculated by shorting one of the ports at a time. This, however, is a major disadvantage of y parameters. Significant damage may occur to your black box if you arbitrarily place a short across the two terminals of a port. Serious damage can also be done to your laboratory grade. A large capacitor may be used as an "ac short," so that measurement of these parameters is feasible at sinusoidal frequencies above zero.

Consider the circuit in Fig. 11.34a, which consists of three admittances in a π network. Shorting port 2 and applying current source $\mathbf{I}_1$, as shown in Fig. 11.34b, allows us to calculate y_{11} and y_{21}.

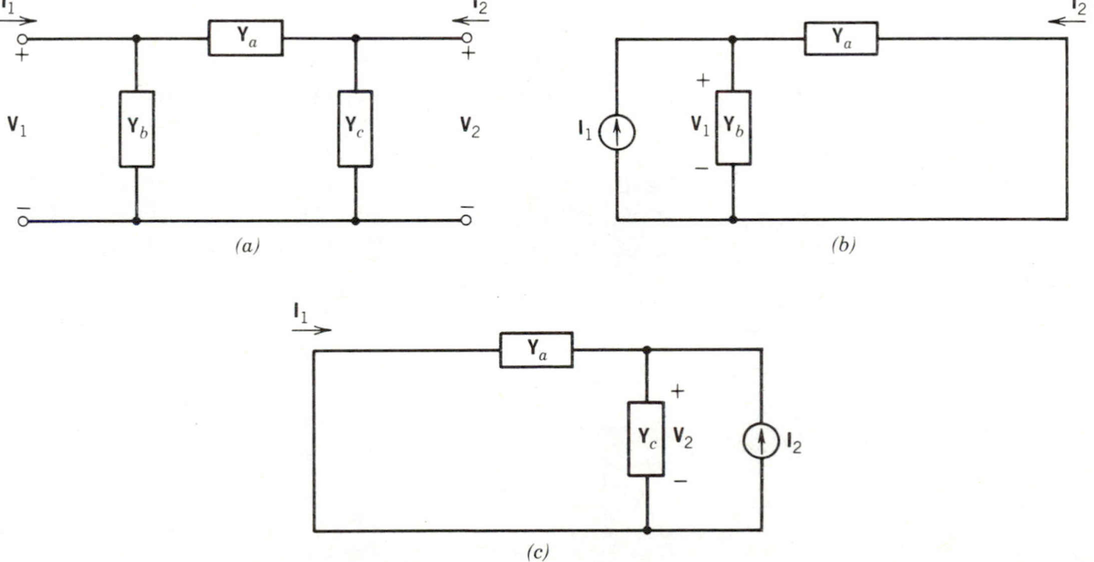

Figure 11.34. A π network.

$$y_{11} = \left.\frac{\mathbf{I}_1}{\mathbf{V}_1}\right|_{\mathbf{V}_2=0} = \mathbf{Y}_a + \mathbf{Y}_b \tag{11.5.3}$$

In the calculation of y_{21} the circuit acts as a current divider, giving

$$\mathbf{I}_2 = \frac{-\mathbf{I}_1\mathbf{Y}_a}{\mathbf{Y}_a + \mathbf{Y}_b}$$

Substituting this into Eq. 11.5.3 gives

$$y_{21} = \left.\frac{\mathbf{I}_2}{\mathbf{V}_1}\right|_{\mathbf{V}_2=0} = -\mathbf{Y}_a$$

The minus sign is a result of our assumed direction for the input currents.

In a similar manner we may calculate y_{22} and y_{12} using Fig. 11.34*c*. The result is given by

$$y_{12} = \left.\frac{\mathbf{I}_1}{\mathbf{V}_2}\right|_{\mathbf{V}_1=0} = -\mathbf{Y}_a$$

$$y_{22} = \left.\frac{\mathbf{I}_2}{\mathbf{V}_2}\right|_{\mathbf{V}_1=0} = \mathbf{Y}_a + \mathbf{Y}_c$$

EXAMPLE 11.5.1 Calculate the y parameters for the network in Fig. 11.35.

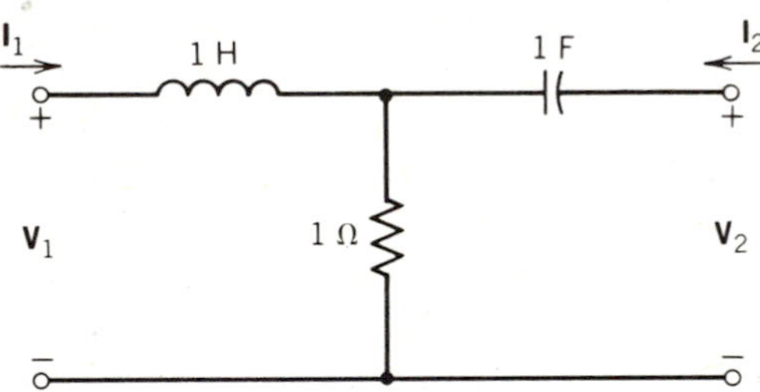

Figure 11.35. Network for Example 11.5.1.

Solution

In Fig. 11.36*a* a voltage source is applied to the left port with $\mathbf{V}_2 = 0$. Notice that admittance values are used. From this diagram we may solve for y_{11} and y_{21}. Direct calculation of the admittance at the left port gives

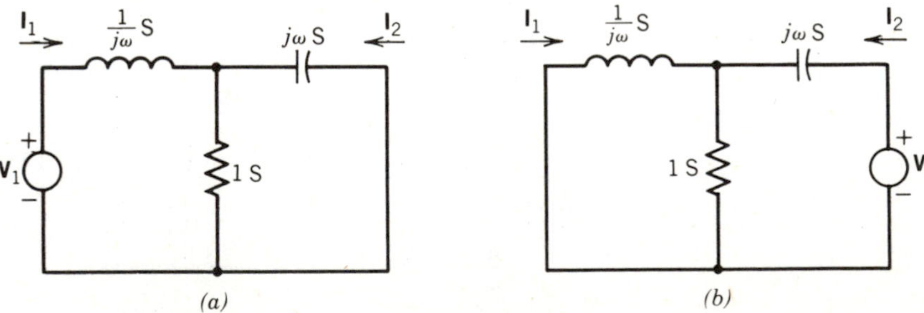

Figure 11.36

$$y_{11} = \frac{\mathbf{I}_1}{\mathbf{V}_1} = \frac{1}{\dfrac{1}{j\omega + 1} + j\omega} = \frac{1 + j\omega}{1 - \omega^2 + j\omega} \tag{11.5.4}$$

Since $\mathbf{I}_2$ is a fraction of $\mathbf{I}_1$ we use current division to obtain the relationship

$$\frac{-\mathbf{I}_2}{\mathbf{I}_1} = \frac{j\omega}{1 + j\omega}$$

Substituting this into Eq. 11.5.4 gives

$$y_{21} = \frac{-j\omega}{1 - \omega^2 + j\omega}$$

In a similar manner we may use Fig. 11.36*b* to calculate y_{12} and y_{22}. The results are given by

$$y_{12} = \frac{-j\omega}{1 - \omega^2 + j\omega}$$

$$y_{22} = \frac{-\omega^2 + j\omega}{1 - \omega^2 + j\omega}$$

Now let us consider the inverse problem of finding a network representation for a given set of y parameters. For reciprocal networks (for which $y_{12} = y_{21}$) the π representation is a convenient model to use. The y parameters for the network in Fig. 11.37*a* are given by

$$\begin{bmatrix} y_{11} & y_{12} \\ y_{21} & y_{22} \end{bmatrix} = \begin{bmatrix} \mathbf{Y}_a + \mathbf{Y}_b & -\mathbf{Y}_a \\ -\mathbf{Y}_a & \mathbf{Y}_a + \mathbf{Y}_c \end{bmatrix} \tag{11.5.5}$$

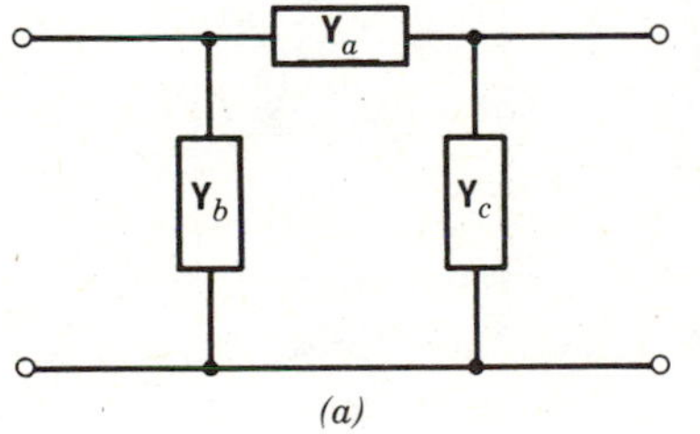

(*a*)

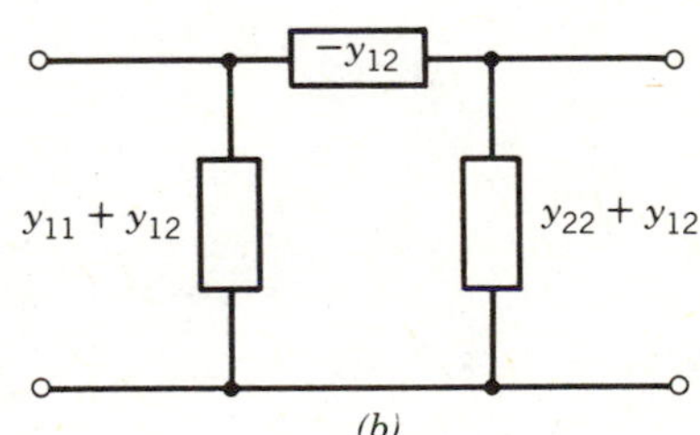

(*b*)

Figure 11.37

Solving these equations for $\mathbf{Y}_a$, $\mathbf{Y}_b$, and $\mathbf{Y}_c$ gives the network in Fig. 11.37*b*, where

$$\mathbf{Y}_a = -y_{12} = -y_{21} \tag{11.5.6a}$$

$$\mathbf{Y}_b = y_{11} + y_{12} \tag{11.5.6b}$$

$$\mathbf{Y}_c = y_{22} + y_{21} \tag{11.5.6c}$$

EXAMPLE 11.5.2 Find a network representation for a set of y parameters given by

$$\begin{bmatrix} 3 + j4\omega & -2 \\ -2 & 5 + j3\omega \end{bmatrix}$$

Solution

Using Eqs. 11.5.5 and 11.5.6 we see by inspection that

$$y_{12} = -\mathbf{Y}_a = -2\text{S}$$

Since $\mathbf{Y}_a + \mathbf{Y}_b = 3 + j4\omega$ we must have

$$\mathbf{Y}_b = 1 + j4\omega$$

giving the 4-F capacitor and 1-S conductor in parallel. Similarly the $\mathbf{Y}_c$ admittance consists of a 3-S conductor in parallel with a 3-F capacitor. The circuit is shown in Fig. 11.38.

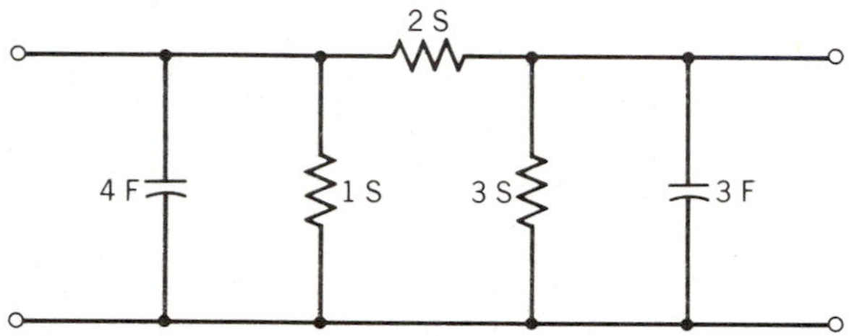

Figure 11.38. Solution to Example 11.5.2.

When the y parameters are not reciprocal (when $y_{12} \neq y_{21}$) then a controlled-source network realization is used. A two controlled-source model is shown in Fig. 11.39, and a one controlled-source model is shown in Fig. 11.40. Notice that the one controlled-source model reduces to the π-network model for $y_{12} = y_{21}$.

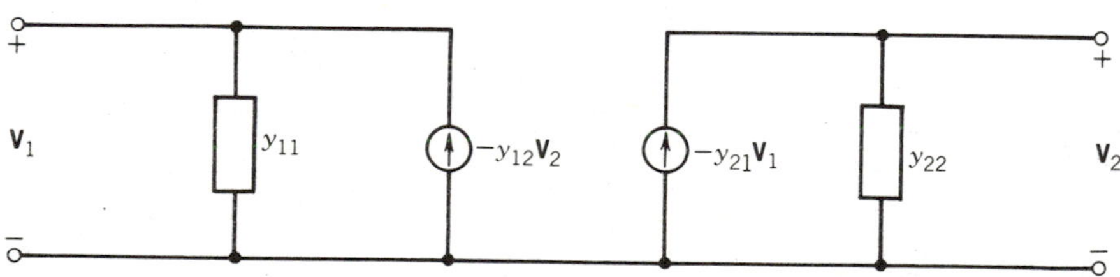

Figure 11.39. A two controlled-source y-parameter model.

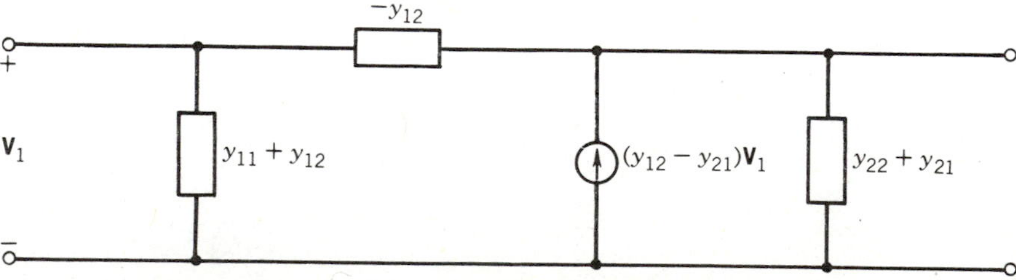

Figure 11.40. A one controlled-source y-parameter model.

LEARNING EVALUATIONS

1. Find the current $\mathbf{I}_2$ in Fig. 11.33 if the y parameters are given by

$$y_{11} = 0.1 \text{ S}$$
$$y_{12} = y_{21} = -0.05 \text{ S}$$
$$y_{22} = 0.2 \text{ S}$$

2. Find a π realization for y parameters given by

$$\begin{bmatrix} y_{11} & y_{12} \\ y_{21} & y_{22} \end{bmatrix} = \begin{bmatrix} 2 + j3\omega & -1 \\ -1 & 3 + j4\omega \end{bmatrix}$$

11.6 Hybrid Parameters

LEARNING OBJECTIVES

After completing this section you should be able to do the following:

1. Define the hybrid parameters for any given two-port network.
2. Use the h parameters for a two-port network to find voltages and currents or voltage gains and current gains through components attached to the two-port network.

There are two sets of hybrid parameters corresponding to case 3 and case 4 in Table 11.1. Case 3, where $\mathbf{I}_1$ and $\mathbf{V}_2$ correspond to the variables $\mathbf{Q}_1$ and $\mathbf{Q}_2$ in Eq. 11.3.1, produces the g parameters. these equations have the form

$$\mathbf{I}_1 = g_{11}\mathbf{V}_1 + g_{12}\mathbf{I}_2 \tag{11.6.1a}$$

$$\mathbf{V}_2 = g_{21}\mathbf{V}_1 + g_{22}\mathbf{I}_2 \tag{11.6.1b}$$

These equations define the g parameters. Keep in mind that all terms are functions of frequency. That is, $\mathbf{I}_1 = \mathbf{I}_1(\omega)$, $g_{11} = g_{11}(\omega)$, and so on. From Eqs. 11.6.1 we see that the g parameters are defined by

$$g_{11} = \left.\frac{\mathbf{I}_1}{\mathbf{V}_1}\right|_{\mathbf{I}_2=0} \tag{11.6.2}$$

$$g_{12} = \left.\frac{\mathbf{I}_1}{\mathbf{I}_2}\right|_{\mathbf{V}_1=0} \tag{11.6.3}$$

$$g_{21} = \left.\frac{\mathbf{V}_2}{\mathbf{V}_1}\right|_{\mathbf{I}_2=0} \tag{11.6.4}$$

$$g_{22} = \left.\frac{\mathbf{V}_2}{\mathbf{I}_2}\right|_{\mathbf{V}_1=0} \tag{11.6.5}$$

These defining equations also indicate the procedure for measuring the g parameters. With port 2 open we may measure g_{11} and g_{21} by applying a source at port 1 and finding the appropriate ratios. Similarly, with port 1 shorted we may measure g_{12} and g_{22}.

The h parameters are defined by identifying $\mathbf{Q}_1$ and $\mathbf{Q}_2$ in Eq. 11.3.1 with $\mathbf{V}_1$ and $\mathbf{I}_2$, as indicated in Table 11.1. These equations have the form

$$\mathbf{V}_1 = h_{11}\mathbf{I}_1 + h_{12}\mathbf{V}_2 \tag{11.6.6a}$$

$$\mathbf{I}_2 = h_{21}\mathbf{I}_1 + h_{22}\mathbf{V}_2 \tag{11.6.6b}$$

The h parameters are defined as

$$h_{11} = \left.\frac{\mathbf{V}_1}{\mathbf{I}_1}\right|_{\mathbf{V}_2=0} \tag{11.6.7}$$

$$h_{12} = \left.\frac{\mathbf{V}_1}{\mathbf{V}_2}\right|_{\mathbf{I}_1=0} \tag{11.6.8}$$

$$h_{21} = \left.\frac{\mathbf{I}_2}{\mathbf{I}_1}\right|_{\mathbf{V}_2=0} \tag{11.6.9}$$

$$h_{22} = \left.\frac{\mathbf{I}_2}{\mathbf{V}_2}\right|_{\mathbf{I}_1=0} \tag{11.6.10}$$

Equation 11.6.7 shows that h_{11} has the units of impedance and is measured by shorting the output terminals of our black box. Since h_{11} relates $\mathbf{V}_1$ to $\mathbf{I}_1$, it is often called the *short circuit input impedance*. h_{12} is measured by opening the input terminals and relates the input voltage to output voltage. This dimensionless parameter is named the *open circuit input/output voltage gain*. The third parameter, h_{21}, relates the output current to input current when the output voltage is shorted and is called the *short circuit current gain*. h_{21} is also dimensionless. The last parameter, h_{22}, is observed to have the units of admittance and is called the *open circuit output admittance*.

Observing Eqs. 11.6.6 also allows us to construct a circuit model. In Eq. 11.6.6a $\mathbf{V}_1$ sees an impedance of h_{11} and a dependent voltage source that is a function of $\mathbf{V}_2$. $\mathbf{I}_2$ sees a dependent current source that is a function of $\mathbf{I}_1$ in parallel with an admittance h_{22}. This circuit model is shown in Fig. 11.41.

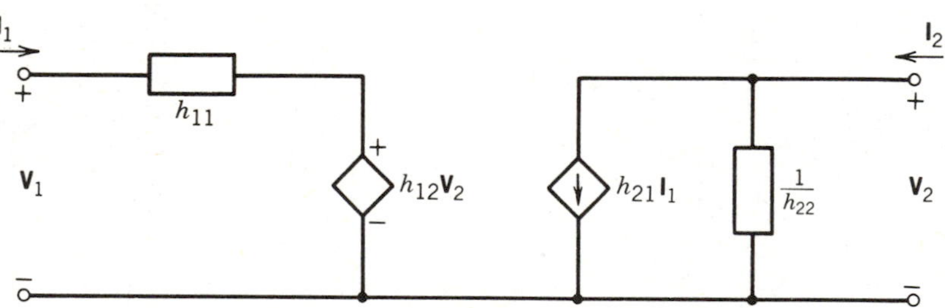

Figure 11.41. Circuit model for the hybrid parameters.

EXAMPLE 11.6.1 A 2N2604 transistor has the following hybrid parameter values when connected in a common base configuration and properly biased:

$$h_{11} = 30\ \Omega, \qquad h_{12} = 10^{-3}, \qquad h_{21} = -0.99, \qquad h_{22} = 1 \times 10^{-6}\ \mathrm{S}$$

A current source is connected to the emitter (input) and a 1-kΩ load resistor is placed across the output (collector). Find the current gain ($\mathbf{I}_L/\mathbf{I}_S$) and determine the effect of varying the value of the load resistor.

Solution

The two-port equivalent circuit is shown in Fig. 11.42 and we desire the current gain $\mathbf{I}_L/\mathbf{I}_S$. By observation,

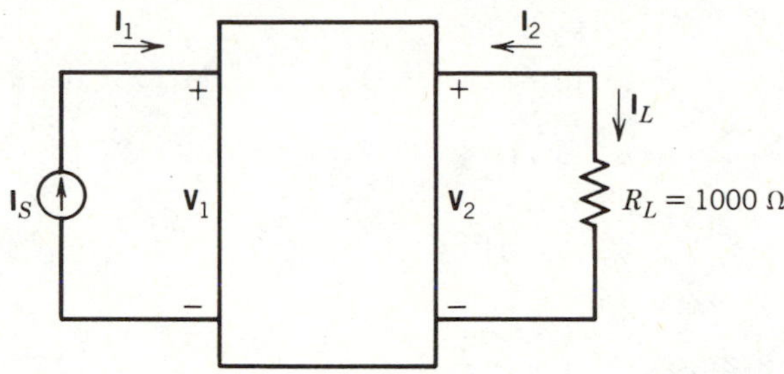

Figure 11.42

$$\mathbf{I}_1 = \mathbf{I}_S \tag{11.6.11}$$

and

$$\mathbf{I}_2 = \frac{-\mathbf{V}_2}{1000} \tag{11.6.12}$$

By definition of the hybrid parameters (Eq. 11.6.6b),

$$\mathbf{I}_2 = h_{21}\mathbf{I}_1 + h_{22}\mathbf{V}_2 \tag{11.6.13}$$

Equations 11.6.11–11.6.13 represent three equations in three unknowns. Solving Eq. 11.6.12 for $\mathbf{V}_2$ gives

$$\mathbf{V}_2 = -1000\mathbf{I}_2 \tag{11.6.14}$$

Putting Eqs. 11.6.11 and 11.6.14 into 11.6.13 yields

$$\mathbf{I}_2 = -0.99\mathbf{I}_s - 0.001\mathbf{I}_2$$

Since $\mathbf{I}_L = -\mathbf{I}_2$ we can calculate our answer as

$$\frac{\mathbf{I}_L}{\mathbf{I}_S} = 0.98901$$

If R_L were 10 Ω instead of 1 kΩ, Eq. 11.4.9 becomes

$$\mathbf{V}_2 = -10\mathbf{I}_2$$

and our gain would be

$$\frac{\mathbf{I}_L}{\mathbf{I}_S} = 0.9801$$

Finally, a value of $R_L = 100$ k Ω results in a gain of 0.90. From this we see clearly one distinctive characteristic of a common base transistor circuit:

the current gain is approximately 1 over a wide range of values of load resistance.

LEARNING EVALUATIONS

1. Find the current $\mathbf{I}_2$ in Fig. 11.33 if the g parameters are given by
$g_{11} = 1$
$g_{12} = 5$
$g_{21} = 10$
$g_{22} = 20$
2. Find the current $\mathbf{I}_2$ in Fig. 11.33 if the h parameters are given by
$h_{11} = 2$
$h_{12} = 20$
$h_{21} = 0.5$
$h_{22} = 0.05$

11.7 *ABCD* Parameters

LEARNING OBJECTIVES

After completing this section you should be able to do the following:

1. Define the $ABCD$ and $\mathscr{A}\mathscr{B}\mathscr{C}\mathscr{D}$ parameters.
2. Find the $ABCD$ parameters for a given network.

The last two entries in Table 11.1 give us the $ABCD$ and $\mathscr{A}\mathscr{B}\mathscr{C}\mathscr{D}$ parameters, respectively. These are also called the transmission and inverse-transmission parameters, respectively. The $ABCD$ parameters are defined by

$$\mathbf{V}_1(\omega) = A(\omega)\mathbf{V}_2(\omega) - B(\omega)\mathbf{I}_2(\omega) \tag{11.7.1a}$$

$$\mathbf{I}_1(\omega) = C(\omega)\mathbf{V}_2(\omega) - D(\omega)\mathbf{I}_2(\omega) \tag{11.7.1b}$$

In all our previous discussion of two-port parameters one term from the input and one term from the output port have appeared on the left of the defining equations. Notice that the input voltage and current are on the left in these equations. Hence each parameter A, B, C, and D defines a relationship between an input variable and an output variable, which provides an explanation for the term transmission parameters. From Eq. 11.7.1 we see that the $ABCD$ parameters are defined by

$$A(\omega) = \left.\frac{\mathbf{V}_1(\omega)}{\mathbf{V}_2(\omega)}\right|_{\mathbf{I}_2(\omega)=0} \tag{11.7.2}$$

$$B(\omega) = -\left.\frac{\mathbf{V}_1(\omega)}{\mathbf{I}_2(\omega)}\right|_{\mathbf{V}_2(\omega)=0} \tag{11.7.3}$$

$$C(\omega) = \left.\frac{\mathbf{I}_1(\omega)}{\mathbf{V}_2(\omega)}\right|_{\mathbf{I}_2(\omega)=0} \tag{11.7.4}$$

$$D(\omega) = -\left.\frac{\mathbf{I}_1(\omega)}{\mathbf{I}_2(\omega)}\right|_{\mathbf{V}_2(\omega)=0} \tag{11.7.5}$$

The negative sign on the $\mathbf{I}_2$ term is present to account for the difference between the convention of defining current at any port into the network, as opposed to the convention of defining the load current out of the network at the output port in a transmission system.

EXAMPLE 11.7.1 Find the transmission parameters for the network in Fig. 11.27.

Solution

First open port 2 so that $\mathbf{I}_2 = 0$ and apply voltage $\mathbf{V}_1$ to the input as shown in Fig. 11.43*a*. Then solve for $\mathbf{V}_2$ and $\mathbf{I}_1$ so that $A(\omega)$ and $C(\omega)$ may be calculated. Since $\mathbf{V}_1$ appears across the resistor we may solve for $\mathbf{V}_2$ by voltage division.

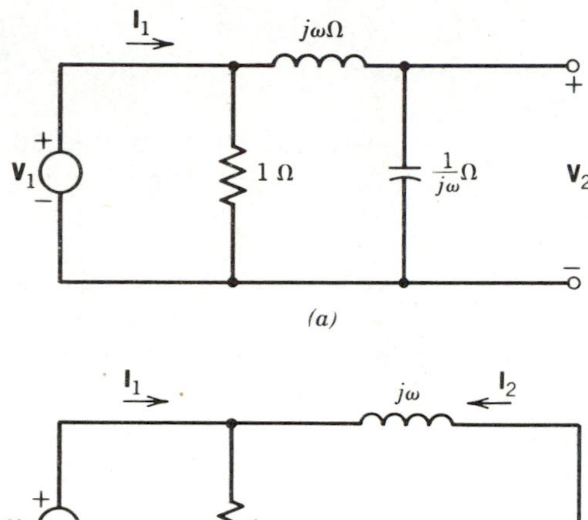

Figure 11.43

$$\frac{\mathbf{V}_2}{\mathbf{V}_1} = \frac{1/j\omega}{j\omega + 1/j\omega} = \frac{1}{1-\omega^2} \tag{11.7.6}$$

Hence $A(\omega) = 1 - \omega^2$. Current $\mathbf{I}_1$ is the sum of the current through the resistor plus the inductor current.

$$\mathbf{I}_1 = \mathbf{V}_1 + \frac{\mathbf{V}_1}{j\omega + 1/j\omega} = \mathbf{V}_1\left(\frac{1-\omega^2 + j\omega}{1-\omega^2}\right)$$

Upon solving for $\mathbf{V}_1$ and substituting into Eq. 11.7.6 we get

$$C(\omega) = 1 - \omega^2 + j\omega$$

The parameters B and D may be calculated using Fig. 11.43*b*, which shows the output terminals shorted. The capacitor may be removed from the circuit in this case, because no voltage will appear across it and no current will flow through it. Since $\mathbf{V}_1$ appears directly across the inductor we may calculate B directly.

$$B(\omega) = -\frac{\mathbf{V}_1}{\mathbf{I}_2} = j\omega$$

Also, the ratio of the two currents may be calculated by current division, giving

$$D(\omega) = -\frac{\mathbf{I}_1}{\mathbf{I}_2} = 1 + j\omega$$

The transmission parameters are particularly suited for calculating load voltage and current if either the input voltage or current is known. Since I_2 and V_2 are related by Ohm's law, only one of the two defining Eqs. 11.7.1 need be used in any calculation. Here is an example to illustrate this point.

EXAMPLE 11.7.2 Suppose a 1-Ω resistor is connected across the output port of the circuit in Fig. 11.27. Find $\mathbf{V}_2$ if the input voltage is 10 V at a frequency $\omega = 1$.

Solution

The network is shown in Fig. 11.44 with the transmission parameters evaluated using the results in Example 11.7.1. Since $\mathbf{V}_2 = -\mathbf{I}_2$ we may substitute into Eq. 11.7.1a.

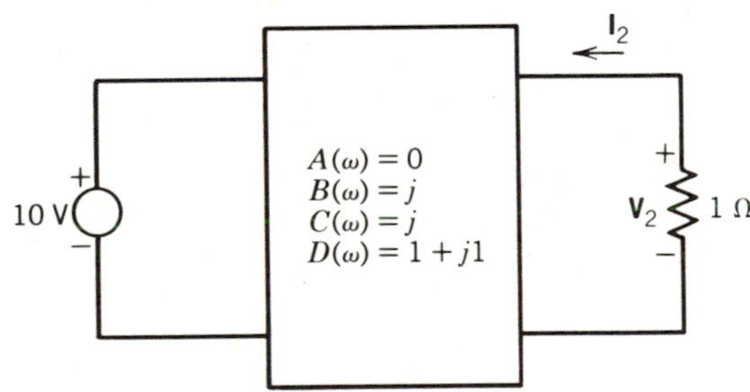

Figure 11.44

$$10 = -j\mathbf{I}_2 = j\mathbf{V}_2 \tag{11.7.7a}$$

or

$$\mathbf{V}_2 = -j10 \text{ V} \tag{11.7.7b}$$

A network model for the transmission parameters consisting of three controlled sources and one impedance is shown in Fig. 11.45. The validity of this model may easily be checked by applying KVL at port 1 and KCL at port 2.

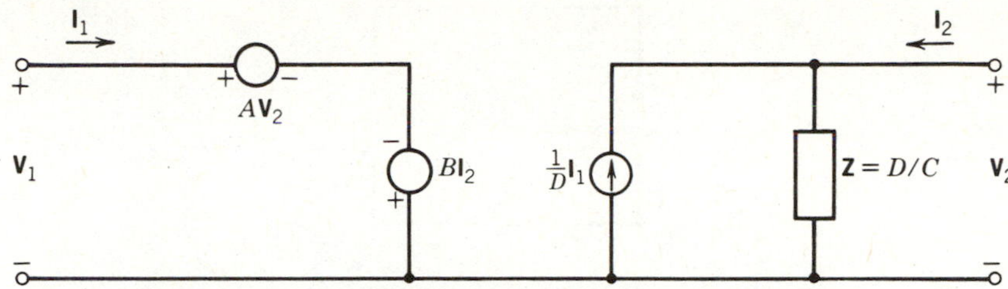

Figure 11.45. A network realization for the *ABCD* parameters.

The inverse-transmission parameters are defined by

$$\mathbf{V}_2(\omega) = \mathscr{A}(\omega)\mathbf{V}_1(\omega) - \mathscr{B}(\omega)\mathbf{I}_1(\omega) \tag{11.7.8a}$$

$$\mathbf{I}_2(\omega) = \mathscr{C}(\omega)\mathbf{V}_1(\omega) - \mathscr{B}(\omega)\mathbf{I}_1(\omega) \tag{11.7.8b}$$

The characteristics of this set of parameters differ little from the *ABCD* parameters. For this reason, and because most applications utilize forward transmission, they are used infrequently.

LEARNING EVALUATION

Find $\mathbf{I}_2$ in Fig. 11.33 if the *ABCD* parameters are given by

$$A = 5$$
$$B = 0.1$$
$$C = 1$$
$$D = 20$$

PROBLEMS

SECTION 11.1

1. A resistance of 10 Ω is placed in series with a variable capacitor, across a sinusoidal 100-V source.

a. What is the maximum current that can flow?

b. Sketch the locus of the current as C is varied, and mark the point corresponding to $C = 3.18\ \mu$F if the source frequency is $f = 5$ kHz.

2. A resistance of 10 Ω is placed in series with a variable inductor, across a sinusoidal 100-V source.

a. What is the maximum current that can flow?

b. Sketch the locus of the current as L is varied, and mark the point corresponding to $L = 0.32$ mH if the source frequency is $f = 5$ kHz.

3. Find the current through the circuit in Fig. 11.46 and the voltage across each element. The applied voltage is $v(t) = 10\cos[12\pi\ (10)^3 t]$.

4. Find the resonant frequency, the bandwidth, and the quality factor for the series *RLC* circuit in Fig. 11.46.

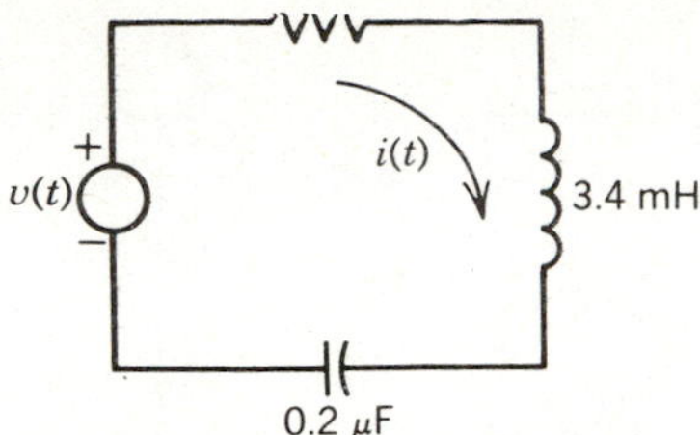

Figure 11.46

5. The power dissipated in the resistor in Fig. 11.46 is a maximum at the resonant frequency. At the 3-dB frequencies this power is down by a factor of $\frac{1}{2}$. Find the upper and lower frequencies where the power is down by $\frac{1}{4}$.
6. As L is varied to produce resonance in a series circuit containing $R = 100\ \Omega$, $C = 10\ \mu$F, and a constant voltage source at $f = 60$ Hz, find the voltage drop across L. Is the voltage drop across L maximum at resonance?
7. Design two series RLC circuits to have a resonant frequency of $5(10)^6$ Hz and a bandwidth of 16 kHz. First use a value of $L = 10^{-3}$ H. Second, use a value of $L = 10^{-4}$ H. Compare the needed values of R for the two designs. Which circuit uses a larger value of R?
8. In Fig. 11.47 the fixed resistor has resistance $R = 10\ \Omega$ and the inductor has reactance $X_L = 10\ \Omega$. The source is sinusoidal with magnitude 10 V and constant frequency. What is the maximum power delivered to R_L as R_L is varied from zero to infinity?

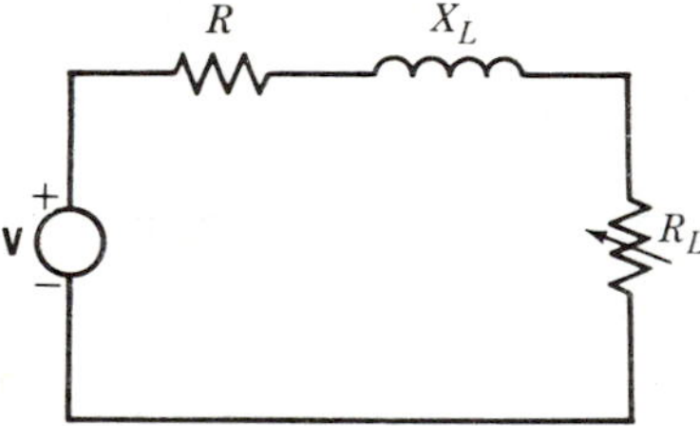

Figure 11.47

SECTION 11.2

9. A resistance of 10 Ω is placed in parallel with a variable capacitor, across a sinusoidal 1-A current source.
   a. What is the maximum voltage across the parallel circuit?
   b. Sketch the locus of the voltage as C is varied and mark the point corresponding to $C = 3.18\ \mu$F if the source frequency is $f = 5$ kHz.
10. A resistance of 10 Ω is placed in parallel with a variable inductor, across a sinusoidal 1-A current source.
   a. What is the maximum voltage across the parallel circuit?

b. Sketch the locus of the voltage as L is varied and mark the point corresponding to $L = 0.32$ mH if the source frequency is $f = 5$ kHz.

11. Find the voltage across the circuit in Fig. 11.48 and the current through each element. The applied current is $i(t) = 0.1 \cos 12\pi(10^3 t)$.

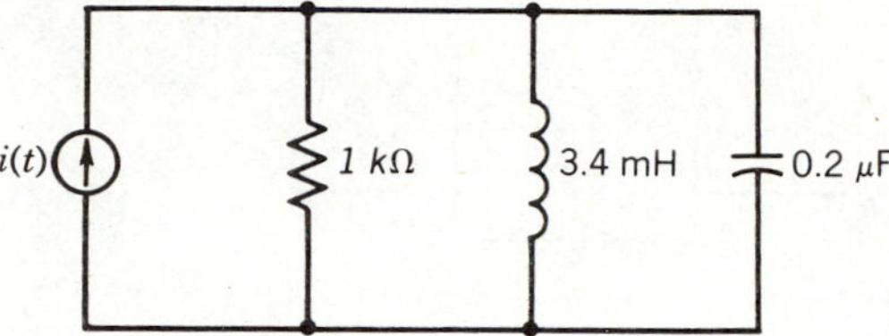

Figure 11.48

12. Find the resonant frequency, the bandwidth, and the quality factor for the parallel *RLC* circuit in Fig. 11.48.

13. The power dissipated in the resistor in Fig. 11.48 is a maximum at the resonant frequency. At the 3-dB frequencies this power is down by a factor of $\frac{1}{2}$. Find the upper and lower frequencies where the power is down by $\frac{1}{4}$.

14. As L is varied to produce resonance in a parallel circuit containing $R = 100$ Ω, $C = 10$ μF, and a constant current source at $f = 60$ Hz, find the current through L. Is the current through L maximum at resonance?

SECTION 11.4

15. Find the z parameters for each two-port network shown in Fig. 11.49.

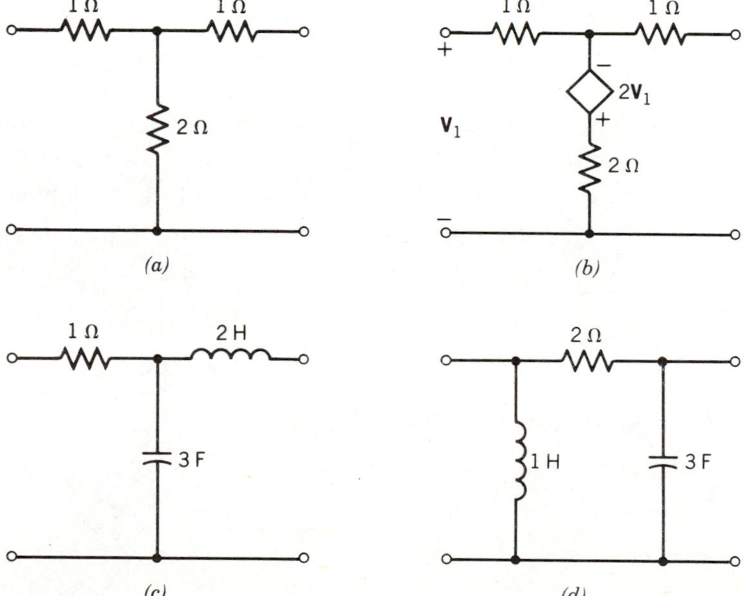

Figure 11.49

16. Find the voltage $\mathbf{V}_2$ in Fig. 11.50 if the z parameters are given by

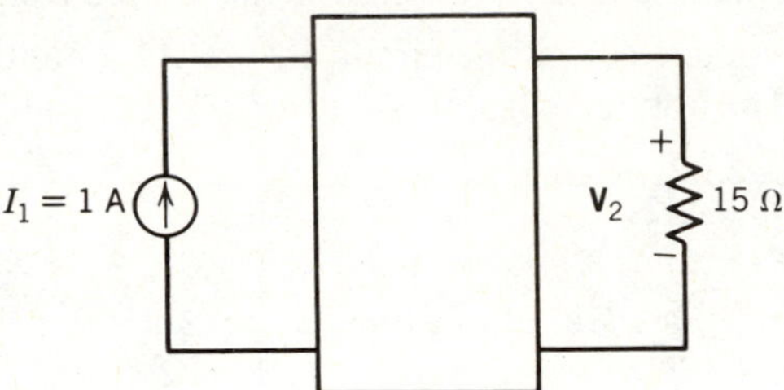

Figure 11.50

$$z_{11} = 20$$
$$z_{12} = z_{21} = 10$$
$$z_{22} = 25$$

17. Find a network that realizes each of the following sets of z parameters.

$$\begin{bmatrix} 3 & 2 \\ 2 & 5 \end{bmatrix} \quad \begin{bmatrix} 2 + \dfrac{4}{j\omega} & \dfrac{1}{j\omega} \\ \dfrac{1}{j\omega} & 3 + \dfrac{2}{j\omega} \end{bmatrix} \quad \begin{bmatrix} 3 & 2 \\ -2 & 5 \end{bmatrix}$$

SECTION 11.5

18. Show that the y parameters may be obtained from the z parameters by

$$y_{11} = \frac{z_{22}}{\Delta}, \qquad y_{12} = -\frac{z_{12}}{\Delta}$$
$$y_{21} = -\frac{z_{21}}{\Delta}, \qquad y_{22} = \frac{z_{11}}{\Delta}$$

where $\Delta = z_{11}z_{22} - z_{12}z_{21}$. Use this result to find the y parameters for each network in Problem 17.

19. Find the y parameters for each network in Problem 15.

20. Find the voltage $\mathbf{V}_2$ in Fig. 11.50 if the y parameters are given by

$$y_{11} = 0.2$$
$$y_{12} = y_{22} = 0.4$$
$$y_{22} = 0.3$$

SECTION 11.6

21. Show that the hybrid parameters may be obtained from the z parameters by

$$h_{11} = \frac{\Delta}{z_{22}} \qquad h_{12} = \frac{z_{12}}{z_{22}}$$
$$h_{21} = -\frac{z_{21}}{z_{22}} \qquad h_{22} = \frac{1}{z_{22}}$$

$$g_{11} = \frac{1}{z_{11}} \qquad g_{12} = -\frac{z_{12}}{z_{11}}$$

$$g_{21} = \frac{z_{21}}{z_{11}} \qquad g_{22} = \frac{\Delta}{z_{11}}$$

where $\Delta = z_{11}z_{22} - z_{12}z_{21}$. Use this to find both sets of hybrid parameters for each network in Problem 17.

22. Find the g parameters for each network in Problem 15.

23. Find the h parameters for each network in Problem 15.

24. Find the voltage $\mathbf{V}_2$ in Fig. 11.50 if the h parameters are given by

$$h_{11} = 1$$
$$h_{12} = 25$$
$$h_{21} = 0.4$$
$$h_{22} = 0.1$$

SECTION 11.7

25. Show that the $ABCD$ parameters may be obtained from the z parameters by

$$A = \frac{z_{11}}{z_{21}} \qquad B = \frac{\Delta}{z_{21}}$$

$$C = \frac{1}{z_{21}} \qquad D = \frac{z_{22}}{z_{21}}$$

where $\Delta = z_{11}z_{22} - z_{12}z_{21}$. Use this to find the $ABCD$ parameters for each network in Problem 17.

26. Find the $ABCD$ parameters for each network in Problem 15.

27. Find the voltage $\mathbf{V}_2$ in Fig. 11.50 if the $ABCD$ parameters are given by

$$A = 4$$
$$B = 0.2$$
$$C = 1$$
$$D = 15$$

LEARNING EVALUATION ANSWERS

Section 11.1

1. See Fig. 11.51.

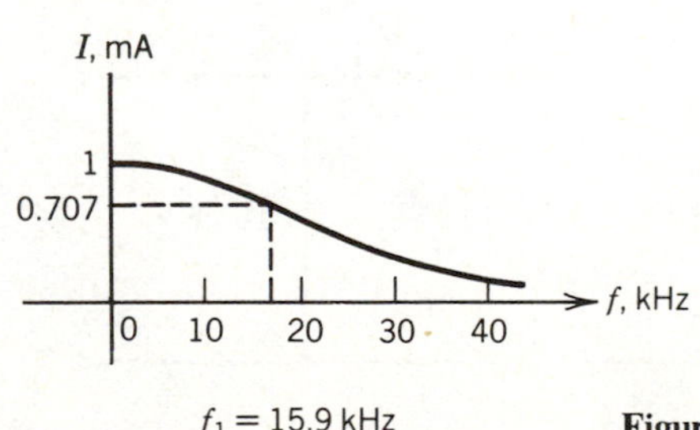

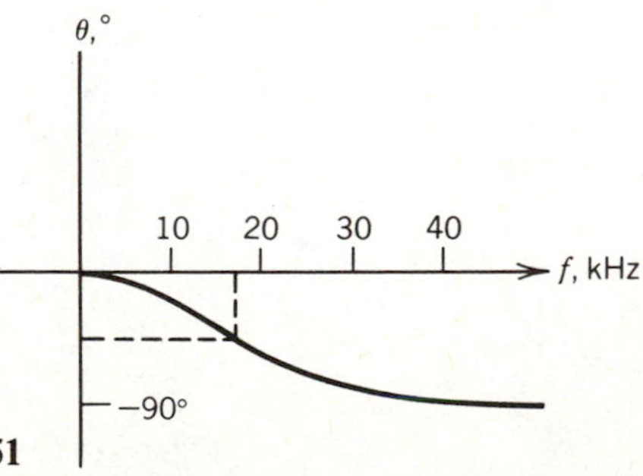

Figure 11.51

2. $fr = 795.8$ Hz
   $Q = 2$
   $B = 398$ Hz

Section 11.2

1. See Fig. 11.52.

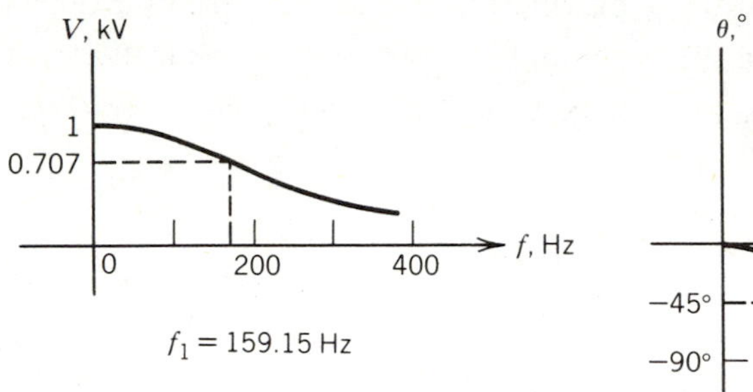

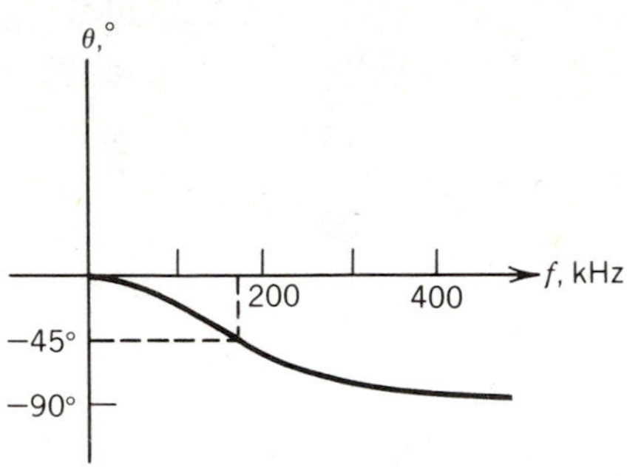

Figure 11.52

2. $fr = 795.8$ Hz
   $Q = 0.5$
   $B = 1591$ Hz

Section 11.3

1. See Fig. 11.53.

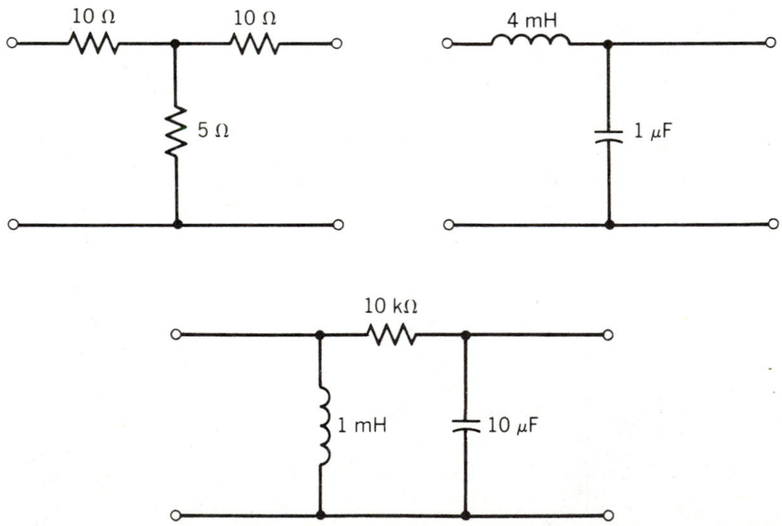

Figure 11.53

2. See Fig. 11.54.

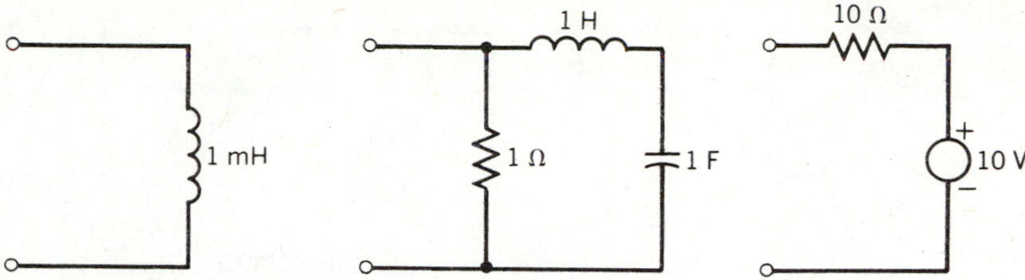

Figure 11.54

Section 11.4

1. $I_2 = -0.125$ A

2. See Fig. 11.55.

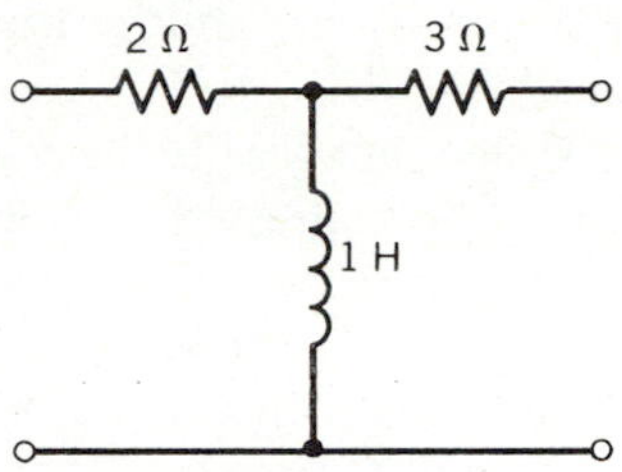

Figure 11.55

Section 11.5

1. $I_2 = -\frac{1}{9}$ A

2. See Fig. 11.56.

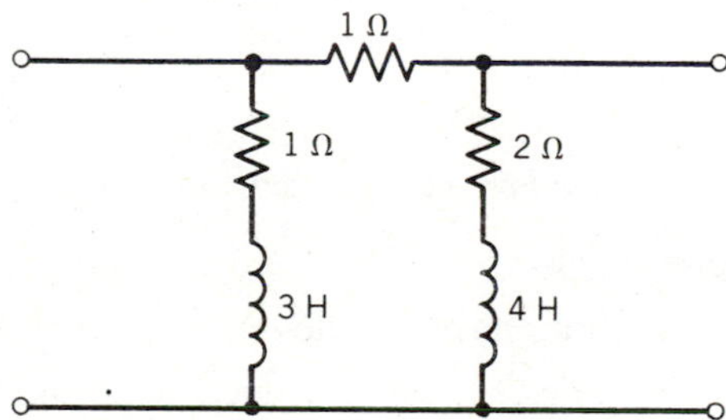

Figure 11.56

Section 11.6

1. $I_2 = 1$ A

2. $I_2 = 0.25$ A

Section 11.7

$I_2 = -\frac{1}{40}$ A

CHAPTER

12 FOURIER TECHNIQUES

This chapter continues and considerably broadens our discussion of frequency analysis. The concepts of linearity and time invariance that were introduced in Chapter 8 will be used extensively here, because frequency analysis is based on the LTI properties. Frequency analysis was used briefly in Chapters 9–11. The concepts of frequency analysis apply equally to discrete-time and continuous-time systems. The names Fourier series, Fourier transform, and Laplace transform are used when continuous-time systems are being analyzed. The terms discrete Fourier transform, fast Fourier transform, and z transform are used in the analysis of discrete-time systems.

The central problem of LTI system analysis is the following: Find that signal $x_0(t)$ such that other signals may be expressed as linear combinations of this basic signal, and such that the system response to $x_0(t)$ can easily be found. If these two conditions are satisfied, the LTI properties can be applied to find the system response to arbitrary input signals. In Chapter 8 the basic signal $x_0(t) = \delta(t)$ was used to form the basic input–output pair $\delta(t)$, $h(t)$. The convolution integral (or summation) could then be applied to find the system response.

This same procedure is used in frequency analysis. The basic input signal $x_0(t)$ is exponential; e^{st} for continuous-time systems, and z^n for discrete-time systems. The response to this exponential signal is called the transfer function, which was introduced in Chapter 9, a term of great importance in system analysis.

12.1 The Fourier Series

LEARNING OBJECTIVES

After completion of this section you should be able to do the following:

1. Define the Fourier series.
2. Describe which functions have and do not have a Fourier series representation.
3. Find the Fourier series, $V(k)$, for any function, $v(t)$, that satisfies the necessary conditions.
4. Invert the Fourier series. That is, given $V(k)$, find $v(t)$.

The Fourier series of a function $v(t)$ is a transform given by the equation

$$V(k) = V(f_k) = \frac{1}{T}\int_{t'}^{t'+T} v(t)e^{-j2\pi f_k t}\,dt \qquad (12.1.1)$$

pictured in Fig. 12.1. This transform operates on a signal $v(t)$ to produce a signal $V(k)$. A typical transform pair is also illustrated in Fig. 12.1. The signal $V(k)$ is

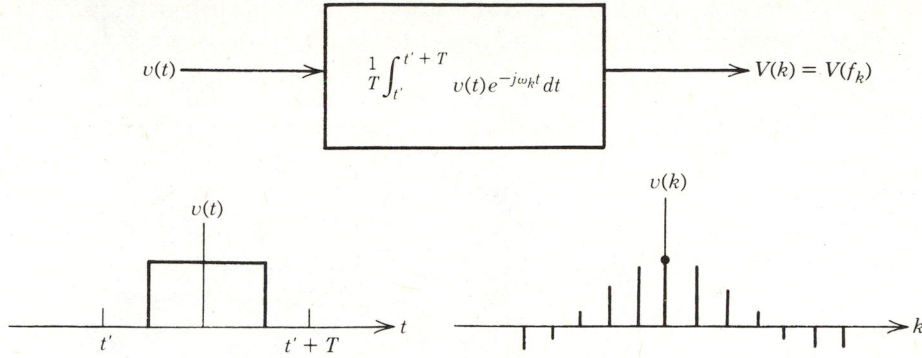

Figure 12.1. The Fourier series.

always discrete. The signal $v(t)$ may be either a continuous-time signal, as shown, or a discrete-time signal. We call it the discrete Fourier series, or discrete Fourier transform, when the signal $v(t)$ is a discrete-time signal. In this section we will concentrate exclusively on the Fourier series as it applies to continuous-time signals. Then in Section 12.2 we will discuss the discrete Fourier transform.

The parameter t can represent any physical quantity or attribute. For example, one application has identified t as space and f or k as wave number. The French mathematician Fourier first used the series in his study of heat, an application that is physically quite different from ours.

We will restrict the input functions to be real valued functions, but there is no inherent reason for this. The Fourier series applies to complex valued functions, just as the inverse transform applies to the complex valued functions in our application. The symbols $V(f_k), V(k), V_k, V(\omega_k)$ are all used interchangeably for a representative element of a Fourier series. The parameters, k, f_k, T, and $\omega_k = 2\pi f_k$ have the following interpretation in our applications. If we expand $v(t)$ in a Fourier series over the interval of length T, then the fundamental frequency f_1 is related to T by $f_1 = 1/T$, and $kf_1 = k/T = f_k$. Also, $\omega_k = 2\pi f_k = 2\pi k/T$.

We described the Fourier series as a transform. A transform is a mathematical operation that changes a function, such as $v(t)$, into another function, in this case $V(f_k)$, by a prescribed operation. There are many useful transforms, some of which are the Laplace, z, Mellin, and Hilbert transforms. All of them may be pictured as a black box whose input is a function, and whose output is another function. The usefulness of these transforms arises because of two factors. First, the properties of a particular transform are useful in certain applications (in our case, the properties of Fourier, Laplace, and z transforms are useful in LTI system analysis). Second, the transform may be inverted. That is, there exists another black box whose input is $V(f_k)$ and whose output is $v(t)$.

The inverse Fourier series operation is

$$v(t) = \sum_{k=-\infty}^{+\infty} V(k)e^{j2\pi f_k t} \tag{12.1.2}$$

Again, this is a transform; it is illustrated in Fig. 12.2. The input is a function $V(f_k) = V_k$, and the output is a function $v(t)$.

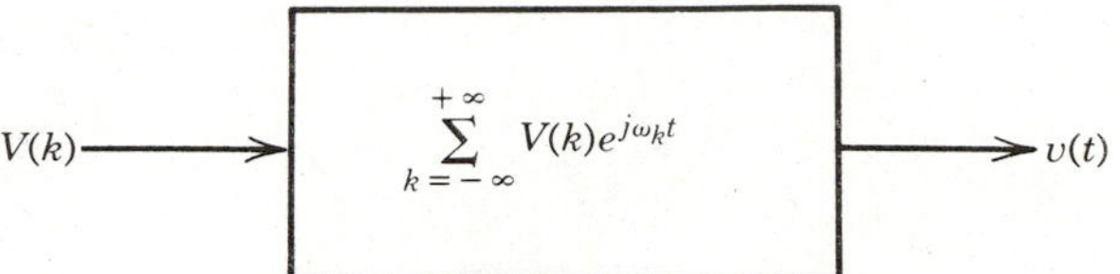

Figure 12.2. The inverse Fourier series.

Which functions have a Fourier series? Look at Eq. 12.1.1. If the magnitude of V_k is to be finite for each k, then we must have

$$\left| \int_{t'}^{t'+T} v(t) e^{-j2\pi f_k t}\, dt \right| < \infty$$

Since the magnitude of $e^{j\theta}$ is unity (for any θ), then it will suffice to have

$$\int_{t'}^{t'+T} |v(t)|\, dt < \infty \tag{12.1.3}$$

This is known as the weak Dirichlet (pronounced der-clay) condition. If, in addition to satisfying Eq. 12.1.3, the function $v(t)$ is finite and has a finite number of maxima and minima in the interval t' to $t' + T$, then the Fourier series converges uniformly.

Any function that has the following three properties is said to satisfy the strong Dirichlet conditions.

1. It satisfies Eq. 12.1.3.
2. It is finite.
3. It has a finite number of maxima and minima in an interval $t' < t < t' + T$.

We will take these three conditions as our criteria to determine whether or not a function possesses a Fourier series. We can make a few observations about these criteria:

1. Any function we can generate in the laboratory has a Fourier series.
2. The strong Dirichlet conditions are only one of several sets of conditions that we could have used. The problem of finding both necessary and sufficient conditions has not yet been solved.
3. Notice that it is not necessary for the function $v(t)$ to be periodic before it has a Fourier series, as you may have assumed if you have had previous experience with the Fourier series.
4. These conditions apply to ordinary functions, but not to δ functions.

EXAMPLE 12.1.1 The periodic square wave $v(t)$ is shown in Fig. 12.3. Find the Fourier coefficients $\{ V_k \}$.

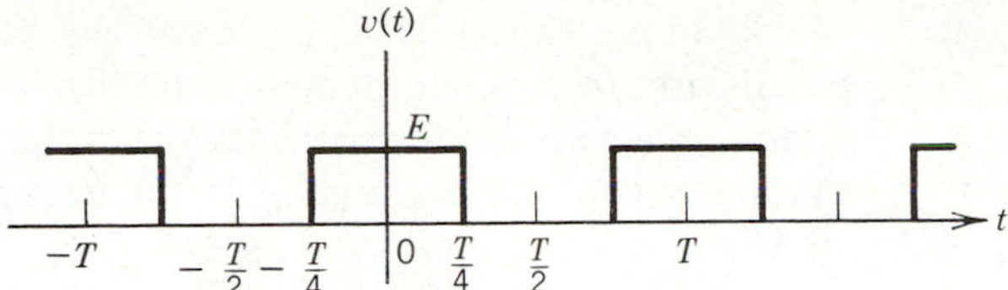

Figure 12.3. Periodic square wave.

Solution

According to Eq. 12.1.1, we have

$$V_k = \frac{1}{T}\int_{-T/2}^{T/2} v(t)e^{-j\omega_k t}\,dt = \frac{1}{T}\int_{-T/4}^{T/4} Ee^{-j\omega_k t}\,dt$$

$$= E\frac{e^{j\omega_k T/4} - e^{-j\omega_k T/4}}{j\omega_k T} \qquad (12.1.4)$$

Now $\omega_k = k\omega_1$ and ω_1 is the fundamental frequency related to T by $f_1 = 1/T$. Therefore, $\omega_k T = k\omega_1 T = k2\pi$ and

$$V_k = E\frac{e^{jk2\pi/4} - e^{-jk2\pi/4}}{j2\pi k} = \frac{E}{2}\,\frac{\sin k\pi/2}{k\pi/2}$$

Figure 12.4 shows a plot of coefficients V_k versus frequency. The dashed wave is the envelope

$$V(f) = \frac{E}{2}\,\frac{\sin \omega T/4}{\omega T/4}$$

[$V(f)$ is derived from Eq. 12.1.4 with $\omega = \omega_k$.] The values of the coefficients V_k are given by the values of $V(f)$ evaluated at $V(f) = V(f_k) = V_k$. Thus the coefficients V_k are functions of frequency. Alternatively, we think of the coefficients V_k simply as numbers, and if we know the fundamental period T, we can reconstruct the original periodic time function from these numbers.

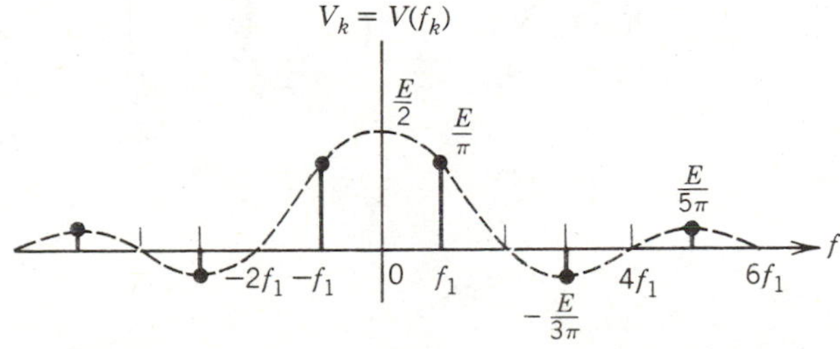

Figure 12.4. The coefficients V_k versus frequency.

Since each of the functions $e^{jk\omega_1 t}$ is periodic, the expansion given in Eq. 12.1.2 always yields a periodic function with period $T = 2\pi/\omega_1$. The expansion of nonperiodic functions, for example, yields periodic functions with period T. For this reason, the Fourier series is most often used on periodic signals.

In Example 12.1.1, each coefficient is a real (not complex) number. This is a special case. In general, the set of numbers $\{V_k\}$ are complex. Therefore a graph of the coefficients versus frequency will usually require either a three-dimensional diagram or two separate graphs. In the next example the coefficients are complex numbers.

EXAMPLE 12.1.2 The periodic square wave of Example 12.1.1 is shifted to the right by $T/4$ (Fig. 12.5). In this case the coefficients $\{G_k\}$ are given by

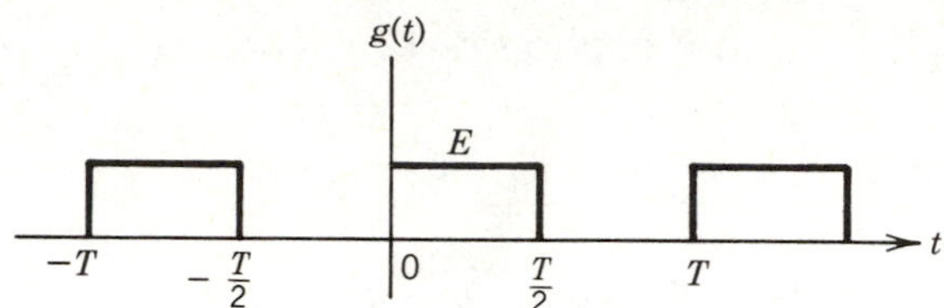

Figure 12.5. $g(t) = v(t - T/4)$.

$$G_k = \frac{1}{T}\int_0^{T/2} Ee^{-j\omega_k t}\,dt = \frac{E}{j\omega_k T}(1 - e^{-j\omega_k T/2})$$

$$= Ee^{-j\omega_k T/4}\frac{e^{j\omega_k T/4} - e^{-j\omega_k T/4}}{j\omega_k T}$$

which, from Eq. 12.1.4, is

$$G_k = V_k e^{-j\omega_k T/4} = V_k e^{-jk\pi/2}$$

A plot of the magnitude and phase of this function of frequency is shown in Fig. 12.6.

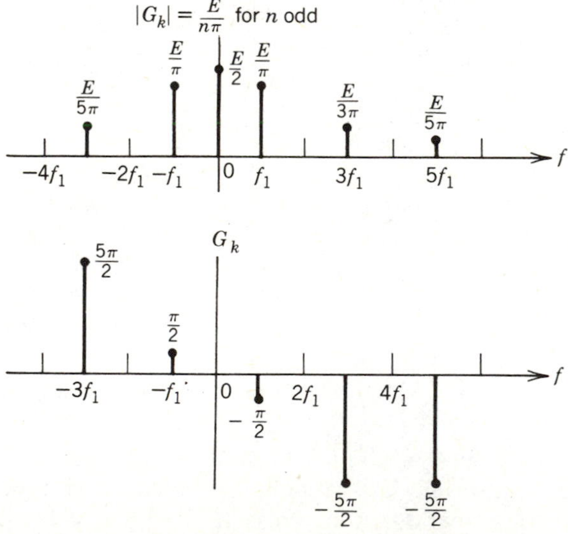

Figure 12.6. Magnitude and phase of G_k.

EXAMPLE 12.1.3 In the previous examples the voltage was on half the period and off half the period. We now examine the more general case of arbitrary pulse width (Fig. 12.7). Find the coefficients W_k.

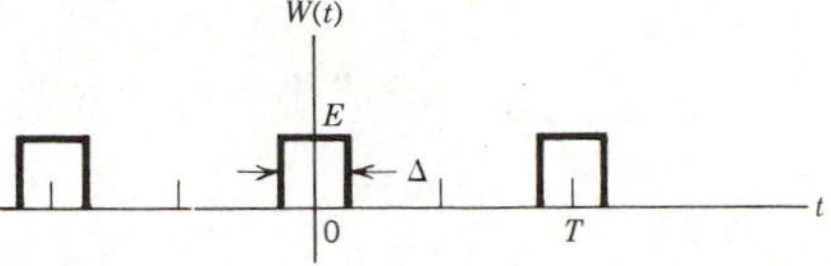

Figure 12.7. Periodic square wave with arbitrary pulse width Δ.

Solution

$$W_k = \frac{1}{T}\int_{-\Delta/2}^{\Delta/2} Ee^{-j\omega_k t}\,dt$$

$$= \frac{\Delta E}{T}\,\frac{\sin k\pi\Delta/T}{k\pi\Delta/T}$$

Figure 12.8 shows the graph of W_k. The spacing between lines is the same as before. This spacing depends on T, the period of the time function. The $(\sin x)/x$ envelope depends on Δ, the pulse width. This $(\sin x)/x$ envelope is characteristic of periodic square pulses. Different pulse shapes results in different envelope shapes.

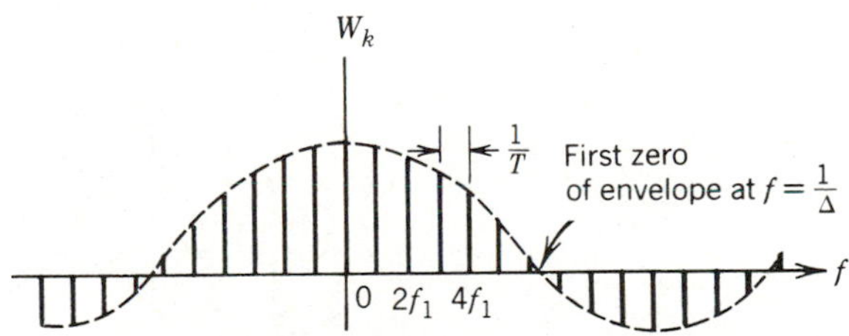

Figure 12.8. The coefficients W_k.

The coefficients are real in Example 12.1.3, just as they were in Example 12.1.1. This is because the time functions are even. An important property of Fourier series is that even time functions result in purely real coefficients, and odd time functions result in purely imaginary coefficients. Other properties are discussed later.

EXAMPLE 12.1.4 Represent the function e^t (Fig. 12.9) over the interval $(0 < t < 1)$ by the exponential Fourier series.

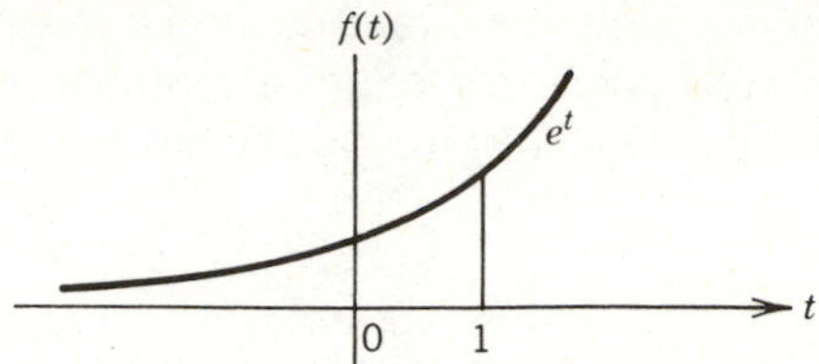

Figure 12.9. The exponential function e^t.

Solution

$$F_n = \frac{1}{T}\int_0^T f(t)e^{-jn\omega_0 t}, \qquad T = 1, \qquad \omega_0 = \frac{2\pi}{T} = 2\pi$$

$$F_n = \int_0^1 e^t e^{-jn\omega_0 t}\,dt = \int_0^1 e^{(1-jn2\pi)t}\,dt$$

$$= \frac{1}{1-jn2\pi} e^{(1-jn2\pi)t}\Big|_0^1 = \frac{e^{(1-jn2\pi)} - 1}{1 - jn2\pi}$$

The first few terms are plotted in Fig. 12.10. The function F_n represents a periodic function with period $T = 1$, although $f(t)$ is not periodic.

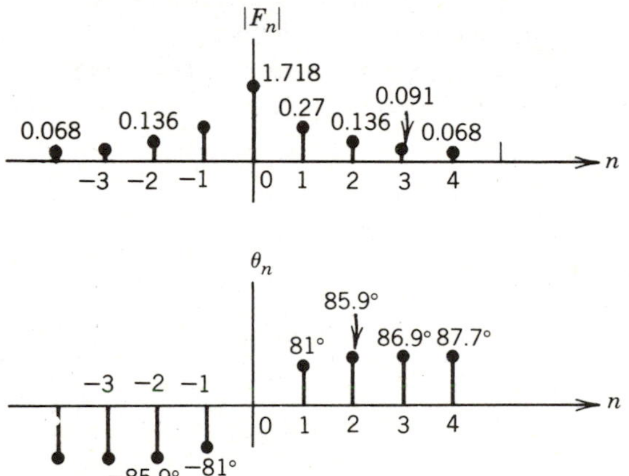

Figure 12.10. The Fourier coefficients of e^t, $0 < t < 1$.

The Trigonometric Fourier Series

Euler's formulas relate complex exponential functions to sine and cosine functions.

$$\cos \omega t = \frac{e^{j\omega t} + e^{-j\omega t}}{2} \tag{12.1.5}$$

$$\sin \omega t = \frac{e^{j\omega t} - e^{-j\omega t}}{2j} \tag{12.1.6}$$

Using these relationships, Eq. 12.1.2 may be rewritten in the form

$$v(t) = \frac{a_0}{2} + \sum_{k=1}^{\infty} (a_k \cos k\omega_1 t + b_k \sin k\omega_1 t) \tag{12.1.7}$$

where

$$a_0 = \frac{2}{T}\int_{t'}^{t'+T} v(t)\, dt \tag{12.1.8a}$$

$$a_k = \frac{2}{T}\int_{t'}^{t'+T} v(t) \cos k\omega_1 t\, dt, \qquad k = 1, 2, 3, \ldots \tag{12.1.8b}$$

$$b_k = \frac{2}{T}\int_{t'}^{t'+T} v(t) \sin k\omega_1 t\, dt, \qquad k = 1, 2, 3, \ldots \tag{12.1.8c}$$

Equation 12.1.8 is equivalent to Eq. 12.1.1, and Eq. 12.1.7 is equivalent to Eq. 12.1.2. That is, Eq. 12.1.8 represents the forward transform, and Eq. 12.1.7 is the inverse transform. Furthermore, the exponential and trigonometric forms of the Fourier series may be derived from each other by using Euler's formulas.

EXAMPLE 12.1.5 Find the trigonometric Fourier series for the periodic square pulse $v(t)$ shown in Fig. 12.3.

Solution

Using Eq. 12.1.8 with $t' = -T/2$ gives

$$a_0 = \frac{2}{T}\int_{-T/2}^{T/2} v(t)\, dt = \frac{2}{T}\int_{-T/4}^{T/4} E\, dt = E$$

$$a_k = \frac{2}{T}\int_{-T/4}^{T/4} E \cos(k2\pi t/T)\, dt = E\frac{\sin(k\pi/2)}{k\pi/2}$$

$$b_k = \frac{2}{T}\int_{-T/4}^{T/4} E \sin(k2\pi t/T)\, dt = 0$$

Therefore $v(t)$ may be expressed as

$$v(t) = \frac{E}{2} + \frac{2E}{\pi}\cos\left(\frac{2\pi t}{T}\right) - \frac{2E}{3\pi}\cos\left(\frac{6\pi t}{T}\right) + \frac{2E}{5\pi}\cos\left(\frac{10\pi t}{T}\right) - \cdots \tag{12.1.9}$$

This solution must be the same as that found in Example 12.1.1. To demonstrate the equality of the two solutions, write

$$v(t) = \cdots + V_{-2}e^{-j4\pi t/T} + V_{-1}e^{-j2\pi t/T} + V_0 + V_1 e^{j2\pi t/T} + V_2 e^{j4\pi t/T} + \cdots$$

Combine terms using Euler's formula. For example,

$$V_{-2}e^{-j4\pi t/T} + V_2 e^{j4\pi t/T} = a_2 \cos\left(\frac{4\pi t}{T}\right)$$

since $a_k = 2V_k = 2V_{-k}$ in this particular case. When this process is continued, every pair of terms in the exponential series whose coefficients are V_k and V_{-k} combine to give a term equal to $a_k \cos(k2\pi t/T)$ in the trigonometric series. Therefore the two solutions are equal.

Both the exponential and trigonometric series have their reason for being. The exponential series is easier to use because there is only one type of term, exponential, while the trigonometric series consists of sine and cosine terms. The trigonometric series has more physical meaning, or at least it is easier to visualize since it contains no complex numbers. To demonstrate, the first few terms of Eq. 12.1.9 are plotted in Fig. 12.11 to show that as more terms are used, the sum approaches $v(t)$.

Inverting the Fourier Series

Here we discuss the process of operating on a sequence V_k by Eq. 12.1.2 to obtain the time function $v(t)$. Which sequences V_k may we substitute into Eq. 12.1.2 to obtain the corresponding $v(t)$? Again, as in our discussion of the forward transform, we require that the sum be finite.

$$|v(t)| = \left| \sum_{n=-\infty}^{+\infty} V_n e^{jn\omega_1 t} \right| < \infty \tag{12.1.10}$$

In practice, any sequence V_k that we obtain from the forward transform by operating on $v(t)$ with Eq. 12.1.1 will satisfy this condition. We will delay further discussion of conditions under which such a series converges until our treatment of the z transform.

What happens if we perform the operations indicated in Fig. 12.12? Here we select an appropriate function $v(t)$ that satisfies the Dirichlet conditions and operate on it by the Fourier series integral to obtain V_k. This sequence is then supplied to the inverse Fourier series to obtain $v'(t)$ [which may be different from $v(t)$].

Although $v'(t)$ and $v(t)$ are equal in the interval of expansion, $t' < t < t' + T$, they are not necessarily equal outside this interval, as our previous discussion should indicate. In fact, the function $v'(t)$ is periodic with period T, regardless of the nature of $v(t)$ outside the interval of expansion. The validity of this statement is demonstrated by the following argument.

First, the V_k terms appear in conjugate pairs, $V_k = V_{-k}^*$. This must be true if the inverse Fourier series is to produce a real function of time, for in the sum

$$v(t) = \sum_{k=-\infty}^{+\infty} V_k e^{jk\omega_1 t} \tag{12.1.11}$$

we have one term given by $V_k e^{jk\omega_1 t}$, and since this is a complex number the conjugate $V_k^* e^{-jk\omega_1 t}$ must also appear so that their sum is real.

Second, the sum of two such terms is a sinusoid of period T/k, as can be shown by applying Euler's formulas relating the sinusoidal and exponential forms. Therefore the sum of all these terms given by Eq. 12.1.11 is periodic with a common period T. This explains the reason that the Fourier series is most often

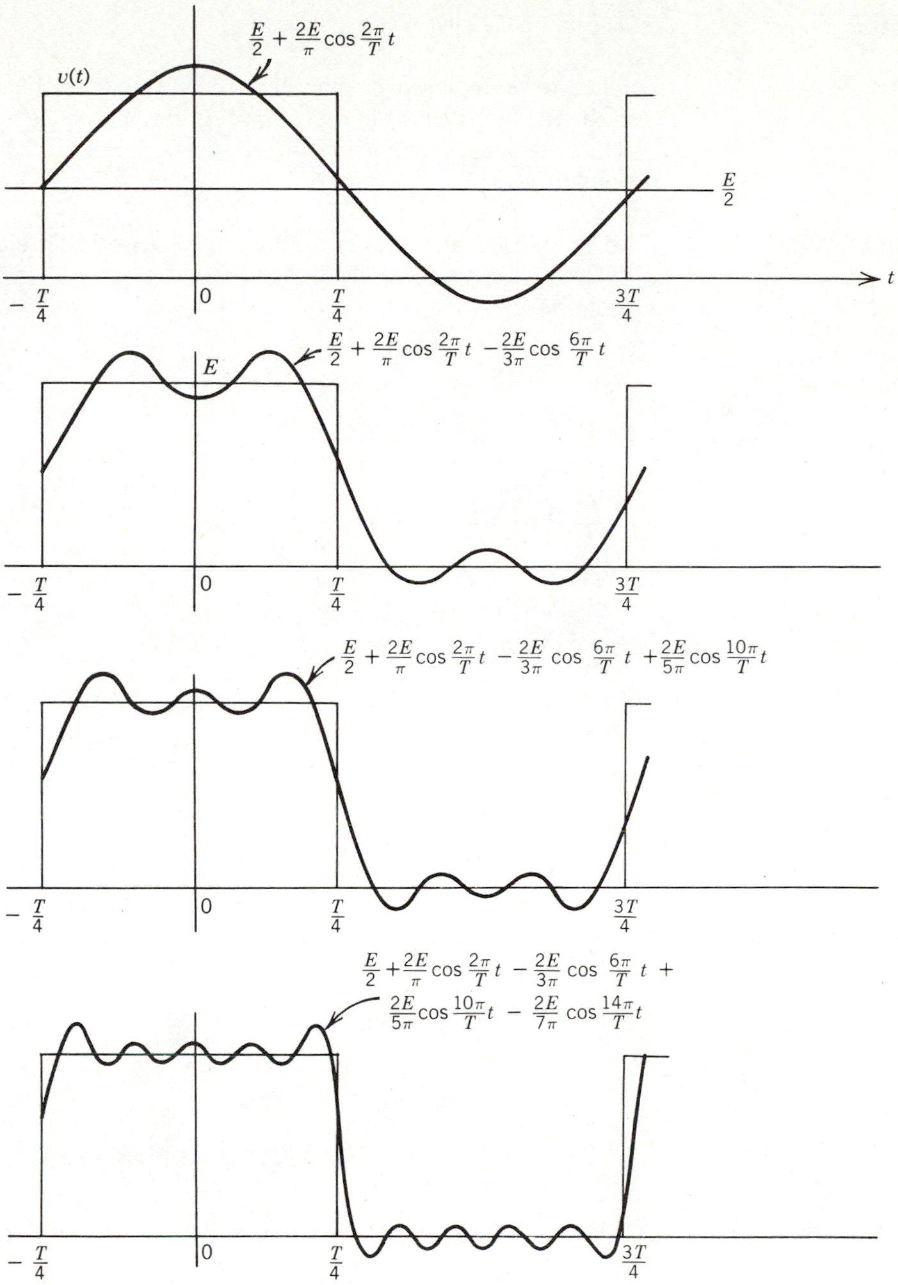

$v(t)$

$\frac{E}{2} + \frac{2E}{\pi} \cos \frac{2\pi}{T} t$

$\frac{E}{2}$

t

$-\frac{T}{4}$ 0 $\frac{T}{4}$ $\frac{3T}{4}$

$\frac{E}{2} + \frac{2E}{\pi} \cos \frac{2\pi}{T} t - \frac{2E}{3\pi} \cos \frac{6\pi}{T} t$

E

$-\frac{T}{4}$ 0 $\frac{T}{4}$ $\frac{3T}{4}$

$\frac{E}{2} + \frac{2E}{\pi} \cos \frac{2\pi}{T} t - \frac{2E}{3\pi} \cos \frac{6\pi}{T} t + \frac{2E}{5\pi} \cos \frac{10\pi}{T} t$

$-\frac{T}{4}$ 0 $\frac{T}{4}$ $\frac{3T}{4}$

$\frac{E}{2} + \frac{2E}{\pi} \cos \frac{2\pi}{T} t - \frac{2E}{3\pi} \cos \frac{6\pi}{T} t +$
$\frac{2E}{5\pi} \cos \frac{10\pi}{T} t - \frac{2E}{7\pi} \cos \frac{14\pi}{T} t$

$-\frac{T}{4}$ 0 $\frac{T}{4}$ $\frac{3T}{4}$

Figure 12.11. Sum of a trigonometric series approaching a square wave.

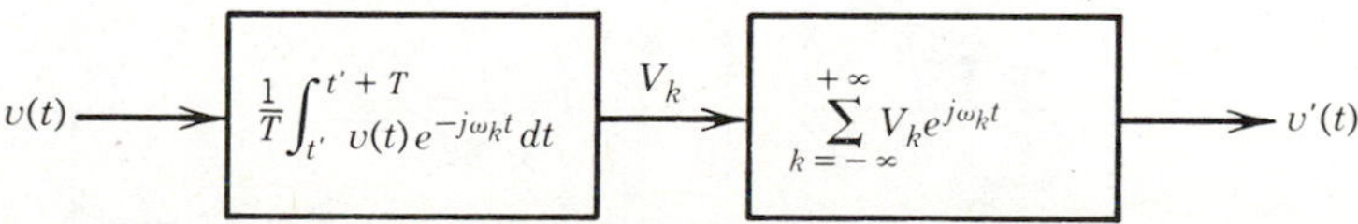

Figure 12.12. The Fourier series followed by the inverse Fourier series.

applied to periodic waveforms. If $v(t)$ is periodic in Fig. 12.12, then $v'(t) = v(t)$ for all t. If not, then $v'(t) = v(t)$ only in the interval $t' < t < t' + T$.

Here are some examples of the application of Eq. 12.1.11 to various sequences to obtain $v(t)$.

EXAMPLE 12.1.6 The amplitude and phase spectrum of a periodic signal $v(t)$ are shown in Fig. 12.13. Write an analytical expression for $v(t)$ in terms of trigonometric functions.

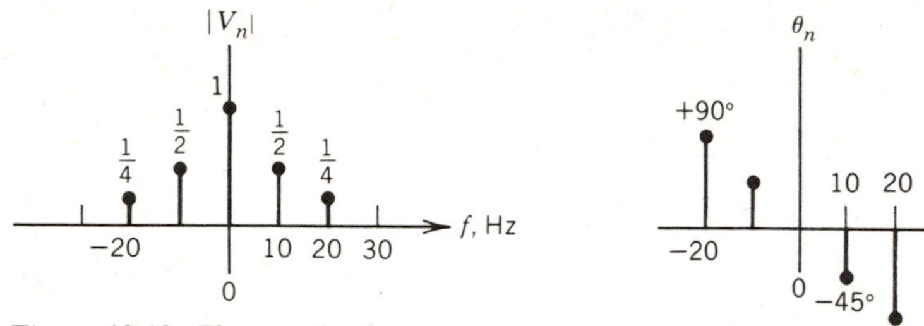

Figure 12.13. The amplitude and phase of a periodic signal.

Solution

The function $v(t)$ is given by

$$v(t) = \sum_{n=-\infty}^{\infty} V_n e^{jn\omega_1 t}, \qquad t' < t < t' + T$$

where $\omega_1 = 2\pi f_1 = 2\pi(10)$. From Fig. 12.13 the V_n coefficients are given by

$$V_{-2} = \tfrac{1}{4}e^{j90^\circ} \qquad V_1 = \tfrac{1}{2}e^{-j45^\circ}$$
$$V_{-1} = \tfrac{1}{2}e^{j45^\circ} \qquad V_2 = \tfrac{1}{4}e^{-j90^\circ}$$
$$V_0 = 1 \qquad V_n = 0, \quad \text{all other } n$$

Therefore $v(t)$ is given by

$$V(t) = \tfrac{1}{4}e^{j90^\circ}e^{-j2\pi(20)t} + \tfrac{1}{2}e^{j45^\circ}e^{-j2\pi(10)t} + 1$$
$$+ \tfrac{1}{2}e^{-j45^\circ}e^{j2\pi(10)t} + \tfrac{1}{4}e^{-j90^\circ}e^{j2\pi(20)t}$$
$$= 1 + \cos[2\pi(10)t - 45^\circ] + \tfrac{1}{2}\cos[2\pi(20)t - 90^\circ]$$

EXAMPLE 12.1.7 Repeat Example 12.1.6 for the spectrum shown in Fig. 12.14.

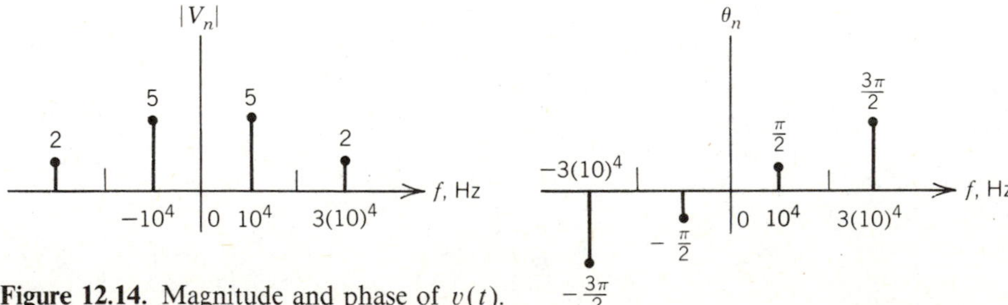

Figure 12.14. Magnitude and phase of $v(t)$.

Solution

$$v(t) = 10\cos\left[2\pi(10)^4 t + \frac{\pi}{2}\right] + 4\cos\left[6\pi(10)^4 t + \frac{3\pi}{2}\right]$$

EXAMPLE 12.1.8 The amplitude and phase spectrum of the periodic signal $f(t)$ is shown in Fig. 12.15. Write an analytical expression for $f(t)$ in terms of trigonometric functions.

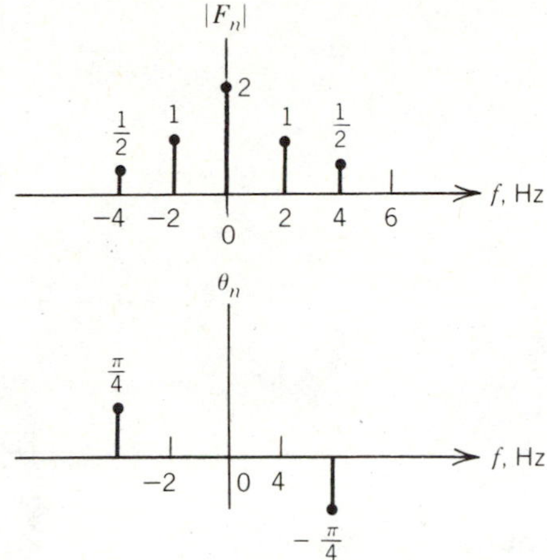

Figure 12.15. Magnitude and phase of $f(t)$.

Solution

$$f(t) = 2 + 2\cos 4\pi t + \cos\left(8\pi t - \frac{\pi}{4}\right)$$

LEARNING EVALUATIONS

1. Write down the forward and inverse Fourier series.
2. List the Dirichlet conditions.
3. Find and plot the exponential Fourier series representation for the time function shown in Fig. 12.16.

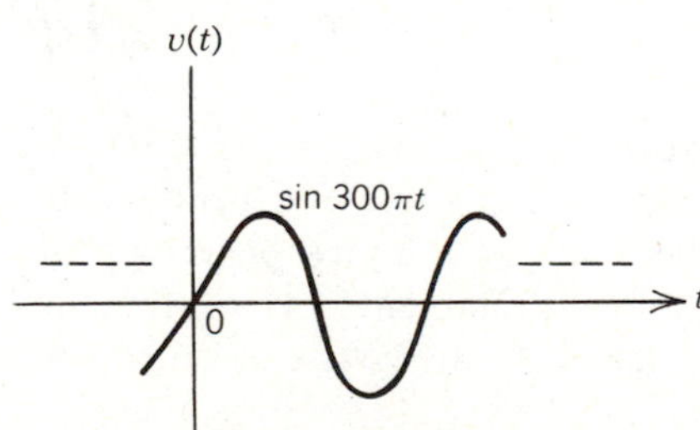

Figure 12.16

4. The amplitude and phase spectrum of the periodic signal $v(t)$ is shown in Fig. 12.17. Write an analytical expression for $v(t)$ in terms of trigonometric functions.

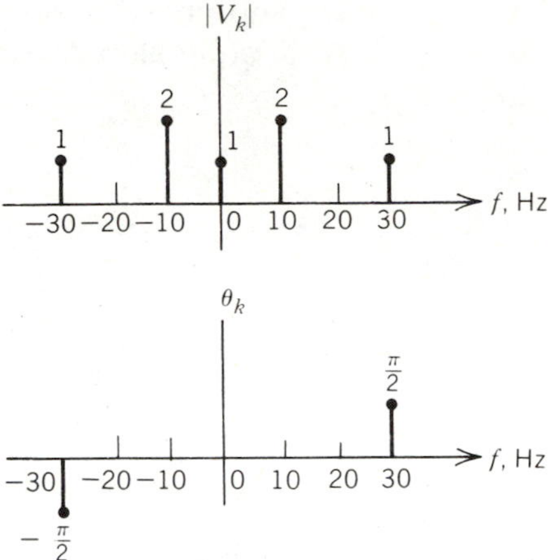

Figure 12.17

12.2 The Discrete Fourier Transform (DFT)

LEARNING OBJECTIVES

After completing this section you should be able to do the following:

1. Evaluate the DFT for a given digital signal.
2. Find the inverse DFT.
3. Use a digital computer to evaluate the DFT.

In Section 12.1 we said that the Fourier series given by Eq. 12.1.1 applied equally to continuous-time and discrete-time functions. We now demonstrate the validity of this statement, and then proceed to investigate some of the uses of the discrete Fourier transform. When Eq. 12.1.1 is applied to a discrete-time function, the resulting transform is called the discrete Fourier transform and abbreviated DFT. Let us begin with Eq. 12.1.1, repeated here for convenience.

$$V_k = \frac{1}{T}\int_{t'}^{t'+T} v(t)e^{-j2\pi/T}\,dt$$

Now suppose $v(t)$ is the discrete-time sequence of samples spaced 1 s apart beginning at $t' = i$ and lasting a total of $T = N$ s as shown in Fig. 12.18. Thus there are N samples beginning at $t = i$ and lasting until $t = i + N - 1$. Each sample is represented as a δ function so that we may apply Eq. 12.1.1.

We claim the V_k are given by the sum

$$V_k = \frac{1}{N}\sum_{n=i}^{i+N-1} v(n)e^{-j2\pi nk/N} \tag{12.2.1}$$

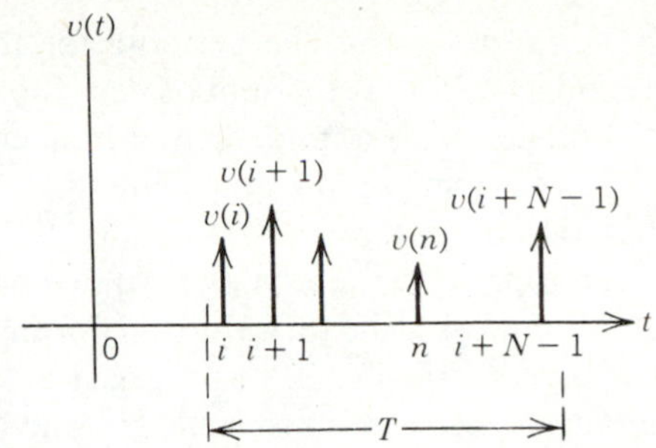

Figure 12.18. A sequence of N samples spaced 1 s apart.

where $v(n)$ is the strength of each δ function. Equation 12.2.1 is called the discrete Fourier transform. To show that Eq. 12.2.1 is equivalent to Eq. 12.1.1 for the discrete-time signal of Fig. 12.18, write $v(t)$ as

$$v(t) = v(i)\delta(t-i) + v(i+1)\delta(t-i-1) + \cdots + v(i+N-1)\delta(t-i-N+1) \tag{12.2.2}$$

Now if Eq. 12.1.1 is applied to a typical term, say $v(n)$, the result is

$$\frac{1}{T}\int_{t'}^{t'+T} v(n)\delta(t-n)e^{-j(2\pi/T)kt}\,dt = \frac{1}{T}v(n)e^{-j(2\pi/T)kn}$$

and since $T = N$ this gives a typical term in Eq. 12.2.1. Integration is a linear operation, so the integral of the sum in Eq. 12.2.2 gives the sum of the integrals to yield Eq. 12.2.1.

The discrete Fourier transform is the sum of a finite number of terms. Therefore, we need pay little attention to the question "Which functions have a DFT?" Every sequence has a DFT if each term $v(n)$ is finite. In practice, this means that every sequence we encounter has a DFT.

The inversion from V_k to $v(n)$ is given by a finite sum of N terms instead of by the infinite sum that was used for continuous-time signals.

$$v(n) = \sum_{k=1}^{N} V_k e^{j2\pi nk/N} \tag{12.2.3}$$

This sum has only N terms because V_k is periodic. That is, for the DFT both $v(n)$ and V_k are periodic. To see this we will use the same argument that was used before in Section 12.1. Look at Fig. 12.19. The result of applying the DFT to $v(n)$

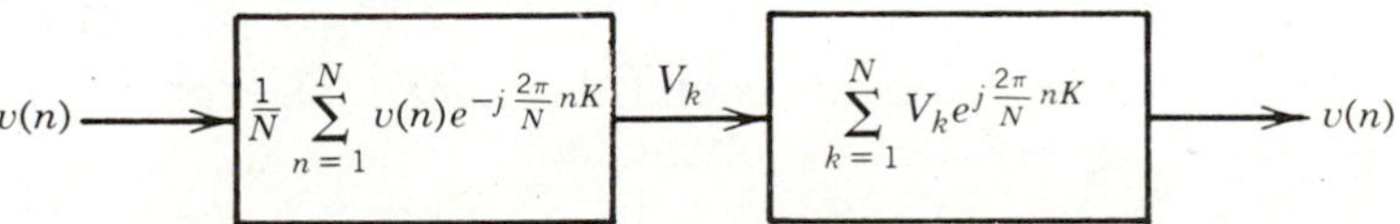

Figure 12.19. The forward and inverse DFT.

is V_k, and V_k must be periodic for the same reason that $v'(t)$ was periodic in Section 12.1. V_k is the sum of periodic exponential functions where each exponential function has a fundamental frequency that is an integral multiple of $1/N$.

Notice that we do not claim that V_k is the sum of periodic sinusoids (real functions), but only that V_k is the sum of periodic exponential functions. Therefore, V_k is ia complex valued sequence. We have chosen to associate the $1/N$ term with the forward transform. Many authors associate the $1/N$ term with the inverse transform. From Fig. 12.19 we see that in going from $v(n)$ to V_k and finally to $v(n)$ we must multiply one of the sums by $1/N$, but it really does not matter which way we do it. Our choice was motivated by the continuous Fourier series with $N = T$ in Eq. 12.1.1. The index of summation may begin at any value. That is, the sum in Eq. 12.2.3 may begin at $k = 0$ and continue to $k = N - 1$. It makes no difference since V_k is periodic.

EXAMPLE 12.2.1 Find the DFT for the periodic signal $v(n)$ shown in Fig. 12.20.

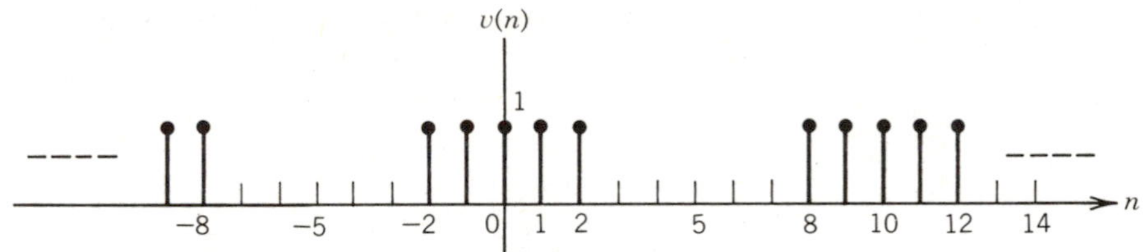

Figure 12.20. The periodic discrete-time function $v(n)$.

Solution

The period is $N = 10$. The summation can begin anywhere and end N steps later. For hand calculation it is easier to begin at $n = -4$ and stop at $n = 5$, giving

$$V(k) = \frac{1}{10} \sum_{n=-2}^{2} (1) e^{-j(2\pi/N)nk}$$

$$= \frac{1}{10} e^{j(2\pi/10)(-2k)} + \frac{1}{10} e^{j(2\pi/10)(-k)} + \frac{1}{10}$$

$$+ \frac{1}{10} e^{j(2\pi/10)k} + \frac{1}{10} e^{j(2\pi/10)(2k)}$$

$$= \frac{1}{5} \cos \frac{4\pi}{10} k + \frac{1}{5} \cos \frac{2\pi}{10} k + \frac{1}{10} \qquad (12.2.4)$$

If this expression is evaluated for each value of k, the function plotted in Fig. 12.21 is obtained.

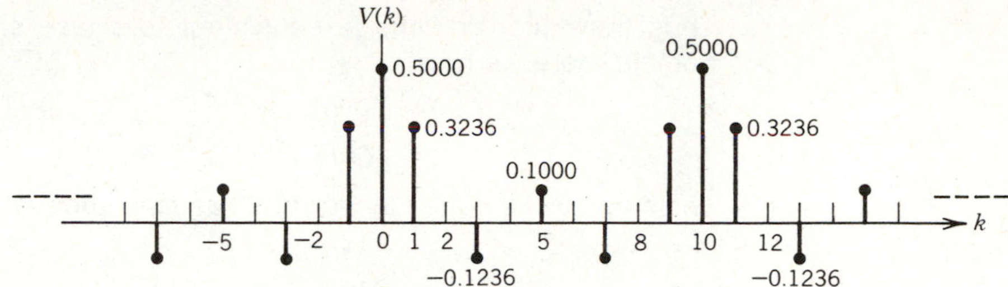

Figure 12.21. The DFT for $v(n)$.

EXAMPLE 12.2.2 Let $g(n) = v(n-2)$ as shown in Fig. 12.22. Find the DFT of $g(n)$.

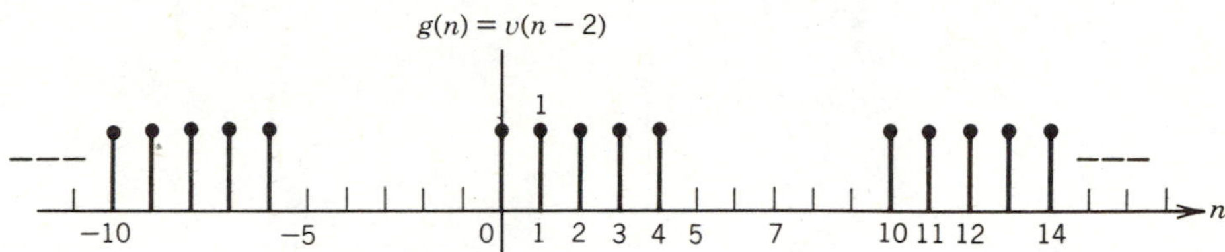

Figure 12.22. The function $g(n) = v(n-2)$.

Solution

$$G(k) = \frac{1}{10}\sum_{n=0}^{4}(1)e^{-j(2\pi/10)nk}$$

$$= \frac{1}{10} + \frac{1}{10}e^{-j(2\pi/10)k} + \frac{1}{10}e^{-j(2\pi/10)(2k)} + \frac{1}{10}e^{-j(2\pi/10)(3k)} + \frac{1}{10}e^{-j(2\pi/10)(4k)}$$

We can multiply $G(k)$ by $e^{j(2\pi/10)2(k)}e^{-j(2\pi/10)(2k)} = 1$ to obtain

$$G(k) = e^{-j(4\pi k/10)}\left[\frac{1}{10}e^{j(2\pi/10)(2k)} + \frac{1}{10}e^{j(2\pi/10)k} + \frac{1}{10} + \frac{1}{10}e^{-j(2\pi/10)k} + \frac{1}{10}e^{-j(2\pi/10)(2k)}\right]$$

$$= e^{-j(4\pi/10)k}\left[\frac{1}{5}\cos\frac{4\pi}{10}k + \frac{1}{5}\cos\frac{2\pi}{10}k + \frac{1}{10}\right]$$

$$= e^{-j(4\pi/10)k}V(k)$$

Thus translation in time is equivalent to phase shift in the frequency domain. Here we have

$$g(n) = v(n-2)$$

$$G(k) = V(k)e^{-j(2\pi k/10)\cdot 2}$$

This function is plotted in Fig. 12.23 as amplitude and phase.

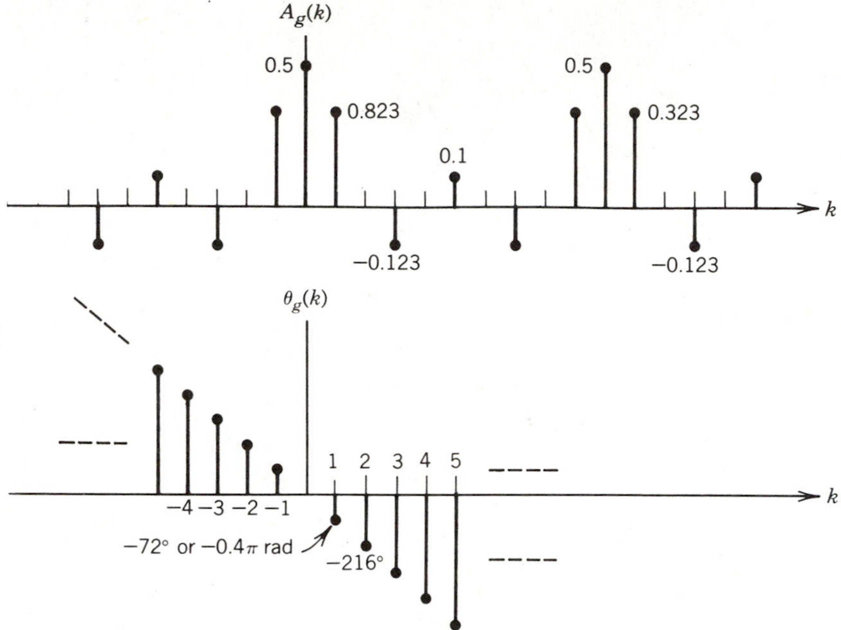

Figure 12.23. The amplitude and phase of $G(k)$.

There is a subtle difference between magnitude and amplitude as we are using these terms. The magnitude of a complex number, say $|G(k)|$, is always positive and real. The amplitude, as plotted in Fig. 12.23, can be either positive or negative and real. This allows our graphs to demonstrate the relationship between $G(k)$ and the function $V(k)$ of Example 12.2.1. If we made each term in our amplitude graph positive, then the phase corresponding to each negative term would be shifted by 180°.

We use the symbol A_g for the amplitude of G, and we use θ_g for the phase. Here $A_g(k) = V(k)$ and the phase is the exponential part of $G(k)$ without the j, that is, $\theta_g(k) = -2(2\pi k/10)$ rad.

Now let us see which features of $V(k)$ remain unchanged for $G(k)$. First, $G(k)$ is periodic with a period of 10, as was $V(k)$. Second, each coefficient is no longer real. Third, the coefficients are no longer symmetrical about $k=5$ and all multiples of 5, but only because of the phase $\theta_g(k)$. If we know the phase, and if we know $A_g(k)$ for $k = 0$, 1, 2, 3, 4, and 5, then we can construct $G(k)$ for all k.

There is no inherent reason to restrict the values of n and k to be integers in the DFT. In fact, it is sometimes useful to use noninteger values for n or k, or perhaps both. Here is an example where the values of n are not integers.

EXAMPLE 12.2.3 The function $v(n)$ in Fig. 12.24 is periodic with period $N = 8$. Values of $v(n)$ occur at integer multiples of $n = 0.5$. Find the DFT.

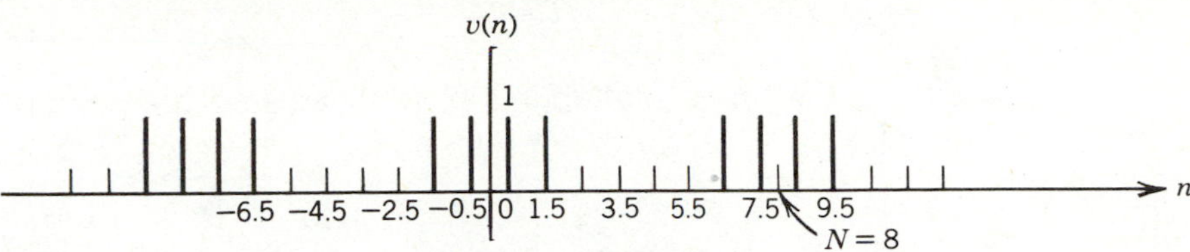

Figure 12.24. The discrete-time function $v(n)$ with values that occur at 0.5, 1.5, 2.5, etc.

Solution

Use Eq. 12.2.1 and sum over any interval of length $N = 8$. If the summation index begins at, say, $n = -4.5$, then

$$V_k = \frac{1}{8} \sum_{n=-4.5}^{2.5} v(n) e^{-j(2\pi/8)(nk)}$$

$$= \frac{1}{8} [e^{-j(2\pi/8)(-1.5k)} + e^{-j(2\pi/8)(-0.5k)} + e^{-j(2\pi/8)(0.5k)} + e^{-j(2\pi/8)(1.5k)}]$$

This expression yields the function shown in Fig. 12.25 for integer values of k. Since $v(n)$ is an even function of n, then V_k is real. Likewise, since $v(n)$ is real, then V_k is an even function of k.

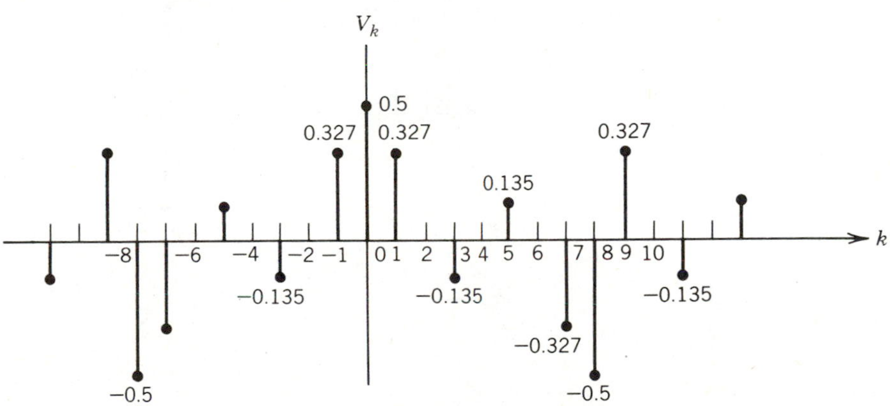

Figure 12.25. The DFT of $v(n)$.

EXAMPLE 12.2.4 Find the DFT of the function $g(n)$ shown in Fig. 12.26.

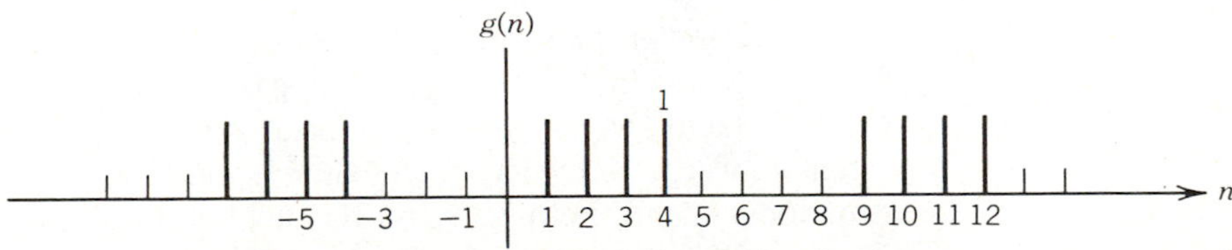

Figure 12.26. The function $g(n) = v(n - 2.5)$.

Solution

Here $g(n) = v(n - 2.5)$, so

$$G(k) = V(k)e^{-j(2k/8)(2.5)}$$

which is plotted in Fig. 12.27 as amplitude and phase.

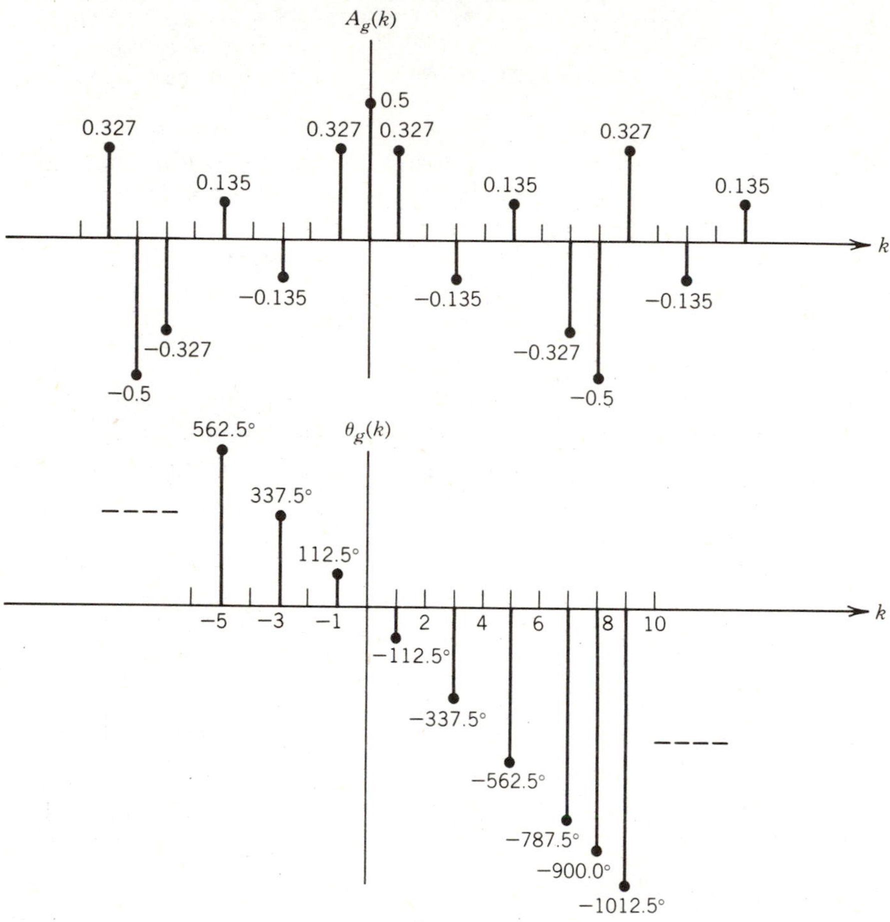

Figure 12.27. Amplitude and phase of $G(k)$.

The Inverse DFT

A comparison of the forward DFT and the inverse DFT given by Eqs. 12.2.1 and 12.2.3 indicates that the only difference is in the sign on the exponent (aside from the $1/N$ factor, which could be associated with either sum). Both sums are finite, ranging over a sequence of N values, and both can handle complex valued functions. That is, both $v(n)$ and V_k may be complex valued functions. Therefore, evaluation of the inverse DFT is identical in principle to the evaluation of the forward DFT. Here is an example.

EXAMPLE 12.2.5 Find the time function $v(n)$ whose transform $V(k)$ is shown in Fig. 12.28. The period is $N = 4$.

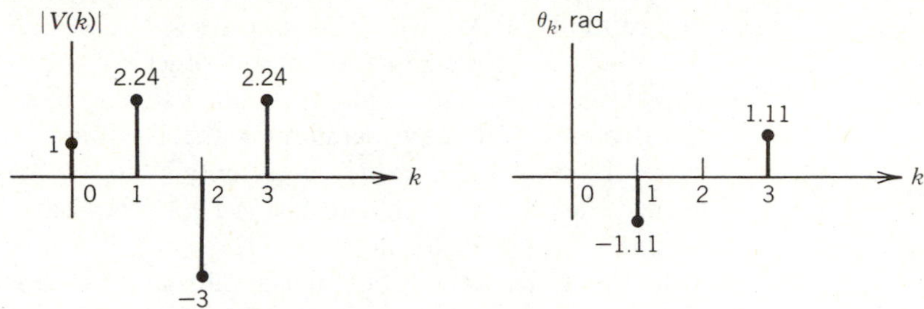

Figure 12.28. $V(k)$ for Example 12.2.5.

Solution

The coefficients $V(k)$ are given in polar coordinates. Thus $V(0) = 1$, $V(1) = 2.24e^{-j1.11}$, $V(2) = -3$, $V(4) = 2.24e^{j1.11}$, where the angles are in radians. Using Eq. 12.2.3 gives

$$v(0) = 1 + 2.24e^{-j1.11} - 3 + 2.24e^{j1.11} = 0$$

$$v(1) = 1 + 2.24e^{-j1.11}e^{j\pi/2} - 3e^{j\pi} + 2.24e^{j1.11}e^{j3\pi/2} = 8$$

Similarly, $v(2) = -4$ and $v(3) = 0$. This time function is shown in Fig. 12.29.

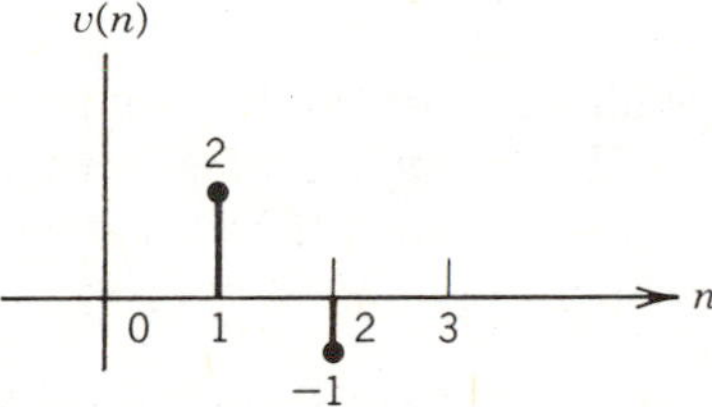

Figure 12.29. $v(n)$ for the $V(k)$ of Fig. 12.28.

The coefficients given in Fig. 12.28 could have been specified in Cartesian coordinates. If they had, their values would be $v(0) = 1$, $V(1) = (1 - j2)$, $V(2) = -3$, and $V(3) = (1 + j2)$. Also the values of $v(n)$ are all real. Of course, any actual discrete-time signal must consist of real values, but the DFT (and the Fourier series) apply equally to complex valued time functions.

Computer Calculation of the DFT

Since the two operations in Fig. 12.19 are identical except for the sign on the exponential and the $1/N$ multiplying factor, one subroutine can be used for computer calculation of either the forward or inverse DFT. The subroutine should be supplied with the number N, the complex (two-dimensional) array V, and the sign on the exponent. The output of the subroutine should be a complex array consisting of the transformed sequence. (In practice, the sign on the exponent need not be supplied, because with some additional processing the same subroutine may be used to calculate both the forward and inverse transform. This will be explained shortly.)

Currently available subroutines exist in most languages for computer calculation of the DFT. We will frame our discussion in FORTRAN, since this language is familiar to most engineers, but any other high-level language may be used. The number of operations required for the calculation appears to be proportional to N^2 if one sets about programming the DFT in a straightforward manner, but currently available routines used only $N \log N$ operations, and are therefore termed fast Fourier transforms (FFT). This saving in machine calculations becomes significant for large N.

To illustrate, let us construct a subroutine in a straightforward manner (N^2 operations) using the definition of the DFT. As a first step, we must be able to calculate the sum

$$\sum_{i=1}^{N} v(i) e^{\pm j(2\pi/N)ik}$$

for a given k. This is accomplished by the following DO loop.

```
  SUM = (0.0, 0.0)
  DO 1 I = 1, N
  E = SIGN * 6.283185307 * FLOAT(I) * FLOAT(K)/FLOAT(N)
  D = CMPLX(0.0, E)
1 SUM = SUM + V(I) * CEXP(D)
```

Now all we must do is compute this sum for each value of k to obtain V_k from $v(n)$. Here is the complete program. The array Y(N) is used for temporary storage so that this subroutine destroys the original array V(N) and replaces it by the transform.

```
  SUBROUTINE DFT (V, N, SIGN)
  COMPLEX V(N), Y(N), SUM, D
  DO 2 K = 1, N
  SUM = (0.0, 0.0)
  DO 1 I = 1, N
  E = SIGN * 6.283185307 * FLOAT(I) * FLOAT(K)/FLOAT(N)
  D = CMPLX (0.0, E)
1 SUM = SUM + V(I) * CEXP(D)
2 Y(K) = SUM
  DO 3 I = 1, N
3 V(I) = Y(I)
  RETURN
  END
```

Notice that there are two nested DO loops, each with a range of N, so the number of required calculations is N^2. Several improvements could be made to this routine, but the nested DO loops seem to be unavoidable. But they only *seem* to be unavoidable. The FFT avoids the necessity for N^2 operations by taking advantage of some properties of exponential functions to reduce the number of required calculations. Here is a simple FFT algorithm as published by Cooley, Lewis, and Welch in the *IEEE Transaction on Education*, March, 1969.

```
      SUBROUTINE FFT(A, M)
      COMPLEX A(1024), U, W, T
      N = 2**M
      NV2 = N/2
      NM1 = N - 1
      J = 1
      DO 7 I = 1, NM1
      IF(I.GE.J)GO TO 5
      T = A(J)
      A(J) = A(I)
      A(I) = T
5     K = NV2
6     IF(K.GE.J)GO TO 7
      J = J - K
      K = K/2
      GO TO 6
7     J = J + K
      PI = 3.14159265358979
      DO 20 L = 1, M
      LE = 2**L
      LE1 = LE/2
      U = (1.0, 0.)
      C = COS(PI/FLOAT(LE1))
      S = SIN(PI/FLOAT(LE1))
      W = CMPLX(C, S)
      DO 20 J = 1, LE1
      DO 10 I = J, N, LE
      IP = I + LE1
      T = A(IP)*U
      A(IP) = A(I) - T
10    A(I) = A(I) + T
20    U = U*U
      RETURN
      END
```

The number M in the FFT routine is related to the number N in the above DFT routine by $N = 2^M$. Hence, this FFT routine may only be used for $N = 2, 4, 8, 16, 32, \ldots,$ or higher powers of 2. The DFT routine listed above may be used for any number of points. The FFT routine is NOT an approximation. If the same value of π is used in each routine, they will give exactly the same results. The FFT algorithm will calculate the inverse DFT as listed. There is no SIGN parameter, but the forward DFT can be computed if some additional data processing is done. Except for the $1/N$ factor, the forward DFT is defined as

$$V_k = \sum_{n=1}^{N} v(n) e^{-j(2\pi/N)nk}$$

Take the complex conjugate to obtain

$$V_k^* = \sum_{n=1}^{N} v^*(n) e^{j(2\pi/N)nk}$$

which is recognized as the inverse DFT of $v^*(n)$. Now if the conjugate is taken once more, the result is

$$V_k = \left[\sum_{n=1}^{N} v^*(n) e^{j(2\pi/N)nk} \right]^*$$

which is the desired output sequence for the forward transform using the FFT algorithm as listed. To find the forward transform, first conjugate the input data $v(n)$ (which is unnecessary if the discrete-time data is real), and then conjugate the output.

LEARNING EVALUATIONS

1. Find and plot the DFT $V(k)$ corresponding to the digital signal $v(n)$ shown in Fig. 12.30. The period is $N = 4$.

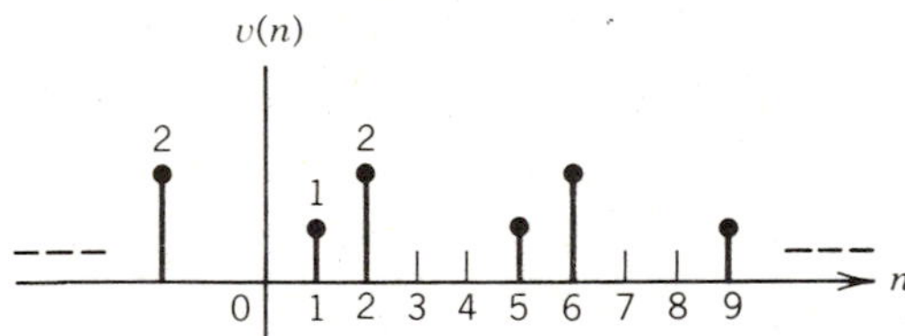

Figure 12.30

2. Find and plot the time function $v(n)$ corresponding to $V(k)$ shown in Fig. 12.31. The period is $N = 4$, and the coefficients are given in rectangular coordinates.

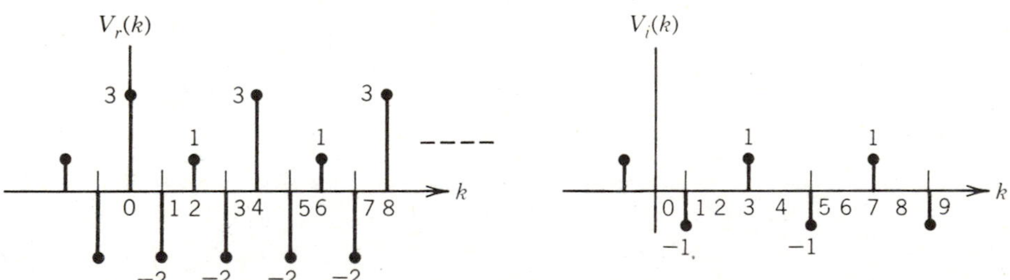

Figure 12.31

3. Write and run a program to evaluate the DFT of the signal shown in Fig. 12.30. You may use either program listed in this section, or write your own DFT subroutine in a language other than FORTRAN.

12.3 Response of LTI Systems by Fourier Series and by DFT

LEARNING OBJECTIVES

After completing this section you should be able to do the following:

1. Find the response of continuous-time systems to periodic forcing functions by the use of Fourier series.
2. Find the response of discrete-time systems to periodic forcing functions by the use of DFT.

In Chapter 8 we found the steady-state response of an LTI system to an eternal exponential signal. We used e^{st}, $-\infty < t < \infty$, for continuous-time systems and z^k, $-\infty < k < \infty$, for discrete-time systems. It is a simple matter to extend this concept to the sum of exponential signals by applying the LTI properties. Then in Sections 12.1 and 12.2 we learned to express a given signal as the sum of exponential signals by the Fourier series. We now combine these two concepts to find the response of LTI systems.

In general, s and z are complex numbers, but the Fourier series uses a special form of the exponential signal. For continuous-time systems the values of frequency lie on the imaginary axis in the complex s plane. For discrete-time systems the values of frequency lie on the unit circle in the complex z plane. The Laplace and z transforms use the entire plane, so the Fourier series is a special case of these more general transforms. It will be worthwhile, however, for us to concentrate first on these special cases.

Continuous-Time System Response

Recall that the steady-state response to an exponential signal $e^{j\omega t}$ is $H(j\omega)e^{j\omega t}$. This is true for any frequency $\omega = \omega_1$. By the LTI properties, if the input signal is the sum of two exponential signals, say $e^{j\omega_1 t} + e^{j\omega_2 t}$, then the response is $H(j\omega_1)e^{j\omega_1 t} + H(j\omega_2)e^{j\omega_2 t}$. Hence, the response to a signal expressed by its Fourier series, say,

$$v(t) = \sum_{n=-\infty}^{+\infty} V_n e^{jn\omega_1 t} \tag{12.3.1}$$

is given by

$$y(t) = \sum_{n=-\infty}^{+\infty} H(jn\omega_1) V_n e^{jn\omega_1 t} \tag{12.3.2}$$

Therefore, the procedure for calculating the steady-state response of continuous-time LTI systems is as follows:

1. Calculate the Fourier coefficients for the input signal $v(t)$. That is, calculate V_n for each frequency $n\omega_1$.
2. Calculate the transfer function $H(j\omega)$.
3. Multiply $H(jn\omega_1)$ by V_n for each n.
4. Use Eq. 12.3.2 to compute the response $y(t)$.

The magnitude $|H(j\omega_n)|$ is called the gain of the system at the frequency $\omega = \omega_n$. This concept applies only to LTI systems, and the gain is a function of frequency.

Discrete-Time System Response

The steady-state response to z^k is $H(z)z^k$ for a discrete-time system. If $z = e^{j\omega}$ then the response to $e^{j\omega k}$ is $H(e^{j\omega})e^{j\omega k}$. If the input signal is the sum of exponential signals,

$$v(k) = \sum_{n=0}^{N-1} V_n e^{jn\omega_1 k} \tag{12.3.3}$$

then, by the LTI properties, the response is given by

$$y(k) = \sum_{n=0}^{N-1} H(e^{jn\omega_1}) V_n e^{jn\omega_1 k} \tag{12.3.4}$$

Therefore, the procedure for calculating the steady-state response of discrete-time LTI systems is as follows:

1. Find the DFT of the input signal.
2. Calculate the transfer function $H(e^{j\omega})$.
3. Multiply $H(e^{jn\omega_1})$ by V_n for each n in the range $0 < n < N - 1$.
4. Use Eq. 12.3.4 to compute the response $y(k)$.

The response of an LTI system may be divided into a steady-state and a transient response. Fourier methods apply only to the steady-state response, because the input signal is assumed to be periodic, and therefore eternal. In order to find the complete response by these methods to a periodic signal applied at $t = 0$, we may do the following:

1. Find the steady-state response as outlined above. To do this, we assume that the signal is eternal.
2. Then find the transient response by methods studied in the solution of differential equations.
3. Add the steady-state solution to the transient response to obtain the complete response, and then solve for the unknown constants from initial conditions.

This section is devoted only to the steady-state response to periodic signals. It should be noted that the Laplace (Chapter 13) and z (Chapter 14) transforms may be used to find the complete response. This is one feature of these transforms that makes them so much more general and powerful than Fourier analysis.

There are some cases, however, where the DFT is more useful than the other transforms. One example is that of image processing. In order to improve the quality of an image (remove blurs, sharpen edges, etc.) or to determine if certain information is contained in an image (for example, an "enemy" tank or a tumor in an x-ray), the image is digitized and methods based on convolution or the DFT (or both) are used. Striking results can be obtained from these methods.

Here are some examples for both continuous- and discrete-time systems. We will use ideal filters in the following examples, so let us pause to introduce this concept.

A filter is a device that passes certain information with little or no attenuation and completely or greatly attenuates other information. Sun glasses are a very familiar type of filter. The filter on a furnace passes air but inhibits certain size particles that may be in the air. In electrical circuits, a filter allows signals of certain frequencies to pass through the filter but inhibits all signals outside the desired frequency range. The filter characteristic for an ideal low-pass filter is plotted in Fig. 12.32*a*, and that for an ideal band pass filter in Fig. 12.32*b*. The name ideal means that the input signal is multiplied by one in the pass band, and multiplied by zero in the rejection band. Thus, the frequency content of the input signal is changed by the filter, and for an ideal filter it is changed in an ideal way. Practical filters cannot accomplish this ideal filtering, because ideal filters are not physically realizable. Use of ideal filters greatly simplifies the mathematics involved in circuit analysis and the results closely approximate the output of good practical filters.

The bandwidth of the ideal low-pass filter in Fig. 12.32*a* is f_1 Hz. On the other hand, the bandwidth of the band pass filter is $2f_1$ Hz. The frequency f_0 is called the center frequency of the band pass filter.

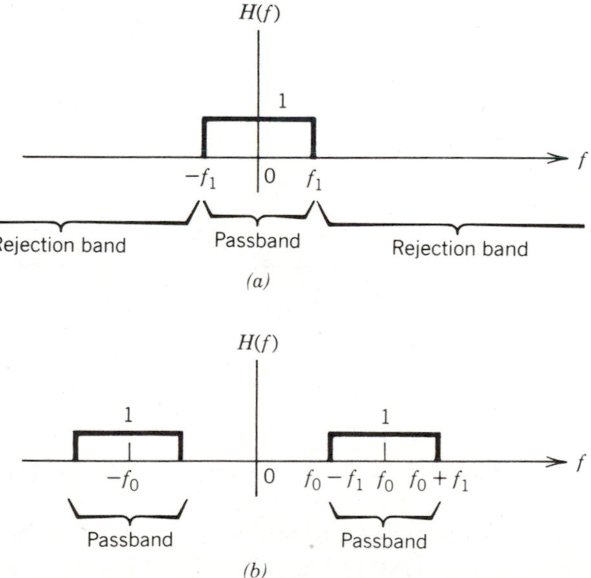

Figure 12.32. Ideal filters. (*a*) Low pass. (*b*) Band pass.

EXAMPLE 12.3.1 The continuous-time signal $f(t)$ shown in Fig. 12.33 is supplied to an ideal low-pass filter with cutoff frequency $f_1 = 18$ Hz. Find the steady-state output $y(t)$.

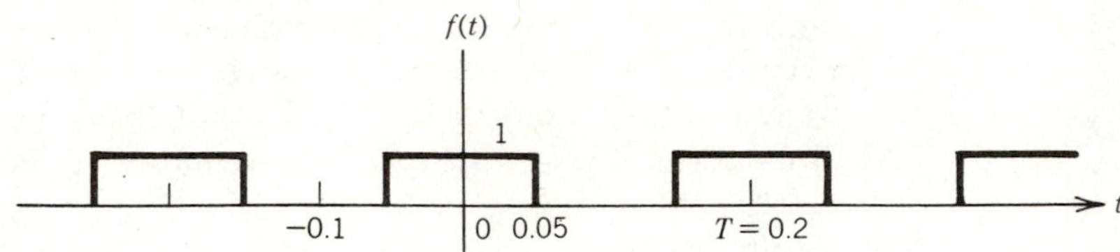

Figure 12.33. The periodic pulse function with period 0.2 s and pulse width 0.1 s.

Solution

Using the definition for Fourier series with $f_1 = 1/T = 5$ Hz, we find

$$F_k = \frac{1}{T}\int_{-T/2}^{T/2} f(t)e^{-jk\omega_1 t}\,dt$$

$$= 5\int_{-0.05}^{0.05} (1)e^{-jk2\pi 5t}\,dt$$

$$= 0.5\left(\frac{\sin 0.5k\pi}{0.5k\pi}\right)$$

The first few terms of this series are plotted in Fig. 12.34, along with the ideal filter characteristic. Therefore, the output $y(t)$ is given by

$$y(t) = -\frac{1}{3\pi}e^{-j3(2\pi)5t} + \frac{1}{\pi}e^{-j2\pi 5t} + \frac{1}{2} + \frac{1}{\pi}e^{j2\pi 5t} - \frac{1}{3\pi}e^{-j5(2\pi)5t}$$

$$= \frac{1}{2} + \frac{2}{\pi}\cos 2\pi 5t - \frac{2}{3\pi}\cos 2\pi 15t$$

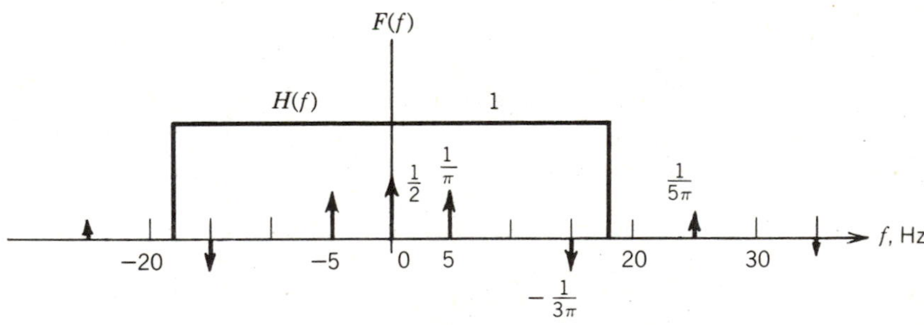

Figure 12.34. The output consists of those signal components between $-18 < f < 18$ Hz.

EXAMPLE 12.3.2 The signal $f(t)$ of the previous example is now supplied to the RC low-pass filter shown in Fig. 12.35. Find the first few terms in the steady-state output $g(t)$.

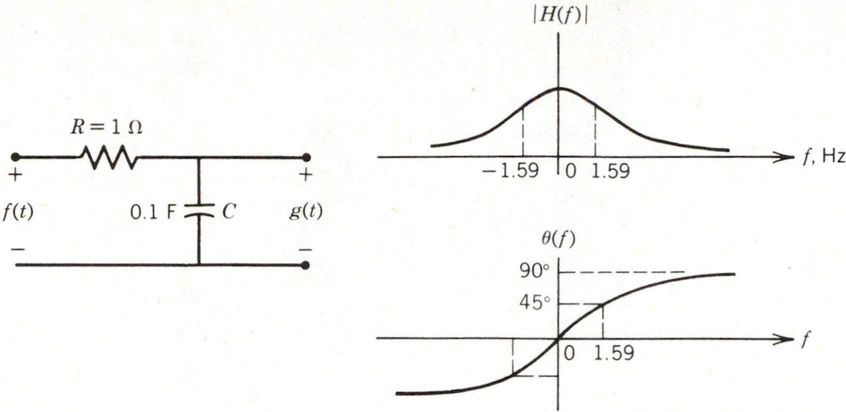

Figure 12.35. An *RC* low-pass filter with cutoff (3 dB) frequency $f_c \simeq 1.59$ Hz.

Solution

The filter characteristic is given by

$$H(f) = \frac{1}{1 + j2\pi f/10}$$

At frequencies $f = 0, 5, 15$ the value of $H(f)$ is shown in Table 12.1, along with the input components and the output components.

TABLE 12.1
EXAMPLE 12.3.2

f	$H(f)$	$F(f)$	$G(f) = H(f)F(f)$
0	1	$\frac{1}{2}$	$\frac{1}{2}$
5	$0.303e^{-j72.34°}$	$1/\pi$	$0.0965e^{-j72.34°}$
−5	$0.303e^{j72.34°}$	$1/\pi$	$0.0965e^{j72.34°}$
15	$0.106e^{-j83.84°}$	$-1/3\pi$	$-0.011e^{-j83.94°}$
−15	$0.106e^{j83.94°}$	$-1/3\pi$	$-0.011e^{j83.94°}$

Thus, the output $g(t)$ is approximately given by

$$\begin{aligned} g(t) &\simeq -0.011e^{-j3(2\pi)5t}e^{j84°} + 0.0965e^{-j2\pi 5t}e^{j72.4°} + \tfrac{1}{2} \\ &\quad + 0.0965e^{j2\pi 5t}e^{-j72.4°} - 0.011e^{j3(2\pi)5t}e^{-j84°} \\ &= \tfrac{1}{2} + 0.193\cos(10\pi t - 72.34°) - 0.022\cos(30\pi t - 83.94°) \end{aligned}$$

Comparing the results of Examples 12.3.1 and 12.3.2, we see that the *RC* circuit is an approximation for an ideal low-pass filter but a fairly poor one. Except for the dc component, the *RC* circuit attenuates each term and also causes a phase shift.

EXAMPLE 12.3.3 The periodic discrete-time signal from Example 12.2.1 (see Figs. 12.20 and 12.21) is applied to an ideal low-pass filter with cutoff frequency at $k = 1.5$. Find the steady-state response $y(k)$.

Solution

The filter is shown in Fig. 12.36. The periodic nature of the filter is due to the fact that discrete-time frequency is measured on the unit circle in the z plane, that is, $z = e^{jk\omega_1}$. Each time $k\omega_1$ increases by 2π, the filter characteristic $H(z)$ repeats itself. Since $\omega_1 = \frac{2\pi}{10}$, the period is 10.

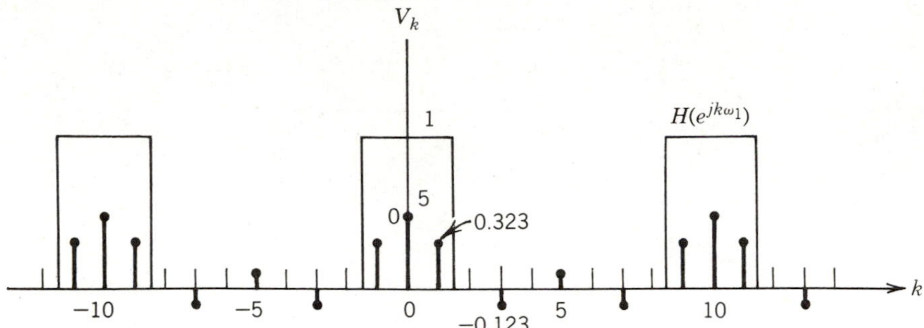

Figure 12.36. The ideal filter characteristic superimposed on the signal V_k.

The output of the filter is given by

$$\begin{aligned} y_k &= 0.5, \qquad k = 0 \\ &= 0.323, \qquad k = +1 \quad \text{and} \quad -1 \\ &= 0 \text{ for all other } k \text{ in one period} \end{aligned}$$

For the inverse DFT, we have

$$\begin{aligned} y(n) &= \sum_{k=-4}^{5} y_k e^{j2\pi nk/10} \\ &= 0.5 + 0.646 \cos \frac{n\pi}{5} \end{aligned} \tag{12.3.5}$$

which is plotted in Fig. 12.37. The ideal filter has changed the periodic square wave into the waveform $y(n)$.

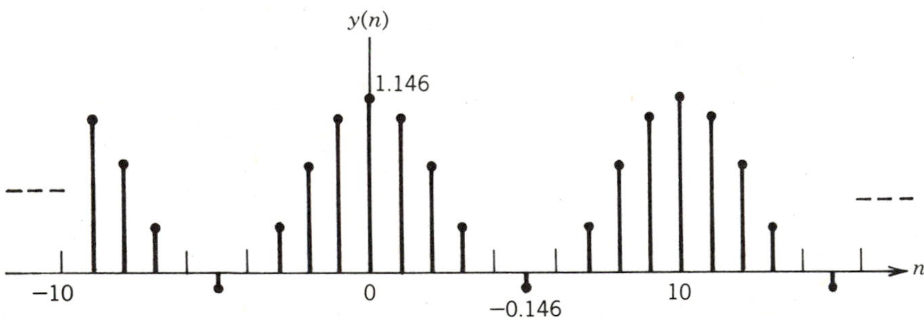

Figure 12.37. The output $y(n)$.

EXAMPLE 12.3.4 Now apply the periodic square wave of Example 12.2.1 to the system whose difference equation model is

$$y(n) = 0.7x(n-1) + 0.3y(n-1)$$

Find the steady-state response $g(n)$.

Solution

We found the transfer function for this system in Example 9.5.5. It is given by

$$H(z) = \frac{0.7}{z - 0.3}$$

In our present application, $z = e^{jk\omega_1}$ where $\omega_1 = 2\pi/10$. Hence,

$$H(e^{jk\pi/5}) = \frac{0.7}{e^{jk\pi/5} - 0.3}$$

The values of $H(k)$, $V(k)$, and $G(k) = H(k)V(k)$ are shown in Table 12.2. We find $g(n)$ by taking the inverse DFT of $G(k)$.

TABLE 12.2
EXAMPLE 12.3.4

k	$H(k) = H(e^{jk\omega_1})$	$V(k)$	$G(k) = H(k)V(k)$
0	1	0.5	0.5
1	$0.9e^{-j857r}$	0.323	$0.29e^{-j.857r}$
2	$0.736e^{-j1.56r}$	0	0
3	$0.619e^{-j2.14r}$	-0.123	$-0.076e^{-j2.14r}$
4	$0.557e^{-j2.65r}$	0	0
5	$0.583e^{-j\pi r}$	0.1	$0.0538e^{-j\pi r}$
6	$0.557e^{j2.65r}$	0	0
7	$0.619e^{j2.14r}$	-0.123	$-0.076e^{j2.14r}$
8	$0.736e^{j1.56r}$	0	0
9	$0.9e^{j0.854r}$	0.323	$0.29e^{j0.857r}$

$$g(n) = \sum_{k=0}^{9} G(k)e^{j2\pi nk/10}$$

$$= 0.5 + 0.0538e^{j(n+1)\pi} + 0.58\cos\left(\frac{n\pi}{5} - 0.857\right)$$

$$-0.152\cos\left(\frac{3n\pi}{5} - 2.14\right)$$

In this example, the phase angles are expressed in radians. The output is shown in Fig. 12.38.

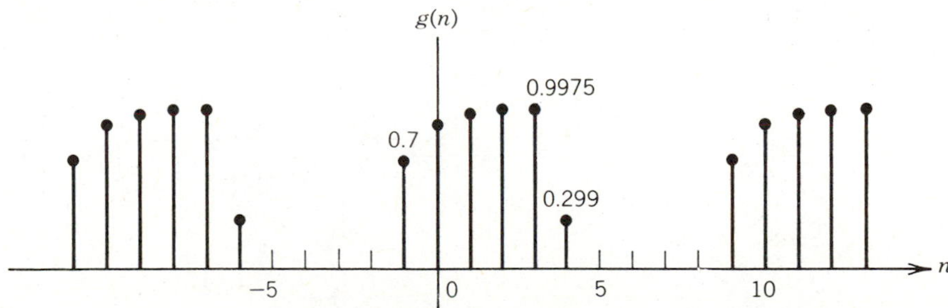

Figure 12.38. The response $g(n)$.

In all our previous examples the input waveform has been an even function of time, so that the corresponding frequency spectrum has been real. We next consider some examples where the frequency spectrum is complex.

EXAMPLE 12.3.5 The periodic square wave $f(t)$ shown in Fig. 12.39 has period $T = 100\ \mu s$. This voltage is applied to an ideal band pass filter with center frequency $f_0 = 50$ kHz and bandwidth $B = 10$ kHz. Find the steady-state output voltage $y(t)$.

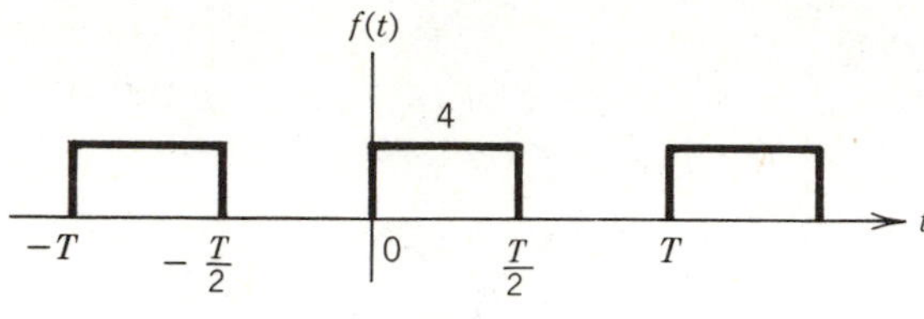

Figure 12.39

Solution

The Fourier series is plotted in Fig. 12.40 as amplitude and phase. That is, we allow the amplitude to be either a positive or negative real number. Notice that this amplitude is the same (identical) to the Fourier series for an even function. Thus shifting the function $f(t)$ along the time axis serves to change only the phase spectrum. This is one of the properties of transforms that we will study later.

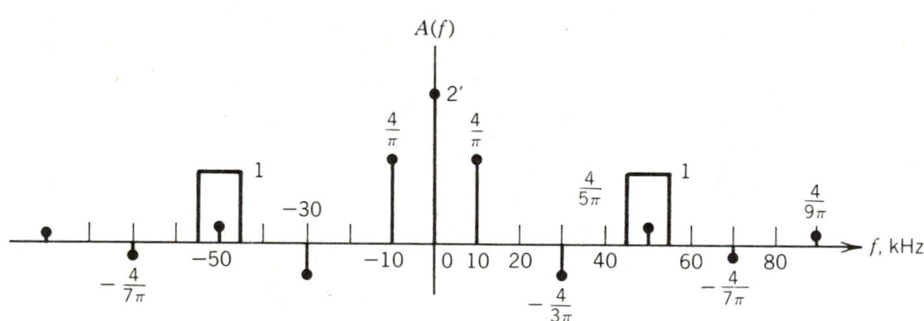

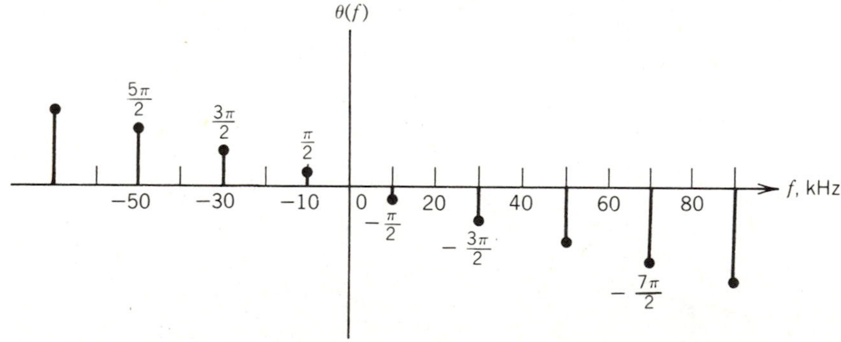

Figure 12.40. Amplitude and phase of $f(t)$ for Example 12.3.5.

The fifth harmonic is the only output of the filter. Therefore the output is given by

$$y(t) = \frac{4}{5\pi} e^{-j\pi 5/2} e^{j2\pi 5(10)^4 t} + \frac{4}{5\pi} e^{j5\pi/2} e^{-j2\pi 5(10)^4 t}$$

$$= \frac{8}{5\pi} \cos\left(10^5 \pi t - \frac{5\pi}{2}\right)$$

EXAMPLE 12.3.6 The physically realizable filter shown in Fig. 12.41 is now used in place of the ideal band pass filter in Example 12.3.5. The resonant circuit has a center frequency $\omega_0 = 1/\sqrt{LC} = 3.16(10)^5$ or $f_0 = \omega_0/2\pi = 50.33$ kHz. The coil has Q_0 at resonance of 31.6. Find $y(t)$.

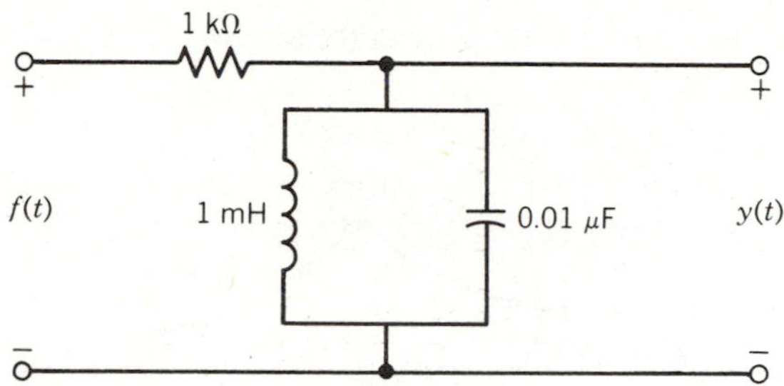

Figure 12.41. A realizable filter. The coil has $Q_0 = 31.6$.

Solution

The quality factor Q_0 is given by

$$Q_0 = \frac{\omega_0 L}{R_s} = \frac{R_p}{\omega_0 L}$$

where R_s is in series with the coil and R_p is in parallel. Therefore, we have either the equivalent circuit in Fig. 12.42*a* or 12.42*b*, which is valid only at resonance.

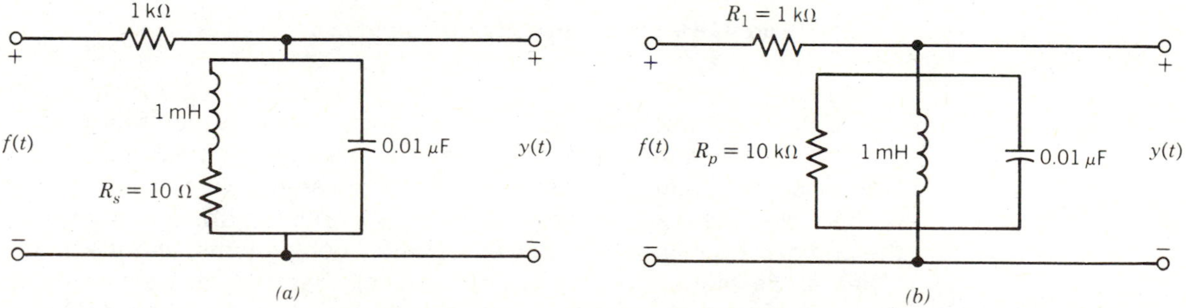

Figure 12.42. Two circuits that are equivalent to Fig. 12.41 at resonance.

Lacking any further information about the behavior of the coil at different frequencies, and also to simplify matters, we will assume the circuit in Fig. 12.42*b* is valid at all frequencies. With Z_p the impedance of the parallel *RLC* branch, we have

$$Z_p(j\omega) = \frac{1}{\frac{1}{R} + \frac{1}{Z_L} + \frac{1}{Z_c}} = \frac{1}{\frac{1}{10^4} + \frac{1}{10^{-3}j\omega} + 10^{-8}j\omega}$$

$$= \frac{10^8 j\omega}{(j\omega)^2 + 10^4 j\omega + 10^{11}}$$

Then the transfer function is

$$H(j\omega) = \frac{Z_p(j\omega)}{R_1 + Z_p(j\omega)} = \frac{10^5 j\omega}{(j\omega)^2 + 1.1(10)^5 j\omega + 10^{11}}$$

This filter characteristic is plotted in Fig. 12.43.

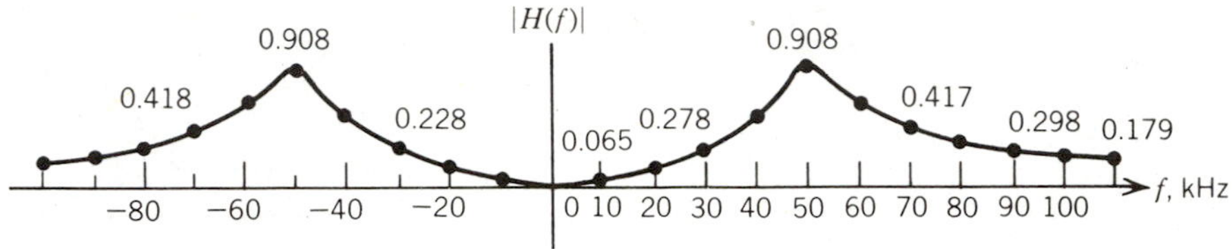

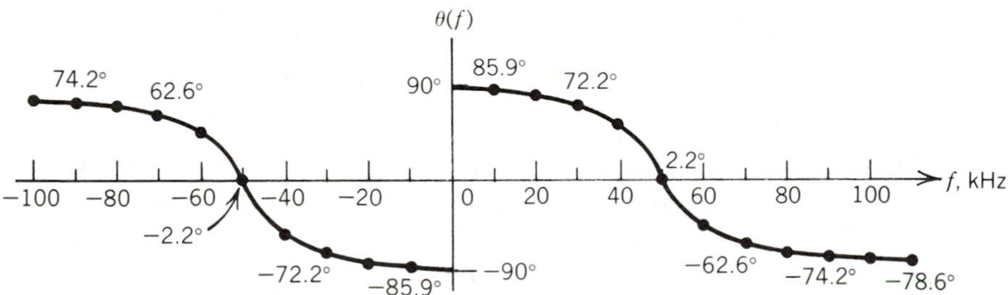

Figure 12.43. Filter characteristic for a realizable band pass filter.

In Table 12.3 we select each input signal frequency component and multiply by the transfer function at that frequency to obtain the output. Only positive frequency terms through the seventh harmonic are shown in the table. The negative frequency components are obtained by taking the complex conjugate of positive frequency components, and the amplitudes at the higher harmonics are so small that we can neglect them. These frequency components are shown in Fig. 12.44. The output is given by

TABLE 12.3
EXAMPLE 12.3.6

f	$F(f)$	$H(f)$	$Y(f)$
10^4	$\frac{4}{\pi}e^{-j\pi/2}$	$0.065e^{j85.9°}$	$0.083e^{-j4.1°}$
$3(10)^4$	$-\frac{4}{3\pi}e^{-j3\pi/2}$	$0.278e^{j72.2°}$	$-0.118e^{-j197.8°}$
$5(10)^4$	$\frac{4}{5\pi}e^{-j5\pi/2}$	$0.908e^{j2.2°}$	$0.231e^{-j447.8°}$
$7(10)^4$	$-\frac{4}{7\pi}e^{-j7\pi/2}$	$0.418e^{-j62.6°}$	$-0.076e^{-j692.6°}$

$$\begin{aligned}
y(t) &= 0.083e^{-j4.1°}e^{j\omega_1 t} + 0.083e^{j4.1°}e^{-j\omega_1 t} \\
&\quad - 0.118e^{-j197.8°}e^{j3\omega_1 t} - 0.118e^{j197.8°}e^{-j3\omega_1 t} \\
&\quad + 0.231e^{-j447.8°}e^{j5\omega_1 t} + 0.231e^{j447.8°}e^{-j5\omega_1 t} \\
&\quad - 0.076e^{-j692.6°}e^{j7\omega_1 t} - 0.076e^{j692.6°}e^{-j7\omega_1 t} \\
&= 0.166\cos(\omega_1 t - 4.1°) - 0.236\cos(3\omega_1 t - 197.8°) \\
&\quad + 0.462\cos(5\omega_1 t - 447.8°) - 0.152\cos(7\omega_1 t - 692.6°)
\end{aligned}$$

where $\omega_1 = 2\pi(10)^4$ rad/s.

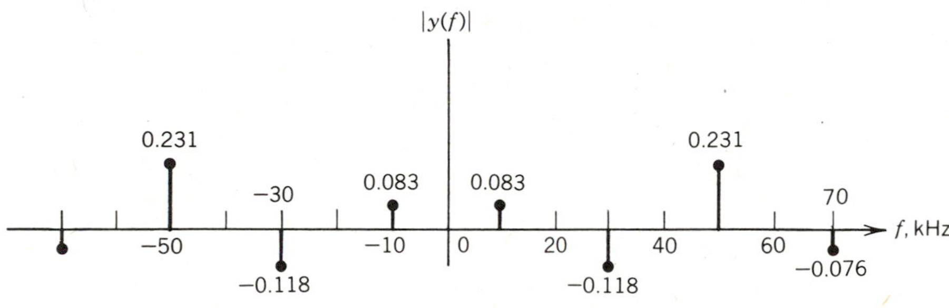

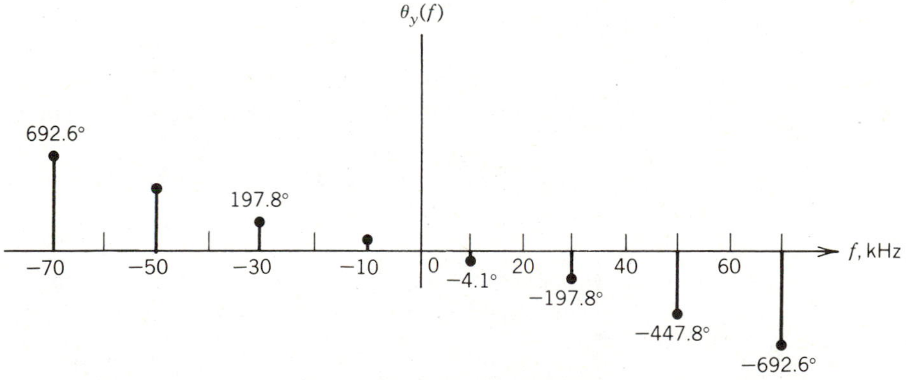

Figure 12.44. The output $Y(f)$.

LEARNING EVALUATIONS

1. The periodic pulse train $v(t)$ shown in Fig. 12.45 is supplied to an ideal low-pass filter with bandwidth $f_c = 300$ Hz. Find the steady-state response $y(t)$.

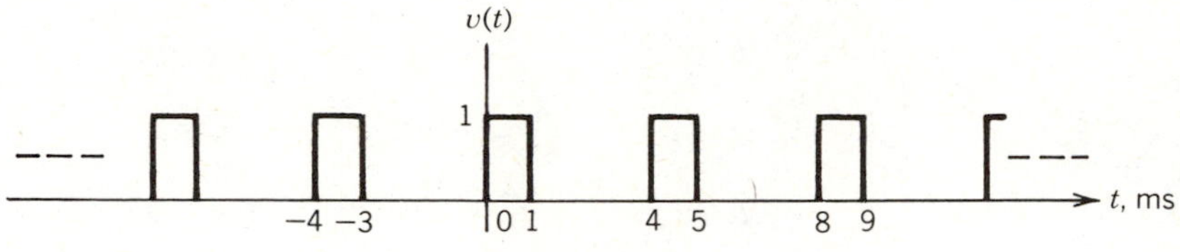

Figure 12.45

2. The periodic digital signal $x(n)$ shown in Fig. 12.46 is supplied to a system whose difference equation model is

$$y(n) = 0.5x(n-1) + 0.2y(n-1)$$

Find the steady-state response $y(n)$.

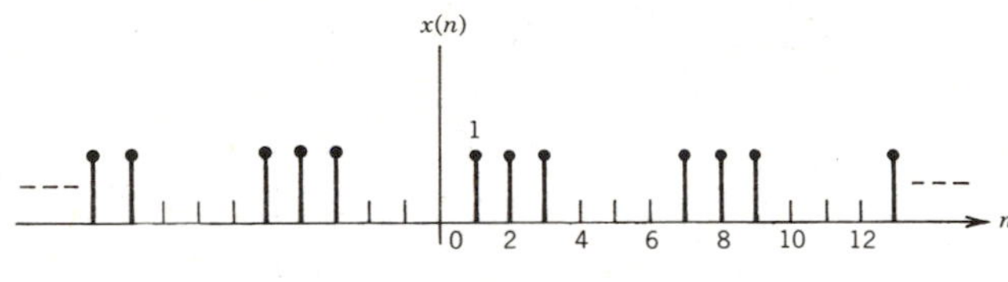

Figure 12.46

12.4 The Fourier Transform

LEARNING OBJECTIVES

After completing this section you should be able to do the following:

1. Define the Fourier transform.
2. Determine whether or not a given function has a Fourier transform.
3. Evaluate the transform for arbitrary functions of time that satisfy the requirements.
4. Evaluate the inverse transform for arbitrary functions of frequency that satisfy the requirements.

All transforms of the type we discuss in this text may be viewed as a black box with an input and an output. The Fourier transform is given by

$$V(f) = \int_{-\infty}^{\infty} v(t)e^{-j\omega t}\,dt \tag{12.4.1}$$

and is shown as the black box in Fig. 12.47. A particular time function $v(t)$ is selected and fed into the black box. What comes out is the corresponding

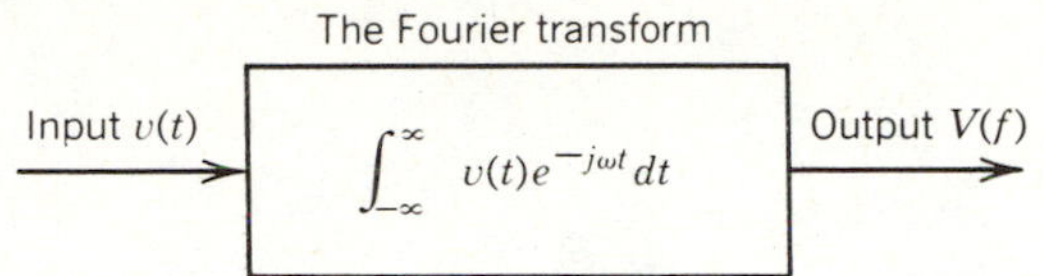

Figure 12.47. The Fourier transform as a black box.

frequency function $V(f)$. Furthermore, this transformation is linear, for it satisfies the additivity and homogeneity properties.

In our application, $v(t)$ will always be identified as a function of time and $V(f)$ will be a function of frequency. That is, the domain of v is a set of numbers that represent values of time, and the domain of V is a set of numbers that represent values for frequency. The parameter f is continuous, in contrast to the parameter f_k for the Fourier series, which is discrete.

The range of both v and V can be a set of complex numbers. Since we will be most often concerned with signals that can be generated in the laboratory, the functions v will, for the most part, be real valued.

The inverse transform is given by

$$v(t) = \int_{-\infty}^{+\infty} V(f)e^{j2\pi ft}\,df \tag{12.4.2}$$

or, equivalently,

$$v(t) = \frac{1}{2\pi}\int_{-\infty}^{+\infty} V(\omega)e^{j\omega t}\,d\omega \tag{12.4.3}$$

Equation 12.4.3 is related to Eq. 12.4.2 by a simple change of variable, $\omega = 2\pi f$.

The inverse Fourier transform is pictured in Fig. 12.48. The input to the black box is selected from a set of frequency functions, and the corresponding output is a member of a set of time functions.

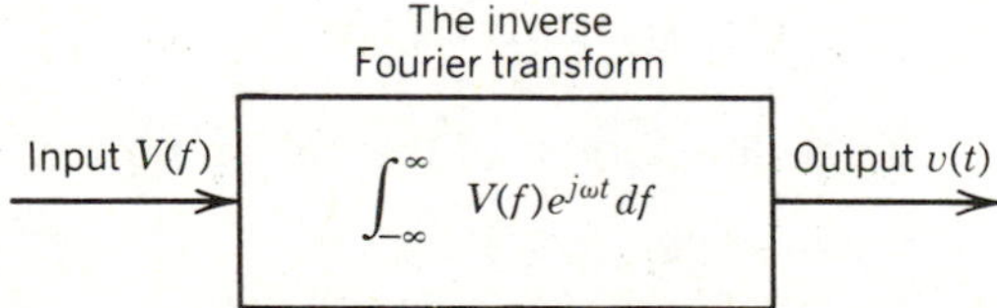

Figure 12.48. The inverse transform as a black box.

Which functions have a Fourier transform? Notice the similarity between the forward and inverse transforms, Eqs. 12.4.1 and 12.4.2. The only difference is the sign of the exponent. Therefore, the conditions that specify the domain of the forward transform are precisely those conditions that specify the domain of the inverse transform. From Eq. 12.4.1 if $|V(f)|$ is to be finite then we must have

$$\begin{aligned}|V(f)| &= \left|\int_{-\infty}^{\infty} v(t)e^{-j\omega t}\,dt\right| \\ &\le \int_{-\infty}^{\infty} |v(t)||e^{-j\omega t}|\,dt < \infty\end{aligned}$$

Since

$$|e^{-j\omega t}| = 1$$

we have

$$\int_{-\infty}^{\infty} |v(t)|\, dt < \infty \qquad (12.4.4)$$

Again, as in Section 12.1, we will use the strong Dirichlet conditions as our criteria for $v(t)$ to possess a Fourier transform. These are:

1. It satisfies Eq. 12.4.4.
2. It is finite.
3. It has a finite number of maxima and minima.

Any function $V(f)$ that satisfies the Dirichlet conditions [with $V(f)$ substituted for $v(t)$ and integrated with respect to the variable f] has a Fourier transform $v(t)$ given by Eq. 12.4.2. As in Section 12.1, the criteria used to specify the domain are not both necessary and sufficient. In fact, this problem has not been solved. Many methods of specifying sufficient conditions on v are available, but no one has found necessary conditions for v to have a Fourier transform.

We now have methods to describe energy signals and periodic power signals in the frequency domain. As yet, we have no such methods to describe nonperiodic power signals. This problem is encountered in random signal theory. An energy signal $v(t)$ satisfies the condition

$$\int_{-\infty}^{\infty} |v(t)|^2\, dt < \infty$$

while a power signal satisfies the condition

$$\lim_{T\to\infty} \frac{1}{2T} \int_{-T}^{T} |v(t)|^2\, dt < \infty$$

Pulse-type signals are energy signals, while periodic signals are power signals.

EXAMPLE 12.4.1 Determine which of the following functions has a Fourier transform.

1. $f(t) = \cos 2\pi t,\ -\infty < t < \infty$
2. $f(t) = \cos 2\pi t,\ 0 < t < 1$
3. The function in Fig. 12.49*a*.
4. The function in Fig. 12.49*b*.

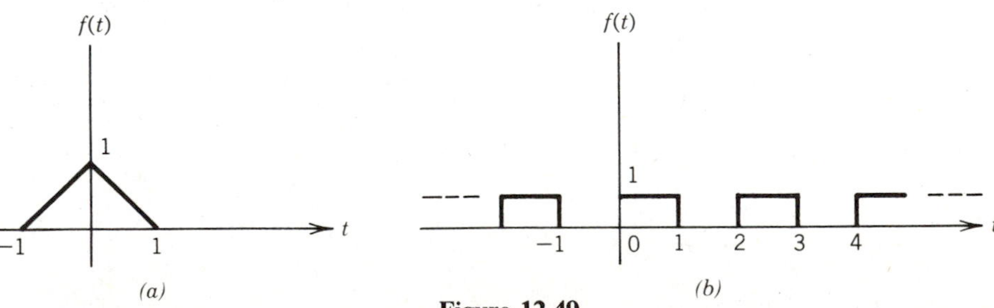

Figure 12.49

Solution

2 and **3**.

Finding the Fourier Transform

In this section we present a series of examples that illustrate the application of Eq. 12.4.1 to various waveforms that meet the criteria for having a Fourier transform.

EXAMPLE 12.4.2 Find the Fourier transform of $f_1(t)$, shown in Fig. 12.50*a*. The function $f_1(t)$ is given by

$$f_1(t) = \begin{matrix} e^{at} & t<0 \\ e^{-at} & t>0 \end{matrix}$$

Solution

$$F_1(f) = \int_{-\infty}^{0} e^{at}e^{-j\omega t}\,dt + \int_{0}^{\infty} e^{-at}e^{-j\omega t}\,dt$$

$$= \frac{1}{a-j\omega} + \frac{1}{a+j\omega} = \frac{2a}{a^2+\omega^2}$$

The solution is shown in Fig. 12.50*b*. Notice that $F_1(f)$ is a real function of frequency. This is true because $f_1(t)$ is even. Generally, the Fourier transform of a time function is a complex function of frequency. Therefore, to graph the function, we will need either a three-dimensional plot or two separate plots.

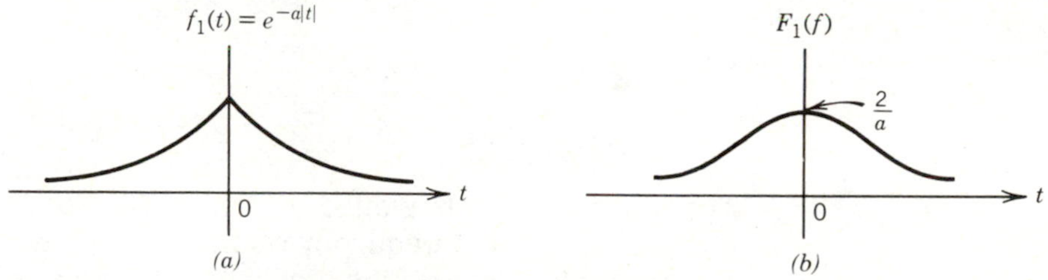

Figure 12.50. An even function of time and its transform.

EXAMPLE 12.4.3 Find the Fourier transform of $f_2(t)$ shown in Fig. 12.51*a*. The function $f_2(t)$ is given by

$$f_2(t) = \begin{matrix} -e^{at} & t<0 \\ e^{-at} & t>0 \end{matrix}$$

Solution

$$F_2(f) = \int_{-\infty}^{0} -e^{at}e^{-j\omega t}\,dt + \int_{0}^{+\infty} e^{-at}e^{-j\omega t}\,dt$$

$$= \frac{-1}{a-j\omega} + \frac{1}{a+j\omega} = \frac{-j2\omega}{a^2+\omega^2}$$

The solution is shown in Fig. 12.51*b*. Since $f_2(t)$ is odd, $F_2(f)$ is an imaginary function of frequency. That is, the real part is zero.

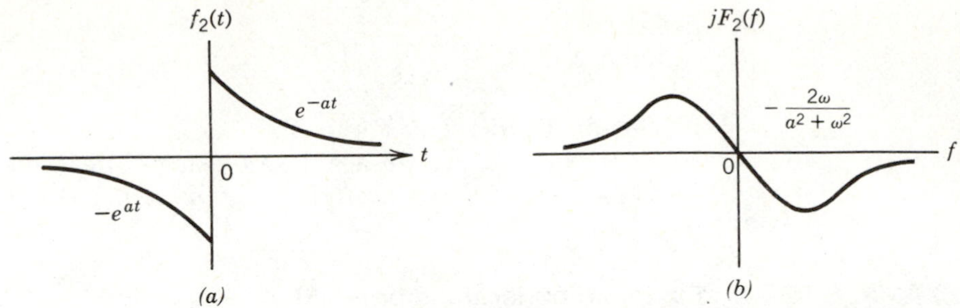

Figure 12.51. An odd function of time and its transform.

We have demonstrated the following property of Fourier transforms with these examples. If $f(t)$ is even, then $F(f)$ is real; if $f(t)$ is odd, then $F(f)$ is imaginary.

EXAMPLE 12.4.4 Find the Fourier transform of the square pulse shown in Fig. 12.52.

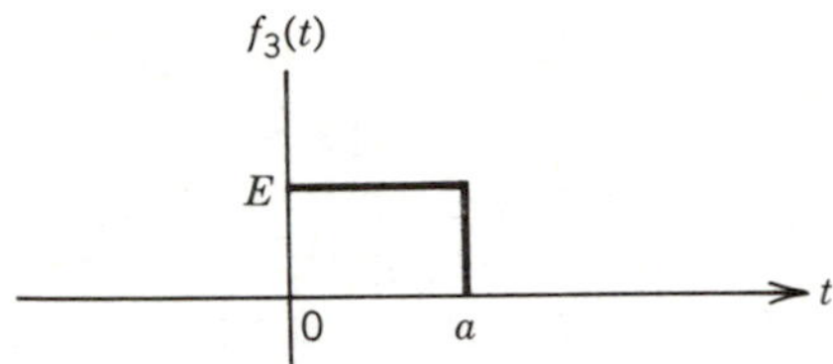

Figure 12.52. The square pulse $f_3(t)$.

Solution

$$F_3(f) = \int_0^a Ee^{-j\omega t}\,dt = \frac{E}{j\omega}(1 - e^{-j\omega a}) = Ea\left(\frac{\sin \omega a/2}{\omega a/2}\right)e^{-j\omega a/2}$$

The function $F_3(f)$ is plotted in Fig. 12.53. Notice that since $F_3(f)$ is a complex function of frequency we need two plots, one for amplitude and one for angle. We write the function $F_3(f)$ in the following form,

$$F_3(f) = A_3(f)e^{j\theta_3(f)}$$

where $A_3(f)$ is the amplitude and $\theta_3(f)$ is the angle.

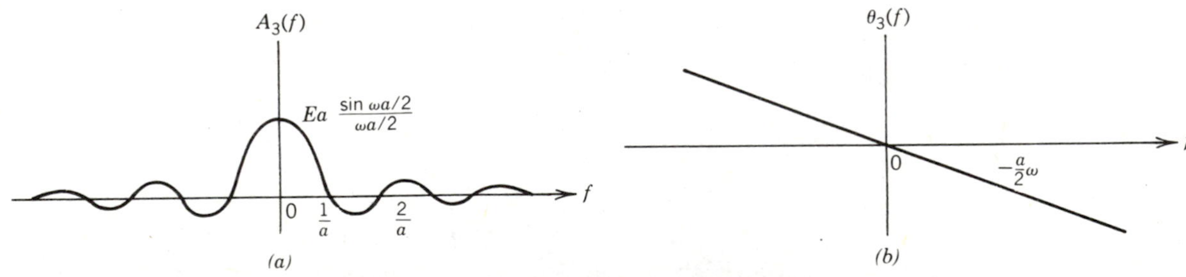

Figure 12.53. The transform of $f_3(t)$ in Fig. 12.52.

EXAMPLE 12.4.5 Find the Fourier transform of the periodic square wave in Fig. 12.54.

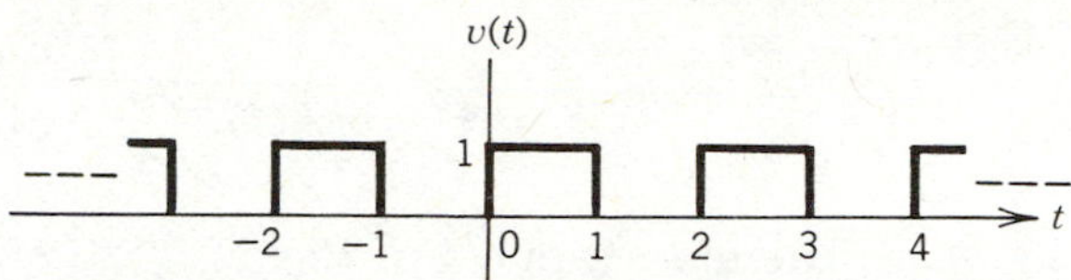

Figure 12.54. A periodic square wave does not satisfy the Dirichlet conditions.

Solution

This waveform does not meet our criteria, so it has no Fourier transform in the usual sense. However, the Fourier transform does exist as distribution; that is, delta functions are present in the transform. We will find it convenient to combine our notation for the Fourier series and Fourier transform, and we will discuss this problem after studying the properties of the transform.

EXAMPLE 12.4.6 Find the Fourier transform of the exponential function shown in Fig. 12.55.

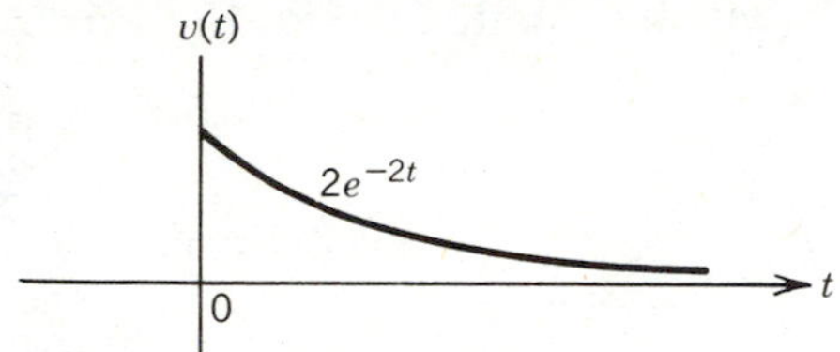

Figure 12.55. The exponential signal $v(t) = 2e^{-2t}$.

Solution

$$V(f) = \frac{2}{2 + j\omega}$$

Now let us reverse the above examples. That is, given the frequency function $F(f)$, find the corresponding time function $f(t)$ by applying the inverse transform. Here are some examples.

EXAMPLE 12.4.7 The Fourier transform of a function $f(t)$ is shown in Fig. 12.56. Find $f(t)$. Here, $F(f) = A(f)e^{j\theta(f)}$.

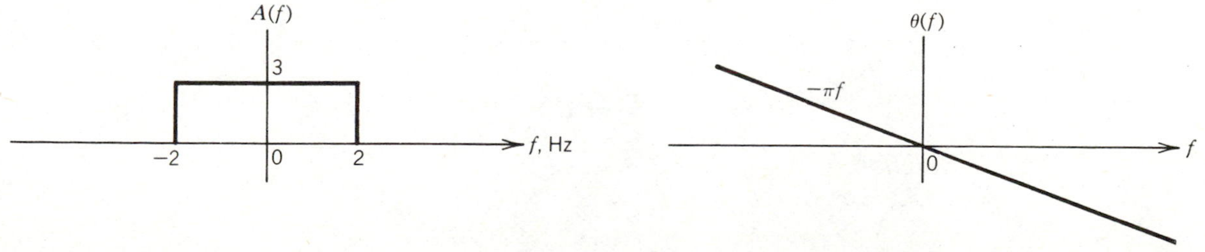

Figure 12.56. The signal $F(f) = A(f)e^{j\theta(f)}$.

Solution

The function $F(f)$ is given by

$$F(f) = A(f)e^{j\theta(f)} = \begin{cases} 0 & f < -2 \\ 3e^{-j\pi f} & -2 < f < 2 \\ 0 & f > 2 \end{cases}$$

Therefore (apply Eq. 12.4.2),

$$\begin{aligned} f(t) &= \int_{-2}^{2} 3e^{-j\pi f}e^{j\omega t}\,df \\ &= 3\int_{-2}^{2} e^{j(2\pi t-\pi)f}\,df \\ &= \frac{3}{j(2\pi t-\pi)}[e^{j(2\pi t-\pi)2} - e^{-j(2\pi t-\pi)2}] \\ &= 12\left[\frac{\sin(4\pi t - 2\pi)}{4\pi t - 2\pi}\right] \end{aligned}$$

which is plotted in Fig. 12.57.

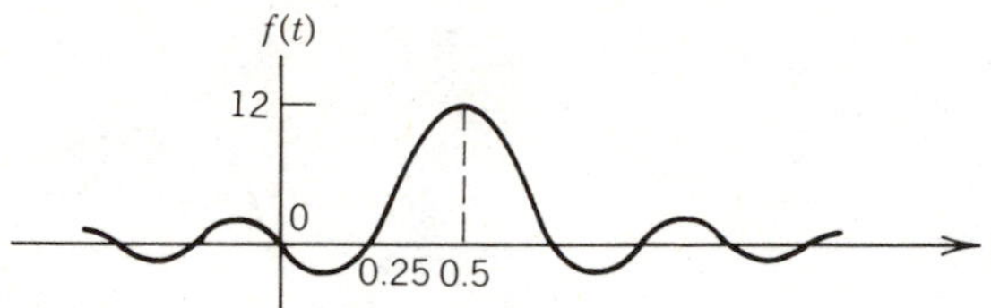

Figure 12.57. The transform of $F(f)$ from Fig. 12.56.

This is an important example because it illustrates why an ideal filter is not physically realizable. Notice that the time function $f(t)$ is not zero over any inverval of time which implies that $f(t) \neq 0$ for $t < 0$. Hence, this cannot represent the impulse response of a physically realizable filter.

EXAMPLE 12.4.8 Find the function $v(t)$ corresponding to the function $V(f)$ in Fig. 12.58.

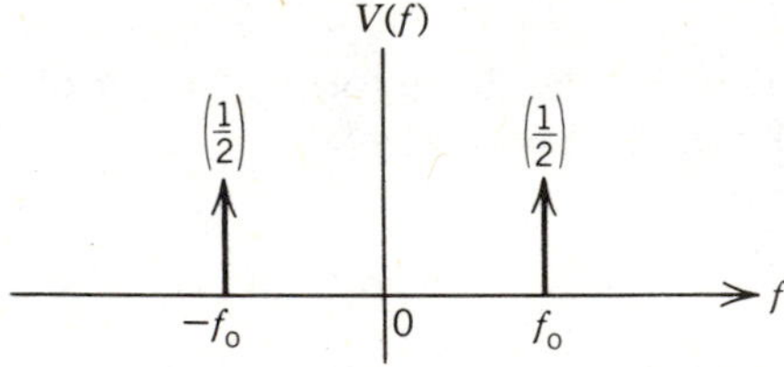

Figure 12.58. The signal $V(f)$.

Solution

Use Eq. 12.4.2.

$$v(t) = \int_{-\infty}^{\infty}\left[\frac{1}{2}\delta(f-f_0) + \frac{1}{2}\delta(f+f_0)\right]e^{j\omega t}\,df$$

$$= \frac{1}{2} e^{j2\pi f_0 t} + \frac{1}{2} e^{-j2\pi f_0 t} = \cos 2\pi f_0 t$$

This is an example of the special case discussed in Example 12.4.5. The transform of the power signal $v(t)$ exists as a distribution.

EXAMPLE 12.4.9 Find the function $y(t)$ corresponding to the function $Y(\omega)$ in Fig. 12.59.

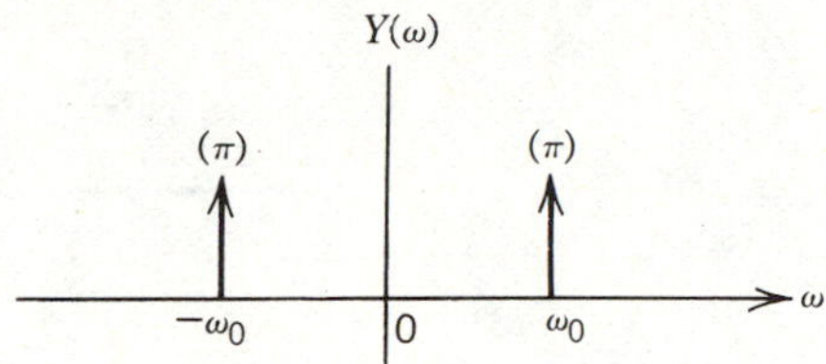

Figure 12.59. The signal $Y(\omega)$.

Solution

Use Eq. 12.4.3.

$$y(t) = \frac{1}{2\pi} \int_{-\infty}^{\infty} [\pi\delta(\omega - \omega_0) + \pi\delta(\omega + \omega_0)] e^{j\omega t} \, d\omega$$

$$= \frac{1}{2} e^{j\omega_0 t} + \frac{1}{2} e^{-j\omega_0 t} = \cos \omega_0 t$$

which is the same as $v(t)$ in Example 12.4.8.

Note that for a given function of frequency (f or ω) we must multiply the area under δ functions by 2π when changing the variable from f in hertz to ω in radians per second.

EXAMPLE 12.4.10 Find $s(t)$ corresponding to $S(\omega)$ shown in Fig. 12.60, where $S(\omega) = A(\omega)e^{j\theta(\omega)}$

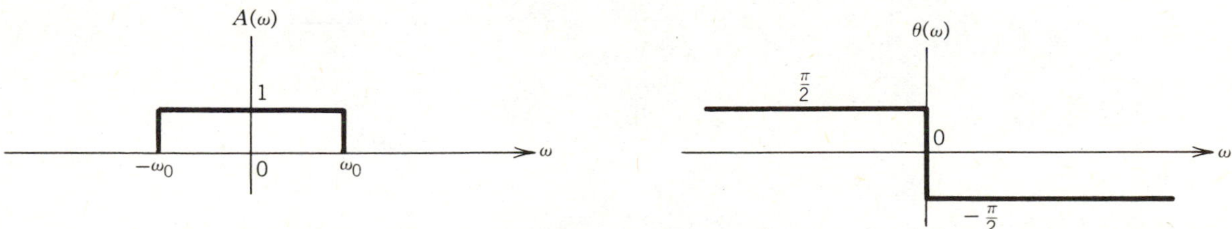

Figure 12.60. The signal $S(\omega) = A(\omega)e^{j\theta(\omega)}$.

Solution

$$s(t) = \frac{1}{2\pi} \int_{-\infty}^{\infty} A(\omega) e^{j\theta(\omega)} e^{j\omega t} \, d\omega$$

$$= \frac{1}{2\pi}\int_{-\omega_0}^{0} e^{j\pi/2}e^{j\omega t}\,d\omega + \frac{1}{2\pi}\int_{0}^{\omega_0} e^{-j\pi/2}e^{j\omega t}\,d\omega$$

$$= \frac{1}{\pi t}\left[\sin\frac{\pi}{2} + \sin\left(\omega_0 t - \frac{\pi}{2}\right)\right]$$

EXAMPLE 12.4.11 Find $v(t)$ corresponding to $V(f)$ shown in Fig. 12.61, where $V(f) = A(f)e^{j\theta(f)}$.

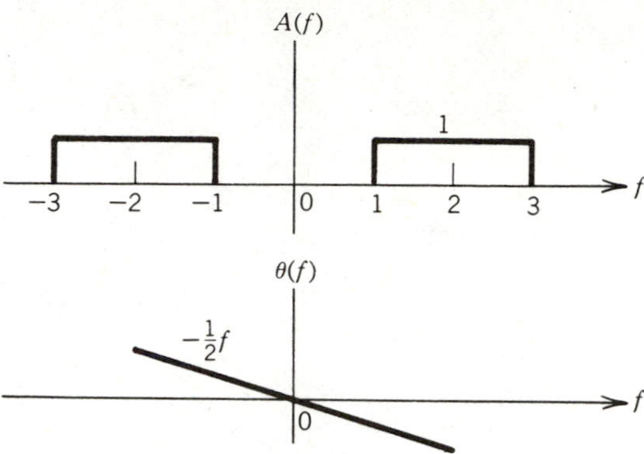

Figure 12.61. The signal $V(f) = A(f)e^{j\theta(F)}$.

Solution

$$v(t) = \frac{1}{j(2\pi t - \frac{1}{2})}\left[e^{-j2\pi(t-1/2)} - e^{-3j(2\pi t-1/2)} + e^{j3(2\pi t-1/2)} - e^{j(2\pi t-1/2)}\right]$$

$$= 4\left[\frac{\sin(2\pi t - \frac{1}{2})}{2\pi t - \frac{1}{2}}\right]\cos\left[2(2\pi t - \frac{1}{2})\right]$$

Table 12.4 consists of eight common Fourier transform pairs and is provided for future use for two reasons. First, it will save labor in evaluating needed

TABLE 12.4
FOURIER TRANSFORM PAIRS

$v(t)$	$V(\omega)$
1. $e^{-at}u(t)$	$\frac{1}{a+j\omega}$
2. $te^{-at}u(t)$	$\frac{1}{(a+j\omega)^2}$
3. $\delta(t)$	1
4. 1	$2\pi\delta(\omega)$
5. $u(t)$	$\pi\delta(\omega) + \frac{1}{j\omega}$
6. $e^{-a\lvert t\rvert}$	$\frac{2a}{a^2+\omega^2}$
7. $\cos\omega_0 t$	$\pi[\delta(\omega+\omega_0) + \delta(\omega-\omega_0)]$
8. $\sin\omega_0 t$	$j\pi[\delta(\omega+\omega_0) - \delta(\omega-\omega_0)]$

transforms. Second, we will have need for some transform pairs that we are not equipped to evaluate—those for which $v(t)$ does not meet the Dirichlet requirements. We will learn to evaluate these pairs in Section 12.5.

LEARNING EVALUATIONS

1. Describe the Dirichlet conditions.
2. Which of the functions shown in Fig. 12.62 satisfy the Dirichlet conditions.

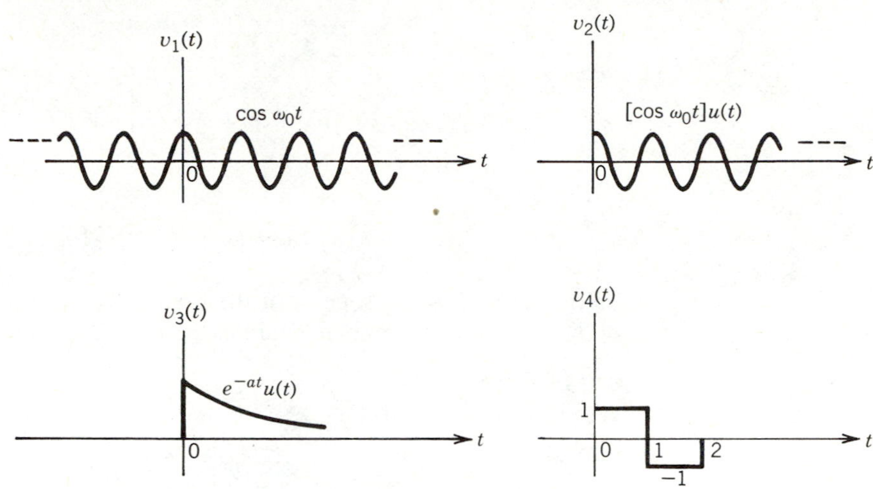

Figure 12.62

3. Find the Fourier transform of $v_3(t)$ shown in Fig. 12.62.
4. Find the function $v(t)$ corresponding to $V(f)$, where

$$V(f) = \begin{matrix} e^{-j2\omega} & -10 < f < 10 \\ 0 & \text{elsewhere} \end{matrix}$$

12.5 Properties of the Fourier Transform

LEARNING OBJECTIVES

After completing this section you should be able to use the properties listed in Eqs. 12.5.1 through 12.5.7 to aid you in finding the transform of various signals.

Some of the more useful properties of the Fourier transform are listed below. The properties of the Fourier series are identical to those for the transform, with appropriate modifications for the discrete nature of the frequency domain. These modifications will be illustrated in some of the examples that follow.

Superposition

$$av_1(t) + bv_2(t) \leftrightarrow aV_1(f) + bV_2(f) \tag{12.5.1}$$

Differentiation

$$\frac{d^n}{dt^n}v(t) \leftrightarrow (j\omega)^n V(f) \tag{12.5.2}$$

Scaling

$$v\left(\frac{t}{a}\right) \leftrightarrow aV(af) \qquad (12.5.3)$$

Delay

$$v(t-t_0) \leftrightarrow V(f)e^{-j\omega t_0} \qquad (12.5.4)$$

Modulation

$$v(t)e^{j\omega_0 t} \leftrightarrow V(f-f_0) \qquad (12.5.5)$$

Convolution

$$\int_{-\infty}^{+\infty} v_1(\lambda)v_2(t-\lambda)\,d\lambda \leftrightarrow V_1(f)V_2(f) \qquad (12.5.6)$$

Multiplication

$$v_1(t)v_2(t) \leftrightarrow \int_{-\infty}^{+\infty} V_1(\lambda)V_2(f-\lambda)\,d\lambda \qquad (12.5.7)$$

Each of these is also a property of the Fourier series if ω_k or f_k is substituted for ω or f in the above formulas. Here is the proof of some of these properties.

Proof of Differentiation

We wish to prove that

$$\frac{d^n}{dt^n}v(t) \leftrightarrow (j\omega)^n V(f)$$

where $v(t)$ is given by

$$v(t) = \int_{-\infty}^{\infty} V(f)e^{j\omega t}\,df \qquad (12.5.8)$$

Take the nth derivative of both sides of Eq. 12.5.8 to obtain

$$\frac{d^n}{dt^n}v(t) = \int_{-\infty}^{\infty} (j\omega)^n V(f)e^{j\omega t}\,df$$

which is recognized as the transform of $(j\omega)^n V(f)$.

Proof of Delay

Begin by writing the transform of a function $g(t)$ as

$$G(f) = \int_{-\infty}^{\infty} g(t)e^{-j\omega t}\,dt$$

Identify $g(t)$ as $g(t) = v(t-t_0)$. Then

$$G(f) = \int_{-\infty}^{\infty} v(t-t_0)e^{-j\omega t}\,dt$$

Now change variable of integration and let $\lambda = t - t_0$.

$$G(f) = \int_{-\infty}^{\infty} v(\lambda)e^{-j\omega(\lambda+t_0)}\,d\lambda$$

$$= e^{-j\omega t_0}\int_{-\infty}^{\infty} v(\lambda)e^{-j\omega\lambda}\,d\lambda = e^{-j\omega t_0}V(f)$$

and the property is proved.

Proof of Convolution

We will do this for the convolution property of the Fourier series, given by

$$\frac{1}{T}\int_{t'}^{t'+T} v_1(\lambda)v_2(t-\lambda)\,d\lambda \leftrightarrow V_1(k)V_2(k) \tag{12.5.9}$$

That is, if $v_1(t)$ and $v_2(t)$ are periodic waveforms, each with period T, then their convolution is given by the left side of Eq. 12.5.9, where we integrate over one period and divide by T. This convolution of $v_1(t)$ with $v_2(t)$ produces another time function $v_3(t)$, given by

$$v_3(t) = \frac{1}{T}\int_{t'}^{t'+T} v_1(\lambda)v_2(t-\lambda)\,d\lambda \tag{12.5.10}$$

We wish to show that the coefficients $V_3(k)$ are given by $V_1(k)V_2(k)$, which also means that

$$v_3(t) = \sum_{k=-\infty}^{+\infty} V_1(k)V_2(k)e^{j\omega_k t}$$

By the definition of the Fourier series, $V_3(k)$ is given by

$$\begin{aligned} V_3(k) &= \frac{1}{T}\int_{t'}^{t'+T} e^{-j\omega_k t}v_3(t)\,dt \\ &= \frac{1}{T}\int_{t'}^{t'+T} e^{-j\omega_k t}\left[\frac{1}{T}\int_{t'}^{t'+T} v_1(\lambda)v_2(t-\lambda)\,d\lambda\right]dt \end{aligned}$$

where we have substituted Eq. 12.5.10 to obtain this last expression. Changing the order of integration, we get

$$V_3(k) = \frac{1}{T}\int_{t'}^{t'+T} v_1(\lambda)\left[\frac{1}{T}\int_{t'}^{t'+T} v_2(t-\lambda)e^{-j\omega_k t}\,dt\right]d\lambda$$

We recognize the term in the brackets as

$$\frac{1}{T}\int_{t'}^{t'+T} v_2(t-\lambda)e^{-j\omega_k t}\,dt = V_2(k)e^{-j\omega_k\lambda}$$

(This is the delay property for the series.) Therefore,

$$V_3(k) = V_2(k)\left[\frac{1}{T}\int_{t'}^{t'+T} v_1(\lambda)e^{-j\omega_k\lambda}\,d\lambda\right] = V_2(k)V_1(k)$$

which was to be proved.

Here are some examples that illustrate the use of these properties in evaluating transforms.

EXAMPLE 12.5.1 Use the differentiation property to find the Fourier transform of the square pulse shown in Fig. 12.63*a*.

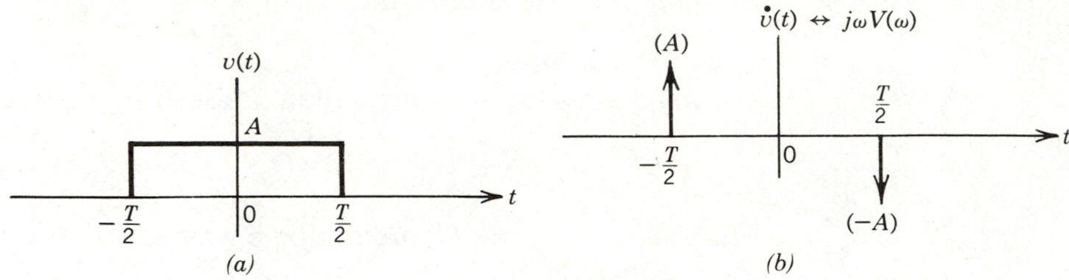

Figure 12.63. The square pulse $v(t)$ and its derivative.

Solution

The derivative $\dot{v}(t)$ is shown in Fig. 12.63*b*. Since the transform of a δ function occurring at t_0 is given by

$$\delta(t - t_0) \leftrightarrow e^{-j\omega t_0} \tag{12.5.11}$$

we have

$$j\omega V(\omega) = Ae^{j\omega T/2} - Ae^{-j\omega T/2}$$

or

$$V(\omega) = \frac{Ae^{j\omega T/2} - Ae^{-j\omega T/2}}{j\omega}$$

$$= AT\left(\frac{\sin \omega T/2}{\omega T/2}\right)$$

EXAMPLE 12.5.2 Use the differentiation property to find the Fourier transform of $f(t)$ in Fig. 12.64*a*.

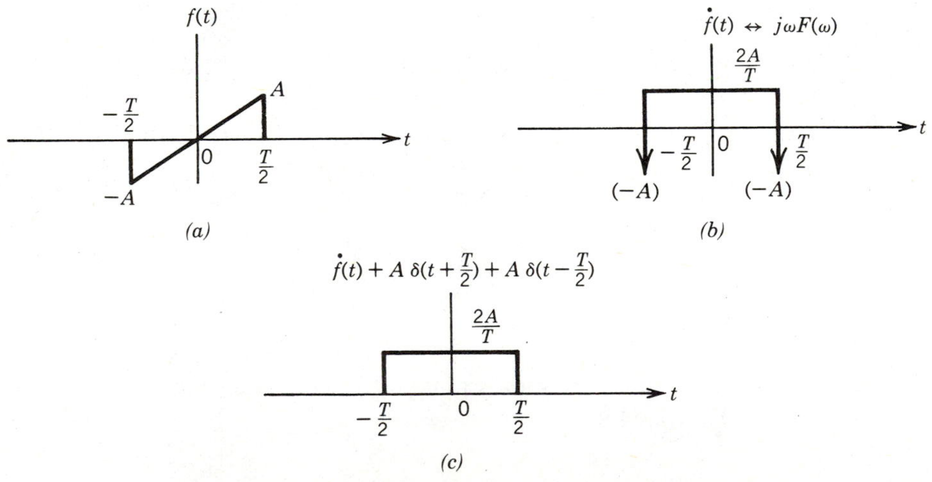

Figure 12.64. Waveforms for Example 12.5.2.

Solution

The derivative $\dot{f}(t)$ is shown in Fig. 12.64*b*, and the derivative minus the two delta functions is shown in Fig. 12.64*c*. Let us call this function $g(t)$ so that

$$g(t) = \dot{f}(t) + A\delta\left(t + \frac{T}{2}\right) + A\delta\left(t - \frac{T}{2}\right)$$

Therefore, $G(f)$ is given by

$$G(f) = j\omega F(f) + Ae^{j\omega T/2} + Ae^{-j\omega T/2}$$

where we have used Eq. 12.5.11 for the last two terms. We just found the transform of a square pulse similar to $g(t)$ in the previous example, so $G(f)$ is given by

$$G(f) = 2A\left(\frac{\sin \omega T/2}{\omega T/2}\right)$$

Equating terms and solving for $F(f)$ gives

$$j\omega F(f) + Ae^{j\omega T/2} + Ae^{-j\omega T/2} = 2A\left(\frac{\sin \omega T/2}{\omega T/2}\right)$$

$$F(f) = \frac{2A}{j\omega}\left(\frac{\sin \omega T/2}{\omega T/2} - \cos\frac{\omega T}{2}\right)$$

EXAMPLE 12.5.3 Use the differentiation property to find the transform of the cosine pulse in Fig. 12.65*a*.

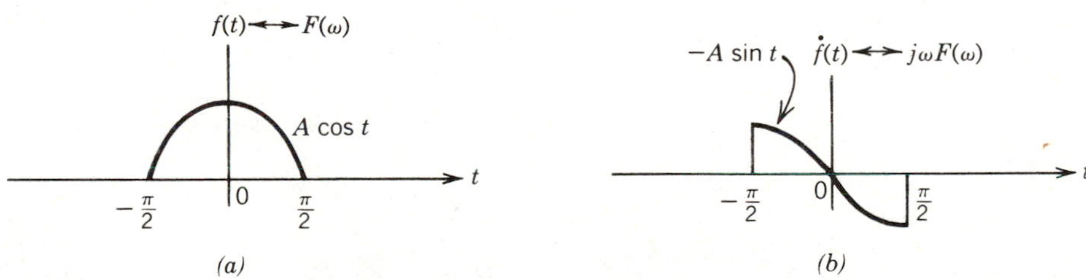

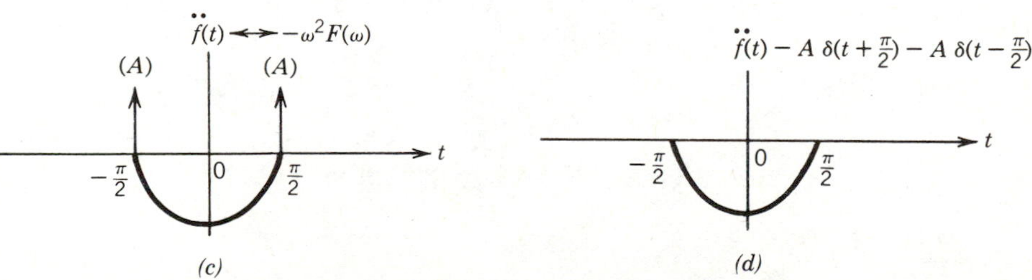

Figure 12.65. Waveforms for Example 12.5.3.

Solution

The second derivative is shown in Fig. 12.65*c*. Notice that this second derivative would be proportional to $f(t)$ if it were not for the two δ functions. In Fig. 12.65*d* we show $\ddot{f}(t)$ minus the two δ functions. Let us call this $g(t)$, so that $g(t)$ is given by

$$g(t) = \ddot{f}(t) - A\delta\left(t + \frac{\pi}{2}\right) - A\delta\left(t - \frac{\pi}{2}\right)$$

Therefore the transform of $g(t)$ is given by

$$G(\omega) = -\omega^2 F(\omega) - Ae^{j\omega\pi/2} - Ae^{-j\omega\pi/2} \tag{12.5.12}$$

But how do we find $G(\omega)$? Since $g(t)$ is related to $f(t)$ by

$$g(t) = -f(t)$$

then $G(\omega)$ must be related to $F(\omega)$ by

$$G(\omega) = -F(\omega)$$

Combining this with Eq. 12.5.12 we have

$$-F(\omega) = -\omega^2 F(\omega) - Ae^{j\omega\pi/2} - Ae^{-j\omega\pi/2}$$

or

$$F(\omega) = \frac{A}{1-\omega^2}\left(e^{j\omega\pi/2} + e^{-j\omega\pi/2}\right)$$

$$= \frac{2A}{1-\omega^2}\cos\left(\frac{\omega\pi}{2}\right)$$

EXAMPLE 12.5.4 Use the delay property and the solution to Example 12.5.1 to find the transform of $g(t)$ shown in Fig. 12.66.

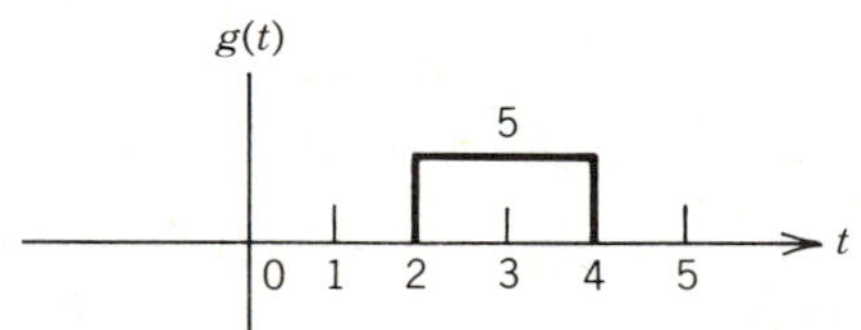

Figure 12.66. Waveform for Example 12.5.4.

Solution

If $T/2 = 1$, the function $v(t)$ in Fig. 12.63 is related to $g(t)$ by $g(t) = v(t-3)$. Since

$$V(\omega) = AT\frac{\sin \omega T/2}{\omega T/2}$$

we apply the delay property to obtain

$$G(\omega) = V(\omega)e^{-j\omega 3}$$

$$= 10\left(\frac{\sin \omega}{\omega}\right)e^{-j\omega 3}$$

EXAMPLE 12.5.5 Use the convolution property and the solution to Example 12.5.1 to find the transform of $f(t)$ shown in Fig. 12.67.

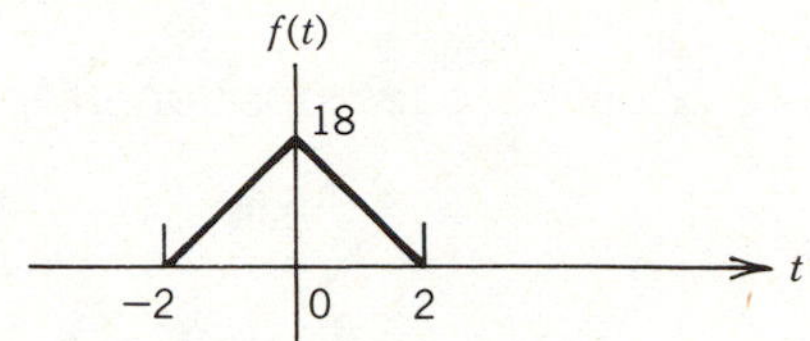

Figure 12.67. Waveform for Example 12.5.5

Solution

The function $v(t)$ in Fig. 12.63 is related to $f(t)$ by convolution,

$$f(t) = v(t) * v(t)$$

where $A = 3$ and $T = 2$ in Fig. 12.63. Therefore by the convolution property

$$F(\omega) = V^2(\omega) = 36\left(\frac{\sin \omega}{\omega}\right)^2$$

The Transform of Power Signals

Although power signals do not satisfy the Dirichlet conditions, the transform of many such signals exist as distribution. This is convenient because we can combine our notation for the Fourier series and transform. We now set about using the properties to derive the transform of several power signals.

EXAMPLE 12.5.6 Find the transform of the eternal exponential signal $g(t) = e^{j\omega_0 t}$.

Solution

The transform of a constant $v(t) = 1$ is $\delta(f) = 2\pi\delta(\omega)$. This is easily seen by direct application of the inverse Fourier transform.

$$v(t) = \int_{-\infty}^{\infty} \delta(f) e^{j2\pi tf} df = 1$$

or

$$v(t) = \frac{1}{2\pi} \int_{-\infty}^{\infty} 2\pi\delta(\omega) e^{jt\omega} d\omega = 1$$

Since the external exponential $g(t)$ is related to $v(t)$ by

$$g(t) = v(t) e^{j\omega_0 t}$$

we apply the modulation property to obtain

$$e^{j\omega_0 t} \leftrightarrow \delta(f - f_0) = 2\pi\delta(\omega - \omega_0) \tag{12.5.13}$$

EXAMPLE 12.5.7 Find the transform of $f(t) = \sin \omega_0 t$.

Solution

Since

$$\sin \omega_0 t = \frac{1}{2j} e^{j\omega_0 t} - \frac{1}{2j} e^{-j\omega_0 t}$$

we apply the superposition property to Eq. 12.5.13 to obtain

$$\sin \omega_0 t \leftrightarrow \frac{1}{2j}\delta(f - f_0) - \frac{1}{2j}\delta(f + f_0) \qquad (12.5.14)$$

EXAMPLE 12.5.8 Find the transform of the signum function $s(t)$ shown in Fig. 12.68.

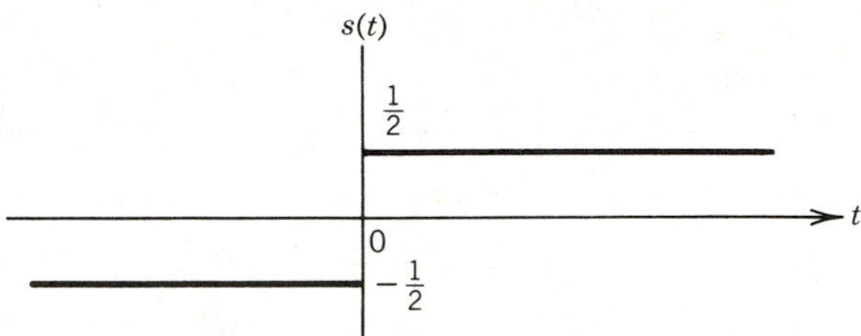

Figure 12.68. The signum function.

Solution

Apply the differentiation property. The derivative $\dot{s}(t) = \delta(t) \leftrightarrow j\omega S(f)$, and by Eq. 12.5.11 the transform of $\delta(t)$ is given by

$$\delta(t) \leftrightarrow 1 = j\omega S(f)$$

Hence

$$S(f) = \frac{1}{j\omega} \qquad (12.5.15)$$

EXAMPLE 12.5.9 Use the superposition property and Eq. 12.5.15 to find the transform of the unit step $u(t)$ shown in Fig. 12.69.

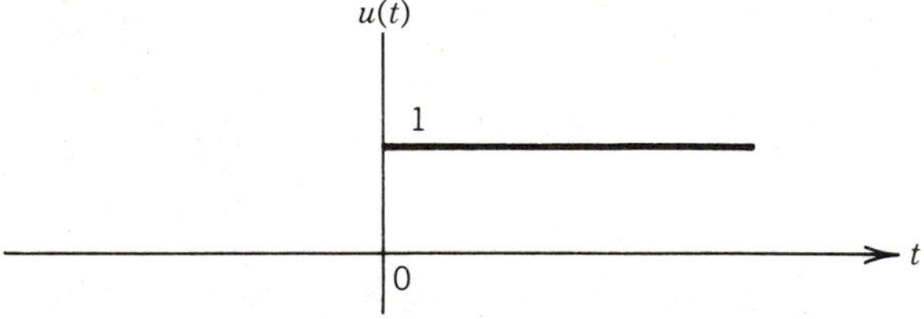

Figure 12.69. The unit step function.

Solution

The unit step $u(t)$ is related to the signum function $s(t)$ by

$$u(t) = s(t) + \frac{1}{2}$$

Hence

$$U(f) = S(f) + \frac{1}{2}\,\delta(f)$$

$$= \frac{1}{j\omega} + \frac{1}{2}\delta(f) \qquad (12.5.16)$$

Note that the derivatives of $u(t)$ and $s(t)$ are equal. We cannot apply the differentiation property directly to find the transform of $u(t)$ since the dc level is lost in taking the derivative.

We can generalize this observation. If you take the derivative of a power signal, the dc level (zero frequency component) is lost. This does not apply to an energy signal, however, because an energy signal has no dc level.

EXAMPLE 12.5.10 Find the transform of $p(t) = (\sin \omega_0 t)u(t)$ shown in Fig. 12.70.

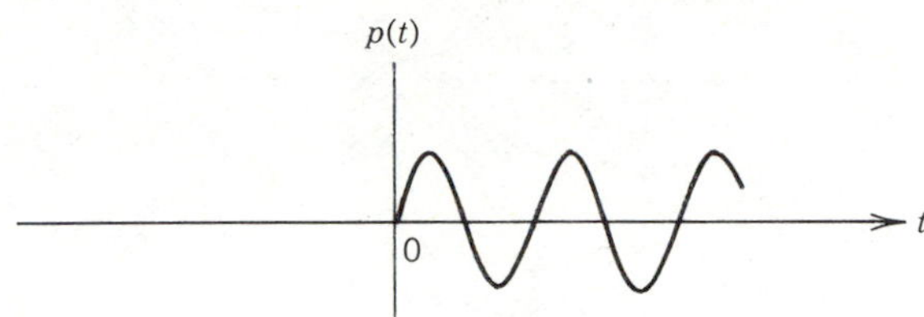

Figure 12.70. $p(t) = \sin(\omega_0 t)u(t)$.

Solution

Apply the multiplication property. Since $p(t)$ is the product of $f(t)$ in Eq. 12.5.14 with $u(t)$ in Eq. 12.5.16 we convolve $F(f)$ with $U(f)$ to obtain

$$P(f) = F(f) * U(f)$$

$$= \frac{1}{4j}\,\delta(f - f_0) - \frac{1}{4j}\,\delta(f + f_0) - \frac{\omega_0}{\omega^2 - \omega_0^2} \qquad (12.5.17)$$

EXAMPLE 12.5.11 Find the transform of $q(t) = (\cos \omega_0 t)u(t)$.

Solution

$$Q(f) = \frac{1}{4}\,\delta(f - f_0) + \frac{1}{4}\,\delta(f + f_0) - \frac{j\omega}{\omega^2 - \omega_0^2} \qquad (12.5.18)$$

The Fourier transform properties are an aid to understanding many practical systems. Example 12.5.12, where we apply several of the above concepts, is an apt illustration of this point.

EXAMPLE 12.5.12 *Amplitude modulation.* Here we will describe how an ordinary AM radio works. The transmitter is essentially a multiplier, where two signals (time functions) are multiplied together. The receiver is also a multiplier followed by a low-pass filter.

Suppose that a voice or music signal $f(t)$ is to be transmitted by amplitude modulation. The signal and its spectrum $F(f)$ are shown in Fig. 12.71*a*. Now if $f(t)$ is multiplied by a carrier waveform $r(t) = \cos \omega_0 t$, which is shown in Fig. 12.71*b*, then the multiplication property of Eq. 12.5.7 can be applied to find the resulting spectrum. The multiplication property indicates that we should convolve $F(f)$ and $R(f)$ to find the spectrum of the transmitted signal $x(t)$. That is, if $s(t) = f(t)r(t)$, then $X(f) = F(f) * R(f)$. This convolution is particularly simple since impulses are involved. The spectrum $F(f)$ is translated f_0 units to the right and left, and multiplied by the impulse area of $\frac{1}{2}$, as indicated in Fig. 12.71*c*.

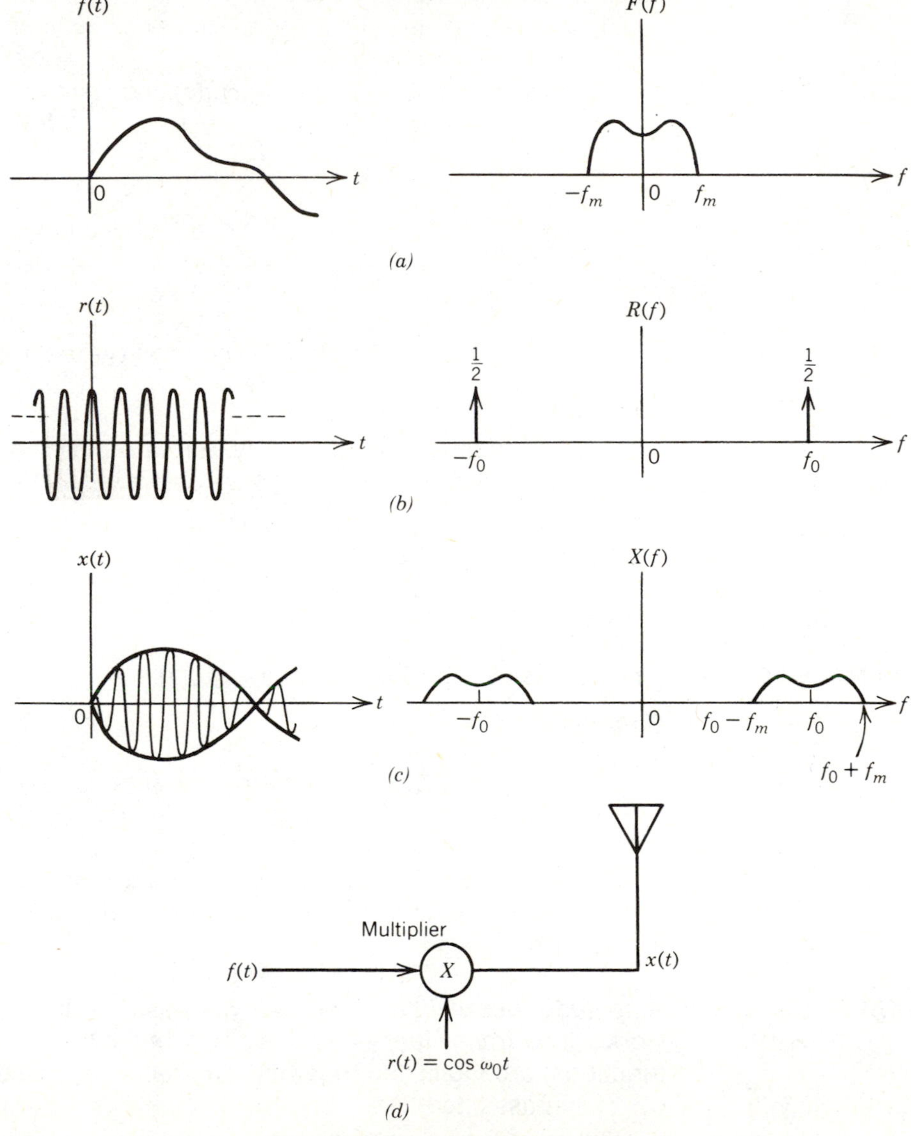

Figure 12.71. Amplitude modulation.

Now the signal x is ready for broadcasting, so it is amplified by a power amplifier stage and fed to the antenna. Aside from the necessary amplifier stages, this transmitter is essentially a product device, as indicated in Fig. 12.71*d*. This in practice is not the way commercial AM broadcasting is accomplished, for the receiver becomes unduly complex and expensive if the transmitter shown in Fig. 12.71 is used.

Commercial AM broadcasting is accomplished by adding the carrier to x, as shown in Fig. 12.72, where

$$\begin{aligned} y(t) &= x(t) + Ar(t) \\ &= f(t)\cos\omega_0 t + A\cos\omega_0 t \\ &= [A + f(t)]\cos\omega_0 t \end{aligned} \tag{12.5.19}$$

The reason for adding the carrier back in is to simplify the receiver, which we now consider.

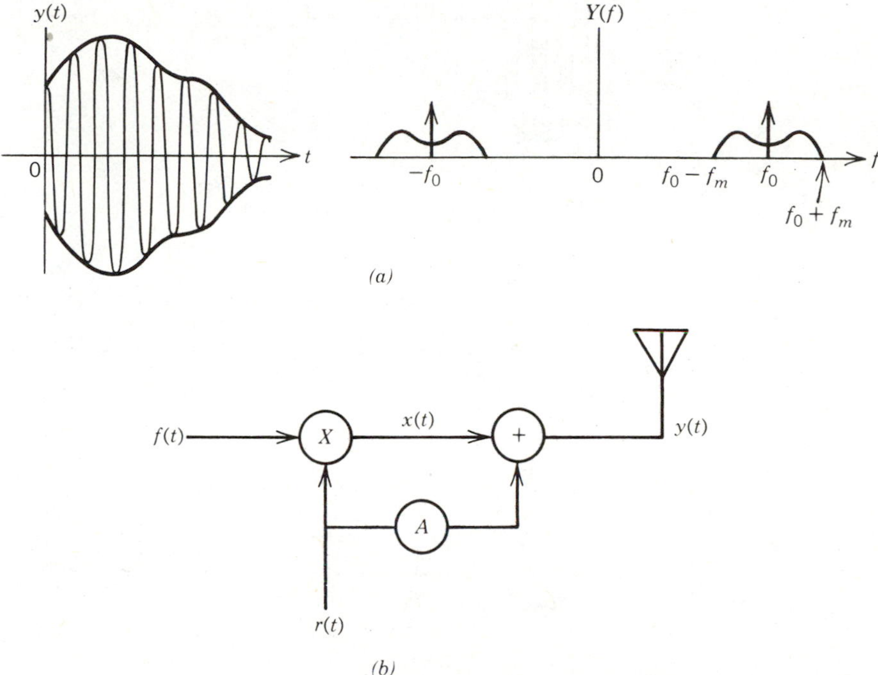

Figure 12.72. Commercial AM broadcasting scheme.

Notice that the envelope of $y(t)$ is proportional to the modulating signal $f(t)$. The central component in the receiver is called an envelope detector, because this component succeeds in extracting the envelope from the received signal $y(t)$. As shown in Fig. 12.73, the envelope detector consists of a diode, an RC low-pass filter, and a coupling capacitor C_c. The diode acts as a switch that is closed when the input $y(t)$ is positive, and one that is open when $y(t)$ is negative. The effect of this switch on $y(t)$ is pictured in Fig. 12.74*a*. This does not provide us with a useful picture, however, for it is to the remainder of Fig. 12.74 that we must look for an explanation of the importance of the diode.

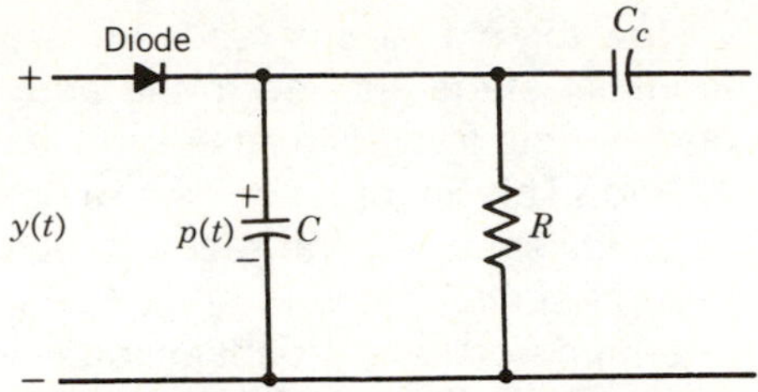

Figure 12.73. An envelope detector circuit.

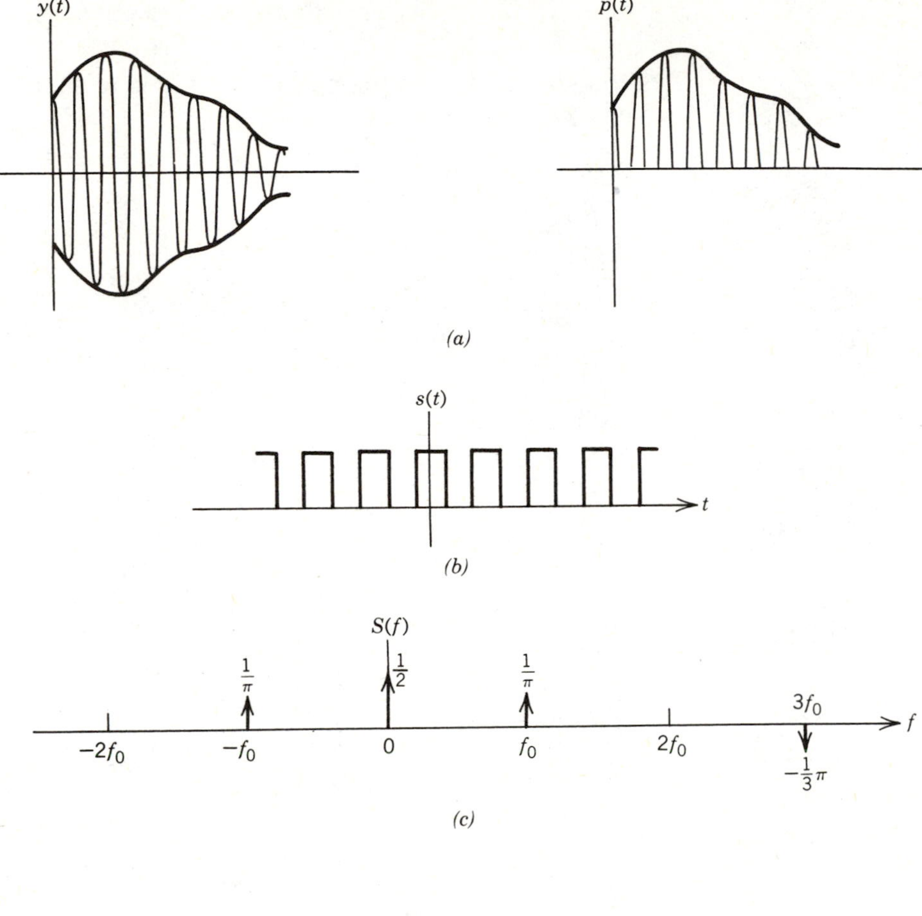

Figure 12.74. AM demodulation.

Mathematically, the diode operation is equivalent to multiplication by the periodic square wave in Fig. 12.74*b*, so long as $A > |f(t)|_{max}$ in Eq. 12.5.19. The transform of the switching function $s(t)$ is shown in Fig. 12.74*c*. Since we are multiplying $s(t)$ by $y(t)$, we convolve $S(f)$ with $Y(f)$. We label this product $p(t)$ and picture $P(f)$ in Fig. 12.74*d*. Notice that $P(f)$ consists of the original spectrum $F(f)$ centered around zero frequency, plus other spectral components every f_0 Hz. The low-pass filter in the envelope detector serves to eliminate these other spectral components, and the coupling capacitor C_c serves to eliminate the dc component (the δ function at zero frequency). Hence, we recover a signal that is proportional to the original modulating signal $f(t)$.

12.6 Response of LTI Systems by Fourier Transform

LEARNING OBJECTIVES

After completing this section you should be able to find the response of a given circuit to an arbitrary input that is Fourier transformable.

The procedures for calculating the response of LTI systems when the signal is expressed by the Fourier series and when the signal is expressed by the Fourier transform are analogous. For the Fourier series, the sum is discrete. For the Fourier transform, the sum is continuous and is given by

$$v(t) = \frac{1}{2\pi}\int_{-\infty}^{\infty} V(\omega)e^{j\omega t}\,d\omega \tag{12.6.1}$$

We know that the system response to a single exponential signal $V(\omega)e^{j\omega t}$ is given by $H(j\omega)V(\omega)e^{j\omega t}$. Therefore, by the LTI properties, the response to a signal expressed by the continuous sum in Eq. 12.6.1 is given by

$$y(t) = \frac{1}{2\pi}\int_{-\infty}^{\infty} H(j\omega)V(\omega)e^{j\omega t}\,d\omega \tag{12.6.2}$$

You should compare these equations to Eqs. 12.3.1 and 12.3.2. The following procedure is analogous to that given in Section 12.3.

1. Calculate the Fourier transform for the input signal $v(t)$. That is, calculate $V(\omega)$.
2. Calculate the transfer function $H(j\omega)$.
3. Multiply the functions $V(\omega)$ and $H(j\omega)$.
4. Use Eq. 12.6.2 to compute the response $y(t)$.

EXAMPLE 12.6.1 An impulse $v(t) = \delta(t)$ is applied to the ideal low-pass filter shown in Fig. 12.75. Calculate the response $y(t)$.

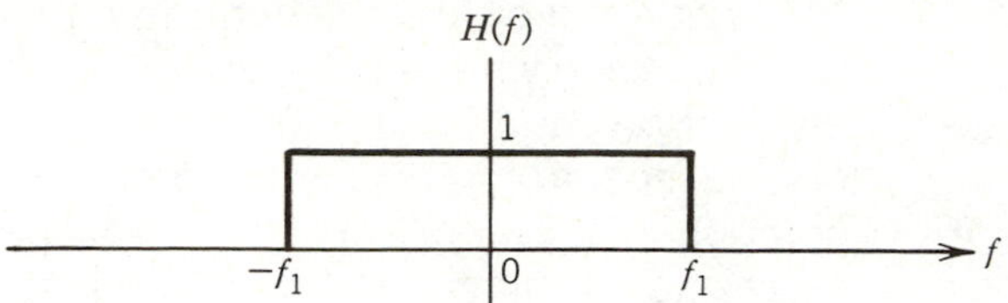

Figure 12.75. An ideal low-pass filter.

Solution

We first calculate $V(\omega)$.

$$V(\omega) = \int_{-\infty}^{\infty} \delta(t) e^{-j\omega t}\, dt = 1$$

Step 2 in the above procedure is provided by Fig. 12.75. For step 3 we have $V(\omega)H(\omega)$ given by

$$V(\omega)H(\omega) = \begin{matrix} 1 & |\omega| < \omega_1 \text{ where } \omega_1 = 2\pi f_1 \\ 0 & \text{elsewhere} \end{matrix}$$

From step 4 we have

$$y(t) = \frac{1}{2\pi} \int_{-\omega_1}^{\omega_1} e^{j\omega t}\, d\omega = \frac{\omega_1}{\pi}\left(\frac{\sin \omega_1 t}{\omega_1 t}\right), \qquad -\infty < t < \infty$$

This response is plotted in Fig. 12.76. Notice that there is a response for $t < 0$. That is, a response occurs before we apply the input. This is the reason that the ideal filter is not physically realizable.

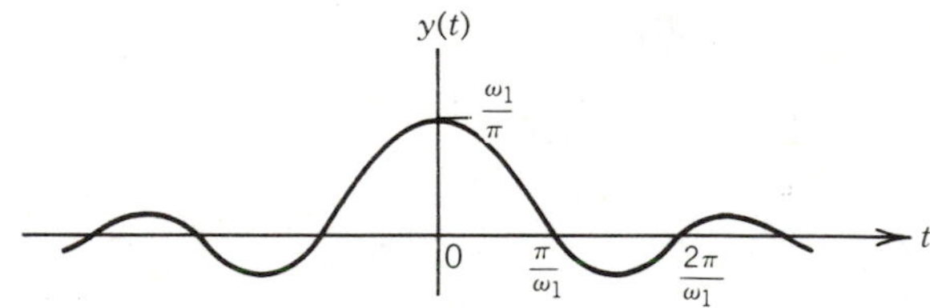

Figure 12.76. The impulse response of the ideal low-pass filter.

EXAMPLE 12.6.2 Find the impulse response of the *RC* low-pass filter shown in Fig. 12.77.

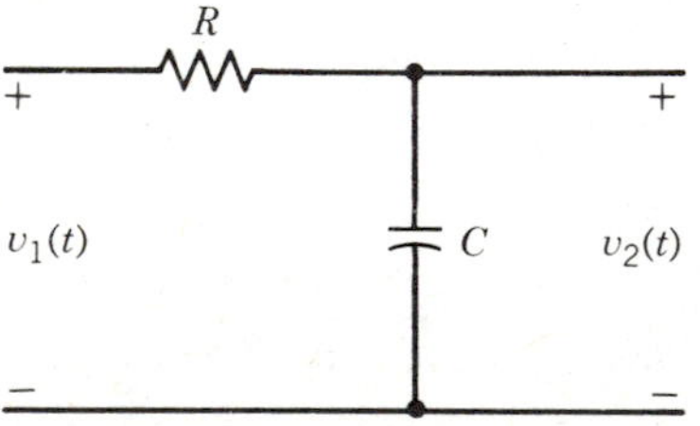

Figure 12.77. An *RC* low-pass filter.

Solution

The transfer function $H(j\omega)$ is given by

$$H(j\omega) = \frac{1/RC}{j\omega + 1/RC}$$

Since the transform of $v_1(t) = \delta(t)$ is given by 1, the output $v_2(t)$ is given by

$$v_2(t) = \frac{1}{2\pi}\int_{-\infty}^{\infty} H(j\omega)e^{j\omega t}\,d\omega$$

$$= \frac{1}{2\pi}\int_{-\infty}^{\infty} \frac{1/RC}{j\omega + 1/RC}e^{j\omega t}\,d\omega$$

We can avoid the difficulty encountered in evaluating this integral by using Table 12.4. From the first entry in the table we find that

$$v_2(t) = \frac{1}{RC}e^{-t/RC}u(t)$$

EXAMPLE 12.6.3 Find the current in the resistor of the circuit shown in Fig. 12.78 if the input voltage is $v(t)$ as shown.

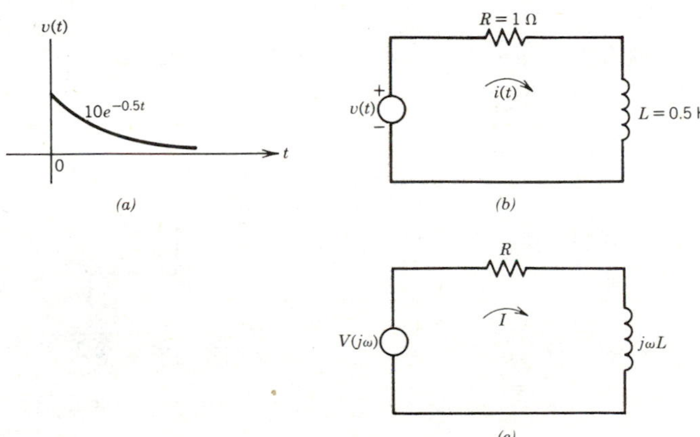

Figure 12.78

Solution

Step 1 of our procedure is to calculate $V(j\omega)$.

$$V(j\omega) = \int_0^{\infty} 10e^{-0.5t}e^{-j\omega t}\,dt = \frac{10}{0.5 + j\omega}$$

To calculate the transfer function (step 2) use the procedure outlined in

Chapter 9. The frequency domain model of the circuit is shown in Fig. 12.78*c*. From this

$$H(j\omega)=\frac{I(j\omega)}{V(j\omega)}=\frac{1}{R+j\omega L}=\frac{1}{1+j0.5\omega}$$

In step 3,

$$Y(j\omega)=V(j\omega)H(j\omega)=\frac{10}{(0.5+j\omega)(1+j0.5\omega)}$$

To accomplish the last step, find the inverse Fourier transform of $Y(j\omega)$. It is easier to use Table 12.4 than Eq. 12.6.2, even though there is no entry in the table that has the same form as $Y(j\omega)$. Partial fraction expansion will be introduced in the next chapter, but for the moment some simple algebra may be used to convince yourself that $Y(j\omega)$ is given by

$$Y(j\omega)=\frac{10}{(0.5+j\omega)(1+j0.5\omega)}=\frac{40/3}{0.5+j\omega}-\frac{20/3}{1+j0.5\omega}$$

Hence

$$y(t)=[\tfrac{40}{3}e^{-0.5t}-\tfrac{40}{3}e^{-2t}]u(t)$$

The Laplace transform is a generalization of the Fourier transform. As such, it has wider applicability to problems of the type that we are illustrating here. The next example illustrates a problem that is difficult to solve by Fourier transforms, but relatively easy to solve using Laplace transforms.

EXAMPLE 12.6.4 Find the current in the circuit in Fig. 12.78 if the input is the square pulse shown in Fig. 12.79.

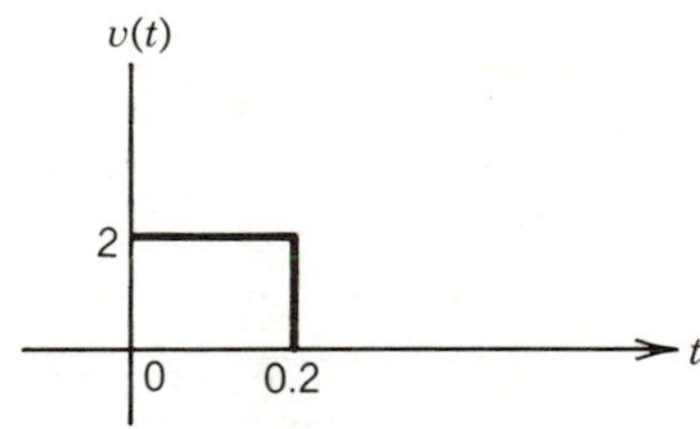

Figure 12.79

Solution

The Fourier transform of the input signal $v(t)$ is given by

$$V(j\omega)=\int_0^{0.2}2e^{-j\omega t}\,dt=0.4\left(\frac{\sin 0.1\omega}{0.1\omega}\right)e^{-j0.1\omega}$$

The product of $V(j\omega)$ and $H(j\omega)$ gives

$$Y(j\omega)=\left(\frac{0.4}{1+j0.5\omega}\right)\left(\frac{\sin 0.1\omega}{0.1\omega}\right)e^{-j0.1\omega}$$

Although the inverse transform of this function can be found, we have no convenient method for doing so. The Laplace transform is so much more convenient when solving problems of this type that we will postpone concluding this example until the next chapter.

Another convenient method for solving the above problem is to use convolution. The Laplace transform method is comparable to convolution in complexity for problems of this type.

LEARNING EVALUATION

Find the current in the *RL* circuit of Fig. 12.80 if the voltage is an impulse $v(t) = \delta(t)$.

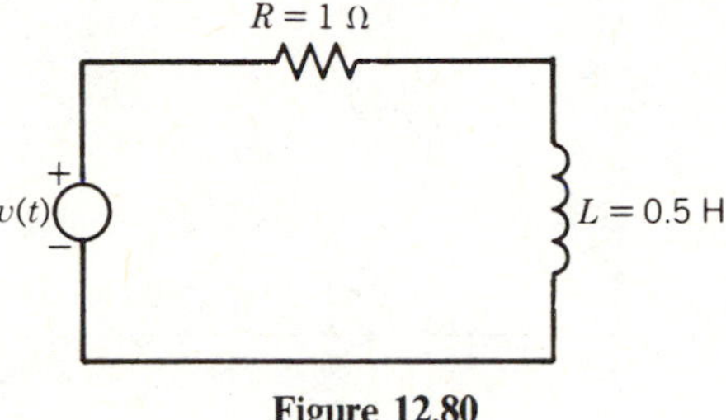

Figure 12.80

PROBLEMS

SECTION 12.1

1. Find and plot the exponential Fourier series for each waveform in Fig. 12.81. The period of each waveform is $T = 10$ s.

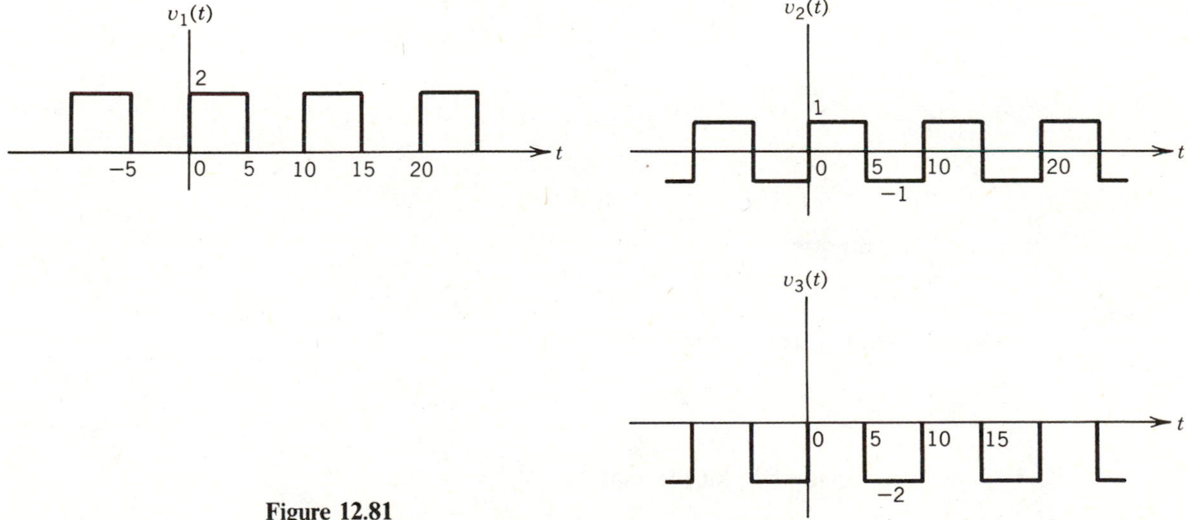

Figure 12.81

2. Find and plot the exponential Fourier series for the signal $v(t) = t$, $0 < t < 1$. The period of expansion is $T = 1$.

3. A periodic function $v(t)$ with period $T = 2$ s has Fourier series coefficients given by

$$V_k = 1, \qquad k = 0, \pm 1, \pm 2$$
$$= 0 \qquad \text{for all other } k$$

   a. Evaluate $v(t)$ at $t = 0$.
   b. Evaluate $v(t)$ at $t = 1$.
   c. Write an equation for $v(t)$ for all t.

4. Find and plot the time function $g(t)$ that is represented by the Fourier series shown in Fig. 12.82.

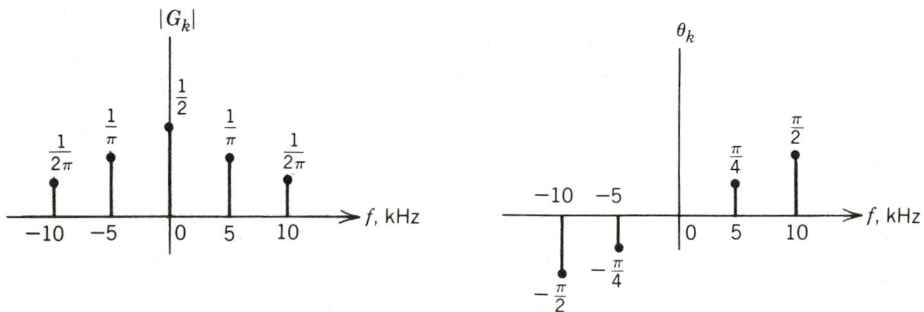

Figure 12.82. The Fourier series for Problem 4.

5. This problem is designed to illustrate convergence of the Fourier series. The series for the periodic function $v(t)$ in Fig. 12.83 is given by

$$v(t) = \sum_{k=-\infty}^{\infty} V_k e^{j2\pi kt}$$

since $T = 1$. The coefficients for $|k| \leq 5$ are given in Table 12.5.

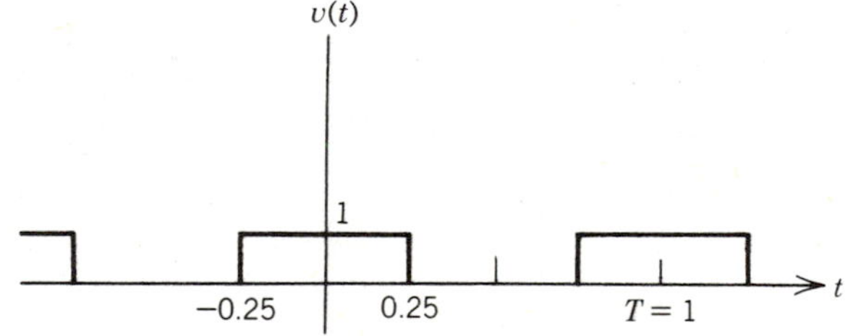

Figure 12.83. The function for Problem 5.

TABLE 12.5

k	V_k	k	V_k
−5	$\frac{1}{5\pi}$	1	$\frac{1}{\pi}$
−4	0	2	0
−3	$\frac{-1}{3\pi}$	3	$\frac{-1}{3\pi}$
−2	0	4	0
−1	$\frac{1}{\pi}$	5	$\frac{1}{5\pi}$
0	0.5		

   a. Find and plot the sum

$$v_1(t) = \sum_{k=-1}^{1} V_k e^{j2\pi kt} = \frac{1}{\pi} e^{-j2\pi t} + \frac{1}{2} + \frac{1}{\pi} e^{j2\pi t}$$

$$= \frac{1}{2} + \frac{2}{\pi} \cos 2\pi t$$

over the interval $-\frac{1}{2} < t < \frac{1}{2}$. Also calculate the mean square error between $v_1(t)$ and $v(t)$ defined by

$$\overline{e_1^2} = \frac{1}{T} \int_0^T [v(t) - v_1(t)]^2 \, dt$$

b. Find and plot the sum

$$v_3(t) = \sum_{k=-3}^{3} V_k e^{j2\pi kt}$$

over the interval $-\frac{1}{2} < t < \frac{1}{2}$. Calculate the mean square error between $v_3(t)$ and $v(t)$.

c. Find and plot the sum

$$v_5(t) = \sum_{k=-5}^{5} V_k e^{j2\pi kt}$$

over the interval $-\frac{1}{2} < t < \frac{1}{2}$. Again find the mean square error between $v_5(t)$ and $v(t)$.

d. What do you speculate the mean square error to be when all the coefficients V_k, $-\infty < k < \infty$ are used in the series?

SECTION 12.2

6. A digital signal $v(n)$ consisting of three values is given by

$$v(0) = 1.5$$
$$v(1) = 0.5$$
$$v(2) = -0.5$$

Find and plot the DFT $V(k)$, $k = 0, 1, 2$. Notice that $V(k)$ is complex valued, and therefore you must use two plots, one for the amplitude and the second for the phase angle, or else one plot for real values and the second for imaginary values. Make both plots and then sketch an equivalent three-dimensional plot.

7. Find the DFT of the signal $x(n)$ where

$$x(n) = n, \quad 0 \le n \le 3$$
$$= 0, \quad 4 \le n \le 7$$

The function is periodic with period $N = 8$.

8. Apply the inverse operation to your answer to Problem 6, and see that you obtain the original signal $v(n)$.

9. Find and plot the inverse DFT of the function plotted in Fig. 12.84. The period is $N = 3$.

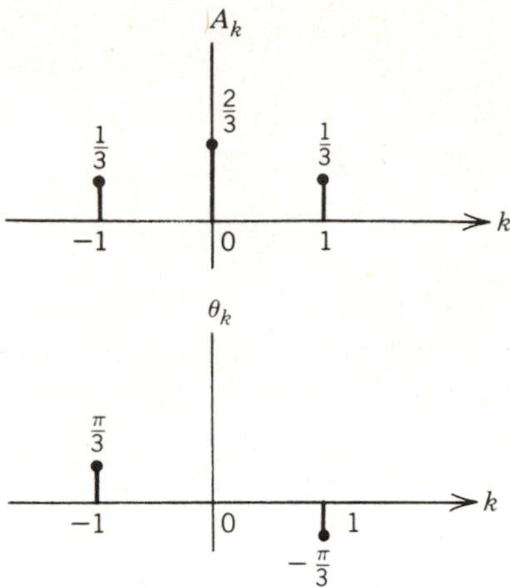

Figure 12.84

SECTION 12.3

10. The signal $v_1(t)$ in Fig. 12.81 is supplied to an ideal low-pass filter with cutoff frequency $f_1 = 0.35$ Hz.

a. Find the steady-state output $y_1(t)$.
b. Repeat for input signal $v_2(t)$ in Fig. 12.81.

11. The signal $v_1(t)$ in Fig. 12.81 is supplied to an RC low-pass filter like that illustrated in Fig. 12.35, except that here the cutoff frequency is 0.35 Hz.

a. Find the approximate steady-state output $y_1(t)$ by calculating the first few terms (through the third harmonic).
b. Repeat for input signal $v_2(t)$ in Fig. 12.81.

12. The signal $v_1(t)$ in Fig. 12.81 is supplied to an ideal band pass filter centered at $f = 0.5$ Hz with a bandwidth of 0.05 Hz. Find and plot the response $y_1(t)$.

13. The signal $v_1(t)$ in Fig. 12.81 is supplied to an ideal band reject filter centered at $f = 0.5$ Hz with a bandwidth of 0.05 Hz. Therefore this filter rejects only the fifth harmonic, passing all others. Find and plot the response $y_1(t)$.

14. The periodic digital signal in Fig. 12.30 has a period of 4. If this signal is supplied to an ideal low-pass filter whose cutoff frequency is $k = 1.5$, find and plot the response $y(n)$.

15. Replace the ideal filter in Problem 14 by the system whose difference equation is given by

$$y(n) = 0.5x(n-1) + 0.5y(n-1)$$

Find and plot the steady-state response $y(n)$.

SECTION 12.4

16. Find the Fourier transform of each pulse waveform shown in Fig. 12.85. Do as little calculation as possible.

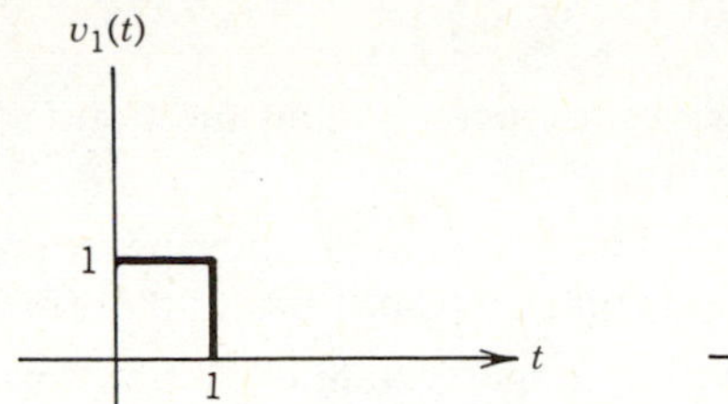

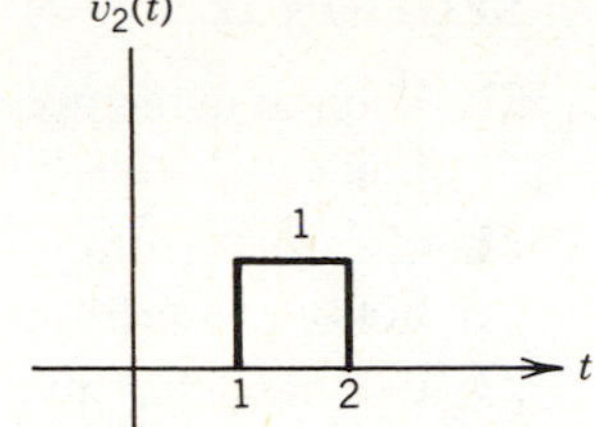

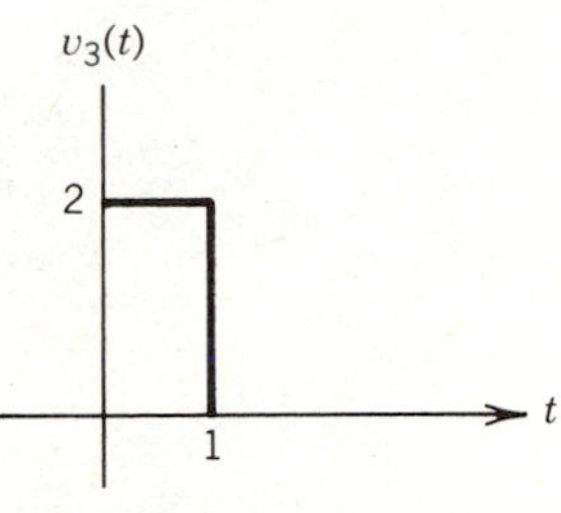

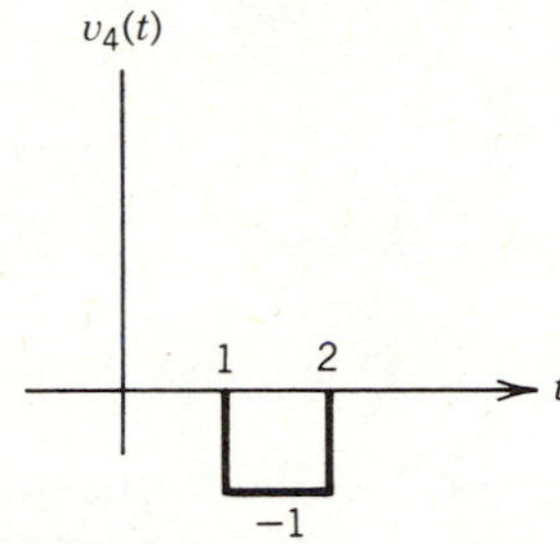

Figure 12.85

17. Notice that $x(t)$ in Fig. 12.86 is the sum of two functions in Fig. 12.85, namely $v_1(t)+v_4(t)$. Given the transforms $V_1(f)$ and $V_4(f)$, can you guess the transform of $x(t)$? Prove your answer by calculating $X(f)$.

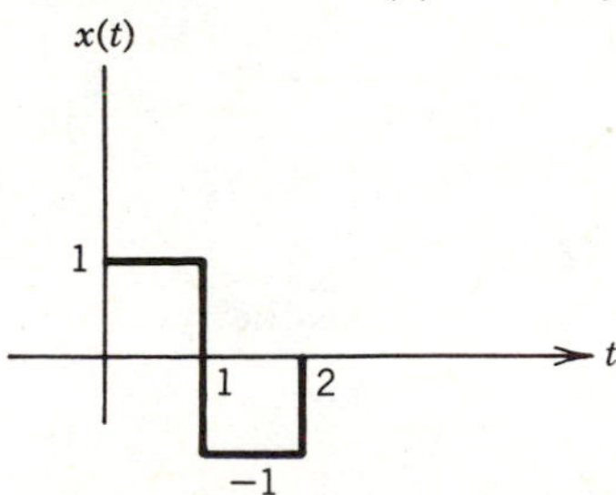

Figure 12.86

18. The Fourier transform of a time function $v(t)$ is given by

$$V(f)=\frac{1}{2j}\delta(f+1)-\frac{1}{2j}\delta(f-1)$$

Find $v(t)$.

19. Find the time function $v(t)$ corresponding to

$$V(f)=\begin{matrix} 5e^{-j2\pi f} & \quad -1<f<1 \\ 0 & \quad \text{elsewhere} \end{matrix}$$

20. Find the time function $v(t)$ corresponding to

$$V(f)=2\frac{\sin\omega}{\omega}e^{-j\omega}, \qquad -\infty<f<\infty$$

Note that you will need to guess at the solution (try a square pulse) and see if your guess is correct.

SECTION 12.5

21. Use the differentiation property to find the Fourier transform of the pulse in Fig. 12.67. (See Example 12.5.5.)

22. Use the results of Example 12.4.4 and the appropriate properties of the Fourier transform to find the transform of $x(t)$ in Fig. 12.86.

23. Use the differentiation property to find the transform of $v(t)$ in Fig. 12.87.

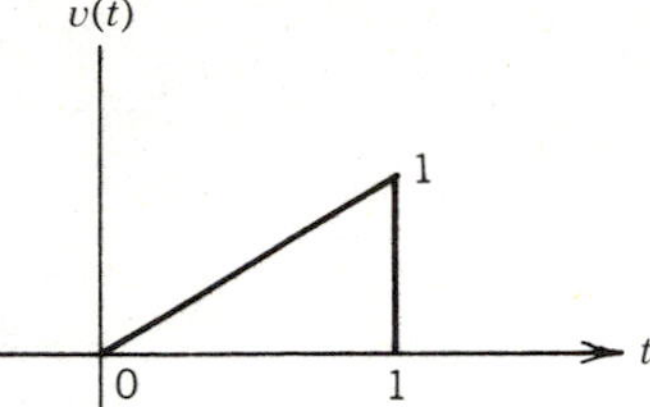

Figure 12.87

24. Use the results of Problem 23 along with the delay and additivity properties to find the Fourier transform of the signal $x(t)$ in Fig. 12.88

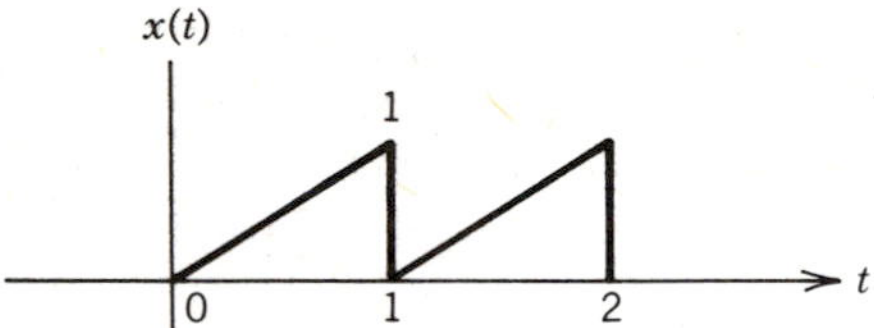

Figure 12.88

25. In the system in Fig. 12.89 $x(t)$ has the spectrum $X(f)$ as shown. Sketch the spectrum for $r(t)$ and $v(t)$.

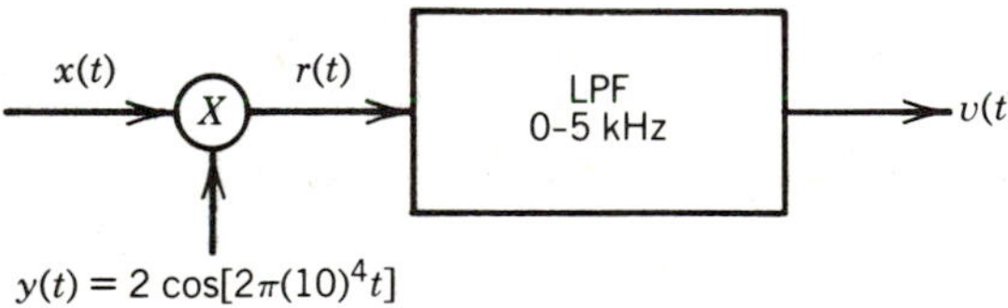

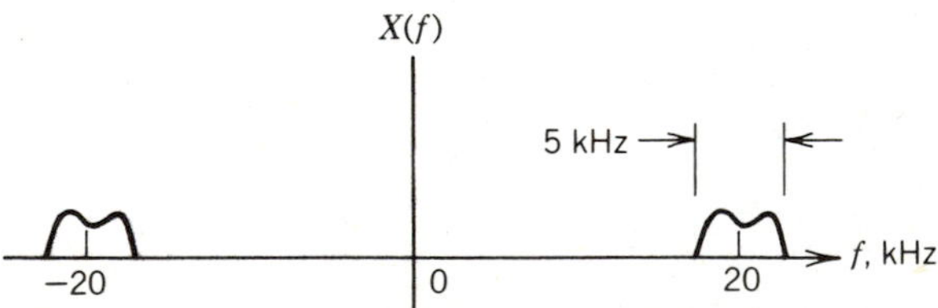

Figure 12.89

SECTION 12.6

26. Use Fourier transforms to find the impulse response of the RL circuit in Fig. 12.90. The input is the voltage $v(t)$ and the response is the current $i(t)$.

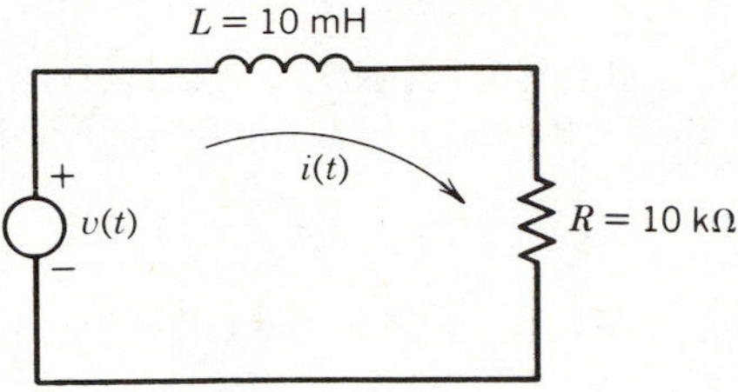

Figure 12.90

27. Use Fourier transforms to find the current in Fig. 12.90 if the exponential signal $v(t)$ in Fig. 12.78 is applied to the circuit.

LEARNING EVALUATION ANSWERS

Section 12.1

1. $V_k = \dfrac{1}{T}\displaystyle\int_{t'}^{t'+T} v(t)e^{-j2\pi f_1 kt}\,dt$

$$v(t) = \sum_{k=-\infty}^{+\infty} V_k e^{j2\pi k f_1 t}$$

2. See page 494.

3. The amplitude plot has impulses with amplitude 0.5 at ± 150 Hz. The phase plot has impulses with amplitude $\pi/2$ at ± 150 Hz.

4. $v(t) = 1 + 4\cos 20\pi t + 2\cos(60\pi t + \pi/2)$.

Section 12.2

1 and 2. Except for the multiplying factor $1/N$, the functions in Figs. 12.30 and 12.31 form a DFT pair.

3. Here is a FORTRAN program for use with our subroutine DFT.

```
   COMPLEX V(4)
   V(1) = (1.0, 0.0)
   V(2) = (2.0, 0.0)
   V(3) = (0.0, 0.0)
   V(4) = (0.0, 0.0)
   CALL DFT(V, 4, -1.0)
   DO 1 I = 1, 4
1  PRINT 10, I, V(I)
10 FORMAT (10X, I5, 2E20.8)
   STOP
   END
```

Section 12.3

1. The Fourier series for $v(t)$ is

$$V_n = \frac{1}{4}\left(\frac{\sin n\pi/4}{n\pi/4}\right)e^{-jn\pi/4}$$

where the fundamental frequency is $f_1 = 250$ Hz. Hence, the signal $y(t)$ out of the filter consists of the dc value plus the first component.

$$y(t) = 0.25 + 0.46\cos\left(1570\,t - \frac{\pi}{4}\right)$$

2. The values of $Y(k)$ are computed in the table. The inverse DFT gives

$$y(n) = \sum_{k=0}^{5} Y(k)e^{jn2\pi k/6}$$

where each value of $Y(k)$ is taken from the table. The values of $y(n)$ for one period are given in the last column.

K	$H(k)$	$X(k)$	$Y(k) = H(k)X(k)$	$y(n)$
0	0.625	0.5	0.313	0.553
1	$0.56e^{-j1.24}$	$0.33e^{-j2\pi/3}$	$0.178e^{-j3.33}$	0.531
2	$0.45e^{-j2.25}$	0	0	1.02
3	-0.4166	$-\frac{1}{6}$	0.06944	1.113
4	$0.45e^{j2.25}$	0	0	1.135
5	$0.56e^{j1.24}$	$0.33e^{j2\pi/3}$	$0.178e^{j3.33}$	0.646

Section 12.4

1. See page 528.
2. $v_3(t)$ and $v_4(t)$.
3. $V_3(f) = \dfrac{1}{a + j2\pi f}, \quad -\infty < f < \infty$
4. $v(t) = 20\dfrac{\sin 20\pi(t-2)}{20\pi(t-2)}, \quad -\infty < t < \infty$

Section 12.6

$i(t) = 2e^{-2t}u(t)$

CHAPTER 13 LAPLACE TRANSFORM

From the outset of the text it has been our goal to develop circuit analysis procedures that apply to general *RLC* circuits with virtually any type of forcing function and initial conditions. We completed the last chapter with the ability to use the Fourier series and the Fourier transform to analyze circuits whose forcing functions met the criteria stated by Eq. 12.4.4.

Unfortunately several signals of direct interest to engineers do not meet this criterion. Examples include the unit step, unit ramp, parabolic, and nondeterministic (random) signals. Further, we cannot incorporate initial conditions into the Fourier transform method.

Fortunately these restrictions can be overcome by adding a "convergence factor," $e^{-\sigma t}$, to the Fourier transform where σ is a positive real number of sufficient magnitude to guarantee that Eq. 12.4.4 is satisfied. This modified Fourier transform can be represented as

$$V(\sigma, \omega) = \int_{-\infty}^{\infty} e^{-\sigma t} v(t) e^{-j\omega t}\, dt$$

$$= \int_{-\infty}^{\infty} v(t) e^{-(\sigma + j\omega)t}\, dt$$

or

$$V(s) = \int_{-\infty}^{\infty} v(t) e^{-st}\, dt$$

where $s = \sigma + j\omega$ as before.

We will find that this is the definition of the Laplace transform. The Laplace transform is also a linear transformation, so all our LTI properties apply. Further, this transform is easily capable of incorporating initial energy storage that may be present in our circuit.

13.1 Definition of Laplace Transform

LEARNING OBJECTIVES

After completing this section you should be able to do the following:

1. Define the Laplace transform and determine which functions have a transform.
2. Evaluate the transform and determine the region of convergence in the s plane for a given time function.
3. Find the time function corresponding to a given frequency function.

There are two forms of the Laplace transform, and we first introduce the bilateral or two-sided form because of its similarity to the Fourier transform. It is given by

$$V(s)=\int_{-\infty}^{+\infty} v(t)e^{-st}\,dt \tag{13.1.1}$$

This transform is pictured in Fig. 13.1.

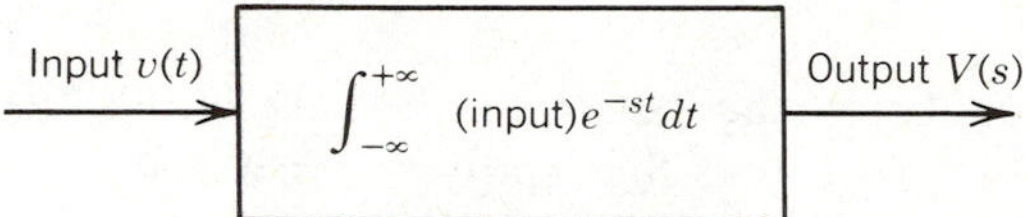

Figure 13.1. The bilateral Laplace transform.

In the Fourier transform the result (output) of the operator was a function $V(f)$ where f was a real variable. The function V was a complex valued function, but the values of f were real. Now the variable s is also allowed to be complex, and we write $s=\sigma+j\omega$.

The inverse Laplace transform is given by

$$v(t)=\frac{1}{2\pi j}\int_{\sigma-j\infty}^{\sigma+j\infty} V(s)e^{st}\,ds \tag{13.1.2}$$

and this transform is pictured in Fig. 13.2. Since s is a complex variable the limits of integration are complex. Otherwise, the Fourier and Laplace transforms are quite similar. In fact, the Laplace transform can be derived from the Fourier transform by a substitution of variable $s=\sigma+j\omega$, although we will not do so here.

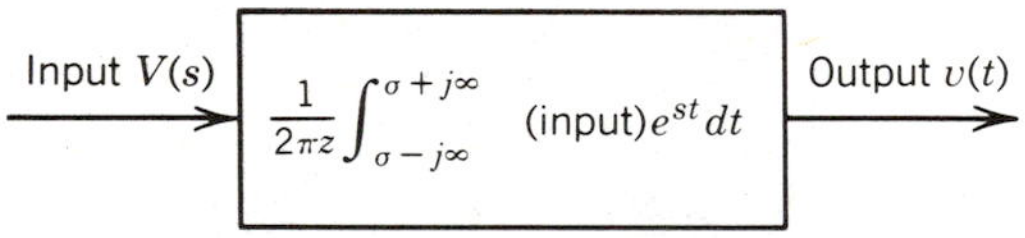

Input $V(s)$
$\frac{1}{2\pi z}\int_{\sigma-j\infty}^{\sigma+j\infty}$ (input)$e^{st}\,dt$
Output $v(t)$

Figure 13.2. The inverse Laplace transform.

The unilateral or one-sided form of the Laplace transform is widely used in circuit theory because most time functions encountered start at $s=0$. This form is given by

$$V(s)=\int_{0}^{\infty} v(t)e^{-st}\,dt \tag{13.1.3}$$

The inverse transform is again given by Eq. 13.1.2. The only difference is that the lower limit is zero in Eq. 13.1.3, instead of $-\infty$ as in Eq. 13.1.1.

Which functions have a Laplace transform? Look at Eq. 13.1.1. If $v(t)$ is to have a Laplace transform then the magnitude of $V(s)$ must be finite. This is written as

$$|V(s)|=\left|\int_{-\infty}^{\infty} v(t)e^{-st}\,dt\right|<+\infty \tag{13.1.4}$$

But since

$$|v(t)e^{-st}| = |v(t)|e^{-\sigma t}$$

the existence of the Laplace transform is guaranteed if

$$\int_{-\infty}^{\infty} |v(t)|e^{-\sigma t}\,dt < +\infty \tag{13.1.5}$$

Obviously the value of $|V(s)|$ in Eq. 13.1.4 depends on the value of σ used in Eq. 13.1.5. Equation 13.1.4 might be satisfied for some values of σ but not for others.

If a value of σ can be found for which $|V(s)|$ is finite, then the Laplace transform of $v(t)$ exists (for that value of σ). Therefore, we are faced with the following problem: Given a function $v(t)$, how do we either (1) determine that $v(t)$ does not have a Laplace transform, or (2) how do we choose the right value of σ to satisfy Eq. 13.1.5?

If there exists a positive finite number M, so that for a real α and β we have

$$|v(t)| \leq \begin{matrix} Me^{\alpha t} & \text{for } t > 0 \\ Me^{\beta t} & \text{for } t < 0 \end{matrix} \tag{13.1.6}$$

then Eq. 13.1.5 is satisfied for any value of σ between α and β. That is, for

$$\alpha < \sigma < \beta \tag{13.1.7}$$

Thus, if we can find values of M, α, β that satisfy the condition of Eq. 13.1.6, then if $\alpha < \beta$, any value of σ between α and β will do.

In order to show that Eq. 13.1.6 defines the domain of the Laplace transform, substitute these terms into Eq. 13.1.4. This gives

$$\begin{aligned} |V(s)| &\leq \left| \int_{-\infty}^{0} Me^{(\beta - s)t}\,dt + \int_{0}^{+\infty} Me^{(\alpha - s)t}\,dt \right| \\ &= \left| \frac{M}{\beta - s} e^{(\beta - s)t} \Big|_{-\infty}^{0} + \frac{M}{\alpha - s} e^{(\alpha - s)t} \Big|_{0}^{\infty} \right| \end{aligned} \tag{13.1.8}$$

and it is evident that each term will be finite so long as σ satisfies Eq. 13.1.7.

We can make the following observations:

1. A positive-time function is one that satisfies the condition $f(t) = 0$, $t < 0$. A negative-time function satisfies the condition $f(t) = 0$, $t > 0$. Now consider Eqs. 13.1.6 and 13.1.7. It is evident that the region of convergence for a positive-time function is to the right of some boundary in the complex plane, that is, $\alpha < \sigma$. For a negative-time function, the region of convergence is to the left of β, that is, $\sigma < \beta$. These regions of convergence are plotted in Fig. 13.3.
2. Notice our original condition for convergence given by Eq. 13.1.4, namely $|V(s)| < \infty$. This means that if the transform of a positive-time function has any poles [values of s for which $V(s) = \infty$] they must lie to the left of $\sigma = \alpha$ in the complex plane. Likewise, any poles of a negative-time function must lie to the right of the $\sigma = \beta$ line.

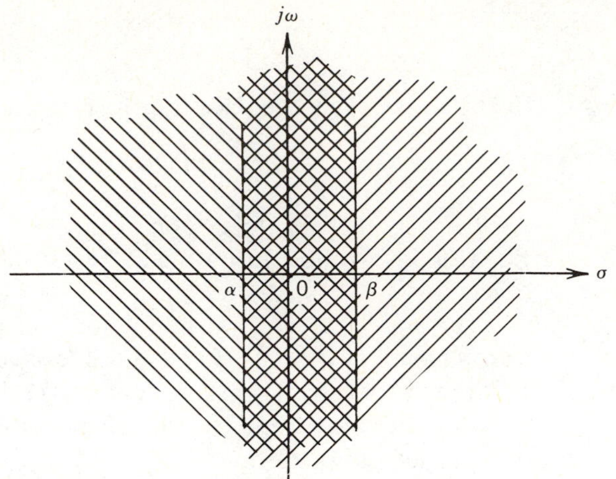

Figure 13.3. Regions of convergence for positive-time functions (to the right of α) and for negative-time functions (to the left of β).

3. If the region of convergence includes the $\sigma = 0$ line, then $v(t)$ has a Fourier transform, and it is found from the bilateral Laplace transform by setting $\sigma = 0$ (or $s = j\omega$).

We will now find the Laplace transform for several functions, along with the region of convergence in the s plane.

EXAMPLE 13.1.1 *Unit step* (Fig. 13.4).

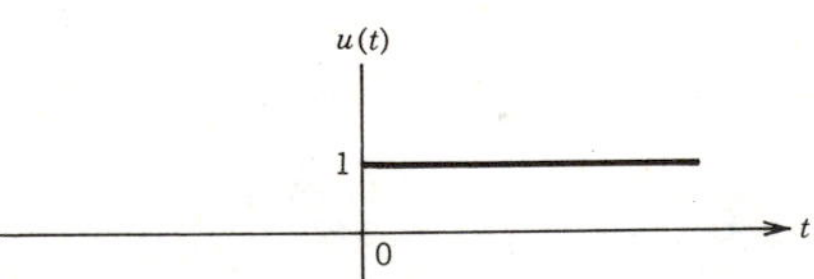

Figure 13.4. The unit step function $u(t)$.

$$v(t) = u(t)$$

$$V(s) = \int_0^{\infty} e^{-st}\,dt = \frac{1}{s} \qquad (13.1.9)$$

The region of convergence is $0 < \sigma$. To find this, set $M = 2$ (say) in Eq. 13.1.6 so that for $t > 0$, any value of $\sigma > 0$ will satisfy the condition

$$Me^{\alpha t} \geq |v(t)|, \qquad t > 0$$

Thus, the minimum value of α is zero. Any value of σ greater than this minimum α is in the region of convergence for the positive-time function. The region of convergence does not include $\sigma = 0$. Therefore, we cannot find the Fourier transform for a unit step by simply setting $s = j\omega$ in Eq. 13.1.9.

EXAMPLE 13.1.2 *Exponential function* (Fig. 13.5).

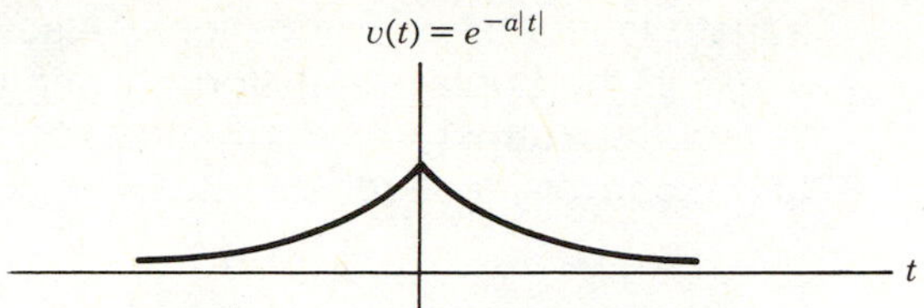

Figure 13.5. An exponential function.

$$v(t) = \begin{matrix} e^{at} & t < 0 \\ e^{-at} & t > 0 \end{matrix}$$

$$v(s) = \int_{-\infty}^{0} e^{at}e^{-st}\,dt + \int_{0}^{+\infty} e^{-at}e^{-st}\,dt$$

$$= \frac{1}{a-s} + \frac{1}{a+s} = \frac{-2a}{s^2 - a^2} \qquad (13.1.10)$$

The region of convergence is given by $-a < \sigma < a$. First consider $t > 0$. For any $M > 1$ in Eq. 13.1.6, we have

$$Me^{\alpha t} > e^{-at}, \qquad t > 0$$

so long as $\alpha > -a$. This condition will not be true for any $\alpha < -a$ for large values of t, no matter how large M is chosen.

Next consider $t < 0$. For any $M > 1$ in Eq. 13.1.6 we have

$$Me^{\beta t} > e^{at}, \qquad t < 0$$

so long as $\beta < a$. Now the region of convergence does include $\sigma = 0$. Therefore the Fourier transform is found from Eq. 13.1.10 by substituting $s = j\omega$.

$$V(\omega) = \frac{2a}{\omega^2 + a^2}$$

EXAMPLE 13.1.3 *Eternal sinusoid* (Fig. 13.6). The Laplace transform of this function does not exist because there is no value of σ for which the Laplace transform converges. Testing this function in Eq. 13.1.6 we find that for $t > 0$ we must have $\sigma > 0$. For $t < 0$ we must have $\sigma < 0$. Thus there is no value of σ for which $|V(s)| < \infty$.

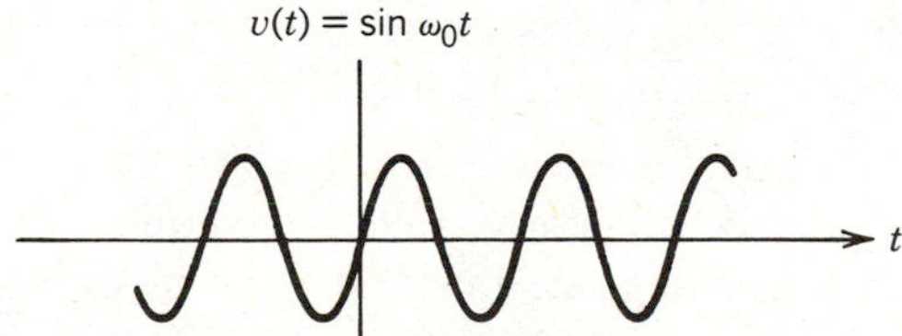

Figure 13.6. The sinusoid $\sin \omega_0 t$, $-\infty < t < +\infty$.

But recall that in Example 12.4.8 the Fourier transform of a cosine waveform was given by two δ functions. The transform of $v(t)$ is

$$V(f) = \frac{1}{2j}\delta(f - f_0) - \frac{1}{2j}\delta(f + f_0)$$

If $\sigma = 0$ is on the boundary of the convergence region, then the Fourier transform contains δ functions. Also notice that simple substitution of $s = j\omega$ in the Laplace transform (when it exists) is not appropriate in order to find the Fourier transform of a time function if $\sigma = 0$ is on the boundary of the convergence region.

EXAMPLE 13.1.4 *Positive-time sinusoid.*

$$v(t) = (\sin \omega_0 t) u(t)$$

$$V(s) = \int_0^{+\infty} \sin \omega_0 t e^{-st}\, dt = \frac{\omega_0}{s^2 + \omega_0^2} \qquad (13.1.11)$$

The region of convergence is $\sigma > 0$.

Note that the Fourier transform cannot be found by substituting $s = j\omega$ into Eq. 13.1.11. Since $\sigma = 0$ is on the boundary, the Fourier transform contains δ functions. It is given by

$$V(\omega) = \frac{\pi}{2j}\left[\delta(\omega - \omega_0) - \delta(\omega + \omega_0)\right] + \frac{\omega_0}{\omega_0^2 - \omega^2}$$

EXAMPLE 13.1.5 Determine the region of convergence for the following functions.

1. $e^{2t}u(t)$
2. $\sin 2\pi t$
3. $e^{-3|t|}$
4. $e^{2t}u(-t)$

Solution

1. $\sigma > 2$
2. None
3. $-3 < \sigma < 3$
4. $\sigma < 2$

The Inverse Transform

All of the above examples illustrated the forward transform. We now discuss the inverse transform, beginning with an illustration of the importance of the region of convergence in finding the inverse transform. The same frequency function $F(s)$ can correspond to two different time functions if the regions of convergence are different. In the following we consider how to select the proper time function for a given region of convergence in the s plane. Here is an example where two different time functions have identical transforms except for the regions of convergence.

EXAMPLE 13.1.6 Find the Laplace transform of $f_1(t)$ and $f_2(t)$ in Fig. 13.7.

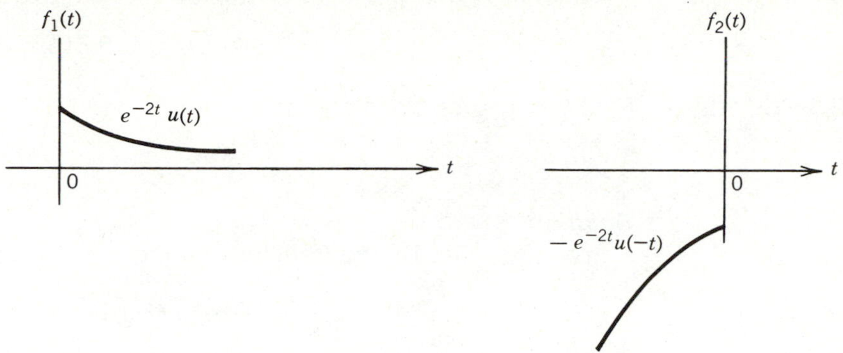

Figure 13.7. Two functions with the same transform except for the regions of convergence.

Solution

$$F_1(s) = \int_0^{\infty} e^{-2t}e^{-st}\,dt = \frac{1}{s+2}, \qquad \sigma > -2$$

$$F_2(s) = \int_{-\infty}^{0} -e^{-2t}e^{-st}\,dt = \frac{1}{s+2}, \qquad \sigma < -2$$

Notice that $F_1(s)$ and $F_2(s)$ are given by the same expression. Only the regions of convergence are different.

We will make frequent use of the following formulas.

$$e^{-at}u(t) \leftrightarrow \frac{1}{s+a}, \qquad \sigma > -a \tag{13.1.12}$$

$$-e^{-at}u(t) \leftrightarrow \frac{1}{s+a}, \qquad \sigma < -a \tag{13.1.13}$$

EXAMPLE 13.1.7 Find the inverse transform of

$$F(s) = \frac{1}{s+1}, \qquad \sigma < -1$$

Solution

From Eq. 13.1.13, we have

$$f(t) = -e^{-t}u(-t)$$

EXAMPLE 13.1.8 Find the inverse transform of

$$F(s) = \frac{1}{s+1} + \frac{1}{s+2}, \qquad -2 < \sigma < -1$$

Solution

Combining the results of Examples 13.1.6 and 13.1.7 gives

$$f(t) = e^{-2t}u(t) - e^{-t}u(-t)$$

EXAMPLE 13.1.9 Find the inverse transform of

$$F(s) = \frac{1}{s+1} + \frac{1}{s+2}, \qquad \sigma > -1$$

Solution

From Eq. 13.1.12, the transform of the first term is

$$\frac{1}{s+1} \leftrightarrow e^{-t}u(t), \qquad \sigma > -1$$

Confusion may arise about the second term because the region of convergence is not $\sigma > -2$, but if $\sigma > -1$ is it not also greater than -2? Therefore, the transform of the second term is

$$\frac{1}{s+2} \leftrightarrow e^{-2t}u(t), \qquad \sigma > -1$$

Combining these results gives

$$f(t) = (e^{-t} + e^{-2t})u(t)$$

EXAMPLE 13.1.10 Find the inverse transform of the following functions.

1. $F_1(s) = \dfrac{1}{s+a}, \qquad \sigma < -a$
2. $F_2(s) = \dfrac{1}{s+1} + \dfrac{1}{s+2}, \qquad \sigma < -2$
3. $F_3(s) = \dfrac{1}{s+a}, \qquad \sigma > -a$
4. $F_4(s) = \dfrac{1}{s}, \qquad \sigma > 0$

Solution

1. $f_1(t) = -e^{-at}u(-t)$
2. $f_2(t) = (-e^{-t} - e^{-2t})u(-t)$
3. $f_3(t) = e^{-at}u(t)$
4. $f_4(t) = u(t)$

For later reference, Table 13.1 lists the more useful Laplace transform pairs.

LEARNING EVALUATIONS

1. Define the bilateral and unilateral Laplace transform
2. Find the Laplace transform for each time function below. Also indicate the region of convergence.

TABLE 13.1
TABLE OF COMMON LAPLACE TRANSFORMS

	$v(t)$	$V(s)$	Range of σ
1.	$\delta(t)$	1	$\infty < \sigma < \infty$
2.	$u(t)$	$\frac{1}{s}$	$\sigma > 0$
3.	$u(-t)$	$\frac{1}{s}$	$\sigma < 0$
4.	$tu(t)$	$\frac{1}{s^2}$	$\sigma > 0$
5.	$-tu(-t)$	$\frac{1}{s^2}$	$\sigma < 0$
6.	$t^n u(t)$	$\frac{n!}{s^{n+1}}$	$\sigma > 0$
7.	$e^{at}u(t)$	$\frac{1}{s-a}$	$\sigma > \mathrm{Re}(a)$
8.	$-e^{at}u(-t)$	$\frac{1}{s-a}$	$\sigma < \mathrm{Re}(a)$
9.	$t^n e^{at}u(t)$	$\frac{n!}{(s-a)^{n+1}}$	$\sigma > \mathrm{Re}(a)$
10.	$(\cos at)u(t)$	$\frac{s}{s^2+a^2}$	$\sigma > 0$
11.	$(\sin at)u(t)$	$\frac{a}{s^2+a^2}$	$\sigma > 0$

a. $v(t) = 10u(t)$
b. $v(t) = tu(t)$
c. $v(t) = 115\cos 377t, \quad -\infty < t < \infty$
d. $v(t) = 100e^{-10|t|}, \quad -\infty < t < \infty$

3. Find the time function corresponding to each frequency function listed below.

a. $F(s) = \frac{1}{s}, \quad \sigma > 0.$
b. $F(s) = \frac{1}{s^2}, \quad \sigma > 0.$
c. $F(s) = \frac{1}{s+2} - \frac{1}{s-2}, \quad -2 < \sigma < 2$

13.2 Partial Fraction Expansion

LEARNING OBJECTIVES

After completing this section you should be able to do the following:

1. Convert any ratio of polynomials into a proper fraction plus a polynomial.
2. Use the methods of equating coefficients, Heaviside's expansion theorem, and substitution of variables for partial fraction expansion.

If $F(s)$ is a ratio of polynomials in s, then partial fraction expansion can be used to simplify the inverse transform operation. We can therefore avoid the complex integration of Eq. 13.1.2 by using a table of transform pairs. Most functions $F(s)$ will not be in the form of one of the transforms in Table 13.1. In this section we explain how to convert a proper fraction into a sum of terms, each of which can be found in Table 13.1.

Suppose that $F(s)$ is expressed as the ratio of polynomials,

$$F(s)=\frac{a_n s^n + a_{n-1}s^{n-1} + \cdots + a_1 s + a_0}{b_m s^m + b_{m-1}s^{m-1} + \cdots + b_1 s + b_0} = \frac{N(s)}{D(s)} \tag{13.2.1}$$

Then n is the order of the numerator $N(s)$, and m is the order of the denominator $D(s)$. We will distinguish two cases, $n \geq m$, and $n < m$. If $n \geq m$ then $F(s)$ is termed an improper fraction. If $n < m$ then the fraction is called proper. If $F(s)$ is improper then it can always be separated into a sum of polynomials in s plus a proper fraction. For example, if

$$F(s)=\frac{s^3+4s^2+6s+6}{s^2+3s+2}$$

then dividing the denominator into the numerator gives

$$F(s)=s+1+\frac{s+4}{s^2+3s+2}$$

Now $F(s)$ is in the form of a polynomial in positive powers of s plus a proper fraction. The following discussion is confined to proper fractions.

We will present three methods for expanding a proper fraction into partial fractions. Any one of these three methods may be the best to use, depending on the situation. They are called equating coefficients, Heaviside's expansion theorem, and substitution of variables. (Of course, there are more than just these three methods.)

Equating Coefficients

Suppose the proper fraction $F(s)$ is given by

$$F(s)=\frac{N(s)}{(s-s_1)(s-s_2)\cdots(s-s_m)} = \frac{A_1}{s-s_1}+\frac{A_2}{s-s_2}+\cdots+\frac{A_m}{s-s_m} \tag{13.2.2}$$

The constants A_i are known as residues. Thus A_i is the residue of the pole at $s=s_i$. We wish to evaluate the residues in Eq. 13.2.1. One approach is to equate coefficients, and this method is illustrated in Example 13.2.1.

EXAMPLE 13.2.1 Find $v(t)$ if $V(s)$ is given by

$$V(s)=\frac{5}{s(s+\frac{1}{2})}, \qquad \sigma>0$$

Solution

There is no transform pair in Table 13.1 that can be used to find $v(t)$. Therefore, we must use partial fraction expansion, where

$$\frac{5}{s(s+\frac{1}{2})}=\frac{A}{s}+\frac{B}{s+\frac{1}{2}}$$

Now if the two terms on the right side of this equation are combined by finding a common denominator, the result is

$$\frac{A}{s}+\frac{B}{s+\frac{1}{2}}=\frac{A(s+\frac{1}{2})+Bs}{s(s+\frac{1}{2})}$$

The constants A and B can be found by equating like coefficients in the numerators. That is, since $A(s+\frac{1}{2})+Bs=5$, set

$$As+Bs=0$$

and

$$\frac{A}{2}=5$$

or

$$A=10, \qquad B=-10$$

Therefore, $V(s)$ is given by

$$V(s)=\frac{10}{s}-\frac{10}{s+\frac{1}{2}}, \qquad \sigma>0$$

Use the table of Laplace transform pairs to obtain

$$v(t)=10(1-e^{-t/2})u(t)$$

EXAMPLE 13.2.2 Find $v(t)$ if $V(s)$ is given by

$$V(s)=\frac{s+1}{s^2+4s+4}, \qquad \sigma<-2$$

Solution

The denominator has repeated roots at $s=-2$. Therefore, the expansion is given by

$$\frac{s+1}{(s+2)^2}=\frac{A}{s+2}+\frac{B}{(s+2)^2}$$

Combining terms on the right gives

$$\frac{A}{s+2}+\frac{B}{(s+2)^2}=\frac{A(s+2)+B}{(s+2)^2}$$

Equating like coefficients in the numerator gives

$$s+1=A(s+2)+B$$

$$As=s \qquad \text{or} \qquad A=1$$

$$1=2A+B \qquad \text{or} \qquad B=-1$$

Therefore, $V(s)$ is given by

$$V(s)=\frac{1}{s+2}-\frac{1}{(s+2)^2}, \qquad \sigma<-2$$

and

$$v(t)=(-e^{-2t}+te^{-2t})u(-t)$$

This procedure produces m simultaneous equations, where m is the order of the denominator. When m is large, either of the next two procedures may be more appropriate.

Heaviside's Expansion Theorem

Instead of producing simultaneous equations, the method due to Oliver Heaviside (1850–1925) produces one equation at a time. Each equation contains only one unknown residue. Consider once again the proper fraction given by Eq. 13.2.2. To evaluate A_1 multiply both sides by $(s-s_1)$.

$$\frac{(s-s_1)N(s)}{(s-s_1)(s-s_2)\cdots(s-s_m)}=A_1+\frac{(s-s_1)A_2}{(s-s_2)}+\cdots+\frac{(s-s_1)A_m}{(s-s_m)} \tag{13.2.3}$$

This equation must hold for every value of s. Notice that if $s=s_1$ then every term on the right side of Eq. 13.2.3 is zero, except for A_1. The left side is not zero because the $(s-s_1)$ terms in the numerator and denominator cancel. Therefore, A_1 may be evaluated as follows:

$$A_1=(s-s_1)\left.\frac{N(s)}{D(s)}\right|_{s=s_1}$$

or, in general

$$A_i=(s-s_i)\left.\frac{N(s)}{D(s)}\right|_{s=s_i} \tag{13.2.4}$$

EXAMPLE 13.2.3 Find $f(t)$ if $F(s)$ is given by

$$F(s)=\frac{s^4+3s^3+4s^2+2s+3}{(s+1)(s+2)(s+3)(s+4)(s+5)}, \qquad -4<\sigma<-3$$

Solution

We write

$$\frac{s^4+3s^3+4s^2+2s+3}{(s+1)(s+2)(s+3)(s+4)(s+5)}=\frac{A_1}{s+1}+\frac{A_2}{s+2}+\frac{A_3}{s+3}+\frac{A_4}{s+4}+\frac{A_5}{s+5}$$

Using Eq. 13.2.4, we have

$$A_1=\left.\frac{s^4+3s^3+4s^2+2s+3}{(s+2)(s+3)(s+4)(s+5)}\right|_{s=-1}=\frac{1}{8}$$

$$A_2 = \left.\frac{s^4+3s^3+4s^2+2s+3}{(s+1)(s+3)(s+4)(s+5)}\right|_{s=-2} = -\frac{7}{6}$$

Similarly,

$$A_3 = \frac{33}{4}, \qquad A_4 = -\frac{41}{2}, \qquad A_5 = \frac{343}{24}$$

Therefore, $F(s)$ is given by

$$F(s) = \frac{1/8}{(s+1)} - \frac{7/6}{(s+2)} + \frac{33/4}{(s+3)} - \frac{41/2}{(s+4)} + \frac{343/24}{(s+5)}, \qquad -4 < \sigma < -3$$

Using Table 13.1 we find

$$f(t) = \left(-\frac{1}{8}e^{-t} + \frac{7}{6}e^{-2t} - \frac{33}{4}e^{-3t}\right)u(-t) + \left(-\frac{41}{2}e^{-4t} + \frac{343}{24}e^{-5t}\right)u(t)$$

An attempt to use the method of equating coefficients in the previous example should convince you that Heaviside's expansion theorem is worthwhile.

EXAMPLE 13.2.4 Find $f(t)$ if $F(s)$ is given by

$$F(s) = \frac{s+2}{s^2+2s+2} = \frac{s+2}{(s+1+j)(s+1-j)}, \qquad \sigma > -1$$

Solution

Expand $F(s)$ in partial fractions to obtain the following form.

$$\frac{s+2}{(s+1+j)(s+1-j)} = \frac{A}{s+1+j} + \frac{B}{s+1-j}$$

Now multiply each term by $s+1+j$ to evaluate the residue A.

$$\frac{s+2}{s+1-j} = A + \frac{(s+1+j)B}{s+1-j}$$

or

$$A = \left.\frac{s+2}{s+1-j}\right|_{s=-1-j} = \frac{1}{2} + \frac{j}{2}$$

We can use the same procedure to evaluate B, or we can recognize that B must be the conjugate of A to obtain

$$B = \frac{1}{2} - \frac{j}{2}$$

Therefore,

$$F(s) = \frac{\frac{1}{2} + (j/2)}{s+1+j} + \frac{\frac{1}{2} - (j/2)}{s+1-j}$$

Use pair 7 from Table 13.1 to obtain

$$f(t) = \left[\left(\frac{1}{2} + \frac{j}{2}\right) e^{-(1+j)t} + \left(\frac{1}{2} - \frac{j}{2}\right) e^{-(1-j)t}\right] u(t)$$
$$= e^{-t}(\cos t + \sin t) u(t)$$

This last term is obtained by using Euler's formula.

In order to extend Heaviside's method for repeated roots, consider

$$F(s) = \frac{s}{(s+1)^2} = \frac{A}{(s+1)^2} + \frac{B}{s+1}$$

Multiply both sides by $(s+1)^2$ to give

$$s = A + (s+1)B \tag{13.2.5}$$

So A can be evaluated by setting $s = -1$. This is Heaviside's method, as already discussed, but we need to modify the procedure to evaluate the coefficient B. Notice that if both sides of Eq. 13.2.5 are differentiated with respect to s the result is

$$1 = B$$

Thus we conclude that differentiation is the key to enable us to evaluate residues of repeated roots. The procedure may be stated as follows. Let

$$\frac{N(s)}{D(s)} = \frac{N(s)}{(s-s_1)^r} = \frac{A_1}{s-s_1} + \frac{A_2}{(s-s_1)^2} + \cdots + \frac{A_n}{(s-s_1)^n} + \cdots + \frac{A_r}{(s-s_1)^r}$$

First multiply both sides by $(s-s_1)^r$.

$$N(s) = (s-s_1)^{r-1} A_1 + \cdots + (s-s_1)^{r-n} A_n + \cdots + A_r$$

To evaluate the residue A_n we differentiate $(r-n)$ times to obtain

$$\frac{d^{r-n}}{ds^{r-n}} N(s) = (r-n)!\, A_n + \text{terms containing } (s-s_1)$$

We now set $s = s_1$ to eliminate the terms containing $(s-s_1)$. Dividing by $(r-n)!$ we obtain the general form given by

$$A_n = \frac{1}{(r-n)!} \frac{d^{r-n}}{ds^{r-n}} N(s)\bigg|_{s=s_1} \tag{13.2.6}$$

This procedure as stated is based on the assumption that $D(s)$ consists entirely of $(s-s_1)^r$. Usually there will be other terms in the denominator, so the form of Eq. 13.2.6 must be changed appropriately. In general, Eq. 13.2.6 should be as follows.

$$A_n = \frac{1}{(r-n)!} \frac{d^{r-n}}{ds^{r-n}} \frac{N(s)}{D(s)} (s-s_1)^r \bigg|_{s=s_1} \tag{13.2.7}$$

EXAMPLE 13.2.5 Find the inverse Laplace transform of $F(s)$ given below using Heaviside's method.

$$F(s) = \frac{4s^3 + 13s^2 + 14s + 6}{(s+1)^3(s+2)}, \qquad \sigma > -1$$

$$= \frac{A_1}{(s+1)^3} + \frac{A_2}{(s+1)^2} + \frac{A_3}{s+1} + \frac{A_4}{s+2} \tag{13.2.8}$$

Solution

In order to solve for A_1 multiply both sides by $(s+1)^3$ to obtain

$$\frac{4s^3 + 13s^2 + 14s + 6}{s+2} = A_1 + (s+1)A_2 + (s+1)^2 A_3 + \frac{(s+1)^3}{s+2} A_4 \tag{13.2.9}$$

Setting $s = -1$ gives

$$A_1 = \left. \frac{4s^3 + 13s^2 + 14s + 6}{(s+2)} \right|_{s=-1} = 1$$

To solve for A_2 differentiate each term in Eq. 13.2.9 with respect to s to obtain

$$\frac{(s+2)[12s^2 + 26s + 14] - [4s^3 + 13s^2 + 14s + 6]}{(s+2)^2}$$

$$= A_2 + 2(s+1)A_3 + \frac{3(s+2)(s+1)^2 - (s+1)^3}{(s+2)^2} A_4$$

If we now set $s = -1$ all terms except A_2 are eliminated on the right side. Therefore,

$$A_2 = \left. \frac{(s+2)(12s^2 + 26s + 14) - (4s^3 + 13s^2 + 14s + 6)}{(s+2)^2} \right|_{s=-1} = -1$$

Take one more derivative to evaluate A_3. We will leave out the details, but the correct answer is $A_3 = 2$. Now to evaluate A_4 return to Eq. 13.2.8, multiply both sides by $(s+2)$, and set $s = -2$. This gives $A_4 = 2$. Thus we have

$$F(s) = \frac{1}{(s+1)^3} - \frac{1}{(s+1)^2} + \frac{2}{s+1} + \frac{2}{s+2}, \qquad \sigma > -1$$

or

$$f(t) = \left(\frac{t^2}{2} e^{-t} - te^{-t} + 2e^{-t} + 2e^{-2t} \right) u(t)$$

Repeated roots are a complicated problem regardless of the method used. It is debatable whether or not our first method (which would require the solution of four simultaneous equations) is easier than the above process of evaluating derivatives. Our third method, presented below, is usually easier to use on problems with repeated roots.

Substitution of Variables

Assume that $F(s)$ has a rth order root at $s=s_1$. Thus, $F(s)$ can be written as

$$F(s)=\frac{N(s)}{D(s)}=\frac{N(s)}{(s-s_1)^r D_1(s)} \tag{13.2.10}$$

where $D_1(s)$ is the polynomial remaining after $(s-s_1)^r$ has been factored from $D(s)$. Now multiply both sides of Eq. 13.2.10 by $(s-s_1)^r$ to obtain

$$F(s)(s-s_1)^r=\frac{N(s)}{D_1(s)} \tag{13.2.11}$$

Make the substitution $p=s-s_1$ in Eq. 13.2.11 to obtain

$$F(s)p^r=\frac{N(p+s_1)}{D_1(p+s_1)}$$

If we now divide $D_1(p+s_1)$ into $N(p+s_1)$ with these polynomials ordered in ascending powers of p, the result can be used to algebraically solve for the r residues of $(s-s_1)^r$.

Instead of presenting the procedure formally we illustrate it with an example. We will again find the residues of $F(s)$ given in Eq. 13.2.8. First multiply both sides by $(s+1)^3$ to obtain

$$F(s)(s+1)^3=\frac{4s^3+13s^2+14s+6}{s+2}$$

Make the substitution $p=s+1$ to obtain

$$F(s)p^3=\frac{4p^3+p^2+1}{p+1}$$

Now divide the numerator by the denominator with both polynomials in ascending powers of p.

$$\begin{array}{r|l}
 & 1-p+\ 2p^2 \\ \hline
1+p & 1+0+p^2\ +4p^3 \\
 & 1+p \\ \hline
 & -p+\ p^2 \\
 & -p-\ p^2 \\ \hline
 & \qquad 2p^2+4p^3 \\
 & \qquad 2p^2+2p^3 \\ \hline
 & \qquad\qquad 2p^3
\end{array}$$

Thus,

$$F(s)p^3 = 1 - p + 2p^2 + \frac{2p^3}{p+1}$$

Dividing both sides by p^3 and substituting $p = s + 1$ gives

$$F(s) = \frac{1}{(s+1)^3} - \frac{1}{(s+1)^2} + \frac{2}{s+1} + \frac{2}{s+2} \tag{13.2.12}$$

which completes the partial fraction expansion. If the last term in Eq. 13.2.12 contained more than the single pole at $s = -2$, we would have to expand this last term to complete the solution.

EXAMPLE 13.2.6 Use the substitution of variables method to find the inverse Laplace transform of

$$F(s) = \frac{3s^2 + 2s + 4}{(s+2)^3}, \qquad \sigma > -2$$

Solution

Multiply both sides by $(s+2)^3$.

$$F(s)(s+2)^3 = 3s^2 + 2s + 4$$

Substitute $p = s + 2$.

$$F(s)p^3 = 3(p-2)^2 + 2(p-2) + 4$$
$$= 12 - 10p + 3p^2$$

or

$$F(s) = \frac{12}{(s+2)^3} - \frac{10}{(s+2)^2} + \frac{3}{s+2}$$

From Table 13.1 we find

$$f(t) = (6t^2e^{-2t} + 10te^{-2t} + 3e^{-2t})u(t)$$

LEARNING EVALUATIONS

1. Convert the following ratio of polynomials into a proper fraction plus a polynomial.

$$F(s) = \frac{s^3 + 4s^2 + 5s + 3}{s^2 + 4s + 5}$$

2. Let $F(s) = \dfrac{3}{s^2 + 5s + 4}$, $\sigma > -1$

Expand $F(s)$ in partial fractions and find $f(t)$.

a. Use the method of equating coefficients.

b. Use Heaviside's expansion theorem.

3. Let $F(s) = \dfrac{2s+1}{(s+1)^2(s+2)}$, $\qquad \sigma > -1$

Expand $F(s)$ in partial fractions and find $f(t)$.

a. Use Heaviside's expansion theorem.
b. Use the method of substitution of variables.

13.3 Properties of the Laplace Transform

LEARNING OBJECTIVES
After completing this section you should be able to do the following:

1. Use the differentiation property for the unilateral transform to solve differential equations.
2. Use the various other properties as an aid in evaluating Laplace transforms.

It is the properties of transforms that make them useful. The properties of the Fourier, Laplace, and z transforms make them useful in system analysis, and that is the reason that they are introduced here. The following is a list of some of the more useful properties of the Laplace transform. These properties apply to either the bilateral or unilateral transform.

Superposition

$$av_1(t) + bv_2(t) \leftrightarrow aV_1(s) + bV_2(s) \qquad (13.3.1)$$

Scaling

$$v\left(\frac{t}{a}\right) \leftrightarrow aV(as) \qquad \text{for real} \quad a > 0 \qquad (13.3.2)$$

Delay

$$v(t - t_0) \leftrightarrow V(s)e^{-st_0} \qquad (13.3.3)$$

Modulation

$$v(t)e^{s_0 t} \leftrightarrow V(s - s_0) \qquad (13.3.4)$$

Convolution

$$\int_{-\infty}^{+\infty} v_1(\lambda)v_2(t-\lambda)\,d\lambda \leftrightarrow V_1(s)V_2(s) \qquad (13.3.5)$$

Multiplication

$$v_1(t)v_2(t) \leftrightarrow \frac{1}{2\pi j}\int_{\sigma-j\infty}^{\sigma+j\infty} V_1(\lambda)V_2(s-\lambda)\,d\lambda \qquad (13.3.6)$$

Here are some examples to illustrate the use of these properties. After presenting several examples, we will discuss two additional properties, the differentiation property and the integration property for the single-sided transforms. We are isolating our discussion of these properties because they apply only to the unilateral transform, and also because they are particularly useful in the solution of differential equations.

EXAMPLE 13.3.1 Find the Laplace transform of the pulse shown in Fig. 13.8.

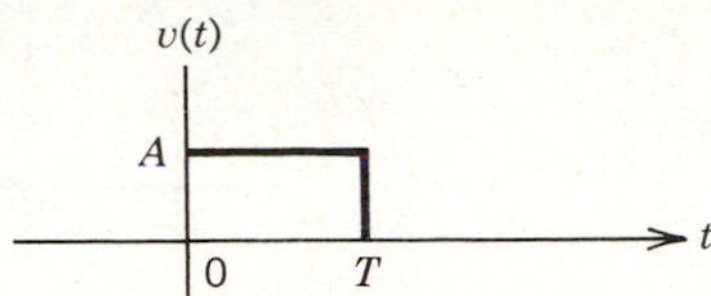

Figure 13.8. The square pulse $v(t)$.

Solution

We could use the definition, Eq. 13.1.1, but in order to illustrate the use of properties we first express $v(t)$ as the sum of two step functions.

$$v(t) = Au(t) - Au(t-T)$$

The delay property, plus Eq. 13.1.9, gives

$$u(t-T) \leftrightarrow \frac{1}{s} e^{-sT}$$

By the superposition property we have

$$V(s) = \frac{A}{s} - \frac{A}{s} e^{-sT}$$

The region of convergence is the entire s plane.

> For any signal that is nonzero only over a finite time interval, the region of convergence is the entire s plane.

EXAMPLE 13.3.2 Find the Laplace transform for the signal $v(t)$ shown in Fig. 13.9.

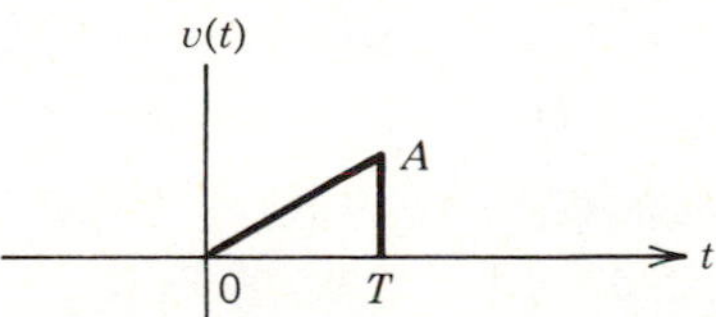

Figure 13.9. A triangular pulse.

Solution

The Laplace transform of a ramp signal $tu(t)$ is given by

$$tu(t) \leftrightarrow \frac{1}{s^2}, \qquad \sigma > 0$$

We can express $v(t)$ as the sum of ramp and step functions as shown in Fig. 13.10. Therefore,

$$V(s) = \frac{A}{T}\left(\frac{1}{s^2} - \frac{1}{s^2} e^{-Ts}\right) - \frac{A}{s} e^{-Ts}, \qquad -\infty < \sigma < \infty$$

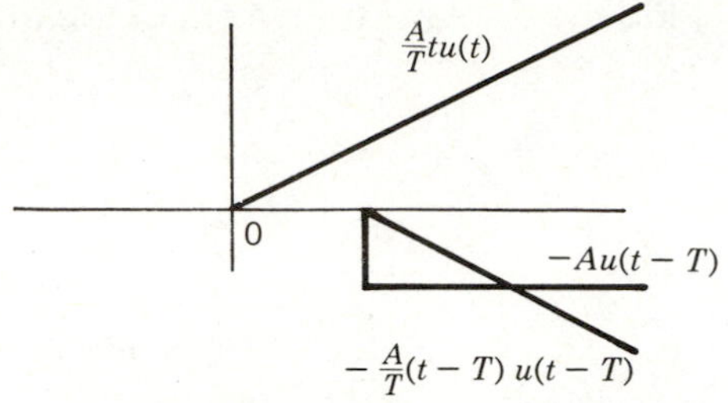

Figure 13.10. The components of $v(t)$.

If we apply Eq. 13.1.1 directly we also obtain

$$V(s)=\int_0^T \frac{A}{T} te^{-st}\,dt=\frac{A}{T}\left\{-\frac{t}{s}e^{-st}\Big|_0^T+\int_0^T \frac{1}{s}e^{-st}\,dt\right\}$$

$$=\frac{A}{T}\left[\frac{1}{s^2}-\frac{1}{s^2}e^{-Ts}\right]-\frac{A}{s}e^{-Ts}$$

The region of convergence is the entire s plane, even though the region of convergence for each ramp and step function in Fig. 13.10 is only a half-plane.

Also since the $\sigma=0$ line is in the region of convergence, we can find the Fourier transform directly by substituting $s=j\omega$ into the expression for $V(s)$.

EXAMPLE 13.3.3 Find the Laplace transform of the following functions.

1. $(t-1)u(t)$
2. $(t-1)u(t-1)$
3. $e^{-a|t|}, \qquad -\infty<t<\infty$
4. $tu(-t)$

Solution

1. $\dfrac{1}{s^2}-\dfrac{1}{s}, \qquad \sigma>0$
2. $\dfrac{1}{s^2}e^{-s}, \qquad \sigma>0$
3. $\dfrac{-2a}{s^2-a^2}, \qquad -a<\sigma<a$
4. $\dfrac{-1}{s^2}, \qquad \sigma<0$

Let us now discuss the use of Laplace transforms as an aid in the solution of differential equations. The differentiation property of the single-sided transform is given by

$$\frac{dv}{dt}\leftrightarrow sV(s)-v(0) \tag{13.3.7}$$

It is this property that makes the single-sided transform so useful in the solution of differential equations, because the initial conditions are included as a matter of course.

The proof of this property is straightforward. By definition

$$\mathscr{L}\left(\frac{dv}{dt}\right) = \int_0^\infty \frac{dv}{dt} e^{-st}\, dt$$

where $\mathscr{L}(\cdot)$ stands for the Laplace transform. Integrating by parts gives

$$\mathscr{L}\left(\frac{dv}{dt}\right) = v(t)e^{-st}\Big|_0^\infty + s\int_0^\infty v(t)e^{-st}\, dt$$

Since s lies in the region of convergence, the quantity $v(t)e^{-st}$ approaches zero at $t = \infty$. Therefore,

$$\mathscr{L}\left(\frac{dv}{dt}\right) = -v(0) + sV(s)$$

For higher derivatives we find, either by repeated application of Eq. 13.3.7, or by repeated integration, that the differentiation property is given by

$$\frac{d^n v}{dt^n} \leftrightarrow s^n V(s) - s^{n-1}v(0) - s^{n-2}\frac{dv}{dt}(0) - \cdots - \frac{d^{n-1}v}{dt^{n-1}}(0) \qquad (13.3.8)$$

The integration property of the unilateral transform is given by

$$\int_{-\infty}^{t} v(\lambda)\, d\lambda = \int_{-\infty}^{0} v(\lambda)\, d\lambda + \int_0^t v(\lambda)\, d\lambda \leftrightarrow \frac{1}{s}v(0) + \frac{1}{s}V(s)$$

The proof of this property is straightforward.

EXAMPLE 13.3.4 Use the differentiation property of single-sided transforms to solve the following differential equation.

$$\frac{d^2y}{dt^2} + 5\frac{dy}{dt} + 4y(t) = 10, \qquad t > 0$$

$$y(0) = 0.2$$

$$\frac{dy}{dt}(0) = 1$$

Solution

From Eq. 13.3.8, the second derivative has the transform

$$\frac{d^2y}{dt^2} \leftrightarrow s^2Y(s) - sy(0) - \frac{dy}{dt}(0)$$

and

$$\frac{dy}{dt} \leftrightarrow sY(s) - y(0)$$

Therefore, we may replace the differential equation by its transform, given by

$$s^2Y(s) - 0.2s - 1 + 5sY(s) - 0.2 + 4Y(s) = \frac{10}{s}$$

Solving for $Y(s)$ gives

$$Y(s) = \frac{0.2s^2 + 1.2s + 10}{s(s+4)(s+1)}$$

$$= \frac{2.5}{s} + \frac{0.7}{s+4} - \frac{3}{s+1}$$

Therefore, the solution to the differential equation is the inverse Laplace transform of $Y(s)$, given by $y(t) = (2.5 + 0.7e^{-4t} - 3e^{-t})u(t)$.

The region of convergence for $Y(s)$ is in the right-half s plane $\sigma > 0$. Since we are using the single-sided transform, and since the differential equation is valid for positive time, the solution to all our differential equations will be a positive-time function.

This powerful technique allows us to solve second order circuits in a straightforward manner. Compare the solution of the second order circuit in Example 13.3.5 with the methods we used earlier in the text.

EXAMPLE 13.3.5 The circuit shown in Fig. 13.11 has the following initial conditions

$$i_L(0) = 0.25 \text{ V}$$
$$v_c(0) = 1 \text{ V}$$

Find $i(t)$ if the forcing function is

$$v_{in}(t) = e^{-2t}u(t)$$

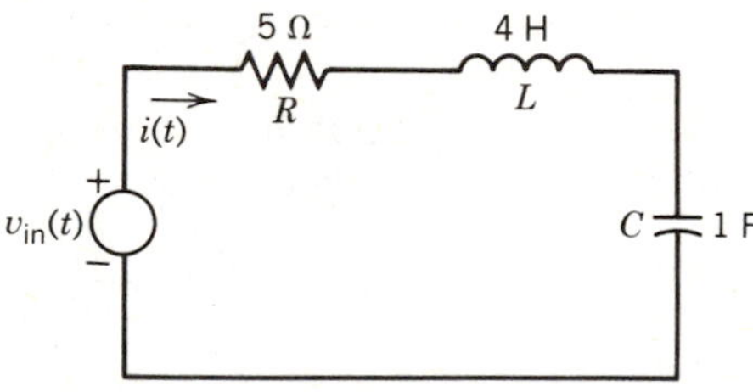

Figure 13.11

Solution

Writing a mesh equation around the circuit gives

$$-v_{in}(t) + Ri(t) + L\frac{di}{dt} + \frac{1}{C}\int i(t)\,dt = 0$$

Transforming this equation using our recently discussed properties yields

$$-\frac{1}{s+2} + 5I(s) + 4sI(s) - 4i(0) + \frac{1}{sC}I(s) + \frac{1}{C}\frac{v_c(0)}{s} = 0$$

Solving for $I(s)$ gives

$$I(s) = \frac{(s^2 + 2s - 2)}{(s+2)(4s+1)(s+1)}$$

Using partial fraction expansion we can obtain

$$I(s) = \frac{-\frac{2}{7}}{s+2} + \frac{-\frac{13}{7}}{4s+1} + \frac{1}{s+1}$$

Then taking the inverse transform gives $i(t)$.

$$i(t) = \left(-\tfrac{2}{7}e^{-2t} - \tfrac{13}{28}e^{-0.25t} + e^{-t}\right)u(t)$$

To check our result, evaluate $i(t)$ for $t=0$:

$$i(0) = -\tfrac{2}{7} - \tfrac{13}{28} + 1 = 0.25$$

which is correct.

LEARNING EVALUATIONS

1. Use the properties listed in Eqs. 13.3.1 through 13.3.6 to find the Laplace transform of the time functions shown in Fig. 13.12.

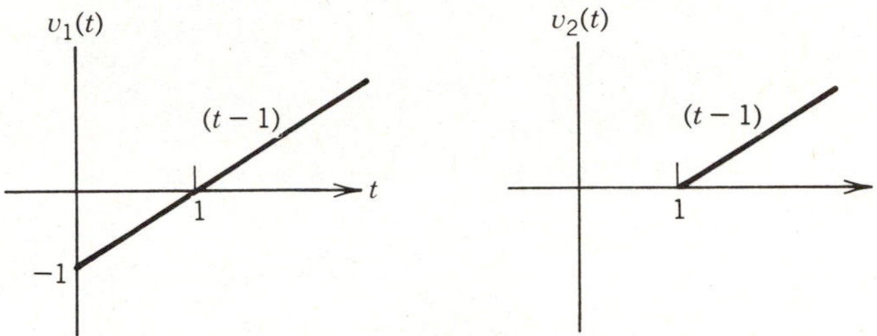

Figure 13.12

2. Solve the following differential equation by Laplace transforms.

$$\frac{d^2y}{dt^2} + 3\frac{dy}{dt} + 2y(t) = 5, \qquad t \geq 0$$

$$\frac{dy}{dt}(0) = 1$$

$$y(0) = -1$$

13.4 The Response by Laplace Transform

LEARNING OBJECTIVES

After completing this section you should be able to find the response of a given circuit to an arbitrary input that is Laplace transformable.

Here we combine the concepts of transfer function and the transform of signals to find the system response. Since the continuous-time transfer function for a stable linear system is the response to e^{st} divided by e^{st}, and since the Laplace transform is a method of expressing the input signal in terms of e^{st}, then we use our input–output pair methods of Chapter 8 to find the system response. The procedure is as follows:

1. Find the LTI system transfer function.

2. Find the Laplace transform of the input signal.
3. Multiply the input by the transfer function.
4. Find the inverse Laplace transform of the product. This time function is the desired response.

EXAMPLE 13.4.1 A unit step voltage is applied to the *RC* low-pass filter shown in Fig. 13.13. Find the response by using Laplace transforms.

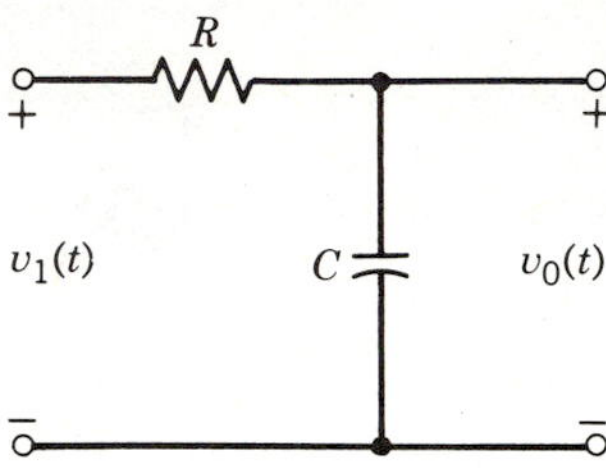

Figure 13.13. Low-pass filter.

Solution

The transfer function is given by

$$H(s) = \frac{V_0(s)}{V_1(s)} = \frac{1/sC}{R + 1/sC} = \frac{1/RC}{s + 1/RC}$$

The Laplace transform of the input signal is

$$u(t) \leftrightarrow \frac{1}{s}$$

The product of the input signal and the transfer function is

$$V_0(s) = V_1(s)H(s) = \frac{1/RC}{s(s + 1/RC)}$$

$$= \frac{1}{s} - \frac{1}{s + 1/RC}$$

Taking the inverse transform of this expression gives the desired response.

$$v_0(t) = (1 - e^{-t/RC})u(t)$$

We stated previously that the Laplace transform had wider application to the problem of finding system response than did the Fourier transform. In Example 12.6.4 we encountered a problem that was difficult at best. Here is the solution to that problem by Laplace transform.

EXAMPLE 13.4.2 Find the current in the circuit of Fig. 13.14 if the input is the square pulse shown.

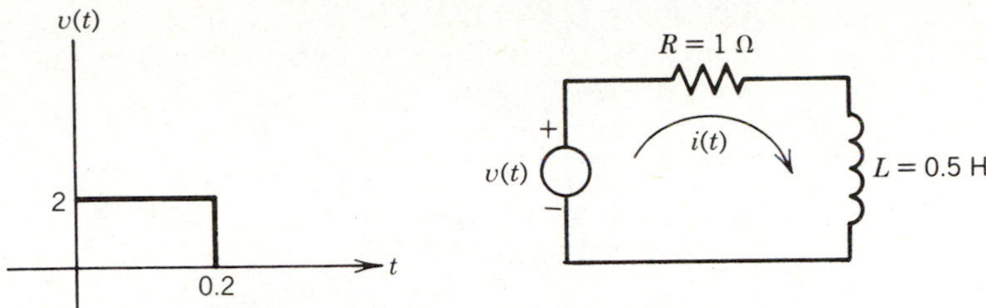

Figure 13.14. Circuit for Example 13.4.2.

Solution

The transfer function is given by

$$H(s) = \frac{I(s)}{V(s)} = \frac{1}{R + sL} = \frac{2}{s + 2}$$

The Laplace transform of the input $v(t)$ is

$$V(s) = \frac{2}{s} - \frac{2}{s}e^{-0.2s}$$

Therefore, the response in the frequency domain is

$$I(s) = V(s)H(s) = \frac{4}{s(s+2)} - \frac{4e^{-0.2s}}{s(s+2)}$$

Expanding the first term in partial fractions gives

$$\frac{4}{s(s+2)} = \frac{2}{s} - \frac{2}{s+2} \leftrightarrow 2(1 - e^{-2t})u(t)$$

The second term in the response is identical to the first term except for the $e^{-0.2s}$ factor. This factor serves to delay the time signal by 0.2 s. Therefore,

$$\frac{4e^{-0.2s}}{s(s+2)} \leftrightarrow 2(1 - e^{-2(t-0.2)})\,u(t - 0.2)$$

Combining these results gives

$$i(t) = 2(1 - e^{-2t})u(t) - 2[1 - e^{-(t-0.2)}]\,u(t - 0.2)$$

The power of the Laplace transform can be appreciated if this example is compared to Example 12.6.4. But while we are making comparisons, it should be noted that convolution is probably a better method to solve this problem than either of the frequency domain approaches.

With the techniques of this chapter added to our bag of tools we can now completely analyze any continuous-time circuit containing resistors, inductors, and capacitors which contains a forcing function that is Laplace transformable. Remember, however, that we always want to choose the least complicated tool that will allow us to perform the required analysis. Although possible, it is not necessary to use Laplace transforms to determine the current through a resistor that is connected in series with a voltage source.

LEARNING EVALUATION

Find the current in the RC circuit shown in Fig. 13.15 if the input is the square pulse shown.

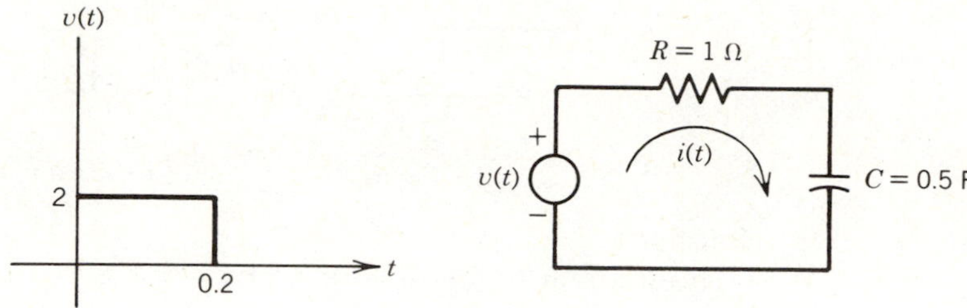

Figure 13.15

PROBLEMS

SECTION 13.1

1. Determine the region of convergence for the bilateral Laplace transform in the complex s plane for each time function below.
   a. $v(t) = 10u(-t)$
   b. $v(t) = e^{-0.8|t|}, \qquad -\infty < t < \infty$
   c. $v(t) = e^{-0.8t}u(t)$
   d. $v(t) = e^{t}u(-t)$
2. Repeat Problem 1 for the following time functions.
   a. $v(t) = (\cos 2\pi t)u(t)$
   b. $v(t) = e^{3t}u(-t)$
   c. $v(t) = tu(t)$
   d. $v(t) = e^{-3t}, \qquad -\infty < t < \infty$
3. Find the Laplace transform of those functions in Problem 1 where possible.
4. Find the Laplace transform of those functions in Problem 2 where possible.
5. Find the inverse Laplace transform of the functions given below.

$$F_1(s) = \frac{1}{s+2}, \qquad \sigma < -2$$

$$F_2(s) = \frac{1}{s+1}, \qquad \sigma > -1$$

$$F_3(s) = \frac{1}{s+1} + \frac{1}{s+2}, \qquad -2 < \sigma < -1$$

$$F_4(s) = \frac{1}{s+1} + \frac{1}{s+2}, \qquad \sigma > -1$$

6. Repeat Problem 5 for the functions given below.

$$F_1(s) = \frac{1}{s+3}, \qquad \sigma > -3$$

$$F_2(s) = \frac{1}{s+3} + \frac{1}{s+2}, \qquad \sigma < -3$$

$$F_3(s) = \frac{1}{s+3} + \frac{1}{s+2}, \qquad -3 < \sigma < -2$$

$$F_4(s) = \frac{1}{s}, \qquad \sigma > 0$$

SECTION 13.2

7. Convert the following ratio of polynomials into a proper fraction plus a polynomial. Then find the inverse Laplace transform.

$$F(s) = \frac{s(s+1)}{s^2+5s+6}, \qquad \sigma > -2$$

8. Expand $V(s)$ in partial fractions and find $v(t)$.

a. Use the method of equating coefficients.
b. Use Heaviside's expansion theorem.
c. Use the method of substitution of variables.

$$V(s) = \frac{s}{(s+3)^2(s+1)}, \qquad \sigma > -1$$

9. Find $g(t)$ corresponding to $G(s)$.

$$G(s) = \frac{s^2+2s+1}{s^2+4s+3}, \qquad -3 < \sigma < -1$$

10. Find $f(t)$ corresponding to $F(s)$.

$$F(s) = \frac{s^2+3s+2}{(s+3)^2(s+4)}, \qquad \sigma > -3$$

SECTION 13.3

11. The function $x(t)$ in Fig. 13.16 can be derived by convolving $v(t)$ with itself. Use the Laplace transform of $v(t)$ as derived in Example 13.3.1 to find the Laplace transform of $x(t)$.

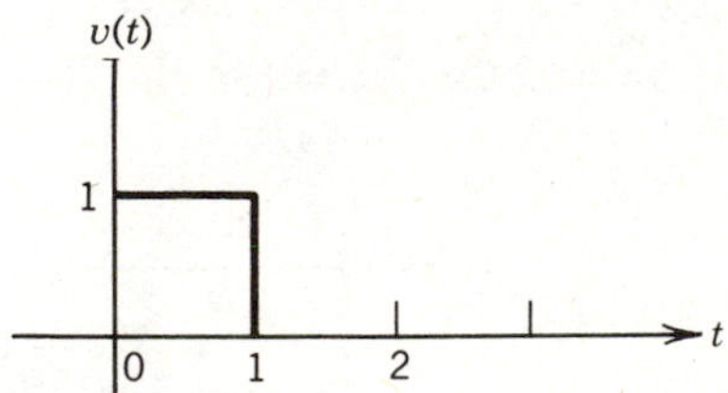

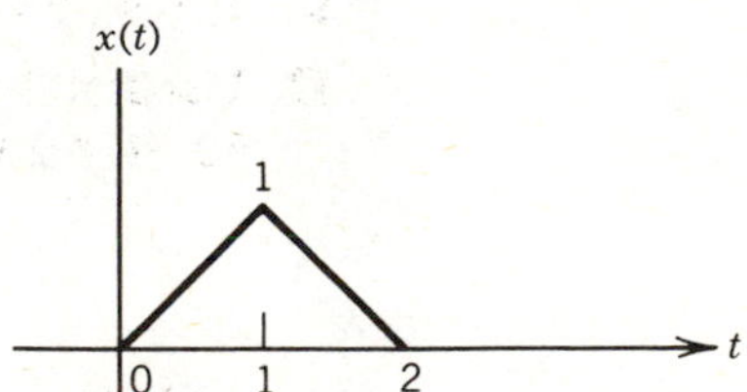

Figure 13.16

12. The transform of a triangular pulse $v(t)$ is derived in Example 13.3.2. Use the delay property to find the transform of $x(t)$ in Fig. 13.17.

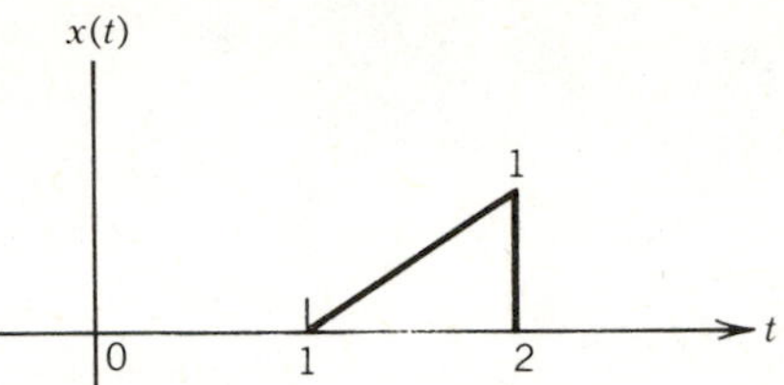

Figure 13.17

13. Write the differential equation for the current in the circuit shown in Fig. 13.18. The initial current in the inductor is zero. Use Laplace transforms to find the current.

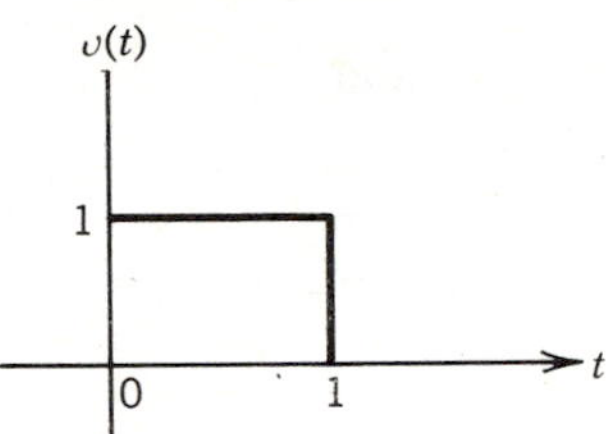

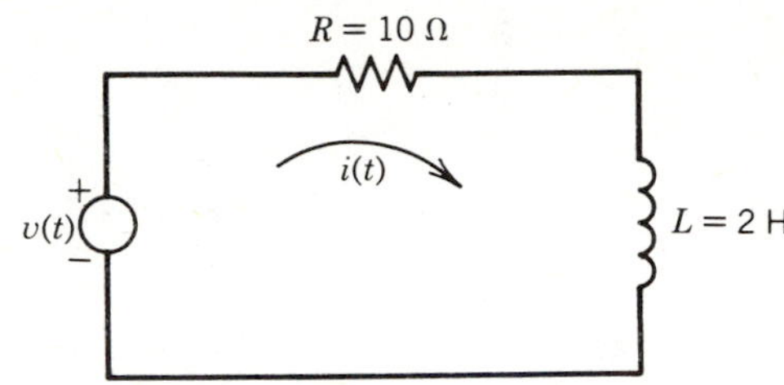

Figure 13.18

14. Repeat Problem 13 if the inductor current at $t = 0$ is 0.2 A.

SECTIONS 13.3 AND 13.4

15. Use Laplace transforms to find the current in the *RC* circuit shown in Fig. 13.19. The initial capacitor voltage is $v_c(0) = 0$.

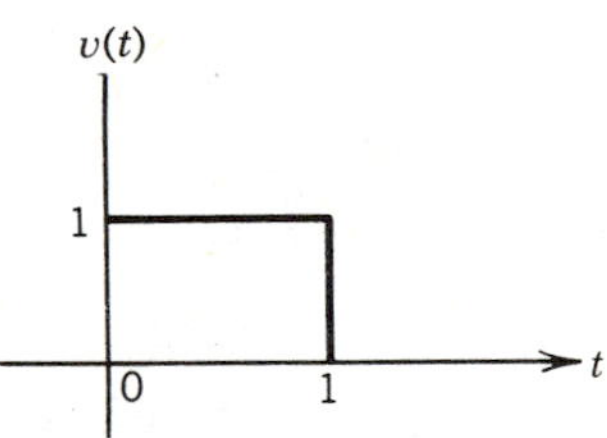

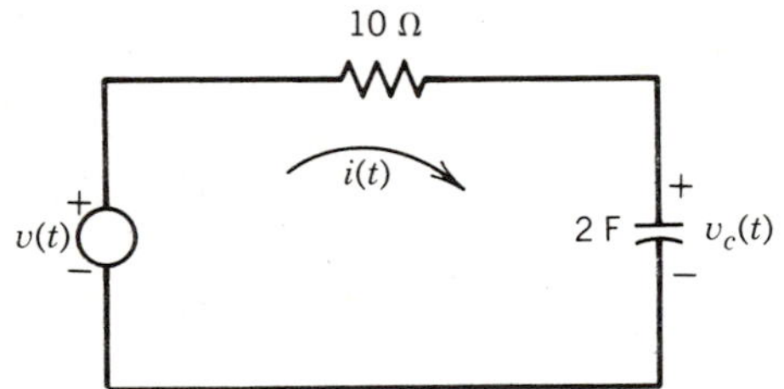

Figure 13.19

16. Repeat Problem 15 if $v_c(0) = 2$ V.

17. Use Laplace transforms to find the current in the *RLC* circuit shown in Fig. 13.20. Assume $i_L(0) = v_c(0) = 0$.

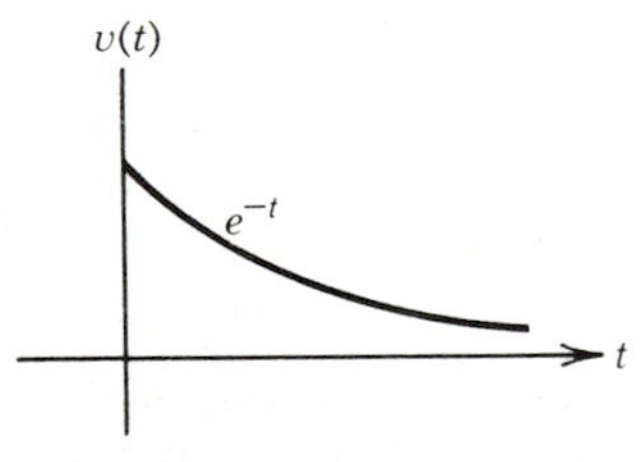

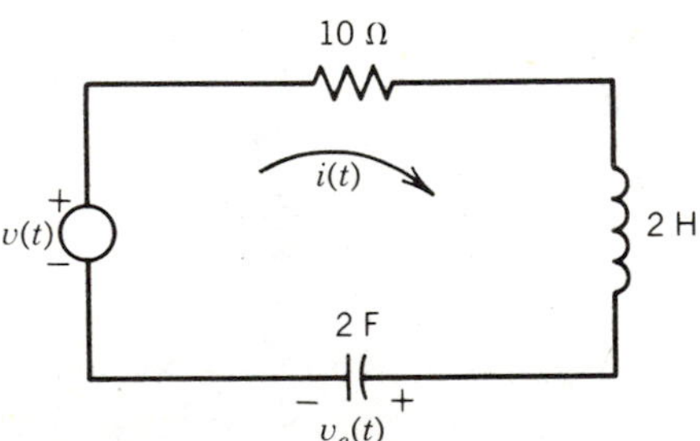

Figure 13.20

18. Repeat Problem 17 if $i_L(0) = 0.2$ A and $v_c(0) = 2$ V.

19. Find the transfer function for the circuit in Fig. 13.21, where the input is $v_1(t)$ and the output is $v_2(t)$. Now use Laplace transforms to find the response to the triangular pulse shown.

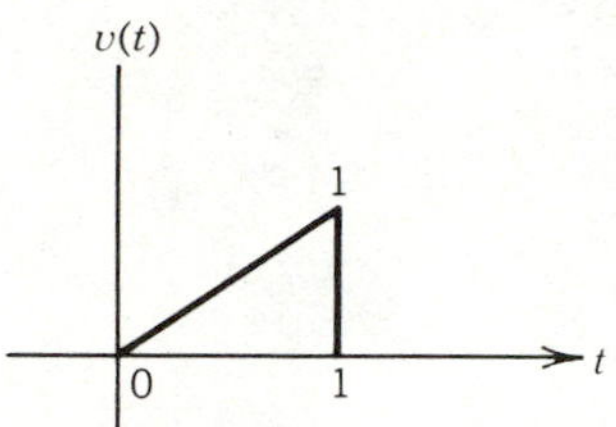

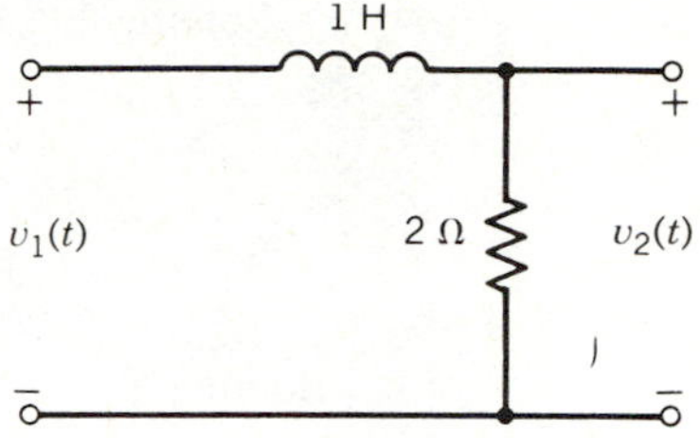

Figure 13.21

LEARNING EVALUATION ANSWERS

Section 13.1

1. See Eqs. 13.1.1 and 13.1.3.

2. **a.** $V(s) = 1/s, \quad \sigma > 0$
b. $V(s) = 1/s^2, \quad \sigma > 0$
c. None.
d. $V(s) = \dfrac{2000}{100 - s^2}, \quad -10 < \sigma < 10$

3. **a.** $f(t) = u(t)$
b. $f(t) = -tu(-t)$
c. $f(t) = e^{-2t}u(t) - e^{2t}u(-t)$

Section 13.2

1. $F(s) = s + \dfrac{3}{s^2 + 4s + 5}$

2. $F(s) = \dfrac{1}{s+1} - \dfrac{1}{s+4}$
$f(t) = (e^{-t} - e^{-4t})u(t)$

3. $F(s) = \dfrac{-1}{(s+1)^2} + \dfrac{3}{s+1} - \dfrac{3}{s+2}$
$f(t) = (te^{-t} + 3e^{-t} - 3e^{-2t})u(t)$

Section 13.3

1. Express $v_1(t)$ as

$$v_1(t) = tu(t) - u(t)$$

so

$$V_1(s) = \frac{1}{s^2} - \frac{1}{s}$$

Since $v_2(t) = (t-1)u(t-1)$, then $V_2(s) = (1/s^2)e^{-s}$.

2. The transform of the differential equation is

$$s^2Y(s) + 3sY(s) + 2Y(s) + s + 2 = \frac{5}{s}$$

Solving for $Y(s)$ gives

$$Y(s) = \frac{-s^2 - 2s + 5}{s(s+2)(s+1)} = \frac{\frac{5}{2}}{s} + \frac{\frac{5}{2}}{s+2} - \frac{6}{s+1}$$

Therefore, $y(t) = (\frac{5}{2} + \frac{5}{2}e^{-2t} - 6e^{-t})u(t)$.

Section 13.4

$V(s) = \dfrac{2}{s} - \dfrac{2}{s}e^{-0.25}$

$H(s) = \dfrac{s}{s+2}$

$I(s) = V(s)H(s) = \dfrac{2}{s+2} - \dfrac{2e^{-0.2s}}{s+2}$

so $i(t) = 2e^{-2t}u(t) - 2e^{-2(t-0.2)}u(t-0.2)$.

CHAPTER

14 z TRANSFORM

14.1 Definition of *z* Transform

LEARNING OBJECTIVES

After completing this section you should be able to do the following:

1. Define the bilateral and unilateral z transforms.
2. Determine the region of convergence in the complex z plane for the transform of a given discrete-time function.

The z transform is an operation performed on a discrete-time sequence $v(n)$ to produce a function $V(z)$, where z is a complex variable. There are two forms of the z transform. The two-sided or bilateral form is given by

$$V(z) = \sum_{n=-\infty}^{+\infty} v(n) z^{-n} \tag{14.1.1}$$

and the single-sided or unilateral form is given by

$$V(z) = \sum_{n=0}^{+\infty} v(n) z^{-n} \tag{14.1.2}$$

The only difference is the lower limit on the summation. If $v(n)=0$ for $n<0$, then the two forms yield identical functions $V(z)$. The two-sided form is pictured in Fig. 14.1.

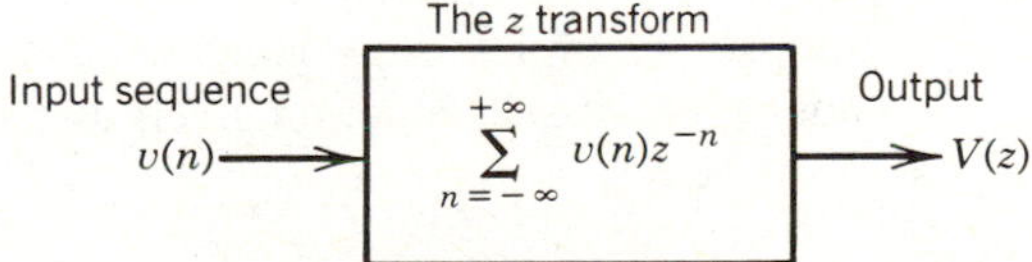

Figure 14.1. The z transform.

The inverse z transform is given by

$$v(n) = \frac{1}{2\pi j} \oint V(z) z^{n-1}\, dz \tag{14.1.3}$$

This is a complex integral because z is a complex variable. The inverse transform is an operation performed on the function $V(z)$ to produce $v(n)$, as pictured in Fig. 14.2.

Which functions have a z transform? Look at Eq. 14.1.1. If $v(n)$ is to have a z transform, then the magnitude of $V(z)$ must be finite.

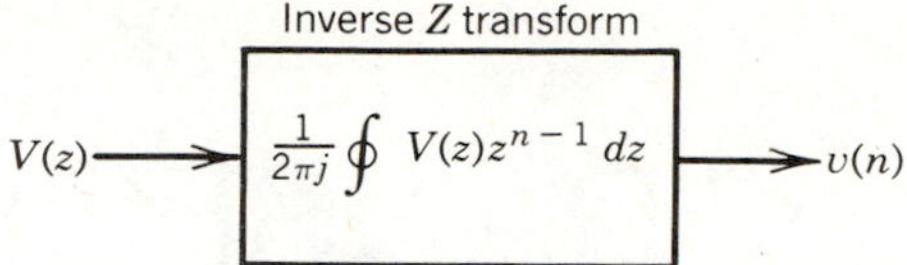

Figure 14.2. The inverse z transform.

$$|V(z)| = \left| \sum_{n=-\infty}^{+\infty} v(n)z^{-n} \right| < +\infty \tag{14.1.4}$$

The set of all z for this inequality to hold is called the region of convergence in the complex plane. Therefore, to determine which functions have a z transform, we must be able to find this region of convergence, if one exists. If not, we must be able to determine that there is no region of convergence, and this means that the function has no z transform.

The z transform is a power series. Both negative and positive powers of z are involved in the two-sided transform of Eq. 14.1.1. Because of this, we need to break Eq. 14.1.1 into two parts, one for positive powers, and one for negative powers, as follows.

$$V(z) = \sum_{n=-\infty}^{-1} v(n)z^{-n} + \sum_{n=0}^{+\infty} v(n)z^{-n}$$

It is true that a series of positive powers of a complex variable converges inside a circle of radius R_1, and a series of negative powers converges outside a circle of radius R_2 in the complex plane. If $R_1 < R_2$ then there is no region of convergence and the function $v(n)$ has no z transform. On the other hand, if $R_1 > R_2$, as we have shown in Fig. 14.3, then the region of convergence is in the ring between the circles of radius R_1 and R_2. Rings of convergence for the z transform correspond to strips of convergence for the Laplace transform.

How do we find these regions of convergence for various $v(n)$? Let us first consider the unit step function given by (see Fig. 14.4a)

$$v(n) = 1, \qquad n \geq 0$$
$$v(n) = 0, \qquad n < 0$$

Equation 14.1.1 then gives

$$V(z) = \sum_{n=0}^{+\infty} z^{-n}$$

which converges were $|z| > 1$.

Next, consider the sequence given by (see Fig. 14.4b)

$$v(n) = 0, \qquad n \geq 0$$
$$v(n) = 1, \qquad n < 0$$

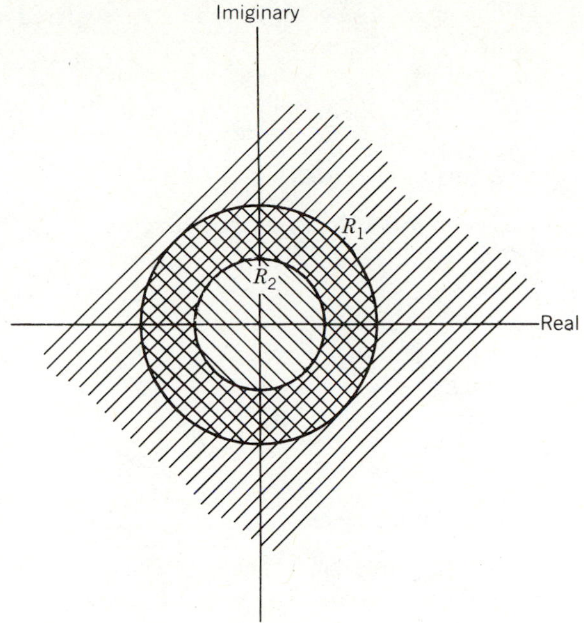

Figure 14.3. Regions of convergence.

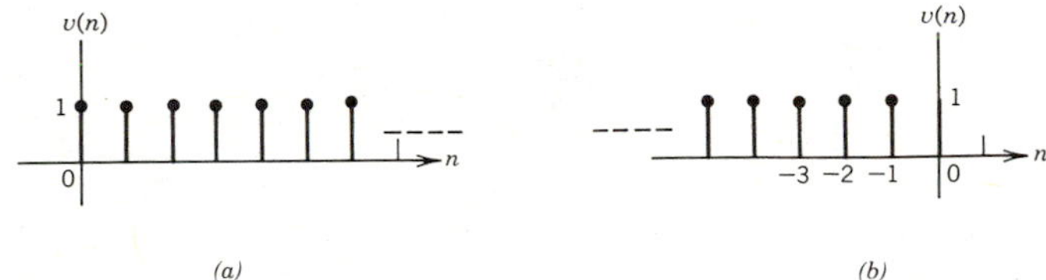

Figure 14.4. (*a*) Discrete step function. (*b*) Negative time function.

Equation 14.1.1 then gives

$$V(z) = \sum_{n=-\infty}^{-1} z^{-n}$$

which converges for $|z| < 1$.

These are just two examples that illustrate a theorem for power series which we now state without proof. We will make extensive use of this theorem.

Theorem 14.1a. Every power series

$$\sum_{n=0}^{+\infty} v(n)(z - z_0)^n$$

converges in a circle of radius R_1 centered at z_0, that is, $|z - z_0| < R_1$. The value of R_1 is given by

$$R_1 = \lim_{n \to \infty} \left| \frac{v(n)}{v(n+1)} \right|$$

if this limit exists.

Theorem 14.1b. Every power series

$$\sum_{n=0}^{\infty} v(n)(z - z_0)^{-n}$$

converges outside a circle or radius R_2 centered at z_0, that is, $|z - z_0| > R_2$. The value of R_2 is given by

$$R_2 = \lim_{n \to \infty} \left| \frac{v(n+1)}{v(n)} \right|$$

if this limit exists.

A discrete-time function where $v(n) = 0$, $n > 0$, is called a negative-time function. We use Theorem 14.1a with $z_0 = 0$ for negative-time functions. When $v(n) = 0$, $n < 0$, then $v(n)$ is called a positive-time function. We use Theorem 14.1b with $z_0 = 0$ for positive-time functions.

EXAMPLE 14.1.1 Find the region of convergence for the series

$$v(n) = \begin{matrix} 2 & n \geq 0 \\ 0 & n < 0 \end{matrix}$$

Solution

From Theorem 14.1b, R_2 is given by

$$R_2 = \lim_{n \to \infty} \left| \frac{2}{2} \right| = 1$$

Therefore, $|z| > 1$ is the region of convergence.

EXAMPLE 14.1.2 Find the region of convergence for

$$v(n) = \begin{matrix} (\frac{1}{2})^n & n \geq 0 \\ 2 & n < 0 \end{matrix}$$

Solution

From Theorem 14.1a R_1 is given by

$$R_1 = \lim_{n \to \infty} \left| \frac{2}{2} \right| = 1$$

and from Theorem 14.1b R_2 is given by

$$R_2 = \lim_{n \to \infty} \frac{(\frac{1}{2})^{n+1}}{(\frac{1}{2})^n} = \frac{1}{2}$$

Therefore, the region of convergence is $\frac{1}{2} < |z| < 1$.

EXAMPLE 14.1.3 Find the region of convergence for

$$v(n) = \begin{matrix} 2^n & n \geq 0 \\ 1 & n < 0 \end{matrix}$$

Solution

Here $R_1 = 1$ and $R_2 = 2$. Since $R_2 > R_1$, there is no region of convergence and $v(n)$ has no z transform.

At this point it is well to make the following observations:

1. Any discrete-time function with finite duration that has a finite value for every n has a z transform. Therefore, any function that we use in a computer has a z transform. The radius $R_2 = 0$ and the radius $R_1 = +\infty$ for a finite length sequence, so the region of convergence is the entire z plane, except for isolated points.
2. Every positive-time function has a region of convergence outside some radius R_2, where possibly $R_2 = \infty$.
3. Every negative-time function has a region of convergence inside some radius R_1, where possibly $R_1 = 0$.
4. It may seem unnecessary to understand how to select functions with z transforms, especially since any signal used in practice has a z transform. But remember that we will use the z transform as a tool, so we must understand the restrictions on its use.

LEARNING EVALUATIONS

1. Define the bilateral and unilateral z transforms.
2. Determine the region of convergence for each time function below.
   a. $v(n) = 10u(n)$
   b. $v(n) = 10(0.5)^n, \qquad -\infty < n < \infty$
   c. $v(n) = 10(0.5)^{|n|}, \qquad -\infty < n < \infty$
   d. $v(n) = u(n) - u(n-10)$

14.2 Evaluating the z Transform

LEARNING OBJECTIVES

After completing this section you should be able to do the following:

1. Find the z transform of a given discrete-time function, and specify the region of convergence.
2. Find the inverse z transform of a given frequency function when the region of convergence is specified.

The purpose of this section is to gain more familiarity with the z transform by evaluating the series

$$V(z) = \sum_{n=-\infty}^{+\infty} v(n) z^{-n} \tag{14.2.1}$$

for various discrete-time functions $v(n)$. Later in this section we will evaluate the inverse z transform for several functions $V(z)$, but here we are concerned only with the forward transform given by Eq. 14.2.1. For finite length sequences $v(n)$, the series 14.2.1 converges over the entire z plane, except possibly at some isolated points. Here is an example.

EXAMPLE 14.2.1 Find the z transform of $v(n)$ consisting of two values as shown in Fig. 14.5.

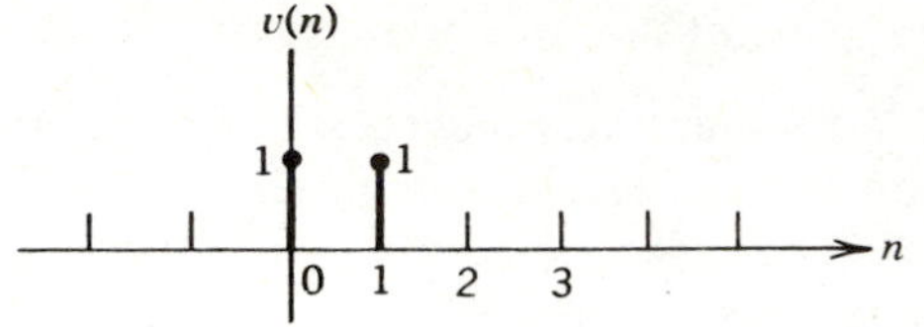

Figure 14.5

Solution

From Eq. 14.2.1,

$$V(z) = \sum_{n=0}^{1} z^{-n} = 1 + \frac{1}{z} = \frac{z+1}{z}, \qquad z \neq 0$$

There is one isolated point, $z = 0$, for which this series does not converge.

The situation is different for infinite length discrete-time functions. Instead of isolated points, there are regions of the z plane where the series does not converge. When evaluating the z transform of infinite length time functions, the regions of convergence in the z plane must be specified.

We will make frequent use of the following identity.

$$\frac{1}{1-x} = 1 + x + x^2 + x^3 + \cdots, \qquad |x| < 1 \tag{14.2.2}$$

which can be shown by long division.

$$\begin{array}{r|l} & 1 + x + x^2 + x^3 + \cdots \\ \hline 1 - x & 1 \\ & \underline{1 - x} \\ & \quad x \\ & \quad \underline{x - x^2} \\ & \qquad x^2 \\ & \qquad \underline{x^2 - x^3} \\ & \qquad\quad x^3 \\ & \qquad\quad \vdots \end{array}$$

Of course this sum must be finite, and this is assured by the condition $|x| < 1$.

EXAMPLE 14.2.2 Find the z transform of the unit step function shown in Fig. 14.6.

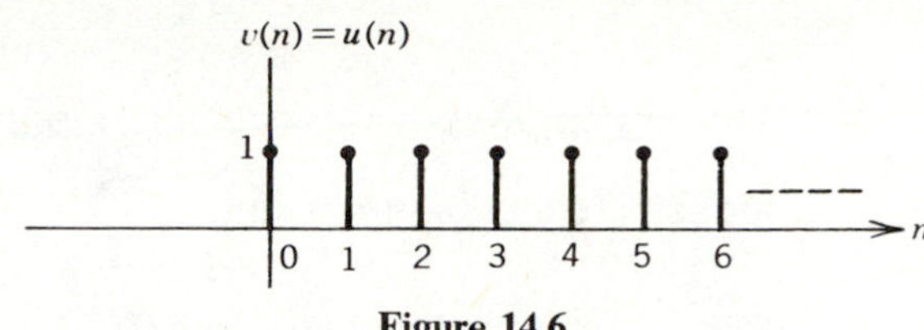

Figure 14.6

Solution

$$V(z) = \sum_{n=0}^{+\infty} z^{-n} = 1 + \frac{1}{z} + (\frac{1}{z})^2 + (\frac{1}{z})^3 + \cdots$$

Using the identity 14.2.2 with $x = 1/z$ gives

$$V(z) = \frac{1}{1 - 1/z} = \frac{z}{z-1}, \qquad |z| > 1 \qquad (14.2.3)$$

Note that the region of convergence is outside the circle with radius 1, which follows from Eq. 14.2.2 with $|1/z| < 1$.

EXAMPLE 14.2.3 Find the z transform of $v(n)$ given by (see Fig. 14.7)

$$v(n) = \begin{matrix} 0.5^n & n \geq 0 \\ 1 & n < 0 \end{matrix}$$

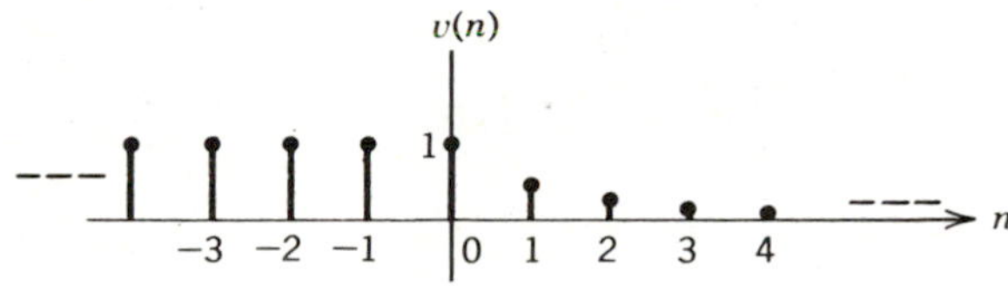

Figure 14.7

Solution

$$V(z) = \sum_{n=-\infty}^{-1} z^{-n} + \sum_{n=0}^{+\infty} (0.5)^n z^{-n}$$

The first summation term is identified with Eq. 14.2.2 with $x = z$.

$$\sum_{n=-\infty}^{-1} z^{-n} = z + z^2 + z^3 + \cdots$$

$$= \frac{1}{1-z} - 1 = \frac{-z}{z-1}, \qquad |z| < 1$$

In a similar manner the second summation is identified with Eq. 14.2.2 if we set $x = 0.5/z$ to obtain

$$\sum_{n=0}^{+\infty} (\frac{0.5}{z})^n = \frac{1}{1 - 0.5/z} = \frac{z}{z - 0.5}, \qquad |z| > 0.5$$

Combining these two results gives

$$V(z) = \frac{-z}{z-1} + \frac{z}{z-0.5}, \qquad 0.5 < |z| < 1$$

The region of convergence is a ring in the z plane.

EXAMPLE 14.2.4 Suppose we sample the signal

$$v(t) = e^{-at}, \qquad t \geq 0$$

every T s. That is, the sampling rate is $1/T$ samples per second. Find the z transform of the resulting sampled signal. (See Fig. 14.8.)

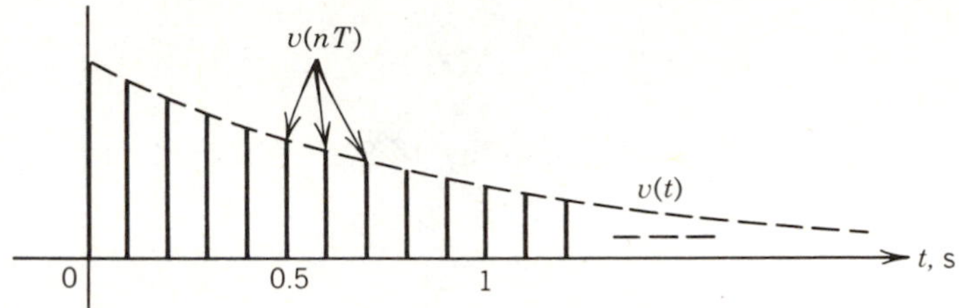

Figure 14.8

Solution

The resulting discrete-time function is given by

$$v(n) = e^{-anT}, \qquad n \geq 0$$

Therefore,

$$\begin{aligned} V(z) &= \sum_{n=0}^{+\infty} e^{-anT} z^{-n} \\ &= 1 + e^{-aT} z^{-1} + e^{-2aT} z^{-2} + \cdots \\ &= \frac{1}{1 - e^{-aT}(z^{-1})} = \frac{z}{z - e^{-aT}}, \qquad |z| > e^{-aT} \end{aligned}$$

If we sample a time function $v(t)$ every T s the result is $v(nT)$. Here are two tables of common z transform pairs. Table 14.1 lists transforms of sequences, while Table 14.2 lists transforms of sampled continuous-time signals. We could deduce one table from the other, but it avoids confusion to have both available.

The Inverse z Transform

We now turn our attention to the task of evaluating the inverse z transform. The region of convergence must be specified or understood in any expression for $V(z)$. The reason for this is illustrated as follows. Consider the two discrete-time functions $v_1(n)$ and $v_2(n)$ shown in Fig. 14.9. If we find the forward transform for $v_1(n)$ we obtain

$$V_1(z) = \sum_{n=0}^{+\infty} \left(\frac{a}{z}\right)^n = 1 + \frac{a}{z} + \left(\frac{a}{z}\right)^2 + \left(\frac{a}{z}\right)^3 + \cdots$$

TABLE 14.1
TRANSFORMS OF COMMON SEQUENCES

	$v(n), n \geq 0$	$V(z)$	Region of Convergence
1	1	$\frac{z}{z-1}$	$1 < \lvert z \rvert$
2	a^n	$\frac{z}{z-a}$	$a < \lvert z \rvert$
3	n	$\frac{z}{(z-1)^2}$	$1 < \lvert z \rvert$
4	na^{n-1}	$\frac{z}{(z-a)^2}$	$a < \lvert z \rvert$
5	$\frac{a^n}{n!}$	$e^{az^{-1}}$	$0 < \lvert z \rvert$
6	$\sin \omega n$	$\frac{z \sin \omega}{z^2 - 2z \cos \omega + 1}$	$1 < \lvert z \rvert$
7	$\cos \omega n$	$\frac{z^2 - z \cos \omega}{z^2 - 2z \cos \omega + 1}$	$1 < \lvert z \rvert$
8	$a^n \cos \pi n$	$\frac{z}{z+a}$	$a < \lvert z \rvert$

TABLE 14.2
TRANSFORMS OF COMMON SAMPLED TIME FUNCTIONS

	$v(t)$	$v(nT), n \geq 0$	$V(z)$	Region of Convergence
1	$u(t)$	1	$\frac{z}{z-1}$	$1 < \lvert z \rvert$
2	$tu(t)$	nT	$\frac{Tz}{(z-1)^2}$	$1 < \lvert z \rvert$
3	$e^{-at}u(t)$	e^{-anT}	$\frac{z}{z-e^{-aT}}$	$e^{-aT} < \lvert z \rvert$
4	$te^{-at}u(t)$	nTe^{-anT}	$\frac{Tz}{(z-e^{-aT})^2}$	$e^{-aT} < \lvert z \rvert$
5	$\sin(\omega t)u(t)$	$\sin(n\omega T)$	$\frac{z \sin \omega T}{z^2 - 2z \cos \omega T + 1}$	$1 < \lvert z \rvert$
6	$\cos(\omega t)u(t)$	$\cos(n\omega T)$	$\frac{z^2 - z \cos \omega T}{Z^2 - 2z \cos \omega T + 1}$	$1 < \lvert z \rvert$
7	$e^{-at}\sin(\omega t)u(t)$	$e^{-anT}\sin(\omega nT)$	$\frac{[e^{-aT} \sin \omega T]z}{z^2 - [2e^{-aT} \cos \omega T]z + e^{-2aT}}$	$e^{-aT} < \lvert z \rvert$
8	$e^{-aT}\cos(\omega t)u(t)$	$e^{-anT}\cos(\omega nT)$	$\frac{[z - e^{-aT} \cos \omega T]z}{z^2 - [2e^{-aT} \cos \omega T]z + e^{-2aT}}$	$e^{-aT} < \lvert z \rvert$

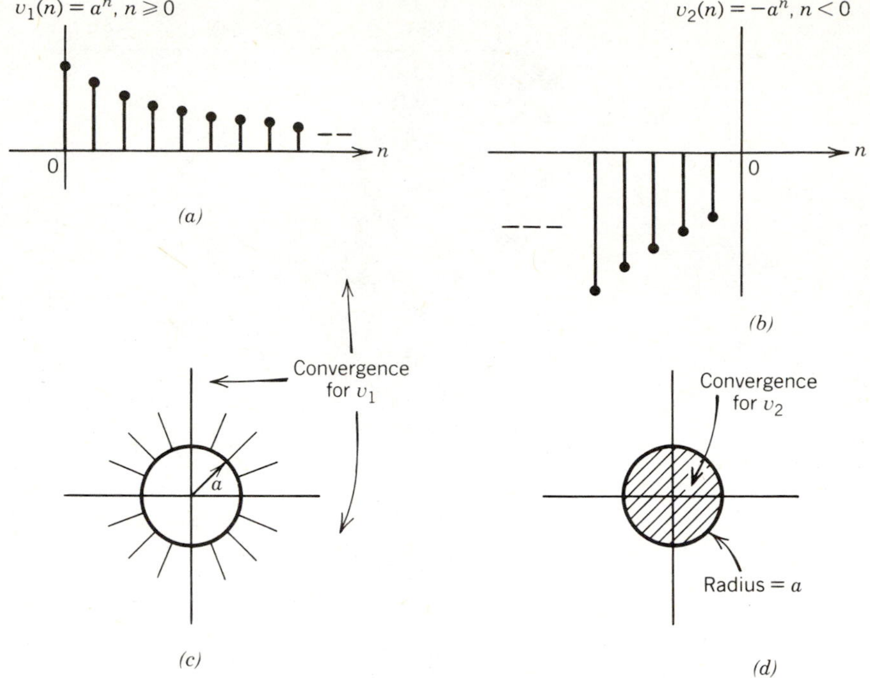

Figure 14.9. Two functions with the same z transform but different regions of convergence.

Comparing to Eq. 14.2.2 we find that $V_1(z)$ is given by

$$V_1(z) = \frac{1}{1 - a/z} = \frac{z}{z - a}, \qquad \left|\frac{a}{z}\right| < 1 \quad \text{or} \quad |z| > a$$

Notice that the region of convergence is $|z| > a$ shown in Fig. 14.9c.

The forward z transform of $v_2(n)$ is

$$V_2(z) = \sum_{n=-\infty}^{-1} -\left(\frac{a}{z}\right)^n = -\left[\frac{z}{a} + \left(\frac{z}{a}\right)^2 + \left(\frac{z}{a}\right)^3 + \cdots\right]$$

Again, comparing to Eq. 14.2.2 we have

$$V_2(z) = \frac{1}{1 - z/a} + 1 = \frac{z}{z - a}, \qquad \left|\frac{z}{a}\right| < 1 \quad \text{or} \quad |z| < a$$

Notice that the expressions for $V_1(z)$ and $V_2(z)$ are identical; only the regions of convergence are different. Therefore, the inverse transform of a function of z depends on the region of convergence as well as on the expression for $V(z)$. We will make frequent use of the following transform pairs.

$$a^n, \quad n \geq 0 \leftrightarrow \frac{z}{z - a}, \qquad |z| > a \tag{14.2.4}$$

$$-a^n, \quad n < 0 \leftrightarrow \frac{z}{z - a}, \qquad |z| < a \tag{14.2.5}$$

It is seen that the region of convergence is a connected region. Also, positive-time functions, for which we may use the single-sided transform, always have their region of convergence outside some contour in the z plane.

The inverse z transform may be evaluated by any one of several different methods. Here are some of them.

1. By use of the definition, Eq. 14.1.3.
2. By referring to a table of transform pairs such as Table 14.1. This is the easiest and most common method, but it requires that the expression for $V(z)$ be one of the forms in the table. Partial fraction expansion is useful here.
3. The Power series method. If the expression for $V(z)$ is available in power series form, we can identify each term as the coefficient of z^{-n} in the power series

$$V(z) = \sum_{n=-\infty}^{+\infty} v(n) z^{-n}$$

We will not discuss the first method, for that requires us to evaluate complex integrals. Here are some examples that illustrate methods 2 and 3.

EXAMPLE 14.2.5 Find the inverse z transform of

$$V(z) = \frac{z(z-1)}{z^2 - 5z + 6} = \frac{z(z-1)}{(z-2)(z-3)}, \qquad |z| > 3$$

Solution by Partial Fraction Expansion

Expand $V(z)/z$ rather than $V(z)$, for we will need z in the numerator of each fraction when using Table 14.1.

$$\frac{V(z)}{z} = \frac{z-1}{(z-2)(z-3)} = \frac{-1}{z-2} + \frac{2}{z-3}$$

$$V(z) = \frac{-z}{z-2} + \frac{2z}{z-3}, \qquad |z| > 3$$

$$v(n) = (-2^n + 2 \cdot 3^n)u(n) \tag{14.2.6}$$

Solution by Power Series

By long division we have

$$\begin{array}{r|l}
 & 1 + 4z^{-1} + 14z^{-2} + 46z^{-3} + \cdots \\
\hline
z^2 - 5z + 6 & z^2 - z \\
 & \underline{z^2 - 5z + 6} \\
 & \quad 4z - 6 \\
 & \quad \underline{4z - 20 + 24z^{-1}} \\
 & \qquad 14 - 24z^{-1} \\
 & \qquad \underline{14 - 70z^{-1} + 84z^{-2}} \\
 & \qquad\qquad 46z^{-1} - 84z^{-2}
\end{array}$$

or $v(0)=1$, $v(1)=4$, $v(2)=14$, $v(3)=46, \ldots$. This result checks with the closed form expression for $v(n)$ above, Eq. 14.2.6. Occasionally we can recognize the series obtained by long division as a familiar closed form, but for the majority of cases this is not true.

EXAMPLE 14.2.6 Find the inverse z transform of

$$V(z)=\frac{-z}{(z-2)(z-1)}, \qquad 1<|z|<2$$

Solution

Expanding $V(z)/z$ in partial fractions gives

$$\frac{V(z)}{z}=\frac{-1}{z-2}+\frac{1}{(z-1)^2}+\frac{1}{z-1}$$

or

$$V(z)=\frac{-z}{z-2}+\frac{z}{(z-1)^2}+\frac{z}{z-1}$$

We can use Table 14.1 for the last two terms, but since the region of convergence is $1<|z|<2$, the first term must correspond to a negative-time function. Using Eq. 14.2.5 we obtain

$$v(n)=\begin{matrix} 2^n & n<0 \\ n+1 & n\geq 0 \end{matrix}$$

LEARNING EVALUATIONS

1. Calculate the z transform and specify the region of convergence for each function listed.

a. $v(n)=n, \qquad 0<n<5$
b. $v(n)=nu(n)$
c. $v(n)=n, \qquad -\infty<n<\infty$

2. Calculate the inverse z transform for each function listed below.

a. $V(z)=\dfrac{z-2}{z}, \qquad z\neq 0$
b. $V(z)=\dfrac{z}{z-2}, \qquad 2<|z|$
c. $V(z)=\dfrac{z}{z-2}, \qquad 2>|z|$

14.3 Properties of *z* Transforms

LEARNING OBJECTIVES

After completing this section you should be able to do the following:

1. Use the z transform to solve difference equations.
2. Use the various properties as an aid in evaluating transforms of discrete-time signals.

Here are some of the more useful properties of the one-sided z transform.

Superposition

$$av_1(n) + bv_2(n) \leftrightarrow aV_1(z) + bV_2(z) \tag{14.3.1}$$

Scaling

$$a^n v(n) \leftrightarrow V\left(\frac{z}{a}\right) \tag{14.3.2}$$

Delay

$$v(n-k) \leftrightarrow z^{-k}V(z) \qquad \text{for} \quad k \geq 0 \tag{14.3.3}$$

Advance

$$v(n+k) \leftrightarrow z^k V(z) - \sum_{j=0}^{k-1} v(j) z^{k-j}, \qquad k \geq 0 \tag{14.3.4}$$

Differentiation

$$nv(n) \leftrightarrow -z\frac{dV(z)}{dz} \tag{14.3.5}$$

Summation

$$\sum_{k=0}^{n} v(k) \leftrightarrow \frac{z}{z-1}V(z) \tag{14.3.6}$$

Convolution

$$\sum_{k=0}^{n} v_1(k) v_2(n-k) \leftrightarrow V_1(z) V_2(z) \tag{14.3.7}$$

Initial Value

$$v(0) = \lim_{z \to \infty} V(z) \tag{14.3.8}$$

Final Value

$$\lim_{n \to \infty} v(n) = \lim_{z \to 1} (z-1) V(z) \tag{14.3.9}$$

Periodic Repetition. If $v(t)$ is periodic (beginning at $n = 0$) with period N,

$$v(n) = \sum_{k=0}^{\infty} v_1(n - kN)$$

then

$$V(z) = \frac{V_1(z)}{1 - z^{-N}} \tag{14.3.10}$$

Keep in mind that these properties are for the one-sided transform. Our primary use of these properties will be in connection with positive-time signals applied to realizable systems, so we will not consider properties of the two-sided transform. Notice that these properties are similar to those for the Laplace transform, as we would expect since the Laplace and z transforms are mathematically similar.

Property 4, advance, is used in the solution of linear difference equations by z transform. A linear difference equation is a relationship between the input $x(n)$, the response $y(n)$, and shifted versions of these functions. If we take the z transform of each term in a difference equation, then the result is an equation in terms of $X(z)$, $Y(z)$, and powers of z. Since $x(n)$ is known, if it is z transformable, then an algebraic expression with only one unknown, namely $Y(z)$, is the result. If the inverse transform of $Y(z)$ can be evaluated, then an explicit expression for $y(n)$ can be obtained. In many instances this method is easier than the classical solution by the method of undetermined coefficients (or any one of several other classical methods that we have not discussed). In many instances, classical methods are easier. Using transforms to solve simple equations is like using a crane to pick up a pebble. Here are some examples.

EXAMPLE 14.3.1 Find the response of the following first order system to a unit step.

$$y(n+1) - 0.2y(n) = u(n)$$
$$y(0) = 1.5$$

Solution

Take the z transform of each term and apply the advance property to $y(n+1)$.

$$y(n+1) \leftrightarrow zY(z) - zy(0)$$

so

$$zY(z) - 1.5z - 0.2Y(z) = \frac{z}{z-1}$$

Solving for $Y(z)$ gives

$$Y(z) = \frac{1.5z}{z-0.2} + \frac{z}{(z-1)(z-0.2)}$$

The last term is expanded in partial fractions to give

$$Y(z) = \frac{1.5z}{z-0.2} + \frac{1.25z}{z-1} - \frac{1.25z}{z-0.2}$$
$$= \frac{1.25z}{z-1} + \frac{0.25z}{z-0.2}$$

Using Table 14.1 to obtain the inverse transform, we have

$$y(n) = 1.25 + 0.25(0.2)^n, \qquad n \geq 0$$

The other properties of the z transform have many uses, and we will indicate some of them here. In Section 9.6 a procedure for finding the transfer function of a discrete-time system was presented without justification. We promised to explain the procedure later, and this explanation is now possible with the aid of the properties.

With x denoting input and y denoting output, the difference equation model for a linear discrete-time system has the form

$$\sum_{i=0}^{k} b_i y(n-i) = \sum_{i=0}^{l} a_i x(n-i) \tag{14.3.11}$$

The algorithm presented in Section 9.6 replaced each $y(n-i)$ term by $z^{-i}Y(z)$, and replaced each $x(n-i)$ term by $z^{-i}X(z)$. The transfer function was then given by the ratio $Y(z)/Z(z)$.

$$H(z) = \frac{Y(z)}{X(z)} = \frac{\sum_{i=0}^{l} a_i z^{-i}}{\sum_{i=0}^{k} b_i z^{-i}} \tag{14.3.12}$$

This procedure is justified by the delay property. It is this property that allows us to replace terms such as $y(n-i)$ by $z^{-i}Y(z)$. That is, if $Y(z)$ is the transform of $y(n)$, then by the delay property $z^{-i}Y(z)$ is the transform of $y(n-i)$.

EXAMPLE 14.3.2 Find the z transform of the signal $v(n)$ shown in Fig. 14.10.

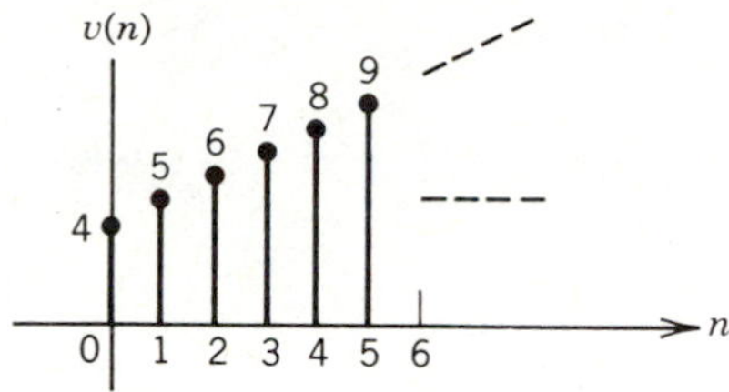

Figure 14.10

Solution

The signal $v(n)$ may be expressed as the sum of a step and a ramp.

$$v(n) = 4u(n) + n$$

By the superposition property, the transform is

$$V(z) = \quad = \frac{4z}{z-1} + \frac{z}{(z-1)^2}$$

$$= \frac{4z^2 - 3z}{(z-1)^2}, \qquad |z| > 1$$

where entries 1 and 3 from Table 14.1 were used to find the transforms of the step and ramp.

Many of the transforms given in Table 14.1 may be derived by using the properties. For example, pair 2 may be derived by applying the scaling property to the unit step transform. Since the transform of $u(n)$ is given by

$$u(n) \leftrightarrow \frac{z}{z-1}$$

then the transform of $a^n u(n)$ is given by

$$a^n u(n) \leftrightarrow \frac{z/a}{z/a - 1} = \frac{z}{z-a}$$

which is the second entry in Table 14.1.

The summation property may be used to derive pair 3 from pair 1 in the table. Since

$$n = \sum_{k=0}^{n} u(k-1)$$

the transform of the ramp must be given by $z/(z-1)$ times the transform of $u(k-1)$. By the delay property, this is $z^{-1}U(z)$. Therefore,

$$nu(n) \leftrightarrow \frac{z}{z-1}\left(z^{-1}\frac{z}{z-1}\right) = \frac{z}{(z-1)^2}$$

The initial value and final value properties allow us to determine these values of the time function if we are given the frequency function.

EXAMPLE 14.3.3 Find the initial value $v(0)$ if we are given $V(z)$ below.

$$V(z) = \frac{z^2(z+1)}{(z-1)(z-0.25)(z-0.75)}$$

Solution

Divide numerator and denominator by z^3 to obtain

$$V(z) = \frac{(1 + 1/z)}{(1 - 1/z)(1 - 0.25/z)(1 - 0.75/z)}$$

Therefore,

$$v(0) = \lim_{z \to \infty} V(z) = 1$$

which is in agreement with the corresponding time function. We can check our work by finding $v(n)$ and substituting $n = 0$. After some algebra we find

$$v(n) = \tfrac{32}{3} + \tfrac{5}{6}(0.25)^n - 10.5(0.75)^n$$

Then $v(0) = 1$ as advertised.

LEARNING EVALUATIONS

1. Use z transforms to solve the following difference equation.

$$y(n+1) - 0.5y(n) = (0.75)^n$$
$$y(0) = 1$$

2. Find the transfer function for the above system.

14.4 System Realizations

LEARNING OBJECTIVES

After completing this section you should be able to do the following:

1. Draw the type 2 realization from the describing difference equation, transfer function, or impulse response.
2. Derive cascade and parallel realization forms from the system equation.

The direct form 1 realization was introduced in Chapter 7. We now have the necessary background to discuss practical realization forms. The direct form 1 realization is not practical because the number of delay units used may be as high as twice the order of the system. The direct form 2 realization uses only as many delays as the order of the system.

To begin our derivation we assume that the transfer function may be expressed as the ratio of two polynomials, $N(z)$ and $D(z)$.

$$H(z) = \frac{N(z)}{D(z)} \tag{14.4.1}$$

Since $H(z)$ is the output divided by the input, we can write

$$Y(z) = H(z)X(z) = \frac{N(z)X(z)}{D(z)} \tag{14.4.2}$$

Define a new variable $W(z)$ as

$$W(z) = \frac{X(z)}{D(z)} \tag{14.4.3}$$

$$Y(z) = N(z)W(z) \tag{14.4.4}$$

EXAMPLE 14.4.1 If $D(z) = 1 - 0.2z^{-1} + z^{-2}$, then from Eq. 14.4.3

$$W(z)(1 - 0.2z^{-1} + z^{-2}) = X(z)$$

or

$$w(n) = x(n) + 0.2w(n-1) - w(n-2)$$

Example 14.4.1 illustrates the application of Eq. 14.4.3 in deriving a difference equation that relates $x(n)$ to $w(n)$. This relationship forms the basis for the direct form 2 realization. In general,

$$w(n) = x(n) - \sum_{i=1}^{k} b_i w(n-i) \tag{14.4.5}$$

Also, from Eq. 14.4.4,

$$y(n) = \sum_{i=0}^{k} a_i w(n-i) \tag{14.4.6}$$

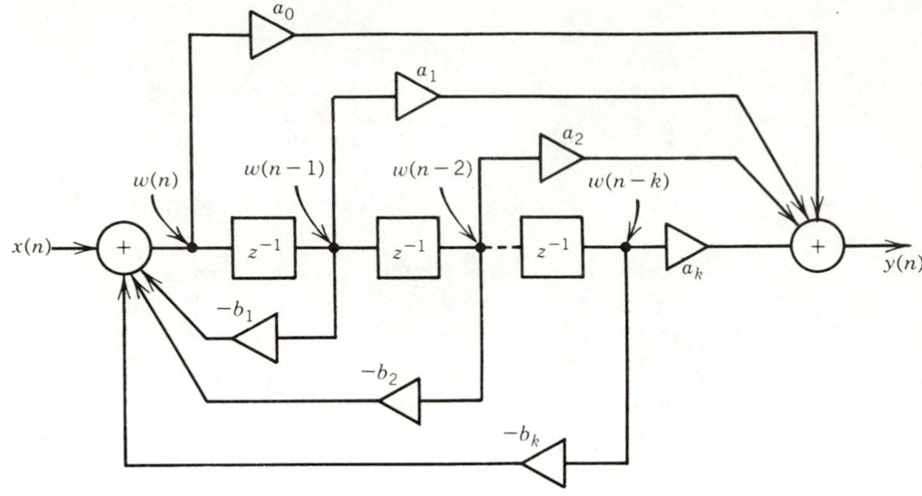

Figure 14.11. The direct form 2 realization.

These two equations describe the system shown in Fig. 14.11, the direct form 2 realization. Equation 14.4.5 states that the output of the first summer in Fig. 14.11, $w(n)$, is the sum of $x(n)$ and all the $-b_i w(n-i)$ terms. Equation 14.4.6 states that the output of the second summer, $y(n)$, is the sum of all the $a_i w(n-i)$ terms. The diagram says the same thing.

EXAMPLE 14.4.2 Find the direct form 2 realization for the system whose transfer function is given by

$$H(z) = \frac{2 + 5z^{-1} + 2z^{-2}}{1 + 0.6z^{-1} - 0.16z^{-2}}$$

Solution

Let

$$W(z) = \frac{X(z)}{1 + 0.62z^{-1} - 0.16z^{-2}}$$

then

$$W(z) = X(z) - 0.6z^{-1}W(z) + 0.16z^{-2}W(z)$$

$$w(n) = x(n) - 0.6w(n-1) + 0.16w(n-2) \qquad (14.4.7)$$

From Eq. 14.4.4,

$$Y(z) = (2 + 5z^{-1} + 2z^{-2})W(z)$$

$$y(n) = 2w(n) + 5w(n-1) + 2w(n-2) \qquad (14.4.8)$$

Equations 14.4.7 and 14.4.8 are used to derive the system shown in Fig. 14.12.

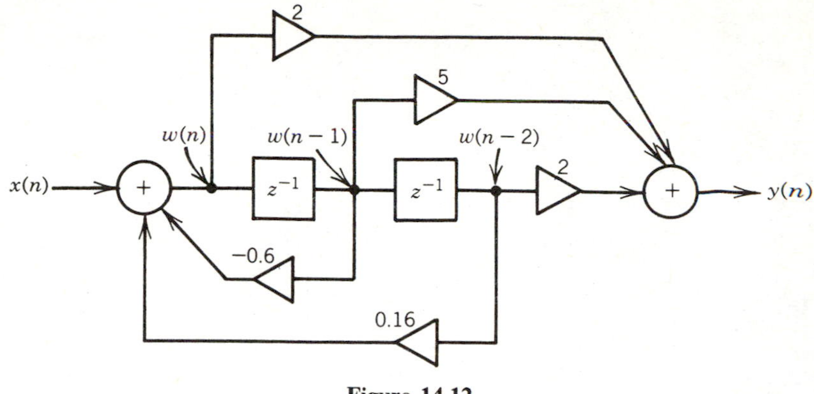

Figure 14.12

Although the direct form 2 realization is effective in reducing the number of delays over the direct form 1 realization, parameter quantization effects remain. The cascade and parallel forms discussed below serve to reduce these errors. In general, there are three sources of such error: quantization of the input signal into discrete levels; accumulation of round-off errors in arithmetic operations, and the representation of coefficients a_i and b_i by a finite number of digits. Of these three, we can reduce the effect of the last two by avoiding systems with sections of high order. This implies that our systems should be designed, not as one big system (Fig. 14.11), but rather as a set of smaller sections that can be combined to provide the overall desired transfer function. We will discuss two methods of accomplishing this, the cascade (or series) form and the parallel form.

The cascade form is obtained by decomposing $H(z)$ into the product of first or second order transfer functions.

$$H(z) = a_0 H_1(z) H_2(z) \dots H_l(z) \tag{14.4.9}$$

First order sections are preferred, but when poles are complex we must combine conjugate pairs to form a second order section with real coefficients. A first order section has the form

$$H_i(z) = \frac{1 + a_{i1}z^{-1}}{1 + b_{i1}z^{-1}} \tag{14.4.10}$$

Second order sections have the form

$$H_i(z) = \frac{1 + a_{i1}z^{-1} + a_{i2}z^{-2}}{1 + b_{i1}z^{-1} + b_{i2}z^{-2}} \tag{14.4.11}$$

These are illustrated in Fig. 14.13.

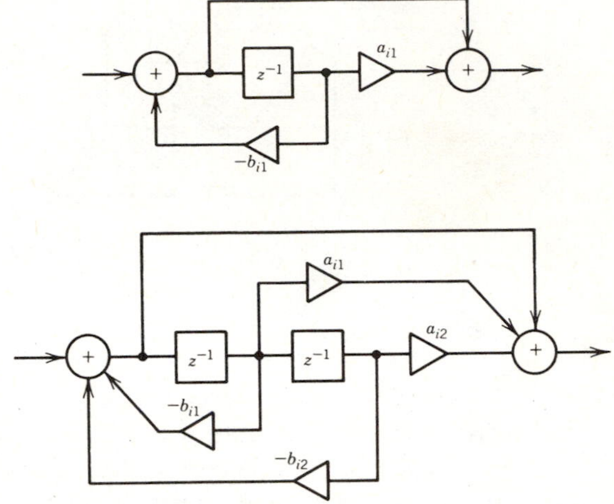

Figure 14.13. Forms used in cascade realizations.

EXAMPLE 14.4.3 Find a cascade realization for the system in the previous example, namely

$$H(z) = \frac{2 + 5z^{-1} + 2z^{-2}}{1 + 0.6z^{-1} - 0.16z^{-2}}$$

Solution

$H(z)$ may be factored into terms given by

$$H(z) = \frac{2(z+2)(z+0.5)}{(z+0.8)(z-0.2)} = \frac{2(1+2z^{-1})(1+0.5z^{-1})}{(1+0.8z^{-1})(1-0.2z^{-1})}$$

If we arbitrarily associate the first term in the numerator with the first term in the denominator we get the diagram in Fig. 14.14.

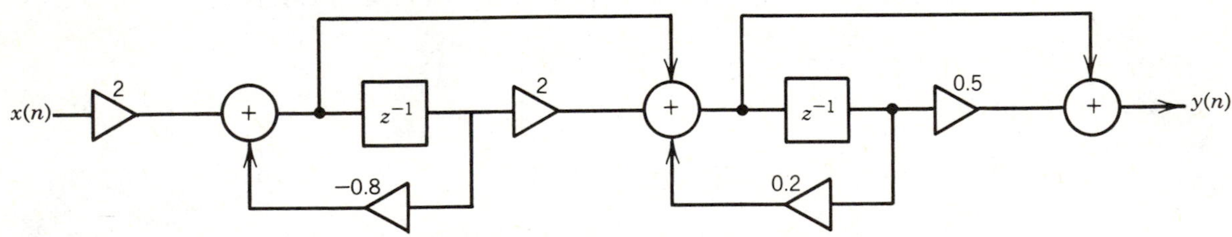

Figure 14.14

The parallel form is obtained by decomposing $H(z)$ into the sum of several terms, each of first or second order, and a constant.

$$H(z) = A + H_1(z) + \cdots + H_l(z) \tag{14.4.12}$$

A first order section has the form

$$H(z) = \frac{a_{i0}}{1 + b_{i1}z^{-1}} \tag{14.4.13}$$

Second order sections have the form

$$H(z) = \frac{a_{i0} + a_{i1}z^{-1}}{1 + b_{i1}z^{-1} + b_{i2}z^{-2}} \tag{14.4.14}$$

These are illustrated in Fig. 14.15.

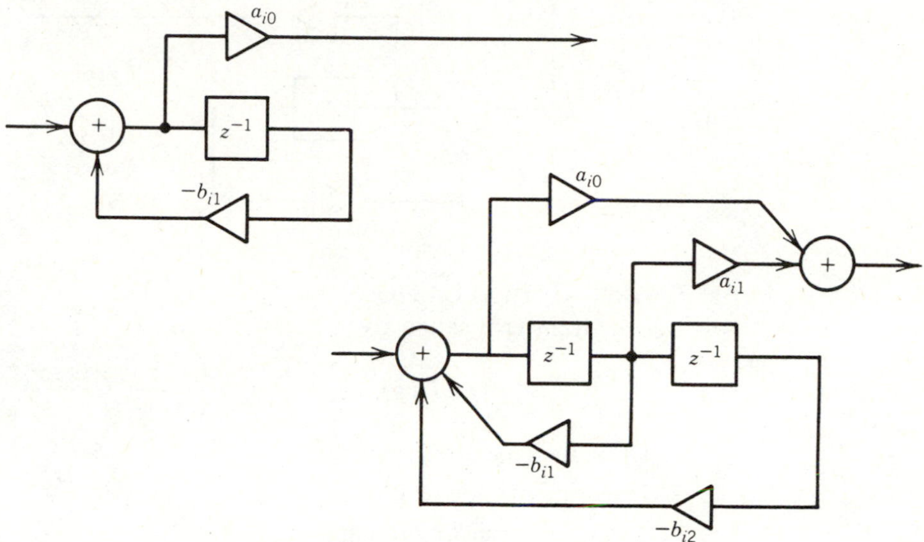

Figure 14.15. Realizations of Eqs. 14.4.13 and 14.4.14.

EXAMPLE 14.4.4 Find a parallel realization of the system in Example 14.4.3.

$$H(z) = \frac{2 + 5z^{-1} + 2z^{-2}}{1 + 0.6z^{-1} - 0.16z^{-2}}$$

Solution

The parallel form is obtained by decomposing $H(z)/z$ by partial fraction expansion, just as we did to find the inverse transform.

$$\frac{H(z)}{z} = \frac{2(z+2)(z+0.5)}{z(z+0.8)(z-0.2)} = \frac{A}{z} + \frac{B}{z+0.8} + \frac{C}{z-0.2}$$

Upon solving for A, B, and C, and after multiplying by z, we obtain

$$\begin{aligned} H(z) &= -12.5 + \frac{0.9z}{z+0.8} + \frac{15.4z}{z-0.2} \\ &= -12.5 + \frac{0.9}{1+0.8z^{-1}} + \frac{15.4}{1-0.2z^{-1}}. \end{aligned}$$

This system is shown in Fig. 14.16.

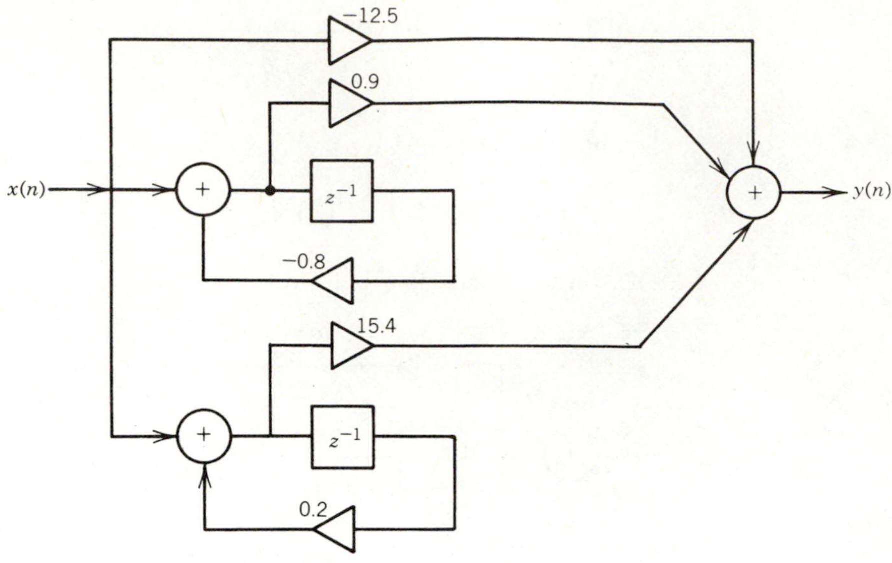

Figure 14.16

LEARNING EVALUATION

For the transfer function given by

$$H(z) = \frac{2 + 3z^{-1} + 1.12z^{-2} + 0.2z^{-3}}{1 + 0.125z^{-3}}$$

1. Draw the type 2 realization.
2. Draw the cascade realization.
3. Draw the parallel representation.

14.5 Response by *z* Transform

LEARNING OBJECTIVES

After completing this section you should be able to do the following:

Find the response of an LTI discrete-time system by the transform method. The z transform serves the same purpose for discrete-time systems that the Laplace transform serves for continuous-time systems. We expressed everything in terms of the signal e^{st} when using Laplace transforms. Here we use z^k. Since the discrete-time transfer function for a stable LTI system is the steady-state response to z^k divided by z^k, then we use our input–output pair methods of Chapter 9 to find the system response. Here is the procedure.

1. Find the LTI system transfer function.
2. Find the z transform of the input signal.
3. Multiply the input by the transfer function.
4. Find the inverse z transform of the product. This time function is the desired response.

EXAMPLE 14.5.1 A unit step is applied to the system described by the difference equation

$$y(n) = 0.3y(n-1) + 0.7x(n)$$

Find the response by z transforms.

Solution

The system transfer function is found by transforming the above difference equation and then finding the ratio $Y(z)/X(z)$.

$$H(z) = \frac{Y(z)}{X(z)} = \frac{0.7z}{z-0.3}$$

The transform of the input is given by

$$u(n) \leftrightarrow \frac{z}{z-1}$$

Therefore, the response as a function of frequency is given by

$$Y(z) = H(z)X(z) = \frac{0.7z^2}{(z-0.3)(z-1)} = \frac{z}{z-1} - \frac{0.3z}{z-0.3}$$

The response as a function of time is therefore

$$y(n) = [1 - 0.3(0.3)^n]u(n)$$

This function is graphed in Fig. 14.17.

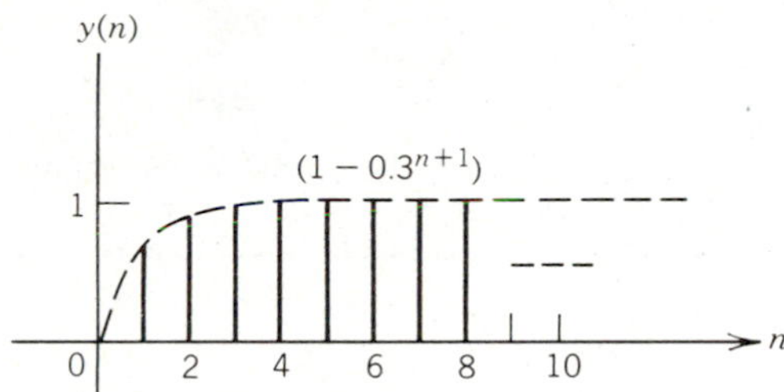

Figure 14.17

EXAMPLE 14.5.2 Find the response of the system shown in Fig. 14.18 to the input signal $x(n)$, where

$$x(n) = 0.9^n, \qquad n > 0$$

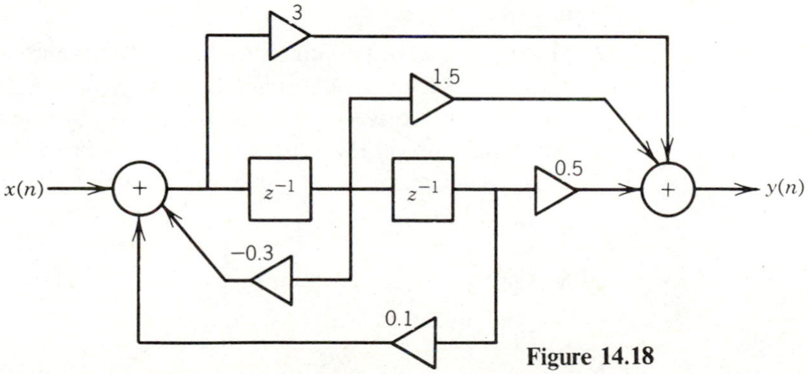

Figure 14.18

Solution

The transfer function obtained from the system diagram is given by (see Example 14.4.2)

$$H(z) = \frac{3 + 1.5z^{-1} + 0.5z^{-2}}{1 + 0.3z^{-1} - 0.1z^{-2}} = \frac{3z^2 + 1.5z + 0.5}{(z + 0.5)(z - 0.2)}$$

The transform of the input signal is

$$X(z) = \frac{z}{z - 0.9}$$

The product $H(z)x(z)$ gives $Y(z)$.

$$Y(z) = \frac{z(3z^2 + 1.5z + 0.5)}{(z - 0.9)(z + 0.5)(z - 0.2)}$$

Expanding $Y(z)/z$ in partial fractions gives

$$Y(z) = \frac{4.36735}{z - 0.9} + \frac{0.51020}{z + 0.5} - \frac{1.87755}{z - 0.2}$$

Multiply by z and find the inverse transform to obtain

$$y(n) = 4.36735(0.9)^n + 0.51020(-0.5)^n - 1.8775(0.2)^n, \qquad n \geq 0$$

LEARNING EVALUATION

Find the response of the system whose impulse response is given by

$$h(n) = \tfrac{14}{9}(0.4)^n - \tfrac{5}{9}(-0.5)^n$$

to a unit step by the transform method.

PROBLEMS

SECTION 14.1

1. Determine the region of convergence in the complex z plane for each time function below.
   a. $v(n) = 10u(-n)$
   b. $v(n) = 0.8^{|n|}, \qquad -\infty < n < \infty$
   c. $v(n) = 0.8^n u(n)$
   d. $v(n) = 2^n u(-n)$
2. Since negative-time functions [those for which $v(n) = 0$ for $n > 0$] have a region of convergence inside a circle of radius R_1, and positive-time functions converge in a region outside some radius R_2, the region of convergence must determine the values of time for which the time function is zero. In each case below determine the time interval for which the time function is zero.
   a. $V(z) = \dfrac{z}{z+1}, \qquad |z| < 1$
   b. $V(z) = \dfrac{z}{z+1}, \qquad |z| > 1$
   c. $V(z)\dfrac{z}{1-z} + \dfrac{z}{z+0.2}, \qquad 0.2 < |z| < 1$

d. $V(z) = \dfrac{z + 0.5}{z}, \qquad z \neq 0$

SECTION 14.2

3. Find the bilateral z transform and specify the region of convergence for each time function given below.

a. $v(n) = 0.5^n, \qquad 0 < n < 5$

b. $v(n) = 0.5^n u(n)$

c. $v(n) = 0.5^n, \qquad -\infty < n < \infty$

4. Find the inverse z transform of $V(z)$ where

a. $V(z) = \dfrac{2}{z^2} + \dfrac{1}{z^5}, \qquad z \neq 0$

b. $V(z) = \dfrac{z}{z - 0.5}, \qquad |z| > 0.5$

c. $V(z) = \dfrac{z}{z - 0.5}, \qquad |z| < 0.5$

5. The impulse response of a discrete-time system is the inverse z transform of the transfer function. For each transfer function below find the system impulse response.

a. $H(z) = \dfrac{z - 2}{(z - 3)(z - 4)}, \qquad |z| > 4$

b. $H(z) = \dfrac{z}{(z - 3)(z - 4)}, \qquad |z| > 4$

6. Find the first four terms in the function $v(n)$ by the power series method if $V(z)$ is given by

$$V(z) = \frac{z - 2}{z^2 + 0.3z - 0.1}, \qquad |z| > 0.5$$

7. Find the inverse transform of

$$V(z) = \frac{z - 1}{(z - 5)(z - 6)}, \qquad 5 < |z| < 6$$

SECTION 14.3

8. Find the initial and final values of the time function $v(n)$ if

$$V(z) = \frac{z}{(z - 1)(z - 0.2)}, \qquad |z| > 1$$

Check your work by finding $v(n)$.

9. According to the differentiation property given in Eq. 14.3.5, if the transform of a unit step $v(n) = u(n)$ is $V(z)$, then the transform of $nu(n)$ should be given by the derivative of $V(z)$ multiplied by $-z$.

a. Show that this is true by differentiating

$$V(z) = \frac{z}{z - 1}$$

b. Now find the z transform of $n^2u(n)$.

10. Use the advance property to solve the following difference equation for $y(n)$ by z transform.

$$y(n+1) - 0.5y(n) = 0.2^n u(n)$$
$$y(0) = 5$$

11. Solve the following difference equation by z transform.

$$y(n) - y(n-1) + \tfrac{3}{16}y(n-2) = 0.5u(n) + 0.5u(n-1)$$
$$y(0) = 0, \qquad y(-1) = 0$$

12. When we discussed the classical solution of difference equations in Section 7.5, we found that special precautions were necessary when the forcing function contained one of the systems natural frequencies. The z transform method automatically handles this situation. To illustrate this, solve by z transform for $y(n)$, where

$$y(n) - y(n-1) + \tfrac{3}{16}y(n-2) = (0.75)^n u(n)$$
$$y(0) = 0, \qquad y(-1) = 0$$

Then refer to Example 7.5.6.

13. In the digital correlator shown in Fig. 14.19, the input is $x(n)$ and the output is $y(n)$. If $s(n)$ is the signal shown, find the response of the correlator if $x(n) = u(n)$.

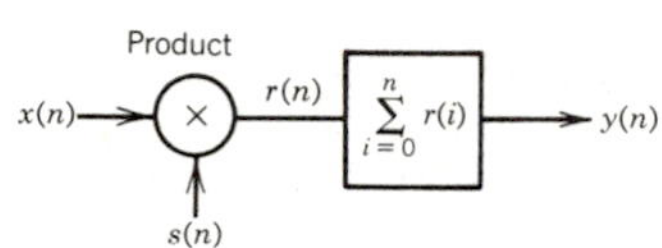

Figure 14.19

SECTION 14.4

14. Find the direct form 2 realization for the system whose transfer function is given by

$$H(z) = \frac{0.8 - 0.4z^{-1}}{1 - z^{-1} + \frac{3}{16}z^{-2}}$$

15. Find the direct form 2 realization for the system whose difference equation is given by

$$y(n) - y(n-1) + \tfrac{3}{16}y(n-2) = 0.8x(n) - 0.4x(n-1)$$

16. Find the direct form realization for the system whose impulse response is given by

$$h(n) = \left[0.4(0.75)^n + 0.4(0.25)^n\right]u(n)$$

17. Find a cascade realization for the system in Problem 14.

18. Find a parallel realization for the system in Problem 14.

19. Manhole covers are round the world over. There is a good reason for this. If the local officials in, say, Laramie, Wyoming (or in Stolkholm Sweden, or in Timbuktu) decide to use something other than round manhole covers in their streets, they soon discover their mistake and revert back to round covers. In a similar vein, it is necessary to use second order sections in some cascade and parallel realizations. Can you explain why

   a. manhole covers are round.
   b. second order sections are sometimes necessary.

20. A system is described by the following difference equation relating input x to output y.

$$y(n)-0.5y(n-1)+0.6y(n-2)=0.2x(n)+0.2x(n-1)$$

   a. Find the transfer function.
   b. Find the impulse response.
   c. Find the type 2 realization.

SECTION 14.5

21. A unit step is applied to the system in Problem 20. Use z transforms to find the response.

22. Find the response of the system in Problem 14 to an exponential input $x(n)=0.9^n u(n)$ by z transforms.

LEARNING EVALUATION ANSWERS

Section 14.1

1. See Eqs. 14.1.1 and 14.1.2.
2. a. $1<|z|$
   b. None.
   c. $0.5<|z|<2$
   d. The entire z plane, except $z=0$.

Section 14.2

1. a. $V(z)=\dfrac{1}{z}+\dfrac{2}{z^2}+\dfrac{3}{z^3}+\dfrac{4}{z^4},\ z\neq 0$
   b. $V(z)=\dfrac{z}{(z-1)^2},\ 1<|z|$
   c. None.
2. a. $v(n)=\begin{matrix}1 & n=0\\ -2 & n=1\\ 0 & \text{elsewhere}\end{matrix}$
   b. $v(n)=2^n u(n)$. See Eq. 14.2.4.
   c. $v(n)=-2^n u(-n)$. See Eq. 14.2.5.

Section 14.3

1. You should obtain the following expression for $Y(z)$.

$$Y(z) = \frac{z}{z-0.5} + \frac{z}{(z-0.5)(z-0.75)}$$

from which $y(n)$ is given by

$$y(n) = 5(0.5)^n - 4(0.75)^n$$

2. $H(z) = \dfrac{1}{z-0.5}$

Section 14.4

1. With

$$Y(z) = \frac{N(z)}{D(z)} X(z) = N(z)W(z)$$

$$Y(z) = (2 + 3z^{-1} + 1.12z^{-2} + 0.12z^{-3})W(z)$$

or

$$\boxed{y(n) = 2w(n) + 3w(n-1) + 1.12w(n-2) + 0.12w(n-3)}$$

Also,

$$X(z) = W(z)D(z) = W(z)(1 + 0.125z^{-3})$$

so

$$\boxed{x(n) = w(n) + 0.125w(n-3)}$$

The two emphasized equations are used to draw the type 2 realization shown in Fig. 14.20.

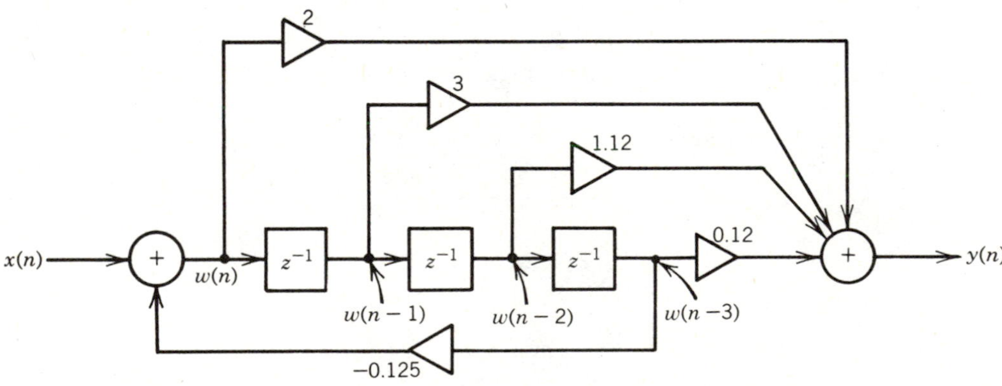

Figure 14.20

2. The numerator and denominator polynomials in $H(z)$ are factored to give

$$H(z)=\frac{2(z+0.2)(z+0.3)(z+0.9)}{(z+0.5)(z^2-0.5z+0.25)}$$

If we arbitrarily associate the $(z+0.2)$ factor with the real pole at $z=0.5$, the cascade realization is expressed by

$$H(z)=2\left(\frac{1+0.2z^{-1}}{1+0.5z^{-1}}\right)\left(\frac{1+1.2z^{-1}+0.27z^{-2}}{1-0.5z^{-1}+0.25z^{-2}}\right)$$

The realization diagram is shown in Fig. 14.21.

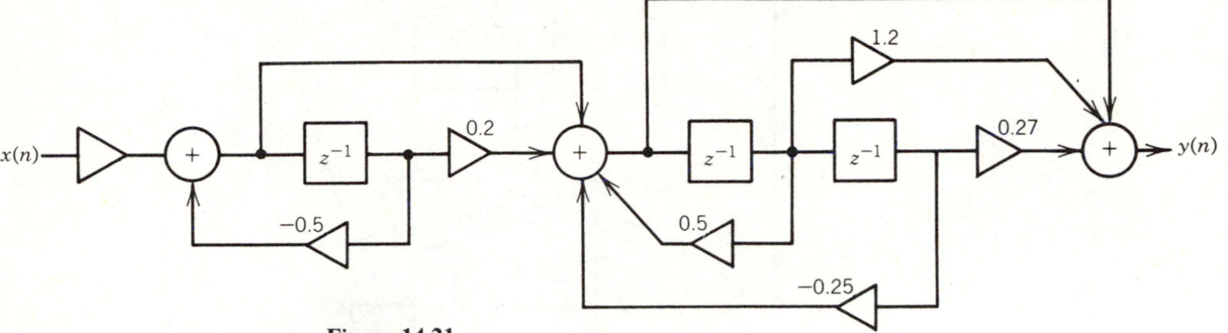

Figure 14.21

3. For the parallel realization we expand $H(z)/z$ in partial fractions to obtain

$$\frac{H(z)}{z}=\frac{2(z+0.2)(z+0.3)(z+0.9)}{z(z+0.5)(z^2-0.5z+0.25)}$$
$$=\frac{0.864}{z}-\frac{0.128}{z+0.5}+\frac{1.264z+2.104}{z^2-0.5z+0.25}$$

Therefore,

$$H(z)=0.864-\frac{0.128}{1+0.5z^{-1}}+\frac{1.264+2.104z^{-1}}{1-0.5z^{-1}+0.25z^{-2}}$$

The system diagram is shown in Fig. 14.22.

Section 14.5

The transfer function is given by

$$H(z)=\frac{\frac{14}{9}z}{z-0.4}-\frac{\frac{5}{9}z}{z+0.5}=\frac{z^2+z}{(z-0.4)(z+0.5)}$$

Therefore

$$Y(z)=H(z)X(z)=\frac{z(z^2+z)}{(z-0.4)(z+0.5)(z-1)}$$

$$\frac{Y(z)}{z} = \frac{2.222}{z-1} - \frac{1.0370}{z-0.4} - \frac{0.1852}{z+0.5}$$

so

$$y(n) = 2.222 - 1.0370(0.4)^n - 0.1852(-0.5)^n, \qquad n > 0$$

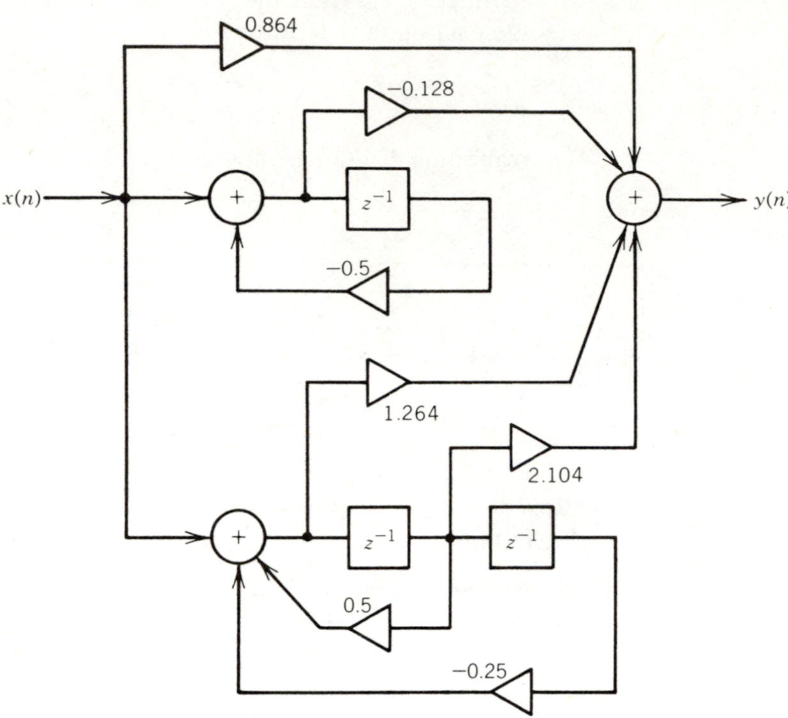

Figure 14.22

APPENDIX

MATRIX SOLUTION TO SIMULTANEOUS EQUATIONS

LEARNING OBJECTIVES

After completing this appendix you should be able to do the following:

1. Define: matrix, vector, matrix elements, square matrix, and determinant.
2. Find the value of determinants.
3. Find the solution to systems of linear independent equations using determinants.

A.1.1 Definitions

A *matrix* is a rectangular array of numbers consisting of n rows and m columns. This matrix is called an n-by-m $(n \times m)$ matrix. Each number in the array is called an *element* of the matrix. For example, an $n \times m$ matrix A is composed of elements a_{ij} and may be written as

$$A = \begin{bmatrix} a_{11} & a_{12} & a_{13} & \cdots & a_{1m} \\ a_{21} & a_{22} & a_{23} & & a_{2m} \\ a_{31} & a_{32} & a_{33} & & a_{3m} \\ \vdots & & & & \\ a_{n1} & a_{n2} & a_{n3} & & a_{nm} \end{bmatrix}$$

where the i subscript indicates the row containing element a_{ij} and the j subscript indicates the column containing the element. Clearly the following inequalities must hold

$$0 \le i \le n$$
$$0 \le j \le m$$

A matrix where the number of rows and number of columns are equal $(n = m)$ is called a *square matrix*. The matrix B is a square matrix:

$$B = \begin{bmatrix} a_{11} & a_{12} & a_{13} \\ a_{21} & a_{22} & a_{23} \\ a_{31} & a_{32} & a_{33} \end{bmatrix}$$

An $n \times 1$ matrix is called a *column vector*. The matrix B above is made up of three column vectors:

$$\begin{bmatrix} a_{11} \\ a_{21} \\ a_{31} \end{bmatrix} \begin{bmatrix} a_{12} \\ a_{22} \\ a_{32} \end{bmatrix} \begin{bmatrix} a_{13} \\ a_{32} \\ a_{33} \end{bmatrix}$$

In general a column vector will contain n terms. Conversely, a $1 \times n$ matrix is called a *row vector*. The matrix B also contains three row vectors.

$$[a_{11} \quad a_{12} \quad a_{13}]$$
$$[a_{21} \quad a_{22} \quad a_{23}]$$
$$[a_{31} \quad a_{32} \quad a_{33}]$$

In general an $n \times m$ matrix contains m column vectors of n elements each and n row vectors of m elements each.

A.1.2 Determinants

A matrix does not have a value. A matrix is an array of numbers, not a number itself. We can however define an operation to yield a number if the matrix is a square matrix. This number we will call the *determinant* of the matrix. If the matrix is $n \times n$, then the determinant is said to be of order n. The determinant of the matrix A is written as $|A|$. For low-order determinants ($n = 2$ or $n = 3$), a "brute force" method can be used. For instance, if

$$A = \begin{bmatrix} a_{11} & a_{12} \\ a_{21} & a_{22} \end{bmatrix}$$

then

$$|A| \triangleq a_{11}a_{22} - a_{12}a_{21}$$

or if

$$A = \begin{bmatrix} a_{11} & a_{12} & a_{13} \\ a_{21} & a_{22} & a_{23} \\ a_{31} & a_{32} & a_{33} \end{bmatrix}$$

then

$$\begin{aligned} |A| = {} & a_{11}a_{22}a_{33} + a_{12}a_{23}a_{31} + a_{13}a_{32}a_{21} \\ & - a_{13}a_{22}a_{31} - a_{23}a_{32}a_{11} - a_{33}a_{21}a_{12} \end{aligned}$$

EXAMPLE A.1.2.1 Let

$$X = \begin{bmatrix} 7 & 8.3 \\ 0.97 & 1.98 \end{bmatrix} \quad \text{and} \quad Y = \begin{bmatrix} 2 & 9 & 1 \\ 4 & 3 & 6 \\ 1 & 5 & 5 \end{bmatrix}$$

Find $|X|$ and $|Y|$.

Solution

$$\begin{aligned} |X| &= (7)(1.98) - (0.97)(8.3) \\ &= 5.809 \\ |Y| &= (2)(3)(5) + (9)(6)(1) + (1)(5)(4) \\ &\quad - (1)(3)(1) - (6)(5)(2) - (5)(4)(9) \\ &= -139 \end{aligned}$$

If the order of the determinant is > 3, the procedure outlined below is to be used. It is sometimes quicker to use this method even for $n = 3$ if some elements of the matrix are zero. Before introducing the new method we must make two more definitions:

Minor: If A is an $n \times n$ matrix then the minor of element a_{ij} is the determinant of the $(n-1) \times (n-1)$ matrix formed when row i and column j are removed from the matrix A.

Cofactor: The cofactor of element a_{ij} is defined as the minor of a_{ij} multiplied by $(-1)^{i+j}$. The cofactor is denoted as A_{ij}.

EXAMPLE A.1.2.2 For the matrices X and Y in Example A.1.2.1 find the minors and cofactors.

Solution

Since X is 2×2, when one row and one column are removed, the remainder is simply an element and the determinant is the value of the element. Therefore, by inspection

$$\text{minor}\,(x_{11}) = 1.98$$

$$\text{minor}\,(x_{12}) = 0.97$$

$$\text{minor}\,(x_{21}) = 8.3$$

$$\text{minor}\,(x_{22}) = 7$$

The cofactors are the minors multiplied by $(-1)^{i+j}$ so

$$X_{11} = (-1)^{1+1} \times \text{minor}\,(a_{11}) = 1.98$$

$$X_{12} = (-1)^{1+2} \times \text{minor}\,(a_{12}) = -0.97$$

$$X_{21} = (-1)^{1+2}(8.3) = -8.3$$

$$X_{22} = (-1)^{2+2}(7) = 7$$

Since Y is 3×3, it contains 9 minors and 9 cofactors. A few of these are:

$$\text{minor}\,(y_{11}) = \begin{vmatrix} 3 & 6 \\ 5 & 5 \end{vmatrix} = -15$$

$$\text{minor}\,(y_{22}) = \begin{vmatrix} 2 & 1 \\ 1 & 5 \end{vmatrix} = 9$$

$$\text{minor}\,(y_{32}) = \begin{vmatrix} 2 & 1 \\ 4 & 6 \end{vmatrix} = 8$$

$$Y_{32} = (-1)^{3+2}(8) = -8$$

$$Y_{21} = (-1)^{2+1} \begin{vmatrix} 9 & 1 \\ 5 & 5 \end{vmatrix} = -40$$

$$Y_{13} = (-1)^{1+3} \begin{vmatrix} 4 & 3 \\ 1 & 5 \end{vmatrix} = 17$$

The determinant of an $n \times n$ matrix is defined as the sum of products formed by choosing any row vector or any column vector and multiplying each element of the chosen vector by its cofactor. Mathematically this is written as

$$|A| = \sum_{i=1}^{n} a_{ij} A_{ij} \qquad \text{for any } j$$

or

$$|A| = \sum_{j=1}^{n} a_{ij} A_{ij} \qquad \text{for any } i$$

We emphasize that *any* vector may be chosen. To illustrate this consider again the matrix Y in Example A.1.2.1. Choose column 1 and evaluate the determinant. By our definition,

$$\begin{aligned} |Y| &= (2)Y_{11} + (4)Y_{21} + (1)Y_{31} \\ &= 2\cdot(-1)^{1+1}\begin{vmatrix} 3 & 6 \\ 5 & 5 \end{vmatrix} + 4\cdot(-40) + 1\cdot(-1)^{3+1}\begin{vmatrix} 9 & 1 \\ 3 & 6 \end{vmatrix} \\ &= 2\cdot(15-30) + (-160) + 1\cdot(54-3) \\ &= -139 \end{aligned}$$

Now choose row 2 and repeat the calculation:

$$\begin{aligned} |Y| &= 4A_{21} + 3A_{22} + 6A_{23} \\ &= 4\cdot(-40) + 3\cdot(-1)^{2+2}\begin{vmatrix} 2 & 1 \\ 1 & 5 \end{vmatrix} + 6\cdot(-1)^{2+3}\begin{vmatrix} 2 & 9 \\ 1 & 5 \end{vmatrix} \\ &= -160 + 3\cdot(10-1) - 6\cdot(10-9) \\ &= -139 \end{aligned}$$

Either approach gives us the same answer we calculated in the example.

To minimize calculations it is usually best to choose the vector containing the greatest number of zeros as our vector for calculating the determinant.

EXAMPLE A.1.2.3 Find the determinant of the matrix D.

$$D = \begin{bmatrix} 1 & 4 & 7 & 2 \\ 6 & 8 & 0 & 3 \\ 2 & 0 & 0 & 1 \\ 0 & 3 & 0 & 5 \end{bmatrix}$$

Solution

Since column 3 contains several zeros, use the column to find $|D|$.

$$|D| = 7\cdot A_{13} = 7\cdot(-1)^{1+3}\begin{vmatrix} 6 & 8 & 3 \\ 2 & 0 & 1 \\ 0 & 3 & 5 \end{vmatrix}$$

Expand the 3×3 determinant about column 1:

$$\begin{aligned} |D| &= 7\cdot(-1)^4\cdot\left[6\cdot(-1)^{1+1}\begin{vmatrix} 0 & 1 \\ 3 & 5 \end{vmatrix} + 2\cdot(-1)^{2+1}\begin{vmatrix} 8 & 3 \\ 3 & 5 \end{vmatrix}\right] \\ &= -560 \end{aligned}$$

A.1.3 Solution to Systems of Linear Independent Equations

The method for using determinants of matrices to solve systems of linear independent equations is straightforward. The method described here is readily adaptable to calculator solution. The set of equations must be arranged in the following form:

$$\begin{gathered}a_{11}x_1 + a_{12}x_2 + a_{13}x_3 + \cdots + a_{1n}x_n = b_1 \\ a_{21}x_1 + a_{22}x_2 + a_{23}x_3 + \cdots + a_{2n}x_n = b_2 \\ \vdots \\ a_{n1}x_1 + a_{n2}x_2 + a_{n3}x_3 + \cdots + a_{nn}x_n = b_n\end{gathered} \tag{A.1.3.1}$$

Step One

Form the matrix of coefficients, A.

$$A = \begin{bmatrix} a_{11} & a_{12} & a_{13} & \cdots & a_{1n} \\ a_{21} & a_{22} & a_{23} & \cdots & a_{2n} \\ \vdots & & & & \\ a_{n1} & a_{n2} & a_{n3} & \cdots & a_{nn} \end{bmatrix}$$

and the vector B

$$B = \begin{bmatrix} b_1 \\ b_2 \\ \vdots \\ b_n \end{bmatrix}$$

Step Two

For a solution for a particular variable, replace the column of coefficients associated with that variable in matrix A with the vector B. That is, if we are solving for x_2 the modified matrix A_n would be

$$A_n = \begin{bmatrix} a_{11} & b_1 & a_{13} & \cdots & a_{1n} \\ a_{21} & b_2 & a_{23} & & a_{2n} \\ \vdots & & & & \\ a_{n1} & b_n & a_{n3} & & a_{nn} \end{bmatrix}$$

Step Three

The variable under investigation, say x_2, then has a value

$$x_2 = \frac{|A_2|}{|A|}$$

Likewise,

$$x_1 = \frac{|A_1|}{|A|}$$

$$\vdots$$

$$x_n = \frac{|A_n|}{|A|}$$

By inspection the denominator need only be calculated once, so in order to solve an nth order set of equations, we must evaluate $n+1$ determinants of order n.

EXAMPLE A.1.3.1 Solve the following set of simultaneous equations:

$$3x_1 + 5x_2 = 21$$
$$5x_1 - 2x_2 = 4$$

Solution

Step 1

$$A = \begin{bmatrix} 3 & 5 \\ 5 & -2 \end{bmatrix} \qquad B = \begin{bmatrix} 21 \\ 4 \end{bmatrix}$$

Step 2

$$A_1 = \begin{bmatrix} 21 & 5 \\ 4 & -2 \end{bmatrix} \qquad A_2 = \begin{bmatrix} 3 & 21 \\ 5 & 4 \end{bmatrix}$$

Step 3

$$x_1 = \frac{|A_1|}{|A|} = \frac{(21)(-2) - (4)(5)}{(3)(-2) - (5)(5)} = \frac{-62}{-31} = 2$$

$$x_2 = \frac{|A_2|}{|A|} = \frac{(3)(4) - (5)(21)}{-31} = \frac{-93}{-31} = 3$$

EXAMPLE A.1.3.2 Solve for x_3 in the following set of equations:

$$3x_1 + 7.2x_2 - 5x_3 = 2$$
$$x_1 - 0 + 4x_3 = 6$$
$$-4.3x_1 + 5x_2 - 1.7x_3 = 0$$

Solution

Step 1

$$A = \begin{bmatrix} 3 & 7.2 & -5 \\ 1 & 0 & 4 \\ -4.3 & 5 & -1.7 \end{bmatrix}$$

$$B = \begin{bmatrix} 2 \\ 6 \\ 0 \end{bmatrix}$$

Step 2

$$A_3 = \begin{bmatrix} 3 & 7.2 & 2 \\ 1 & 0 & 6 \\ -4.3 & 5 & 0 \end{bmatrix}$$

Step 3

$$x_3 = \frac{|A_3|}{|A|} = \frac{-265.76}{-196.6} = 1.35$$

LEARNING EVALUATIONS

1. Which of the matrices are square matrices?

a. $\begin{bmatrix} 7 & 18 \\ 3 & 5 \end{bmatrix}$

b. $\begin{bmatrix} 5.1 & 0 & 8.8 \\ -7.7 & 3.1 & 6.7 \\ 45 & 2 & 8 \end{bmatrix}$

c. $\begin{bmatrix} 1 & 7 & 9 \\ 0 & 5 & 12 \\ 0 & 0 & 2 \end{bmatrix}$

d. $\begin{bmatrix} 16 \\ -8 \\ -3 \\ -5 \end{bmatrix}$

2. Find the determinant of the square matrices in Problem 1.

3. Solve the systems of equations below using matrices.

a. $4x_1 + x_2 = 6$
$-2x_1 + 7x_2 = 5$

b. $7.3y_1 - 55.2y_2 - 14.9y_3 = 33.8$
$23.8y_1 + 48.7y_2 - 12y_3 = 18$
$-5y_1 + 22y_2 + 57.3y_3 = 29.9$

LEARNING EVALUATION ANSWERS

1. **a, b, c**

2. **a.** -19
b. -1304.98
c. 10

3. **a.** $x_1 = 37/30, x_2 = 32/30$
b. $y_1 = 2.363,\ y_2 = -0.554,\ y_3 = 0.941$

APPENDIX

2 COMPLEX NUMBERS

LEARNING OBJECTIVES

After completing this appendix you should be able to do the following:

1. Perform binary operations on complex numbers. That is, be able to add, subtract, multiply, and divide using complex numbers.
2. Find powers, roots, and logarithms of complex numbers.
3. Use Euler's formulas to derive trigonometric relationships.

A.2.1 What Are Complex Numbers?

Suppose we have an equation such as

$$x - 1 = 0$$

The value of x that makes this a true statement is $x = 1$. Next, suppose our equation is a quadratic.

$$x^2 - 3x + 2 = 0$$

We discover by substitution that there are two values, namely $x = 1$ and $x = 2$, that satisfy this equation. The difficulty appears when we try to find a real number x that satisfies the following equation.

$$x^2 - 2x + 2 = 0$$

There are no real numbers that we can use for x to make this a true statement. However, if $x = 1 + \sqrt{-1}$, or if $x = 1 - \sqrt{-1}$, then the above equation is satisfied. In electrical engineering the symbol j is used for $\sqrt{-1}$, instead of the symbol i that is used by everyone else. The reason for this, of course, is that i is firmly entrenched as the symbol for current. Therefore, the two solutions of the above quadratic equation are written as

$$x = 1 + j1$$
$$x = 1 - j1$$

It is an unfortunate happenstance that the square root of a negative number is called "imaginary." The cause of this misfortune is rooted in history. The first brave soul to write down on paper the square root of a negative number was the sixteenth-century Italian mathematician Girolamo Cardano. He did so with due apology, stating that the thing is meaningless, fictitious, and imaginary. Some 200 years later the famous Swiss mathematician Leonhard Euler published a book on algebra in which the symbol i was introduced for $\sqrt{-1}$. Although he demonstrated many uses for complex numbers, he also apologized for using "impossible" and "imaginary" numbers. It was not until two amateur mathematicians named Wessel and Argand gave a geometric interpretation to complex numbers that they gained wide acceptance. Caspar Wessel, a Norweigan, gave the first geometric interpretation in 1799. His work received no attention, and the geometric interpretation was rediscovered by Jean Robert Argand, a Parisian bookeeper, in 1806. Their thinking in developing this interpretation probably went somewhat along the lines of the following argument.

We are accustomed to the real number line. A horizontal line is drawn, and two points are marked on it. These points are labeled 0 and 1, as shown in Fig. A.2.1. Once this basic length is established, all other real numbers may be associated with a unique point on the line. It was Wessel's and Argand's insight to construct a vertical axis through the origin, and to associate the imaginary number system with points on this vertical axis, as shown in Fig. A.2.2.

Figure A.2.1. The real number line.

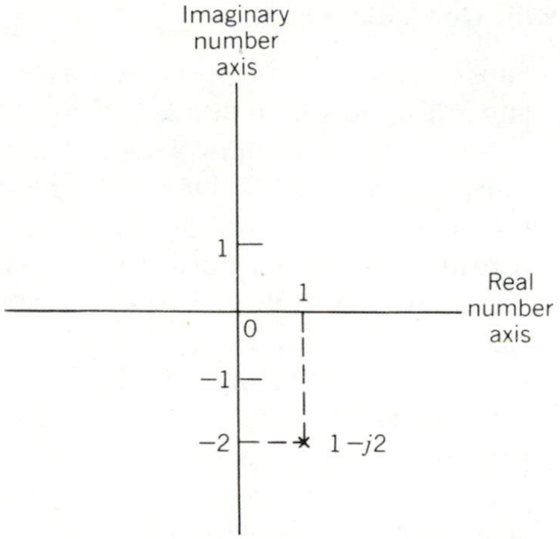

Figure A.2.2. The complex plane (also called an Argand diagram).

The complex number system may now be associated with the complex plane. A unique point in the plane is associated with each complex number. For example, the number $1 - j2$ is associated with a point in the plane found by measuring off one unit along the positive real axis, followed by measuring off two units in the negative vertical direction, as illustrated in the diagram.

Happily the representation of complex numbers as points in a plane has many uses. For instance, a real number a is represented by a point on the real axis. Now if this real number is multiplied by $\sqrt{-1}$, the number becomes ja, and is represented by a point on the vertical axis. This corresponds to a 90° rotation in the counterclockwise direction. This is a general feature of the diagram. When any complex number is multiplied by $\sqrt{-1}$, its representation in the Argand diagram is rotated 90° in the ccw direction. The binary operations of addition, multiplication, subtraction, and division between complex numbers are consistent with the diagram, as well as the operations of root extraction and exponentiation. Another use for the Argand diagram is in interpreting complex numbers in polar coordinates. This makes for easier computation when multiplying, dividing, finding roots, and so on.

A German mathematician, Leopold Kronecker, said, "God gave us the counting numbers, all else is the invention of man." Here is an example from Isaac Asimov[1] that illustrates the meaning of this statement, and also illustrates that complex numbers are no more abstract than the real numbers. In refuting an argument that mathematicians dealt with "illusory" and "imaginary" numbers, he asked his adversary to hand him one-half piece of chalk. Naturally, the fellow complied, breaking a piece of chalk into two pieces. But Asimov argued that he was receiving one piece of chalk, not one-half of a piece. His point was that the number one-half is an abstraction, and one could argue against the existence of such numbers in the real world, just as his adversary was arguing against the existence of imaginary numbers.

A.2.2 Operations with Complex Numbers

Suppose for a moment that you have just invented the complex number system, and it is up to you to decide how to add, subtract, multiply, divide, and find roots, powers, and logarithms of these numbers. You will want to define these operations in such a way that if the imaginary part of all numbers is zero in a particular problem, all results will be consistent with all these operations performed on the real number system. Furthermore, it would be nice if most of the properties of operations on real numbers also apply to complex number operations. For example, $x + y = y + x$ in the real numbers, and you should strive to satisfy this property when defining addition for complex numbers. How should you proceed?

Quite naturally, you would first attempt to use the same methods and definitions for complex numbers that are used for the reals. And things would work out just fine when this is done, so the following procedures should be familiar. Some small difficulty is encountered when finding powers, roots, and logarithms, because of the differences between real and complex numbers, but this will present no real problem.

Polar and Rectangular Form

A complex number x can be expressed in either of two forms,

$$\text{polar form:} \qquad x = Ae^{j\theta}$$
$$\text{rectangular form:} \qquad x = a + jb$$

where the components are related as shown in Fig. A.2.3. That is,

$$A = a^2 + b^2 \qquad a = A\cos\theta$$
$$\theta = \tan^{-1}\left(\frac{b}{a}\right) \qquad b = A\sin\theta$$

[1]Isaac Asimov, *Asimov on Numbers*, Pocket Books, New York, 1977, p. 115.

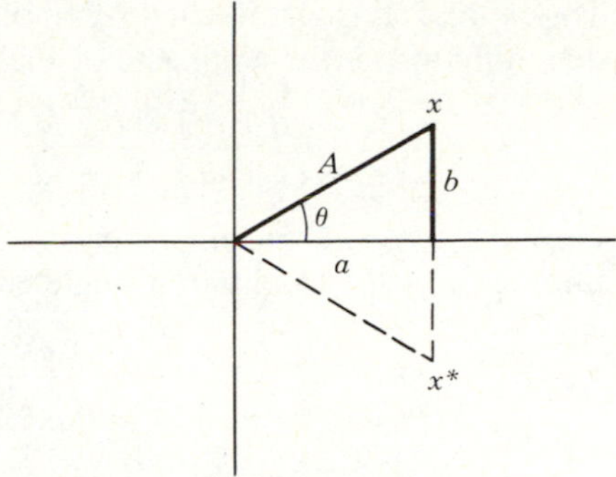

Figure A.2.3

Conjugation

The conjugate of a complex number x, written x^*, is found by replacing each j by $-j$. In polar form, if $x = Ae^{j\theta}$ then $x^* = Ae^{-j\theta}$. In rectangular form, if $x = a + jb$ then $x^* = a - jb$. This is also illustrated in Fig. A.2.3.

We will use the following notation in our subsequent discussion of binary operations on complex numbers. Two numbers x and y are expressed in both rectangular and polar coordinates as follows:

$$x = a + jb = A_1 e^{j\theta_1}$$

$$y = c + jd = A_2 e^{j\theta_2}$$

Equality

In rectangular form, $x = y$ if and only if (iff) $a = c$ and $b = d$. In polar form, $x = y$ iff $A_1 = A_2$ and $\theta_1 = \theta_2$.

Addition and Subtraction

$$x + y = (a + c) + j(b + d)$$
$$x - y = (a - c) + j(b - d)$$

Note: Addition and subtraction should be performed in rectangular coordinates.

Multiplication

In rectangular form,

$$x \cdot y = (a + jb)(c + jd) = (ac - bd) + j(ad + bc)$$

In polar form,

$$x \cdot y = (A_1 e^{j\theta_1})(A_2 e^{j\theta_2}) = A_1 A_2 e^{j(\theta_1 + \theta_2)}$$

Division

In rectangular form,

$$\frac{x}{y} = \frac{a + jb}{c + jd}$$

To express this as a single complex number, first multiply numerator and denominator by the conjugate of the denominator

$$\frac{(a+jb)}{(c+jd)}\cdot\frac{(c-jd)}{(c-jd)}=\frac{(ac+bd)+j(bc-ad)}{c^2+d^2}$$

Now the real and imaginary parts of the numerator may be divided by the real number c^2+d^2 to obtain a single complex number. In polar coordinates,

$$\frac{x}{y}=\frac{A_1e^{j\theta_1}}{A_2e^{j\theta_2}}=\left(\frac{A_1}{A_2}\right)e^{j(\theta_1-\theta_2)}$$

EXAMPLE A.2.2.1 Let $x=1+j2=2.236e^{j1.107}$ and $y=2-j3=3.605e^{-j0.983}$. The angles are in radians. Addition, subtraction, multiplication, and division are now illustrated for these two numbers.

$$x+y=(1+j2)+(2-j3)=3-j1$$
$$x-y=(1+j2)-(2-j3)=-1+j5$$

In rectangular coordinates, the product is

$$x\cdot y=(1+j2)(2-j3)=(2+6)+j(4-3)$$
$$=8+j1=8.062e^{j0.124}$$

In polar coordinates,

$$x\cdot y=(2.236e^{j1.107})(3.605e^{-j0.983})=8.062e^{j0.124}$$

Division in rectangular coordinates is given by

$$\frac{x}{y}=\frac{1+j2}{2-j3}\cdot\frac{(2+j3)}{(2+j3)}=\frac{-4+j7}{4+9}$$
$$=-\frac{4}{13}+j\frac{7}{13}=0.620e^{j2.090}$$

In polar coordinates,

$$\frac{x}{y}=\frac{2.236e^{j1.107}}{3.605e^{-j0.983}}=0.620e^{j2.090}$$

Powers

Squares and higher powers of complex numbers are found by repeated multiplication. The square of a real number is real and positive. The square of an imaginary number is real and negative. The square of a complex number is, in general, complex.

Roots

Every positive real number has two square roots. The square roots of 4 are $+2$ and -2. What about cube roots? We know that $8^{1/3}$ is 2, but are there three cube roots of 8? The answer is yes, and the other two are complex numbers. In general, there are n roots of every (real or complex) number x, that is, for

$$y=x^{1/n}$$

there are n numbers y that satisfy the equation.

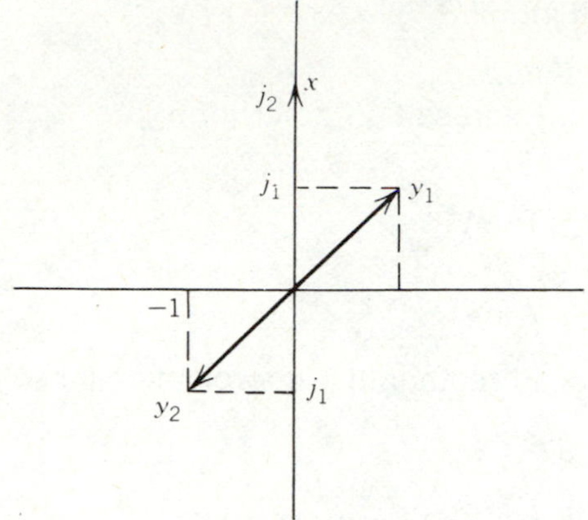

Figure A.2.4

Consider the situation shown in Fig. A.2.4. There are three complex numbers shown in the Argand diagram, $y_1 = 1 + j1$, $y_2 = -1 - j1$, and $x = j2$. Notice that

$$y_1^2 = (1 + j1)^2 = j2$$

and

$$y_2^2 = (-1 - j1)^2 = j2$$

Hence, y_1 and y_2 are the square roots of x. Can we discover a general rule or procedure for finding these two roots?

First express x in polar form.

$$x = 2e^{j\pi/2}$$

Then notice that

$$y_1 = \sqrt{2}\, e^{j\pi/4}$$

and

$$y_2 = \sqrt{2}\, e^{j(\pi/4 + \pi)}$$

It appears that if x is expressed in polar form,

$$x = Ae^{j\theta}$$

then

$$\sqrt{x} = \sqrt{A}\, e^{j\theta/2}$$

and

$$x = Ae^{j(\theta + 2\pi/2)}$$

This rule can be generalized. To find the nth root of x, first express x in polar form

$$x = Ae^{j\theta} \qquad \text{where} \qquad \theta < 2\pi \text{ rad}$$

Then the n roots of x all have magnitude $A^{1/n}$, and the angles are equally spaced around the complex plane with the first angle equal to θ/n. Thus the angles can be found by the formula $(\theta + k2\pi)/n$ where $k = 0, 1, \ldots, n - 1$.

EXAMPLE A.2.2.2 Find $8^{1/3}$.

Solution

Express 8 in polar form,

$$x = 8e^{j0}$$

Then

$$8^{1/3} = 8^{1/3}e^{j(0+k2\pi)/3}$$
$$= 2e^{j0}, 2e^{j2\pi/3}, 2e^{j4\pi/3}$$

In rectangular coordinates, these roots are

$$2 + j0, \qquad -1 + j\sqrt{3}, \qquad -1 - j\sqrt{3}$$

EXAMPLE A.2.2.3 Find the fourth roots of

$$x = 625e^{j120°}$$

Solution

$$y_1 = (625)^{1/4}e^{j120°/4} = 5e^{j30°}$$
$$y_2 = (625)^{1/4}e^{j(120+360)/4} = 5e^{j120°}$$
$$y_3 = (625)^{1/4}e^{j(120+720/4)} = 5e^{j210°}$$
$$y_4 = (625)^{1/4}e^{j(120+1080/4)} = 5e^{j300°}$$

Logarithms of Complex Numbers

Can we find the logarithm of a negative number? The answer is no if the number system we are using is the reals, but the answer is yes if we are using the complex number system. Express the complex number in polar form,

$$x = Ae^{j\theta} \qquad \text{where} \qquad \theta < 2\pi \text{ rad}$$

Then the natural logarithm of x is

$$\ln(x) = \ln(A) + j\theta$$

Thus the logarithm of -1 is

$$\ln(1e^{j\pi}) = \ln(1) + j\pi = j\pi$$

Note: The angle must be expressed in radians when finding logarithms.

Euler's Formulas

Consider the relationship between polar and rectangular coordinates,

$$x = Ae^{j\theta} = a + jb$$

where $a = A\cos\theta$ and $b = A\sin\theta$. If these values of a and b are substituted we have

$$Ae^{j\theta} = A\cos\theta + jA\sin\theta$$

or

$$e^{j} = \cos\theta + j\sin\theta \tag{A.2.2.1}$$

Also, using the definition of complex conjugate,

$$e^{-j\theta} = \cos\theta - j\sin\theta \tag{A.2.2.2}$$

Equations A.2.2.1 and A.2.2.2 are known as Euler's formulas. If Eqs. A.2.2.1 and A.2.2.2 are solved for $\cos\theta$ and $\sin\theta$, the result is

$$\cos\theta = \frac{e^{j\theta} + e^{-j\theta}}{2} \tag{A.2.2.3}$$

$$\sin\theta = \frac{e^{j\theta} - e^{-j\theta}}{2j} \tag{A.2.2.4}$$

which expresses Euler's formulas in an alternate form. These formulas are valuable because they express a fundamental relationship between trigonometric and exponential forms.

EXAMPLE A.2.2.4 Use Euler's formulas to derive the expression

$$\sin^2\theta = 0.5 - 0.5\cos 2\theta$$

Solution

$$\sin^2\theta = \left(\frac{e^{j\theta} - e^{-j\theta}}{2j}\right)^2 = \frac{2 - e^{j2\theta} - e^{-j2\theta}}{4}$$

$$= \frac{1}{2} - \frac{1}{2}\cos 2\theta$$

LEARNING EVALUATIONS

1. Let $x = 3 + j4$ and $y = 8e^{j135°}$. Find

- **a.** $x + y$
- **b.** $x - y$
- **c.** $x \cdot y$
- **d.** x/y

2. Find x^2, $y^{1/3}$, and $\ln(y)$.

3. Use Euler's formulas to derive the identity

$$\sin\theta\cos\theta = 0.5\sin 2\theta$$

LEARNING EVALUATION ANSWERS

1. a. $x + y = (3 + j4) + (-5.657 + j5.657)$
$= -2.657 + j9.657$

b. $x - y = 8.657 - j1.657$

c. $x \cdot y = (5e^{j53.13°})(8e^{j135°}) = 40e^{j188.13°}$

d. $\frac{x}{y} = \frac{5}{8}e^{j(53.13° - 135°)} = 0.625e^{-j81.87°}$

2. $x^2 = (5e^{j53.13^\circ})^2 = 25e^{j106.26^\circ}$

$$y^{1/3} = 8^{1/3}e^{j(135 + k360/3)}, \; k = 0, 1, 2$$
$$= 2e^{j45^\circ}, \; 2e^{j165^\circ}, \; 2e^{-j75^\circ}$$
$$\ln(y) = \ln(8e^{j2.356\text{rad}}) = 2.079 + j2.356$$

3. $$\sin\theta\cos\theta = \left(\frac{e^{j\theta} - e^{-j\theta}}{2j}\right)\left(\frac{e^{j\theta} + e^{-j\theta}}{2}\right)$$
$$= \frac{1}{2}\left(\frac{e^{j2\theta} - e^{-j2\theta}}{2j}\right) = \frac{1}{2}\sin 2\theta$$

INDEX